PHYSICS RESEARCH AND TECHNOLOGY

GEOMAGNETOSPHERE AND COUPLING PHENOMENA

VOLUME I

SOLAR WIND/IMF COUPLING WITH GEOMAGNETOSPHERE/IONOSPHERE

PHYSICS RESEARCH AND TECHNOLOGY

Additional books in this series can be found on Nova's website
under the Series tab.

Additional e-books in this series can be found on Nova's website
under the eBooks tab.

Physics Research and Technology

Geomagnetosphere and Coupling Phenomena

Volume I

Solar Wind/IMF Coupling with Geomagnetosphere/Ionosphere

Lev I. Dorman

Library of Congress Cataloging-in-Publication Data

ISBN: 978-1-53610-564-3

Published by Nova Science Publishers, Inc. † New York

Volume I is dedicated to the memory of my late parents Isaac Dorman (1884–1954) and Eva Dorman (née Globman), 1894–1958, who helped our family survive the famine known as the Holodomer by relocating us in 1932 from Ukraine to Moscow. In the photo, which was taken in 1934 while we lived in Moscow, part of our family is shown from left to right: Musja (Maria), who was 14 years old; our mother Eva, who was 40 years old; Mirele (Mara, Marina), who was 10 years old; Leibele (Lev), who was 5 years old; our father Isaac who was 50 years old; and Zuss who was 18 years old. Fligele (born in 1924) unfortunately passed away a year later, and Abraham (born in 1914) accompanied my maternal great grandparents to Palestine in 1925. Consequently, I never met Abraham until 1989, where I found him in Paris following Gorbachev's political movement known as the perestroika.

CONTENTS

PREFACE TO VOLUME I

Let me start from short autobiographical notes. I was born at 1-st May 1929 in Ukraine as last child, "mizinecel", in many children family of Bratslav's rabbi in Vinnitsa Region. In 1932, during great "Golodomor", our family transferred to Moscow. In 1941 part of family with youngest children was evacuated to Magnitogorsk, where in 1942 - 1944 I worked as carpenter and turner in the great Metallurgy Plant. In 1944 we returned to Moscow, where I in 1945 start to learn in Lomonosov's Moscow State University (MSU). After graduation in December 1950 MSU (Nuclear Physics Division, the Cathedra of Theoretical Physics), my supervisor Professor D.I. Blokhintsev planned for me, as a winner of a Red (Honora) Diploma, to continue my education as PhD-student in his very secret object in the framework of Atomic Problem. To my regret, the KGB withheld permission, and I, together with a lot of other Jewish students who graduate Nuclear Physics Divisions of Moscow and Leningrad Universities and Institutes, were faced with a real prospect of being without work according our education. It was our good fortune that after some time there was being brought the new, Cosmic Ray Project and we were all directed to work in the frame of this Project, which was organized and headed by Prof. S.N. Vernov and Prof. N.V. Pushkov (Director of IZMIRAN); Prof. E.L. Feinberg headed the theoretical part of the Project. Within the framework of this Project there was organized in former Soviet Union in 1951–1952 a wide network of CR stations equipped with a Compton type of large ASC-1 (volume 1000 liters) and ASC-2 (volume 50 liters) ionization chambers developed and constructed in USSR. At that time many experimental results on CR time variations were obtained, but they were very considerably affected by meteorological effects and by meson-nuclear cascade in the atmosphere. Therefore it was not possible to make reasonable transformation from observed CR time variations in the atmosphere and underground to the variations expected in space. To solve this problem, I developed in 1951-1952 a full theory of cosmic ray meteorological effects and a special method of coupling functions between primary and secondary CR variations. All these important results received a grief "full secret". Only from 1954 it becomes possible results on CR variations to publish in open scientific literature, and from 1955 – to take part (mostly by presentation of papers) in International Cosmic Ray Conferences. At the beginning of International Geophysical Year (IGY), in July 1957 was published in Moscow my first book (Dorman, M1957), and at beginning of 1958 – in English in USA (Dorman, M1958). In 1958 under the auspices of the International CR Commission the Committee of CR Meteorological Effects was organized, and I became its first Chairman.

In 1957 I was invited to work on special problems in Magnetic Laboratory of the Ac. of Sci. of USSR and then in I.V. Kurchatov Institute of Atomic Energy, where I founded and became a Head of MHD Laboratory. In parallel I also worked in Moscow State University during about 30 years as Professor at CR and Space Research Cathedra and gave lectures on CR Origin and Time Variations, and for students of Astronomy Division – lectures on Nuclear and Elementary Particle Physics. I also gave lectures on MHD, Plasma Physics and CR in many Universities of USSR (Alma Ata, Irkutsk, Nalchik, and others). As Vice-President of All-Union Section of Cosmic Ray variations (founded in 1955), I took active part in organization the Soviet net of CR stations to the IGY (International Geophysical Year, 1957-1958): we equipped all soviet stations in USSR and in Antarctica with standard cubic and semi-cubic muon telescopes and with neutron monitors of IGY (or Simpson's) type. Then, in connection with preparing for the IQSY (the International Quiet Sun Year, 1964-1965), the soviet net of CR stations was extended about two times and they were equipped with neutron super-monitors of IQSY type (with an effective surface about 10 times bigger than the previous monitor of IGY type).

In 1965 I returned back to IZMIRAN, and founded the Cosmic Ray Department with three Laboratories. For the 30 years, I was a Head of this Department, which became in USSR and then in Russia as the Center of scientific CR research in geophysical and astrophysical aspects. Our Department supported the work of all Soviet and social countries CR stations and undertook the entire work of Soviet CR stations in Antarctica. In cooperation with SibIZMIRAN we organized many CR expeditions inside USSR and in the Arctic Ocean, as well as in Pacific, Atlantic, Indian, and Southern Oceans on the ships "Academician Kurchatov", "Kislovodsk" and others (expeditions were equipped with a neutron super-monitor of IQSY type, with multi-directional muon telescope, with radio-balloon CR measurements in the troposphere and stratosphere). Very important data were obtained about coupling functions, integral multiplicities, and on the planetary distributions of CR intensity and cut-off rigidities.

At the end of the 1960s, in cooperation with my former student from Moscow State University V. Yanke, the theory of CR meteorological effects was generalized to take into account the spectrum and angular distribution of muons at the decay of charged pions, and Coulomb scattering of muons during their propagation in the atmosphere (reviewed in Dorman, M1972, M2004). I proposed and developed the spectrographic method for separation of observed CR variations (corrected on meteorological effects) on variations of geomagnetic and extra-terrestrial origin (Chapter 3 in Dorman, M2004). It became possible based on CR data determine the change of cut-off rigidity and from this the structure of magnetospheric currents and their time variations during large geomagnetic storms (these results reviewed in detail in Dorman, M2009). Simultaneously it became possible, based on CR data, to investigate in detail the variation of the CR spectrum in space, outside the Earth's magnetosphere. This method was then generalized and developed in two directions. The first - by also considering CR meteorological effects as being unknown (the so called generalized spectrographic method), allowed, based only on CR data, to determine simultaneously and separately of each class of CR variations: atmospheric, geomagnetic, and extra-terrestrial. The second, by taking into account CR anisotropy (the so called global spectrographic method), allowed, based only on the CR data of many CR stations (about 40–50) corrected for meteorological effects, to determine simultaneously of the change of cut-off rigidities on our planet and the CR distribution function in space. These methods we consider in details in Dorman (M2004, M2009, M2010).

From 1955 I took part in all International Cosmic Ray Conferences by presenting original papers, as well as: Invited Lectures at 1959 in Moscow and at 1965 in London (in 1965 it was read by Prof. E.L. Feinberg), Rapporteur Papers (at 1969 in Budapest and at 1987 in Moscow), Highlight Paper (at 1999 in Salt Lake City), but I was able to go abroad only in 1966-1969 (thanks to my chief in Kurchatov's Institute of Atomic Energy, Vice-President of Academy of Science of USSR, Academician M.D. Millionshchikov) and then from 1988, thanks to Gorbachev's "perestroika".

In 1961-1963 I worked hard on my second monograph Dorman (M1963a), where is a special Chapter devoted to research on behavior of energetic particles in different space magnetic traps (and, particularly, in traps formed by magnetic field of dipole type). It was a time when only recently was discovered very important for physics of the Earth's magnetosphere phenomenon – radiation belts filled by energetic particles of very high intensity. In those times I had a lot of discussions on radiation belts, on cosmic ray interactions with atmosphere and formation of albedo and plasma processes in the Earth's magnetosphere with S.N. Vernov, A.E. Chudakov, V.P. Shabansky, B.A. Tverskoy and many others from Moscow State University, Physical Lebedev Institute, and IZMIRAN. The formation of Geomagnetosphere and its changes are caused by internal (main) magnetic field and by its interaction with solar wind, shock waves, coronal mass ejections (CME), and other perturbations in the solar wind. This leads to the formation of specific plasmas and energetic processes inside Geomagnetosphere, including the generation and trapping of energetic particles (which can be called as Magnetospheric Cosmic Rays – MCR, in analogy with Solar Cosmic Rays – SCR and Galactic Cosmic Rays – GCR), the formation of radiation belts, ring current and other electrical currents inside the magnetosphere, formation of substorms and magnetic storms, auroras and energetic particle precipitation into atmosphere and escaping into interplanetary space (low energetic local magnetospheric cosmic rays)

For research coordination in this new important field of science the previous Section "Cosmic Ray Variations" was reformatted in Academy of Science of USSR at beginning of 1960s to Section "Cosmic Ray Variations and Radiation Belts" (President S.N. Vernov, Vice-President L.I. Dorman, and from 1982 L.I. Dorman became President of this Section). In 1969 and 1987 I was invited to be rapporter on International Cosmic Ray Conferences (held, correspondingly, in Budapest and in Moscow) on cosmic ray geophysical effects, cosmogenic nuclides, and radiation belts. During preparation of rapporter reports and corresponding review papers (Dorman, 1969, 1987), I had a lot of discussions with many authors on problems of radiation belts and plasma processes in Geomagnetosphere. Recently I had fruitful discussions on these problems with experts in this field of science from Moscow State University, Institute of Space Research, and IZMIRAN.

From my opinion, the importance and actuality of the Geomagnetosphere research are determined mainly by following three factors:

1. The Geomagnetosphere is the nearest giant natural Laboratory, where is possible by many satellites, rockets, balloons, and ground observations investigate in details many different plasmas and energetic processes in space caused finally by interaction of solar wind plasmas and its perturbations (Coronal Mass Ejections - CME, Interplanetary Shock Waves – ISW, Interplanetary Interaction Regions – IIR) with

frozen in Interplanetary Magnetic Fields (IMF) with the rotated main geomagnetic field. This interaction leads to dynamic transformation magnetic fields in Geomagnetosphere, generation and trapping high energy particles, generation of many types instabilities and electromagnetic radiations. These processes are in principle similar to processes in Magnetospheres of other planets and their moons, in the atmosphere of the Sun and other stars, in interplanetary and in interstellar space, in many different astrophysical objects, i.e., the development of this research is important for fundamental Space and Astrophysical science.

2. In the modern time the Technology, Economics, Navigation, TV, Radio wave propagation, Internet, Military aspects, and the life of people on our planet are strong connected with the work of many satellites, moving inside the Geomagnetosphere. Plasmas and energetic processes in the Geomagnetosphere influenced on the satellites work and often lead to satellite malfunctions up to full destroying work of their electronics – satellite became 'dead' (Insurance companies paid many millions dollars for loosing satellites). We hope that detail research of plasmas and energetic processes in the Geomagnetosphere will help to develop methods of forecasting dangerous situation for satellites on different orbits and to decrease the risk of satellite malfunctions and loosing satellites. It means that this research have also very important practical application.
3. The interaction of Coronal Mass Ejections, Interplanetary Shock Waves, and Interplanetary Interaction Regions with the Geomagnetosphere leads (as result of plasmas and energetic processes in the Geomagnetosphere) to generation big magnetic storms accompanied with Forbush decreases and precursory effects in cosmic ray intensity. These big magnetic storms are dangerous not only for satellites, but also on the Earth's surface for technology, communications, train and car accidents, people health (e.g., increasing frequency of infarct myocardial and brain strokes). Investigations of causes of magnetic storms can help to develop methods of their forecasting (including also cosmic ray data) and decreasing the level of magnetic storms hazards. Therefore, the other practical application of this research is connected with the problem of space weather influence on the technology, communications, transportation, and people health on the Earth in dependence of altitude and latitude.

Although the problem of MCR (Magnetospheric Cosmic Rays – energetic particles inside the Earth's magnetosphere) began to be developed far later than that for Galactic CR at supernovae, in interstellar space, and other remote objects and even that for Solar CR on the Sun (Dorman and Dorman, M2014), in present a mechanisms of energetic particles generation, interactions, propagation, and acceleration in the Geomagnetosphere seems to be the most completely studied both experimentally and theoretically. On the other hand, the above-mentioned mechanisms in various regions of the Universe, in different objects have, undoubtedly, some general features (Dorman, M2006). Hence, it is hardly to doubt that the progress in the knowledge of these processes in the Geomagnetosphere will be of great importance for the development of our knowledge concerning the generation, interactions, propagation, and acceleration processes in many types of remote objects in the Universe. Let us note that the main results on the Geomagnetosphere, on its coupling with space and atmospheric processes, on plasmas and energetic magnetospheric processes were obtained at

50s, 60s, and 70s, which became classical, and sufficiently developed later; see monographs and books, collected papers discussed on different scientific Meetings: Alfvén, M1961; Chamberlain, M1961; Alfvén and Fälthammar, M1963; Dorman (M1957, M1963a,b), Williams and Mead (Eds.), M1969; Dorman et al., M1971, M1972a,b; Dorman and Kozin, M1983, Alpert et al., M1964; Akasofu, M1968, M1977; Akasofu and Chapman, M1972; Akasofu and Kan (Eds.), M1981; Baranov and Krasnobaev, M1977; Berezhko et al., M1988; Bleeker et al., M2002; Chamberlain and Hunten, M1987; Chapman and Bartels, M1962; Corovillano et al. (Eds.), M1968; Daglis (Ed.), M2001; Demidovich et al., M1962; Dorman, M1975a,b; Gombosi, M1998; Helliwell, M1965; Hess, Ml972; Isaev and Pudovkin, M1972; Jones, M1974; Kennel, M1995; Lemaire, M1985; Lemaire and Gringauz, M1998; Lemaire et al. (Eds), M1997; Lyatsky, M1978; Lyons and Williams, M1984; Mayaud (Ed.), M1980; McCormac (Ed.), M1970, M1972; Nishida, M1978; Pudovkin and Semenov, M1985; Roederer, M1970; Schulz and Lanzerotti, M1974; Schunk and Nagy, M2000; Sergeev and Tsyganenko, M1980; Shabansky, M1972; Störmer, M1955; Tsurutani et al. (Eds.), M1997; Tverskoy, M1968; Velinov et al., M1974; Hayakawa and Fujinawa (Eds.), M1994; Hayakawa and Molchanov (Eds.), M2002; Pulinets and Boyarchuk, M2004; Walker, M2005; Walt, M1994; Dorman, M2004, M2006, M2009, M2010, M2017a,b; Dorman and Dorman, M2014. Different aspects of the plasmas-energetic processes in the Geomagnetosphere/Ionosphere were discussed in many review papers: Hulot et al. (2010a,b), Thébault et al. (2010), Vasyliūnas (2011), Olsen et al. (2010), Donadini et al. (2010), Aubert et al. (2010), Amit et al. (2010), Finlay et al. (2010), Siscoe et al. (2011), Sterenborg et al. (2011), Shen et al. (2012), Gjerloev (2012), Valdivia et al. (2013), Sokoloff et al. (2014). Important review papers with a lot of original results were published by Russian scientists in Panasyuk (Ed.), M2007 and Zeleny and Veselovsky (Eds.), M2008: Alexeev and Kalegaev (2007, 2008), Kovtyukh (2007a,b), Kuznetsov and Tverskaya (2007a,b), Lazutin (2007), Vaisberg et al. (2008), Malova and Zeleny (2008), Kotova et al. (2008), Kozelov et al. (2008), Deminov (2008), Petrukovich (Ed., 2008).

After publication at beginning of 2017 review-book "Plasmas and Energetic Processes in Geomagnetosphere", I decided little change the structure-organization of the book: instead of many Volumes of one book, to publish several books and each will be contain not more than three Volumes.

The review-book **Plasmas and Energetic Processes in Geomagnetosphere** includes two Volumes:

Volume I. Internal and External Sources, Structure, and Main Properties of Geomagnetosphere. Published by Nova Science Publishers, New York: Dorman (M2017a). It includes six Chapters:

Chapter 1. Internal and External Sources of Geomagnetosphere, Inverse Periods and Secular Variations, Structure, Instabilities, Geomagnetic Indexes, Energy Transfer from Macro to Micro Objects, Magnetospheric Cosmic Rays, Effects before Earthquakes and Tsunami, Radio and CR Tomography.

Chapter 2. Foreshock and Bow Shock.

Chapter 3. Magnetopause/Plasmapause.

Chapter 4. Plasmasphere.

Chapter 5. Cusps.

Chapter 6. Magnetotail.

Volume II. Magnetospheric Sheets, Reconnections, Particle Acceleration, and Substorms. Published in 2017 in the Nova Science Publishers, New York: Dorman (M2017b). It includes seven Chapters:

Chapter 1. Plasmas/Magnetic and Current Sheets: Main Properties, Structure, Expansion, and Particle Transport.

Chapter 2. Magnetospheric Sheets: Turbulence and Fluctuations, Instabilities, Storm Time Variations, Fast Flows, Temperature Anisotropies, and Magnetic Field Structure.

Chapter 3. Magnetospheric Sheets: Bifurcation and Flapping, Pressure Gradients, Influence of Solar Wind and IMF Conditions, ULF, EUV, Auroral Arc Waves, Dipolarization and Antidipolarization Fronts, Parallel Electric Fields, Plasmoids and Jets.

Chapter 4. Substorms: Main Properties, Onset and Expansion Phases.

Chapter 5. Reconnections and Particle Acceleration during Substorms, Statistical Studies.

Chapter 6. Modeling and Simulation of Magnetospheric Substorms.

Chapter 7. Substorms: Energetics, Relation with Magnetic Storms, Plasmas-Energetic Processes in the Ionosphere, Pulsations, Turbulence, Plasma Bubbles, New Substorm Index.

The present review-book **Geomagnetosphere and Coupling Phenomena** reflects the development of research phenomena on the Sun, in the Interplanetary Space, inside the Earth, and in low Atmosphere which coupling with Geomagnetosphere, Ionosphere, and different parts of neutral Earth's atmosphere. It consists from two Volumes.

Volume I. Solar Wind/IMF Coupling with Geomagnetosphere/Ionosphere (is in press in the Nova Science Publishers, New York). This Volume includes six Chapters:

Chapter 1. Coupling and Energy Transfer: Time Variations and Main Problems

Chapter 2. Coupling of Solar Wind and IMF with Geomagnetosphere/Ionosphere

Chapter 3. Efficiency of Solar Wind-Magnetosphere/Ionosphere Coupling for Solar Wind Pressure Variations

Chapter 4. Coupling Solar Wind and IMF with High Latitude Magnetospheric and Auroral Activity

Chapter 5. Coupling of Interplanetary Shock Waves, Coronal Mass Ejections, Corotating Interaction Regions, and other Solar Wind Discontinuities with Geomagnetosphere

Chapter 6. Coupling of Geomagnetosphere-Ionosphere system with Processes in Space and in Atmosphere

Volume II. Coupling of the Geomagnetosphere/Ionosphere/Atmosphere System with Processes Underground, in Low Atmosphere, on the Sun, and in Space (is in press in the Nova Science Publishers, New York). This Volume includes also six Chapters:

Chapter 1. Solar activity and Solar Wind/IMF coupling with Geomagnetosphere-Ionosphere system.

Chapter 2. Coupling Geomagnetosphere–Ionosphere system with Thermosphere.

Chapter 3. Solar activity and Solar Wind/IMF coupling with Geomagnetosphere-Ionosphere-Thermosphere-Mesosphere-Stratosphere-Troposphere system.

Chapter 4. Influence of geomagnetic storms and polar substorms on Ionosphere, Thermosphere, Mesosphere, Stratosphere, and Troposphere.

Chapter 5. Influence of ICME, SIR, HSS, Solar Flares, and Solar Eclipses on the Magnetosphere-Ionosphere-Atmosphere System.

Chapter 6. Coupling of Magnetosphere/Ionosphere/Atmosphere System with Processes Underground and in Low Atmosphere

The review-book **Geomagnetosphere as Space Magnetic Trap** includes two Volumes:
Volume I. Geomagnetosphere's Trap for Energetic Particles
Chapter 1. The Earth's Magnetosphere as a Space Magnetic Trap
Chapter 2. Particle Acceleration in the Geomagnetosphere's Trap
Chapter 3. Auroras and the Magnetospheric-Ionospheric Acceleration Processes
Chapter 4. Inner and Outer Radiation Belts, Trapping and Quasi-Trapping of Anomalous, Solar, and Galactic CR
Chapter 5. Isotopes of He and H in Radiation Belts
Chapter 6. Protons, Antiprotons, and Heavy Energetic Ions in Radiation Belts; Precipitation into Atmosphere and Escaping into Interplanetary Space
Volume II. Geomagnetosphere's Trap for Waves/Pulsations
Chapter 1. Electromagnetic Radiation and Alfvén Waves in Geomagnetosphere's Trap
Chapter 2. Ultra-Low Frequency (ULF) Electromagnetic Waves/Pulsations in Geomagnetosphere's Trap
Chapter 3. Whistlers, Localized Electromagnetic Waves and Cyclotron Interactions in Geomagnetosphere's Trap
Chapter 4. EMIC, ion/electron gyro-frequencies, and ELF (chorus, magnetosonic mode, and hiss) emissions in Geomagnetosphere's Trap
Chapter 5. WEDS, LW, LF, VLF, EN, SSW, MLR, ESW, MF/HF, KH, QTDW, and BEN Emission/Noise in Geomagnetosphere's Trap

The book **Radiation Hazards in Geomagnetosphere from Cosmic Rays.** This book is in preparation with coauthors – my former PhD students/colleagues from Cosmic Ray Department of IZMIRAN (Russia), and colleagues from Israel. It includes three Volumes:
Volume I. Space-Time Distribution of Cosmic Ray Cutoff Rigidities in Geomagnetosphere and low Atmosphere up to 2050.
Volume II. Radiation Hazards for Space-Crafts and Air-Crafts from Galactic Cosmic Rays.
Volume III. Radiation Hazards for Space-Crafts and Air-Crafts during Solar Energetic Particle Events and possibilities for forecasting.

The review-book **Geomagnetosphere: Currents, Winds, and Internal Sources of Geomagnetic Storms** (in preparation).
The review-book **Geomagnetic Storms: Space Sources, Hazards for Satellites, Technology, and People Health, Possibilities of Forecasting** (in preparation).

The detail information on Chapters 1 – 6 of the present Volume I "Solar Wind/IMF Coupling with Geomagnetosphere/Ionosphere" is given in Contents. At the end of each Volume there are References for Monographs and Books (in the text corresponding references marked by letter M before the year). For each Chapter there are separate lists of References. For the convenience of the reader, at the end of each Volume there are put a Subject Index and Author Index. I shall be grateful for any comments, suggestions, and reprints which can be useful for possible next Edition of the book; they may be sent by e-mail (lid010529@gmail.com, lid@physics.technion.ac.il).

Lev I. Dorman,
June 2017, Qazrin, Moscow, Princeton

REFERENCES

Alexeev I.I. and V.V. Kalegaev, 2007. "Magnetosphere of the Earth", in *Model of Cosmos*, Vol. 1 (Ed. M.I. Panasyuk), Moscow, 417-455. In Russian.

Alexeev I.I. and V.V. Kalegaev, 2008. "Magnetic field and main current systems in magnetosphere", in *Plasmas Heliogeophysics*, Vol. 1 (Eds. L.M. Zeleny and I.S. Veselovsky). Physmatlit, Moscow, 422-434 (in Russian).

Amit H., R. Leonhardt, and J. Wicht, 2010. "Polarity Reversals from Paleomagnetic Observations and Numerical Dynamo Simulations", *Space Sci. Rev.*, **155**, 293–335. DOI 10.1007/s11214-010-9695-2.

Aubert J., J.A. Tarduno, and C.L. Johnson, 2010. "Observations and Models of the Long-Term Evolution of Earth's Magnetic Field", *Space Sci. Rev.*, **155**, 337–370. DOI 10.1007/s11214-010-9684-5.

Deminov M.G., 2008. "Ionosphere of the Earth", in *Plasmas Heliogeophysics*, Vol. 2 (Eds. L.M. Zeleny and I.S. Veselovsky). Physmatlit, Moscow, 92-163 (in Russian).

Donadini F., M. Korte, and C. Constable, 2010. "Millennial Variations of the Geomagnetic Field: from Data Recovery to Field Reconstruction", *Space Sci. Rev.*, **155**, 219–246. DOI 10.1007/s11214-010-9662-y.

Dorman L.I., 1969. "Geophysical effects and properties of the various components of the cosmic radiation in the atmosphere (Rapporteur Talk)". *Proc. of 11-th Intern. Cosmic Ray Conf.*, Budapest, Volume of Invited Papers and Rapporteur Talks, pp. 381-444.

Dorman L.I., 1987. "Geomagnetic and atmospheric effects in primary and secondary cosmic rays; cosmogeneous nuclei (rapporteur paper)". *Proc. 20-th Intern. Cosmic Ray Conf.*, Moscow, Vol. 8, pp 186-237.

Eastwood J.P., H. Hietala, G. Toth, T. D. Phan, and M. Fujimoto, 2015. "What Controls the Structure and Dynamics of Earth's Magnetosphere?", *Space Sci. Rev.*, **188**, No. 1-4, 251-286.

Finlay C.C., M. Dumberry, A. Chulliat, and M.A. Pais, 2010. "Short Timescale Core Dynamics: Theory and Observations", *Space Sci. Rev.*, **155**, 177–218. DOI 10.1007/s11214-010-9691-6.

Gillet N., V. Lesur, N. Olsen, 2010. "Geomagnetic Core Field Secular", *Space Sci. Rev.*, **155**, 129–145. DOI 10.1007/s11214-009-9586-6.

Goldstein J. and D.J. McComas, 2013. "Five Years of Stereo Magnetospheric Imaging by TWINS", *Space Sci. Rev.*, **180**, 39–70.

Gubbins D., 2010. "Terrestrial Magnetism: Historical Perspectives and Future Prospects "Baggage We Carry with Us", *Space Sci. Rev.*, **155**, 9–27.

Hulot G., A. Balogh, U.R. Christensen, C.G. Constable, M. Mandea, and N. Olsen, 2010a. "The Earth's Magnetic Field in the Space Age: An Introduction to Terrestrial Magnetism", *Space Sci. Rev.*, **155**, 1–7.

Hulot G., C.C. Finlay, C.G. Constable, N. Olsen, M. Mandea, 2010b. "The Magnetic Field of Planct Earth", *Space Sci Rev*, **152**, 159–222.

Johnson J.R., S. Wing, and P.A. Delamere, 2014. "Kelvin Helmholtz Instability in Planetary Magnetospheres", *Space Sci. Rev.*, **184**, No. 1-4, 1-31.

Kotova G.A., A.S. Leonovich, B.A. Mazur, A.S. Kovtyukh, M.I. Panasyuk, V.Yu. Trakhtengerts, and A.G. Demekhov, 2008. "Inner magnetosphere", in *Plasmas Heliogeophysics*, Vol. 1 (Eds. L.M. Zeleny and I.S. Veselovsky), Physmatlit, Moscow, 484-569 (in Russian).

Kovtyukh A.S., 2007a. "Outer plasmas magnetosphere's covers", in *Model of Cosmos*, Vol. 1 (Ed. M.I. Panasyuk), Moscow, 456-481. In Russian.

Kovtyukh A.S., 2007b. "Ring current", in *Model of Cosmos*, Vol. 1 (Ed. M.I. Panasyuk), Moscow, 482-517. In Russian.

Kozelov B.V., V.A. Pilipenko, and V.Yu. Trakhtengerts, 2008. "Ionosphere-magnetosphere action and physics of auroras phenomena", in *Plasmas Heliogeophysics*, Vol. 1 (Eds. L.M. Zeleny and I.S. Veselovsky), Physmatlit, Moscow, 569-586 (in Russian).

Kuznetsov S.N. and L.V. Tverskaya, 2007a. "Radiation belts", in *Model of Cosmos*, Vol. 1 (Ed. M.I. Panasyuk), Moscow, 518-546. In Russian.

Kuznetsov S.N. and L.V. Tverskaya, 2007b. "Cosmic rays penetrating into magnetosphere", in *Model of Cosmos*, Vol. 1 (Ed. M.I. Panasyuk), Moscow, 579-591. In Russian.

Lazutin L.L., 2007. "Auroral magnetosphere", in *Model of Cosmos*, Vol. 1 (Ed. M.I. Panasyuk), Moscow, 547-578. In Russian.

Li X. and M.A. Temerin, 2001. "The electron radiation belt", *Space Sci. Rev.*, **95**, No. 1-2, 569-580.

Malova H.V. and L.M. Zeleny, 2008. "Structure and dynamics of magnetosphere's tail", in *Plasmas Heliogeophysics*, Vol. 1 (Eds. L.M. Zeleny and I.S. Veselovsky). Physmatlit, Moscow, 434-483 (in Russian).

Mandea M., R. Holme, A. Pais, K. Pinheiro, A. Jackson, and G. Verbanac, 2010. "Geomagnetic Jerks: Rapid Core Field Variations and Core Dynamics", *Space Sci. Rev.*, **155**, 147–175 DOI 10.1007/s11214-010-9663-x.

Matzka J., A. Chulliat, M. Mandea, C.C. Finlay, and E. Qamili, 2010. "Geomagnetic Observations for Main Field Studies: From Ground to Space", *Space Sci. Rev.*, **155**, 29–64. DOI 10.1007/s11214-010-9693-4.

Olsen N., K.-H. Glassmeier, and X. Jia, 2009. "Separation of the Magnetic Field into External and Internal Parts", *Space Sci. Rev.*, **152**, No. 1-4, 135-157.

Olsen N., G. Hulot, and T.J. Sabaka, 2010. "Measuring the Earth's Magnetic Field from Space: Concepts of Past, Present and Future Missions", *Space Sci. Rev.*, **155**, No. 1.65-93. DOI 10.1007/s11214-010-9676-5.

Petrukovich A.A. et al., 2008. "Solar-terrestrial relations and space weather" in *Plasmas Heliogeophysics*, Vol. 2 (Eds. L.M. Zeleny and I.S. Veselovsky). Physmatlit, Moscow, 175-257 (in Russian).

Shabansky V.P., 1968. "Magnetospheric processes and related geophysical phenomena", *Space Sci. Rev.*, **8**, No. 3, 366-454.

Sokoloff D.D., R.A. Stepanov, and P.G. Frick, 2014. "Dynamos: from an astrophysical model to laboratory experiments", *Physics-Uspekhi*, **57**, No. 3, 292-311. Russian Text: *Uspekhi Fizicheskikh Nauk*, **184**, No. 3, 313-335.

Thébault E., M. Purucker, K.A. Whaler, B. Langlais, and T.J. Sabaka, 2010. "The Magnetic Field of the Earth's Lithosphere", *Space Sci Rev*, **155**, 95–127.

Vaisberg O.L., V.N. Smirnov, G.N. Zastenker, S.P. Savin, and M.I. Verigin, 2008. "Interaction of solar wind with outer magnetosphere of the Earth", in *Plasmas Heliogeophysics*, Vol. 1 (Eds. L.M. Zeleny and I.S. Veselovsky). Physmatlit, Moscow, 378-422 (in Russian).

Valdivia J.A., J. Rogan, V. Muñoz, B. A. Toledo, and M. Stepanova, 2013. "The magnetosphere as a complex system", *Adv. in Space Research*, **51**, No. 10, 1934-1941.

Vasyliūnas V.M., 2011. "Physics of Magnetospheric Variability", *Space Sci. Rev*, **158**, No. 1, 91-118.

ACKNOWLEDGMENTS

It is our great pleasure to cordially thank:

my Teachers Professor, Academician Eugenie Feinberg (1912-2005) in former USSR and Professor, Minister of Science Yuval Ne'eman (1925-2006) in Israel;

authors of papers and monographs, and editors of books, collected papers and reports, reflected and discussed in this Volume;

Elsevier, Springer, Editors of ICRC Proceedings, COSPAR, and others for possibility to use figures and tables;

my former students who became colleagues and friends – for many years of collaboration and interesting discussions – M.V. Alania, R.G. Aslamazashvili, V.Kh. Babayan, M. Bagdasariyan, L. Baisultanova, V. Bednaghevsky, A.V. Belov, A. Bishara, D. Blenaru, Ya.L. Blokh, A.M. Chkhetia, L. Churunova, T.V. Dzhapiashvili, E.A. Eroshenko, S. Fisher, E.T. Franzus, L. Granitskij, R.T. Gushchina, O.I. Inozemtseva, K. Iskra, N.S. Kaminer, V.L. Karpov, M.E. Katz, T.V. Kebuladse, Kh. Khamirzov, Z. Kobilinsky, V.K. Koiava, E.V. Kolomeets, V.G. Koridse, V. Korotkov, V.A. Kovalenko, Yu.Ya. Krestyannikov, T.M. Krupitskaja, A.E. Kuzmicheva, A.I. Kuzmin, I.Ya. Libin, A.A. Luzov, N.P. Milovidova, L.I. Miroshnichenko, Yu.I. Okulov, I.A. Pimenov, L.V. Raichenko, L.E. Rishe, A.B. Rodionov, O.G. Rogava, A. Samir Debish, V.S. Satsuk, A.V. Sergeev, A.A. Shadov, B. Shakhov, L.Kh. Shatashvili, G.Sh. Shkhalakhov, V.Kh. Shogenov, V.S. Smirnov, M.A. Soliman, F.A. Starkov, M.I. Tyasto, V.V. Viskov, V.G. Yanke, K.F. Yudakhin, A.G. Zusmanovich;

for many years support of our research in the former USSR and in Russia Federation – A.E. Chudakov, G.B.Khristiansen, V.D. Kuznetsov, V.V. Migulin, M.D. Millionshikov, V.N. Oraevsky, N.V. Pushkov, I.V. Rakobolskaya, S.N. Vernov, G.T. Zatsepin;

for interesting discussions and fruitful collaboration – H.S. Ahluwalia, T.M. Aleksanyan, V.V. Alexeenko, I.V. Alexeev, H. Alfvén, E.E. Antonova, W.I. Axford, J.H. Allen, E. Bagge, G.A. Bazilevskaja, G. Bella, N.P. Ben'kova, M. Bercovitch, E.G. Berezhko, V.S. Berezinsky, J.W. Bieber, R.C. Binford, S.P. Burlatskaya, G. Cini Castagnoli, A.N. Charakhchyan, T.N. Charakhchyan, A. Chilingarian, J. Clem, E. Cliver, H. Coffey, J.W. Cronin, I. Daglis, E. Daibog, A. Dar, R. Davis, Jr., H. Debrunner, V.A. Dergachev, V.A. Dogiel, A.Z. Dolginov,

S.S. Dolginov, L.O.C. Drury, M. Duldig, V.M. Dvornikov, D. Eichler, E. Etzion, Yu.I. Fedorov, P. Ferrando, E.O. Fluckiger, V. Fomichev, M. Galli, A.M. Galper, Yu.I. Galperin, E.S. Glokova, N.L. Grigorov, O.N. Gulinsky, A.V. Gurevich, S.R. Habbal, J.E. Humble, N. Iucci, I.P. Ivanenko, G.S. Ivanov-Kholodny, R. Kallenbach, V.S. Kirsanov, G.E. Kocharov, I.D. Kozin, O.N. Kryakunova, G.F. Krymsky, K. Kudela, L.V. Kurnosova, V. Kuznetsov, O. Kuzetsova, A.A. Lagutin, A. Laor, A.K. Lavrukhina, Yu.I. Logachev, C. Lopate, H. Mavromichalaki, K.G. McCracken, B. Mendoza, M.A. Mogilevsky, I. Moskalenko, Y. Muraki, M. Murat, K. Mursula, V.S. Murzin, N. Nachkebia, K. Nagashima, G.M. Nikolsky, S.I. Nikolsky, V. Obridko, M.I. Panasyuk, J. Pap, E.N. Parker, M. Parisi, S.B. Pikelner, L.P. Pitaevsky, M.K.W. Pohl, A. Polyakov, M.S. Potgieter, C. Price, N.G. Ptitsyna, V.S. Ptuskin, P. Pullkinen, L.A. Pustil'nik, R. Pyle, A.I. Rez, S.I. Rogovaya, I.L. Rozental, S. Sakakibara, N. Sanchez, V. Sarabhai, I.A. Savenko, K. Scherer, V. Sdobnov, V.B. Semikoz, V.P. Shabansky, Yu.G. Shafer, G.V. Shafer, M.M. Shapiro, P.I. Shavrin, M.A. Shea, I.S. Shklovsky, Ya. Shwarzman, B.I. Silkin, J.A. Simpson, G.V. Skripin, D.F. Smart, A. Somogyi, T. Stanev, M. Stehlic, A. Sternlieb, P.H. Stoker, M. Storini, Yu.I. Stozhkov, A. Struminsky, A.K. Svirzhevskaya, S.I. Syrovatsky, P.J. Tanskanen, A.G. Tarkhov, I. Transky, E. Troitskaja, V.A. Troitskaya, B.A. Tverskoy, I.G. Usoskin, J.F. Valdes-Galicia, E.V. Vashenyuk, P. Velinov, D. Venkatesan, S.N. Vernov, E.S. Vernova, I.S. Veselovsky, G. Villoresi, V.P. Vizgin, T. Watanabe, J. Wdowczyk, J.P. Wefel, G. Wibberenz, A.W. Wolfendale, V. Yakhot, G. Yom Din, A.K. Yukhimuk, N.L. Zangrilli, G.T. Zatsepin, L.M. Zeleny, G.B. Zhdanov, V.N. Zirakashvili, I.G. Zukerman;

for constant support and the kind-hearted atmosphere during our education and long way in science–parents Isaac Meerovich Dorman (1884-1954) and Eva Markovna Dorman (Globman) (1894-1958), my wife Irina, parents in low Olga Ivanovna Zamsha and Vitaly Lazarevich Ginzburg (1916-2009), our daughters Maria and Victoria, sisters and brothers Manja (Musja Dorman) Tiraspolskaya (1920-2010), Marina (Mara Dorman) Pustil'nik (1924-2014), Abraham (Dorman) Argov (1914-2003), and Zuss Dorman (1916-1958), our son-in-law Michael Petrov, grand-children Elizabeth and Gregory; relatives in Israel – cousins Michal, David, Shlomo, Dickla, and nephews Raja, Ada, Lev, Dan, Dalia, Shlomo; our good friends in Russia Lena and Volodja, Lena and Misha, Natasha, Lena and Tanja, in Israel Grisha and Valeria, and in USA Alla and Abraham, Rita and Zjama, Ella and Milja, Ljuba from Princeton and Ljuba from New York, Nellja, Aron and Natasha, Valentina and Marik, Ljusja and Misha.

for great help and collaboration in the period of working in Israel which made possible to continue the scientific research – Yuval Ne'eman, Abraham Sternlieb, Uri Dai, Aby Har-Even, Isaac Ben Israel, Zvi Kaplan, Lev Pustil'nik, Igor Zukerman, Michael Murat, Alexei Zusmanovich, Lev Pitaevsky, David Eichler, Sami Bar-Lev, Avi Gurevich, Nunzio Iucci, Giorgio Villoresi, Mario Parisi, Marisa Storini, John A. Simpson, W.I. Axford, Arnold W. Wolfendale, Victor Yakhot, Alexander Polyakov, Doraswamy Venkatesan, Harjit Ahluwalia, Jose F. Valdes-Galicia, Yasushi Muraki, Marc Duldig, Heleni Mavromichalaki, Anatoly Belov, Vladimir Ptuskin, Victor Yanke, Eugenia Eroshenko, Raisa Gushchina, Natalie Ptitsyna, Marta Tyasto, Olga Kryakunova;

for great help in preparing full references – Igor Zukerman;

for great help in checking and improving English in this Volume – Fatima Keshtova;

The work of the Israeli-Italian Emilio Ségre Observatory is supported by the Collaboration of Tel Aviv University (ISRAEL) and "Uniroma Tre" University and IFSN/CNR (ITALY) – our great gratitude for foundation and supporting of this collaboration Yuval Ne'eman, Nunzio Iucci, Giorgio Villoresi, Marisa Storini, Isaac Ben Israel, Mario Parisi, Lev Pustil'nik, Abraham Sternlieb, Uri Dai, and Igor Zukerman.

ABBREVIATIONS AND NOTATIONS

3DVAR – 3 Dimensional VARiational techniques
AACGM - Altitude-Adjusted Corrected GeoMagnetic
ACE - Advanced Composition Explorer
ACR- Anomaly Cosmic Rays
ADFs - AntiDipolarization Fronts
ADP - Axial Double Probe instrument
AE - Auroral Electrojet index
AEEF - Anti-Earthward Electron Fluxes
AIC - Alfvén-Ion Cyclotron waves
AKR - Auroral Kilometric Radiation
AL - AuroraL electrojet index
ALU - modified AL index
AMIE - Assimilative Mapping of Ionospheric Electrodynamics
AMPERE - Active Magnetosphere and Planetary Electrodynamics Response Experiment
AMPTE - Active Magnetospheric Particle Tracer Explorers
AMPTE/CCE - Active Magnetospheric Particle Tracer Explorers/Charge Composition Explorer
AMR - Adaptive Mesh Refinement
ANNM - Artificial Neural Network Method
AoLP - Angle of Linear Polarization
AP - Auroral Power
AP - Auroral Precipitation
APES - Acute Precipitating Electron Spectrometer
ARs - Active Regions
ARM - Atmospheric Research Model
ARMA - AutoRegressive Moving-Average
ARTEMIS - Acceleration, Reconnection, Turbulence, and Electrodynamics of the Moon's Interaction with the Sun
ASC - All-Sky Camera
ASI - All-Sky Imager
ASK - Auroral Structure and Kinetics
ATL - Alfvén Transition Layer
ATV - All-sky TV camera

AW - Alfvén Waves
AWFCs - Auroral Westward Flow Channels
BARREL - Balloon Array for RBSP Relativistic Electron Losses
BAS - British Antarctic Survey radiation belt model
BATS-R-US - Block Adaptive Tree Solar wind Roe Upwind Scheme
BBEs - BroadBand Electrons
BBFs - Bursty Bulk Flows
BCS - Bifurcated Current Sheet
BEN - Broadband Electrostatic Waves
BGC - Bernstein–Green–Cruskal disturbance
BGS - British Geological Survey
BGK - Bernstein-Greene-Kruskal, or trapped particle mode
BHE - Broadband Hiss-like Emission
BI - Buneman Instability
BL - Boundary Layer
BPDs - Broad Plasma Decreases
BRI - Balanced Reconnection Interval
BTF - Balanced Tail Flux
CA - Clock Angles
CARISMA - Canadian Array for Real-time Investigations of Magnetic Activity
CC - Correlation Coefficient
C-C - Cross-Correlation
CCA - Canonical Correlation Analysis
CCE - Charge Composition Explorer
CCMC - Community Coordinated Modeling Center
CD - Contact Discontinuity
CDA - Current Driven Alfvén waves
CDC - Cusp Diamagnetic Cavity
CDD -Current Disruption/Dipolarization
CDF - Cumulative Distribution Function
CDPS - Cold-Dense Plasma Sheet
CEJ - Counter ElectroJet
CEP - Cusp Energetic Particles
CERTO - Coherent Electromagnetic Radio TOmography
CGRs - Concentrated Generator Regions
CHAMP - CHAllenging Minisatellite Payload
CHs - Coronal Holes
CIDs - Coseismic Ionospheric Disturbances
CIF - Cleft Ion Fountain
CIMI - Comprehensive Inner Magnetosphere-Ionosphere model
CIR - Corotating Interaction Region
CIRA - COSPAR International Reference Atmosphere
CIS - Comprehensive Ion Spectrometer
CIS - Cluster Ion Spectrometer
CISM - Center for Integrated Space Weather Modeling
CIT - Computerized Ionospheric Tomography

CLRs - Concentrated Load Regions
CMAM - Cologne Middle Atmosphere Model
CME - Coronal Mass Ejection
CMIT Coupled Magnetosphere-Ionosphere-Thermosphere model
CML - Coupled-Map Lattice
CNA - Cosmic Noise Absorption
C/NOFS - Communication/Navigation Outage Forecasting System
CODE - Center for Orbit Determination in Europe
CODIF - COmposition and DIstribution Function analyzer
COSMIC - Constellation Observing System for Meteorology, Ionosphere, and Climate
CPC - Central Polar Cap
CPCP - Cross-Polar Cap Potential
CPM - Cold-Point Mesopause
CPMN - Circum-pan Pacific Magnetometer Network
CPS - Central Plasma Sheet
CR - Cosmic Rays
CRB - Convection Reversal Boundary
CRCM - Comprehensive Ring Current Model
CRRES - Combined Release and Radiation Effects Satellite
CS - Current Sheet
CSKI - Current Sheet Kink Instability
CTIM - Coupled Thermosphere Ionosphere Model
CTIPE - Coupled Thermosphere-Ionosphere-Plasmasphere Electrodynamics (simulations)
CVW - Christmas Valley West radar
DAZ - Diffuse Auroral precipitation Zone
DBM - Drift Ballooning Mode
DC-ULF - Direct Current of UltraLow Frequency
DD - Directional Discontinuity
DD - Directly Driven
DD -Density Depletions
DDEF - Disturbance Dynamo E-Field
Ddyn - Disturbance dynamo
DE - Dynamics Explorer
DE - Density Enhancements
DEMETER - Detection of Electro-Magnetic Waves Transmitted from Earthquake Regions
DF - Dipolarization Front
DFBs - Dipolarizing Flux Bundles
DFC - Distance–Frequency Characteristics
DGCPM - Dynamic Global Core Plasma Model
DKI - Drift-Kink Instability
DL - Disturbances Low
DL - Double Layer
DLPT - D Layer Preparation Time
DMSHM - Daily Mean Spherical Harmonic Models
DMSP - Defense Meteorological Satellite Program
DNL - Distant-tail Neutral Line

DP - Disturbances Polar
DPD - Decrease – Peak– Decrease
D-RAP - D-Region Absorption Model
DS - Double Star
DSP - Digital Signal Processor
DT -directly transmitted
ECDI - Electron Cyclotron Drift Instability
ECE - Expansion – Compression – Expansion
ECH - Electron Cyclotron Harmonic wave
eCMAM - extended Canadian Middle Atmosphere Model
ECRs - Energy Conversion Regions
ECS - Electron Current Sheet
ECS - Equivalent Current Systems
ECT - Energetic particle, Composition, and Thermal plasma
EDI - Electron Drift Instrument
EDR - Electron Diffusion Region
FDTD - Finite-Difference Time-Domain model
EEB - Energetic Electron Burst
EEF - Energetic Electron Flux
EEHS - Echo Extreme Horizontal drift Speed
EEJ - Equatorial ElectroJet
EEJ - Eastward ElectroJet currents
EEP - Energetic Electron Precipitation
EER - Energetic Electron Rate
EF - Earthward Flow
EF - Electric Field
EFW - Electric Field and Wave
EHA - Equatorial Height Anomaly
EHs - Electron Holes
EIA - Equatorial Ionospheric Anomaly
EIS - Energetic Ion Spectrometer
EISCAT - European Incoherent Scatter radar
ELF - Extremely Low Frequency
EMEC - ElectroMagnetic Electron Cyclotron wave
EMFISIS - Electric and Magnetic Field Instrument and Integrated Science
EMHD - Electron MagnetoHydroDynamic
EMIC - ElectroMagnetic Ion Cyclotron wave
ENAs - Energetic Neutral Atoms
ENVISAT - European Environmental Satellite
EOF - Empirical Orthogonal Function
EPA - Equatorial Pitch Angle
EPB - Equatorial Plasma Bubble
EPD - Energetic Particle Detector
EPDs - Equatorial Plasma Depletions
EPE - Electron Precipitation Event
EPIC - Energetic Particles and Ion Composition

EPOP - Enhanced Polar Outflow Probe satellite
EPP - Energetic Particle Precipitation
EPT - Energetic Particle Telescope
EPTA - Equatorial Plasma Temperature Anomaly
EQ - EarthQuake
ER - Energy Release
ERM - Engineering Radiation Monitor
ERR - Error Reduction Ratio analysis
ES - ElectroStatic
ESA - European Space Agency
ESA - ElectroStatic Analyzer
ESF - Equatorial Spread F layer
ESOI – Emilio Ségre Observatory, Israel
ESR - EISCAT Svalbard Radar
ESWs - Electrostatic Solitary Waves
ETA - Equatorial Thermosphere Anomaly
ETSI - Electron Two-Stream Instability
EUV - Extreme UltraViolet
FAB - Field-Aligned Beam population
FACs - Field-Aligned Currents
FAI - Fast Auroral Imager
FAIs - Field-Aligned Irregularity structures
FAST - Fast Auroral SnapshoT
FBs - Foreshock Bubbles
fBm - fractional Brownian motion
FCa - Foreshock Cavities
FEE - Forbidden Electron Enhancement
FEEPS - Fly's Eye Energetic Particle Spectrometer
FFSs - Fast Forward Shocks
FFT - Fast Fourier Transform
FGM - Fluxgate Magnetometer
FIR - Finite-Impulse-Response
FLC - Field Line Curvature
FLIP - Field Line Interhemispheric Model
fLm - fractional Lévy motion
FLR - Field Line Resonance
FLR - Finite Larmor Radius
FMR - Fast Mode Resonance
FMS - Fast MagnetoSonic waves
FOV - Field Of View
FPC - Flux in the Polar Cap
FPI - Fast Plasma Investigation
FPR - Flux Pileup Region
FR - Flux Ropes
FRSs - Fast Reverse Shocks
FS - Fast Shock

FS - Forward Shock
FSH - ForeSHock
FTD - Fast Type Discontinuities
FTE - Flux Transfer Event
FTE crater – crater Flux Transfer Event
FUV – Far Ultra-Violet
GA - Geomagnetic Activity
GAIA - Ground-to-topside model of the Atmosphere and Ionosphere for Aeronomy
GAIM - Global Assimilation of Ionospheric Measurements
GAIM-FP - GAIM–Full Physics
GAIM-GM - GAIM–Gauss Markov
GCR – Galactic Cosmic Rays
GE - Grabbe-Eastman model
GEC - Global Electron Content
GEC - Global Electric Circuit
GEM - Geospace Environment Modeling
GEO – satellites on GEosynchronous Orbit
GIC - Geomagnetically Induced Currents
GIMs - Global Ionospheric Maps
GITM - Global Ionosphere Thermosphere Model
GITMST–Geomagnetosphere-Ionosphere-Thermosphere-Mesosphere-Stratosphere-Troposphere system
GLD - Global Lightning Dataset
GLOW - GLobal airglOW model
GLOWfast - GLobal airglOW fast model
GM - GeoMagnetosphere
GML – GeoMagnetic Latitude
GMR - General Magnetic Reconnection
GNSSs - Global Navigation Satellite Systems
GOCE - Gravity field and steady state Ocean Circulation Explorer
GOES - Geostationary Operational Environmental Satellites
GPB - GyroPhase Bunched population
GPID - Global Plasmasphere Ionosphere Density
GPS - Global Positioning System
GRA - Generic Residue Analysis
GRACE - Gravity Recovery and Climate Experiment
GSs - Geomagnetic Storms
GSE - Geocentric Solar Ecliptic
GSK - Generalized SemiKinetic model
GSM - Geocentric Solar Magnetospheric coordinate system
GSM TIP - Global Self-consistent Model of the Thermosphere, Ionosphere and Protonosphere
GTO - Geostationary Transfer Orbit
GUVI - Global UltraViolet Imager
GW - Gravity Waves
HAARP - High frequency Active Auroral Research Program

HAMMONIA - Hamburg Model of the Neutral and Ionized Atmosphere
HASDM - High Accuracy Satellite Drag Model
HEE- Hot Electron Enhancement
HEEA - High Energy Electron Analyser
HEIDI - Hot Electron and Ion Drift Integrator model
HENA - High Energy Neutral Atom imager
HF - High Frequency
HFA - Hot Flow Anomaly
HIA - Hot Ion Analyzer
HILDCAA - High-Intensity, Long-Duration, Continuous AE Activity events
HIRB - HIgh-Resolution Bubble model
HMB - Heppner-Maynard Boundary
HOPE - Helium Oxygen Proton Electron instrument
HP - auroral Hemispheric Power
HPCA - Hot Plasma Composition Analyzers
HPR - High-Pressure Region
HPS Heliospheric Plasma Sheet
HR - Harang Reversal
HSJs - High-Speed Jets
HSSs - High-speed Solar wind Streams
HWM - Horizontal Wind Model
I - Ionosphere
IAGA - International Association of Geomagnetism and Aeronomy
IAP - Instrument d'Analyse du Plasma
IAW - Inertial Alfvén Wave
IBEX - Interstellar Boundary Explorer
IBs - Isotropic Boundaries
IBIPS - Inner Boundary of Ion Plasma Sheet
IC - Independent Component analyses méthode
ICMEs - Interplanetary Coronal Mass Ejections
ICRH - Ion Cyclotron Resonance Heating
IDS - Inter-Disciplinary Scientist team
IEC - Ionospheric Electron Content
IEF - Interplanetary Electric Field
IES - Imaging Electron Spectrometer
IGS - International GNSS Services
IL - Invariant Latitude
IM - Interplanetary Medium
IMAGE - Imager for Magnetopause-to-Aurora Global Exploration
IMAGE - International Monitor for Auroral Geomagnetic Effects network
IMAGE/WIC - Wideband Imaging Camera on board IMAGE
IMC - Inner Magnetosphere Coupling
IMF - Interplanetary Magnetic Field
IMM – Ionosphere-Magnetosphere Model
IMPTAM - Inner Magnetosphere Particle Transport and Acceleration Model
IP - Ionospheric Potential

IPIM - IRAP Plasmasphere-Ionosphere Model
IPM - Ionosphere-Plasmasphere Model
IPS - InterPlanetary Shock
IPS - Imaging Proton Spectrometer
IRI - International Reference Ionosphere model
IS - Intermediate Shock
ISR - Incoherent Scatter Radar
ISS - International Space Station
ISWs - Interplanetary Shock Waves
IT – Ionosphere/Thermosphere system
ITM – Ionosphere/Thermosphere/Mesosphere system
KAW - Kinetic Alfvén Wave
KBI - Kinetic Ballooning Instabilities
KC - Kilometric Continuum
KH - Kelvin-Helmholtz
KHI - Kelvin–Helmholtz Instability
KHVs - Kelvin-Helmholtz Vortices
KHWs - Kelvin-Helmholtz Waves
LAEs - Low-Altitude Emissions
LANL - Los Alamos National Laboratory
LAP – Low Atmospheric Processes
LAU - Large Amplitude Undulations
LB - Lower Band
LBP - Local Binary Pattern
LB-to-PS - LoBe-to-Plasma Sheet
LCE - Local Cooling Emission
LD - Logarithmic Diagram
LDE - Long Duration Event
LEEA - Low Energy Electron Analyzer
LEMMS - Low-Energy Magnetospheric Measurement System
LENA - Low Energy Neutral Atom
LEPA - Low Energy Plasma Analyser
LF – Low Frequency waves (0.01-2 Hz)
LFM - Lyon-Fedder-Mobarry
LLBL - Low-Latitude Boundary Layer
LMC - Linear Mode Conversion
LMD - Linear Magnetic Decrease
LOFs - Large-scale Open solar Fields
LOS - Line Of Sight
LPAW - Longitudinally Propagating Arc Wave
LPM - Low-Pressure Magnetosheath
LPP - Location of PlasmaPause
LSK - Large-Scale Kinetic
LSWS - Large-Scale Wave Structure
LTE - Local Thermodynamic Equilibrium
LTEI - Lower Thermosphere and E region Ionosphere

LTPT - Liouville Theorem Particle Tracer
LWs - Langmuir Waves
MagEIS - Magnetic Electron Ion Spectrometer
MBI - MHD Bernoulli Integral
MCs - Magnetic Clouds
MCLs - Magnetic Cloud-Like structures
MCR – Magnetospheric Cosmic Rays
MCSs - Multiple Cusp Signatures
MD - Magnetic Decrease
MDM - Midnight Density Maximum
MEP - Monitor of Electrons and Protons
MERRA - Modern-Era Retrospective Analysis for Research and Application
MF - Medium Frequency
MFD - Magnetic Flux Depletion
MFER - Magnetic Field Enhancement Region
MFLs - Magnetic Field Lines
MFR - Magnetic Flux Rope
MFS - Magnetosheath Filamentary Structures
MGWs - Magnetic Gradient Waves
MHD - MagnetoHydroDynamic
M-I - Magnetosphere-Ionosphere
MI - Main Impulse
MICA - Magnetosphere-Ionosphere Coupling in the Alfvén resonator sounding rocket
MIE - Magnetic Impulse Events
MIECO - Midlatitude Ionosphere Electrodynamics COupling
MIPAS - Michelson Interferometer for Passive Atmospheric Sounding
M-IT - Magnetosphere-Ionosphere/Thermosphere coupling
ML - Magnetic Loop
MLAT - Magnetic LATitude
MLT - Magnetic Local Time
MLT - Mesosphere/Lower Thermosphere
MM - Mirror Mode
MMFs - Midlatitude Magnetic Fluctuations
MMS - Magnetospheric MultiScale mission
MP - MagnetoPause
MPA - Magnetospheric Plasma Analyzer
MPB - Midlatitude Positive Bay index
MPBL - MagnetoPause Boundary Layer
MR - Magnetic Reconnection
MS - MagnetoSonic waves
MSEs - Magnetopause Surface Eigenmodes
MSF - Mid-latitude Spread F
MSH - MagnetoSHeath
MSIS - Mass-Spectrometer Incoherent Scatter
MSNA – Midlatitude Summer Nighttime Anomaly
MSP - Meridian Scanning Photometer

MST- Mesosphere-Stratosphere-Troposphere
MSTIDs - Medium-Scale Traveling Ionospheric Disturbances
MTM - Midnight Temperature Maximum
MTSI - Modified Two Stream Instability
MVA - Minimum Variance Analysis
MWA - Murchison Widefield Array radio telescope
MZK - Modified Zakharov–Kuznetsov equation
NARMAX - Nonlinear AutoRegressive Moving Average with eXogenous inputs
NATION - North American Thermosphere Ionosphere Observing Network
NCAR - National Center for Atmospheric Research
NCAR TIE-GCM - National Center for Atmospheric Research Thermosphere-Ionosphere-Electrodynamics General Circulation Model
NECM - Near Earth Current Meander
NEIALs - Naturally Enhanced Ion-Acoustic Lines
NENL - Near-Earth Neutral Line
NEPS - Near-Earth Plasma Sheet
NETR - Near-Earth Transition Region
NFPI - Narrow-field Fabry-Perot Interferometer
NFTEs - Nightside Flux Transfer Events
NGO - Near-Geosynchronous Onset
NH - North Hemisphere
NIF - Normal Incidence Frame
NIMF - Northward IMF
NMP - North Magnetic Pole
NOC - Natural Orthogonal Components
NPC - Nightside Polar Cap
NS - Northward-then-Southward
NTC - NonThermal Continuum radiation
NWA - Nighttime Winter Anomaly
OCB - Open-Closed field line Boundary
OIS - Oblique Ionospheric Sounding
OmF - Observation-minus-Forecast
OMNI - Operating Missions as a Node on the Internet
OP - OVATION Prime
OpenGGCM-CTIM - Open Geospace General Circulation Model–Coupled Thermosphere Ionosphere Model
OpenGGCM/RCM - Open Geospace General Circulation Model/Rice Convection Model
OPS - Outer Plasma Sheet
OT - Outer cusp Throat
PACs - Pulsating Auroral Channels
PAD - Pitch Angle Distribution
PADIE - Pitch Angle and Energy Diffusion of Ions and Electrons code
PAMELA - Payload for Antimatter Matter Exploration and Light-nuclei Astrophysics
PBIs - Poleward Boundary Intensifications
PBL - Plasmasphere Boundary Layer
PBL - Poleward Boundary of Luminosity

PC - Principal Component analysis method
PC - Polar Cap index
PCA - Principal Component Analysis
PCB - Polar Cap Boundary
PCP - Polar Cap Potential
P_d - solar wind dynamic pressure
PDF-Probability Distribution Function
PE - Prediction Efficiency
PEACE - Plasma Electron And Current Experiment
PEC - Plasmaspheric Electron Content
PFISR - Poker Flat Incoherent Scatter Radar
PH - Plasmaspheric Hiss
PIC - Particle-In-Cell
PIXIE - Polar Ionospheric X-Ray Imaging Experiment
PJ - Polarization Jet
PLHR - Power Line Harmonic Radiation
P_m - solar wind magnetic pressure
PM - Polar Magnetic index
PMAF - Poleward Moving Auroral Form
PMRAFs - Poleward Moving Radar Aurora Forms
PMSEs - Polar Mesospheric Summer Echoes
POD - Precise Orbit Determination
POES - Polar-Orbiting Environmental Satellites system
PPA - Patchy Pulsating Aurora
PPBL - PlasmaPause Boundary Layer
PPEF - Prompt Penetration Electric Field
PPMLR - Piecewise Parabolic Method with a Lagrangian Remap
PRC - Partial Ring Current
PRE - Pre-Reversal Enhancement
PRI - Preliminary Reverse Impulse
PS - Plasma Sheet
PsA - Pulsating Auroras
PSBL - Plasma Sheet Boundary Layer
PSD - Phase Space Density
PSR - Plasma Sheet Reconnection
PSW - dynamic Pressure of Solar Wind
PTP - Plasmapause Test Particle
PVI - Partial Variance of Increments
PVR - Plasmaspheric Virtual Resonance model
PVW - Polar Vortex Weakening
PWs - Planetary Waves
PWI - Plasma Wave Instrument
QP - Quasi-Periodic emission 0.5–4 kHz
QTDWs - Quasi Two Day Waves
QW - Quarter Wave
RadFETs - Radiation-sensing Field Effect Transistors

RAPID - Research with Adaptive Particle Imaging Detectors
RBE - Radiation Belt Environment model
RBN - Reverse Beacon Network
RBSP - Radiation Belt Storm Probes
RBSPICE - Radiation Belt Storm Probes Ion Composition Experiment
RC - Ring Current
RCM - Rice Convection Model
RCM-E - Rice Convection Model-Equilibrium version
RCPO - Ring Current/Plasmasphere Overlap region
RDs - Rotational Discontinuities
R_E - Earth's radius
REPT -Relativistic Electron-Proton Telescope
RFT - Rapid Flux Transport events
RG - Renormalization Group
RGSs - Recurrent Geomagnetic Storms
RIM - Ridley Ionosphere Model
RIP - Root-Integrated Power
RJF - Reconnection Jet Front
R-M effect - Russell-McPherron effect
RM - Resonant Moments method
RO - Radio Occultation
ROCC - Rank Order Correlation Coefficient analysis
ROTI - Rate Of Total electron content change Index
RPI - Radio Plasma Imager
RPS - Relativistic Proton Spectrometer
RSs - Reverse Shocks
RWs - Reverse Waves
SAA - South Atlantic Anomaly
SABER - Sounding of the Atmosphere using Broadband Emission Radiometry
SAIDs - SubAuroral Ion Drifts
SAKR - Striated Auroral Kilometric Radiation
SAMNET - Subauroral Magnetometer Network
SAMPLEX - Solar, Anomalous, and Magnetospheric Particle Explorer
SANAE - South African National Antarctic Expedition
SANE - SubAuroral Nonthermal radio Emission
SAPS - SubAuroral Polarization Streams
SAR - Stable Auroral Red arcs
SAWs - Shear Alfvén Waves
SC - Solar Cycle
SC - Sudden Commencement
SCHA - Spherical Cap Harmonic Analysis
SCIAMACHY - Scanning Imaging Absorption Spectrometer for Atmospheric CHartographY
SCM - Search Coil Magnetometer
SCR – Solar Cosmic Rays
SCW - Substorm Current Wedge
SCW2L - Substorm Current Wedge two-Loop

SDA - Shock Drift Acceleration
SDI - Scanning Doppler Imagers
SDP - Soft Diffuse Precipitation
SDP - Spin-plane Double Probe instrument
SDRE - Sustained Dayside Reconnection Events
SD-WACCM - Whole Atmosphere Community Climate Model using Specified Dynamics
SEC - Stagnant Exterior Cusp
SED - Storm-Enhanced Density
SEEs - Stimulated Electromagnetic Emissions
SEGMA - South European GeoMagnetic Array
SEPs – Solar Energetic Particles
SEW - Slow Expansion Wave
SFR - Sweep Frequency Receiver
SH - South Hemisphere
SHFAs - Spontaneous Hot Flow Anomalies
SI - Sudden Impulse
SI - Stream Interfaces
SIR - Stream Interaction Regions
SITL - Scientist-In-The-Loop
SLAMS - Short Large Amplitude Magnetic Structures
SMC - Steady Magnetospheric Convection
SMC - Solar Magnetic Cloud
SMF_2 - Satellite Model of F_2 layer
SMI - Solar wind–Magnetosphere–Ionosphere coupling
SML - SuperMag aL auroral index
SMP - South Magnetic Pole
SN - Southward-then-Northward
SO - Substorm Onset
SO - Self-Organization
SOC - Self-Organized Criticality
SOPHIE - Substorm Onsets and PHases from Indices of the Electrojet
SOHO - Solar and Heliospheric Observatory
SOHO/SEM - Solar and Heliospheric Observatory/Solar EUV Monitor
SOLID - SOLar Irradiance Data exploitation
SPE - SuperPosed Epoch
SR - Schumann Resonance
SR - Synchrotron Radiation
SRC - Symmetric Ring Current
SS - Slow Shock
SSA - shock surfing acceleration
SSC - Storm Sudden Commencement
SSI - Solar Spectral Irradiance
SSW - Sudden Stratosphere Warming
STAFF - Spatio-Temporal Analysis of Field Fluctuations
STD - Slow Type Discontinuities
STEC - Slant TEC

STICS - SupraThermal Ion Composition Spectrometer
STP - Solar Terrestrial Probe program
SuperDARN - Super Dual Auroral Radar Network
SV - Secular Variation
SVRM - Support Vector Regression Machine
SW - Solar Wind
SWCX - Solar Wind Charge eXchange
SWDs - Solar Wind Discontinuities
SWEPAM - Solar Wind Electron Proton Alpha Monitor
SWIMR - Surface Wave Induced Magnetic Reconnection
SWMF - Space Weather Modeling Framework
TaD - TSM-assisted Digisonde
TADs - Traveling Atmospheric Disturbances
TBL - Turbulent Boundary Layer
TCD - Tail Current Disruption
TCL - Thin Current Layer
TCR - Traveling Compression Region
TCR - Terrestrial Continuum Radiation
TCS - Thin Current Sheet
TCVs - Traveling Convection Vortices
TD - Tangential Discontinuity
TDS - Time Domain Structures
TEC - Total Electron Content
TECU - Total Electron Content Unit (10^{16} el/m^2)
TF - Tailward Flow
TGCM - Thermospheric General Circulation Model
THEMIS - Time History of Events and Macroscale Interactions during Substorms
THEMIS - THermal EMission Imaging System
TIDs - Traveling Ionospheric Disturbances
TIEGCM - Thermosphere Ionosphere Electrodynamic General Circulation Model
TIFP - Transient Ion Foreshock Phenomena
TIGER - Tasman Geospace Environment Radar
TIGER - Thermospheric Ionospheric GEospheric Research
TIM - Tucumán Ionospheric Model
TIMED - Thermosphere Ionosphere Mesosphere Energetics and Dynamics satellite
TIME-GCM - Thermosphere-Ionosphere-Mesosphere-Electrodynamics General Circulation Model
TLR - Tail Lobe Reconnection
TLS - THz Limb Sounder
TMT - Theory and Modeling Team
TOI - Tongue Of Ionization
TPE - Thermospheric Potential Energy
TR - Transition Region
TS - Turbulent Sheet
TVD - Total Variation-Diminishing
TWINS - Two Wide-angle Imaging Neutral-atom Spectrometers

UB - Upper Band
UFKW - UltraFast Kelvin Wave
UL - UnLoading
ULF – Ultra Low Frequency waves (10-100 mHz)
UMM - Universal Multifractal Model
UPAFs - Upstream-Propagating Alfvénic Fluctuations
UPAWs - Upstream-Propagating Alfvén Waves
UT - Universal Time
UVI - UltraViolet Imager
UWFB - ULF Wave Foreshock Boundary
$V_{AL,asym}$ - asymmetric Alfvén speed
VASIMR - Variable Specific Impulse Magnetoplasma Rocket
VCSHs - Vary-Chap Scale Heights
VDFs - Velocity Distribution Functions
VER - Volume Emission Rate
VERB - Versatile Electron Radiation Belt code
VLF - Very Low Frequency
VSH - Vector Spherical Harmonic
VTEC - Vertical Total Electron Content
WACCMX+TIMEGCM - Whole Atmosphere Community Climate Model eXtended version plus Thermosphere-Ionosphere-Mesosphere-Electrodynamics General Circulation Model
WAM+GIP - Whole Atmosphere Model plus Global Ionosphere Plasmasphere model
WBD - Wide-Band Data
WDC - World Data Center
WEJ - Westward ElectroJet currents
WFC - WaveForm Capture
WHI - Whole Heliosphere Interval
WI - Whistler Instability
WSA - Wang-Sheeley-Arge method
WSA - Weddell Sea Anomaly
WTS - Westward Traveling Surge
XUV - soft X-ray/ UV-radiation
α_0 - equatorial pitch angle
ρ_e - electron Larmor radius
ρ_i - ion Larmor radius
Ω_{ce} - equatorial electron gyrofrequency
Ω_{ci} - equatorial ion gyrofrequency
λ_i - ion inertial length

Chapter 1

COUPLING AND ENERGY TRANSFER: TIME VARIATIONS AND MAIN PROBLEMS

1.1. THE ENERGY TRANSFER FROM SOLAR WIND TO THE EARTH'S MAGNETOSPHERE AND PROBLEMS OF SOLAR WIND-MAGNETOSPHERE-IONOSPHERE-THERMOSPHERE COUPLING

1.1.1. The Energy Transferred from Solar Wind to the Internal Part of the Earth's Magnetosphere-Ionosphere System

The energy of the magnetospheric acceleration processes is eventually derived from the solar wind energy transferred to the magnetosphere. It is of interest from this viewpoint to consider the expression obtained by Atkinson (1978) for the flux of the energy transferred by solar wind to the internal part of the Earth's magnetosphere-ionosphere system:

$$W = 10^9 \Phi\left(\rho_o u_r^2\right)^{1/3} + \sigma_d \Phi^2 + 70\Phi^{13/7}, \tag{1.1.1}$$

where Φ is the potential difference across the polar cap; ρ_o and u_r are the density and the off-Sun component of the solar wind plasma velocity in the region ahead of bow shock; σ_d is the dayside electric conduction. It has been established that only three of the seven major current systems contribute to the energy transfer to the magnetospheric inside, namely the magnetotail current system and two current systems flowing along the force lines at the polar edges of the aurora zone. An acceptable model of the processes occurring inside the magnetosphere has been developed which permits the energy flux to be estimated. Consideration has been given to the plasma streaming through bow shock, magnetic field merging, force line reconnection and the magnetotail neutral line, the inner magnetosphere shield by the Alfvén layer against plasma convection, the night-side pressure balance, and the particle precipitation. The intensity and orientation of the interplanetary magnetic field and the direction and velocity of the solar wind movement have been shown to be the determinant parameters in the energy transfer to the magnetosphere inside through the plasma layer on the geo-magnetospheric night side. The solar-wind concentration, speed, and dynamical pressure together with frozen in interplanetary magnetic field (IMF) and induced interplanetary electrical field are essential parameters determined the variations of magnetosphere and processes of energy transfer.

1.1.2. Comparison of the Solar Wind Energy Input to the Magnetosphere Measured by Wind and Interball-1

As outlined Petrukovich et al. (2001), timely solar wind measurements are indispensable for space weather forecasts and magnetospheric studies, but solar wind variations detected by a distant spacecraft might be different from those actually hitting Earth's magnetosphere. To determine how important these differences can be for geophysical applications, Petrukovich et al. (2001) compared energy input to the magnetosphere which was simultaneously measured by the Wind and Interball-1 spacecraft at various distances from the Earth. The percentage of equal (with differences less than 15%) measurements was found to increase from 30% at energies associated with small substorms to 100% for storm-level energies. The degree of the spacecraft separation along the X_{GSE} coordinate and in the YZ_{GSE} plane appeared to be of minor importance within the limits of Wind and Interball-1 orbits.

1.1.3. Multiscale Coupling in Planetary Magnetospheres

Russell (2002) noted that processes in planetary magnetospheres occur on a variety of scales. On the largest scales are the plasma circulations induced in the magnetospheric plasma externally by the solar wind interaction or internally by processes such as mass-loading of the Jovian magnetosphere by the moon Io. These large-scale processes are influenced by small-scale processes such as particle scattering by waves, gyro motion, and charge exchange. It is not always clear which of two coupled processes are in control. For example, many believe that the onset of reconnection requires a microscale instability but it is possible that external forcing leads to the magnetic configuration at which reconnection eventually occurs, and it is only the existence of the appropriate magnetic configuration over a sufficiently large region that is required to cause the macroscopic flows. Russell (2002) reviews such coupling in the magnetospheres of the Earth and Jupiter.

1.1.4. Bimodal Nature of Solar Wind–Magnetosphere–Ionosphere–Thermosphere Coupling

As outlined Siscoe and Siebert (2006), the solar wind-magnetosphere-ionosphere-thermosphere system has two ways to transfer the force that the solar wind exerts at the magnetopause to the Earth: (1) a dipole interaction mode and (2) a thermospheric drag mode. The paper of Siscoe and Siebert (2006) uses MHD simulations to discuss both modes comparatively. The dipole interaction mode is relatively well known and so is used as a model against which to contrast the less well known thermospheric drag mode. The thermospheric drag mode itself consists of two distinct mechanisms, a direct mechanism in which the solar wind pulls on the thermosphere and (here discussed for the first time) an indirect mechanism involving an interaction between the region 1 current system and the geomagnetic dipole which increases the drag on the thermosphere nearly an order of magnitude over the direct drag mechanism. This additional force appears to be necessary to account for observed high-latitude thermospheric winds that correlate with IMF B_z.

1.1.5. Cross-Scale Coupling within Rolled-up MHD-Scale Vortices and Its Effect on Large Scale Plasma Mixing across the Magnetospheric Boundary

Fujimoto et al. (2006) noted that Kelvin-Helmholtz Instability (KHI) is an MHD-scale instability that grows in a velocity shear layer such as the low-latitude boundary layer of the magnetosphere. KHI is driven unstable when a velocity shear is strong enough to overcome the stabilization effect of magnetic field. When the shear is significantly strong, vortices in the nonlinear stage of KHI is so rolled-up as to situate magnetospheric plasma outward of the magnetosheath plasma and vice versa. The big question is if such highly rolled-up vortices contribute significantly to the plasma transport across the boundary and to the filling of the plasma sheet by cool magnetosheath component, which is observed under northward IMF condition. Fujimoto et al. (2006) discussed recent results from two-fluid simulations of MHD-scale KHI with finite electron inertia taken into account. The results indicate that there is coupling between the MHD-scale dynamics and electron-scale dynamics in the rolled-up stage of the vortices. While the details differ depending on the initial magnetic geometry, the general conclusion is that there is significant modification of the MHD-scale vortex flow pattern via coupling to the micro-physics. The kick-back from the parasitic micro-physics enhances highly the potential for large-scale plasma mixing of the parent MHD-scale vortices, which is prohibited by definition in ideal-MHD. Fujimoto et al. (2006) also review their recent 3-D MHD simulation results indicating that KHI vortex can indeed roll-up in the magnetotail-flank situation despite the strong stabilization by the lobe magnetic field. These results encouraged to search for evidence of rolled-up vortices in the Cluster formation flying observations. A nice event was found during northward IMF interval. This interval is when the plasma transport via large scale reconnection becomes less efficient. The finding supports the argument that KHI is playing some role in transporting solar wind into the magnetosphere when the normal mode of transport cannot dominate.

1.1.6. Space- and Ground-Based Investigations of Solar Wind–Magnetosphere–Ionosphere Coupling

Milan et al. (2006) provides a brief review of the understanding of the coupled solar wind-magnetosphere-ionosphere system that can be attained from observations of the size of the ionospheric polar caps from space and the ground. These measurements allow the occurrence and rate of dayside and nightside reconnection to be deduced. The former can be correlated with upstream IMF observations to give an estimate of the effective length of the dayside reconnection X-line. The latter allows a preliminary statistical study of flux closure during substorms and other, smaller nightside reconnection events. Obtained results are illustrated by Figure 1.1.1.

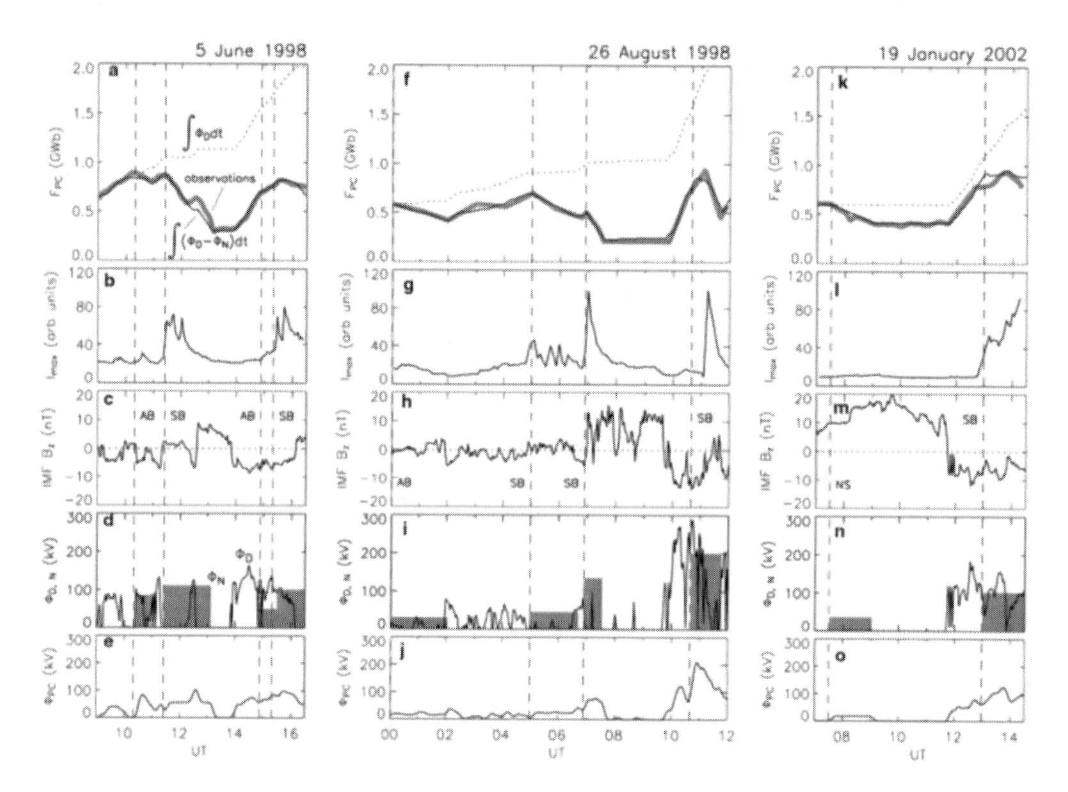

Figure 1.1.1. Measurements related to the investigation of dayside and nightside reconnection for three events on 5 June 1998 (panels a–e), 26 August 1998 (panels f–j), and 19 January 2002 (panels k–o). Vertical dashed lines (labelled in panels c, h, m) indicate the onsets of substorms and auroral brightenings. Panels a, f, k - observed variation in polar cap flux (grey), predicted polar cap flux, including and excluding the contribution from nightside reconnection (solid and dotted, respectively). Panels b, g, l - maximum auroral intensity on the nightside, Imax. Panels c, h, m - IMF B_z lagged to the magnetopause. Panels d, i, n - deduced day - and nightside reconnection rates, UD and UN. Panels e, j, o - cross polar cap potential. According to Milan et al. (2006). Permission from COSPAR.

1.1.7. Magnetospheric Feedback in Solar Wind Energy Transfer

As outlined Palmroth et al. (2010), the solar wind kinetic energy fueling all dynamical processes within the near-Earth space is extracted in a dynamo process at the magnetopause. This direct energy transfer from the solar wind into the magnetosphere depends on the orientation of the IMF as well as other solar wind parameters, such as the IMF magnitude and solar wind velocity. Using the GUMICS-4 MHD simulation, Palmroth et al. (2010) find that the energy input from the solar wind into the magnetosphere depends on this direct driving as well as the magnetopause magnetic properties and their time history in such a way that the energy transfer can continue even after the direct driving conditions turned unfavorable. Such a hysteresis effect introduces discrepancies between the energy input proxies and the energy input measured from GUMICS-4, especially after strong driving, although otherwise the simulation energy input captures the system dynamics. For the cause of the effect, we propose a simple feedback mechanism based on magnetic flux accumulation in the tail lobes. By ideal MHD theory, the energy conversion at the magnetopause is proportional to the product of normal and tangential magnetic fields, the magnetic stress. During large magnetic flux accumulation, the tangential field at the magnetopause strengthens, enhancing the local instantaneous energy conversion and transfer. Simulations of Palmroth et al. (2010) show that this mechanism supports the energy transfer even under weak driving followed by favorable solar wind conditions and transfer up to 50% more power than without the feedback.

1.1.8. Timing of Changes in the Solar Wind Energy Input in Relation to Ionospheric Response

Pulkkinen et al. (2010a) investigate four consecutive substorms using observations and global MHD simulation code GUMICS-4. The solar wind energy input is estimated by the epsilon parameter and evaluated from the simulation by direct integration of the energy transfer through the magnetopause. The ionospheric dissipation is estimated using the AE index and also evaluated by computing the Joule heating from both hemispheres in the simulation. Obtained results show that there is a clear and repeatable time delay between changes in the solar wind parameters and the energy input through the magnetopause. In the simulation, the Joule heating is quite directly driven by the energy input through the magnetopause, without any significant delays. The AE index shows two distinct responses: when the energy input through the magnetopause increases, the AE index responds with a delay; this can be interpreted as the growth phase. When the energy input decreases, the AE decreases simultaneously, without any significant delay. This indicates that after the onset, the energy input to the ionosphere is directly controlled by the energy input through the magnetopause. The results suggest that the recovery phase timing is determined by the solar wind parameters (with a delay associated with the magnetopause processes), while the onset timing is associated with internal magnetotail and/or magnetosphere-ionosphere coupling processes.

1.1.9. Global Energy Transfer during a Magnetospheric Field Line Resonance

According to Hartinger et al. (2011), field line resonances (FLRs) are important for transferring energy from fast mode waves to shear Alfvén waves in the Earth's magnetosphere. Using simultaneous multi-satellite observations from THEMIS and the IMAGE ground magnetometer array, Hartinger et al. (2011) report on the transfer of energy from compressional magnetopause undulations through an FLR to the ionosphere. Energy diversion from the magnetosphere to the ionosphere took place at the FLR: it was found net energy flux there to have comparable values in the radial and the field-aligned directions. The field-aligned energy flux, when mapped to the ionosphere, was 0.70 mW/m^2 and consistent with the inferred Joule dissipation rate at that time. IMAGE's regional monitoring of wave activity reveals that the temporal evolution of the FLR wave power and energy transfer were correlated with the amplitude profile of magnetopause undulations, confirming these waves to be the FLR driver.

1.1.10. Solar Wind Energy Input during Prolonged, Intense Northward IMF: A New Coupling Function

Sudden energy release (ER) events in the midnight sector auroral zone during intense (B > 10 nT), long-duration (T > 3 h), northward IMF magnetic clouds (MCs) during solar cycle 23 (SC23) have been examined in detail by Du et al. (2011). The MCs with northward-then-southward (NS) IMFs were analyzed separately from MCs with southward-then-northward (SN) configurations. It is found that there is a lack of ER/substorms during the N field intervals of NS clouds. In sharp contrast, ER events do occur during the N field portions of SN MCs. From the above two results it is reasonable to conclude that the latter ER events represent residual energy remaining from the preceding S portions of the SN MCs. Du et al. (2011) derive a new solar wind–magnetosphere coupling function during northward IMFs:

$$E_{NIMF} = \alpha\, N^{-1/12}\, V^{7/3}\, B^{1/2} + \beta\, V\, |Dst_{\min}|. \tag{1.1.2}$$

The first term on the right-hand side of the equation represents the energy input via "viscous interaction," and the second term indicates the residual energy stored in the magnetotail. It is empirically found that the magnetotail/magnetosphere/ionosphere can store energy for a maximum of ~4 h before it has dissipated away. This concept is defining one for ER/substorm energy storage. The proposed scenario indicates that the rate of solar wind energy injection into the magnetotail/magnetosphere/ionosphere for storage determines the potential form of energy release into the magnetosphere/ionosphere. This may be more important to understand solar wind–magnetosphere coupling than the dissipation mechanism itself (in understanding the form of the release).

1.1.11. Anomalously Low Geomagnetic Energy Inputs during 2008 Solar Minimum

As noted Deng et al. (2012), the record-low thermospheric density during the last solar minimum has been reported and it has been mainly explained as the consequence of the anomalously low solar extreme ultraviolet (EUV) irradiance. Deng et al. (2012) examined the variation of the energy budget to the Earth's upper atmosphere during last solar cycle from both solar EUV irradiance and geomagnetic energy, including Joule heating and particle precipitation. The global integrated solar EUV power was calculated from the EUV flux model for aeronomic calculations (EUVAC) driven by the MgII index. The annual average solar power in 2008 was 33 GW lower than that in 1996. The decrease of the global integrated geomagnetic energy from 1996 to 2008 was close to 29 GW including 13 GW for Joule heating and 16 GW for particle precipitation from NOAA Polar-Orbiting Environmental Satellites (POES) measurements. Although the estimate of the solar EUV power and geomagnetic energy vary from model to model, the reduction of the geomagnetic energy was comparable to the solar EUV power. The Thermosphere Ionosphere Electrodynamic General Circulation Model (TIEGCM) simulations indicate that the solar irradiance and geomagnetic energy variations account for 3/4 and 1/4 of the total neutral density decrease in 2008, respectively.

1.1.12. Deep Solar Activity Minimum 2007-2009: Solar Wind Properties and Major Effects on the Terrestrial Magnetosphere

Farrugia et al. (2012) discuss the temporal variations and frequency distributions of solar wind and IMF parameters during the solar minimum of 2007-2009 from measurements returned by the IMPACT and PLASTIC instruments on STEREO-A. It was found that the density and total field strength were significantly weaker than in the previous minimum. The Alfvén Mach number was higher than typical. This reflects the weakness of MHD forces, and has a direct effect on the solar wind-magnetosphere interactions. Farrugia et al. (2012) discuss two major aspects that this weak solar activity had on the magnetosphere, using data from Wind and ground-based observations: **i)** the dayside contribution to the cross-polar cap potential (CPCP), and **ii)** the shapes of the magnetopause and bow shock. For i) it was found a low interplanetary electric field of 1.3±0.9 mV/m and a CPCP of 37.3±20.2 kV. The auroral activity is closely correlated to the prevalent stream-stream interactions. Farrugia et al. (2012) suggest that the Alfvén wave trains in the fast streams and Kelvin-Helmholtz instability were the predominant agents mediating the transfer of solar wind momentum and energy to the magnetosphere during this three-year period. For ii) it was determined 328 magnetopause and 271 bow shock crossings made by Geotail, Cluster 1, and the THEMIS B and C spacecraft during a three-month interval when the daily averages of the magnetic and kinetic energy densities attained their lowest value during the three years under survey. We use the same numerical approach as in Fairfield (1971) empirical model and compare obtained results with three magnetopause models. The stand-off distance of the subsolar magnetopause and bow shock were 11.8 R_E and 14.35 R_E, respectively. When comparing with Fairfield (1971) classic result, we find that the subsolar magnetosheath is thinner by ~1 R_E. This is mainly due to the low dynamic pressure which results in a sunward shift of the magnetopause. The magnetopause is more flared than in Fairfield (1971) model. By contrast the bow shock is less flared, and the latter is the result of weaker MHD forces.

1.1.13. Seasonal and Diurnal Variation of Geomagnetic Activity: Russell-Mcpherron Effect during Different IMF Polarity and/or Extreme Solar Wind Conditions

As outlined Zhao and Zong (2012), the Russell-McPherron (R-M) effect is one of the most prevailing hypotheses accounting for semiannual variation of geomagnetic activity. To validate the R-M effect and investigate the difference of geomagnetic activity variation under different IMF polarity and during extreme solar wind conditions (interplanetary shock), Zhao and Zong (2012) have analyzed 42 years IMF and geomagnetic indices data and 1270 SSC (storm sudden commencement) events from the year 1968 to 2010 by defining the R-M effect with positive/negative IMF polarity (IMF away/toward the Sun). The results obtained in this study have shown that the response of geomagnetic activity to the R-M effect with positive/negative IMF polarity are rather profound: the geomagnetic activity is much more intense around fall equinox when the direction of IMF is away the Sun, while much more intense around spring equinox when the direction of IMF is toward the Sun. The seasonal and diurnal variation of geomagnetic activity after SSCs can be attributed to both R-M effect and the equinoctial hypothesis; the R-M effect explains most part of variance of southward IMF, while the equinoctial hypothesis explains similar variance of ring current injection and geomagnetic indices as the R-M effect. However, the R-M effect with positive/negative IMF polarity explains the difference between SSCs with positive/negative IMF B_y accurately, while the equinoctial hypothesis cannot explain such difference at the spring and fall equinoxes. Thus, the R-M effect with positive/negative IMF polarity is more reasonable to explain seasonal and diurnal variation of geomagnetic activity under extreme solar wind conditions.

1.1.14. Solar Activity Dependence of Total Electron Content Derived from GPS Observations over Mbarara

According to Adewale et al. (2012), vertical total electron content (VTEC) observed at Mbarara (geographic co-ordinates: 0.60°S, 30.74°E; geomagnetic coordinates: 10.22°S, 102.36°E), Uganda, for the period 2001-2009 have been used to study the diurnal, seasonal and solar activity variations. The daily values of the 10.7 cm radio flux (F10.7) and sunspot number (R) were used to represent Solar Extreme Ultraviolet Variability (EUV). VTEC is generally higher during high solar activity period for all the seasons and increases from 0600 h LT and reaches its maximum value within 1400 h-1500 h LT. All analyzed linear and quadratic fits demonstrate positive VTEC-F10.7 and positive VTEC-R correlation, with all fits at 0000 h and 1400 h LT being significant with a confidence level of 95% when both linear and quadratic models are used. All the fits at 0600 h LT are insignificant with a confidence level of 95%. Generally, over Mbarara, quadratic fit shows that VTEC saturates during all seasons for F10.7 more than 200 units and R more than 150 units. The result of this study can be used to improve the International Reference Ionosphere (IRI) prediction of TEC around the equatorial region of the African sector. As note Adewale et al. (2012), the software calculates slant TEC (STEC) using GPS observables from the observation data. The VTEC is derived from STEC by using equation:

$$\mathrm{VTEC} = [\mathrm{STEC} - (b_R + b_S)]/S(E), \tag{1.1.3}$$

where b_R is the interfrequency differential receiver biases and b_S the interfrequency differential satellite biases. The mapping function $S(E)$ employed is given by

$$S(E) = \frac{1}{cosZ} = \left\{1 - \left(\frac{R_E \times cos(E)}{R_E + h_s}\right)^2\right\}^{-0.5}, \qquad (1.1.4)$$

where Z - zenith angle of the satellite as seen from the observing station, R_E - radius of the Earth, E - the elevation angle in radians, and h_s - the altitude of the thin layer above the surface of the Earth (taken as 350 km). Some obtained results are illustrated by Table 1.1.1 and Figures 1.1.2 - 1.1.5.

Table 1.1.1. Correlation coefficients between solar indices and VTEC. According to Adewale et al. (2012). Permission from COSPAR

	Solar flux (F10.7)			Sunspot number (R)		
	0 h LT	06 h LT	14 h LT	0 h LT	6 h LT	14 h LT
DECSOLS	0.72	0.30	0.92	0.67	0.27	0.86
MAREQUI	0.88	0.36	0.94	0.87	0.36	0.91
JUNSOLS	0.78	−0.09	0.79	0.77	−0.07	0.78
SEPEQUI	0.70	0.24	0.85	0.67	0.25	0.82

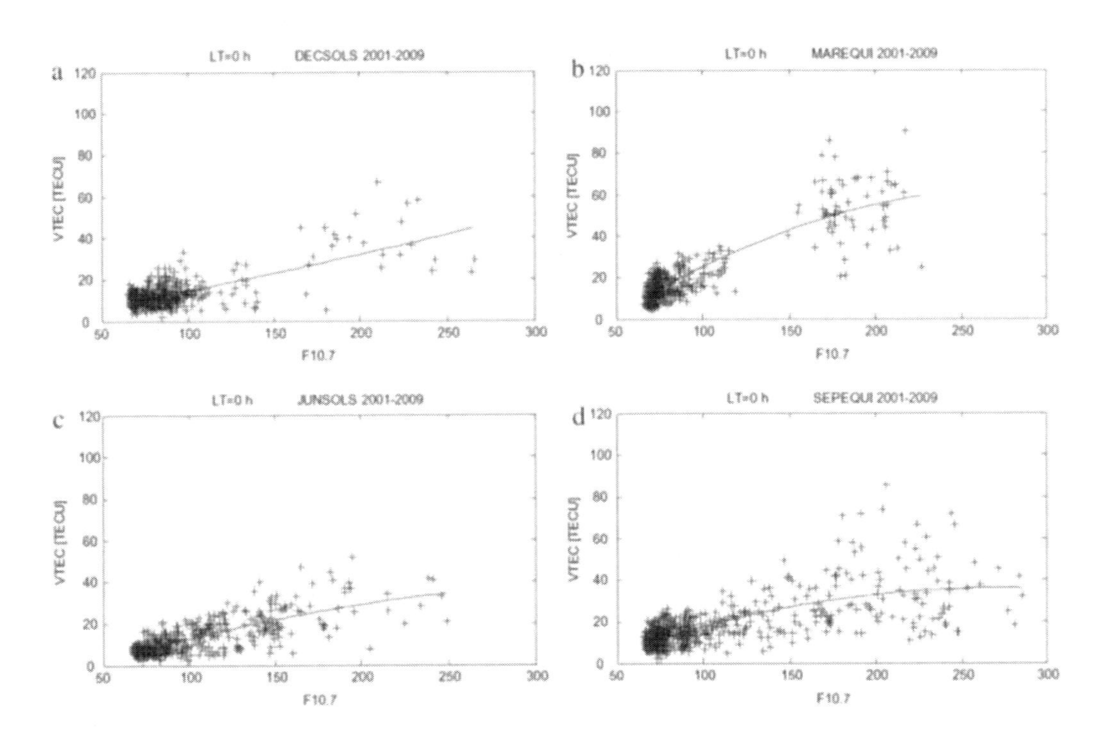

Figure 1.1.2. Variation of VTEC at 0000 h LT with solar flux (F10.7) for December (DECSOLS), March (MAREQUI), June (JUNSOLS), and September (SEPEQUI). According to Adewale et al. (2012). Permission from COSPAR.

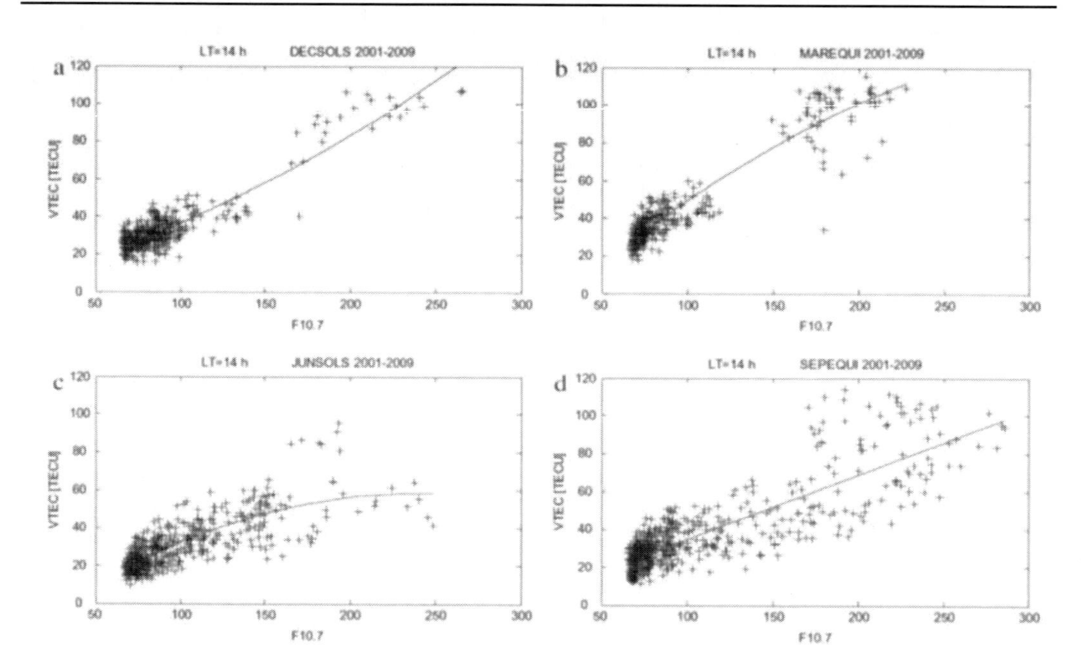

Figure 1.1.3. Variation of VTEC at 1400 h LT with solar flux (F10.7) for DECSOLS, MAREQUI, JUNSOLS, and SEPEQUI. According to Adewale et al. (2012). Permission from COSPAR.

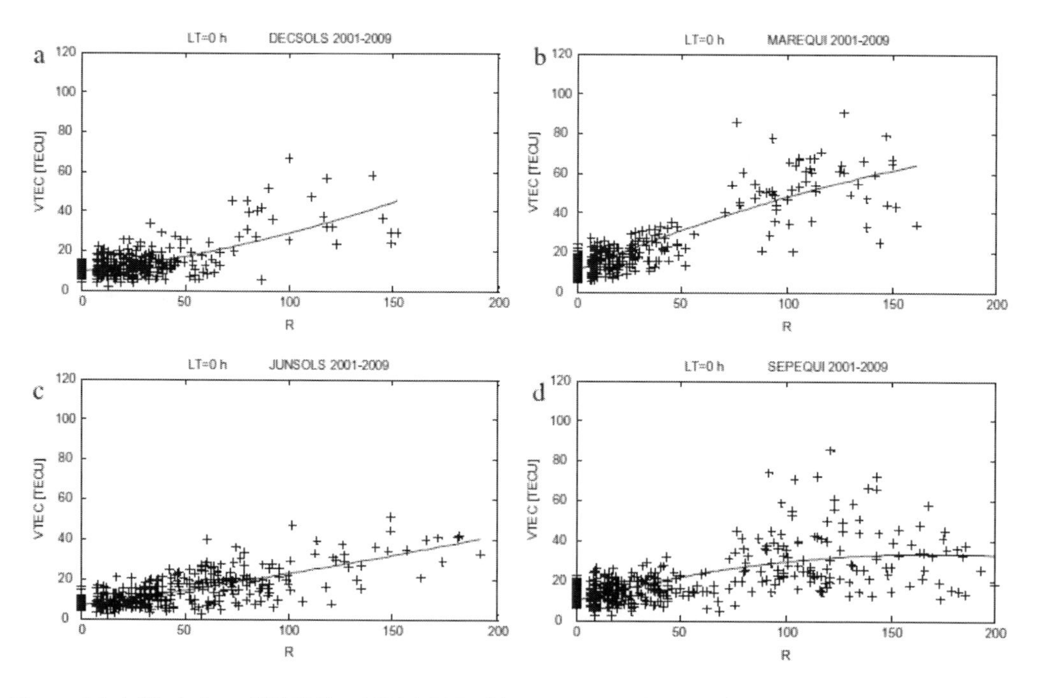

Figure 1.1.4. Variation of VTEC at 0000 h LT with sunspot number (R) for December (DECSOLS), March (MAREQUI), June (JUNSOLS), and September (SEPEQUI). According to Adewale et al. (2012). Permission from COSPAR.

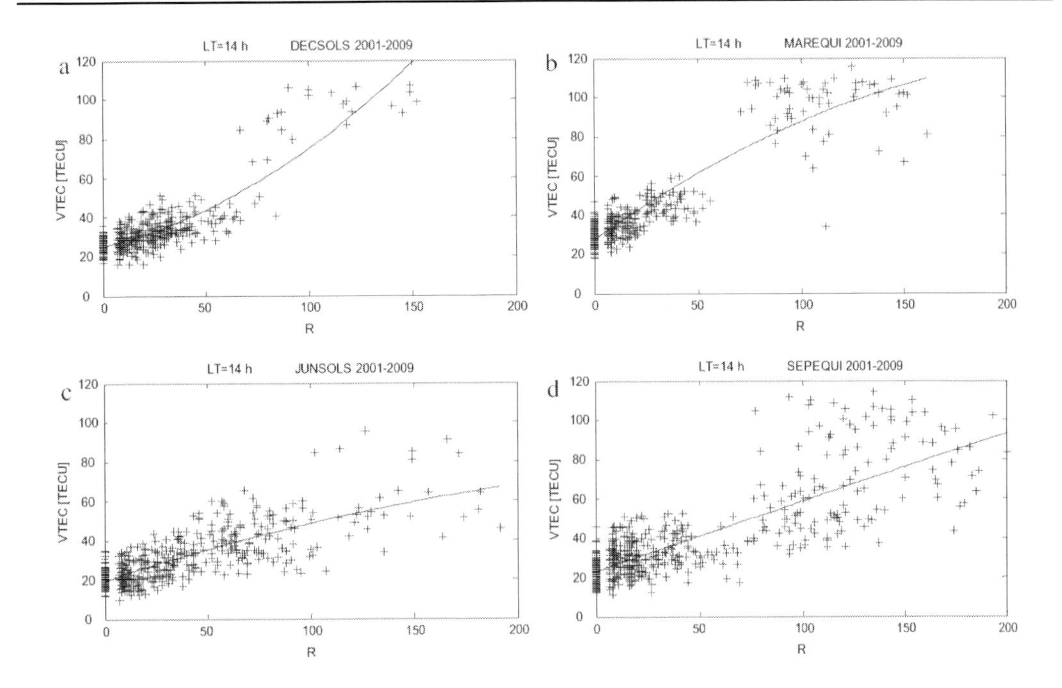

Figure 1.1.5. Variation of VTEC at 1400 h LT with sunspot number (R) for DECSOLS, MAREQUI, JUNSOLS, and SEPEQUI. According to Adewale et al. (2012). Permission from COSPAR.

1.1.15. Changes in Solar Wind–Magnetosphere Coupling with Solar Cycle, Season, and Time Relative to Stream Interfaces

As outlined McPherron et al. (2013), geomagnetic activity depends on a variety of factors including solar zenith angle, solar UV, strength of the IMF, speed and density of the solar wind, orientation of the Earth's dipole, distance of the Earth from Sun, occurrence of CMEs and CIRs, and possibly other parameters. McPherron et al. (2013) have investigated some of these using state-dependent linear prediction filters. For a given state a prediction filter transforms a coupling function such as rectified solar wind electric field (VBs) to an output like the auroral electrojet index (AL). The area of this filter calculated from the sum of the filter coefficients measures the strength of the coupling. When the input and output are steady for a time longer than the duration of the filter the ratio of output to input is equal to this area. It was found coupling strength defined in this way for Es=VBs to AL (and AU) is weakest at solar maximum and strongest at solar minimum. AL coupling displays a semiannual variation being weakest at the solstices and strongest at the equinoxes. AU coupling has only an annual variation being strongest at summer solstice. AL and AU coupling also vary with time relative to a stream interface. Es coupling is weaker after the interface, but ULF coupling is stronger. Total prediction efficiency remains about constant at the interface. The change in coupling strength with the solar cycle can be explained as an effect of more frequent saturation of the polar cap potential causing a smaller ratio of AL to Es. Stronger AL coupling at the equinoxes possibly indicates some process that makes magnetic reconnection less efficient when the dipole axis is tilted along the Earth–Sun line. Strong AU coupling at summer solstice is likely due to high conductivity in northern summer. Coupling changes at a stream interface are correlated with the presence of strong wave activity in ground and satellite measurements and may be an artifact of the method by which solar wind data are propagated.

1.1.16. Energy Coupling during the August 2011 Magnetic Storm

Huang et al. (2013) present results from an analysis of high-latitude ionosphere-thermosphere (IT) coupling to the solar wind during a moderate magnetic storm which occurred on 5–6 August 2011. During the storm, a multipoint set of observations of the ionosphere and thermosphere was available. Huang et al. (2013) make use of ionospheric measurements of electromagnetic and particle energy made by the Defense Meteorological Satellite Program and neutral densities measured by the Gravity Recovery and Climate Experiment satellite to infer (1) the energy budget and (2) timing of the energy transfer process during the storm. It was conclude that the primary location for energy input to the IT system may be the extremely high latitude region. Huang et al. (2013) suggest that the total energy available to the IT system is not completely captured either by observation or empirical models.

1.1.17. Energy Transfer across the Magnetopause for Northward and Southward Interplanetary Magnetic Fields

A three-dimensional adaptive MHD model is used by Lu et al. (2013) to examine the energy flow from the solar wind to the magnetosphere. Using the model, Lu et al. (2013) directly compute fluxes of mechanical and electromagnetic energy across the magnetopause surface. For northward IMF, most of the energy flux inflow occurs near the polar cusps on magnetopause. The viscous interaction leads the carrying energy plasma enter into high latitudes of the tail magnetopause and then divert to low-latitude regions tangentially, where the plasma gets cooler and denser near the flanks of plasma sheet. For southward IMF, the largest electromagnetic energy input into the magnetosphere occurs at the tail lobe behind the cusps, and largest mechanical energy input occurs at near-equatorial dayside magnetopause. Under southward IMF conditions, mechanical energy transfer is enhanced at the flanks of magnetopause in response to increased IMF magnitude, while more electromagnetic energy input can be identified as increasing solar wind density. Obtained results suggest that the mechanisms proposed to energy transfer are mainly due to reconnection and viscous interaction processes for northward IMF. For southward IMF, reconnection is the dominant factor in energy transfer. If the electromagnetic energy coupling between the solar wind and the magnetosphere can be interpreted as a proxy for the reconnection efficiency, the average efficiency during northward IMF is about 20% of that for southward IMF.

1.1.18. Solar Cycle Variations in Polar Cap Area Measured by the Superdarn Radars

Imber et al. (2013) present a long-term study, from January 1996 to August 2012, of the latitude of the Heppner-Maynard Boundary (HMB) measured at midnight using the northern hemisphere Super Dual Auroral Radar Network (SuperDARN). The HMB represents the equatorward extent of ionospheric convection and is used in this study as a measure of the global magnetospheric dynamics. It was found that the yearly distribution of HMB latitudes is single peaked at 64° magnetic latitude for the majority of the 17 year interval. During 2003, the envelope of the distribution shifts to lower latitudes and a second peak in the distribution is observed at 61°. The solar wind-magnetosphere coupling function derived by Milan et al. (2012) suggests that the solar wind driving during this year was significantly higher than during the rest of the 17 year interval. In contrast, during the period 2008–2011, HMB distribution shifts to higher latitudes, and a second peak in the distribution is again observed, this time at 68° magnetic latitude. This time interval corresponds to a period of extremely low solar wind driving during the recent extreme solar minimum. This is the first long-term study of the polar cap area and the results demonstrate that there is a close relationship between the solar activity cycle and the area of the polar cap on a large-scale, statistical basis.

1.1.19. PC Index as a Proxy of the Solar Wind Energy That Entered into the Magnetosphere

As outlined Troshichev et al. (2014), the Polar Cap (PC) index has been approved by the International Association of Geomagnetism and Aeronomy (IAGA XXII Assembly, Merida, Mexico, 2013) as a new index of magnetic activity. The PC index can be considered to be a proxy of the solar wind energy that enters the magnetosphere. This distinguishes PC from AL and Dst indices that are more related to the dissipation of energy through auroral currents or storage of energy in the ring current during magnetic substorms or storms. The association of the PC index with the direct coupling of the solar wind energy into the magnetosphere is based upon analysis of the relationship of PC with parameters in the solar wind, on the one hand, and correlation between the time series of PC and the AL index (substorm development), on the other hand. This paper (the first of a series) provides the results of statistical investigations that demonstrate a strong correlation between the behavior of PC and the development of magnetic substorms. Substorms are classified as isolated and expanded. Troshichev et al. (2014) found that (1) substorms are preceded by growth in the PC index, (2) sudden substorm expansion onsets are related to “leap” or “reverse” signatures in the PC index which are indicative of a sharp increase in the PC growth rate, (3) substorms start to develop when PC exceeds a threshold level 1.5 ± 0.5 mV/m irrespective of the length of the substorm growth phase, and (4) there is a linear relation between the intensity of substorms and PC for all substorm events.

1.1.20. The Solar Cycle Variation of Plasma Parameters in Equatorial and Mid Latitudinal Areas during 2005–2010

Based on the ISL data detected by DEMETER satellite, the solar cycle variation in electron density (Ne) and electron temperature (Te) were studied by Zhang et al. (2014) separately in local daytime 10:30 and nighttime 22:30 during 2005–2010 in the 23rd/24th solar cycles. The semi-annual, annual periods and decreasing trend with the descending solar activity were clearly revealed in Ne. At middle and high latitudes, there exhibited phase shift and even reversed annual variation over Southern and Northern hemisphere, and the annual variation amplitudes were asymmetrical at both hemispheres in local daytime.

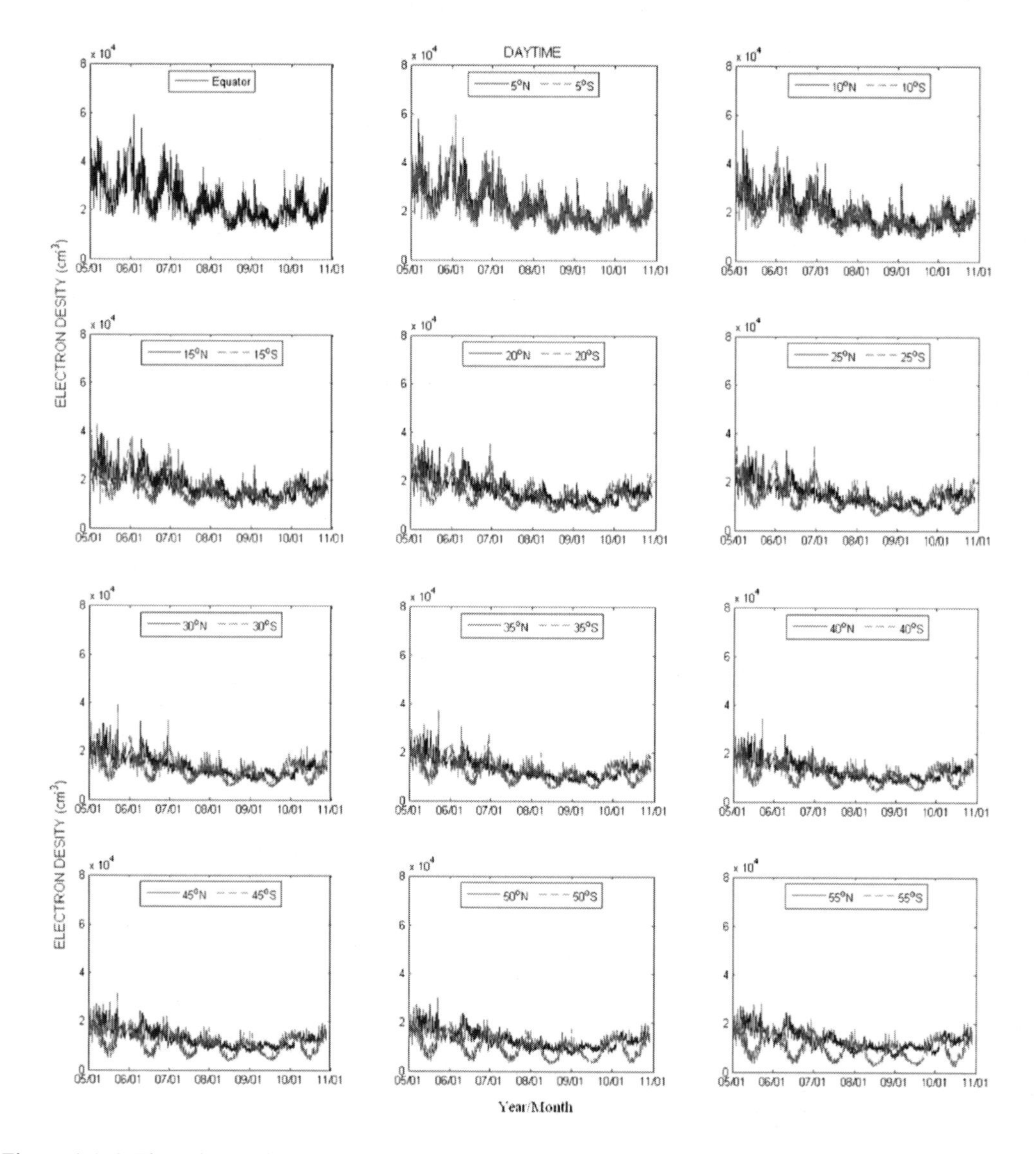

Figure 1.1.6. The solar cycle variations of electron density in local daytime during 2005–2010 (the latitudes were marked at upper middle in each image). According to Zhang et al. (2014). Permission from COSPAR.

In local nighttime, the annual variations of Ne at south and north hemispheres were symmetrical at same latitudes, but the annual variation amplitudes at different latitudes differed largely, showing obviously zonal features. As for Te, the phase shift in annual variations was not as apparent as Ne with the increase of latitudes at Southern and Northern hemisphere in local daytime. While in local nighttime the reversed annual variations of Te were shown at low latitudinal areas, not at high latitudes as those in Ne. The correlation study on Ne and Te illustrated that, in local daytime, Ne and Te showed strong negative correlation at equator and low latitudes, but during the solar minimum years the correlation between Ne and Te changed to be positive at 25–30° latitudes in March 2009. The correlation coefficient CC between Ne and Te also showed semi-annual periodical variations during 2005–2010. While in local nighttime, Ne and Te exhibited relatively weak positive correlation with CC being about 0.6 at low latitudes, however no correlation beyond latitudes of 25° was obtained. Obtained results are illustrated by Figures 1.1.7 – 1.1.11.

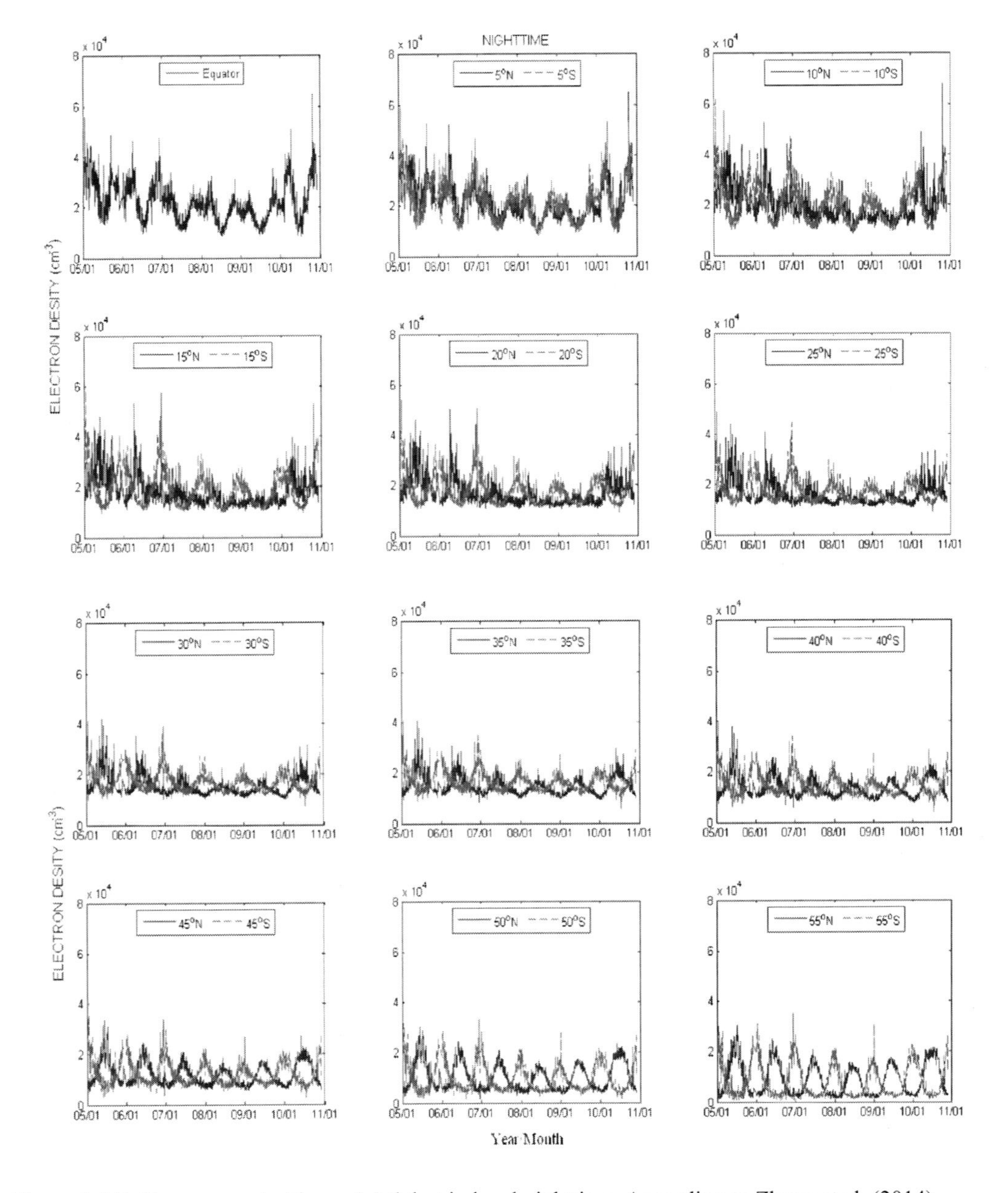

Figure 1.1.7. The same as in Figure 1.1.6, but in local nighttime. According to Zhang et al. (2014). Permission from COSPAR.

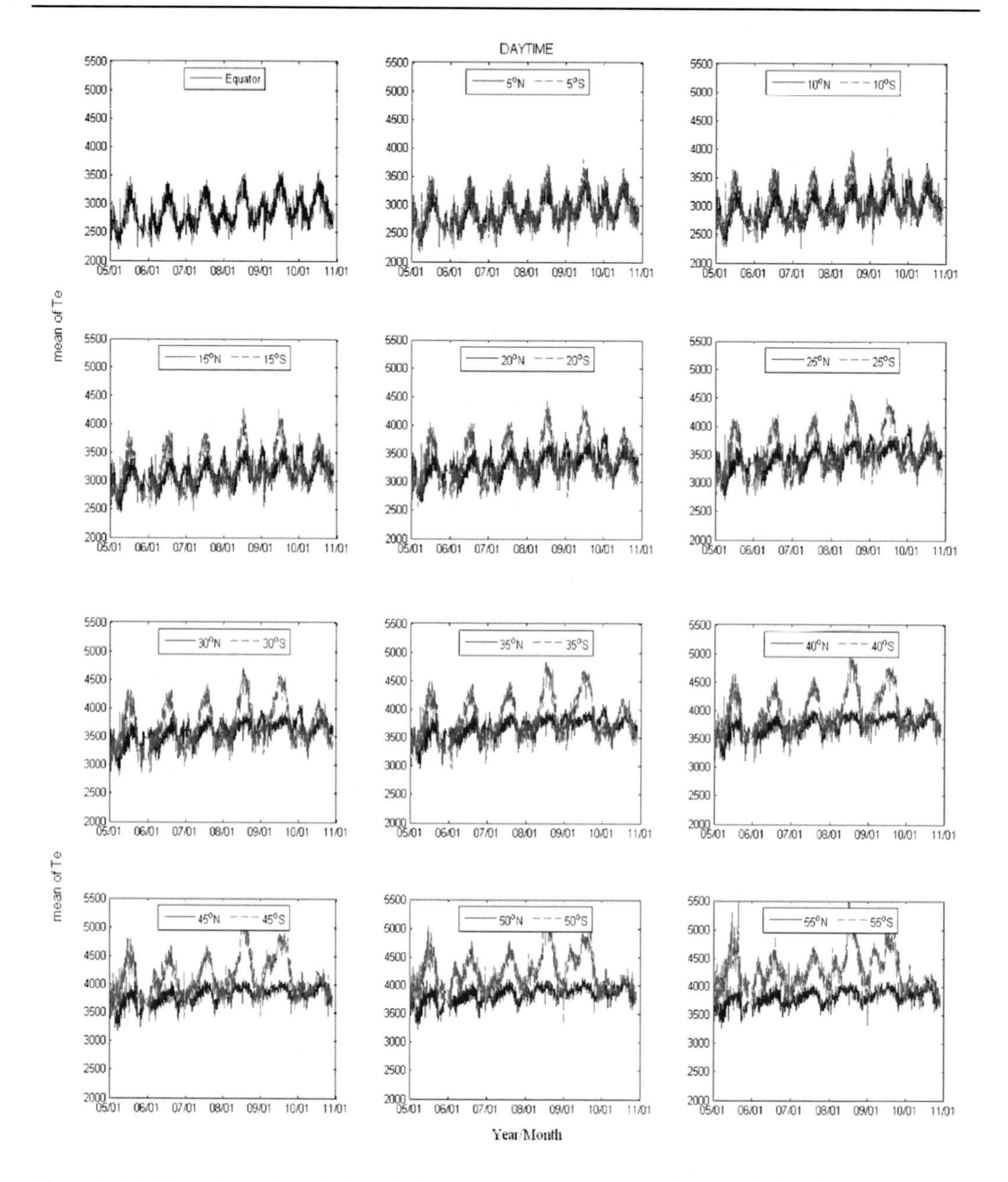

Figure 1.1.8. The solar cycle variation of electron temperature in local daytime during 2005–2010. According to Zhang et al. (2014). Permission from COSPAR.

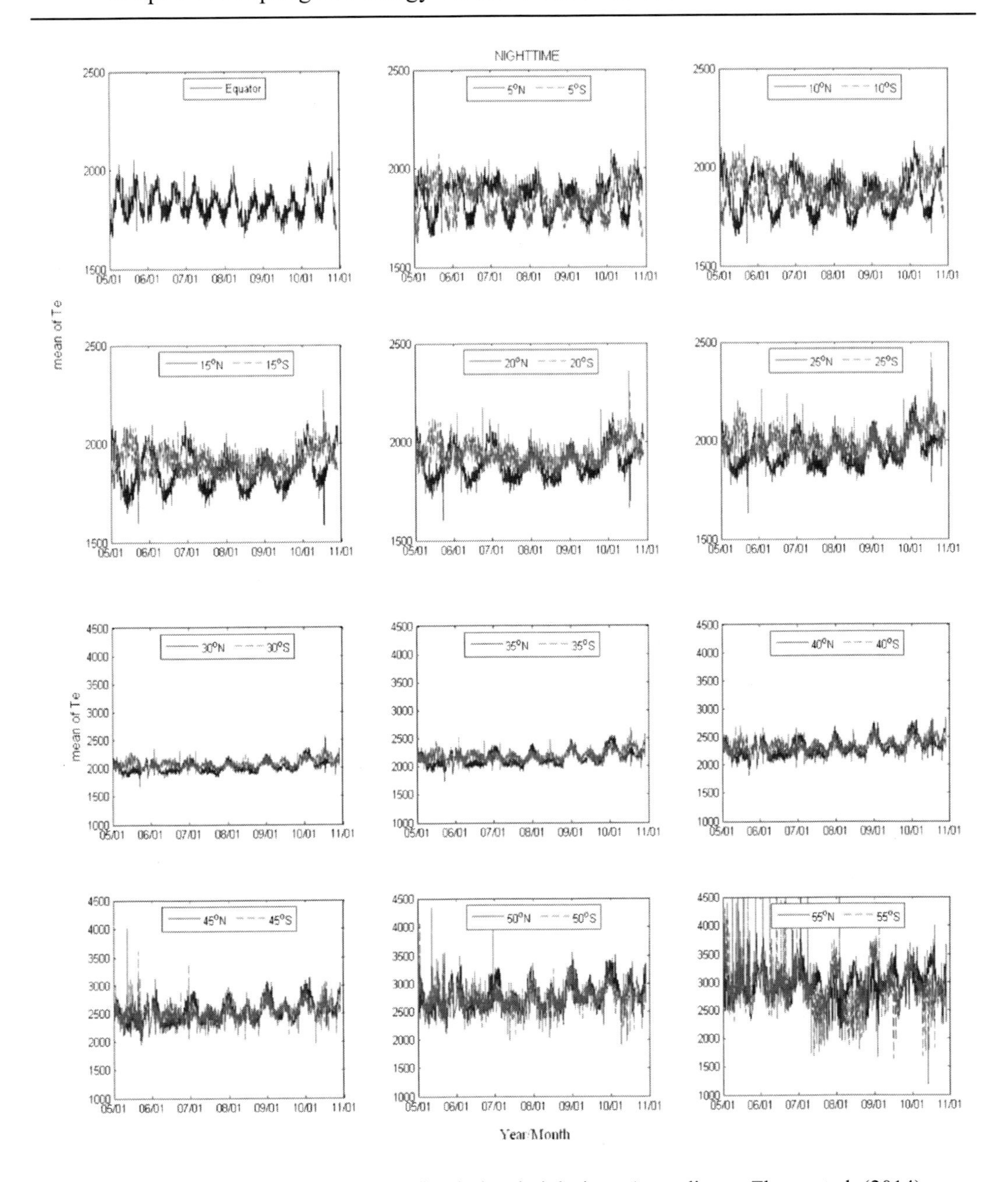

Figure 1.1.9. The same as in Figure 1.1.8, but in local nighttime. According to Zhang et al. (2014). Permission from COSPAR.

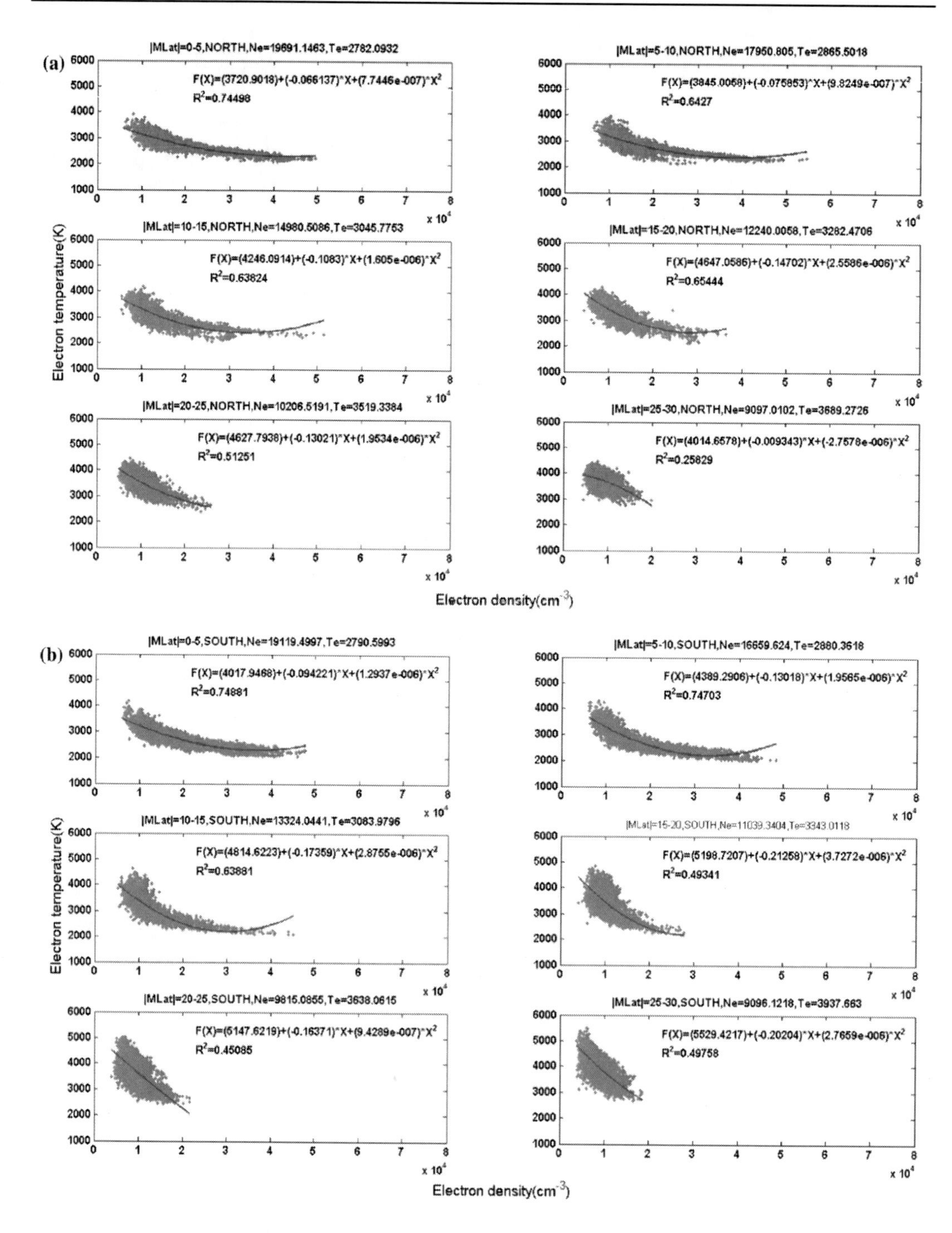

Figure 1.1.10. The correlation between Ne (X-axis) and Te (Y-axis) in local daytime in September 2008 at different latitudes (a: |0–5|°; b: |5–10|°; c: |10–15|°, d: |15–20|°; e: |20–25|°; f: |25–30|°). According to Zhang et al. (2014). Permission from COSPAR.

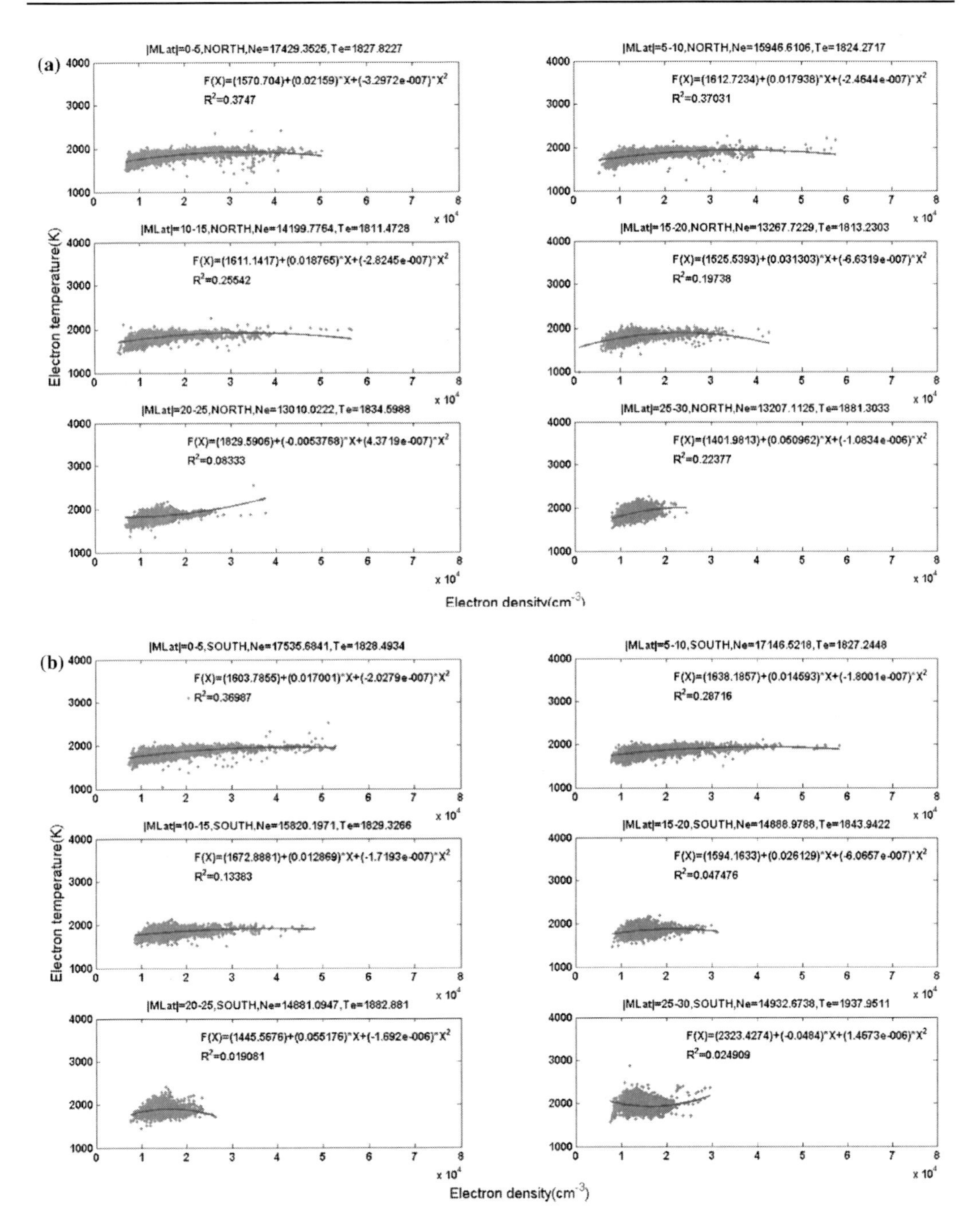

Figure 1.1.11. The same as in Figure 1.1.10, but in local nighttime. According to Zhang et al. (2014). Permission from COSPAR.

1.1.21. Global and Comprehensive Analysis of the Inner Magnetosphere as a Coupled System

As outlined Shprits and Spasojevic (2015), the Earth's inner magnetosphere is home to a number of meteorological, communication, and observational satellites. Understanding the dynamics of this region of space closest to the Earth advances our knowledge of the physics of plasmas in space and has practical applications since inner magnetospheric particle populations provide a hazardous environment for Earth-orbiting satellites and may be also hazardous to humans in space. Recent advances in global modeling of the inner magnetospheric processes along with observations from multiple vantage points clearly demonstrated that different plasma populations, electromagnetic fields, and plasma waves cannot be considered in isolation. Modeling and observations have also shown that the dynamics of the inner magnetosphere is closely coupled to both the solar wind and ionosphere. Understanding the needs of the space weather customers, synergy between different space weather-related missions, interactions between the bulk plasma population, energetic particles, relativistic particles, ultrarelativistic particle populations and fields, and magnetosphere-ionosphere coupling, solar wind-magnetosphere coupling were subjects of discussion during the recent Inner Magnetosphere Coupling (IMC) workshop held in March 2015 at University of California, Los Angeles.

1.2. Solar Wind–Magnetosphere–Ionosphere Coupling: Diurnal, Semi-Annual, and Solar Cycle Variations

1.2.1. The Matter and Short History of the Solar Wind–Magnetosphere–Ionosphere (SMI) Coupling Problem

Using the *Km* index, the F10.7 index, and solar wind data from 1965 to 1996, Nagatsuma (2006) has examined the variations for the efficiency of solar wind–magnetosphere–ionosphere (SMI) coupling. He note that since geomagnetic activity is a manifestation of magnetospheric dynamics, the research of this phenomenon has a long history. It is well known that geomagnetic activity shows the diurnal and semiannual variations. The cause of these variations consists of two effects. One is the periodical change of the solar wind parameters due to a variation of the geometrical condition between the solar wind and the Earth's magnetosphere. The other is the periodical change of the solar wind–magnetosphere–ionosphere (SMI) coupling efficiency. One of the major former effects is the Russell and McPherron (R-M) effect (Russell and McPherron, 1973). Since the R-M effect explained the enhancement of the north-south component of the IMF due to the geometrical relationship between the Earth's dipole axis and the Parker spiral plane, it has been believed that this is the major cause of the semiannual variation of geomagnetic activity. The axial effect, which is defined by the annual variations of the solar wind parameter caused by the annual change of heliographic latitude, also belongs to this category (Cortie, 1912). On the other hand, Cliver et al. (2000) pointed out the importance of the Equinoctial/McIntosh effect. This effect is based on a hypothesis that the maximum of geomagnetic activity appears when the angle between the Earth's dipole axis and the Sun-Earth line is 90º (McIntosh, 1959). Cliver et al. (2000) showed that the efficiency of the solar wind– magnetosphere coupling has a semiannual modulation. They suggested that the change of the coupling efficiency might cause the equinoctial effect. Several studies have pointed out the possible cause for the modulation of the SMI coupling related to the equinoctial effect. Murayama et al. (1980) studied the tilt angle dependence of the solar wind–magnetosphere coupling efficiency using the AL index and solar wind data. They discussed the possible change of the reconnection rate depending on the tilt angle if reconnection process occurs near the polar cusp. Crooker and Siscoe (1986) proposed that the change of the dipole tilt angle might control the magnetopause shape and reconnection geometry. Lyatsky et al. (2001) and Newell et al. (2002) proposed that the diurnal and semiannual variations in the total ionospheric conductivity in the nightside auroral oval cause the variations in geomagnetic activity. Their interpretation is that the geomagnetic activity maximize when no conducting path exists in the ionosphere to complete the circuit of the current system required by SMI coupling.

Nagatsuma (2004) have shown that the total Pedersen conductivity in the northern and southern polar caps controls the saturation of the cross-polar cap potential. The result of Nagatsuma (2004) suggested that the efficiency of SMI coupling depends on the total Pedersen conductivity. Since the total Pedersen conductivity shows the diurnal and semiannual variations, these variations can produce the diurnal and semiannual variations of geomagnetic activity. If the SMI coupling is controlled by the ionospheric conductivity, the coupling efficiency should have solar activity dependence. Nakai and Kamide (1999) found that the solar activity dependence of the coupling efficiency between solar wind parameters and the *AL* index. Using the long-term data sets of geomagnetic indices, the solar wind parameter, and solar radio flux, Nagatsuma (2006) examine the diurnal, semiannual, and solar activity-dependent variations of SMI coupling. Further, he show that these variations depend on the variations of the total Pedersen conductivity in the northern and southern polar caps.

1.2.2. The Method of Analysis and Basis of Using Data on SMI Coupling Variations

To examine the diurnal, semiannual, and solar activity dependent variations of SMI coupling, Nagatsuma (2006) perform a statistical analysis using the geomagnetic *Km* index from 1965 to 1996 (the *Km* index is a logarithmic scale of the *am* index which is a mid-latitude range index based on maximum excursions of the horizontal or declination components of the geomagnetic field over a 3-hour interval after subtracting the regular variation; the *am* index is based on a set of mid-latitude, sub-auroral stations optimally positioned in both latitude and longitude to yield an index free of local time effects, see Mayaud, M1980). Nagatsuma (2006) also used the OMNI 2 database for the solar wind data and the F10.7 index for the proxy of the solar EUV flux during the same period. Since the mid-latitude geomagnetic field variations are mainly caused by the magnetospheric convection, the *am* and *Km* indices are expected to be good measures of the magnetospheric convection. In the case of the auroral latitude index, such as the *AU* and the *AL* indices, the magnitude of the indices are also modified by the enhancement of the conductivity caused by the auroral particle precipitation.

1.2.3. Diurnal and Semiannual Variations of Total Pedersen Conductivity in Northern and Southern Polar Caps

Nagatsuma (2006) introduced a new empirical conductivity model to examine the solar activity dependence. Since the ionospheric conductivity is controlled mainly by the flux of the solar EUV, it is important to consider the solar activity variations of the EUV intensity for the present analysis using the long-term data. For this purpose, Nagatsuma (2006) modified the conductivity model proposed by Moen and Brekke (1993). Although their empirical model is the latest model based on the IS radar observations, they does not take into account the solar EUV effect at the ionospheric altitude when the solar zenith angle is more than 90 degrees. In the present study, we use the new empirical model based on Moen and Brekke (1993) improved by incorporating this effect. Nagatsuma (2004) have shown the importance of this effect. In the new model, the Pedersen conductivity (Σ_P) is described as

$$\Sigma_P = S_a^{0.5}(1.2\cos\chi + 0.1736), \qquad (1.2.1)$$

where S_a is the solar radio flux (F10.7) and χ is the solar zenith angle. The comparison between the models of Senior (1980), Moen and Brekke (1993) and Nagatsuma (2006) shoes that models Moen and Brekke (1993) and Nagatsuma (2006) give similar values when the solar zenith angle is less than 85° and the saturation of the cross polar cap potential (CPCP) is dependent on the total Pedersen conductivity calculated from the sum of the Pedersen conductivities in the polar caps of both hemispheres (Nagatsuma, 2004). The total Pedersen conductivity in Nagatsuma (2006) is calculated from the following equation:

$$\Sigma_{Ptotal} = \Sigma_{PN} + \Sigma_{PS} = S_a^{0.5}\{1.2(\cos\chi_N + \cos\chi_S) + 0.3472\}, \qquad (1.2.2)$$

where χ_N and χ_S are the solar zenith angles at the northern and southern corrected geomagnetic (CGM) poles, respectively. In this calculation, Nagatsuma (2006) assume that the ionospheric conductivity within the polar cap is represented by the solar zenith angle at the CGM pole. Hence the total Pedersen conductivity is a function of the solar radio flux and the total solar zenith angle defined by $\cos\chi_N + \cos\chi_S$. The diurnal and semiannual variations of the total Pedersen conductivity in the northern and southern polar caps was estimated with Nagatsuma's (2006) on the basis of empirical conductivity model. The conditions of Sa = 70 are used for calculating Σ_{Ptotal}. The total Pedersen conductivity tends to decrease in the equinoxes and tends to increase in the solstices. On the basis of the different assumption, Benkevitch et al. (2002) and Newell et al. (2002) showed a similar type of the diurnal and semiannual variations of the total Pedersen conductivity.

1.2.4. Nonlinear Development of the Cross Polar Cap Potential

According to Nagatsuma (2006), the nonlinear development of the CPCP can be estimated from the analytical formula of the Hill model introduced by Siscoe et al. (2002).

The concept of this formula is based on that the enhancement of the region 1 currents regulates the development of magnetospheric convection. The formulas are as follows:

$$\Phi_{PC} = 57.6 E_m P_{SW}^{1/3} / \left(P_{SW}^{1/2} + 0.0125 \xi \Sigma_P E_m \right), \ \xi = 4.45 - 1.08 \log \Sigma_P, \qquad (1.2.3)$$

where Φ_{PC} is the CPCP, $E_m = V_{sw} B_T \sin^2(\theta/2)$ is the merging electric field, introduced by Kan and Lee (1979), where V_{sw} is the solar wind speed, B_T is the projection of the IMF on the solar magnetospheric y-z plane, and θ is a clock angle, P_{SW} is the dynamic pressure, and ξ is a geometrical factor. In Eq. 1.2.3, the Earth's dipole moment is normalized as 1. These formulas were used for studying the saturation of the CPCP. Nagatsuma (2006) emphasizes that these also shows the variations of the SMI coupling efficiency. The Eq. 1.2.3 was modified in Nagatsuma (2006) as,

$$E_m = P_{SW}^{1/2} \Phi_{PC} / \left(57.6 P_{SW}^{1/3} - 0.0125 \xi \Sigma_P \Phi_{PC} \right). \qquad (1.2.4)$$

On the basis of Eq. 1.2.4, it is clear that the intensity of E_m that produces the same magnitude of the CPCP changes depending on the total Pedersen conductivity. This means that the efficiency of SMI coupling can change depending on the total Pedersen conductivity. To examine the efficiency of SMI coupling, Nagatsuma (2006) have analyzed the variations of the merging electric field that produces the same magnitude of geomagnetic activity as a function of the total solar zenith angle ($\cos\chi_N + \cos\chi_S$), which represents the diurnal and semiannual variation of the total Pedersen conductivity. This approach can eliminate the effect from the diurnal and semiannual variations of the solar wind parameter itself. The solar wind parameter of E_m is selected for the coupling parameter with K_m index. Since E_m is one of the best coupling parameter for magnetospheric convection (e.g., Burke et al., 1999), Nagatsuma (2006) compare the K_m index for every 3 hours with the average of 1-hour value of E_m during every 3-hour period. The value of E_m is estimated with the solar wind data from the OMNI-2 database. The two cases for the low ($75 < S_a < 125$) and high ($175 < S_a < 225$) solar activity are analyzed.

1.2.5. Connection with Equinoctial Effect

Nagatsuma (2006) has shown that the total Pedersen conductivity depends from the SMI coupling efficiency. As shown in Figure 1.2.2, the total Pedersen conductivity shows the diurnal and semiannual variations related to the equinoctial effect. This means that the variation of total Pedersen conductivity can explain the diurnal and semiannual variations of geomagnetic activity related to the equinoctial effect. The solar activity dependence of SMI coupling is another important finding of Nagatsuma's (2006) study. According to Nagatsuma (2006), the efficiency of SMI coupling is low during high solar activity and high during low solar activity. Using the AL index, Nakai and Kamide (1999) studied the solar cycle dependence of the relationship between the solar wind parameters and AL. They showed that the values of AL for fixed values of B_zV^2 are greater at the solar minimum than at the solar maximum. If to assume that the origin of geomagnetic activities represented by AL and Km are the same, their results are consistent with Nagatsuma's (2006) study. The solar cycle dependence shown by Nakai and Kamide (1999) is well explained by the total Pedersen conductivity dependence of SMI coupling shown by the Nagatsuma's (2006) study. Nagatsuma (2006) noted that several studies suggested that the efficiency of the dayside merging could change depending on the tilt angle of the Earth's dipole (Murayama et al., 1980; Crooker and Siscoe, 1986; Russell et al., 2003). These studies predicted the dependence on the IMF intensity. However, the development of the CPCP estimated from the PCN index did not show such dependence (Nagatsuma, 2002). This scenario cannot explain the variations of the SMI coupling efficiency associated with solar activity. Lyatsky et al. (2001) and Newell et al. (2002) suggested that geomagnetic disturbances are maximized when the nightside auroral zones of both hemispheres are in darkness, because no conducting path exists in the ionosphere to complete the circuit of the current system required by SMI coupling. They estimated the diurnal and semiannual variations of the total Pedersen conductivity in the night side sector. The trend of these variations was correlated with the diurnal and semiannual variations of the am index. Morioka et al. (2003) showed that the auroral kilometric radiation (AKR), which is the manifestation for the acceleration of auroral particles, disappears during the initial and main phase of geomagnetic storms, which means high geomagnetic activity. Their result implies that the relationship between the geomagnetic activity and the precipitation of auroral particles are not as simple as stated by Lyatsky et al. (2001) and Newell et al. (2002). According to Barth et al. (2004), maximum auroral particle precipitation is seen in the winter hemisphere, and the total precipitating particle flux reaches its maximum in solstice, not in equinox. Auroral activities possibly contribute to the enhancement of geomagnetic activity in equinox as a secondary effect.

1.2.6. Effect of Ionospheric Conductivity at Sub-Auroral Latitudes

Nagatsuma (2006) outlined the important result of Ebihara et al. (2004), showed that the ring current dynamics is controlled by the ionospheric conductivity at sub-auroral latitudes. They suggested that the diurnal, semiannual, and solar cycle variations of the Dst index are possibly produced by this effect. This effect might play some role for the variations of geomagnetic activity at mid-latitude, especially for the case of high level geomagnetic activity. The effect proposed by Ebihara et al. (2004) shows the opposite characteristics of the Siscoe-Hill model. Ebihara et al. (2004) showed that the increase of Pedersen conductivity increases the intensity of the ring current during storm time. This means that the magnetic field variations produced by the ring current increases with increasing Pedersen conductivity in sub-auroral latitude. The results of diurnal and semiannual variations of Dst show the same sense for those of Km, since the diurnal and semiannual variations of Pedersen conductivity at sub-auroral latitude are opposite sense for those at northern and southern polar caps. However, the solar cycle variation of Dst predicted from Ebihara et al. (2004) shows the opposite sense for those of Km.

1.2.7. Possible Interpretation of the Statistical Result

Nagatsuma's (2006) statistical result can be quantitatively interpreted using the Siscoe-Hill model even if will be used a simple model of Pedersen conductivity. One to one correspondence between *Km* and CPCP are reproduced from the tilt angle dependence of the relationship between *Km* and *Em*. This result can explain the semiannual variations of the *Km* index. Further; the value of CPCP estimated from Nagatsuma's (2006) interpretation is comparable with those from previous studies. These quantitative agreements strongly suggest the validity of this interpretation. Since the high latitude side of the region 1 current system is connected to the ionospheric Pedersen current flowing in the polar cap region, the Pedersen current in the polar cap controls the total intensity of the region 1 current system in the high latitude side. The total intensity of this current system regulates the SMI coupling process (Siscoe et al., 2004). This suggests that the polar cap region has an important role in magnetospheric dynamics.

1.2.8. Summary of Main Nagatsuma's Results

Nagatsuma (2006) came to following conclusions:

(i) Using the *Km* index, F10.7, and OMNI-2 solar wind data, it was confirmed that the efficiency of SMI coupling has diurnal and semiannual variations.

(ii) It was found that the coupling efficiency during the period of high solar activity tends to be lower than during low solar activity. This suggests that the efficiency of SMI coupling has a dependence on the solar activity.

(iii) These variations correspond to the variations of the total Pedersen conductivity in the northern and southern polar caps.

(iv) It was found that the variations of the coupling efficiency for fixed *Km* value as a function of the total Pedersen conductivity can be reproduced when the empirical conductivity model is applied in the Siscoe-Hill model (Siscoe et al., 2004).

(v) The described interpretation can explain the diurnal and semiannual variations of geomagnetic activities related to the equinoctial effect, which has not quantitatively been explained in other studies.

(vi) The solar activity dependence of geomagnetic activity is also explained by this idea.

(vii) Obtained result strongly suggests that the ionospheric conductivity plays a significant role in the SMI coupling process.

1.3. Longitudinal and Seasonal Variations in Plasmaspheric Electron Density: Implications for Electron Precipitation

1.3.1. The Matter and Short History of the Longitudinal and Seasonal Variations Problem

As noted Clilverd et al. (2007), the plasmasphere is a region of low-energy ('cold', i.e., T_e ~1 eV) plasma surrounding the Earth, and extending out to L ~2–6 depending on geomagnetic latitude, geomagnetic disturbance levels, and on local time. It is primarily made up of electrons and protons that have diffusively migrated from the underlying ionosphere. Overlapping the plasmasphere are regions of high-energy ('hot', i.e., T_e ~1 MeV) plasma known as the radiation belts. Low-frequency radio waves propagating within the plasmasphere can interact with the high-energy radiation belt particles, changing their energy spectra and causing them to precipitate into the Earth's upper atmosphere, driving chemical changes (e.g., Rozanov et al., 2005). Variability in the background conditions of the plasmasphere is one of the factors in determining the efficiency of wave-particle interactions (e.g., Horne et al., 2003), thus influencing the resultant particle precipitation into the atmosphere. Clilverd et al. (2007) use CRRES satellite measurements of 'cold' plasmaspheric equatorial electron density to investigate the longitudinal and annual variations in density in the range L = 2.5–5.0, and assess the effect on the rate of 'hot' electron precipitation from the overlapping outer radiation belt. The annual variation in equatorial plasmaspheric electron density (N_{eq}) has been observed previously. The first observations were made using natural whistler signals, typically at either American or European longitudes (e.g., Helliwell, M1965; Park et al., 1978; Tarcsai et al., 1988). In these cases N_{eq} showed a maximum in December and a minimum in June, with December larger by a factor of between 1.5 and 3.0 at L = 1.5–2.5, depending on longitude. The annual variation in Neq has been modeled with a view to reproducing the observations and understanding the underlying physical processes responsible. Some models reproduced the December/June annual variation at American longitudes, and then made predictions regarding the effect at other longitudes. Early work by Rasmussen and Shunk (1990) showed a N_{eq} maximum in June rather than December as is actually observed, probably because of the centered dipole model used. Modeling work undertaken by Rippeth et al. (1991), which included a tilted offset dipole in the model, was better able to reproduce the observations at L = 2.5 at American longitudes. Guiter et al. (1995) modeled plasmaspheric densities at L = 2 and found that N_{eq} was 1.5 times higher in December than in June for 300°E longitude. At 120°E longitude the L = 2 N_{eq} was predicted to be higher in June than December by a factor of 1.2. The underlying mechanism driving the annual variation was considered to be variations in ionospheric O^+.

As outlined Clilverd et al. (2007), further modeling using the field line interhemispheric model (FLIP) indicated that the annual variation at American and Australian longitude sectors were likely to be 6 months out of phase (Richards et al., 2000). This work concluded that plasmaspheric thermal structure, not ionospheric density, should play a key role in producing the annual variation at solar minimum. A new approach using dynamical diffusive equilibrium, called the global plasmasphere ionosphere density model (GPID), was able to reproduce the observed seasonal variations in Neq at L = 2.5 during solar maximum, but not at solar minimum (Webb and Essex, 2001). To maintain charge neutrality an annual variation in ion concentration would be anticipated. Berube et al. (2003) used data from a pair of magnetometers at L = 1.74 in the MEASURE array (American longitudes) to determine the plasmapsheric equatorial mass density. They observed an annual variation in mass density with December densities 2–3 times higher than in June. This suggests that the mass densities vary in a similar way to the electron densities. Although at L < 2, the annual variation in field line resonance frequencies is due to the influence of O^+ in the underlying ionosphere changing the Alfvèn speed profile along those flux tubes (Waters et al., 1994).

1.3.2. Determination of Electron Densities in the Plasmasphere

In Clilverd et al. (2007) electron number densities are derived from wave data provided by the Plasma Wave Experiment on board the Combined Release and Radiation Effects Satellite (CRRES). This satellite, which was launched on 25 July 1990, operated in a highly elliptical geosynchronous transfer orbit with a perigee of 305 km, an apogee of 35,768 km and an inclination of 18°. The orbital period was approximately 10 h, and the initial apogee was at a magnetic local time (MLT) of 08:00 MLT. The magnetic local time of apogee decreased at a rate of approximately 1.3 h per month until the satellite failed on 11 October 1991, when its apogee was at about 14:00 MLT. The satellite swept through the plasmasphere on average approximately 5 times per day for almost 15 months. The Plasma Wave Experiment provided measurements of electric fields from 5.6 Hz to 400 kHz, using a 100 m tip-to-tip long wire antenna, with a dynamic range covering a factor of at least 10^5 in amplitude (Anderson et al., 1992).

1.3.3. Determination of Ion Densities in the Plasmasphere

The ion mass densities presented in Clilverd et al. (2007) were calculated using field-line resonant frequencies (FLRs) measured from pairs of ground-based magnetometers, following the analytical expressions described by Taylor and Walker (1984) and Walker et al. (1992). These assume decoupled toroidal mode oscillations and yield essentially identical results to the models described by Orr and Matthew (1971). Techniques for the detection of FLRs were summarized by Menk et al. (1999) and Menk et al. (2000). When examining data from latitudinally separated magnetometers, the resonant frequency is identified by the peak in H-component cross-power and cross-phase, and a unity crossing in H-component power ratio, approximately midway between the stations. Where only one station is available the resonance is indicated by a peak in the power ratio H/D and a rapid change in polarization, that is, in the phase between the H and D components.

1.3.4. Longitudinal and Seasonal Variations in Plasmaspheric Densities

The density variation with longitude from CRRES for December and June for L = 2.5–5.0 was obtained by Clilverd et al. (2007). The densities at these longitudes in December were so large (>2000 el. cm^{-3}) that the upper hybrid frequency could not be determined at all times, and data from a higher range of L-shells (L = 2.7–3.3) were used and linearly extrapolated to L = 2.5. However, the same extrapolation technique used on the L = 2.5 June data, and some of the other panels (L = 3.0–4.0) reproduced the December data to within 5%. This extrapolation was make for the L = 2.5 December data primarily to allow comparison with previous work, it does not materially affect any of the below conclusions.

1.3.5. Implications for Electron Precipitation

As noted Clilverd et al. (2007), the background electron density in the plasmasphere plays a key role in determining the resonant energy of wave-particle interactions. Clilverd et al. (2007) investigate the influence that the annual variation in electron density will have on pitch-angle scattering of energetic electrons into the loss cone, out of the radiation belts, and subsequent precipitation into the atmosphere. Meredith et al. (2006) calculated loss timescales for pitch angle scattering by plasmaspheric hiss using the PADIE code (Glauert and Horne, 2005) with wave properties based on CRRES observations. The determination of the diffusion coefficients requires knowledge of the distribution of the wave power spectral density with frequency and wave normal angle, together with the ratio f_{pe}/f_{ce}, wave mode, and the number of resonances. The ratio f_{pe}/f_{ce} is dependent on the background electron density, and the background magnetic field.

1.3.6. Summary of Main Results on the Longitudinal and Seasonal Variations in Electron Density and Precipitation

Clilverd et al. (2007) summarize main obtained results as following:

(i) There are used CRRES measurements of plasmaspheric equatorial electron density at solar maximum to investigate the longitudinal and annual variation in density in the range L = 2.5–5.0.

(ii) It is found that the largest annual variation occurs at American longitudes (–60°E), and that no annual variation occurs at Asian longitudes (+100°E). These findings are in agreement with Clilverd et al. (1991). The underlying cause is due to the influence of a tilted-offset dipole geomagnetic field. At American longitudes there is the largest discrepancy between geomagnetic latitude and geographic latitude. This leads to substantial annual variations in ionospheric plasma density, which map up into the plasmasphere as a consequence of diffusive equilibrium.

(iii) Plasmaspheric electron density is larger in December than in June in the region covering –180°E to +20°E. Elsewhere the ratio of December to June is very close to 1.0. The annual variation also differs with L-shell. At American longitudes (–60°E), and possibly at New Zealand longitudes, the maximum December/June ratio is at L = 2.5–3.5, with a value of 2.7 at American longitudes at solar maximum. At L = 4.5 and above the annual variation disappears, possibly because the plasmasphere is not in diffusive equilibrium with the ionosphere at these high L-shells, or the inclusion of non-saturated electron density values from CRRES observations. The lowest electron density values for a given L-shell occur at American longitudes. This is particularly clear for the lower L-shells, although apparent as far out as L = 4.5. These values occur in June. Clearly, the plasmasphere is strongly controlled by the configuration of the Earth's magnetic field and the annual variations in the F2 regions that are in diffusive equilibrium with it.

(iv) Ion densities also show significant annual variations. There are similar longitudinal characteristics in the case of IMAGE EUV He^+ measurements taken during June and December 2001. However, there are as yet unexplained differences in atomic mass density measurements calculated using the magnetometer cross-phase technique, where European values are significantly higher than those at American longitudes and require a large correction factor for the ion composition.

(v) Calculations of the effect of changing plasmaspheric density on wave-particle interactions with plasmaspheric hiss indicate that the depletion of the plasmasphere at American longitudes in June results in a harder energy spectrum of electrons being precipitated into the atmosphere at those longitudes than anywhere else. Conversely, the softest energy spectrum occurs at the same longitudes in December. Little variation in precipitation energy spectrum is likely at Asian longitudes due to the absence of any significant annual variation in plasmaspheric density.

1.3.7. Investigating energetic electron precipitation through combining ground-based and balloon observations

A detailed comparison is undertaken by Clilverd et al. (2017) of the energetic electron spectra and fluxes of two precipitation events that were observed in 18/19 January 2013. A novel but powerful technique of combining simultaneous ground-based subionospheric radio wave data and riometer absorption measurements with X-ray fluxes from a Balloon Array for Relativistic Radiation-belt Electron Losses (BARREL) balloon is used for the first time as an example of the analysis procedure. The two precipitation events are observed by all three instruments, and the relative timing is used to provide information/insight into the spatial extent and evolution of the precipitation regions. The two regions were found to be moving westward with drift periods of 5–11 h and with longitudinal dimensions of ~20° and ~70° (1.5–3.5 h of magnetic local time). The electron precipitation spectra during the events can be best represented by a peaked energy spectrum, with the peak in flux occurring at ~1–1.2 MeV. This suggests that the radiation belt loss mechanism occurring is an energy-selective process, rather than one that precipitates the ambient trapped population. As outlined Clilverd et al. (2017), the motion, size, and energy spectra of the patches are consistent with electromagnetic ion cyclotron-induced electron precipitation driven by injected 10–100 keV protons. Radio wave modeling calculations applying the balloon-based fluxes were used for the first time and successfully reproduced the ground-based subionospheric radio wave and riometer observations, thus finding strong agreement between the observations and the BARREL measurements.

1.3.8. North-south asymmetries in cold plasma density in the magnetotail lobes: Cluster observations

Haaland et al. (2017) present observations of cold (0–70 eV) plasma density in the magnetotail lobes. The observations and results are based on 16 years of Cluster observation of spacecraft potential measurements converted into local plasma densities. Measurements from all four Cluster spacecraft have been used, and the survey indicates a persistent asymmetry in lobe density, with consistently higher cold plasma densities in the northern lobe. External influences, such as daily and seasonal variations in the Earth's tilt angle, can introduce temporary north-south asymmetries through asymmetric ionization of the two hemispheres. Likewise, external drivers, such as the orientation of the interplanetary magnetic field can set up additional spatial asymmetries in outflow and lobe filling. As note Haaland et al. (2017), the persistent asymmetry reported in this paper is also influenced by these external factors but is mainly caused by differences in magnetic field configuration in the Northern and Southern Hemisphere ionospheres.

1.4. Magnetospheric Convection during Intermediate Driving: Sawtooth Events and Steady Convection Intervals as Seen in Lyon-Fedder-Mobarry Global MHD Simulations

1.4.1. The Basic Data for Events 18 April 2002 and 3–4 February 1998

Goodrich et al. (2007) have used global MHD simulations to investigate the magnetospheric response to steady solar wind for two events. The event of 18 April 2002 is characterized by periodic particle injections and magnetic field dipolarizations, or sawtooth activity, at geosynchronous orbit, while 3–4 February 1998 is a period of steady magnetospheric convection (SMC). In their simulations Goodrich et al. (2007) find for both events that a general system of convection develops, characterized in the magnetotail by large-scale sunward flows driven by reconnection in region of 30–45 R_E downstream. These flows in general divert around the inner magnetosphere to the dawn and dusk flanks and then converge toward the dayside magnetopause.

1.4.2. MHD Simulations for SMC Event February 3, 1998

In the SMC event the convection system is formed by quasi-steady reconnection in the midtail, which drives steady earthward flows. These flows divert to the flanks, leaving the inner magnetosphere undisturbed. Goodrich et al. (2007) find the difference in the magnitude of IMF B_Z is unlikely to account for difference in activity in the two events, as a simulation of the SMC event with increased $|B_Z|$ produced qualitatively the same steady convection. Solar wind density variations are shown to control the average mass transport but have no correlation with the flow channels responsible for the inner magnetospheric activity in the simulations.

1.4.3. MHD Simulations for Sawtooth Event April 18, 2002

In the sawtooth event of 18 April 2002 Goodrich et al. (2007) find that reconnection is intermittent and patchy, resulting in flow bursts which on average produce the general convection pattern indicated, but a fraction of them penetrate into the inner magnetosphere and are associated with observed plasma injections and field dipolarizations.

1.4.4. Summary of Main Results on the Magnetospheric Convection during Intermediate Driving

Goodrich et al. (2007) have explored the response of the magnetosphere during steady solar wind conditions with moderately strong southward B_Z and come to the following conclusions:

(i) The steady driving leads to a large-scale convection pattern, driven by reconnection in the magnetotail, which convects magnetic flux from the tail to balance reconnection at the dayside magnetopause.

(ii) This reconnection can be intermittent and patchy, which produces strong but intermittent flows that periodically reach the inner magnetosphere.

(iii) These flows cause plasma injections and field dipolarizations such as observed in geosynchronous orbit in sawtooth events. However, most of the flows are diverted to the flanks creating the large-scale convection pattern.

(iv) The tail reconnection can also be quasi-steady and stretch across the tail, leading to a coherent quasi-steady convection system with little disturbance in the inner magnetosphere, as observed in steady magnetospheric convection events.

(v) The magnitude of the solar wind driving does not appear to determine alone whether the magnetotail can attain steady convection. It is a factor, but not the only factor, in determining the radial distance to the reconnection region; a more steady plasma sheet seems to allow reconnection closer to the Earth than a more dynamic one.

(vi) The solar wind density variations appear to be related to the average mass transport (through control of average density) in the plasma sheet, but these variations have no one-to-one relationship with the inner magnetosphere substorm-like dipolarizations that are created by flow channels penetrating to the inner magnetosphere.

1.5. Solar Wind-Magnetosphere Coupling and the Ionospheric and Reconnection Potentials of the Earth: Results from Global MHD Simulations

1.5.1. The Matter and Short History of the Problem on the Ionospheric and Reconnection Potentials

According to Hu et al. (2007), the reconnection potential along the reconnection line provides a quantitative description of magnetic reconnection that dominates the solar wind-magnetosphere coupling; to properly determine the reconnection potential is a first and crucial step in deeply understanding magnetospheric processes. As noted Hu et al. (2007), magnetic reconnection has been considered in the literature to be the main means for the dynamic coupling between the solar wind and the Earth's magnetosphere, and the Earth's ionosphere plays an important role in the coupling (Raeder, 2003). The reconnection is quantitatively described by the tangential electric field at the reconnection line; the associated potential along this line, Φ_R, is referred to as the reconnection potential. The electric field describes the reconnection rate of unit length of reconnection line, or the local strength of magnetic reconnection, whereas the total reconnection potential drop, or in brief, the reconnection voltage, represents the reconnection rate. In the frame of ideal MHD, Φ_R is impressed via equipotential magnetic field lines onto the ionosphere, so the potential in the ionosphere, Φ, has a one-to-one correspondence to Φ_R. Consequently, the so-called transpolar potential in the polar cap, defined as the difference between the positive and negative potential peaks, was naturally taken by some authors as a measure for the reconnection rate (e.g., Fedder et al., 1995). Nevertheless, because of numerical diffusion inherent in any algorithm used in global MHD simulations of the solar wind-magnetosphere-ionosphere system, such a one-to-one correspondence is somewhat distorted, so the transpolar potential differs from, generally smaller than, the reconnection rate. Incidentally, numerical diffusion has been often invoked in the literature in order to reflect real magnetic reconnections, which are bound to occur in the magnetopause and magnetotail. As pointed out by Fedder et al. (1995, 2002), numerical diffusion tends to make numerical reconnection important only where oppositely directed components of the magnetic field are being forced together.

As outlined Hu et al. (2007), the global magnetic merging process is virtually independent of the grid cell size and of the dissipation model and only depends on the solar wind parameters and the ionospheric load. Therefore, the role of numerical diffusion is rather positive in global MHD simulations of the solar wind-magnetosphere-ionosphere system, and the parallel electric field created by numerical diffusion is thus physically meaningful. Nevertheless, a good algorithm should have numerical diffusion that is confined in the reconnection region and is negligible otherwise. Efforts were made in directly calculating the reconnection voltage and the reconnection potential. In discussions of the transpolar potential saturation, (2002a,b) neglected the influence of the ionospheric conductance and presented an empirical formula for the reconnection voltage, which they called 'available magnetospheric convection potential' (denoted by Φ_m in their paper). As pointed out by Merkin et al. (2003, 2005), however, the ionospheric conductance has an important effect on the reconnection voltage. For the due southward IMF case, they traced two magnetic field lines, which pass through the ends of the reconnection line from the ionosphere to the solar wind, and integrated parallel electric field along them. The potential difference between the two field lines in the solar wind was then referred to as the reconnection voltage (named 'reconnection potential' in their paper). This estimate forms an upper bound for the reconnection voltage as the authors indicated. Then it was found that both the transpolar potential and the reconnection voltage vary with the ionospheric conductance and tend to saturate for large solar wind electric field. Siscoe et al. (2001) calculated the reconnection potential along the reconnection line instead of the mere reconnection voltage. The method was still an integration of parallel electric field along magnetic field lines that touch the reconnection line, but the contribution of $\mathbf{v} \times \mathbf{B}$ was incorporated. Analytically, such a contribution should be zero, but Siscoe et al. (2001) argued that a staggered mesh used in the algorithm that centers different MHD variables at different spatial locations leads to a nonzero contribution of the $\mathbf{v} \times \mathbf{B}$ term to the parallel electric field. Physically, it represents the turbulent dissipation to magnetic reconnection as they claimed. Using this method, they obtained the reconnection potential for the due duskward IMF case and found a large parallel potential drop along the field lines that connect the ionosphere and the null points. In addition, based on the detailed distribution of the reconnection potential along the reconnection line, they concluded that magnetopause reconnection becomes more antiparallel-like for northward IMF and more component-like for southward IMF.

As outlined Hu et al. (2007), to acquire a quantitative knowledge on the magnetospheric dynamics from global MHD simulations, the first and crucial step is to calculate the reconnection potential. As mentioned above, efforts were made in finding reasonable ways to implement this step although it suffers uncertainties associated with numerical diffusion. The paper of Hu et al. (2007) is a continuation of these efforts. To avoid complexity, they take the simplest case with due southward IMF, in which the reconnection line is simply a zero contour of B_z in the equatorial plane.

1.5.2. The Numerical PPMLR Method for Global MHD Simulations

A so-called piecewise parabolic method with a Lagrangian remap (PPMLR)-MHD algorithm was developed by Hu et al. (2005) specially for global simulations of solar wind-magnetosphere-ionosphere system. It is an extension of the Lagrangian version of the piecewise parabolic method developed by Colella and Woodward (1984) to MHD. In the PPMLR-MHD algorithm, all MHD-dependent variables are defined at the zone centers as volume averages, and their spatial distributions are obtained by interpolation which is piecewise continuous, with a parabolic profile in each zone. The algorithm consists of three steps. First, a characteristic method similar to that proposed by Dai and Woodward (1995) is used to calculate the local values of the dependent variables at the zone edges and at the half-time point in-between t_n and t_{n+1} ($t_{n+1} - t_n = \Delta t$ is the time step length). The results are then used to calculate the effective fluxes. Next, using the fluxes obtained and starting from the difference approximation of the Lagrangian conservation laws, it was update all dependent variables to t_{n+1} in the Lagrangian coordinates. Finally, the results are remapped back onto the fixed Eulerian grid through solving the corresponding advection equations. The PPMLR-MHD algorithm has a formal accuracy of the third order in space and the second order in time, and a low numerical dissipation, and it can capture a shock within two numerical zones without any appreciable overshot and undershot.

1.5.3. Three Methods to Calculate the Reconnection Potential

Hu et al. (2007) proposes three methods to calculate the reconnection potential based on data obtained by global MHD simulations of the Earth's magnetosphere-ionosphere system. The simulations are limited to the due southward IMF case for simplicity. The three methods are all based on the line integration of electric field and differ in the choice of integration path: radial rays in the equatorial plane for method 1, last closed magnetic field lines for method 2, and IMF lines nearby the reconnection line for method 3.

1.5.4. Comparison of Results from Three Methods

Hu et al. (2007) compare the reconnection potentials Φ_{R1}, Φ_{R2}, and Φ_{R3} as functions of λ (longitude along the reconnection line, $\lambda = 0$ stands for the subsolar point), for six cases with different solar wind speeds and Pedersen conductances calculated by the three methods described in Section 1.5.3. Also was calculated $\Phi(\lambda)$ – the ionospheric potential at the foot point of the closed field line originating from a point on the reconnection line at λ, and Φ_{R1N} and Φ_{R3N}, the contributions of numerical diffusion to Φ_{R1} and Φ_{R3}, as functions of λ. Note that the profiles of Φ_{R3} terminate at a certain longitude, beyond which the IMF lines have not reached the undisturbed solar wind yet in the numerical box.

1.5.5. Summary of Main Results of Global MHD Simulations of the MI System

Hu et al. (2007) summarize obtained results and came to following conclusions: (i) It was proposed three methods to calculate the reconnection potential along the reconnection line based on data obtained by global MHD simulations of the Earth's magnetosphere-ionosphere system. (ii) It was briefly discusses the difference between the transpolar potential and the reconnection voltage caused by the parallel electric field. (iii) The simulations are made with the use of the PPMLR-MHD algorithm developed by Hu et al. (2005), limited to the due southward IMF case. (iv) The three methods are all based on the line integration of electric field and differ in the choice of the reference potential and integration path. (v) The effect of numerical diffusion of the algorithm is included approximately by an equivalent uniform resistivity. (vi) Method 1 takes the intersection between the inner boundary ($r = 3R_E$) and the equatorial plane as the zero point of potential and radial rays that connect the inner boundary and the reconnection line as the integration path. (vii) Method 2 takes the ionospheric potential as the reference and the last closed field lines as the integration path. (viii) Method 3 takes the potential in the undisturbed solar wind as the reference and the IMF lines that pass by the reconnection line as the integration path, and it works only for points connected by IMF lines to the undisturbed solar wind. (ix) For a properly selected numerical resistivity ($\eta_N = 0.003$), the three methods give answers of the reconnection potential, which are reasonably close to each other. (x) On the sunward side, the contribution of numerical diffusion to electric field is essentially negligible in methods 1 and 3, so one needs only to consider the contribution of the convectional electric field. This makes the answers of the reconnection potential on the sunward side more reliable than the tailward side. (xi) The fact that method 2 also gives reasonable answers means that the parallel electric field along the last closed field lines causes the difference between the ionospheric and the reconnection potentials and between the transpolar potential and the reconnection voltage as well. (xii) The ratio between the transpolar potential and the reconnection voltage decreases as the ionospheric conductance increases. (xiii) For steady cases, the curl of E should be zero, so in principle, one can define a potential as a function of position for a steady solar wind-magnetosphere system based on line integration of E. (xiv) Nevertheless, if data come from global MHD simulations, there exists a key problem concerning the uncertainty associated with numerical diffusion of the algorithm. (xv) Fortunately, for due southward IMF cases there may use a single uniform resistivity to represent the effect of numerical diffusion so that the line integral of E along certain closed paths nearby the reconnection line is approximately zero. (xvi) Consequently, there may choose any part of these paths with one end nearby the reconnection line and may integrate the electric field along it to get the reconnection potential. (xvii) In this sense, the methods proposed in Hu et al. (2007) are probably useful for general cases with arbitrary IMF orientations provided that a similar numerical resistivity exists and makes the line integral of E approximately vanish along closed paths associated with last closed field lines, nearby IMF lines, and straight lines connecting the reconnection line with the equator of the inner boundary.

1.6. Solar Wind Electric Field Driving of Magnetospheric Activity: Is it Velocity or Magnetic Field?

1.6.1. The Matter and Short History of the Problem: What Is the Main Cause of Magnetospheric Activity Driving?

Pulkkinen et al. (2007b) tried to solve this problem, formulated in the title of their paper and this Section. For this the Lyon-Fedder-Mobarry global MHD simulation code is used to examine a period of steady magnetospheric convection driven by a moderately southward IMF and a steady and quite low solar wind speed. Two other runs were performed to test the effects of increasing solar wind driving: one with 50% increase in the IMF magnitude and one with 50% increase in the solar wind speed. It was found that larger IMF magnitude leads to nightside reconnection closer to the Earth but the steady state of the magnetotail is not changed. On the other hand, higher solar wind speed enhances Earth-ward mass and Poynting flux transport as well as their variability. The increase in the solar wind E_Y is the same for both runs, but the E_Y in the magnetosheath is larger in the run in which the speed is enhanced, which is associated with the higher magnetotail activity in that simulation.

Pulkkinen et al. (2007b) note that the solar wind and IMF impinging on the Earth's magnetosphere can power a variety of dynamic events during which the magnetosphere processes a portion of the solar wind energy either episodically or in a quasi-continuous fashion. The most typical response during moderate levels of driving solar wind electric field ($\mathbf{E} = -\mathbf{V} \times \mathbf{B}$) is a magnetospheric substorm (Baker et al., 1996), while higher values of the electric field drive full-fledged magnetic storms (Gonzalez et al., 1994). Recently Goodrich et al. (2007) studied two examples of intermediate levels of the driving solar wind electric field using the Lyon-Fedder-Mobarry (LFM) global MHD simulation code (Lyon et al., 2004). The first event was a sawtooth event during which strong, quasiperiodic substorm-like activity (Reeves et al., 2004) was observed. This event was compared with a steady magnetospheric convection (SMC) event, which characteristically showed enhanced activity but no substorm-associated tail reconfigurations (Sergeev et al., 1996a). In the simulation, both events were powered by a large-scale nightside reconnection region in the mid-tail. Goodrich et al. (2007) argued that the higher magnetospheric activity during the sawtooth event was caused by the faster solar wind speed rather than the larger solar wind electric field. This result was further justified by a test run where the SMC event was rerun with a 50% higher IMF magnitude.

Pulkkinen et al. (2007b) revisit the issue of the factors controlling the dynamic state of the magnetosphere through further analysis of the February 3–4, 2004 SMC event. They examine the results of three simulations: one (SMC-run) driven by the observed solar wind and IMF parameters, and two in which these parameters were altered to increase the solar wind E by 50%, either by increasing the IMF B_Z (B-run) or the solar wind speed V_{SW} (V-run).

The V-run, driven with a 50% higher solar wind speed, was run in Pulkkinen et al. (2007b); the SMC and B-run simulations were originally described by Goodrich et al. (2007). Pulkkinen et al. (2007b) find that, despite having equal enhanced driving electric fields, the B-run and V-run simulations exhibit quite different large-scale dynamic states of the magnetosphere.

1.6.2. Simulation Runs

As outlined Pulkkinen et al. (2007b), the LFM global MHD code solves the ideal MHD equations in a box that extends from 30 R_E in the sunward direction to –300 R_E tailward of the Earth, and ±100 R_E in the perpendicular dimensions. The highest spatial resolution in the code is obtained in the inner magnetosphere, plasma sheet, and boundaries where the gradients can be expected to be largest. The upstream boundary uses measured or artificial solar wind and IMF values as boundary conditions, while supersonic outflow conditions are applied at the other boundaries. All results in Pulkkinen et al. (2007b) are given in GSM coordinates. It was drive three simulation runs. The SMC-run uses the measured solar wind and IMF during the February 3–4, 1998 SMC-event as input. The B-run uses otherwise the observed parameters, but the IMF the B_Z is increased by 50%. For the V-run the solar wind speed was increased by 50%. Both of these changes to the original conditions lead to a 50% increase of the driving solar wind E_Y, while only the V-run shows a change in dynamic pressure. The simulation setup for each run is identical to that described by Goodrich et al. (2007).

According to Pulkkinen et al. (2007b), comparison of the three runs shows that the B-run produces the most stable magnetotail: the average Earthward mass transport (mnV_X) and Poynting flux ($S_X = (\mathbf{E} \times \mathbf{B})_X/\mu_0$) as well as their variances are smallest in that run. The V-run shows that increasing solar wind speed increases the average transport as well as its temporal variability and the spatial variances. This indicates that the tail is more active and spatially and temporally structured during that run than during either of the two other runs. The blue curves in the top panels show the lobe magnetic field defined as the maximum of B_X in the northern lobe. The lobe field is largest in the V-run indicating that the cross-tail current is strongest as well. Limiting the averaging area to a fixed region in the tail instead of the entire closed plasma sheet region would not change the results.

As outlined Pulkkinen et al. (2007b), all runs demonstrate that the perpendicular field is small and unchanging in the tail lobes. The B-run shows only very low variance of B_{YZ} in the plasma sheet, which demonstrates that the tail field is both uniform in space and unchanging in time. The V-run shows very strongly structured signatures of high temporal variances in B_{YZ}, indicating frequent occurrence of flow bursts and/or dipolarization events. The magnetosheath is seen as a region of enhanced electric field encircling the magnetopause. It is clear that the magnetosheath field is lowest for the SMC-run, intermediate for the B-run, and largest for the V-run. In the V-run, the increased solar wind speed drapes the magnetic field lines around the magnetopause, which creates a larger magnetosheath electric field.

1.6.3. Discussion and Summary of Main Results of the Global MHD Simulation

Pulkkinen et al. (2007b) discuss and summarize main obtained results as following:

(i) The three LFM simulation runs indicate that the dynamic state of the magnetosphere is not only dependent on the driving electric field but also on its constituents. Under relatively steady driving conditions, larger solar wind speed leads to a more dynamic magnetosphere while larger IMF magnitude leads to steady but more intense nightside reconnection closer to the Earth.

(ii) There are quantified the temporal and spatial structure of the magnetotail in a global MHD simulation by first taking spatial averages over the closed plasma sheet and computing their variances. Similarly, it was obtain temporal averages by averaging over an extended interval of time at each gridpoint separately. These averages and their standard deviations allow to demonstrate quantitatively the characteristic transport and driver properties in the magnetosphere.

(iii) A substantial number of earlier works address solar wind–magnetosphere coupling and the solar wind parameters that most affect the coupling function. Akasofu (1981) summarizes their work, which gives the coupling in the form $\varepsilon = 10^7 l_0^2 V B^2 \sin^4(\theta/2)$, where $\theta = \arctan(B_Y/B_Z)$ is the IMF clock angle and $l_0 = 7\ R_E$ a scaling parameter. Bargatze et al. (1985) use nonlinear filtering techniques and obtain best coupling for the electric field. Most recently, Newell et al. (2006) re-examine a variety of coupling functions and arrive at a best-fit function of the form $V^2 B \sin^4(\theta/2)^{2/3}$. They interpret this function to represent the rate at which magnetic flux is opened at the magnetopause. In this function, the solar wind speed appears squared while the magnetic field is only linear (different from ε and its derivatives), consistent with Pulkkinen et al. (2007b) conclusion that V is stronger driver than B_Z.

(iv) Pulkkinen et al. (2007b) have treated the primary parameters used to solve the MHD equations, the solar wind velocity and magnetic field. While Vasyliunas (2005) points out the secondary nature of the electric field as a derived parameter, in the MHD approximation **E** and **V** × **B** are synonymous. However, as the effects are seen throughout the magnetosphere, Pulkkinen et al. (2007b) do not think that non-MHD terms in the plasma transport equations cause the differences in the magnetospheric dynamics in the examined cases.

(v) Lopez et al. (2004) propose that the magnetospheric activity may be related to the compression ratio at the bow shock. Indeed, changing the speed increases the compression ratio from slightly above 2 to slightly above 3. However, the electric field is preserved across the shock. This, of course, is the reason why the upstream values can be used in computing the various coupling parameters involving the electric field in its various forms. Additional simulation experiments are needed to fully resolve the effects of the shock compression ratio.

(vi) Event studies (Sergeev et al., 1996b) clearly demonstrate that the SMC events occur predominantly during low solar wind speed (~400 km/s) and intermediate values of the southward IMF (a few nT). On the other hand, sawtooth events characteristically show lower IMF magnitude but higher solar wind speed than other stormtime activity, but higher IMF and higher solar wind speed than the SMC events (Pulkkinen et al., 2007a). As the sawtooth events are examples of extremely dynamic and variable magnetosphere, these observational results are qualitatively consistent with the simulation results shown here.

(vii) The global MHD simulations allow to study the role of each driving solar wind parameter individually. It's clearly illustrates that for otherwise exactly the same driving parameters, increasing the solar wind speed changes the magnetospheric response from a steadily convecting state to highly variable both in space and time. Furthermore, even though the increase in the driving electric field is the same, increasing the IMF magnitude does not change the dynamic state of the system, it rather tends to stabilize the magnetotail. These results, supported by the observational evidence, allow to conclude that the solar wind speed is the main factor controlling the state of the magnetotail under otherwise similar (and relatively steady) driving conditions. As discussed above, this study is limited to relatively low Mach numbers. However, a statistical study by Pulkkinen et al. (2007a) suggests that the result holds also for high Mach numbers, as they obtain a similar result by examining the AE-response as a function of V and B for both high and low Mach number regime. Furthermore, complementary results by Laitinen et al. (2007) using the GUMICS-4 global MHD simulation demonstrate that velocity controls the efficiency of energy input from the solar wind into the magnetosphere, while the density plays a very minor role in that process.

1.7. Dependence of the Power Consumed by the Magnetosphere on the Solar Wind Parameters

1.7.1. Short History and the Matter of Problem of the Power Consumed by the Magnetosphere

As noted Ponomarev et al. (2009), the process of energy transfer from the solar wind to the Earth's magnetosphere or, more specifically, to the convecting magnetospheric plasma seems to be rather complex. A bow shock front is the main transformer of the solar wind kinetic energy into the electromagnetic energy (Ponomarev et al., 2006a,b). The intensity of the tangential component of the solar wind's magnetic field and the plasma density increase several times when the front is crossed. Therefore, a bow shock front is also a current sheet. It can be shown that this sheet carries a divergent current; i.e., the front generates the current. Since the plasma with the magnetic field crosses the front, the electric field is generated in the coordinate system of the front (Ponomarev et al., 2006a,b). Thus, the bow shock front is the source of the electric power. This electric power is distributed between two consumers: the transition layer (TL, or the magnetosheath) and the magnetosphere proper. It should be noted that TL can only conditionally be classified as a consumer because it can operate as a generator in a certain regime (Ponomarev et al., 2000). A potential difference exists between the bow shock front and the magnetosphere. This difference is uniquely related to the flow velocity of the TL plasma (because the TL magnetic field depends on the solar wind's magnetic field). Thus, the magnetopause potential is functionally related to the solar wind's parameters. The power consumed by the magnetosphere is spent for the work of the compressor and consists of the active and reactive power. The active power compensates the loss in the ionosphere (mainly ohmic loss), and the reactive power is returned to the compressor (Sedykh and Ponomarev, 2002). Most likely, the generator power at the bow shock front is independent of the power consumed by the magnetosphere; however, an 'energy depot', where the power produced by the bow shock front but not consumed by the magnetosphere can be released, becomes necessary in this case an. The magnetosheath can play this role. In what follows Ponomarev et al. (2009) will mark all parameters of solar wind, TL, and the magnetosphere with subscripts 0, 1, and 2, respectively. The task of the Ponomarev et al. (2009) work is to formalize the above qualitative considerations.

In Ponomarev et al. (2009) the results of the previous studies, where the expressions were obtained for the electric current, which is generated at the bow shock front and is closed through the magnetosphere, and for the magnetopause potential as a function of such solar wind parameters as the plasma density and velocity and the IMF intensity, are used. The power (W) consumed by the magnetosphere is equal to the Poynting vector flux **S** through the magnetopause. According to the special case of the Poynting theorem applied by Heikkila (1997) to the magnetosphere, the energy flux can be expressed in terms of the electric potential (the integration is carried out over the entire surface of the magnetosphere). As a result, the required dependence, which is quadratic with respect to the IMF B_z component, has been obtained for W. It is discussed why the magnetosphere is energy-isolated at the northward IMF B_z component despite this.

1.7.2. The Problem Statement

Ponomarev et al. (2009) consider the bow shock front as a paraboloid of revolution, whose axis coincides with the X axis of the solar–magnetosphere system of coordinates. The equation of the parabola generatrix is

$$r = \left(y_g/2\right)\cos\left(\varphi/2\right), \tag{1.7.1}$$

where r is the radius vector of a point of the generatrix parabola, $y_g/2$ is the distance from the nose point to the focus, and φ is the angle between the radius vector and X axis. In this representation the magnetopause is described by the same equation but with a different focal distance of the nose point, $y_m/2$. Assume that the solar wind velocity (V_0) is directed along the X axis, and IMF in solar wind is $\mathbf{B}_0 = (B_{0x}, B_{0y}, B_{0z})$ or (in unit vectors) $\mathbf{B}_0 = (b_x, b_y, b_z)$. It is convenient to use the parabolic system of coordinates for description (Madelung, M1960). In this system the position of a point in space is defined as follows. Assume that two conjugate hyperbolas $u(y_u, \varphi)$ and $v(y_v, \varphi^*)$ are specified on the plane passing through the axis of the paraboloid of revolution. Linear parameters y_v have the same meaning as y_g in (Ponomarev et al., 2006a). From the condition of conjugacy it follows that parabolas have a common focus and $\varphi + \varphi^* = \pi$. Assume that the plane is chosen such that the point with coordinates to be found lies on it. Obviously, the coordinates of the point in the plane can be determined by intersection of two parabolas (i.e., by numbers y_u and y_v), and the position of the plane can be characterized by the angle ψ between the plane and the Z axis. In other words, the position of each point can be given by three numbers, y_u, y_v, and ψ. Since $y_v = y_u \tan^2(\varphi/2)$, coordinate y_v can be replaced by coordinate φ. The last representation is especially convenient. Here are some relationships that will be used below:

$$u^2 = r + x,\ r = \left(x^2 + y^2 + z^2\right)^{1/2} = \left(x^2 + \rho^2\right)^{1/2},\ y = uv\sin\psi,\ v^2 = r - x,\ \rho = \left(y^2 + z^2\right)^{1/2},$$
$$z = uv\cos\psi,\ \ \psi = \arctan(y/z),\ \ \rho = uvr = \left(u^2 + v^2\right)/2,\ \ x = \left(u^2 - v^2\right)/2, \tag{1.7.2.}$$

Ponomarev et al. (2009) express the components of vector $\mathbf{B}_0$ on the paraboloid surface $u(y_g, \varphi, \psi)$ in terms of unit vectors of the Cartesian coordinate system b_x, b_y, b_z, where $b_x^2 + b_y^2 + b_z^2 = 1$:

$$B_{0u} = B_0\left[\left(1 - b_x^2\right)^{1/2}\cos(\psi - \psi_0)\sin(\varphi/2) + b_x\cos(\varphi/2)\right],$$
$$B_{0v} = B_0\left[\left(1 - b_x^2\right)^{1/2}\sin(\psi - \psi_0)\cos(\varphi/2) - b_x\sin(\varphi/2)\right],\ \ B_{0\psi} = -B_0\left(1 - b_x^2\right)^{1/2}\sin(\psi - \psi_0). \tag{1.7.3}$$

In the parabolic system of coordinates the electric field directed tangentially with respect to the surface of paraboloid u is

$$E_u = -\left(B_\psi V_v\right)/c. \tag{1.7.4}$$

Taking into account that the element of a parabola length

$$dl = \left(y_g/2\right)\cos^2(\varphi/2)d\varphi, \tag{1.7.5}$$

Ponomarev et al. (2009) find the potential on the paraboloid surface by integrating E_u:

$$\Phi_g = -(B_0V_0/c)y_g\left(1 - b_x^2\right)^{1/2}\tan(\varphi/2)\sin(\psi - \psi_0). \tag{1.7.6}$$

At the front crossing, the plasma density experiences a jump $\rho_1 = \chi\rho_0$, where $\chi = (\gamma + 1)/(\gamma - 1)$, and γ is the adiabatic exponent ($\gamma = 5/3$ for a monoatomic gas). Thus, $\chi = 4$ for a monoatomic gas. As was noted above, the tangential component of the magnetic field also experiences a jump at the bow shock front crossing, $B_{1t} = \sigma B_{0t}$. Parameter σ is connected to χ by the relationship:

$$\chi - 1 = \left(A^2 - 1\right)(\sigma - 1)/\left(A^2 + (\sigma - 1)\right), \tag{1.7.7}$$

where $A = A_0\cos(\varphi/2)$, A_0 is the Mach–Alfvén number for solar wind. Since $A_0 >> 1$, $\sigma \approx \chi$ to an accuracy not less than 7%, if $\cos(\varphi/2)$ is greater than 0.5. Now it can be find the density of the current perpendicular to the bow shock front:

$$j_{1v} = (c/4\pi)\mathrm{curl}_v B_1 = -\left(\frac{cB_0}{\pi y_g}\right)\left(1-b_x^2\right)^{1/2}\cos^3\left(\frac{\varphi}{2}\right)\sin(\psi-\psi_0)\left(\frac{\partial\sigma/\partial\varphi}{\sin\varphi}\right). \quad (1.7.8)$$

The expression for the rotor in the parabolic system of coordinates is given in (Madelung, M1960). The $\sigma(\varphi)$ dependence can be approximated by the formula

$$\sigma = 0.5[(\sigma_0+1)+(\sigma_0-1)\cos\varphi]. \quad (1.7.9)$$

In such a case, Eq. 1.7.8 will take the form

$$j_{1v} = -\left(\frac{cB_0}{\pi y_g}\right)\left(1-b_x^2\right)^{1/2}\cos^3\left(\frac{\varphi}{2}\right)\sin(\psi-\psi_0)(\sigma_0-1). \quad (1.7.10)$$

The equation for the potential in the parabolic system of coordinates can be written as $\Delta\Phi = 0$, or

$$\partial^2\Phi/\partial u^2 + \partial\Phi/u\partial u - \partial^2\Phi/u^2\partial\psi^2 + \partial^2\Phi/\partial v^2 + \partial^2\Phi/v\partial v^2 - \partial^2\Phi/v^2\partial\psi^2 = 0 \quad (1.7.11)$$

From the set of solutions Eq. 1.7.10 Ponomarev et al. (2009) select such solutions that satisfy the conditions:

$$\Phi = [A_{1m}uv + \Phi_{2m}J_1(ku)I_1(kv)]\sin(\psi-\psi_0). \quad (1.7.12)$$

Equation (1.7.12) on the magnetopause $u = (y_m)^{1/2}$ takes the form

$$\Phi_m = \left[A_{1m}y_m\tan(\varphi/2) + \Phi_{2m}J_1\left(ky_m^{1/2}\right)I_1\left(ky_m^{1/2}\tan(\varphi/2)\right)\right]\sin(\psi-\psi_0). \quad (1.7.13)$$

Ponomarev et al. (2009) tried to relate the magnetopause potential Φ_m to the potential of the bow shock front Φ_g. Note that the velocity of the plasma motion in the magnetosheath directly depends on the potential difference and on the known intensity of the magnetic field in the gap between the bow shock front and magnetopause. Under stationary conditions, the consumption of plasma in TL can be estimated because its supply from solar wind is known. Indeed, the entire solar wind that crosses the bow shock front will flow in the gap between the magnetopause and the mentioned front, if the magnetopause is impermeable for external plasma. Strictly speaking, this is not the case; however, the estimates demonstrate that the flux from TL into the magnetosphere is several times as small as the flux from solar wind into TL. In such a case,

$$n_0V_0\pi\rho_g^2 = 2\pi\int n_1V_x\rho d\rho. \quad (1.7.14)$$

In addition, since the velocity along the TL axis is related to its projection onto the X axis by the obvious condition $V_u = V_x \sin(\varphi/2)$, then

$$V_u = \left(V_0 \sin(\varphi/2) y_g^2\right) / \left(2\sigma y_m \left(y_g - y_m\right)\right), \tag{1.7.15}$$

hence

$$\left(\Phi_g - \Phi_m\right)/D = V_u B \psi / c, \tag{1.7.16}$$

where

$$D = \left(y_g - y_m\right)/\left(2\cos(\varphi/2)\right), \tag{1.7.17}$$

As a result, Ponomarev et al. (2009) obtained

$$\left(\Phi_g - \Phi_m\right) = -\left(V_0 B_0 / c\right)\left(1 - b_x^2\right)\sin\left(\psi - \psi_0\right)\tan(\varphi/2)\left(y_g^2 / 4\sigma y_m\right). \tag{1.7.18}$$

Comparing Eqs. 1.7.6, 1.7.18, and the first term in Eq. 1.7.12, may be seen that the compared potentials differ in only the constant, if we neglect a weak dependence of σ on φ. In such a case, the boundary conditions on the magnetopause are satisfied if *ku* is set equal to the first root $J_1(z_1) = 0$, where $z_1 = ku = 3.8317$. Then the second term in Eq. 1.7.13 will be zero throughout the model magnetopause, and $k = z_1 / y_m^{1/2}$. From Eqs 1.7.6 and 1.7.18 can be obtained

$$\Phi_m = -\left(\frac{V_0 B_0}{c}\right)\left(1 - b_x^2\right)\sin\left(\psi - \psi_0\right) y_g \tan\left(\frac{\varphi}{2}\right)\left(1 - \frac{y_g}{4\sigma y_m}\right) = \Phi_g \left(1 - \frac{y_g}{4\sigma y_m}\right). \tag{1.7.19}$$

It can be seen from 1.7.19 that the magnetopause potential is proportional to the bow shock potential and differs from it by 7–12%. In what follows that Ponomarev et al. (2009) assume for convenience that factor

$$K = \left\{1 - \left(y_g / 2y_m\right)\left[\left(\sigma_0 + 1\right) - \left(\sigma_0 - 1\right)\cos\varphi\right]\right\} \tag{1.7.20}$$

is the constant (0.88), which corresponds to the value of σ at the dawn–dusk meridian. After this can be obtain the expression for the potential in the entire magnetosphere:

$$\Phi(r, \varphi, \psi) = \left[A_{1m} r \sin\varphi + \Phi_{2m} J_1\left(q\cos\left(\frac{\varphi}{2}\right)\right) I_1\left(q\sin\left(\frac{\varphi}{2}\right)\right)\right]\sin\left(\psi - \psi_0\right), \tag{1.7.21}$$

where

$$A_{1m} = -(V_0 B_0/c)(1-b_x^2)^{1/2} K, \quad q = 0.38317(2r/y_m)^{1/2}. \qquad (1.7.22)$$

To estimate the electromagnetic energy flux into the magnetosphere, Ponomarev et al. (2009) use the representation proposed by Heikkila (1997) for the density of electromagnetic energy flux through the surface:

$$\mathbf{S} = \Phi\mathbf{j} - (c/4\pi)\mathrm{curl}(\Phi\mathbf{B}). \qquad (1.7.23)$$

It is obvious that the integral of **S** over a closed surface is zero if the surface is equipotential. It is clear that div**S** = $-E\mathbf{j}$, i.e., the energy flux consists of two parts: potential $\mathbf{S}_1 = \Phi\mathbf{j}$ and vortical $\mathbf{S}_2 = (c/4\pi)\mathrm{curl}(\Phi\mathbf{B})$. In this case Ponomarev et al. (2009) are interested only in the potential part. Thus, there is known the potential of the magnetopause. The problem of the inflowing current is more complicated. The current density is known only under the bow shock arch and can be quite different at the magnetopause because part of the current can close in TL. It is still unclear how a 'current channel' is formed in collisionless plasma. In such a medium, the current flows along pressure isolines. Thus, to change the current channel, its should change the plasma pressure distribution. If the condition of a constant total pressure is satisfied, this means that the magnetic field is displaced from the region where plasma is accumulated. Then currents appear as a result of the diamagnetic effect. The magnetic field can be forced out by plasma in the process of convection, when plasma moves toward a stronger magnetic field. In this case the velocity of plasma pressure motion is evidently equal to the convection velocity. However, a different situation is possible when the plasma and magnetic pressures change together. Then the region with an increased magnetic pressure moves at the velocity of fast magnetic sound. The accompanying current moves at the same velocity. Thus, in the magnetosphere the timescale of the electric field formation is about L/V_A, and that of the electric current is L/V_c, where V_A and V_c are the Alfvén and convective velocities, respectively. In the magnetosheath both velocities are of the same order of magnitude.

1.7.3. Dependence of the Power Consumed by the Magnetosphere on the Solar Wind Parameters

Now, according to Ponomarev et al. (2009) it can be easily write the general expression for the power consumed by the magnetosphere:

$$W = A_{1m} j^0 y_m^2 \int_0^{\varphi_c}\int_0^{2\pi} \sin^2(\psi - \psi_0)\tan(\varphi/2)d\psi d\varphi, \qquad (1.7.24)$$

where

$$j^0 = (cB_0/\pi y_g)(1-b_x^2)^{1/2}(\sigma_0 - 1). \qquad (1.7.25)$$

The first integral in Eq. 1.7.24 over ψ is taken from 0 to 2π; the second integral, from 0 to some limiting value φ_c. This limiting value is determined from the consideration that the energy from very distant regions of the geomagnetotail cannot participate in the organization of a substorm. It simply has no time to approach the Earth. Assuming that the substorm process duration is ~5 × 10³ s and the mean convection velocity is 25 km/s, it can be find that plasma will have time to arrive from a distance of about 20 Earth's radii by the end of the process. This distance corresponds to the angle φ_c = 137°. According to Ponomarev et al. (2009), after the integration of Eq. 1.7.24 it will be

$$W = -\left(V_0 B_0^2 y_m^2 / y_g\right) K\left(1 - b_x^2\right)\left(\sigma_0 - 1\right)\left(\tan\left(\frac{\varphi_c}{2}\right) - \frac{\varphi_c}{2}\right). \quad (1.7.26)$$

At a solar wind speed of ~5×10⁷ cm/s, IMF intensity of ~10⁻⁴ T, $b_x^2 = 2/3$, and previously accepted values of the remaining quantities, it was obtained in Ponomarev et al. (2009) that W = 2.4 × 10¹⁸ erg/s, which corresponds to the accepted value for the power of a medium-strength substorm. However, Ponomarev et al. (2009) noted that it was assumed in this case that the density of the extraneous current is $j = j_{1v}$; i.e., it was admitted that the current generated at the bow shock front is closed through the body of the magnetosphere. equation 1.7.26 indicates that the energy flux across the magnetopause does not depend on the IMF sign; however, it is well known that the magnetospheric activity considerably decreases if the B_z component sign changes from negative (southward) to positive. It can be seen two reasons why this can take place. First, the current in the magnetosphere cannot quickly reverse. This requires time on the order of at least 30–60 min (see the above comments on the timescale of formation of the current and field); therefore, at the first instant the extraneous current with the new direction simply does not penetrate into the magnetosphere. Second, the reversed electric field in the magnetosphere is much less efficient. This is related to the fact that only the first term in Eq. 1.7.21 for the potential changes its sign. Some idea about the effect of the B_z sign on the distribution of the fields along the dawn–dusk line is also given in the Table 1.7.1. The first line in Table 1.7.1 lists the relative values of the A_{1m} coefficient (see Eq. 1.7.21), which approximately correspond to B_z in nanoteslas; the second line contains mean values of the anticonvection field responsible for the velocity of the antisunward plasma flow; finally, the third line presents the values of the convection field.

Table 1.7.1. Relationship between A_{1m} (corresponds to B_z in nT), anticonvection electric field E_{ak}, and convection electric field E_k. From Ponomarev et al. (2009)

A_{1m}	3.0	2.0	1.5	1.0	0.5	0.0	−0.5	−1.0	−1.5	−2.0	−3.0
E_{ak}	−4.0	−3.75	−3.63	−3.5	−3.35	−3.25	−3.1	−3.0	−2.85	−2.72	−2.5
E_k	0.35	0.85	1.1	1.35	1.6	1.8	2.1	2.35	2.6	2.85	3.35

A power density of magnetospheric process depends on the rate of magnetospheric plasma compression by the Ampere force (Ponomarev, M1985):

$$\mathbf{V}\,\mathrm{grad}\, p = E_k j. \quad (1.7.27)$$

Therefore, small magnitudes of the convection field imply a low intensity of substorm-type magnetospheric processes. Thus, the total energy consumption by the magnetosphere decreases for the northward IMF because of a sharp decrease in the extraneous current through the magnetosphere soon after the northward reversal of the magnetic field and in the convection field, which increases with increasing the northward B_z component. Then the density of the extraneous current j_{2v} will increase during 30–60 min, but the convection electric field will correspond to the E_k current value as before; therefore, magnetospheric activity in the auroral zone and equatorward of this zone will remain low in any case. Since the electric field is large in the anticonvection region, which is projected on the polar cap, activity in the cap will on the contrary be increased. This pattern qualitatively corresponds to the observations (Pudovkin et al., M1977).

1.7.4. Main Results and Discussion

Based on the above considerations, Ponomarev et al. (2009) present the following pattern of supply of the magnetosphere with the electric power generated at the bow shock front.

(i) The magnetosphere constantly includes the magnetospheric electric currents related to the existing distribution of a gas pressure, 'diamagnetic' currents, and natural electric fields associated with the existing convection. These currents are described by the second term in Eq. 1.7.21. Due to the loss in the ionosphere, these currents and fields cannot exist without 'pumping' of energy from outside (Sedykh and Ponomarev, 2002). In this case the natural currents and pressure gradients will evidently decrease.

(ii) A bow shock is the generator of the electric current and electric field, which produce an energy flux into the magnetosphere, sufficient to compensate the magnetospheric loss. This energy flux depends on the solar wind parameters and the capability of the magnetosphere to conduct the extraneous current. Under stationary conditions, precisely a decrease in the natural current is immediately compensated by the extraneous current; therefore, the pressure gradient remains unchanged.

(iii) The power sign is independent of the IMF sign, and the power is always directed into the magnetosphere. The power magnitude is different and is implemented in different zones of the magnetosphere depending on the IMF sign. When B_z is negative, the convection electric field is larger and the field of anticonvection is smaller than in the case when IMF of the same magnitude is positive. Therefore, the energy is dissipated mainly in the auroral zone in the former case and in the polar cap in the latter case. The electric current caused by a bow shock, which also changes its sign during the B_z reversal, cannot first enter the magnetosphere because it still included the pressure gradient corresponding to the previous current value. Only after a lapse of time, when a new convection system reorganizes pressure, the current can enter the magnetosphere in a new direction. This process was considered only qualitatively; therefore, alternative solutions are possible.

1.8. Magnetospheric Modes and Solar Wind Energy Coupling Efficiency

1.8.1. The Matter and Short History of the Problem on the Solar Wind Energy Coupling Efficiency

As outlined Pulkkinen et al. (2010), the solar wind–magnetosphere interaction is a complex set of processes driven by variations of the solar wind plasma parameters and the IMF. The magnetospheric activity that follows has been empirically categorized to several distinct event classes, and efforts have been made to identify the drivers that are associated with each activity type. As a rule of thumb, substorms follow a period of southward IMF longer than about 30 min, and a magnetic storm is caused by southward IMF duration longer than about 3 h (Gonzalez et al., 1994). Earlier work on coupling functions have clearly identified the role of southward IMF and solar wind speed in driving magnetospheric activity. The most often used coupling functions are the Y component of the solar wind electric field (Burton et al., 1975)

$$E_Y = -(\mathbf{V}\times\mathbf{B})_Y \tag{1.8.1}$$

and the epsilon parameter defined as (Akasofu, 1981)

$$\varepsilon = (4\pi/\mu_0)l_0^2 VB^2 \sin^4(\theta/2), \tag{1.8.2}$$

where the IMF clock angle is given by $\theta = \tan^{-1}(B_Y/B_Z)$ and $l_0 = 7R_E$ is an empirical scaling parameter. While the former can be related to the electric field inside the "convecting and reconnecting" magnetosphere (Dungey, 1961), the latter is essentially the Poynting flux incident at the magnetopause. More recently, several more complex coupling functions involving V, B, and θ have been devised using correlation analysis techniques (e.g., Newell et al., 2007, and references therein). These works have led to a firm association of the energy transfer with the magnetic reconnection process (e.g., Siscoe et al., 2001).

Pulkkinen et al. (2010) noted that in the large scale, global MHD simulations can be used to trace the temporal evolution of the fields and plasmas in the magnetosphere. However, the real magnetosphere is a much more complex system than that represented by ideal MHD simulations; major limitations of the simulations arise from the MHD approximation of the plasma physics, incomplete resolution both in some parts of the magnetosphere and in the ionosphere, simplified description of the ionospheric physics, and highly simplified description of the magnetosphere-ionosphere coupling. Both the plasma physics description and grid resolution issues limit the ability of MHD simulations to describe the details of the reconnection process within the diffusion region itself, but comparisons of simulation runs of actual events with proxy indices and in situ measurements show quite comparable temporal behavior in the observations and in the simulation (Palmroth et al., 2004). Therefore, Pulkkinen et al. (2010) assume that the temporal evolution and the characteristic behavior of the energy transfer and dissipation processes driven by reconnection can be quite well reproduced in the simulation magnetosphere.

1.8.2. Main Obtained Results and Discussion on the Solar Wind Energy Coupling Efficiency

Using observations and two different global MHD simulations, Pulkkinen et al. (2010) demonstrate that the solar wind speed controls the magnetospheric response such that the higher the speed, the more dynamic and irregular is the magnetospheric response. For similar level of driving solar wind electric field, the magnetospheric modes can be organized in terms of speed: Low speed produces steady convection events, intermediate speeds result in periodic sawtooth oscillations, and high speeds drive large geomagnetic storms. Pulkkinen et al. (2010) show that the control parameter of energy transfer and coupling is the electric field along the large-scale X line. They demonstrate using global MHD simulations that for slowly varying IMF, the reconnection line is tilted approximately by an angle $\theta/2$, where θ is the IMF clock angle. Then, for clock angles away from northward, the magnetospheric energy entry and response scale with the electric field along the reconnection line (E_{PAR}), rather than the traditionally used E_Y. If we define the energy coupling efficiency as response/E_{PAR}, we can show it to be independent of the IMF clock angle and only weakly dependent on the solar wind dynamic pressure. These results demonstrate the ability of the localized reconnection line to control the energy input through the entire magnetopause.

1.8.3. Magnetosheath Control of Solar Wind-Magnetosphere Coupling Efficiency

Pulkkinen et al. (2016) examine the role of the magnetosheath in solar wind-magnetosphere-ionosphere coupling using the THEMIS plasma and magnetic field observations in the magnetosheath together with OMNI solar wind data and auroral electrojet recordings from the IMAGE magnetometer chain. Pulkkinen et al. (2016) demonstrate that the electric field and Poynting flux reaching the magnetopause are not linear functions of the electric field and Poynting flux observed in the solar wind: the electric field and Poynting flux at the magnetopause during higher driving conditions are lower than those predicted from a linear function. It is also shown that the Poynting flux normal to the magnetopause is linearly correlated with the directly driven part of the auroral electrojets in the ionosphere. This indicates that the energy entering the magnetosphere in the form of the Poynting flux is directly responsible for driving the electrojets. Furthermore, Pulkkinen et al. (2016) argue that the polar cap potential saturation discussed in the literature is associated with the way solar wind plasma gets processed during the bow shock crossing and motion within the magnetosheath.

1.9. Magnetospheric Energy Budget during Huge Geomagnetic Activity

1.9.1. The Matter and Short History of the Problem on the Magnetospheric Energy Budget

Rosenqvist et al. (2006) note that the energy transfer process into the Earth magnetosphere is highly dependent on the orientation of the IMF. Most of the energy transfer is suggested to be a consequence of a magnetic reconnection process (Dungey, 1961). The energy input to the magnetosphere can be estimated by empirical coupling functions derived from solar wind measurements: see Gonzalez (1990) for a review. One of the most widely used energy input functions is the so-called ε parameter (U_ε) of Akasofu (Perreault and Akasofu, 1978; Akasofu, 1979, 1981). The U_ε parameter is given in the SI units as

$$U_\varepsilon(W) = \frac{4\pi}{\mu_o} vB^2 \sin^4(\theta/2) l_o^2, \tag{1.9.1}$$

where θ is the clock-angle of the IMF orientation, in the GSM coordinate system,

$$\theta = \begin{cases} \arctan(|B_y|/|B_z|) & if \quad B_z > 0, \\ 180^\circ - \arctan(|B_y|/|B_z|) & if \quad B_z < 0. \end{cases} \tag{1.9.2}$$

The factor $\frac{4\pi}{\mu_o} vB^2$ corresponds to 4π times the absolute value of the Poynting flux vector $(\mathbf{E}\times\mathbf{B})/\mu_o$, where $\mathbf{E} = -\mathbf{v}\times\mathbf{B}$. The factor l_0 denotes the linear dimension of an 'effective cross-sectional area' of the magnetosphere determined empirically to l_0= 7 R_E (Perreault and Akasofu, 1978). It was scaled to numerically correspond to the estimated energy output in the magnetosphere and the physical dimension of power for the energy input rate. They assumed the main energy sinks in the magnetosphere to be the ring current, auroral Joule dissipation, and particle precipitation in the ionosphere. The dominating energy sink was considered to be the ring current (60%), but this has not been confirmed by recent studies that rather account Joule heating as the main dissipation mechanism (e.g., Knipp et al., 1998). Since then the importance of energy sinks in the form of plasmoids and plasma sheet heating have been identified and found to be comparable to the energy dissipated in the ionosphere during substorms (Slavin et al., 1993; Ieda et al., 1998). A revised version of the U_ε parameter suggests that the scaling factor should be l_0= 10 r_E to account for such substorm-related tail energy sinks (Koskinen and Tanskanen, 2002) or else the meaning of the U_ε parameter should be interpreted as an energy input parameter to the inner magnetosphere only.

Rosenqvist et al. (2006) use the original value of l_0= 7 R_E, as they do not take into account substorm-related tail energy sinks. If magnetic reconnection is assumed to take place at the nose of the magnetosphere, the actual energy conversion from the magnetosheath to the magnetotail in the MHD picture is due to tangential stress at the magnetopause (e.g., Siscoe and Cummings, 1969; Siscoe and Crooker, 1974). As the merged field lines are dragged down-tail along the high-latitude magnetopause, they exert a tangential force on the solar wind. Beyond approximately the cusp region, this tangential force starts to act against the solar wind flow, extracting the kinetic energy from the solar wind and converting it to magnetic energy in the magnetosphere. Maxwell's magnetic stress tensor (e.g., Jackson, M1975) is given by

$$T_{ij} = \frac{1}{\mu_o}\left(B_i B_j - \frac{1}{2}\delta_{ij} B_k B_k \right), \quad (1.9.3)$$

where δ_{ij} is the Kronecker delta and $\mu_0 = 4\pi 10^{-7}$ Vs/Am. The force exerted on the plasma at the magnetopause is thus

$$\vec{F} = \int_V \vec{\nabla} \cdot \vec{T} dr \quad (1.9.4)$$

over the magnetopause volume V. This force is in fact the magnetopause dynamo responsible for breaking the incoming solar wind flow. This gives rise to an instantaneous power per unit volume given by

$$u(W/m^3) = \vec{f} \cdot \vec{v}, \quad (1.9.5)$$

where $\vec{f}$ is the force per unit volume and $\vec{v}$ is the solar wind velocity. Furthermore, the magnetopause volume can be expressed as dr = dAdn, where dn is along the normal to the magnetopause surface element dA. Thus the total power transferred through the surface A of the magnetopause is given by

$$U(W) = \int_A dA \int_n \left(\vec{\nabla} \cdot \vec{T}\right) \cdot \vec{v}_t dn, \quad (1.9.6)$$

where $\vec{v}_t$ is the magnetosheath velocity tangential to the magnetopause (neglect the normal component as it was assumed pressure balance across the magnetopause). This is the total power through the area on the magnetopause where the energy transfer occurs. A portion of this power is later dissipated in the ionosphere in the form of Joule heating and auroral precipitation, to the ring current and tail processes such as plasmoid release and plasma sheet heating. The energy partitioning between different energy sinks has remained one of the key questions in magnetospheric physics perhaps owing to the complexity and simultaneous occurrence of these processes.

Østgaard and Tanskanen (2004) have investigated the energetics of both isolated and storm-time substorms based on parametrized empirical relations. They found that the Joule heating during both types of substorms is the dominant sink in the magnetosphere-ionosphere system and that particle precipitation energy flux is about half as much. Furthermore, Østgaard et al. (2002a) found average contributions of the total time-integrated energy dissipation over the duration of a substorm to be 15% (ring current), 56% (Joule heat), and 29% (particle precipitation). In this study the rate of energy deposition from particle precipitation in the entire range of energies (0.1–100 keV) important for the ionosphere was based on remote sensing techniques in both the UV and X-ray spectrum. The ring current energy increase and Joule heating rate were estimated by empirical methods. Both of these studies find that the total energy dissipated through these sinks often exceeded the estimated total energy input by the Akasofu U_ε parameter.

Another study for the entire period of a geomagnetic storm initiated by a corotating interaction region with a duration of 10 days found similar contributions of the total time-integrated energy dissipation (17%, 60%, and 23%) (Knipp et al., 1998). This study was based on low-energy particle measurements from several polar orbiting satellites, whereas the Joule heating rate was determined by the assimilative mapping of ionospheric electrodynamic (AMIE) technique. The energy dissipation from the ring current was estimated by the Dessler-Parker-Sckopke relation (see Eq. 1.9.9 below). By comparison, Lu et al. (1998) studied the global magnetospheric energy deposition, using a wide range of satellite and ground-based data and utilizing the AMIE technique together with empirical estimations for the ring current energy injection, during a magnetic cloud event. They found that during the 2 day storm the average dissipation rates corresponded to 30%, 47.5%, and 22.5% for ring current, Joule heat, and particle heat, respectively. This added up to almost 90% of the average solar wind energy transfer rate as represented by the U_{ε} parameter. Pulkkinen et al. (2002) found that the combined energy dissipation through the ring current and Joule heating during a strong storm is about a third of the total energy input (see Figures 1.9.1 - 1.9.2).

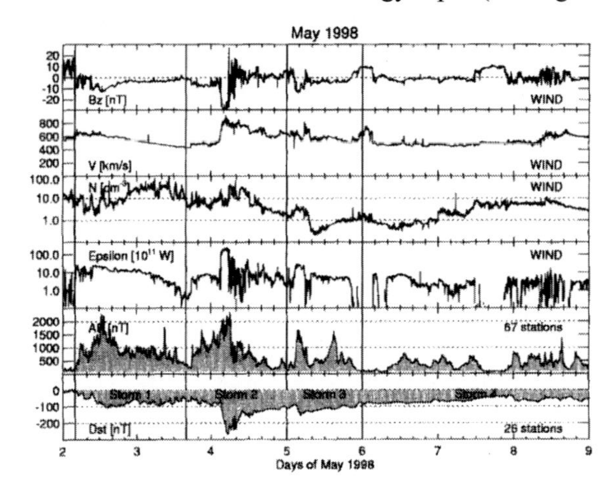

Figure 1.9.1. Solar wind drivers: IMF B_Z, solar wind velocity v, solar wind density n, and the ε parameter determined from the WIND measurements. The bottom panels show the AE and Dst indices; the vertical lines separate the four intervals selected for study. According to Pulkkinen et al. (2002). Permission from COSPAR.

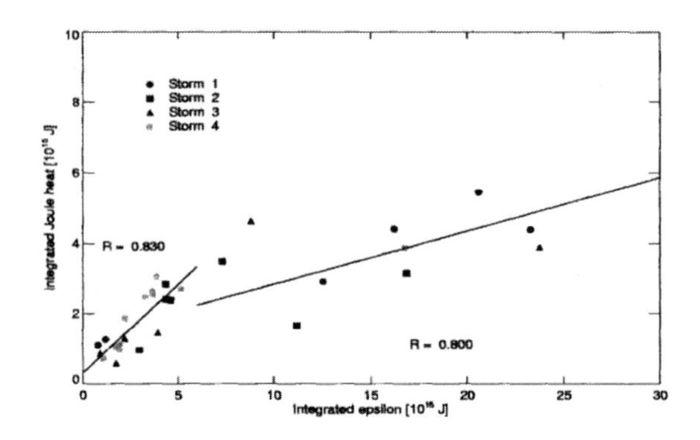

Figure 1.9.2. Scatter plot of the integrated energy input vs. the integrated energy dissipation in Joule heating. According to Pulkkinen et al. (2002). Permission from COSPAR.

In the study of Pulkkinen et al. (2002) the ring current energy was determined both with spacecraft observations and empirical methods while the Joule heating rate was based on the Ahn et al. (1983) parameterized results. In a study of 839 substorms during 1997 (solar minimum) and 1999 (solar maximum), based entirely on the Ahn et al. (1983) parameterized results, it was found that on average the Northern Hemisphere Joule heating accounts for 30% of solar wind energy input (Tanskanen et al., 2002). Assuming that one can multiply this by two to account for both hemispheres, this corresponds to a total dissipation ratio of 60%. It should be noted that in the above studies the effects of energetic electrons (~20–100 keV) is often neglected. On the basis of Polar Ionospheric X-Ray Imaging Experiment (PIXIE) and Ultraviolet Imager (UVI) observations from the Polar satellite, Aksnes et al. (2004, 2006) find that this may result in a reduction of the estimated Joule heating rate by 20% on a global scale. They find that during a severe storm the Joule heating still represents the largest energy form and contributes to 1.67 times more than the energy flux by precipitating particles (although for substorms occurring during non-storm or minor storm periods, the auroral particle precipitation seem to be of increased importance for the energy budget). A summary of a part of these findings can be found in Table 1.9.1.

The ionospheric energy dissipation in the form of particle and Joule heating can be determined locally from, e.g., radar or satellite measurements. Furthermore, the ring current energy can be directly obtained from in situ energetic particle measurements assuming a constant ring current volume. However, since neither of these processes can be directly monitored continuously several empirical relations have been developed.

Table 1.9.1. Summary of the relative contributions to the total energy dissipation from: ring current dissipation, Joule heating, and particle precipitation heating (during different kinds of geomagnetic activity). According to Rosenqvist et al. (2006)

Activity	Reference	Ring current, %	Joule heating, %	Particle heating, %
Substorm	Østgaard and Tanskanen (2004)	15	56	29
CIR storm	Knipp et al. (1998)	17	60	23
Magnetic cloud storm	Lu et al. (1998)	30	47.5	22.5

The ionospheric energy dissipation has been shown to be related to the strengths of the auroral electojets. For example, the Ahn et al. (1983b) formulas converts the westward electrojet index AL (in nT), to Joule and accelerated particle dissipation

$$U_{J,Ahn}(W) = 3\times10^{8}\,AL, \quad U_{A,Ahn}(W) = 0.8\times10^{8}\,AL\,. \tag{1.9.7}$$

All events on which this proxy is based on took place during summer and the ionospheric dissipation rates were determined based on magnetometer data and empirical conductivity models. Note that a factor two should be added to these formulas to account for both hemispheres. However, it should be noted that the summer hemisphere accounts for about 60% of the total Joule heating dissipation, while the winter hemisphere only dissipation about

40% (Østgaard et al., 2002a). Empirical relations that accounts for this summer and winter asymmetry are the Østgaard et al. (2002a,b) formulas,

$$U_{J,Øst}(W) = (0.54AE + 1.8)\times 10^9,\ U_{A,Øst}(W) = 2\times\left(4.4\sqrt{AL} - 7.6\right)\times 10^9, \tag{1.9.8}$$

where AE is the auroral electrojet index. The formula for the Joule heating was obtained from a combination of past results based on seasonal consideration while the particle precipitation was found from global UV and X-ray remote sensing as mentioned above.

The ring current energy can be estimated using the ground-based Dst index associated with the total energy in the ring current particles (Dessler and Parker, 1959; Sckopke, 1966) via the so-called Dessler-Parker-Sckopke (DPS) relation,

$$\frac{Dst^*}{B_0} = \frac{\Delta B_{RC}}{B_0} = -\frac{2}{3}\frac{W_{RC}}{W_{mag}}, \tag{1.9.9}$$

where $W_{mag} = (4\pi/3\mu_o)B_0^2 r_E^3$ is the energy in the Earth's dipole field above the surface, W_{RC} is the total energy of the ring current particles, and B_0 is the surface magnetic field at the equator. The Dst index measures also current systems other than the ring current, such as the dayside magnetopause current, magnetotail currents, and ground-induced currents. Thus their contribution should be removed from the index before the DPS relation can be used to derive the ring current energy (Turner et al., 2001). The magnetopause current can be removed using a suggested correction by Burton et al. (1975):

$$Dst^* = Dst - b\sqrt{p_{sw}} + C, \tag{1.9.10}$$

where p_{sw} is the solar wind dynamic pressure in nP. The values of b and C have been calculated by O'Brien and McPherron (2000) to be b = 7.26nT(nPa)$^{-1/2}$ and C = 11.0 nT. Furthermore, Turner et al. (2001) concluded that the magnetotail currents account for about 25% of the measured Dst and that the removal of ground-induced current contribution requires a reduction of 21% of measured Dst.

1.9.2. The Energy Budget during the Huge Geomagnetic Disturbances within the Intense Geomagnetic Storm on 30 October 2003

Rosenqvist et al. (2006) note that up until now we are not aware of any attempts of determining the solar wind energy transfer directly by observational means, partly due to the difficulties of correct estimation of boundary properties. However, with the multi-spacecraft mission Cluster there are improved techniques of making accurate magnetopause boundary normal and velocity determinations with both minimum variance and timing methods. Furthermore, the current density can be determined with the curlometer technique. Rosenqvist et al. (2006) examine the energy budget during the huge geomagnetic disturbances above Scandinavia within the intense geomagnetic storm on 30 October 2003. These substorm-like disturbances were triggered by pressure pulses in the solar wind. Thus the Cluster spacecraft located at the flank of the magnetospheric tail were passed several times by an inward and consecutively outward moving magnetopause in close relation to the disturbances (Rosenqvist et al., 2005).

1.9.3. Summary of Main Obtained by Rosenqvist et al. (2006) Results on the Research of Magnetospheric Energy Budget

Main obtained results are summarized by Rosenqvist et al. (2006) as following:

(i) It was made the first observational estimate of the local solar wind power input utilizing the four-spacecraft Cluster observations across the magnetopause in close relation to extreme substorm activity, and obtained a local value of input energy ~0.5 mW/m^2. However, about half of the contribution to this local power input comes from local pressure gradients tangential to the magnetopause. Thus only ~0.25 mW/m^2 due to tangential stress at the magnetopause can be considered in a global extrapolation of energy input.

(ii) The local value was extrapolated in order to obtain a quantitative global power input into the inner magnetosphere using a simple model of the magnetosphere and of the spatial variations of the local power input. The global power input was found to be between 17 and 40 TW depending on the distance to the onset of reconnection in the near-Earth tail. Spacecraft observations of the distance to the near-Earth tail suggest that the lower bound of 17 TW is the most likely value. According to the well-known U_{ε} parameter, this value is estimated to be 37 TW which is more than two times larger than the lower value found with the observational method.

(iii) The EISCAT radar monitored the ionosphere during this sequence of events and local ionospheric energy dissipation rates could be estimated based on these observations. The Joule heating rate obtained by EISCAT was found to be in reasonable agreement with results based on AMIE modeling during this sequence of events. AMIE, in addition, provides a global distribution of the Ohmic heating.

(iv) On the basis of the observational global power estimates it was found that about 30% of the total power input from the solar wind into the inner magnetosphere is dissipated via Joule heating in the ionosphere. This percentage is in rather good agreement to corresponding ratios found in previous studies (~50–60%) but might suggest that the global Cluster estimate is somewhat overestimated.

(v) In comparison, global electrojet index-based empirical estimations of the Joule heating deposition in the ionosphere predict a ratio of only 3% of the total solar wind energy input obtained by the U_{ε} parameter when pushed to the extreme circumstances of this storm period. In absolute terms, it was found that the observational values of the power output via Joule heating are much larger than the empirical estimates. The AMIE technique observes a value about four times larger than the corresponding empirical proxies based on electrojet *AE* and *AL* indices.

(vi) In summary, Rosenqvist et al. (2006) find observationally that the power that enters the inner magnetosphere is balanced to a rather large fraction (30%) by the Joule heating deposition in the ionosphere. On the other hand, the energy input estimated by the empirical

U_{ε} parameter is more than a factor two larger than the Cluster observation and the electrojet index based proxies for Joule heating is at least a factor four smaller than AMIE and EISCAT observations. This extreme combination shows that empirical estimates predict an extremely low ratio of the total energy input deposited in the ionosphere as Joule heating (only 3%). This suggests that empirical proxies are an insufficient tool for energy budget estimates during storm and substorms of this extreme character.

(vii) Rosenqvist et al. (2006) plan to further examine the observed discrepancy between observation and empirical estimates taking into account the intensity of the geomagnetic activity. Empirical relations should be revised in order to account for the magnitude of the storm/substorm.

1.10. The Plasmapause Formation Seen from Meridian Perspective by KAGUYA

According to Murakami et al. (2016), observations by the extreme ultraviolet (EUV) imager on board the IMAGE spacecraft revealed that the formation of a sharp plasmapause occurs in the postmidnight sector soon (<1 h) after the convection enhancement. These results cannot be explained simply by the conventional theory of the plasmapause formation that the plasmapause coincides with the last closed equipotential of the convection electric field superposed on the Earth's corotation electric field. However, due to the limitation that the EUV imager provides information on only the azimuthal distribution of the plasmapause, the formation mechanism still remains an open issue. Now global images of the plasmasphere from meridian perspective become available, thanks to the telescope of extreme ultraviolet (TEX) instrument on board the KAGUYA spacecraft. Murakami et al. (2016) studied the plasmapause formation mechanism by analyzing the sequential TEX images of an erosion event during the geomagnetic disturbance (Kp = 5) on 1–2 May 2008. The temporal evolution of the plasmapause locations at postmidnight observed by TEX agreed with those predicted by the dynamic simulations based on the interchange mechanism. Furthermore, the He^{+} column density in the nightside plasmasphere decreased by ~30% only at the low latitudes (<20°) during the enhanced convection period. This suggests that the plasmapause formation occurs first near the equatorial region during a geomagnetic disturbance, and it agrees with the plasmapause formation mechanism based on the interchange instability. Although Murakami et al. (2016) cannot conclude exclusively for the interchange mechanism, this is the first study to present the plasmapause formation viewed from the meridian perspective.

1.11. Partitioning of Integrated Energy Fluxes in Four Tail Reconnection Events Observed by Cluster

Tyler et al. (2016) present the partitioning of integrated energy flux from four tail reconnection events observed by Cluster, focusing on the relative contributions of Poynting flux, electron, H^{+} and O^{+} enthalpy, and kinetic energy flux in the tailward and earthward directions in order to study temporal and spatial features of each event. Tyler et al. (2016)

further subdivide the Poynting flux into three frequency bands to examine the possible structures and waves that contribute most significantly to the total Poynting flux from the reconnection region. Obtained results indicate that H^+ enthalpy flux is often dominant, but O^+ enthalpy, electron enthalpy, Poynting flux, and H^+ kinetic energy flux can contribute significant or greater total energy flux depending on spacecraft location with respect the current sheet, flow direction, temporal scale, and local conditions. Tyler et al. (2016) observe integrated H^+ enthalpy fluxes that differ by factors of 3–4 between satellites, even over ion inertial length scales. There are observed strong differences in behavior between H^+ and O^+ enthalpy fluxes in all events, highlighting the importance of species-specific energization mechanisms. It is find tailward-earthward asymmetry in H^+ enthalpy flux, possibly indicative of the influence of the closed earthward boundary of the magnetotail system. Frequency filtering of the Poynting flux shows that current sheet surface waves and structures on the timescale of current sheet flapping contribute significantly, while large-scale structure contributions are relatively small. Tyler et al. (2016) observe that the direction and behavior of the Poynting flux differs between bands, indicating that the observed flux originates from multiple distinct sources or processes.

1.12. Hall Effect Control of Magnetotail Dawn-Dusk-Asymmetry: A Three-Dimensional Global Hybrid Simulation

As outlined Lu et al. (2016), magnetotail reconnection and related phenomena (e.g., flux ropes, dipolarizing flux bundles, flow bursts, and particle injections) occur more frequently on the duskside than on the dawnside. Because this asymmetry can directly result in dawn-dusk asymmetric space weather effects, uncovering its physical origin is important for better understanding, modeling, and prediction of the space weather phenomena. However, the cause of this pervasive asymmetry is unclear. Using three-dimensional global hybrid simulations, Lu et al. (2016) demonstrate that the Hall physics in the magnetotail current sheet is responsible for the asymmetry. The current sheet thins progressively under enhanced global convection; when its thickness reaches ion kinetic scales, some ions are decoupled from the magnetized electrons (the Hall effect). The resultant Hall electric field E_z is directed toward the neutral plane. The Hall effect is stronger (grows faster) on the duskside; i.e., more ions become unmagnetized there and do not comove with the magnetized dawnward $E_z \times B_x$ drifting electrons, thus creating a larger additional cross-tail current intensity j_y (in addition to the diamagnetic current) on the duskside, compared to the dawnside. The stronger Hall effect strength on the duskside is controlled by the higher ion temperature, thinner current sheet, and smaller normal magnetic field B_z there. These asymmetric current sheet properties are in turn controlled by two competing processes that correspond to the Hall effect: (1) the dawnward $E \times B$ drift of the magnetic flux and magnetized ions and electrons and (2) the transient motion of the unmagnetized ions which do not execute $E \times B$ drift.

1.13. Void Structure of O^+ Ions in the Inner Magnetosphere Observed by the Van Allen Probes

According to Nakayama et al. (2016), the Van Allen Probes Helium Oxygen Proton Electron instrument observed a new type of enhancement of O^+ ions in the inner magnetosphere during substorms. As the satellite moved outward in the premidnight sector, the flux of the O+ ions with energy ~10 keV appeared first in the energy-time spectrograms. Then, the enhancement of the flux spread toward high and low energies. The enhanced flux of the O^+ ions with the highest energy remained, whereas the flux of the ions with lower energy vanished near apogee, forming what we call the void structure. The structure cannot be found in the H^+ spectrogram. Nakayama et al. (2016) studied the generation mechanism of this structure by using numerical simulation. They traced the trajectories of O^+ ions in the electric and magnetic fields from the global magnetohydrodynamics simulation and calculated the flux of O^+ ions in the inner magnetosphere in accordance with the Liouville theorem. The simulated spectrograms are well consistent with the ones observed by Van Allen Probes. Nakayama et al. (2016) suggest the following processes: (1) when magnetic reconnection starts, an intensive equatorward and tailward plasma flow appears in the plasma lobe. (2) the flow transports plasma from the lobe to the plasma sheet where the radius of curvature of the magnetic field line is small. (3) the intensive dawn-dusk electric field transports the O^+ ions earthward and accelerates them nonadiabatically to an energy threshold; (4) the void structure appears at energies below the threshold.

1.14. Three-Scale Structure of Diffusion Region in the Presence of Cold Ions

As note Divin et al. (2016), kinetic simulations and spacecraft observations typically display the two-scale structure of collisionless diffusion region (DR), with electron and ion demagnetization scales governing the spatial extent of the DR. Recent in situ observations of the nightside magnetosphere, as well as investigation of magnetic reconnection events at the Earth's magnetopause, discovered the presence of a population of cold (tens of eV) ions of ionospheric origin. Divin et al. (2016) present two-dimensional particle-in-cell simulations of collisionless magnetic reconnection in multicomponent plasma with ions consisting of hot and cold populations. They show that a new cold ion diffusion region scale is introduced in between that of hot ions and electrons. Demagnetization scale of cold ion population is several times (~4–8) larger than the initial cold ion gyroradius. Cold ions are accelerated and thermalized during magnetic reconnection and form ion beams moving with velocities close to the Alfvén velocity.

1.15. Intermittent Energy Dissipation by Turbulent Reconnection

As outlined Fu et al. (2017), magnetic reconnection—the process responsible for many explosive phenomena in both nature and laboratory—is efficient at dissipating magnetic energy into particle energy. To date, exactly how this dissipation happens remains unclear, owing to the scarcity of multipoint measurements of the "diffusion region" at the sub-ion scale. Fu et al. (2017) report such a measurement by Cluster—four spacecraft with separation of 1/5 ion scale. They discover numerous current filaments and magnetic nulls inside the diffusion region of magnetic reconnection, with the strongest currents appearing at spiral nulls (O-lines) and the separatrices. Inside each current filament, kinetic-scale turbulence is significantly increased and the energy dissipation, $E' \times j$, is 100 times larger than the typical value. At the jet reversal point, where radial nulls (X-lines) are detected, the current, turbulence, and energy dissipations are surprisingly small. All these features clearly demonstrate that energy dissipation in magnetic reconnection occurs at O-lines but not X-lines.

1.16. Extremely Field-Aligned Cool Electrons in the Dayside Outer Magnetosphere

According to Mozer et al. (2017), for 200 days in 2016 while THEMIS-D was in the dayside, equatorial magnetosphere, its electron energy coverage was modified such that the first 15 energy steps covered the range of 1–30 eV and 16 steps covered energies to 30 keV. These measurements were free of backgrounds from photoelectrons, secondaries, or ionospheric plasma plumes. Three energy bands of electrons were observed: cold electrons having energies below 1 eV (plasmaspheric plumes measured by the spacecraft potential); cool electrons, defined as electrons having energies of 1–25 eV; and hot electrons are having energies of 25 eV to 30 keV. The cool electron fluxes at fixed radial distances varied by an order of magnitude from one orbit to the next. These fluxes often increased with increasing radial distance, suggesting an external source. They were extremely field aligned, having pitch angle ratios (flux at 0–20° and 160–180° divided by the flux at 80–100°) greater than 100. Evidence is presented that they resulted from cusp electrons moving from open to closed magnetospheric field lines due to their $\boldsymbol{E} \times \boldsymbol{B}/B^2$ drift. They constituted the majority of the electron energy density at such times and places. As outlined Mozer et al. (2017), these electrons were not associated with magnetopause reconnection because they were not observed at the magnetopause, but they were observed as far as 3 R_E inside of it. Their occurrence probability in the outer magnetosphere was ~50% in June and ~10% in September, suggesting a dayside source attributed to the tilt of the northern cusp toward the Sun during the summer.

1.17. Electron Currents Supporting the Near-Earth Magnetotail during Current Sheet Thinning

As outlined Artemyev et al. (2017), formation of intense, thin current sheets (i.e., current sheet thinning) is a critical process for magnetospheric substorms, but the kinetic physics of this process remains poorly understood. Using a triangular configuration of the three THEMIS spacecraft at the end of 2015 Artemyev et al. (2017) investigate field-aligned and transverse currents in the magnetotail current sheet around 12 R_E downtail. Combining the curlometer technique with direct measurements of ion and electron velocities, Artemyev et al. (2017) demonstrate that intense, thin current sheets supported by strong electron currents form in this region. Electron field-aligned currents maximize near the neutral plane $B_x \sim 0$, attaining magnitudes of ~ 20 nA/m^2. Carried by hot (>1 keV) electrons, they generate strong magnetic shear, which contributes up to 20% of the vertical (along the normal direction to the equatorial plane) pressure balance. Electron transverse currents, on the other hand, are carried by the curvature drift of anisotropic, colder (<1 keV) electrons and gradually increase during the current sheet thinning. In the events under consideration the thinning process was abruptly terminated by earthward reconnection fronts which have been previously associated with tail reconnection further downtail. It is likely that the thin current sheet properties described herein are similar to conditions further downtail and are linked to the loss of stability and onset of reconnection there. The findings of Artemyev et al. (2017) are likely applicable to thin current sheets in other geophysical and astrophysical settings.

1.18. Climatology Characterization of Equatorial Plasma Bubbles Using GPS Data

The climatology of equatorial plasma bubbles (EPBs) for the period 1998–2008 was studied by Magdaleno et al. (2017) using slant total electron content (sTEC) derived from global positioning system (GPS) data. The sTEC values were calculated from data measured at 67 International GNSS Service (IGS) stations distributed worldwide around the geomagnetic equator and embracing the region of the ionospheric equatorial anomaly (IEA). EPBs and their characteristics were obtained using the Ionospheric Bubble Seeker (IBS) application, which detects and distinguishes sTEC depletions associated with EPBs. This technique bases its analysis on the time variation of the sTEC and on the population variance of this time variation. IBS finds an EPB by default when an sTEC depletion is greater than 5 TEC units (TECu). The analysis of the spatial behavior shows that the largest rate of EPB takes place at the equator and in the South America-Africa sector, while their occurrence decreases as the distance from the magnetic equator increases. The depth and duration of the sTEC depletions also maximize at the equator and in the South America-Africa sector and weaken departing from the equator. The results of the temporal analysis for the data of the IGS stations located in AREQ, NKLG, IISC, and GUAM indicate that the greatest rate of EPB occurrence is observed for high solar activity.

1.19. Modeling the Daytime Energy Balance of the Topside Ionosphere at Middle Latitudes

According to Hsu and Heelis (2017), recently reported measurements from the Defense Meteorological Satellite Program (DMSP) indicate that the O^+ temperature in the topside ionosphere is dependent on the fractional H^+ density. This finding indicates that the mass–dependent energy exchange rate between O^+ and H^+ plays an important role in the thermal balance of the topside ionosphere. Hsu and Heelis (2017) utilize the SAMI2 model to retrieve both T_{H+} and T_{O+} and verify the previously observed dependence of ion temperature on ion composition. The model shows that in the topside at middle latitudes when a single ion is dominant, O^+ or H^+ is heated by electron collisions and cooled by conduction as expected. However, as outlined Hsu and Heelis (2017), in the intervening altitude region where both O^+ and H^+ are present, O+ is heated by collisions with H^+ and cooled by conduction, while H^+ is heated by collisions with electrons and cooled by collisions with O^+.

1.20. Some Problems of Identifying Types of Large-Scale Solar Wind and Their Role in the Physics of the Magnetosphere

The paper of Yermolaev et al. (2017) discusses the errors in analyzing solar-terrestrial relationships, which result from either disregarding the types of interplanetary drivers in studying the magnetosphere response on their effect or from the incorrect identification of the type of these drivers. In particular, it has been shown that the absence of selection between the Sheath and ICME (the study of so-called CME-induced storms, i.e., magnetic storms generated by CME) leads to errors in the studies of interplanetary conditions of magnetic storm generation, because the statistical analysis has shown that, in the Sheath + ICME sequences, the largest number of storm onsets fell on the Sheath, and the largest number of storms maxima fell at the end of the Sheath and the beginning of the ICME. That is, the situation is observed most frequently when at least the larger part of the main phase of storm generation falls on the Sheath and, in reality, Sheath-induced storms are observed. In addition, Yermolaev et al. (2017) consider several cases in which magnetic storms were generated by corotating interaction regions, whereas the authors attribute them to CME.

References

Adewale A.O., E.O. Oyeyemi, and J. Olwendo, 2012. "Solar activity dependence of total electron content derived from GPS observations over Mbarara", *Adv. Space Res.*, **50**, No. 4, 415-426.

Ahn B.-H., S.-I. Akasofu, and Y. Kamide, 1983. "The joule heat production rate and the particle energy injection rate as a function of the geomagnetic indices ae and al", *J. Geophys. Res.*, **88**, No. A8, 6275-6287.

Akasofu S.-I., 1979. "Energy coupling between the solar wind and the magnetosphere", *Planet Space Sci.*, **27**, No. 4, 425-431.

Akasofu S.-I., 1981. Energy coupling between the solar wind and the magnetosphere, Space Sci. Rev., 28, 121–190.

Aksnes A., J. Stadsnes, G. Lu, N. Østgaard, R.R. Vondrak, D.L. Detrick, T.J. Rosenberg, G.A. Germany, and M. Schultz, 2004. "Effects of energetic electrons on the electrodynamics in the ionosphere", *Ann. Geophys.*, **22**, No. 2, 475-496.

Aksnes A., J. Stadsnes, N. Østgaard, G.A. Germany, K. Oksavik, R.R. Vondrak, A. Brekke, and U.P. Lovhaug, 2006. "Height profiles of the ionospheric electron density derived using space-based remote sensing of UV and X-ray emissions, and EISCAT radar data: A ground-truth experiment", *J. Geophys. Res.*, **111**, No. A2, A02301, doi: 10.1029/2005JA011331, 1-11.

Anderson R.R., D.A. Gurnett, and D.I. Odem, 1992. "CRRES Plasma-Wave Experiment, *J. Spacecr. Rockets*, **29**, No. 4, 570-573.

Artemyev A.V., V. Angelopoulos, J. Liu, and A. Runov, 2017, "Electron currents supporting the near-Earthmagnetotail during current sheet thinning", *J. Geophys. Res. Lett.*, **44**, 5–11, doi:10.1002/2016GL072011.

Atkinson G., 1978. "Energy flow and closue of current systems in the magnetosphere", *J. Geophys. Res.*, **A83**, No.3, 1089-1103.

Baker D.N., T.I. Pulkkinen, V. Angelopoulos, W. Baumjohann, and R.L. McPherron, 1996.. "The neutral line model of substorms: Past results and present view", *J. Geophys. Res.*, **101**, No. A6, 12975-13010.

Bargatze L.F., D.N. Baker, R.L. McPherron, and E.W. Hones, 1985. "Magnetospheric impulse response for many levels of geomagnetic activity", *J. Geophys. Res.*, **90**, No. A7, 6387-6394.

Barth C.A., D.N. Baker, and S.M. Baileym, 2004. "Seasonal variation of auroral electron precipitation", *Geophys. Res. Lett.*, **31**, No. 4, L04809, doi:10.1029/2003GL018892, 1-4).

Benkevitch L.V., W.B. Lyatsky, A.V. Koustov, G.J. Sofko, and A.M. Hamza, 2002. "Substorm onset times as derived from geomagnetic indices", *Geophys. Res. Lett.*, **29**, No. 10, 1496, doi:10.1029/2001GL014386, 134-1-4.

Berube D., M.B. Moldwin, and J.M. Weygand, 2003. "An automated method for the detection of field line resonance frequencies using ground magnetometer techniques", *J. Geophys. Res.*, **108**, No. A9, 1348, doi:10.1029/2002JA009737, SMP10-1-6.

Burke W.J., D.R. Weimer, and N.C. Maynard, 1999. "Geoeffective interplanetary scale sizes derived from regression analysis of polar cap potentials", *J. Geophys. Res.*, **104**, No. A5, 9989-9994.

Burton R.K., R.L. McPherron, and C.T. Russell, 1975. "An empirical relationship between interplanetary conditions and *Dst*", *J. Geophys. Res.*, **80**, No. 31, 4204-4214.

Clilverd M.A., A.J. Smith, and N.R. Thomson, 1991. "The annual variation in quiet time plasmaspheric electron density determined from whistler mode group delays", *Planet. Space Sci.*, **39**, No. 7, 1059-1067.

Clilverd M.A., N.P. Meredith, R.B. Horne, S.A. Glauert, R.R. Anderson, N.R. Thomson, F.W. Menk, and B.R. Sandel, 2007. "Longitudinal and seasonal variations in plasmaspheric electron density: Implications for electron precipitation", *J. Geophys. Res.*, **112**, A11210, doi:10.1029/2007JA012416, 1-10.

Clilverd M.A., C.J. Rodger, M. McCarthy, R. Millan, L.W. Blum, N. Cobbett, J.B. Brundell, D. Danskin, and A.J. Halford, 2017. "Investigating energetic electron precipitation through combining ground-based and balloon observations", *J. Geophys. Res. Space Physics,* **122**, 534–546, doi:10.1002/2016JA022812.

Cliver E.W., Y. Kamide, and A.G. Ling, 2000. "Mountains versus valleys: Semiannual variation of geomagnetic activity", *J. Geophys. Res.*, **105**, No. A2, 2413-2424.

Colella P. and P.R. Woodward, 1984. "The piecewise parabolic method (PPM) for gas-dynamical simulations, *J. Comput. Phys.*, **54**, No. 1, 174-201.

Cortie A.L., 1912. “Sunspots and terrestrial magnetic phenomena, 1898–1911: The cause of the annual variation in magnetic disturbances”, *Mon. Not. R. Astron. Soc.*, **73**, 52-60.

Crooker N.U. and G.L. Siscoe, 1986. “On the limits of energy transfer through dayside merging”, *J. Geophys. Res.*, **91**, No. A12, 13393-13397.

Dai W. and P.R. Woodward, 1995. "A simple Riemann solver and high-order Godunov schemes for hyperbolic systems of conservation laws", *J. Comput. Phys.*, **121**, No. 1, 51-65.

Deng Y., Y. Huang, S.C. Solomon, L. Qian, D.J. Knipp, D.R. Weimer, and J.-S. Wang, 2012. “Anomalously low geomagnetic energy inputs during 2008 solar minimum”, *J. Geophys. Res.*, **117**, No. A9, A09307, doi:10.1029/2012JA018039, 1-9.

Dessler A.J. and E.N. Parker, 1959. “Hydromagnetic theory of geomagnetic storms”, *J. Geophys. Res.*, **64**, No. 12, 2239-2252.

Divin A., Y.V. Khotyaintsev, A. Vaivads, M. André, S. Toledo-Redondo, S. Markidis, and G. Lapenta, 2016, “Three-scale structure of diffusion region in the presence of cold ions”, *J. Geophys. Res. Space Physics*, **121**, 12,001–12,013, doi:10.1002/2016JA023606.

Du A.M., B.T. Tsurutani, and W. Sun, 2011. “Solar wind energy input during prolonged, intense northward interplanetary magnetic fields: A new coupling function”, *J. Geophys. Res.*, **116**, No. A12, A12215, doi:10.1029/2011JA016718, 1-14.

Dungey J.W., 1961. “Interplanetary magnetic field and the auroral zones”, *Phys. Rev. Lett.*, **6**, No. 2, 47-48.

Ebihara Y., M.-C. Fok, R.A. Wolf, T.J. Immel, and T.E. Moore, 2004. “Influence of ionosphere conductivity on the ring current”, *J. Geophys. Res.*, **109**, No. A8, A08205, doi:10.1029/2003JA010351, 1-14.

Fairfield D., 1971. "Average and unusual locations of the Earth’s magnetopause and bow shock", *J. Geophys. Res.*, **76**, No. 28, 6700-6716.

Farrugia C.J., B. Harris, M. Leitner, C. Möstl, A.B. Galvin, K.D.C. Simunac, R.B. Torbert, M.B. Temmer, A.M. Veronig, N.V. Erkaev, A. Szabo, K.W. Ogilvie, J.G. Luhmann, and V.A. Osherovich, 2012. “Deep Solar Activity Minimum 2007 - 2009: Solar Wind Properties and Major Effects on the Terrestrial Magnetosphere”, *Solar Phys.*, **281**, No. 1, 461-489.

Fedder J.A., J.G. Lyon, S.P. Slinker, and C.M. Mobarry, 1995. "Topological structure of the magnetotail as a function of interplanetary magnetic field direction", *J. Geophys. Res.*, **100**, No. A3, 3613-3621.

Fedder J.A., S.P. Slinker, J.G. Lyon, and C.T. Russell, 2002. “Flux transfer events in global numerical simulations of the magnetosphere”, *J. Geophys. Res.*, **107**, No. A5, 1048, doi:10.1029/2001JA000025, SMP1-1-11.

Fu H.S., A. Vaivads, Y.V. Khotyaintsev, M. André, J.B. Cao, V. Olshevsky, J.P. Eastwood, and A. Retinò, 2017, “Intermittent energy dissipation by turbulent reconnection”, *J. Geophys. Res. Lett.*, **44**, 37–43, doi:10.1002/2016GL071787.

Fujimoto M., T.K.M. Nakamura, and H. Hasegawa, 2006. “Cross-scale coupling within rolled-up MHD-scale vortices and its effect on large scale plasma mixing across the magnetospheric boundary”, *Space Science Reviews*, **122**, 3–18.

Glauert S.A. and R.B. Horne, 2005. “Calculation of pitch angle and energy diffusion coefficients with the PADIE code”, *J. Geophys. Res.*, **110**, No. A4, A04206, doi:10.1029/2004JA010851, 1-15.

Gonzalez W.D., 1990. “A unified view of solar wind-magnetosphere coupling function”, *Planet. Space Sci.*, **38**, No. 5, 627-632.

Gonzalez W.D., J.A. Joselyn, Y. Kamide, H.W. Kroehl, G. Rostoker, B.T. Tsurutani, and V.M. Vasyliunas, 1994. “What is a geomagnetic storm?”, *J. Geophys. Res.*, **99**, No. A4, 5771-5792.

Goodrich C.C., T.I. Pulkkinen, J.G. Lyon, and V.G. Merkin, 2007. "Magnetospheric convection during intermediate driving: Sawtooth events and steady convection intervals as seen in Lyon-Fedder-Mobarry global MHD simulations", *J. Geophys. Res.*, **112**, A08201, doi:10.1029/2006JA012155, 1-12.

Guiter S.M., C.E. Rasmussen, T.I. Gombosi, J.J. Sojka, and R.W. Shunk, 1995. “What is the source of observed annual variations in plasmaspheric density”, *J. Geophys. Res.*, **100**, No. A5, 8013-8020.

Haaland S., B. Lybekk, L. Maes, K. Laundal, A. Pedersen, P. Tenfjord, A. Ohma, N. Østgaard, J. Reistad, and K. Snekvik, 2017. “North-south asymmetries in cold plasma density in the magnetotail lobes: Cluster observations*”, J. Geophys. Res. Space Physics*, **122**, No. 1, 136–149, doi:10.1002/2016JA023404.

Hartinger M., V. Angelopoulos, M.B. Moldwin, K.- H. Glassmeier, and Y. Nishimura, 2011. “Global energy transfer during a magnetospheric field line resonance”, *Geophys. Res. Lett.*, **38**, No. 12, L12101, doi:10.1029/2011GL047846, 1-6.

Heikkila W.J., 1997. “Interpretation of Recent AMPTE Data at the Magnetopause,” *J. Geophys. Res.*, **102**, No. A2, 2115–2124.

Horne R.B., S.A. Glauert, and R.M. Thorne, 2003. “Resonant diffusion of radiation belt electrons by whistler-mode chorus”, *Geophys. Res. Lett.*, **30** No. 9, 1493, doi:10.1029/2003GL016963, 46-1-4.

Hsu C.-T., and R.A. Heelis, 2017. “Modeling the daytime energy balance of the topside ionosphere at middle latitudes”, *J. Geophys. Res. Space Physics*, **122**, 5733–5742, doi:10.1002/2017JA024112.

Hu Y.-Q., X.C. Guo, G.-Q. Li, C. Wang, and Z.-H. Huang, 2005. "Oscillation of quasi-steady Earth’s magnetosphere", *Chin. Phys. Lett.*, **22**, No. 10, 2723-2726.

Hu Y.Q., X.C. Guo, and C. Wang, 2007. "On the ionospheric and reconnection potentials of the earth: Results from global MHD simulations", *J. Geophys. Res.*, **112**, A07215, doi:10.1029/2006JA012145, 1-8.

Huang C.Y., Y.-J. Su, E.K. Sutton, D.R. Weimer, and R.L. Davidson, 2014. “Energy coupling during the August 2011 magnetic storm”, *J. Geophys. Res. Space Physics*, **119**, No. 2, doi:10.1002/2013JA019297, 1219-1232.

Ieda A., S. Machida, T. Mukai, Y. Saito, T. Yamamoto, A. Nishida, T. Terasawa, and S. Kokubun, 1998. “Statistical analysis on the plasmoid evolution with geotail observations”, *J. Geophys. Res.*, **103**, No. A3, 4453-4465.

Imber S.M., S.E. Milan, and M. Lester, 2013. “Solar cycle variations in polar cap area measured by the superDARN radars”, *J. Geophys. Res. Space Physics*, **118**, No. 10, doi:10.1002/jgra.50509, 6188-6196.

Knipp D.J., B.A. Emery, M. Engebretson, X. Li, A.H. McAllister, T. Mukai, S. Kokubun, G.D. Reeves, D. Evans, T. Obara, X. Pi, T. Rosenberg, A. Weatherwax, M.G. McHarg, F. Chun, K. Mosely, M. Codrescu, L. Lanzerotti, F.J. Rich, J. Sharber, and P. Wilkinson, 1998. “An overview of the early November 1993 geomagnetic storm”, *J. Geophys. Res.*, **103**, No. 11, 26197-26220.

Koskinen Hannu E.J. and Eija I. Tanskanen, 2002. “Magnetospheric energy budget and the epsilon parameter”, *J. Geophys. Res.*, **107**, No. A11, 1415, doi:10.1029/2002JA009283, SMP42-1-10.

Laitinen T.V., M. Palmroth, T.I. Pulkkinen, P. Janhunen, and H.E.J. Koskinen, 2007. “Continuous reconnection line and pressure-dependent energy conversion on the magnetopause in a global MHD model”, *J. Geophys. Res.*, **112**, No. A11, A11201, doi:10.1029/2007JA012352, 1-13.

Lopez R.E., M. Wiltberger, S. Hernandez, and J.G. Lyon, 2004. "Solar wind density control of energy transfer to the magnetosphere", *Geophys. Res. Lett.,* **31**, No. 8, L08804, doi:10.1029/2003GL018780, 1-4.

Lu G., D.N. Baker, R.L. McPherron, C.J. Farrugia, D. Lummerzheim, J.M. Ruohoniemi, F.J. Rich, D.S. Evans, R.P. Lepping, M. Brittnacher, L.X. Li, R. Greenwald, G. Soflo, J. Villain, M. Lester, J. Thayer, T. Moretto, D. Milling, O. Troshichev, L.A. Zaitzev, V. Odintzov, L.G. Makarov, and K. Hayashi, 1998. “Global energy deposition during the January 1997 magnetic cloud event”, *J. Geophys. Res.*, **103**, No. A6, 11685-11694.

Lu J.Y., H. Jing, Z.Q. Liu, K. Kabin, and Y. Jiang, 2013. “Energy transfer across the magnetopause for northward and southward interplanetary magnetic fields”, *J. Geophys. Res., Space Phys.*, **118**, No. 5, 2021-2033.

Lu S., Y. Lin, V. Angelopoulos, A.V. Artemyev, P.L. Pritchett, Q. Lu, and X.Y. Wang, 2016, “Hall effect control of magnetotail dawn-duskasymmetry: A three-dimensional global hybrid simulation”, *J. Geophys. Res. Space Physics*, **121**, 11,882–11,895, doi:10.1002/2016JA023325.

Lyatsky W., P.T. Newell, and A. Hamza, 2001. “Solar illumination as cause of the equinoctial preference for geomagnetic activity”, *Geophys. Res. Lett.*, **28**, No. 12, 2353-2356.

Lyon J.G., J.A. Fedder, and C.M. Mobarry, 2004. “The Lyon–Fedder–Mobarry (LFM) global MHD magnetospheric simulation code”, *J. Atmos. Sol. Terr. Phys.*, **66**, No. 15-16, 13333-13350.

Magdaleno S., M. Herraiz, D. Altadill and B.A. de la Morena, 2017, “Climatology characterization of equatorial plasma bubbles using GPS data”, *J. Space Weather Space Clim.*, **7**, A3-1-12, doi:10.1051/swsc/2016039.

McIntosh D.H., 1959. “On the annual variation of magnetic disturbance”, *Philos. Trans. R. Soc. London*, Ser. A, **251**, No. 1001, 525-552.

McPherron R.L., D.N. Baker, T.I. Pulkkinen, T.-S. Hsu, J. Kissinger, and X. Chu, 2013. “Changes in solar wind–magnetosphere coupling with solar cycle, season, and time relative to stream interfaces”, *J. Atmos. Sol. Terr. Phys.*, **99**, 1-13.

Menk F.W., D. Orr, M.A. Clilverd, A.J. Smith, C.L. Waters, D.K. Milling, and B.J. Fraser, 1999. "Monitoring spatial and temporal variations in the dayside plasmasphere using geomagnetic field line resonances", *J. Geophys. Res.*, **104**, No. A9, 19955-19969.

Menk F.W., C.L. Waters, and B.J. Fraser, 2000. “Field line resonances and waveguide modes at low latitudes: 1. Observations”, *J. Geophys. Res.*, **105**, No. A4, 7747-7761.

Merkin V.G., K. Papadopoulos, G. Milikh, A.S. Sharma, X. Shao, J. Lyon, and C. Goodrich, 2003. “Effects of the solar wind electric field and ionospheric conductance on the cross polar cap potential: results of global MHD modeling”, *Geophys. Res. Lett.*, **30**, No. 23, 2180, doi:10.1029/2003GL017903, SSC1-1-5.

Merkin V.G., A.S. Sharma, K. Papadopoulos, G. Milikh, J. Lyon, and C. Goodrich, 2005. "Global MHD simulations of the strongly driven magnetosphere: Modeling of the transpolar potential saturation", *J. Geophys. Res.*, **110**, No. A9, A09203, doi:10.1029/2004JA010993, 1-10.

Milan S.E., J.A. Wild, A. Grocott, and N.C. Draper, 2006. "Space- and ground-based investigations of solar wind magnetosphere ionosphere coupling", *Adv. Space Res.*, **38**, No. 8, 1671-1677.

Moen J. and A. Brekke, 1993. "The solar flux influence on quiet time conductances in the auroral ionosphere", *Geophys. Res. Lett.*, **20**, No. 10, 971-974.

Morioka A., Y. Miyoshi, T. Seki, F. Tsuchiya, H. Misawa, H. Oya, H. Matsumoto, K. Hashimoto, T. Mukai, K. Yumoto, and T. Nagatsuma, 2003. "AKR disappearance during magnetic storms", *J. Geophys. Res.*, **108**, No. A6, 1226, doi:10.1029/2002JA009796., SMP3-1-9.

Mozer F.S., O.A. Agapitov, V. Angelopoulos, A. Hull, D. Larson, S. Lejosne, and J.P. McFadden, 2017, "Extremely field-aligned cool electrons in the dayside outer magnetosphere", *J. Geophys. Res. Lett.*, **44**, 44–51, doi:10.1002/2016GL072054.

Murakami G., K. Yoshioka, A. Yamazaki, Y. Nishimura, I. Yoshikawa, and M. Fujimoto, 2016. "The plasmapause formation seen from meridian perspective by KAGUYA", *J. Geophys. Res. Space Physics*, **121**, 11,973–11,984, doi:10.1002/2016JA023377.

Murayama T., A. Takao, H. Nakai, and K. Hakamada, 1980. "Empirical formula to relate the auroral electrojet intensity with interplanetary parameters", *Planet. Space Sci.*, **28**, No. 8, 803-813.

Nagatsuma T., 2002. "Saturation of polar cap potential by intense solar wind electric fields", *Geophys. Res. Lett.*, **29**, No. 10, 1422, doi:10.1029/2001GL014202, 62-1-4.

Nagatsuma T., 2004. "Conductivity dependence of cross-polar potential saturation", *J. Geophys. Res.*, **109**, No. A4, A04210, doi:10.1029/2003JA010286, 1-7.

Nagatsuma T., 2006. "Diurnal, semiannual, and solar cycle variations of solar wind–magnetosphere–ionosphere coupling", *J. Geophys. Res.*, **111**, No. A9, A09202, doi:10.1029/2005JA011122, 1-6.

Nakai H. and Y. Kamide, 1999. "Solar cycle variations in the storm-substorm relationship", *J. Geophys. Res.*, **104**, No. A10, 22695-22700.

Nakayama Y., Y. Ebihara, S. Ohtani, M. Gkioulidou, K. Takahashi, L.M. Kistler, and T. Tanaka, 2016, "Void structure of O+ ions in the inner magnetosphere observed by the Van Allen Probes", *J. Geophys. Res. Space Physics*, **121**, 11,698–11,713, doi:10.1002/2016JA023013.

Newell P.T., T. Sotirelis, K. Liou, C.-I. Meng, and F.J. Rich, 2006. "Cusp latitude and the optimal solar wind coupling function", *J. Geophys. Res.*, **111**, A09207, doi:10.1029/2006JA011731, 1-11.

Newell P.T., T. Sotirelis, K. Liou, C.-I. Meng, and F.J. Rich, 2007. "A nearly universal solar wind-magnetosphere coupling function inferred from 10 magnetospheric state variables", *J. Geophys. Res.*, **112**, A01206, doi:10.1029/2006JA012015, 1-16.

O'Brien T.P. and R.L. McPherron, 2000. "An empirical phase space analysis of ring current dynamics: Solar wind control of injection and decay", *J. Geophys. Res.*, **105**, No. A4, 7707-7719.

Orr D. and J.A.D. Matthew, 1971. "The variation of geomagnetic micropulsation periods with latitude and the plasmapause", *Planet. Space Sci.*, **19**, No. 8, 897-905.

Østgaard N. and E. Tanskanen, 2004. "Energetics of isolated and stormtime substorms", in *Disturbances in Geospace: The Storm-Substorm Relationship*, Geophys. Monogr. Ser.,

142, edited by A.S. Sharma, Y. Kamide, and G.S. Lakhina, AGU, Washington, D.C., 169-184.

Østgaard N., G. Germany, J. Stadsnes, and R.R. Vondrak, 2002a. "Energy analysis of substorms based on remote sensing techniques, solar wind measurements, and geomagnetic indices", *J. Geophys. Res.*, **107**, No. A9, 1233, doi:10.1029/2001JA002002, SMP9-1-14.

Østgaard N., R.R. Vondrak, J.W. Gjerloev, and G. Germany, 2002b. "A relation between the energy deposition by electron precipitation and geomagnetic indices during substorms", *J. Geophys. Res.*, **107**, No. A9, 1246, doi:10.1029/2001JA002003, SMP16-1-7.

Palmroth M., P. Janhunen, T.I. Pulkkinen, and H.E.J. Koskinen, 2004. "Ionospheric energy input as a function of solar wind parameters: Global MHD simulation results", *Ann. Geophys.*, **22**, No. 2, 549–566.

Palmroth M., H.E.J. Koskinen, T.I. Pulkkinen, P.K. Toivanen, P. Janhunen, S.E. Milan, and M. Lester, 2010. "Magnetospheric feedback in solar wind energy transfer", *J. Geophys. Res.*, **115**, No. A5, A00I10, doi:10.1029/2010JA015746, 1-14.

Park C.G., D.L. Carpenter, and D.B. Wiggin, 1978. "Electron density in the plasmasphere: whistler data on solar cycle, annual and diurnal variations", *J. Geophys. Res.*, **83**, No. A7, 3137-3144.

Perreault P. and S.-I. Akasofu, 1978. "A study of geomagnetic storms", *Geophys. J. R. Astron. Soc. (UK).*, **54**, No. 3, 547-583.

Petrukovich A.A., S.I. Klimov, A. Lazarus, and R.P. Lepping, 2001. "Comparison of the solar wind energy input to the magnetosphere measured by Wind and Interball-1", *J. of Atmospheric and Solar-Terrestrial Physics*, **63,** 1643–1647.

Ponomarev E.A., V.D. Urbanovich, and E.I. Nemtsova, 2000. "On the Excitation Mechanism of Magnetospheric Convection by the Solar Wind", in Proceedings of the 5th International Conference on Substorms, Held 16-20 May, 2000, at the Congress Centre of the Arctic and Antarctic Research Institute, St. Petersburg, Russia. Edited by A. Wilson. European Space Agency, ESA SP-443, 2000, 553-555.

Ponomarev E.A., P.A. Sedykh, and V.D. Urbanovich, 2006a. "Bow Shock as a Power Source for Magnetospheric Processes", *J. Atmos. Solar–Terr. Phys.* **68**, No. 6, 685–690.

Ponomarev E.A., P.A. Sedykh, and V.D. Urbanovich, 2006b. "Generation of Electric Field in the Magnetosphere Caused by Processes in the Bow Shock Region", *J. Atmos. Solar–Terr. Phys.* **68**, No. 6, 679–684.

Ponomarev E.A., P.A. Sedykh, and V.D. Urbanovich, 2009. "Dependence of the Power Consumed by the Magnetosphere on the Solar Wind Parameters", *Geomagnetism and Aeronomy*, **49**, No. 7 (Special Issue 1), 970–974.

Pulkkinen T.I., N.Y. Ganushkina, E.I. Kallio, G. Lu, D.N. Baker, N.E. Turner, T.A. Fritz, J.F. Fennell, and J. Roeder, 2002. "Energy dissipation during a geomagnetic storm: May 1998", *Adv. Space. Res.*, **30**, No. 10, 2231-2240.

Pulkkinen T.I., N. Partamies, R.L. McPherron, M. Henderson, G.D. Reeves, M.F. Thomsen, and H.J. Singer, 2007a. "Comparative statistical analysis of storm time activations and sawtooth events", *J. Geophys. Res.*, **112**, A01205, doi:10.1029/2006JA012024.

Pulkkinen T.I., C.C. Goodrich, and J.G. Lyon, 2007b. "Solar wind electric field driving of magnetospheric activity: Is it velocity or magnetic field?", *Geophys. Res. Lett.*, **34**, L21101, doi:10.1029/2007GL031011, 1-4.

Pulkkinen, T.I., M. Palmroth, P. Janhunen, H.E.J. Koskinen, D.J. McComas, and C.W. Smith, 2010a. "Timing of changes in the solar wind energy input in relation to ionospheric response", *J. Geophys. Res.*, **115**, No. A5, A00I09, doi:10.1029/2010JA015764, 1-9.

Pulkkinen T.I., M. Palmroth, H.E.J. Koskinen, T.V. Laitinen, C.C. Goodrich, V.G. Merkin, and J.G. Lyon, 2010b. "Magnetospheric modes and solar wind energy coupling efficiency", *J. Geophys. Res.*, **115**, A03207, doi:10.1029/2009JA014737.

Pulkkinen T.I., A.P. Dimmock, A. Lakka, A. Osmane, E. Kilpua, M. Myllys, E.I. Tanskanen, and A. Viljanen, 2016. "Magnetosheath control of solar wind–magnetosphere coupling efficiency", *J. Geophys. Res. Space Physics*, **121**, 8728–8739, doi:10.1002/2016JA023011.

Raeder J., 2003. "Global Magnetohydrodynamics - A Tutorial", in *Space Plasma Simulations*, Lecture Notes in Physics, **615**, edited by J. Buchner, C.T. Dum, and M. Scholer, Springer-Verlag, Berlin, Heidelberg, 212-246.

Rasmussen C.E. and R.W. Shunk, 1990. "A three-dimensional time-dependent model of the plasmasphere", *J. Geophys. Res.*, **95**, No. A5, 6133-6144, 6125.

Reeves G., M.C. Henderson, R.M. Skoug, M.F. Thomsen, J.E. Borovsky, H.O. Funsten, P. C. Son Brandt, D.J. Mitchell, J.-M. Jahn, C.J. Pollock, D.J. McComas, and S.B. Mende, 2004. "Image, polar, and geosynchronous observations of substorm and ring current ion injection", in *Disturbances in Geospace: The Storm-Substorm Relationship*, Geophys. Monogr. Ser., **142**, edited by A.S. Sharma, Y. Kamide, and G.S. Lakhina, AGU, Washington, D.C., 91-101.

Richards P.G., T. Chang, and R.H. Comfort, 2000. "On the causes of the annual variation in the plasmaspheric electron density", *J. Geophys. Res.*, or *J. Atmos. Terr. Phys.*, **62**, No. 10, 935-946.

Rippeth Y., R.J. Moffett, and G.J. Bailey, 1991. "Model plasmasphere calculations for *L*-values near 2.5 at the longitude of Argentine Islands, Antarctica", *J. Atmos. Terr. Phys.*, **53**, No. 6-7, 551-555.

Rosenqvist L., H.J. Opgenoorth, S. Buchert, O. Amm, I. McCrea, and C. Lathuillere, 2005. "Extreme solar-terrestrial events of October 2003: High latitude and cluster observations of the large geomagnetic disturbances on October 30", *J. Geophys. Res.*, **110**, No. A9, A09S23, doi:10.1029/2004JA010927, 1-12.

Rosenqvist L., S. Buchert, H. Opgenoorth, A. Vaivads, and G. Lu, 2006. "Manetospheric energy budget during huge geomagnetic activity using Cluster and ground-based data", *J. Geophys. Res.*, **111**, No. A10, A10211, doi:10.1029/2006JA011608, 1-14.

Rozanov E., L. Callis, M. Schlesinger, F. Yang, N. Andronova, and V. Zubov, 2005. "Atmospheric response to NO_y source due to energetic electron precipitation", *Geophys. Res. Lett.*, **32**, No. 14, L14811, doi:10.1029/2005GL023041, 1.

Russell C.T., 2002. "Multiscale coupling in planetary magnetospheres", *Adv. Space Research*, **30**, No. 12, 2647-2656.

Russell C.T. and R.L. McPherron, 1973. "Semiannual variations of geomagnetic activity", *J. Geophys. Res.*, **78**, No. 1, 92-108.

Russell C.T., Y.L. Wang, and J. Raeder, 2003. "Possible dipole tilt dependence of dayside magnetopause reconnection", *Geophys. Res. Lett.*, **30**, No. 18, 1937, doi:10.1029/2003GL017725, SSC5-1-4.

Sckopke N., 1966. "A general relation between the energy of trapped particles and the disturbance field near the Earth", *J. Geophys. Res.*, **71**, No. 13, 3125-3130.

Sedykh P.A. and E.A. Ponomarev, 2002. "Magnetosphere–Ionosphere Coupling in the Region of Auroral Electro-jets", *Geomagn. Aeron.* 42, No. 5, 613–618.

Senior C., 1980. "Les conductivitie ionospheriques et leur role dans la convection magnetospherique, Une etude experimentale et theorique", *Diplome de docteur de 3e cycle,* Univ. Pierre et Marie Curie, Paris.

Sergeev V., V. Angelopoulos, J.T. Gosling, C.A. Cattell, and C.T. Russell, 1996a. "Detection of localized, plasma-depleted flux tubes or bubbles in the midtail plasma sheet", *J. Geophys. Res.*, **101**, No. A5, 10817-10826.

Sergeev V.A., T.I. Pulkkinen, and R.J. Pellinen, 1996b. "Steady magnetospheric convection: A review of recent results", *Space Sci. Rev.*, **75**, No. 3-4, 551-604.

Shprits Y.Y. and M. Spasojevic, 2015. "Global and Comprehensive Analysis of the Inner Magnetosphere as a Coupled System: Physical Understanding and Applications", *SpaceWeather*, **13**, No. 9, doi:10.1002/2015SW001295, 533–535.

Siscoe G. and N. Crooker, 1974. "A theoretical relation between Dst and the solar wind merging electric field", *Geophys. Res. Lett.*, **1**, No. 1, 17-19.

Siscoe G. and W.D. Cummings, 1969. "On the cause of geomagnetic bays", *Planet. Space Sci.*, **17**, No. 10, 1795-1802.

Siscoe G.L. and K.D. Siebert, 2006. "Bimodal nature of solar wind–magnetosphere–ionosphere–thermosphere coupling", *J. of Atmospheric and Solar-Terrestrial Physics*, **68**, 911–920.

Siscoe G.L., G.M. Erickson, B.U.O. Sonnerup, N.C. Maynard, K.D. Siebert, D.R. Weimer, and W.W. White, 2001, "Global role of $E_{\|}$ in magnetopause reconnection: An explicit demonstration", *J. Geophys. Res.*, **106**, No. A7, 13015–13022.

Siscoe G.L., G.M. Erickson, B.U.Ö. Sonnerup, N.C. Maynard, J.A. Schoendorf, K.D. Siebert, D.R. Weimer, W.W. White, and G.R. Wilson, 2002. "Hill model of transpolar potential saturation: Comparisons with MHD simulations", *J. Geophys. Res.*, **107**, No. A6, 1075, doi:10.1029/2001JA000109, SMP8-1-8.

Siscoe G., J. Raeder, and A.J. Ridley, 2004. "Transpolar potential saturation models compared", *J. Geophys. Res.*, **109**, No. A9, A09203, doi:10.1029/2003JA010318, 1-10.

Slavin J.A., M.F. Smith, E.L. Mazur, D.N. Baker, E.W.J. Hones, T. Iyemori, and E.W. Greenstadt, 1993. "Isee 3 observations of traveling compression regions in the Earth's magnetotail", *J. Geophys. Res.*, **98**, No. A9, 15425-15446.

Sotirelis T., J.P. Skura, C.-I. Meng, and W. Lyatsky, 2002. "Ultraviolet insolation drives seasonal and diurnal space weather variations", *J. Geophys. Res.*, **107**, No. A10, 1305, doi:10.1029/2001JA000296, SMP15-1-12.

Tanskanen E., T.I. Pulkkinen, H.E.J. Koskinen, and J.A. Slavin, 2002. "Substorm energy budget during low and high solar activity: 1997 and 1999 compared", *J. Geophys. Res.*, **107**, No. A6, 1086, doi:10.1029/2001JA900153, SMP15-1-11.

Tarcsai G., P. Szemeredy, and L. Hegymegi, 1988. "Average electron density profiles in the plasmasphere between L=1.4 and 3.2 deduced from whistlers", *J. Atmos. Terr. Phys.*, **50**, No. 7, 607-611.

Taylor J.P.H. and A.D.M. Walker, 1984. "Accurate approximate formulae for toroidal standing hydromagnetic oscillations in a dipolar geomagnetic field", *Planet. Space Sci.*, **32**, No. 9, 1119-1124.

Troshichev O.A., N.A. Podorozhkina, D.A. Sormakov, and A.S. Janzhura, 2014. "PC index as a proxy of the solar wind energy that entered into the magnetosphere: Development of magnetic substorms", *J. Geophys. Res. Space Physics*, **119**, No. 8, doi:10.1002/2014JA019940, 6521-6540.

Turner N.E., D.N. Baker, T.I. Pulkkinen, J.L. Roeder, J.F. Fennell, and V.K. Jordanova, 2001. "Energy content in the stormtime ring current", *J. Geophys. Res.*, **106**, No. A9, 19149-19156.

Tyler E., C. Cattell, S. Thaller, J. Wygant, C. Gurgiolo, M. Goldstein, and C. Mouikis, 2016, "Partitioning of integrated energy fluxes in four tail reconnection events observed by Cluster", *J. Geophys. Res. Space Physics*, **121**, 11,798–11,825, doi:10.1002/2016JA023330.

Vasyliunas V.M., 2005. "Relation between magnetic fields and electric currents in plasmas", *Ann. Geophys.*, **23**, No. 7, 2589-2597.

Walker A.D.M., J.M. Ruohoniemi, K.B. Baker, R.A. Greenwald, and J.C. Samson, 1992. "Spatial and temporal behaviour of ULF pulsations observed by the Goose Bay HF radar", *J. Geophys. Res.*, **97**, No. A8, 12187-12202.

Waters C.L., F.W. Menk, and B.J. Faser, 1994. "Low latitude geomagnetic field line resonance; experiment and modeling", *J. Geophys. Res.*, **99**, No. A9, 17547-17558.

Webb P.A. and E.A. Essex, 2001. "A dynamic diffusive equilibrium model of the ion densities along plasmaspheric magnetic flux tubes", *J. Atmos. Solar Terr. Phys.*, **63**, No. 11, 1249-1260.

Yermolaev Y.I., I.G. Lodkina, N.S. Nikolaeva, M.Y. Yermolaev, and M.O. Riazantseva, 2017. "Some Problems of Identifying Types of Large-Scale Solar Wind and Their Role in the Physics of the Magnetosphere", *Cosmic Research*, **55**, No. 3, 178–189. Original Russian Text published in *Kosmicheskie Issledovaniya*, **55**, No. 3, 189–200.

Zhang X., X. Shen, J. Liu, Z. Zeren, L. Yao, X. Ouyang, S. Zhao, G. Yuan, and J. Qian, 2014. "The solar cycle variation of plasma parameters in equatorial and mid latitudinal areas during 2005–2010", *Adv. Space Res.*, **54**, No. 3, 306-319.

Zhao H. and Q.-G. Zong, 2012. "Seasonal and diurnal variation of geomagnetic activity: Russell-McPherron effect during different IMF polarity and/or extreme solar wind conditions", *J. Geophys. Res.*, **117**, No. A11, A11222, doi:10.1029/2012JA017845, 1-15.

Chapter 2

COUPLING OF SOLAR WIND/IMF WITH GEOMAGNETOSPHERE/IONOSPHERE

2.1. SOLAR WIND–MAGNETOSPHERE COUPLING FUNCTION AND FORECASTING

2.1.1. Search of Nearly Universal Solar Wind-Magnetosphere Coupling Function

Newell et al. (2007) developed a nearly universal solar wind-magnetosphere coupling function inferred from several magnetospheric and solar wind state variables. They investigated whether one or a few coupling functions can represent best the interaction between the solar wind and the magnetosphere over a wide variety of magnetospheric activity. Ten variables which characterize the state of the magnetosphere were studied. Five indices from ground-based magnetometers were selected, namely *Dst*, *Kp*, *AE*, *AU*, and *AL*, and five space age indices, namely auroral power, Polar UVI measurements of auroral power, DMSP measurements of cusp latitude (i.e., $\sin(\Lambda_c)$), and the nightside multi-keV ion precipitation boundary (b2i), magnetotail inclination angle from NOAA GOES-8, and the polar cap size as inferred from SuperDARN convection reversal boundaries. These indices were correlated with more than 20 candidate solar wind-magnetosphere coupling functions. The data were downloaded from NOAA's Space Data Center for the years 1984–2005, which is divided into two data runs, one from 1984 to 1994 and the other from 1995 to 2005. The main obtained results are summarized by Newell et al. (2007) as following:

(i) Only one coupling function from checked 20 functions, correlate best with 9 out of 10 indices of magnetospheric activity (with correlation coefficients about 0.7–0.8). This coupling function represents the rate magnetic flux which is opened at the magnetopause:

$$d\Phi_{MP}/dt = v^{4/3} B_T^{2/3} \sin^{8/3}(\theta_c/2), \qquad (2.1.1)$$

where v is the solar wind velocity, B_T – magnetic field amplitude, θ_c is the IMF clock angle.

(ii) Only *Dst* index (one from 10) does not correlate with $d\Phi_{MP}/dt$, but has the best correlation over two solar cycles with $p^{1/2} d\Phi_{MP}/dt$, where p is the dynamic pressure of solar wind (with correlation coefficient 0.87).

(iii) The apparent physical interpretation is that the rate of merging on the dayside magnetopause is the single largest correlate for most magnetospheric activity.

(iv) The most commonly encountered coupling functions in general use are products of the solar wind electric field multiplied by a snippet extracted from the merging rate, with other factors discarded.

(v) Commonly used coupling function $E_{KL} = vB_T \sin^2(\theta_c/2)$ appends an estimate of the fractional merging rate as a function of magnetic shear to the electric field, while not including such factors as the strength of the magnetic field at the magnetopause, or the length of the merging line.

(vi) In contrast, the work of Siscoe and Huang (1985), Lockwood et al. (1990), and Cowley and Lockwood (1992) has demonstrated how the merging rate can explain ionospheric convection without the need to reference the solar wind electric field. In opinion of Newell et al. (2007), the latter is thus superfluous.

(vii) Although these ideas have been most extensively applied to the ionosphere, Newell's et al. (2007) results suggest that they apply to the entire magnetosphere too. Indeed, Newell et al. (2007) have chosen to drop consideration of the solar wind electric field and estimate the four factors which go into the rate magnetic flux are opened at the magnetopause. These are the rate field lines are convected toward the magnetopause (v), the percentage of field lines which subsequently merge ($\sin^{8/3}(\theta_c/2)$), the strength of the magnetic field (B_T), and the length of the merging line ($(B_{MP}/B_T)^{1/3}$).

2.1.2. A Forecasting Model of the Magnetosphere Driven by an Optimal Solar Wind Coupling Function

A new empirical magnetospheric magnetic field model is described by Tsyganenko and Andreeva (2015), driven by interplanetary parameters including a coupling function by Newell et al. (2007) – see previous Section 2.1.1, termed henceforth as '*N* index'. The model uses data from Polar, Geotail, Cluster, and THEMIS satellites, obtained in 1995–2013 at distances 3–60 R_E. The model magnetopause is based on Lin et al. (2010) boundary driven by the solar wind pressure, IMF B_Z, and the geodipole tilt. The model field includes contributions from the symmetric ring current (SRC), partial ring current (PRC) with associated Region 2 field-aligned currents (R2 FAC), tail, Region 1 (R1) FAC, and a penetrated IMF. Increase in the N index results in progressively larger magnitudes of all the field sources, the most dramatic and virtually linear growth being found for the PRC and R1 FAC. The solar wind dynamic pressure P_{dyn} affects the model magnetotail current in proportion to the factor $[P_{dyn}/\langle P_{dyn}\rangle]^{\xi}$, where the exponent ξ on the order of 0.4–0.6 steadily decreases with increasing N index. The PRC peaks near midnight at $N \sim 0$ but turns duskward with growing N. At ionospheric altitudes, both R1 and R2 FAC expand equatorward with growing N and P_{dyn}, and the R2 zone rotates westward. Larger values of N result in a more efficient penetration of the IMF into the magnetosphere and larger magnetic flux connection across the magnetopause. Growing dipole tilt is accompanied by a persistent and significant decrease of the total current in all magnetospheric field sources.

2.2. Propagation and Modification of Interplanetary Shock in the Geomagnetosphere

2.2.1. Propagation of Interplanetary Shock through the Bow Shock, Magnetosheath, and Magnetopause

Samsonov et al. (2006) make detail numerical MHD simulation of propagation of interplanetary shock (IPS) through the bow shock, magnetosheath, and magnetopause. They note that the IPS is one of main causes of large geomagnetic disturbances in the Earth's magnetosphere (e.g., Tsurutani and Gonzalez, 1997). The IPS is a fast MHD shock moving from the Sun and it is often driven by the coronal mass ejection. Three steps can be determined at the starting phase of interaction between the IPS and the magnetosphere: the interaction of the IPS with the bow shock, the propagation of the IPS through the magnetosheath, and finally the interaction of the IPS with the magnetopause. The interactions of the IPS with the bow shock and magnetopause were studied theoretically in several early papers using the Rankine-Hugoniot conditions. It was shown (e.g., Ivanov, 1964; Dryer et al., 1967; Shen and Dryer, 1972; Grib et al., 1979) that the interaction of the IPS with the bow shock (which is a reversed fast shock) results in two fast shocks, forward and reversed, and the contact discontinuity located between these shocks in a one-dimensional (1-D) case without the magnetic field or with magnetic field directed perpendicular to the solar wind velocity and to the shock normals. Comparison between the MHD theory and ISEE data was made by Zhuang et al. (1981). In particular, they demonstrated that all discontinuities (including the bow shock) moved with velocities predicted by the Rankine-Hugoniot conditions after the interaction.

Taking the angle between the IMF and the solar wind velocity equal to 45°, Grib (1982) found by a similar analysis that the interaction of the IPS and bow shock results in an ensemble consisting of a forward fast shock (FS), a forward slow expansion wave (SEW), a contact discontinuity (CD), a reversed slow shock (SS), and finally, a reversed fast shock (i.e., the modified bow shock). It was noted that the SEW and SS are weak in comparison with three other discontinuities. A more general 2-D case for arbitrary angles between the two initial shock fronts and the IMF direction was considered by Pushkar et al. (1991). It was found that the obtained combination of discontinuities resulted from the FS-FS interaction may change in dependence on all input parameters. In particular, the interaction between the IPS and the bow shock in the subsolar region and at the flanks may create different combinations of discontinuities. A number of discontinuities in the resulted ensemble may vary from three up to seven. Note that the last corresponds to a general solution of the Riemann problem (e.g., Jeffrey and Taniuti, M1964).

Samsonov et al. (2006) outlined that the theoretical study of Grib (1982) was confirmed by the numerical 1-D MHD simulations of Yan and Lee (1996), who considered a similar case with the angle between the magnetic field and flow velocity equal to 45° in the initial conditions upstream of the bow shock. They found that the interaction of two fast shocks results in the same five discontinuities: FS, SEW, CD, SS, and the modified bow shock.

The third step of the IPS–magnetosphere interaction mentioned above is the interaction with the magnetopause. An increase of the solar wind dynamic pressure after the IPS causes an earthward magnetopause motion. According to opinion of Samsonov et al. (2006), only global 3-D MHD models of solar wind–magnetosphere coupling can take this motion into account in a self-consistent manner. Samsonov et al. (2006) outlined that using the Rankine-Hugoniot conditions, Grib (1971, 1972), Grib et al. (1979) found that the interaction between a FS and the magnetopause considered as a tangential discontinuity (TD) results in a fast rarefaction wave propagating toward the bow shock. Wu (2003a) obtained the same results from the 2-D MHD simulation. Grib et al. (1979) also predicted that the interaction of this rarefaction wave with the bow shock leads to another reflected rarefaction wave propagating toward the magnetopause. The interaction of this secondary rarefaction wave with the magnetopause causes an outward magnetopause motion that follows its initial compression. One may assume that this chain of wave transformations can repeat many times. Namely, the outward magnetopause motion may in turn result in a reversed fast shock moving to the bow shock which would reflect from the bow shock and so on. However, it is unclear whether this picture is realistic or not because the waves may change during the magnetosheath crossings. Note that Grib et al. (1979) have assumed that the waves propagate through the magnetosheath without any modification. Samsonov et al. (2006) note that Spreiter and Stahara (1994) studied the propagation of the IPS through the magnetosheath using the hydrodynamic model and found that the shape of the shock front in the magnetosheath remains nearly planar during its propagation to the flanks. This result was also demonstrated in the laboratory experiments by Dryer et al. (1967) and by Dryer (1973). Recently, Koval et al. (2005) have shown that the shock front in the magnetosheath may be inclined and this inclination results in a delay of shock arrival to a spacecraft orbiting in the inner magnetosheath. They support their observations by results of MHD modeling. However, Koval et al. (2005) have simulated a model event with weak jumps of parameters across the IPS assuming that the magnetopause does not respond to variations of the solar wind dynamic pressure.

In paper of Samsonov et al. (2006), the propagation of an IPS from the supersonic solar wind to the magnetopause is studied using numerical results of the local MHD magnetosheath model. The IPS front is assumed to be perpendicular to the solar wind velocity and the angle between the IMF and velocity vectors is equal to 45°. They show that this 3-D calculation repeats all features that have been discussed above, i.e., the interaction of an IPS with the bow shock results to inward bow shock motion and generates new discontinuities, and the IPS is slowed down in the magnetosheath. In order to investigate the influence of the variations of the magnetopause shape and its location after the IPS arrival on profiles of magnetosheath parameters, Samsonov et al. (2006) have simulated three cases: (**1**) stationary magnetopause that does not respond to changes of the upstream pressure, (**2**) magnetopause that moves inward with a speed obtained from the pressure balance, and (**3**) magnetopause moves with a speed estimated from magnetopause models. The three-dimensional magnetosheath numerical model of Samsonov and Pudovkin (2000) and Samsonov and Hubert (2004) has been modified by Samsonov (2006) in order to extend the numerical field to the magnetosheath flanks. They simulate the interaction of the supersonic solar wind with a parabolic obstacle using the nonstationary MHD equations. Two coordinate systems have been applied: the parabolic coordinates in the main part of the numerical box, and the spherical coordinates near the subsolar region in order to avoid the singularity of the parabolic coordinates at the Sun-Earth line. The two coordinate systems intersect and the values from inner points of one coordinate system have been used to determine the boundary conditions for the other system.

As the initial conditions, typical supersonic solar wind parameters have been taken at the external boundaries: $\rho_0 = 5$ cm^{-3}, $B_0 = 5$ nT, $T_0 = 2.4\times10^5$ K, $V_0 = 400$ km/s. The solar wind velocity points along the Sun-Earth line, while the magnetic field is in the direction of 45° away from the Sun-Earth line in the XY_{GSM} plane. The bow shock forms inside the numerical field self-consistently in course of the simulations. The shock front is assumed to be a plane perpendicular to the Sun-Earth line. Since the outer boundary of the numerical box is curved, the time varying boundary conditions are imposed in order to describe explicitly propagation of the IPS in the solar wind. The inner boundary coincides with the magnetopause under initial conditions. Assuming no magnetopause magnetic reconnection in this case, Samsonov et al. (2006) impose conditions $B_n = 0$ and $V_n = 0$ (the subscript n stands for the direction normal to the boundary). When the IPS approaches the magnetopause, it would move earthward due to increase of the total pressure in the magnetosheath side. This inward motion can be taken into account in the local magnetosheath model by introduction of an inward velocity, V_n, through the boundary. The boundary itself is at rest in the Earth's frame and thus it represents a surface somewhere in the magnetosheath after magnetopause compression and an amount of plasma should penetrate through it.

The pressure balance conditions would be satisfied at the magnetopause (at least, in the case without magnetic reconnection). Using the model, it is possible to calculate how the total pressure varies on the internal boundary during the propagation of the IPS. The normal component V_n can be determined from a pressure balance:

$$\rho V_n^2/2 = P'_{new} - P'_{old}, \tag{2.2.1}$$

where $p' = p+B^2/8\pi$ is the total (thermal and magnetic) pressure at the boundary, the index old means values taken from the initial conditions, and the index new means values taken in the same grid points after the shock reached the magnetopause. This relation has been found from the equation of motion using an assumption that the total pressure inside the magnetosphere does not change when the magnetopause moves inward. It seems that such assumption results in overestimation of the normal velocity at the boundary, and the velocity obtained from this relation represents an upper limit of the normal velocity. The numerical results of Samsonov et al. (2006) show that new discontinuities form after interaction between the forward IPS and the bow shock. When the launched IPS crossed the bow shock, a new discontinuity in addition to the fast forward shock is seen in numerical results downstream of the bow shock. This discontinuity is characterized by an increase of the magnetic field intensity and plasma density accompanied with a decrease of the temperature, and with no visible change of the velocity.

Samsonov et al. (2006) found that even the increased spatial resolution in the simulations is not sufficient to reproduce the separated discontinuities, although the transition from their one-dimensional to our three-dimensional modeling may play a role too. Their estimations show that the slow wave velocity at the Sun-Earth line decreases from 10 km/s just downstream of the bow shock to nearly zero in the vicinity of the magnetopause. This change of the wave velocity corresponds to decrease of the radial magnetic field component near the magnetopause due to draping effect. Since the flow velocity in the same region is larger than 200 km/s (after the forward IPS), the slow shock velocity in the plasma frame does not exceed a few percents of the flow velocity. It means that the distance between the SEW (or SS) and the CD during their propagation toward the magnetopause would be close to the grid spacing of the simulations. The time of propagation of the forward IPS through the subsolar magnetosheath is about one minute. Taking this time as a characteristic timescale, Samsonov et al. (2006) get an average time interval between the SEW (SS) and the CD equal to a few seconds. It seems to be larger at the magnetosheath flanks, nevertheless, it is hardly possible to identify a particular discontinuity in spacecraft data as well as in the three-dimensional numerical simulations. Thus it may be observed as a discontinuity of a complicated type SEW-CD-SS. Samsonov et al. (2006) note that the bow shock begins to move toward the magnetopause with a speed of about 1 R_E per minute immediately after the IPS crossing. This inward motion agrees with predictions of the MHD theory and with the observations of Zhuang et al. (1981). The bow shock velocity can be determined from the Rankine-Hugoniot conditions using a combination of upstream and downstream parameters, therefore it may vary in time due to following changes of the magnetosheath parameters. In turn, this may cause distortions of the bow shock shape.

Samsonov et al. (2006) note that prior to the description of the propagation of the IPS inside the magnetosheath from the subsolar region to the flanks, it would be important to discuss the interaction between the IPS and magnetopause because the character of this interaction influences strongly magnetosheath parameters. Any reasonable magnetopause model predicts that the magnetopause moves inward after the interaction with the IPS. Two questions are interesting: deepness of this magnetopause trip and the reaction of the magnetosheath on this displacement. First question can be answered comparing results of different magnetopause models. Samsonov et al. (2006) include the magnetopause motion into their local numerical model to answer the second question. The extent of the magnetopause shift in the dayside region has been estimated using the relations for position of the subsolar magnetopause taken from the models of Shue et al. (1998), Pudovkin et al. (1998), and using the condition obtained from the pressure balance for a planar magnetopause. Before the shock launch, these models predict the position of the subsolar point at $r = 10.9, 10.2, 10.4\ R_E$, respectively. After the shock, they predict $r = 8.6, 7.8, 8.0\ R_E$. Although the models predict different positions of the subsolar point before the shock, they give almost the same displacement due to the shock crossing: 2.3 R_E for the first model, and 2.4 R_E for two others. Note that these estimations are made for $B_z = 0$, consistently with Samsonov's et al. (2006) model shock.

Samsonov et al. (2006) came to following conclusions:

(i) The propagation of an interplanetary shock through the bow shock and magnetosheath has been simulated using a local 3-D MHD model. An artificial planar IPS perpendicular to the Sun-Earth line has been studied.

(ii) Interaction of an IPS with the bow shock creates a new discontinuity propagating through the magnetosheath toward the magnetopause with the plasma flow velocity. This discontinuity is characterized by an increase of the density, a decrease of the temperature, and a weak increase of the magnetic field intensity. This is consistent with a combination of the slow expansion wave, contact discontinuity, and slow shock obtained previously in the 1-D MHD model of Yan and Lee (1996), and in the theoretical analysis of Grib (1982).

(iii) The bow shock begins its inward motion immediately after the interaction with the IPS. The region of shifted bow shock follows the IPS front from the subsolar part to the flanks. In the subsolar region, the displacement of the bow shock for typical artificial IPS may exceed 2 r_E. Similar values have been obtained for displacement of the subsolar magnetopause using magnetopause models of Shue et al. (1998) and Pudovkin et al. (1998) and the magnetopause pressure balance condition.

(iv) Both a fast reversed shock and a fast expansion wave may result from interaction of the IPS with the magnetopause depending on boundary conditions of the model. The appearance of the reflected fast expansion wave propagating toward the bow shock agrees with the prediction made by using the Rankine-Hugoniot conditions (e.g., Grib et al., 1979).

(v) Samsonov's et al. (2006) simulations confirm a deceleration of the IPS front in the magnetosheath described earlier by Koval et al. (2005). The delay of arrival of the IPS at the magnetosheath flanks depends on the position inside the magnetosheath and generally increases from the outer to inner magnetosheath.

2.2.2. Modification of Interplanetary Shocks Near the Bow Shock and through the Magnetosheath

Ŝafránková et al. (2007) investigate the modification of interplanetary shocks (IPS) near the bow shock and through the magnetosheath. They continued the research started in Samsonov et al. (2006), which was described in Section 1.10.2. Ŝafránková et al. (2007) noted that the interaction of IPS with Earth's bow shock and their modification through the magnetosheath and magnetopause have been a subject of a quasi-steady gasdynamic numerical modeling (with convected magnetic field) by Spreiter and Stahara (1992). Using the Rankine-Hugoniot conditions, Dryer et al. (1967), Ivanov (1964), Dryer (1973), Shen and Dryer (1972), and Grib et al. (1979) have shown that the interaction of the IPS with the bow shock (which is a fast reversed shock) creates three discontinuities: the fast reverse shock (the original bow shock), fast forward shock (the original IPS) and a contact discontinuity between these shocks in a one-dimensional hydrodynamic case. A comparison between the 3-D MHD simulations and experimental observations was prepared by Zhuang et al. (1981) and the authors concluded that all observed discontinuities moved with velocities corresponding to the Rankine-Hugoniot conditions. In their MHD estimations, Grib (1982) and Pushkar et al. (1991) found that the interaction of the IPS and bow shocks results in a train of different kinds of discontinuities and the number of these discontinuities changes with the distance from the Sun-Earth line (including a complete sequence of seven forward and reverse MHD discontinuities at the magnetospheric flanks). These conclusions were confirmed by the 1-D simulation studies of Yan and Lee (1996) that found five discontinuities and a modified bow shock along the Sun-Earth line. On the other hand, as Zhuang et al. (1981) noted, the propagation of IPS within the magnetosheath is propagation through an inhomogeneous medium to the obstacle: the magnetopause. Analysis of the Rankine-Hugoniot conditions (Grib, 1972; Grib et al., 1979; Wu, 2003a) shows that the interaction between the fast shock and the magnetopause (defined as a tangential discontinuity) results in a rarefaction wave propagating toward the bow shock. Zhuang et al. (1981) indicated that this wave reflected from the magnetopause. In the magnetosheath, the predicted discontinuity shape remains nearly planar in the gasdynamic model (Spreiter and Stahara, 1994). This prediction is supported by observations reported by Szabo (2005) and Russell et al. (2000). On the other hand, deviations from the planarity assumption have been also reported (e.g., Russell et al., 1983; Ŝafránková et al., 1998; Szabo et al., 2001). A series of papers by Koval et al. (2005, 2006a, 2006b) demonstrates that the shock front in the magnetosheath is inclined and this inclination causes a delay of the shock arrival to the magnetopause (see Figures 2.2.1-2.2.3). A nonplanar shock propagation through the magnetosheath derived from experimental observations has been confirmed by two numerical MHD simulations (Koval et al., 2006b).

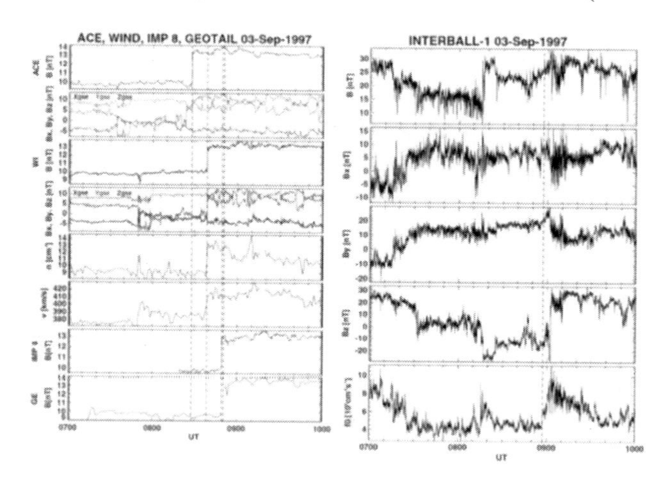

Figure 2.2.1. Observations of an IP shock by WIND (WI), ACE, and IMP 8 in the solar wind (left panels) and a corresponding disturbance observed by INTERBALL-1 in the magnetosheath (right panels). Left panels: solar wind number density and speed measured by WIND and IMF measured by WIND, ACE, and IMP 8; right panels: magnetosheath magnetic field, ion anti-sunward flux, proton number density and speed, ion and electron spectra. According to Koval et al. (2006a). Permission from COSPAR.

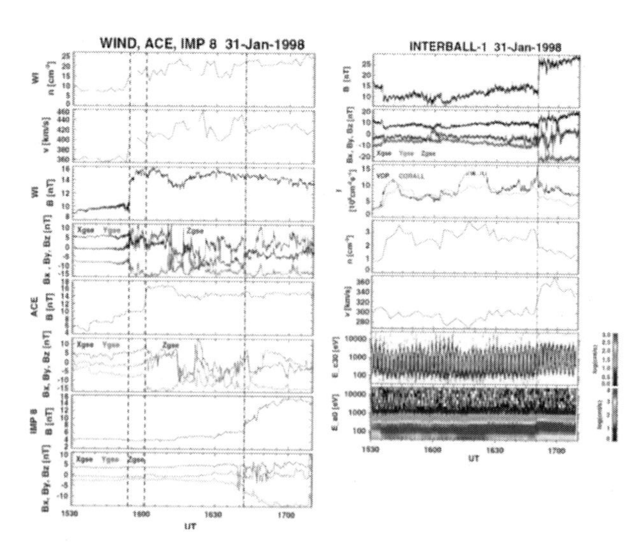

Figure 2.2.2. Observations of an IP shock by ACE, WIND (WI), IMP 8, and GEOTAIL (GE) in the solar wind (left panels) and a corresponding disturbance observed by INTERBALL-1 in the magnetosheath (right panels). Left panels: IMF measured by ACE, WIND, IMP 8, and GEOTAIL and solar wind number density and speed measured by WIND. Right panels: magnetosheath magnetic field and ion flux in anti-sunward direction. According to Koval et al. (2006a). Permission from COSPAR.

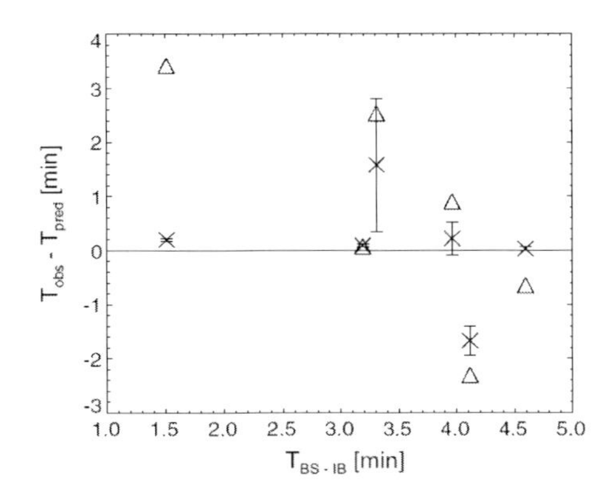

Figure 2.2.3. Differences between observed and predicted times of a shock arrival to the INTERBALL-1 location in the magnetosheath as a function of initial IP shock time of propagation from the Earth's bow shock to INTERBALL-1. Time differences that correspond to the shock parameters obtained from four spacecraft timing are denoted by crosses; time differences that correspond to the shock parameters obtained from Rankine–Hugoniot conservation equations are denoted by triangles. According to Koval et al. (2006a). Permission from COSPAR.

Ŝafránková et al. (2007) outlined, that Samsonov et al. (2006) simulates the propagation of an IPS from the supersonic solar wind to the magnetopause using a local 3-D MHD magnetosheath model (see above, Section 1.2). An IPS is assumed to be perpendicular to the solar wind velocity and the angle between IMF and velocity vectors is equal to 45°. In their 3-D simulations, the interaction of an IPS with the bow shock results to an inward bow shock motion, it generates new discontinuities propagating through the magnetosheath toward the magnetopause with the plasma flow velocity. Moreover, their numerical model allows them to simulate three different types of magnetopause response: (1) a stationary magnetopause without a response to changes of the upstream pressure, (2) a magnetopause that moves inward with a speed derived from pressure balance, and (3) a magnetopause that moves with a speed estimated from magnetopause location models. In Ŝafránková et al. (2007) are presented a series of observations of the interaction of fast forward IPS with the bow shock and their propagation within the magnetosheath and it was compared the experimental observations with predictions of different versions of the aforementioned 3-D MHD model of Samsonov et al. (2006). In the initial conditions, Ŝafránková et al. (2007) take typical solar wind parameters near 1 AU (B_1 = 5 nT, n_1 = 5 cm^{-3}, T_1 = 2 × 10^5 K, v_1 = 400 km/s) with the angle between the IMF and solar wind velocity equal to 45° and B_Z = 0. Then, Ŝafránková et al. (2007) impose an artificial fast forward MHD shock with jump conditions satisfying the Rankine-Hugoniot relations (B_2/B_1 = 2.25, v_2/v_1 = 1.31, n_2/n_1 = 2.84, v_{IPS} = 594 km/s, M_A = 4). Hereafter, the suffixes 1 and 2 stand for upstream and down-stream parameters of the IPS, respectively. Ŝafránková et al. (2007) have analyzed a few observations of IPS in the magnetosheath (28 July 1996, 4 March and 6 August 1998, and 11 February 2000). This analysis was supported by MHD modeling. Attention was devoted to global features of the IPS–bow shock–magnetopause interaction as well as to the structure of the IPS front in the magnetosheath. The results of this analysis can be written as follows:

(i) The IPS passage into the magnetosheath causes the inward motion of the bow shock. This motion is a result of changes of magnetosheath parameters downstream of the IPS.

These observations are in an agreement with the predictions based on the Rankine- Hugoniot conditions (Grib et al., 1979; Zhuang et al., 1981).

(ii) The inward bow shock displacement is followed by an outward motion. A combination of these two motions results in creation of an indentation of the bow surface that moves along the bow shock together with the IPS. The deepness of this indentation should be units of r_E because its motion is very often recorded as two bow shock crossings separated by 1–5 min.

(iii) The outward bow shock motion is probably a consequence of the IPS interaction with an obstacle. This obstacle can be the magnetopause or other boundary in the inner magnetosphere.

(iv) The interaction of the IPS with the bow shock and magnetopause frequently results in two discontinuities. The time lag between them is of the order of 1 min. The first usually exhibits features of the original fast forward shock. The second discontinuity is characterized by a density increase and a decrease of the temperature. The velocity exhibits no or very small change across the discontinuity. The magnetic field magnitude remains nearly unchanged in some cases but in other cases, sharp and large increases are observed.

(v) The discontinuities are products of the IPS–bow shock interaction and the observed patterns are generally consistent with the MHD magnetosheath model of Samsonov et al. (2006).

(vi) It was not able to identify more than one discontinuity following the IPS in the magnetosheath. The train of discontinuities suggested by Grib (1982) or Yan and Lee (1996) probably could not be observed because, as Samsonov et al. (2006) noted, their velocities are very similar and thus they cannot be distinguished in experimental data.

(vii) The experimental data are generally consistent with the Samsonov et al. (2006) model but several points are still waiting for explanation: 1. Why does the model not reproduce the observed splitting of the IPS at the flanks? 2. What is the source of the outward bow shock motion? 3. What is the role of the IMF orientation in the IPS interaction with the bow shock and magnetopause?

(viii) The answers to these questions require further effort in modeling and careful analysis of new observations. Of course, many attempts to answer these and related questions were made in course of years. For example, Shen and Dryer (1972) and Grib et al. (1979) suggested an oscillatory motion of the magnetopause because its interaction with the IPS could explain observed outward motion of the bow shock. However, these calculations were 1-D and thus they can be probably applied to the subsolar region but discussed observations were made on the flanks.

2.3. Solar Wind and IMF Control Processes in Geomagnetosphere

2.3.1. Time Delay of IMF Penetration into Earth's Magnetotail

As outlined Rong et al. (2015), many previous studies have demonstrated that the IMF can control the magnetospheric dynamics. Immediate magnetospheric responses to the external IMF have been assumed for a long time. The specific processes by which IMF penetrates into magnetosphere, however, are actually unclear. Solving this issue will help to accurately interpret the time sequence of magnetospheric activities (e.g., substorm and tail plasmoids) exerted by IMF. With two carefully selected cases, Rong et al. (2015) found that the penetration of IMF into magnetotail is actually delayed by 1–1.5 h, which significantly lags behind the magnetotail response to the solar wind dynamic pressure. The delayed time appears to vary with different auroral convection intensity, which may suggest that IMF penetration in the magnetotail is controlled considerably by the dayside reconnection. Several unfavorable cases demonstrate that the penetration lag time is more clearly identified when storm/substorm activities are not involved.

2.3.2. Solar Wind Control of Plasma Density in the Plasma Sheet

As outlined Nagata et al. (2007), the plasma sheet is a key region to understand the mass and energy transport in the magnetosphere. Among the properties of the plasma sheet, the plasma number density is important for space weather research through the mass transport into the inner magnetosphere. However, prediction of the plasma sheet number density in response to solar wind conditions is more difficult compared to that of other properties such as plasma pressure and temperature (Tsyganenko and Mukai, 2003) due to the complex source, transport, acceleration/heating, and loss (STAHL) processes (Williams, 1997). Thus empirical modeling of plasma sheet number density, which gives constrains for numerical simulation, is a major challenge of plasma sheet research. There are two known solar wind parameters that correlate with the plasma sheet number density. One is the solar wind number density and the other is the north-south component of IMF B_Z. The plasma sheet number density has positive correlation with the solar wind number density in the power law form (Borovsky et al., 1998). IMF B_Z is important to understand the variation of plasma sheet number density under the same level of solar wind number density condition: southward IMF condition results in hot and tenuous plasma sheet and northward IMF condition leads to cool and dense plasma sheet (Terasawa et al., 1997). The time lag between variations of solar wind and plasma sheet is also crucial to understand the transport path of magnetosheath particles into and through the plasma sheet (Terasawa et al., 1997; Borovsky et al., 1998). While the effect of IMF B_Z as well as its duration has been investigated based on remote and in situ observations (Wing and Newell, 2002; Wing et al., 2005; Wang et al., 2006), the effect of solar wind number density was averaged out in those studies. The dependence on solar wind number density was considered by Tsyganenko and Mukai (2003) by developing a global analytic function for the plasma sheet number density that depends on the solar wind number density, velocity, and IMF B_Z. In their study, solar wind parameters averaged over a 30-min interval prior to the plasma sheet observation were used. Hence the time lag between plasma sheet and solar wind parameters was not fully included because the plasma sheet has the time lag of several hours to IMF B_Z (Terasawa et al., 1997). According to Nagata et al. (2007), to improve our knowledge about the solar wind control of plasma sheet number density, it is necessary to consider effects of solar wind number density, IMF B_Z, and time lag simultaneously.

Observations of solar wind and plasma sheet from January 1995 to December 2002 were utilized in the Nagata et al. (2007) study. Solar wind data were provided by the ACE and

WIND spacecraft. WIND data were used before 4 February 1998 and ACE data were used afterward. WIND data were used when the position of WIND was located at the upstream of the bow shock and not far from Sun-Earth line (X_{GSE} > 20 R_E and $|Y_{GSE}|$ < 50 R_E). Plasma parameters were from the SWEPAM (McComas et al., 1998) instrument onboard the ACE and the SWE (Ogilvie et al., 1995) instrument on board the WIND. Magnetic field data were from the MFI (Lepping et al., 1995) instrument onboard the WIND and the MAG (Smith et al., 1998) instrument on board the ACE. All data were time shifted to the position of the Earth by the observed solar wind velocity on the simple linear advection assumption. Solar wind data were averaged over 10-min intervals. Plasma sheet data were from GEOTAIL measurements. Plasma parameters were obtained by the LEP-EAI (Mukai et al., 1994) and the EPIC-STICS (Williams et al., 1994) instruments. The energy range from 20 eV/q to 270 keV/q was covered by these instruments. All particles detected by the LEP-EAI were assumed to be protons. Magnetic field data were from the MGF (Kokubun et al., 1994) instrument. Nagata et al. (2007) used the Geocentric Solar Wind (GSW) coordinate system to represent the GEOTAIL observations. The GSW coordinate system is similar to the GSM coordinate system except that the direction of X axis is antiparallel to the direction of solar wind flow. Investigation of Nagata et al. (2007) was limited to the nightside (X < 0 R_E) and the radial distance of 9 to 30 R_E from the Earth.

Plasma sheet and solar wind observations are fitted in Nagata et al. (2007) to a model equation to derive the distributions of parameters which represent the dependences on solar wind parameters. The model equation was determined on the assumption that the plasma sheet number density depends on the solar wind number density in power law form (Borovsky et al., 1998) and on IMF B_z in exponential form (Terasawa et al., 1997):

$$N_{ps} = N_0\left(\langle N_{sw}\rangle/5\right)^{\eta}\exp\left(\alpha\langle B_z\rangle\right), \tag{2.3.1}$$

where N_{ps} is the plasma sheet number density, $\langle N_{sw}\rangle$ and $\langle B_z\rangle$ are the solar wind number density and IMF B_Z averaged from t_1 hours to t_2 hours prior to each plasma sheet observation (t_1 = 1, 2, . . ., 10 and t_2 = 0, 1, . . ., t_1 − 1). $\langle N_{sw}\rangle$ is normalized by a typical value of 5/cm^3. The parameter N_0 is interpreted as the plasma number density under the condition of $\langle N_{sw}\rangle$ = 5/cm^3 and IMF $\langle B_z\rangle$ = 0 nT. Dependences of plasma sheet number density on solar wind number density and IMF $\langle B_z\rangle$ are represented by the parameters η and α, respectively. Taking logarithm of Eq. 2.3.1, Nagata et al. (2007) used the following equation for the least square fit:

$$\log N_{ps} = \log N_0 + \eta\log\left(\langle N_{sw}\rangle/5\right) + \alpha\langle B_z\rangle . \tag{2.3.2}$$

Nagata et al. (2007) summarized obtained results as following:

(i) On the basis of long-term observations of the solar wind and plasma sheet, it was examined the dependence of the near-Earth plasma sheet number density on the solar wind number density and IMF B_Z including time lags.

(ii) It was found that the plasma number density in the plasma sheet depends on the solar wind number density and the north-south component of IMF B_Z with the time lag of several hours.

(iii) Two-dimensional distributions of response time lags to IMF B_Z and dependence on the solar wind number density under both northward and southward IMF dominant conditions have been revealed for the first time.

(iv) It were examined such dependences as functions of (X, Y) coordinates in the near-Earth plasma sheet by fitting observations of plasma sheet and solar wind to an empirical model described by Eqs. 2.3.1 and 2.3.2.

(v) It was explored shortest and longest response time lags to IMF B_Z by optimizing the correlation coefficient of fit.

(vi) Analyses were conducted separately for northward and southward IMF dominant conditions.

(vii) The dependence of plasma sheet number density on solar wind number density is stronger in the near-tail region ($r > 20\ R_E$) under the southward IMF dominant condition.

(viii) The dependence has weak dawn-dusk asymmetry under the southward IMF dominant condition, whereas it is stronger in the dusk flank under the northward IMF dominant condition.

(ix) The dependence on IMF B_Z is globally positive under the northward IMF dominant condition, whereas the dependence is negative in the near-Earth premidnight region under the southward IMF dominant condition.

(x) Both shortest and longest time lags increase from the mid-tail to the near-Earth premidnight region under the southward IMF dominant condition. On the other hand, the shortest (longest) time lag increases antisunward (sunward) along flanks under the northward IMF dominant condition.

(xi) Obtained qualitative features are consistent with previously proposed mechanisms for the entry and transport of magnetosheath particles: mantle/PSBL entry and subsequent E×B and gradient-curvature drift transport under the southward IMF dominant condition, and poleward-of-cusp reconnection and Kelvin-Helmholtz diffusive entries in the flank region under the northward IMF dominant condition.

(xii) Obtained quantitative features provide observational constrains for numerical simulations which model the mass transport from magnetosheath and ionosphere into the near-Earth plasma sheet.

2.3.3. Temperature versus Density Plots and Their Relation to the LLBL Formation under Southward and Northward IMF

Němeček et al. (2015) report a clear dependence of a basic structure of two sublayers of the low-latitude boundary layer (LLBL) on northward and southward IMF orientations. Regardless of different processes responsible for a formation and evolution of the entire LLBL, the outer part of the LLBL is significantly influenced by the sign of the IMF B_Z component. Under northward IMF conditions this layer is present, whereas it is missing during a southward pointing IMF. This behavior can be understood in terms of a motion of reconnection spots due to the changes of the orientation of the magnetosheath magnetic field in the vicinity of the magnetopause. Analyzed case and statistical studies demonstrate that the changes of the LLBL structure can be observed in the subsolar region as well as on flanks near the dawn-dusk meridian. Moreover, the study emphasizes a role of magnetosheath magnetic field variations on the boundary layer formation.

2.3.4. Solar Wind Energy Input to the Magnetosheath and at the Magnetopause

Using THEMIS observations, Pulkkinen et al. (2015) show that the efficiency of the energy entry through the magnetopause as measured by the Poynting vector normal component depends on the combination of the solar wind speed and the southward component of the IMF: most efficient energy transfer occurs when the IMF B_Z is only moderately negative, and the solar wind speed is high. This means that for the same level of solar wind driver parameters (electric field, epsilon, or other), different combinations of V and B_Z will produce different driving at the magnetopause. The effect is strongest for low to moderate driving conditions, while the influence is smaller for the intense space weather events.

2.3.5. Coupling Parameters, Driving Magnetospheric Activity under Northward IMF Conditions

As noted Pulkkinen et al. (2007), the dependence of geomagnetic activity on the solar wind and IMF parameters has been widely studied and several coupling parameters have been suggested and used in the literature. Most of these involve the solar wind electric field ($E = -V \times B$, where V is the solar wind velocity and B the IMF vector) and some function of the IMF clock angle $\theta = \arctan(B_Y/B_Z)$. The clock-angle dependence, often in the form of $\varepsilon = 10^7 l_0^2 V B^2 \sin^4(\theta/2)$ (where $l_0 = 7\ R_E$ is a scaling parameter), brings an improvement over the pure E_Y component in that it recognizes the energy input into the system even when B_Z is northward if there is a sufficiently large IMF B_Y present (e.g., Akasofu, 1981). Most statistical studies have found that for northward IMF, there is no significant correlation between the activity parameters and the driver intensity (Newell et al., 2006; Lyatsky et al., 2007). Pulkkinen et al. (2007) return to the question of the amount of energy input and the level of magnetospheric activity as a function of the driving solar wind speed and IMF for northward IMF conditions. They show that different combinations of the parameters yielding the same values for the solar wind electric field can produce different levels of activity. These results are used to discuss the roles of the various parameters in the energy coupling between the solar wind and the magnetosphere.

As outlined Pulkkinen et al. (2007), GUMICS-4 is a global MHD simulation (Janhunen, 1996) that solves the ideal MHD equations in a region from 32 R_E upstream of the Earth to ~224 R_E tailward of the Earth and ±64 R_E in the directions perpendicular to the Sun-Earth line. Solar wind density, temperature, velocity, and the IMF components are given as input at the sunward edge of the simulation box, outflow conditions are applied at the other boundaries. The MHD regime is coupled to an electrostatic ionospheric simulation, which solves the ionospheric potential using precipitation and field-aligned currents from the MHD simulation, and returns the potential as a boundary condition for the inner boundary of the magnetospheric part at 3.7 R_E. More details of the simulation setup can be found in work by Janhunen (1996) and Palmroth et al. (2003). Laitinen et al. (2007) report on two sets of runs that they used to examine magnetopause reconnection and energy transfer efficiency dependence on the driving IMF and solar wind. They performed four 'pressure runs', where the pressure was changed from 1 to 10 nPa by either increasing the solar wind speed or by increasing the solar wind density, for constant northward and southward IMF separately ($B = 10$ nT, $\theta = 22.5°$ and $157.5°$, respectively). In the 'IMF rotation runs' the IMF clock angle was rotated full 360° over a period of six hours using two values of solar wind dynamic pressure (2 and 8 nPa) and two values of the IMF magnitude (5 and 10 nT). They found that the reconnection power (characterized as the integral of the Poynting flux divergence over the entire dayside magnetopause and magnetopause thickness of ± 1.5 R_E) decreases for increasingly negative E_Y for the IMF rotation runs while it increases for increasingly negative E_Y for the pressure runs (see Laitinen et al. (2007) for more details on the Poynting flux divergence and reconnection at the magnetopause). The energy input is evaluated through direct integration of the total energy flux vector normal component at the magnetopause (Palmroth et al., 2003). The energy is mostly in the form of Poynting flux, mechanical energy transfer is under 10% of the total energy. For southward IMF (positive E_Y) both sets of runs result in a roughly linear dependence of the energy input on the driving electric field. On the other hand, the behavior for northward IMF (negative E_Y) is opposite for the two sets of runs: When the velocity is changed, the energy input increases for increasing speed (and $-E_Y$). During the IMF rotations, the energy input decreases when B_Z (and $-E_Y$) increase. Thus, the northward IMF behavior seems to depend on how the change in E_Y occurs. These results led Pulkkinen et al. (2007) to re-examine statistical data of the driver – response relationship to demonstrate that the simulation results can indeed be recovered using observational data analysis.

As mentioned Pulkkinen et al. (2007), the OMNI 2 database consists of solar wind and IMF measurements tagged with magnetospheric activity parameters from years 1963 to 2006 at 1-hour resolution. All solar wind and IMF measurements have been propagated from the spacecraft location to the bow shock nose. The data are freely available from http://omniweb.gsfc.nasa.gov/. The entire data set contains 118000 measurements. Pulkkinen et al. (2007) discuss and summarize main obtained results as following:

(i) This study has two main results: First, the magnetospheric activity dependence on the solar wind and IMF parameters is qualitatively and quantitatively different for northward and southward IMF conditions. Secondly, it can explain the northward IMF results by quantitative evaluation of the amount of energy input into the magnetosphere using the global GUMICS-4 MHD simulation.

(ii) In the OMNI data set, during southward IMF the activity scales roughly with the solar wind electric field. However, there is a deviation from pure E_Y-dependence such that large speed and low B_Z magnitude drive larger activity than low speed with higher B_Z magnitude, even if the E_Y for both cases is the same. The same result can be obtained for the ε-parameter by using $B^2\sin^4(\theta/2)$ instead of B_Z (not shown). The results shown here use the AE index as a global measure – using Kp, which is a good indicator of magnetospheric convection

(Thomsen, 2004), yields identical results (not shown). In conclusion, the solar wind speed is a more efficient activity driver than (southward) IMF.

(iii) For the northward IMF conditions, the activity increases for increasing solar wind speed regardless of the IMF clock angle or B_Z-value. On the other hand, the activity shows little or no variability as IMF B_Z changes if the speed is held constant, while there is a clear dependence on the clock angle with weakest activity for purely northward IMF. This suggests that the IMF orientation is more important than its magnitude. Taken together, during northward IMF, the magnetospheric activity is not a simple function of the driving E_Y.

(iv) For southward IMF, global MHD simulation results using the GUMICS-4 code show that the energy input into the magnetosphere is almost linearly dependent on the driving solar wind E_Y (Palmroth et al., 2003; Laitinen et al., 2007). The energy input increases linearly when the solar wind EY is increased both when the E_Y is increased by rotating the IMF clock angle and when the increase is associated with an increase in the solar wind speed. This result is qualitatively consistent with the OMNI data set.

(v) During northward IMF, the simulation shows opposite behavior for increasingly negative EY associated with larger solar wind speed and more northward direction of the IMF. The energy input is weakest for small solar wind speed and purely northward IMF. These results are also consistent with the OMNI data set analysis results.

(vi) The agreement of the simulation results of energy input into the magnetosphere and the statistical observational results for both southward and northward IMF demonstrates two things: (1) the energy input through the magnetopause into the magnetosphere is the most important parameter controlling the level of magnetospheric activity; (2) the amount of energy input is most dependent on the magnitude of the solar wind speed, then the IMF orientation, and least dependent on the IMF magnitude. The results highlight the importance of the solar wind speed in driving magnetospheric activity and point out that the qualitatively different response during southward and northward IMF orientation needs to be accounted for in statistical analyses.

(vii) Laitinen et al. (2007) examine the reconnection power ($\nabla \cdot S < 0$) and dynamo power ($\nabla \cdot S > 0$) at the magnetopause surface separately. They show that while the solar wind speed has an effect on both, changes in the solar wind density only affect the dynamo power, while the reconnection power remains unchanged. Note that this result concerns the total integrals over the surface; changing density changes the size of the magnetopause and the reconnection power surface density in a way that maintains the integral constant. Thus, the effects caused by pressure on the energy transfer processes also depend on whether the pressure change is caused by a change in the solar wind density or in the speed, and the solar wind speed is more effective in changing the energy transfer rate.

(viii) During southward IMF, the electric field maps along open field lines to an ionospheric potential. During northward IMF, the open field lines are created by lobe reconnection and map to the ionospheric reverse cells during four-cell convection patterns. These different responses may be associated with the differences in the energy transfer properties. However, it is still an open question whether the reconnection process itself is different at the dayside and behind the cusps, or whether the coupling to the ionosphere is different. While the observational results provide no clear answer to this, the simulation results hint that the reconnection and energy transfer processes are indeed different in their dependence on the solar wind driver. This is clearly an area that warrants further study.

(ix) Newell et al. (2007) examine various coupling functions for a variety of magnetospheric activity parameters and arrive at a best-fit function averaged for all parameters of the form $V^2 B \sin^4(\theta/2)^{3/2}$, which they interpret to be the rate magnetic flux is opened at the magnetopause. All functions of such form merge positive B_Z values with those with small B or small V (i.e., all weak driving cases), and therefore it is impossible to separate the different factors. However, that function may reflect the observations better in that the solar wind speed appears squared while the magnetic field is only linear, consistent with conclusion of Pulkkinen et al. (2007b) that V is a stronger driver than B_Z.

(x) The results in Pulkkinen et al. (2007b) provide an explanation of the different functional behavior of the activity on the driving solar wind parameters. Simultaneously, the good agreement of the observations with the simulation results improves their confidence in the capability of global MHD simulations to represent the energy transfer and dissipation processes in the magnetosphere. The multiple parameters and processes involved in the energy transfer highlight the complexity of the coupling and the importance of examining each of the effects individually.

2.3.6. Sources, Transport, and Distributions of Plasma Sheet Ions and Electrons in Dependence on Interplanetary Parameters under Northward IMF

Wang et al. (2007) have investigated the Geotail data statistically to understand the particle sources, transport, and spatial distributions of the plasma sheet ions and electrons of different energies during northward IMF, and their dependences on the solar wind density (N_{SW}), the solar wind speed (V_{SW}), and the magnitude of the northward IMF BZ ($|B_{Z,IMF}|$). As outlined Wang et al. (2007b), the plasma sheet is colder and denser during northward IMF (N IMF) than southward IMF (S IMF) (e.g., Terasawa et al., 1997; Wing and Newell, 2002). Despite geomagnetic activity being low during N IMF periods, the more abundant cold plasma that builds up during N IMF can provide more source particles for the inner magnetosphere than under S IMF and lead to a stronger ring current and geomagnetic disturbances after a sudden southward IMF (Thomsen et al., 2003; Lavraud et al., 2006b). Thus the N IMF plasma sheet can play a crucial pre-conditioning role in the development of storms and substorms, and related geomagnetic activity during S IMF. Therefore it is important to understand the distributions of N IMF plasma sheet plasma, the particle sources and transport that result in these distributions, and how the distributions depend on different interplanetary parameters.

As noted Wang et al. (2007), the spatial structure of the plasma sheet plasma is determined by the locations and strengths of the different particle sources, and the transport paths of particles of different energies. The major particle sources for the plasma sheet are believed to be the mantle particles entering through the distant tail and the magnetosheath particles entering through the flank magnetopause. The particles from the mantle, after energization through current sheet crossing (Lyons and Speiser, 1982), have higher temperatures than the particles from the flank magnetosheath. The major large-scale transport of the plasma sheet plasma consists of energy-independent electric drift and energy dependent magnetic drift (Spence and Kivelson, 1993; Wang et al., 2006). Other likely transport processes during N IMF include diffusion (Terasawa et al., 1997) and fast flows (e.g., Garner et al., 2003; Kaufmann et al., 2004), however, their contributions are not well understood.

Wang et al. (2007) mentioned that ion entry from the flanks appears to be more efficient during northward IMF than southward IMF (e.g., Fujimoto et al., 1998). Previous studies (e.g., Borovsky et al., 1998; Tsyganenko and Mukai, 2003) showed that the plasma sheet ion density is highly correlated with the solar wind density and that the plasma sheet ion temperature is highly correlated with the solar wind velocity. Terasawa et al. (1997) showed that the plasma sheet becomes denser and colder with higher northward IMF B_Z. The spatial structure of the plasma sheet is shown to be very different under different IMF B_Z conditions (Wing and Newell, 2002; Wang et al., 2006). These previous studies only focused on ions, thus our understanding of the electron plasma sheet is still limited. Furthermore, these studies mainly focused on the changes in plasma moments, which can be affected by both the changes in particle sources and magnetic field configurations. Therefore investigations of phase space density is necessary to understand if the dependences of the particle sources to the plasma sheet on the interplanetary parameters can account for the observed solar wind-plasma sheet correlations.

Wang et al. (2007) use Geotail data in aberrated GSM coordinates (with the aberration angle determined by one hour averaged solar wind velocity) from 1 January 1995 to 31 December 2005 and in the area $|Y| \leq 22.5\ R_E$ and $0 \leq X \leq -30\ R_E$ (the smallest radial distance covered by Geotail is ~8 R_E). Plasma data from two instruments aboard Geotail are used: the ion and electron data from the Low Energy Particle (LEP) instrument (Mukai et al., 1994) that covers the ion energy range from 21 eV/q to 44 keV/q and the electron energy range from 43 eV to 41 keV, and the proton data from the Energetic Particles and Ion Composition (EPIC) instrument (Williams et al., 1994) that covers the energy range from 46 keV to 3005 keV. Magnetic field data is from the magnetic field (MGF) experiment (Kokubun et al., 1994). The ion density, temperature, and pressure are a summation of the LEP and EPIC moments. The ion and electron count data in four directions (sunward, duskward, anti-sunward, dawnward) are used in the study of phase space density and are used to obtain the electron moments. One min averages of the plasma and magnetic field data are used.

Wang et al. (2007) used also the solar wind and IMF data mainly from Wind. The arrival time of the IMF at the subsolar bow shock at (X = 17, Y = 0, Z = 0) R_E is determined by calculating the minimum variance direction using the minimum variance analysis technique (Weimer et al., 2003). During times when the Wind data is not available or the propagated Wind data is not reliable due to the Wind's position (when Wind is more than ~50 R_E off the Sun- Earth), Wang et al. (2007) used the solar wind and IMF data from ACE, which is available after February 1998. Wang et al. (2007) summarize main obtained results as following:

(i) There are investigated statistically how, during northward IMF periods, the changes in the solar wind density, solar wind speed, or the magnitude of the northward IMF B_Z affect the equatorial distributions of the ion and electron plasma sheet measured by Geotail.

(ii) There are also investigated if the current understanding of the particle sources and transport can account for the observed plasma sheet distributions.

(iii) For both ions and electrons, the overall N_{PS}/N_{SW} is clearly higher when $|B_{Z,IMF}|$ is higher and N_{PS}/N_{SW} in the mid-tail is higher when V_{SW} is lower, suggesting more efficient solar wind entry from the flanks when the $|B_{Z,IMF}|$ is higher or V_{SW} is lower.

(iv) The density (temperature) is higher (lower) near the flanks and lower (higher) near midnight under all conditions.

(v) The N_{ps} becomes higher as the T_{ps} becomes lower, indicating the increase of particles is larger in the low energy than the high-energy range.

(vi) There are fitted the particle energy spectrum with a two components kappa distribution to separate the plasma into a cold and a hot population, and to better estimate the cold population below the instrument's lower cutoff energy.

(vii) For both ions and electrons, the contribution to the density from the cold population (<2 keV ions and <0.1 keV electrons) increases and become more dominant toward the flanks.

(viii) In the near-Earth plasma sheet, the main contribution to the number density is from the hot population. These are consistent with the results from the DMSP observations (Wing et al., 2005).

(ix) The different mixtures of the cold and hot populations in different regions are consistent with Wang et al. (2007) understanding of the particle sources with colder particles coming from the flanks and hotter particles from the distant tail.

(x) The overall phase space density f at low energy increases with increasing $|B_{Z,IMF}|$. The f near the dawn flank and the f in the mid-tail near midnight region are significantly higher with decreasing V_{SW}, while the f near the dusk flank does not change as significantly. This suggests the solar wind entry through the dawn flank becomes more efficient when V_{SW} is lower.

(xi) The f above thermal energy range in the pre-midnight sector is higher (lower) than that in the post-midnight sector for ions (electrons), and dawndusk asymmetries become stronger with increasing energy.

(xii) There are estimated electric and magnetic drift paths of ions and electrons versus different first adiabatic invariants from the observations. The comparisons with the distributions of constant f show that the electric and magnetic drifts are the major transport for particles of thermal and high energy.

(xiii) However, the electric and magnetic drift alone cannot explain the distributions of low energy particles, which peak along the flanks and decrease significantly toward midnight.

(xiv) That the particle source at the flanks sets up a strong gradient in the phase space density for low energy particles and observed long period perturbations in plasma drift flows suggests that the condition is suitable for diffusion to bring these cold particles from the flanks to midnight to account for the observed phase space density in the near-midnight region.

(xv) The 0.3 to 6 keV ions and electrons in the plasma sheet are important sources for the ring current population and the difference between these sources under different interplanetary conditions during N IMF periods can be an order of magnitude, suggesting the resulting ring current can create significantly different storms or substorms.

2.3.7. The Plasmapause Response to the Southward Turning of the IMF Derived from Sequential EUV Images

Murakami et al. (2007) statistically examined the plasmapause response to the southward turning of the IMF using sequential global images of the plasmasphere, in order to understand how the convection electric field propagates to the inner magnetosphere. The extreme ultraviolet (EUV) imager on the Imager for Magnetopause-to-Aurora Global Exploration satellite clearly observed inward motion of the plasmapause driven by the southward turning of the IMF. They surveyed the EUV data in the 2000–2001 period and found 16 events. It was noted that decades of observation show that the radial location of the plasmapause generally moves inward during geomagnetic disturbance periods. The inward motion is explained by a longstanding hypothesis that the dynamics of the plasmasphere is controlled by the superposition of the corotation and convection electric field (Nishida, 1966). The enhanced convection electric field triggered by the southward turning of the IMF forms new $\mathbf{E} \times \mathbf{B}$ drift trajectories, and plasmas on the plasmapause move inward along them (Chappell et al., 1970). Some other mechanisms explaining the inward motion of the plasmapause during geomagnetic disturbance periods have been proposed. Lemaire (1974, M1985) proposed that the centrifugal force drives the plasma upwards and then produces a sharp density gradient along the magnetic field lines tangent to the zero parallel force (ZPF) surface. According to this mechanism, after an increase in the level of magnetic activity the ZPF surface shifts inward and the plasmasphere is peeled off in the post midnight sector where the convection velocity is maximum (Lemaire and Gringauz, M1998). Carpenter and Lemaire (1997) discussed the loss of plasmas through the flow from the plasmasphere into the underlying ionosphere during periods of the enhanced convection. They indicated the possibility of the mechanism involving dumping of plasmas into the ionosphere based on the whistler data.

As outlined Murakami et al. (2007), although it is not completely understood how the convection electric field governing the dynamics mentioned above propagates to the inner magnetosphere, a number of observational studies have examined the ionospheric response to the IMF. For example, the ionospheric convection responds to the IMF quickly (~10 min) (Ridley et al., 1998) and almost simultaneously at the whole ionosphere (Kikuchi et al., 1996; Ruohoniemi and Greenwald, 1998; Ridley et al., 1998; Lu et al., 2002). Furthermore, using the ground magnetometers, Hashimoto et al. (2002) examined the response time of the convection in the inner magnetosphere to the enhancement of the ionospheric convection. Their study suggests that the convection electric field propagates from the dayside magnetopause to the nightside inner magnetosphere through the ionosphere.

As mentioned Murakami et al. (2007), with regard to experimental work on the plasmasphere, the conventional extreme ultraviolet (EUV) photometric experiments from an inside-out view, which had been done intensively in the 1970s, suggested that the Earth's plasmasphere would be imaged by He II (30.4 nm) emission (e.g., Johnson et al., 1971). In the 1990s, the remotesensing method using the EUV emission became a powerful tool to provide global perspectives on the plasmasphere dynamics (Williams et al., 1992). The fundamental technology to detect the EUV emission began with the He II emission through a rocket experiment (Yoshikawa et al., 1997). The two-dimensional He II imaging of the terrestrial plasmasphere from its outside was first done by the Planet-B (Nozomi) spacecraft

(Nakamura et al., 2000). Yoshikawa et al. (2000a, 2000b, 2001) first presented static EUV images of the plasmasphere and the inner magnetosphere. Recent advances in satellite-based imaging techniques have made it possible to routinely obtain full global images of the plasmasphere. The EUV instrument on the Imager for Magnetopause-to-Aurora Global Exploration (IMAGE) satellite gave us complete sequential pictures (one image/10 min) (Burch et al., 2001a,b; Sandel et al., 2000, 2001). Using these sequential images, Goldstein et al. (2003a, 2003b, 2004, 2005) and Spasojević et al. (2003) identified the correlation between the inward motion of the plasmapause and the southward turning of the IMF for specific events. They found that the timescale of the plasmapause response to the IMF is about 30 min in their case studies. Furthermore, Pierrard and Cabrera (2005, 2006) presented the simulations of the plasmapause formation and compared the predicted plasmapause positions with the EUV observations. They remarked the correlation between the inward motion of the plasmapause and the southward turning of the IMF in numerical simulations. However, Pierrard and Cabrera (2005, 2006) did not discuss the timescale of the plasmapause response to the IMF. Murakami et al. (2007) statistically examine the plasmapause response to the southward turning of the IMF in order to understand the propagation mechanism of the convection electric field in the inner magnetosphere. They survey the EUV data set in the period of May 2000–December 2001 obtained by the IMAGE satellite (available at http://euv.lpl.arizona.edu/euv/), and find the events showing the clear inward motion of the nightside plasmapause. Finally, Murakami et al. (2007) deduce the timescale of the plasmapause response to the southward turning of the IMF. Summarizing obtained results, Murakami et al. (2007) came to following conclusions:

(i) Using the sequential images of the plasmasphere observed by the EUV instrument on the IMAGE satellite, it was examined the plasmapause dynamics.

(ii) It was shown that the plasmapause response to the southward turning of the IMF takes 10–30 min, and average 18 min.

(iii) This result is consistent with the timescale derived from the ionospheric observations.

(iv) Therefore it indicates that the electric field penetrates from the magnetopause to the inner magnetosphere through the ionosphere.

(v) However, due to the limited EUV data, it was found only 16 events in the search.

(vi) Future observation of the plasmasphere by the SELENE satellite from the lunar orbit will give not only much more sequential images with higher time resolution but also a deeper understanding of the plasmasphere dynamics.

2.3.8. Global View of Dayside Magnetic Reconnection with the Dusk-Dawn IMF Orientation

On the basis of Double Star/TC-1 and Cluster data Pu et al. (2007) investigate both component reconnection and anti-parallel reconnection occur at the magnetopause when the IMF is predominantly dawnward. The occurrence of these different features under these very similar IMF conditions are further confirmed by a statistical study of 290 fast flows measured in both the low and high latitude magnetopause boundary layers. The directions of these fast flows suggest a possible S-shaped configuration of the reconnection X-line under such a dawnward dominated IMF orientation. It was noted that based on a Polar observation during a cusp crossing, Trattner et al. (2004) postulated that it is likely that both reconnection scenarios, anti-parallel (Crooker, 1979; Luhmann et al., 1984) and component (Sonnerup, 1974; Cowley, 1976; Sonnerup et al., 1981) reconnection are likely to occur simultaneously at the magnetopause. In addition, Moore et al. (2002) found in a model study that magnetic reconnection appeared along an S-shaped X-line extending from low-latitudes to high-latitudes across the dayside magnetopause. Cooling et al. (2001) and Eriksson et al. (2004) also showed similar S-shaped-magnetic reconnection regions. Dorelli et al. (2007) showed in MHD simulation that magnetic reconnection with an S-shaped separator line at the dayside magnetopause displayed features of the processes of both anti-parallel and component magnetic reconnection. Thus, further observational studies are especially needed to confirm these results.

As outlined Pu et al. (2007), an indirect way to locate the magnetic reconnection sites at the magnetopause is to study the global pattern of magnetic reconnection generated fast flows in the magnetopause boundary layer, where the ‘fast flows’ are referred to ‘jets’ viewed from the frame of reference where the background flow is at rest. The coordinated observations of Double Star and the Cluster spacecraft array provide a good opportunity to conduct such a study across a large range of latitudes. Pu et al. (2007) have surveyed 2 years (2004–2005) of fast flow measurements from instruments onboard Cluster and TC-1, the equatorial spacecraft of the Double Star Mission, to obtain a global view of possible X-lines under different IMF conditions. Pu et al. (2007) focus on the events in which the IMF lies predominantly in the dusk-dawn direction in order to better reveal a distinct tilted X-line orientation. Pu et al. (2007) studied 290 fast flow events measured by TC-1 and Cluster with the –Y dominated IMF. The main results are as follows:

(i) In agreement with the component magnetic reconnection scenario, fast flows in the region near the subsolar point are directed mainly north-east or south-west.

(ii) Consistent with either the anti-parallel or the component magnetic reconnection hypotheses, the majority of the fast flows at the high-latitude magnetopause and outer cusps are primarily dawnward and duskward, but having a wide spread of directions into the Z direction.

(iii) Possible X-lines could be statistically constructed with a tilt angle of ~155° in/near the subsolar region and of ~100° and ~86.9° respectively in the northern and southern hemispheres, with the assumption that the effects of the magnetosheath flow might not be significant on average in the statistical sense with a large enough amount of samples.

(iv) The above results suggest that both anti-parallel and component magnetic reconnection may occur at the magnetopause under the same IMF conditions. A global S-shaped configuration of the possible X-line, similar to that of Moore et al. (2002), could be drawn by interconnecting the possible X-line in/near the subsolar region and those at high-latitudes.

2.3.9. Northward IMF Plasma Sheet Entropies

As outlined Johnson and Wing (2009), in the study of the transport of plasma into and within the plasma sheet, it is often useful to consider what parameters are conserved and under what conditions. Conserved quantities are particularly useful to help simplify modeling of a system. Conversely, breaking of conservational invariants may be an indicator that a more complex model is required. One fundamental conserved quantity is entropy, which is conserved when the plasma response is adiabatic. The entropy per unit mass in an ideal, isotropic plasma is

$$\sigma = C\log\left(p^{1/\gamma}/\rho\right), \tag{2.3.3}$$

where C is a constant, p is the plasma pressure, ρ is mass density, and γ is the polytropic index (Birn et al., 2006b). For an entropy-conserving process

$$s = p/\rho^{\gamma} \tag{2.3.4}$$

is equivalently conserved and is commonly referred to as the **specific entropy**. In the ideal MHD description, the pressure response is typically prescribed by an adiabatic equation of state such that $ds/dt = 0$. The specific entropy, s or σ, should be rigorously conserved in an ideal MHD fluid element and is an intensive variable, in that it does not depend on system size. In a multifluid description, the specific entropy for each species should also be conserved as long as the pressure is basically isotropic and there is no heat exchange between species, heat fluxes, or other dissipative processes. Johnson and Wing (2009) noted that the pressure equation also implies an integral constraint on the total mass content of a flux tube,

$$M = \int \rho\, dl/B\,, \tag{2.3.5}$$

and total content of a flux tube (for single ion species),

$$N = \int n\, dl/B\,, \tag{2.3.6}$$

which must be proportional to

$$S = \int p^{1/\gamma}\, dl/B\,, \tag{2.3.7}$$

where dl is an infinitesimal length along **B** and B is the magnetic field strength. Although not precisely the entropy of the flux tube, this quantity S is entropy like in nature (Birn et al., 2006b), and Johnson and Wing (2009) refer to it as the **total entropy**. The total entropy is also an important parameter to identify when the interchange instability is operable. Nonconservation of specific entropy is expected to occur when nonadiabatic processes such as wave-particle interactions, heat flux, and dissipation are operating and when the fluid element loses its integrity, as occurs when hot particles and cold particles move differently (e.g., cold particles move with **E × B** convection and hot particles move by curvature and gradient drift). Nonconservation of specific entropy is normally an indicator that an MHD or multifluid description is inadequate and kinetic or dissipative processes should be included in the dynamical evolution of the system. As noted Johnson and Wing (2009), in the case of a single ion species, $p/\rho^{\gamma} \propto p/n^{\gamma}$. If s is conserved along the convection path, then, in the p – n space, plasma sheet ion p and n should follow lines having constant slopes of γ, termed adiabats. Borovsky et al. (1998) shows that statistically this seems to be the case from midtail region (X = –22.5 R_E) to geosynchronous orbit (X = –6.6 R_E) in their p – n plots (see their Plate 1). Borovsky et al. (1998) finds that γ= 1.52, which is close to 5/3, albeit the scatters are quite large. A similar result was obtained by Baumjohann (1993) and Baumjohann and Paschmann (1989), which report that $\gamma \sim$ 1.69, but also with large scatters in the p – n plot. Where s is conserved, it should be expected that contours of constant s are related to drift paths. However, in the case of energy-dependent drifts such as the curvature and gradient drifts, the fluid element can be compromised where cold particles move with **E × B** and hot particles move according to curvature and gradient drift. However, it is still possible to consider conservation of entropy for particles in a given energy range, which would be conserved along their drift paths. As outlined Johnson and Wing (2009), non-conservation of the total entropy can also occur where non-adiabatic processes are operating, but because it is an integral constraint for a particular flux tube, a localized process that changes the specific entropy in a small region of the flux tube may not significantly affect the total entropy. However, if the localized process leads to field-line reconfiguration (reconnection), then the change of total entropy is more related to the change of flux tube volume than to changes in specific entropy within that volume. In this way, reconnection processes can significantly affect total entropy while introducing little change in the specific entropy as shown theoretically and observationally for substorms (Birn et al., 2004, 2006a,b; Wing et al., 2007; Wing and Johnson, 2009). Johnson and Wing (2009), consider the entropy of plasma in the plasma sheet for the northward IMF configuration. Because the plasma sheet electron contribution to the total pressure is much smaller, roughly 15% (e.g., Baumjohann et al., 1989; Spence et al., 1989), than the ion pressure, it is possible to probe the plasma sheet entropy considering only ion data. In the plasma sheet, ions have been observed to be nearly isotropic irrespective of magnetic activity (Kistler et al., 1992; Huang and Frank, 1994; Spence et al., 1989) because of strong pitch angle scattering (e.g., Lyons and Speiser, 1982) and long drift time compared with the bounce period (e.g., Wolf, 1983). Fortunately, the isotropy of the ions makes it possible to remotely probe the ion distributions by observing only precipitating particles in the ionosphere. Moreover, because the pressure is observed to be isotropic, the total entropy can be simplified when the system is near equilibrium because the pressure is constant along field lines and $S \sim p^{1/\gamma}V$ (where V is flux tube volume) is conserved. Interestingly, Usadi et al. (1996) notes that even when the particle motion violates both the first and the second adiabatic invariant, S is still conserved. The value for γ can range from 0 to ∞, depending on the process involved. For example, for an isobaric (constant pressure) system, γ = 0; for an isothermal (constant temperature) system, γ = 1; for an isometric (constant density) system, $\gamma = \infty$; and for an adiabatic (no heat loss) system and monoatomic gas, $\gamma = 5/3$. The parameter $S = p^{1/\gamma}V$ is an extensive parameter that is proportional to system size, i.e., flux tube volume V.

Johnson and Wing (2009) examine the specific entropy profiles in the plasma sheet under northward IMF configuration. In particular, they consider the entropy of all of the observed plasma as well as the entropy of plasma measured in two energy ranges corresponding to cold, dense plasma and nominal hot plasma sheet material. Their investigation shows that the entropy profiles are significantly different for these two energy ranges, indicating that the two populations evolve according to different dynamics.

In Johnson and Wing (2009) the plasma sheet pressure, temperature, and density profiles inferred from DMSP observations are used to investigate northward interplanetary magnetic field plasma sheet specific entropy s (see above, Eq. 2.13.2). Profiles of s for hot and cold populations are considered. The hot ion population s profile suggests a duskward heat flux that is consistent with the curvature and gradient drift. In contrast, s of the cold population is approximately conserved in the X direction but has a strong gradient toward the midnight meridian. The total entropy S (see above, Eq. 2.13.5) and s under various entry mechanisms are estimated for comparisons with observed values. The cold population s is higher than that of the magnetosheath, suggesting the entry process may heat the ions. Cusp reconnection by itself may not increase the magnetosheath ion s, but cusp reconnection with kinetic Alfvén waves may provide the necessary heating. Localized reconnection in Kelvin-Helmholtz vortices may increase s if the plasma expands non-adiabatically into the large magnetospheric flux tube volume. Kinetic Alfvén waves may lead to diffusion and heating consistent with the observations. The cold population s increases by a factor of 5 from the flanks to the midnight meridian, which provides a constraint for transport mechanism within the plasma sheet. A simple calculation is performed to demonstrate that the spatial gradients of s in the plasma sheet could result from the temporal dependence of transport processes in the magnetopause boundary layer.

2.3.10. Reverse Convection Potential Saturation during Northward IMF under Various Driving Conditions

As outlined Wilder et al. (2009), the primary mechanism whereby energy and momentum are coupled to the magnetosphere from the solar wind is the process of magnetic reconnection on the dayside of the Earth (Dungey, 1961). When two regions of plasma containing anti-parallel magnetic fields collide at high-enough dynamic pressure, it will cause the magnetic fields to merge and shear in an X-shaped formation. These newly sheared magnetic field lines will experience high Maxwell stress and be pulled to the night-side of Earth, where they will reconnect and split in the tail. This reconnection in turn drives convection in the polar ionosphere. Another mechanism described as a viscous-like interaction is also thought to couple energy to the magnetosphere from the solar wind, but at a rate which is small compared to reconnection (Axford and Hines, 1961). During geomagnetically quiet periods when the driving of the magnetosphere by reconnection is negligible, there continues to be a background convection pattern driven by this viscous interaction. The electric potential across the magnetosphere during quiet periods has been estimated to be 20 kilovolts (Papitashvili et al., 1981). Wilder et al. (2009) noted that the magnetospheric convection that results from a combination of magnetic reconnection and viscous interactions maps along magnetic field lines into the high-latitude ionosphere to generate plasma convection and an associated polar electric potential (Dungey, 1961; Axford and Hines, 1961). When the IMF points southward and reconnection occurs at the sub-solar point, the result is a two-cell convection pattern with anti-sunward flow at the highest latitudes. The cross polar cap potential, FPC, associated with this pattern is an important metric of the energy flow through the coupled magnetosphere-ionosphere system. During instances of strongly southward IMF, the polar cap potential as a function of solar wind electric field has been observed to experience nonlinear 'saturation' (Siscoe et al., 2002b; Shepherd et al., 2002; Hairston et al., 2005). When the IMF is northward, reconnection occurs poleward of the cusp on lobe field lines in the Earth's magnetosphere (Dungey, 1963; Watanabe et al., 2005; Dorelli et al., 2007). The resulting convection pattern consists of four cells. Two cells contain anti-sunward flow at higher latitudes and sunward return flow at low latitudes caused by viscous interactions. The other two are reverse convection cells at high latitude on the dayside that exist as a result of lobe reconnection (Crooker, 1992). The reverse convection cells are thought to be driven by high-latitude field aligned currents termed NBZ (Iijima et al., 1984). NBZ refers to the fact that the field aligned currents form when the IMF B_Z is directed northward.

Wilder et al. (2008) used SuperDARN Doppler velocity measurements to demonstrate that the electric potential across the vortices driven by the NBZ Currents, FRC, also exhibits nonlinear saturation behavior. The method used was to create bins of events where the solar wind electric field stayed stable within a minimum and maximum value for a minimum of 40 min. For each solar wind electric field bin, SuperDARN Doppler velocity vectors from each event were then filtered in a manner which generated the most likely convection pattern. A spherical harmonic fit was then applied to determine the electric potential pattern for each solar wind electric field bin. The potential across the reverse convection cells saturated at approximately 20 kV (Wilder et al., 2008). While Wilder et al. (2008) demonstrated that the NBZ cells exhibit saturation behavior, the cause of this saturation was not investigated. For southward IMF, studies have shown the effect of various driving parameters such as dynamic pressure and Alfvénic Mach number on the saturation of the polar cap potential (Siscoe et al., 2002a; Ridley, 2007).

As noted Wilder et al. (2009), there have not yet been any studies on the response of the four cell pattern formed under northward IMF to various driving conditions. Also, there is a difference in conductivity between the winter and summer hemispheres because of the lack of photoionization in the winter, so separating out events of strong northward IMF by season will shed light on the response of the four cell pattern to variations in ionospheric conductivity (Fujii et al., 1981). The purpose of Wilder et al. (2009) study is to build upon the Wilder et al.'s (2008) results and investigate the effects of various solar wind driving conditions as well as ionospheric conductivity on the saturation potential and electric field of the NBZ cells. A clearer picture of the relationship between the saturation of FPC and FRC will also be demonstrated. Wilder et al. (2009) expand on the results presented by Wilder et al. (2008), showing the saturation of the reverse convection electric field under northward IMF, as well as the dependence of FRC on Alfvènic Mach number and solar wind β. Wilder et al. (2009) demonstrate that there is a clear seasonal dependence of the reverse convection potential pattern, and that the reverse convection potential still saturates even when the

Alfvènic Mach number is nominal. Also, there is a clear dependence of the reverse convection potential under strong northward IMF driving on solar wind plasma β.

Wilder et al. (2009) report the results of an investigation of the reverse convection potentials in the dayside high-latitude ionosphere during periods of steady northward IMF. While it has been demonstrated that the reverse convection potentials exhibit saturation behavior, an investigation into how the saturated reverse convection cells respond to various driving conditions has not yet been performed. Wilder et al. (2009) use the OMNI database from 1998 to 2007 to search for events in solar wind data propagated to the bow shock when the IMF is northward and when various driving parameters are stable for more than 40 min. Then they use bin-averaged SuperDARN Doppler radar velocity data to apply a spherical harmonic fit to calculate the average potential pattern for various driving conditions. Results show that the saturated reverse convection potential during northward IMF is substantially lower than during southward IMF (as expected), but the strength of the saturation electric field at a particular location is comparable during northward IMF to what it is during southward IMF. Seasonal dependence and dependence on the Alfvènic Mach number and solar wind β are also investigated. It appears that the reverse convection potential demonstrates the opposite of what is expected for seasonal dependence, and that the Alfvènic Mach number is not the fundamental driver of its saturation.

2.3.11. Interplanetary Magnetic Field–Geomagnetic Field Coupling and Vertical Variance Index

As noted Abraham et al. (2010), strong correlations have been established between ground-based magnetometer observations and F region vertical plasma drifts by Anderson et al. (2002, 2004). Prompt penetration effects of magnetospheric electric fields have been analyzed by Fejer and Scherliess (1995, 1998) using radar and satellite observations. Subsequently, an empirical model for F region plasma drift was developed by Scherliess and Fejer (1999). Instead of a direct attempt to unravel the factors instigating a magnetic storm, the investigation of Abraham et al. (2010) focuses on the normal, mostly quiet time interaction between the solar wind and the geomagnetic field without considering any storm scenario. The IMF and horizontal components of Earth's magnetic field consist of transient variations. An acceptable relationship that yields the geomagnetic field strength observed at a station for a certain value of the incident IMF has not yet been realized. Abraham et al. (2010) introduce a new mathematical index called the vertical variance (VV) index, which can exactly quantify the amount of temporal variations in any time series. The index being of a daily nature, aberrations from quiet time *Sq* variations are minimal. The analysis of interplanetary and observed geomagnetic field variations using the VV index immediately yields empirical relations between the two. It is also observed that the elucidated relationship is specific to each ground-based station and hence can be used for categorizing observatories on that basis. Overall, there is also scope for analyzing storm-type events as an aberration from the normal construct. According to Abraham et al. (2010), the development of the VV index followed from the attempts of the authors to precisely determine the degree of disturbance in various time series data sets. Although there are different indices that are used to gradate temporal variations, many of them are categorizing averaging techniques and are highly specific either in determination or in application (e.g., *K* and *AE*). An important index of more general nature worth mentioning in this context is the interhourly variability (IHV) index developed by Svalgaard and Cliver (2007), which ascertains the hourly fluctuations in geomagnetic field data after eliminating *Sq* variations by not including daylight hours. The VV index is intended for assignment to a time-dependent function varying within specified temporal end points, an exact numerical reckoning that expresses the amount of fluctuations in the data. The versatility of the VV index lies in the fact that it can be applied to both IMF and geomagnetic data, thus enabling a comparison of the contained fluctuations. Defining the VV index for any data set consisting of values $x(t)$, x_i, and x_{i+1}, which are adjacent data values of x corresponding to adjacent time values t_i and t_{i+1}, is find as following:

$$\text{VV index} = n^{-1/2}\left(\sum_{i=1}^{n}\frac{\left(x_{i+1}-x_i\right)^2}{\left(x_{i+1}-x_i\right)^2+\left(t_{i+1}-t_i\right)^2}\right)^{1/2}, \quad (2.3.8)$$

where

$$n = \left(T_2 - T_1\right)/\left(t_{i+1} - t_i\right) \quad (2.3.9)$$

and T_2 and T_1 are the extreme time limits of the period for which the index is being determined. In order to understand the behavior of the index, it may be noted that if $x(t)$ is plotted against t, considering unit change along the time axis, the numerator in summation appears as the vertical change in the plot, while the denominator in the summation appears as the hypotenuse. The gauge of the vertical change gives the name to the index, while the denominator serves to normalize the index.

As noted Abraham et al. (2010), the solar wind impacting at the geomagnetopause contains transient variations in the embedded IMF. These disturbances are mirrored in the horizontal geomagnetic field measured at the Huancayo and Ascension Island geomagnetic stations. Investigation of Abraham et al. (2010) attempts to relate the microtemporal fluctuations in the IMF with the horizontal component of the geomagnetic field by means of a newly developed daily vertical variance (VV) index, which permits the quantification of such pulsations. A linear relationship is established between the disturbances observed in the interplanetary and terrestrial fields on a daily basis.

2.3.12. Three-Dimensional Hybrid Simulation of Magnetosheath Reconnection under Northward and Southward IMF

As noted Pang et al. (2010), solar wind directional discontinuities are abrupt rotations of the heliospheric magnetic field across layers that are observed approximately every 30 min near 1 AU (Burlaga et al., 1977) and thus are a common phenomenon in space. Primarily, two types of directional discontinuities are often present in the solar wind: (1) rotational discontinuities, or large-amplitude Alfvén waves, which possess a finite normal component of the magnetic field, and (2) tangential discontinuities, which are static magnetic boundaries or current sheets without a normal field component. These discontinuities usually are considered to be associated with many physical processes occurring in the Earth's magnetosphere and ionosphere. They are believed to cause rapid equatorward movement of aurora and sudden changes of the *H*-component geomagnetic field at high-latitude geomagnetic stations (Øieroset et al., 1996). Northward turning of the IMF, which could also be viewed as a solar wind discontinuity, has often been referred to as a trigger of the substorm expansion onset (Lyons et al., 1997). The formation of a strong geomagnetic storm is found to be favored by an extended period of southward IMF that is preceded by an earlier interval of northward field or very weak B_Z (Thomsen et al., 2003). Magnetic reconnection (Dungey, 1961) is believed to be a fundamental plasma process in a current sheet, which converts the magnetic energy into particle energy, leading to heating and abrupt acceleration of both ions and electrons (Krauss-Varban and Omidi, 1995; Lin and Swift, 1996; Fu et al., 2006; Pritchett, 2006). Numerous burst events in space plasmas have been confirmed by in situ observations to be associated with magnetic reconnection. Reconnection at the Earth's magnetopause (Sonnerup et al., 1995) provides an efficient mechanism for transfer of solar wind mass, momentum, and energy into the magnetosphere. Evidence of near-Earth and distant magnetotail reconnection has also been reported (e.g., Øieroset et al., 2001; Borg et al., 2005). In the solar wind, some reconnection exhausts have been discovered (Gosling et al., 2005), with a large-scale reconnection X line extending at least hundreds of Earth radii (Phan et al., 2006a). In situ observations have also provided evidence of reconnection in a turbulent plasma downstream of the bow shock (Retin et al., 2007). All these observations suggest that magnetic reconnection is a universal process and plays an important role in physical processes in the coupled Sun–Earth plasma system.

As outlined Pang et al. (2010), MHD simulation using the Integrated Space Weather Prediction Model (ISM) has predicted the merging of interplanetary magnetic field lines in the magnetosheath across opposite sides of a directional discontinuity from the solar wind (Maynard et al., 2002). The result of a 2-D hybrid simulation (Lin, 1997) has also suggested the magnetosheath reconnection as a consequence of the compression of a solar wind current sheet at the bow shock. Such a prediction has recently been confirmed by observations of Cluster spacecraft (Phan et al., 2007). When Cluster 1 was located at a magnetosheath location of (6.7, 2.0, −9.5) R_E in the GSE coordinate system, a current sheet with a width of 525 km or 10 local ion skin depths was found to be accompanied by enhanced flow, density, and ion temperature. Correspondingly, a rather wide current sheet (about 3 r_E or 264 local ion skin depths) was also observed in the solar wind by ACE satellite, and an examination indicated that the thinner one was due to the wider sheet being compressed when propagating into the magnetosheath. How such a wide current sheet is narrowed so significantly to initiate the reconnection when it interacts with the bow shock and magnetopause remains an important issue of study. In Phan et al.'s (2007) case, the magnetic field changes orientation from the north to south across the solar wind directional discontinuity. No dayside magnetopause reconnection is expected under the northward IMF before the current sheet arrives. Such a case under an initially northward IMF was recently investigated by Omidi et al. (2009) through a 2-D hybrid simulation study of magnetosheath reconnection caused by the interaction among TD, bow shock, and magnetopause. Both quasi-steady reconnection and time-independent reconnection were obtained because of different initial widths of TD. According to Pang et al. (2010), a related question to ask is: What if the solar wind condition is different so that the IMF changes direction from the south to north? In this situation, reconnection could also occur between field lines of the IMF and magnetosphere, on the earthward side of the transmitted TD. The two simultaneous reconnection events at the magnetopause and in the magnetosheath could compete with each other to consume the magnetic field energy sandwiched by them, which may affect the magnetosheath reconnection. Meanwhile, the paraboloidal-like configuration of the bow shock and magnetopause on the dayside is expected to introduce 3-D features to the reconnection process.

In paper of Pang et al. (2010) are presented a 3-D global hybrid simulation of the interaction between an interplanetary directional TD and the dayside bow shock and magnetopause. Hybrid simulations have been used to address the large-scale structure and associated ion dynamics in magnetic reconnection (e.g., Krauss-Varban and Omidi, 1995; Lin and Swift, 1996; Lottermoser et al., 1998; Lin, 2001; Omidi and Sibeck, 2007). The focus of the Pang et al. (2010) investigation is on the magnetic field structure and ion particle signatures associated with the magnetosheath reconnection, under different conditions of the IMF direction change across the solar wind TD. Differences between the magnetosheath merging processes corresponding to the north-to-south and south-to-north field rotations are addressed. The evolution of the transmitted TD with different initial widths of TD is also studied. Hereinafter, Pang et al. (2010) will use the terms “merging layer” and “reconnection current sheet” for the current layer in which the magnetosheath reconnection has been initiated and “transmitted TD” and “transmitted current sheet or layer” to describe the transmitted part of the initial tangential discontinuity in general.

The 3-D global hybrid code utilized in Pang et al. (2010), is similar to that used by Lin and Wang (2005) for the dayside bow shock–magnetosphere system. The detailed code description can be found by Swift (1996). In the hybrid model, ions (protons) are treated as fully kinetic particles and electrons are treated as a massless fluid. The simulation domain contains the dayside plasma regions, with GSE X > 0, within the geocentric distance $4\ R_E < r < 25\ R_E$. A spherical coordinate system is applied in the calculation, in which the polar axis is chosen as the GSE Y axis. The polar boundaries of the domain are set at the 20° and 160° polar angles to avoid singular coordinate lines along the Y axis. The dayside geomagnetic polar regions, around the Z axis, are kept in the domain. The initial condition includes a dipole geomagnetic field plus a mirror dipole in $r < 10\ R_E$ and a uniform solar wind in $r > 10\ R_E$, in which the initial IMF can be assumed in an arbitrary direction. The mirror dipole is used in the initial setup to speed up the evolution process of the solar wind–magnetosphere interaction without losing the key physical features. The bow shock, magnetosheath, and magnetopause are formed by the interaction between the solar wind and the dipole field. In Pang et al. (2010) a 3-D global hybrid simulation is carried out for the generation and structure of magnetic reconnection in the magnetosheath because of the interaction of an interplanetary tangential discontinuity (TD) with the bow shock and magnetosphere. Runs are performed for solar wind TDs possessing different polarization senses of magnetic field (north-to-south or south-to-north from the leading to trailing side of the incident TD) and initial half-widths. Two-step compression processes are shown in the transmitted TD, including a ‘shock compression’, as the TD passes through the shock followed by a subsequent ‘convective compression’ while the TD is moving in the magnetosheath toward the magnetopause. In cases with a relatively thin solar wind TD, 3-D patchy reconnection is initiated in the transmitted TD, forming magnetosheath flux ropes. Differences between these flux ropes and those due to magnetopause reconnection are discussed. Multiple components of ion particles are present in the velocity distribution in the magnetosheath merging, accompanied by ion heating. For cases with a relatively wide initial TD, a dominant single X line appears in the subsolar magnetosheath after the transmitted TD is narrowed through the two-step compression process. Specifically, in the cases with a south-to-north field rotation across an incident thin TD, the magnetosheath flux ropes could re-reconnect with the closed geomagnetic field lines to generate a closed field line region with mixed magnetosheath and magnetospheric plasmas, which may contribute to the transport of solar wind plasma into the magnetospheric boundary layer.

2.3.13. Magnetic Merging Line and Reconnection Voltage versus IMF Clock Angle: Results from Global MHD Simulations

As outlined Hu et al. (2009), magnetic reconnection has been considered to be the main means for the dynamic coupling between the solar wind and the Earth's magnetosphere (Raeder, 2003). A vast amount of studies were made to seek a quantitative relation between the magnetopause reconnection and the solar wind magnetosphere coupling. The reconnection rate was often measured by an effective electric field E_{r}, and then the reconnection voltage is estimated via multiplying E_{r} by an effective length l_0 of the magnetic merging line on the sunward side. Efforts were made to express E_{r} and l_0 in terms of the solar wind and IMF parameters, and the resultant formulas are more or less empirical. For instance, l_0 was estimated from either a pressure balance between the solar wind and the magnetosphere (Ridley, 2005) or a simple inverse proportionality to the cube root of B_{IMF} (Newell et al., 2007). As far as E_r is concerned, Sonnerup (1974) derived an expression based on the component reconnection theory:

$$E_r = kv_A B_{IMF} \sin(\theta_{IMF}/2), \tag{2.3.10}$$

where k is the magnetic merging rate, B_{IMF} is the IMF strength, θ_{IMF} is the IMF clock angle, and v_{A} is the Alfvén speed based on B_{IMF}. Kan and Lee (1979) recognized a previously overlooked geometrical relationship between the reconnection electric field and the magnetic field, and derived

$$E_r = v_{sw} B_{IMF} \sin^2(\theta_{IMF}/2), \tag{2.3.11}$$

which does not depend on the merging rate and is referred to as the Kan-Lee electric field in the literature. The advantage of this formula is to directly express the reconnection electric field in terms of the solar wind and IMF parameters, independent of the MHD) steady reconnection theories. They further calculated the power associated with the electric field, and showed that it is approximately proportional to the energy coupling function proposed by Perreault and Akasofu (1978). In the derivations of E_{r} mentioned above, the magnetic fields on the two sides of the magnetopause are assumed equal in magnitude but different in direction. For asymmetric antiparallel reconnection in which the magnetic fields on the two sides of the magnetopause are different in magnitude, Cassak and Shay (2007) derived a scaling law for the reconnection rate, which expresses the reconnection rate in terms of the local plasma parameters near and the geometry of the inflow and outflow to the reconnection site.

Hu et al. (2009) outlined that Borovsky (2008) further applied the Cassak-Shay formula to the dayside reconnection of the magnetosphere and expressed the local parameters near the reconnection site and thus the reconnection rate in terms of upstream solar wind parameters. Using global MHD simulations of the solar wind-magnetosphere-ionosphere (SMI) system with local high resistivity on the nose of the magnetopause, Borovsky et al. (2008) calculated the reconnection electric field over there and compared it with the prediction by the Cassak-Shay formula, and concluded that this formula successfully describes the reconnection rate on the dayside magnetopause for due southward IMF cases. Moreover, they found that the reconnection itself does not significantly modify the local parameters near the nose of the magnetopause and the flow pattern of the magnetosheath. Observables, such as the ionosphere transpolar potential and some indices characterizing the state of the magnetosphere, were used to test various coupling functions which involve the reconnection electric field (cf. Shepherd, 2007; Newell et al., 2007).

Hu et al. (2009) noted that global MHD simulations of the SMI system open a valuable way for quantitative studies of the magnetopause reconnection and its relation to the solar wind-magnetosphere coupling. The reconnection is described by the tangential electric field at the magnetic merging line, defined as the separator between the IMF lines and the geomagnetic closed field lines; the electric potential along this line is referred to as reconnection potential hereinafter. The electric field derived directly from the reconnection potential describes the reconnection rate of unit length of merging line, or the local strength of magnetic reconnection, whereas the total reconnection potential drop, or in brief, the reconnection voltage, represents the total reconnection rate. In order to achieve a quantitative description of the reconnection by global MHD simulations, one has to determine the merging line and calculate the electric potential distribution along it. If the IMF is due southward, the merging line is simply a zero contour of magnetic field strength in the equatorial plane. A magnetic field superposed by a dipole field and a uniform field of arbitrary orientation was often used to mimic the topology of the Earth's magnetospheric magnetic field, and its merging line is simply a great circle of a sphere (Yeh, 1976). For real magnetospheric magnetic fields, the merging line should be a closed space curve which encircles the Earth. Attempts were made to find the merging line of quasi-steady magnetospheric magnetic fields obtained by global MHD simulations. Fedder et al. (1995a) identified the merging line with the intersection between the magnetopause and the topological separatrix between the IMF lines and the inmost open field lines that connect to the ionosphere. They showed the IMF lines for different IMF clock angles in their Plates 1 and 2, from which one can see the merging line by the naked eye. Nevertheless, they did not display the merging line explicitly, and only the angle between the topological separatrix and the IMF was measured. For a precisely duskward IMF case, Siscoe et al. (2001) did find the sunward part of the merging line explicitly, which was identified with the middle part of the last closed field line that comes closest to the two magnetic nulls. Using a similar method, Dorelli et al. (2007) located the merging line for a generic northward IMF case with an IMF clock angle of 45°.

Hu et al. (2009) noted that recently, multispacecraft in situ observations were used to infer the global geometry of the magnetic merging line, or X line as often termed in the literature (see a brief review by Paschmann, 2008). For instance, Phan et al. (2006b) used the Geotail and Wind data during stable dawnward dominated IMF to infer the presence of a tilted X line hinged near the subsolar point and spanning the entire dayside magnetopause. On the basis of a statistical study of 290 fast flow events measured by Double Star/TC-1 in low latitudes and Cluster in high latitudes, Pu et al. (2007) suggested that a possible S-shaped X line exists for generic dawnward IMF cases. The configuration of the merging line inferred from these observations is consistent with the prediction from the component reconnection hypothesis. Once the merging line is found for a specific magnetospheric magnetic field, the next step is to calculate the tangential electric field along it so as to determine the distribution of the reconnection rate and its integration along the merging line, i.e., the reconnection voltage. Fedder et al. (1991, 1995) determined the reconnection voltage from projections of the ionospheric potential contours on the open-closed field boundary, in which the field lines were tacitly assumed to be equipotential. As pointed out by Kan and Lee (1979), Siscoe et al. (2001) and Hu et al. (2007), however, a parallel electric field exists along the field lines that pass through the reconnection site, so the mapped ionospheric potential cannot be used to directly measure the reconnection potential. Siscoe et al. (2001) calculated the reconnection potential along the merging line instead of the mere reconnection voltage. The method was an integration of parallel electric field along magnetic field lines that touch the merging line. Using this method, they obtained the reconnection potential for a due duskward IMF case, and found a large parallel potential drop along the field lines that connect the ionosphere with the null points and a dip of electric field profile along the merging line across the equator. Hu et al. (2007) proposed three methods, which are all based on the curvilinear integration of electric field and differ in the choice of the reference potential and integration path. For due southward IMF cases, they found the differences among the answers via different methods to be within 20% when an equivalent numerical resistivity is carefully selected. In particular, one of these methods takes radial rays that connect the inner boundary ($r = 3\ R_E$, taken to be the unit of length hereinafter) with the merging line as the integration path, and the potential at the inner boundary as the reference potential, which is set to be the potential at the footpoints in the ionosphere of the Earth's dipole field lines that pass through the inner boundary. It happens that the obtained reconnection potential is largely determined by the convectional electric field $\mathbf{E} = -\mathbf{v} \times \mathbf{B}$ in the magnetosphere. Therefore, the knotty problem of appropriate choice of an equivalent numerical resistivity was circumvented. For due southward IMF cases, they found that the reconnection voltage shows a nearly linear relation to the total ionospheric region 1 current as the ionospheric transpolar potential does. Hu et al. (2009) propose a method to quantitatively determine the geometry of the whole merging line, which is applicable for arbitrary IMF clock angles. The radial ray integration method proposed by Hu et al. (2007) is then used to calculate the reconnection potential along the merging line. The magnetic field and the electric field are also evaluated along the merging line, and a preliminary analysis is presented on the physical meaning of the results with emphasis on the distribution of the reconnection rate along the merging line and the reconnection voltage, and their dependence on the IMF clock angle.

In Hu et al. (2009) the PPMLR-MHD code developed by Hu et al. (2005, 2007) is used to make global MHD simulations of the SMI system. It is an extension of the Lagrangian version of the piecewise parabolic method (PPMLR) developed by Colella and Woodward (1984) to MHD. Hu et al. (2009) point out especially that in smooth flow regions, this code has a very low numerical dissipation that is one order of magnitude smaller than ordinary codes (cf. Colella and Woodward, 1984). All numerical runs are made under the following simplifying assumptions: (1) the solar wind is along the Sun–Earth line with a speed of 400 km/s, (2) the solar wind number density and thermal pressure are fixed to be 5 cm^{-3} and 0.0126 nPa, respectively, (3) the IMF lies in the plane perpendicular to the Sun–Earth line, 10 nT in strength, with a clock angle of θ_{IMF} that is adjustable, and (4) the ionosphere is assumed uniform with a fixed Pedersen conductance of 5 S and a zero Hall conductance. Simulation continues for more than 5 h in physical time for each run until a quasi-steady state is reached for the SMI system. The solution domain is taken to be $-300\ R_E \leq x \leq 30\ R_E$ and $-150\ R_E \leq y, z \leq 150\ R_E$ in the GSE coordinates. It is discretized into 160 × 162 × 162 grid points: a uniform mesh is laid out in the near Earth domain of $-10\ R_E \leq x, y, z \leq 10\ R_E$, 0.4 R_E in spacing, and the grid spacing outside increases according to a geometrical series of common ratio 1.05 along each axis. An inner boundary of radius 3 R_E is set for the magnetosphere in order to avoid the complexity associated with the plasmasphere and strong magnetic field.

Hu et al. (2009) propose diagnosis methods to trace the magnetic merging line and to calculate the electric potential along it for Earth's magnetospheric magnetic fields obtained by global MHD simulations of the solar wind-magnetosphere-ionosphere (SMI) system. The points with minimum magnetic field strength along last closed magnetic field lines and properly selected closed field lines are combined to trace the whole merging line, and the radial ray integration of convectional electric field is used to calculate the electric potential on the merging line. The diagnosis methods are then applied to magnetospheric magnetic fields associated with different IMF clock angles, and a preliminary analysis is presented on the clock angle response of the geometry of the merging line and the associated reconnection voltage. The merging line is found to be similar in geometry to that of the compound field superposed by the Earth's dipole field and the IMF, whereas the reconnection voltage is approximately fitted by $\sin^{3/2}(\theta_{IMF}/2)$ for its response to the IMF clock angle θ_{IMF}. The ionospheric transpolar potential and the voltage along the polar cap boundary show different dependence from that of the reconnection voltage, so it is not justified to take them as substitutes for the reconnection voltage. The length of the sunward merging line between the two peaks of reconnection potential shows a nonmonotonic variation in response to θ_{IMF}, peaked at $\theta_{IMF} = 90°$, so it is also not justified to take electric fields along the merging line, however defined they may be, to characterize the total reconnection rate and the coupling strength between the solar wind and the magnetosphere. The reconnection nearby magnetic nulls closest to the subsolar point is found to be negligible, which gives support to the component reconnection hypothesis for dayside reconnection of quasi-steady states of the SMI system.

2.3.14. Particle Injections Observed at the Morning Sector as a Response to IMF Turning

Kozelova and Kozelov (2015) report a detailed case study of substorm on January 6, 2008 in the interval 13–15 UT using data from four THEMIS satellites located in the morning sector magnetosphere eastward the onset location. The substorm of interest presents the ground-based magnetic disturbance consisted from the large-scale pulsations (4–5 min) superposed on the substorm bay. One can distinguish at least two significant activations at different spatial regions. First activation, which follow after a short-living burst of the IMF B_Z, developed westward of second one, which was sequent after the northward turning IMF

B_Z. Kozelova and Kozelov (2015) show the existence of fast magnetosonic mode after second activation. This mode was observed at 7.5 R_E in the morning sector at region of transition from dipole to tail-like configuration of the magnetic field. The increase of z-component of the magnetic field observed in magnetosphere during the non-diamagnetic structure is interpreted as an enhancement of westward ring (or partial-ring) current at closer to Earth distances. The appearance of the sub-keV plasma at ~5.8 R_E (used as a tracer of substorm injection) supports this supposition. Obtained results are illustrated in Figures 2.3.1-2.3.9.

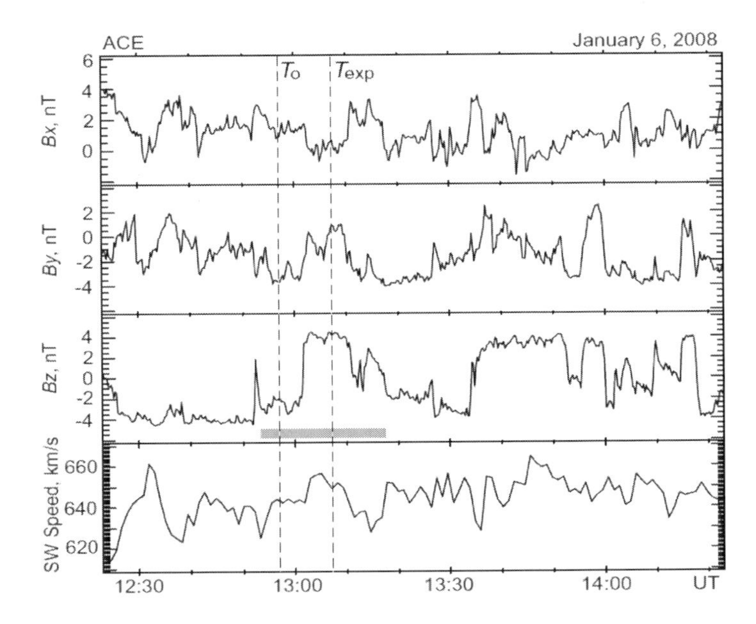

Figure 2.3.1. Data from ACE satellite on January 6, 2008. According to Kozelova and Kozelov (2015). Permission from COSPAR.

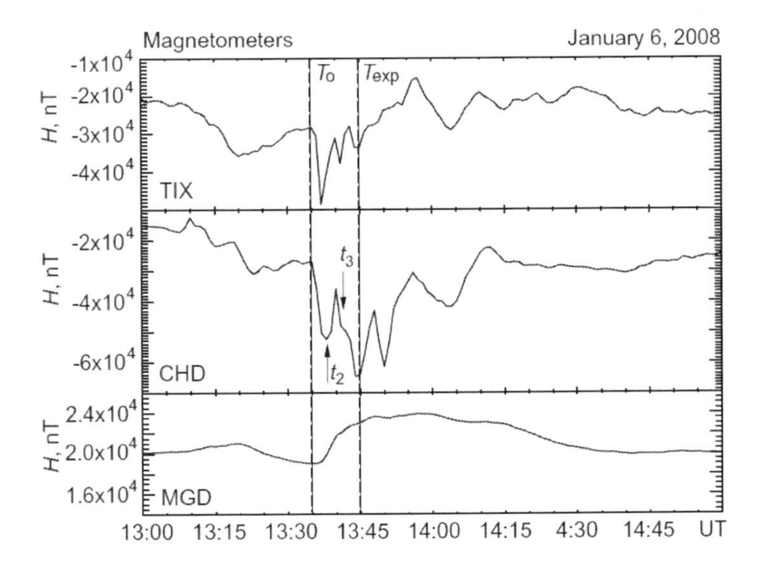

Figure 2.3.2. Substorm on January 6, 2008. The H component from ground magnetometers on the 210 Magnetic Meridian (MM). According to Kozelova and Kozelov (2015). Permission from COSPAR.

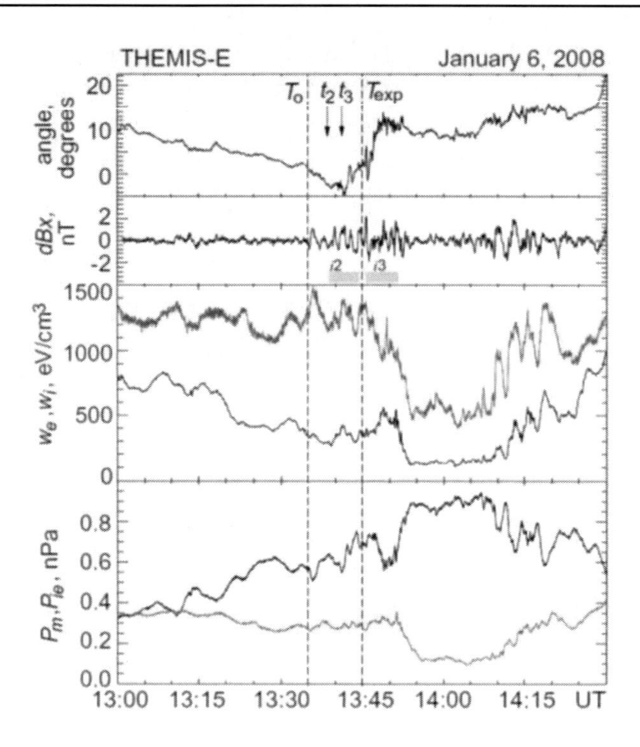

Figure 2.3.3. Data from THEMIS-E (THE) satellite. From top to bottom: (i) inclination angle of the magnetic field relative to the XY plane; (ii) Pi2-like fluctuations of the magnetic field; () ionWi (red line) and electronWe (black line) energy density for low energy ESA particles; (iv) the total pressure of all particles Pie (red line) and pressure of magnetic field Pm (black line). According to Kozelova and Kozelov (2015). Permission from COSPAR.

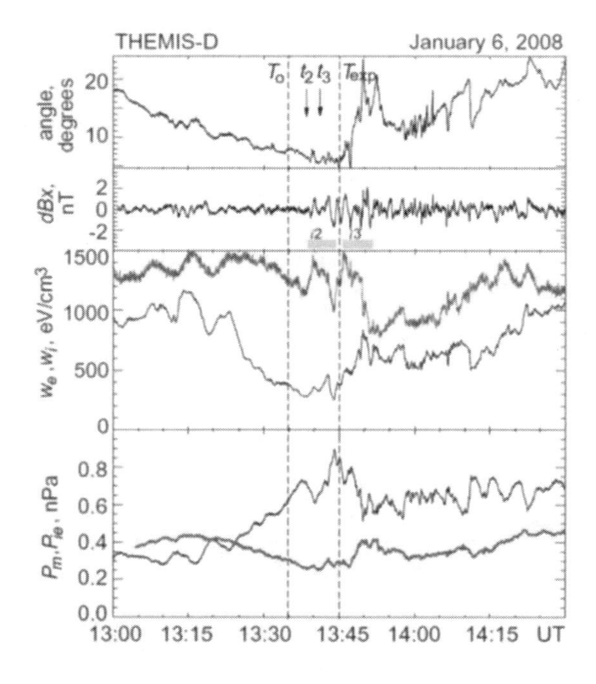

Figure 2.3.4. The same as in Figure 2.3.3, but data from THD satellite. According to Kozelova and Kozelov (2015). Permission from COSPAR.

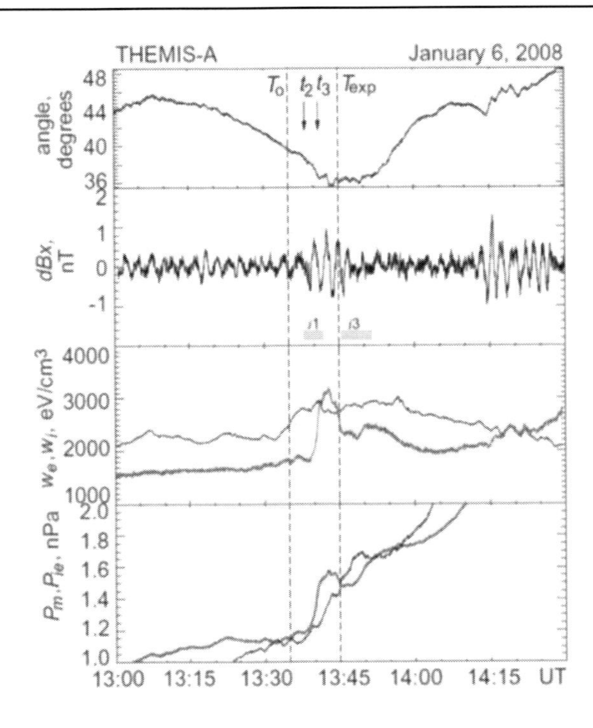

Figure 2.3.5. The same as in Figure 2.3.3, but data from THA satellite. According to Kozelova and Kozelov (2015). Permission from COSPAR.

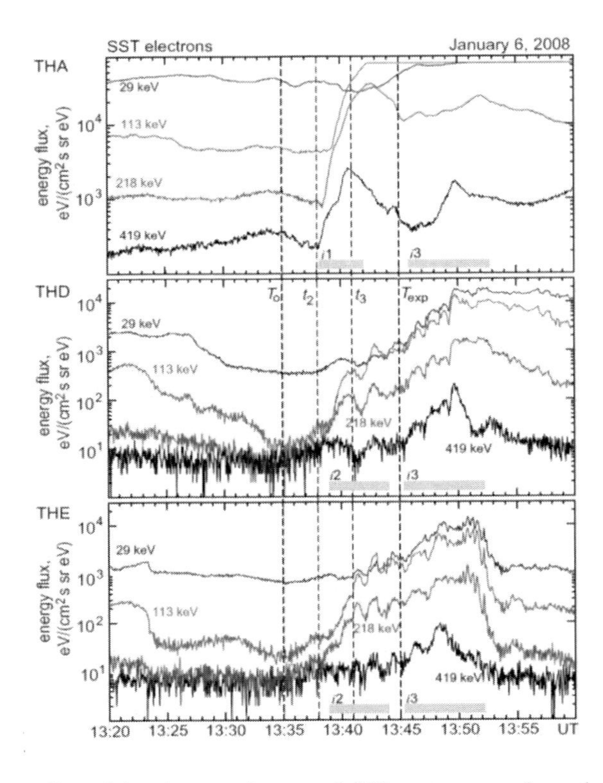

Figure 2.3.6. The energy flux of the electrons in several different energy channels observed by THA, THD and THE satellites during the substorm on January 6, 2008. According to Kozelova and Kozelov (2015). Permission from COSPAR.

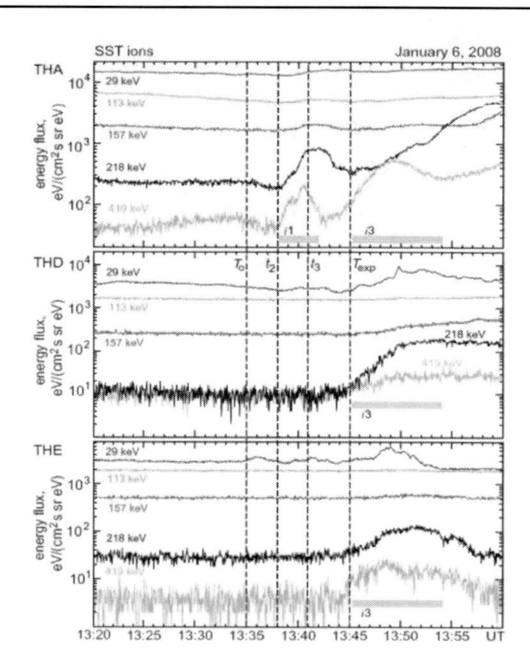

Figure 2.3.7. The energy fluxes of the ions in several different energy channels observed by THA, THD and THE satellites during the substorm on January 6, 2008. According to Kozelova and Kozelov (2015). Permission from COSPAR.

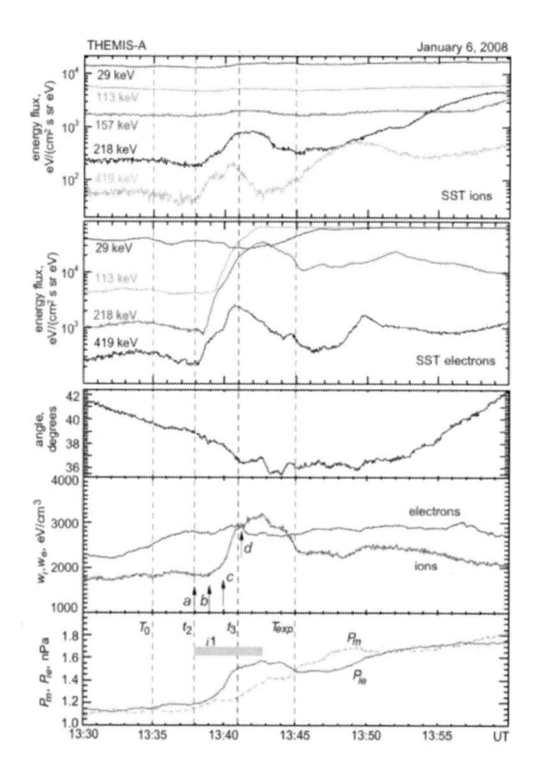

Figure 2.3.8. Particle and magnetic field data during the interval 13:30–14:00 UT from THA. The arrows indicate the moments discussed in Kozelova and Kozelov (2015). According to Kozelova and Kozelov (2015). Permission from COSPAR.

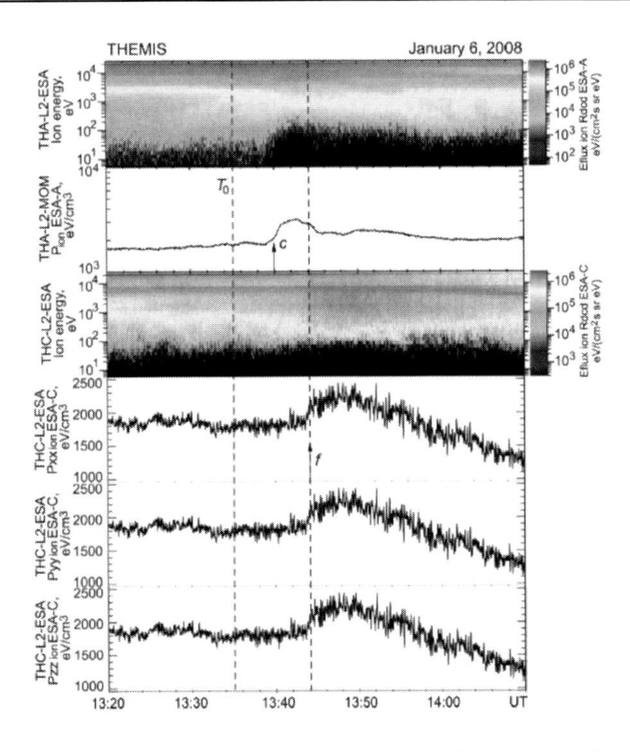

Figure 2.3.9. Low energy ESA ions the interval 13:20–14:10 UT from THA and THC on January 6, 2008. From top to bottom: **(i)** ion energy–time spectrogram at THA; **(ii)** ion energy density at THA; **(iii)** ion energy–time spectrogram at THC; **(iv)** – **(vi)** three components of ion pressure tensor at THC. The arrows indicate the moments discussed in Kozelova and Kozelov (2015). According to Kozelova and Kozelov (2015). Permission from COSPAR.

2.3.15. Impact of Solar Wind ULF BZ Fluctuations on Geomagnetic Activity for Viscous Timescales during Strongly Northward and Southward IMF

Osmane et al. (2015) analyze more than 17 years of OMNI data to statistically quantify the impact of IMF B_Z fluctuations on *AL* by using higher-order moments in the *AL*-distribution as a proxy. For strongly southward interplanetary magnetic field (IMF), the *AL* distribution function is characterized by a decrease of the skewness, a shift of its peak from −30 nT to −200 nT, and a broadening of the distribution core. During northward IMF, the distribution of *AL* is characterized by a significant reduction of the standard deviation and weight in the tail. Following this characterization of *AL* for southward and northward IMF, Osmane et al. (2015) show that IMF fluctuations enhance the driving on timescales smaller than those of substorms by shifting the peak of the probability distribution function by more than 150 nT during southward IMF, and by narrowing the distribution function by a factor of 2 during northward IMF. For both southward and northward IMF, it's demonstrated that high power fluctuations in B_Z systematically result in a greater level of activity on timescales consistent with viscous processes. Obtained results provide additional quantitative evidence of the role of the solar wind fluctuations in geomagnetic activity. The methodology presented also provides a framework to characterize short timescale magnetospheric dynamics taking place on the order of viscous timescales $\tau \ll 1$ hour.

2.3.16. How the IMF B_Y Induces a B_Y Component in the Closed Magnetosphere and How it Leads to Asymmetric Currents and Convection Patterns in the two Hemispheres

Tenfjord et al. (2015) used the Lyon-Fedder-Mobarry global MHD model to study the effects of the IMF B_Y component on the coupling between the solar wind and magnetosphere-ionosphere system. When the IMF reconnects with the terrestrial magnetic field with IMF $B_Y \neq 0$, flux transport is asymmetrically distributed between the two hemispheres. We describe how B_Y is induced in the closed magnetosphere on both the dayside and nightside and present the governing equations. The magnetosphere imposes asymmetric forces on the ionosphere, and the effects on the ionospheric flow are characterized by distorted convection cell patterns, often referred to as "banana" and "orange" cell patterns. The flux asymmetrically added to the lobes results in a nonuniform induced B_Y in the closed magnetosphere. By including the dynamics of the system, Tenfjord et al. (2015) introduce a mechanism that predicts asymmetric Birkeland currents at conjugate foot points. Asymmetric Birkeland currents are created as a consequence of y directed tension contained in the return flow. Associated with these currents, Tenfjord et al. (2015) expect fast localized ionospheric azimuthal flows present in one hemisphere but not necessarily in the other. It's also present current density measurements from Active Magnetosphere and Planetary Electrodynamics Response Experiment that are consistent with this picture. Tenfjord et al. (2015) argue that the induced B_Y produces asymmetrical Birkeland currents as a consequence of asymmetric stress balance between the hemispheres. Such an asymmetry will also lead to asymmetrical foot points and asymmetries in the azimuthal flow in the ionosphere.

2.3.17. Simulations of the Earth's Magnetosphere Embedded in Sub-Alfvénic Solar Wind on 24 and 25 May 2002

According to Chané et al. (2015), during 24 and 25 May 2002, the solar wind conditions at Earth's orbit were very unusual: the density was extremely low (below 0.1/cc) and, as a result, the flow was subfast and sub-Alfvénic (the Alfvén Mach number was as low as 0.4 in the rest frame of the Earth). Consequently, the Earth's bow shock disappeared and two Alfvén wings formed on the flanks of the magnetosphere. These two long structures (estimated extension of 600 R_E for this event) affect the incoming plasma as follows: the velocity is reduced and the magnetic field rotates. In the present study, global MHD simulations of the magnetosphere are performed for such upstream solar wind conditions. The simulations show how the magnetosphere configuration dramatically changes when the sub-Alfvénic solar wind reaches the magnetosphere: the dayside magnetopause expands up to 20 R_E, and on the nightside the position of the last closed magnetic field line diminishes to 20 R_E. As a result the closed magnetic field line region becomes very symmetric. The open field line configuration also changes such that field lines emanating from the Northern Hemisphere all point in the direction of the dawn Alfvén wing (around 8:00 LT), while the field lines from the Southern Hemisphere all point in direction of the other wing (around 22:00 LT). During the formation of the Alfvén wings, the tail lobes completely disappeared and the auroral activity greatly diminished, i.e., the magnetosphere becomes geomagnetically quiet.

2.3.18. Peculiarities of Magnetic Barrier Formation for Southward and Northward Directions of the IMF

As note Slivka et al. (2015), the magnetic barrier is a region which is formed in the course of the solar wind flow around the Earth's magnetosphere and is characterized by depleted plasma and an enhanced magnitude of the magnetic field. There is a general point of view that the magnetic barrier can persist only for the northward IMF, whereas it is practically absent if the IMF is southward directed. Slivka et al. (2015) studied the presence of the magnetic barrier for low-shear (predominantly northward IMF) and high-shear (predominantly southward IMF) magnetopause conditions and analyzed 74 events of low-latitude dayside magnetopause crossings by the THEMIS satellites. It's used the superposed epoch analysis to study variations of magnetic field and plasma parameters near the magnetopause. Results of Phan et al. (1994) that the magnetic barrier is well pronounced for low-shear magnetopauses and practically absent for high-shear events have been confirmed by present analysis. However, 5 from 49 high-shear events show clear signatures of a magnetic barrier. Hence, Slivka et al. (2015) conclude that the magnetic barrier is formed in the same manner for all direction of IMF but for high-shear events it is destroyed by reconnection much faster than it accumulates. Ratio of the events with and without magnetic barrier signatures (5/44) is approximately equal to the ratio of characteristic times of disruption due to reconnection (1–2 min) and the formation of a magnetic barrier (10 min). The reduction of the magnetic field due to reconnection can be used to estimate reconnection rate which turns out to be rather high of the order of 0.3 in 19 from 31 cases.

2.3.19. Asymmetrical Response of Dayside Ion Precipitation to a Large Rotation of the IMF

Berchem et al. (2016) have carried out global MHD simulations together with large-scale kinetic simulations to investigate the response of the dayside magnetospheric ion precipitation to a large rotation (135°) of the IMF. The study uses simplified global MHD model and idealized solar wind conditions where the IMF rotates smoothly from a southward toward a northward direction ($B_x = 0$) to clearly identify the effects of the impact of the discontinuity on the magnetopause. Results of the global simulations reveal that a strong north-south asymmetry develops in the pattern of precipitating ions during the interaction of the IMF rotation with the magnetopause. For a counterclockwise IMF rotation from its original southward direction ($B_y < 0$), a spot of high-energy particle injections occurs in the Northern Hemisphere but not in the Southern Hemisphere. The spot moves poleward and dawnward as the interacting field rotates. In that case, reconnection is found close to the poleward edge of the northern cusp, while it occurs farther tailward in the Southern Hemisphere. Tracing magnetic field lines shows an asymmetry in the tilt of the cusps and indicates that the draping and subsequent double reconnection of newly opened field lines from the Southern Hemisphere over the dayside magnetosphere cause the symmetry breaking. The reverse north-south asymmetry is found for a clockwise IMF rotation from its original southward direction ($B_y > 0$). Trends observed in the ion dispersions predicted from the simulations are in good agreement with Cluster observations of the midaltitude northern cusp, which motivated the study.

2.3.20. MLT Dependence in the Relationship between Plasmapause, Solar Wind, and Geomagnetic Activity Based on CRRES: 1990–1991

Using the database of CRRES in situ observations of the plasmapause crossings, Bandić et al. (2016) develop linear and more complex plasmapause models parametrized by (a) solar wind parameters V (solar wind velocity), BV (where B is the magnitude of the IMF), and $d\Phi_{mp}/dt$ (which combines different physical mechanisms which run magnetospheric activity), and (b) geomagnetic indices Dst, Ap, and AE. The complex models are built by including a first harmonic in magnetic local time (MLT). The method of Bandić et al. (2016) based on the cross-correlation analyses provides not only the plasmapause shape for different levels of geomagnetic activity but additionally yields the information of the delays in the MLT response of the plasmapause. All models based on both solar wind parameters and geomagnetic indices indicate the maximal plasmapause extension in the postdusk side at high geomagnetic activity. The decrease in the convection electric field places the bulge toward midnight. These results are compared and discussed in regard to past works. The study of Bandić et al. (2016) shows that the time delays in the plasmapause response are a function of MLT and suggests that the plasmapause is formed by the mechanism of interchange instability motion. Bandić et al. (2016) observed that any change quickly propagates across dawn to noon, and then at lower rate toward midnight. The results further indicate that the instability may propagate much faster during solar maximum than around solar minimum. This study contributes to the determination of the MLT dependence of the plasmapause and to constrain physical mechanism by which the plasmapause is formed.

2.3.21. The Integrated Dayside Merging Rate is Controlled Primarily by the Solar Wind

In Lopez (2016) an argument is presented to support the view that the rate of merging between the geomagnetic field and the interplanetary magnetic field across the dayside magnetosphere is controlled primarily by the solar wind parameters. The controlling parameters are the solar wind electric field in the Earth's frame of reference and the solar wind magnetosonic fast mode Mach number. These factors control to first order the total amount of magnetic flux that is carried by the magnetosheath flow to the dayside merging region. Lopez (2016) argues that the global dayside merging rate is governed by the amount of flux that is delivered to the dayside merging line by the magnetosheath flow. The ionospheric conductance also plays an important role by modulating the shape of the magnetospheric obstacle around which the magnetosheath flow is deflected. The local conditions at the magnetopause, especially changes in magnetospheric plasma density will affect the local reconnection rate, but not the global dayside merging rate because to change the global merging rate the entire pattern of magnetosheath flow must be changed. The conceptual model presented here can explain how dayside merging depends on solar wind values, including both linear and nonlinear dependencies, through the application of a single, unifying perspective, without the need for ad hoc-mechanisms that limit the dayside merging rate.

2.3.22. Temporal Evolutions of the Solar Wind Conditions at 1 AU Prior to the Near-Earth X Lines in the Tail: Superposed Epoch Analysis

Utilizing conjunction observations of the Geotail and ACE satellites from 1998 to 2005, Zhang et al. (2016a) investigated the temporal evolutions of the solar wind conditions prior to the formation of X lines in the near-Earth magnetotail. It is first show the statistical properties of Bz, By, density, and velocity of the solar wind related to the 374 tail X line events. A superposed epoch analysis is performed to study the temporal evolutions of the solar wind conditions 5 h prior to the tail X lines. The solar wind conditions for tail X lines during southward IMF (SW-IMF) and northward IMF (NW-IMF) are analyzed. The main results are as follows: (1) For events classified as SW-IMF, near-Earth X line observations in the magnetosphere are preceded by ~2 h intervals of southward IMF; (2) for events classified as NW-IMF, the northward IMF orientation preceding near-Earth X line observations lasts ~ 40 min.

2.3.23. Bursty Bulk Flows at Different Magnetospheric Activity Levels: Dependence on IMF Conditions

Based on concurrent observations of the ACE and Geotail satellites from 1998 to 2005, Zhang et al. (2016b) statistically analyzed and compared the earthward bursty bulk flows (BBFs) with local positive Bz under different IMF conditions. Four different magnetospheric activity levels (MALs), including quiet times and substorm growth/expansion/recovery phases, are considered. The properties of the BBFs, including their ion temperature (T), Vx component, x component of the energy flux density (Qx), and the solar wind dawn-dusk electric field Ey (observed at ~1 AU), are analyzed. Main observations include the following: (1) BBF tends to have less penetration distance for northward IMF (NW-IMF) than for southward IMF (SW-IMF). Inward of 15 R_E the BBFs for SW-IMF are dominant. Few BBFs for NW-IMF occur within 15 R_E. (2) The occurrence probabilities of the BBFs at each MAL depend highly on the orientations of the IMF. During quiet times, the BBFs for NW-IMF are dominant. Reversely, during the growth and expansion phases of a substorm, the BBFs for SW-IMF are dominant. (3) The strengths of the BBF have significant evolution with substorm development. For SW-IMF condition, the strengths of the BBFs are the lowest for quiet times. The strength of the BBFs tends to increase during the growth phase and reaches to the strongest value during the expansion phase, then, decays during the recovery phase. For NW-IMF condition, the strengths of the BBFs evolve with the substorm development in a similar way as for SW-IMF condition. (4) For SW-IMF, the solar wind Ey evolves with the substorm development in a similar way to the strength of the BBFs. However, no clear evolution is found for NW-IMF. (5) The strengths of the BBF Qx and solar wind Ey are closely related. Both tend to be stronger for growth phase than for quite time, reach the strongest for expansion phase, and then decay for recovery phase. It appears that to trigger a substorm, the strength of the BBFs should achieve energy thresholds with values different for NW-IMF and SW-IMF.

2.3.24. An Empirical RBF Model of the Magnetosphere Parameterized by Interplanetary and Ground-Based Drivers

As outlined Tsyganenko and Andreeva (2016), in their paper of Andreeva and Tsyganenko (2016), a novel method was proposed to model the magnetosphere directly from spacecraft data, with no a priori knowledge nor ad hoc assumptions about the geometry of the magnetic field sources. The idea was to split the field into the toroidal and poloidal parts and then expand each part into a weighted sum of radial basis functions (RBF). Tsyganenko and Andreeva (2016) take the next step forward by having developed a full-fledged model of the near magnetosphere, based on a multiyear set of space magnetometer data (1995–2015) and driven by ground-based and interplanetary input parameters. The model consolidates the largest ever amount of data and has been found to provide the best ever merit parameters, in terms of both the overall RMS residual field and record-high correlation coefficients between the observed and model field components. By experimenting with different combinations of input parameters and their time-averaging intervals, Tsyganenko and Andreeva (2016) found the best so far results to be given by the ram pressure Pd, SYM-H, and N-index by Newell et al. (2007). In addition, the IMF By has also been included as a model driver, with a goal to more accurately represent the IMF penetration effects. The model faithfully reproduces both externally and internally induced variations in the global distribution of the geomagnetic field and electric currents. Stronger solar wind driving results in a deepening of the equatorial field depression and a dramatic increase of its dawn-dusk asymmetry. The Earth's dipole tilt causes a consistent deformation of the magnetotail current sheet and a significant north-south asymmetry of the polar cusp depressions on the dayside.

2.3.25. Sq Solar Variation at Medea Observatory (Algeria), from 2008 to 2011

The paper of Anad et al. (2016) presents the regular variations of terrestrial magnetic field recorded by a new magnetic Observatory Medea, Algeria (geographic latitude: 36.85°N, geographic longitude: 2.93°E, geomagnetic latitude: 27.98°N, geomagnetic longitude: 77.7°E) during 2008–2011. The diurnal and seasonal variations of the solar quiet (Sq) variations are analyzed. The results show differences in the diurnal pattern of the northward-component Sq variation (SqX) at different seasons. The seasonal variation of SqX is similar in different years. The diurnal pattern of SqX from July through September cannot be explained by an equivalent current system that is symmetric about the noon time sector. The observations indicate that the major axis of the elliptic current system is tilted towards the equator in the morning hours during those months. The diurnal pattern of SqY indicates southward currents in the morning and northward currents in the afternoon, except during February–March 2009 when there is apparently no southward current during the morning. For the other months, the observations indicate that the maximum northward current intensity in the afternoon tends to be greater than the maximum southward current intensity in the morning. This is because of the UT variation of the Sq current system. That is, the pattern and strength of the Sq current system are different when SqY is measured in the morning around 8 UT and in the afternoon around 14 UT. The amplitude of these extreme varies linearly with the solar cycle. For the SqY component, the changes in the morning maximum have an annual variation while that of the afternoon minimum has a semi-annual variation. These variations are attributed to seasonal variations in the ionospheric E-region conductivity and atmospheric tidal winds. The field-aligned currents can also contribute to the seasonal variation of SqY. As outlined Anad et al. (2016), the two-dimensional approach used in this article does not allow to quantitatively determine their influences. Some obtained results are illustrated by Figures 2.3.10 – 2.3.13.

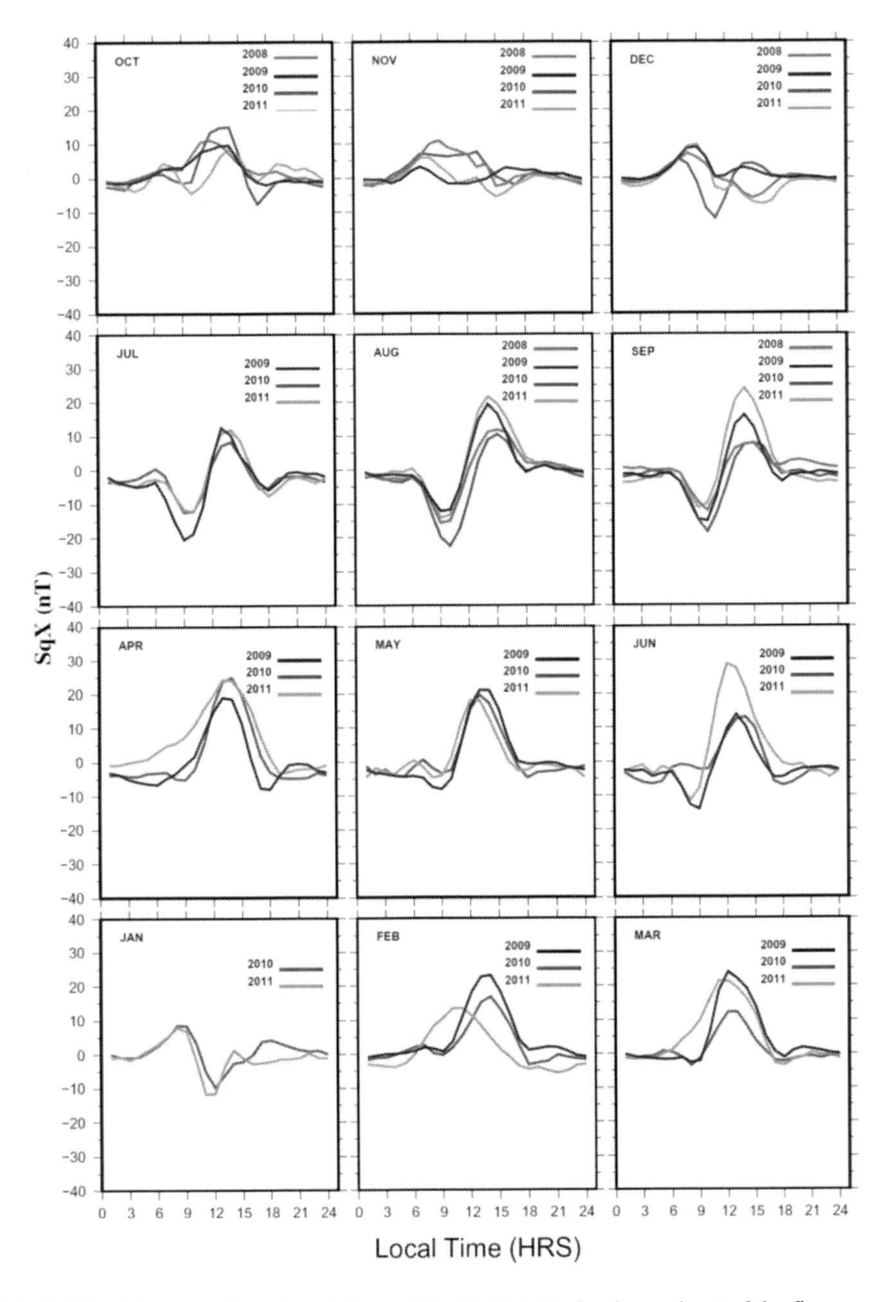

Figure 2.3.10. Monthly mean diurnal variations of the SqX at Medea for each set of the five international quiet days of each month from August 2008 to December 2011. According to Anad et al. (2016). Permission from COSPAR.

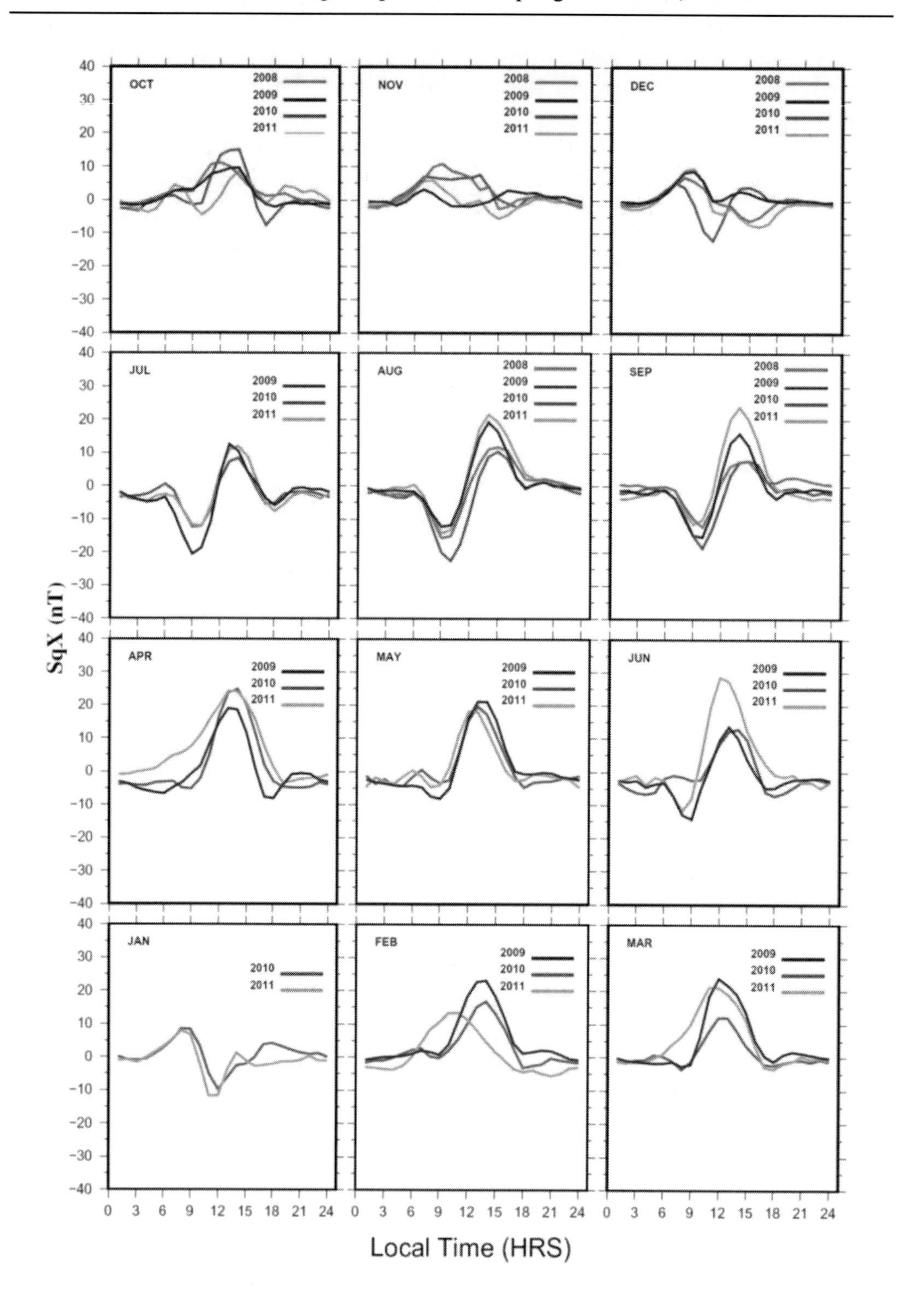

Figure 2.3.11. Monthly mean diurnal variations of the SqY at Medea for each set of the five international quiet days of each month from August 2008 to December 2011. According to Anad et al. (2016). Permission from COSPAR.

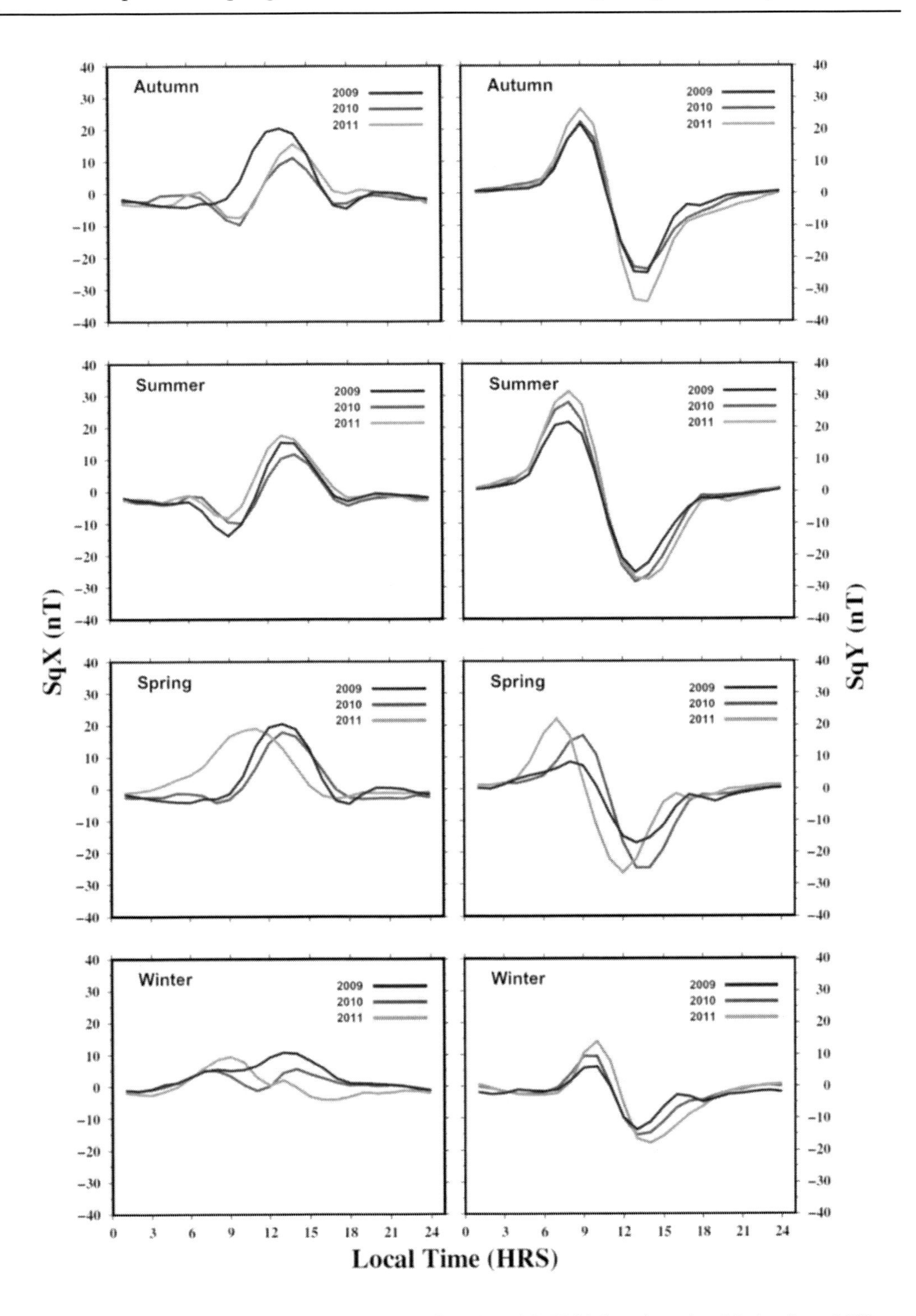

Figure 2.3.12. Seasonal variations of SqX (left column) and SqY (right column) at Medea from 2009 to 2011. From top to bottom each row represents the corresponding season for each year 2009, 2010 and 2011 respectively. According to Anad et al. (2016). Permission from COSPAR.

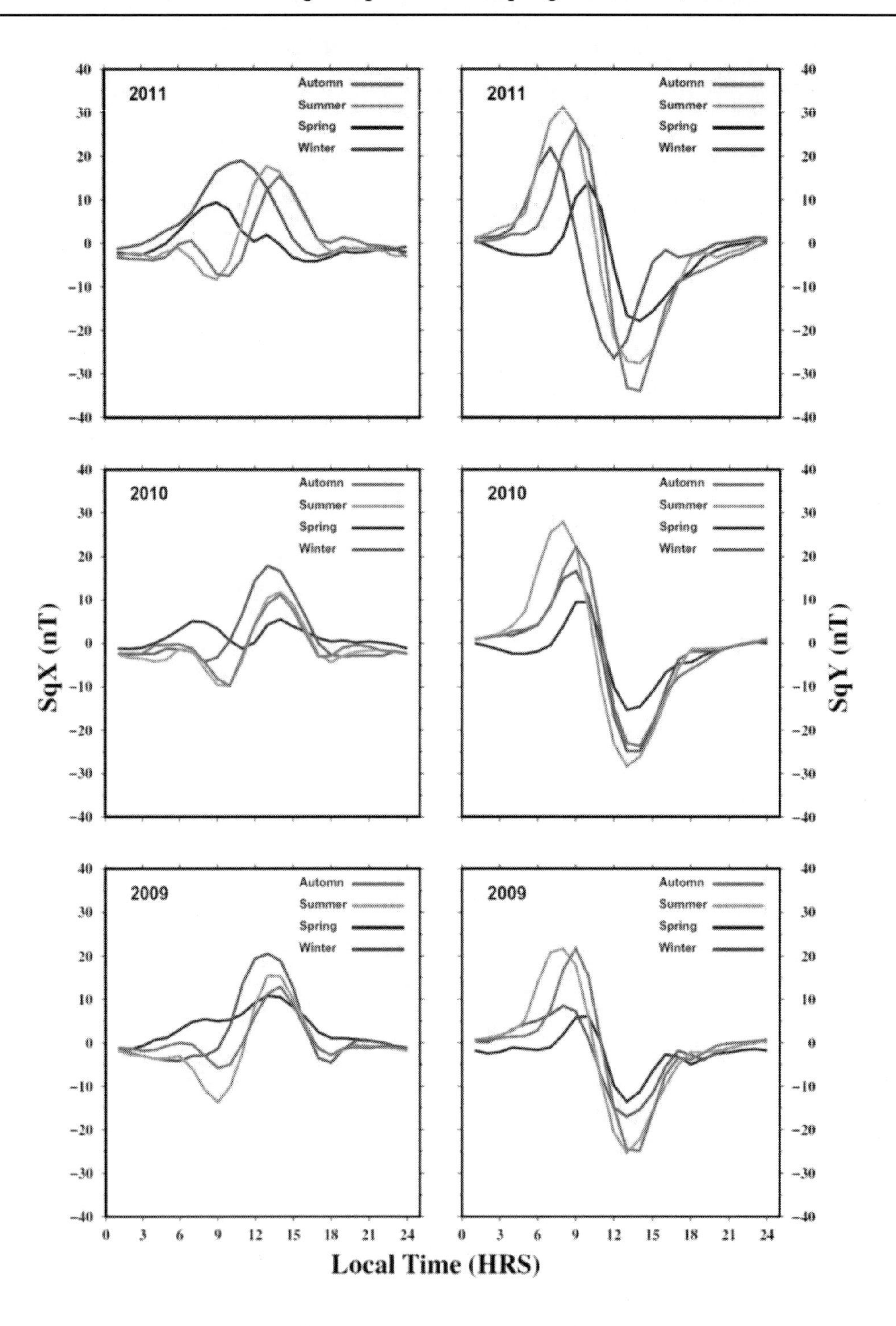

Figure 2.3.13. Seasonal variations of SqX (left column) and SqY (right column) from 2009 to 2011. From top to bottom each panel shows the four seasons of the year. Each figure shows 4 curves corresponding to the variation of either SqX or SqY over the 4 seasons (Winter, Spring, Summer and Autumn). From top to bottom each row represent the seasonal variation across each individual year. According to Anad et al. (2016). Permission from COSPAR.

2.3.26. Seasonal Variation of the Sq Focus Position during 2006–2010.

In Vichare et al. (2017), the perception of the seasonal variation of the Sq focus position is re-examined during low solar activity period (2006–2010). Equivalent current vectors are plotted for each geomagnetic quiet day (Ap ≤ 5), using diurnal variations of H and D components measured at the magnetic observatories located in a narrow longitudinal belt of the Indo-Russian region. On the formation of well-defined Sq current loop, the information about the Sq focus is extracted by identifying a pair of neighboring stations with opposite zonal currents and nearby local times with opposite meridional currents. Thus, the method employed here is different from the methods used in earlier studies. Prominent seasonal variations in the Sq focus latitude, as well as in the local time of Sq focus, are observed. It is observed that the Sq focus is located at ~30 deg in March equinox, but it moves to lower latitudes in the month of September. In winter, it shows large variability and also the formation of clear Sq current loop is less frequent. The local time of Sq focus is at ~12 LT in March and shifts to ~10 LT during September. It is clearly evident from the present analysis that the March and September equinoxes behave differently. The dominance of DE3 and semidiurnal waves in the September equinox could be the reason for the observed disparity. Some obtained In Vichare et al. (2017) results are illustrated by Figures 2.3.14 – 2.3.16.

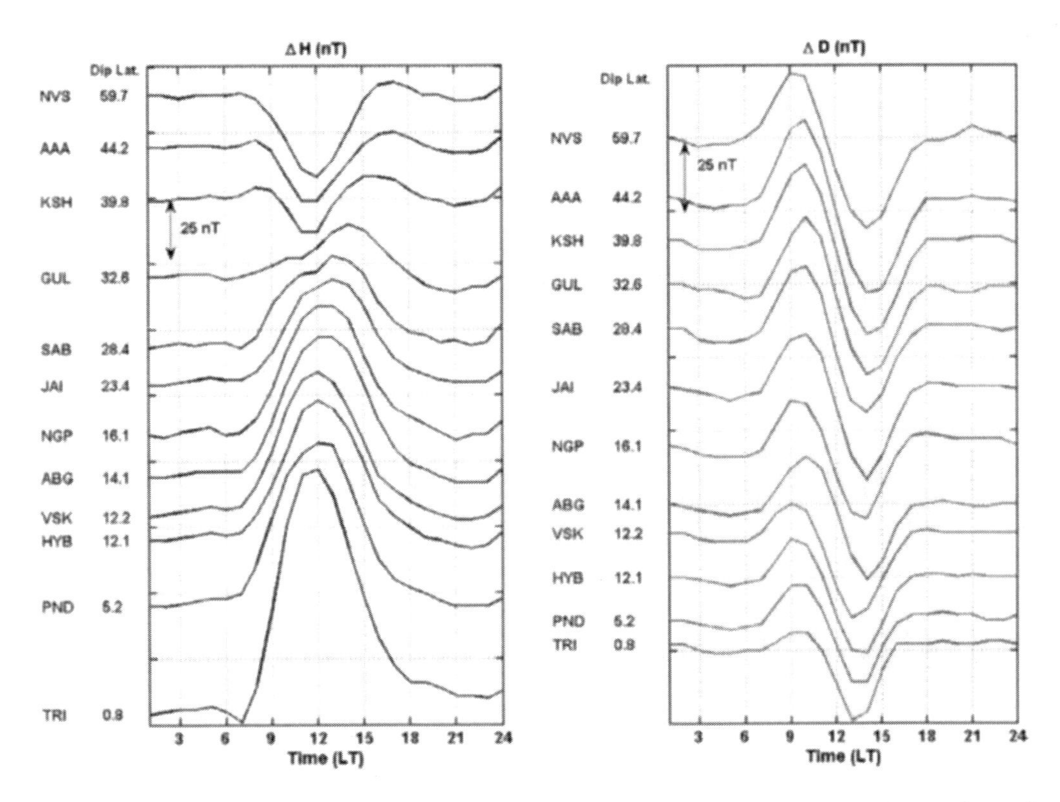

Figure 2.3.14. Sq daily variations in H and D components during 20 March 2007 (Ap = 1) at Indian and Russian observatories. According to Vichare et al. (2017). Permission from COSPAR.

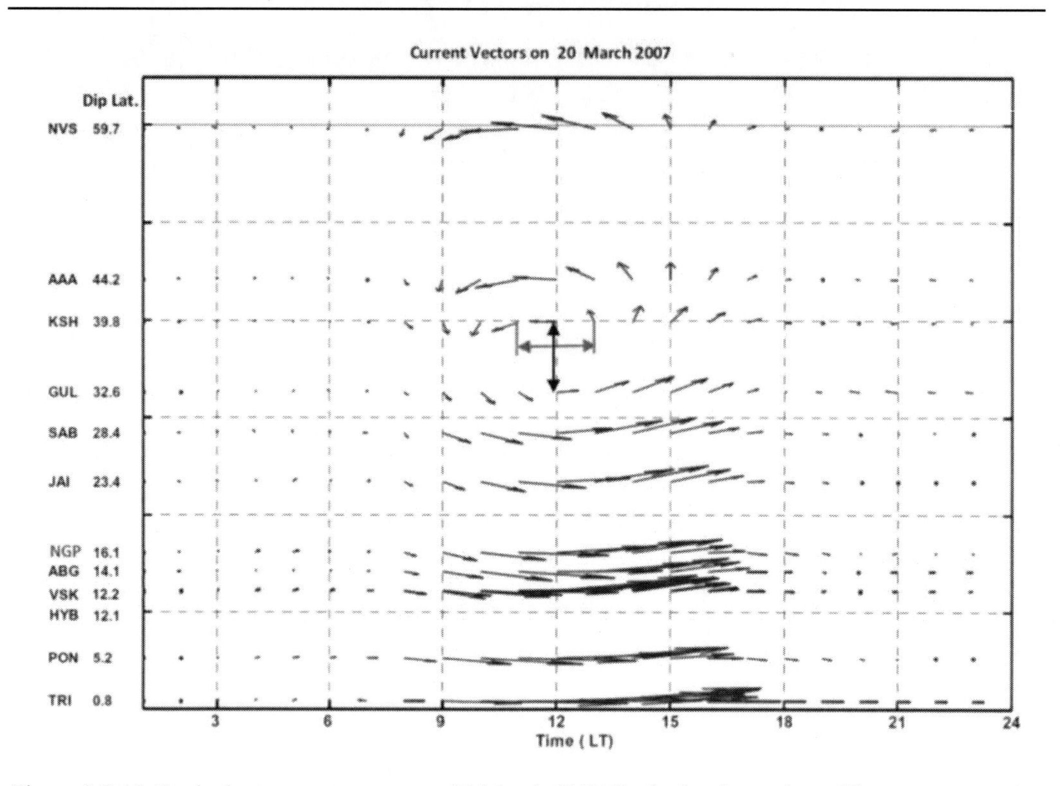

Figure 2.3.15. Equivalent current vectors on 20 March 2007 displaying formation of Sq current vortex. According to Vichare et al. (2017). Permission from COSPAR.

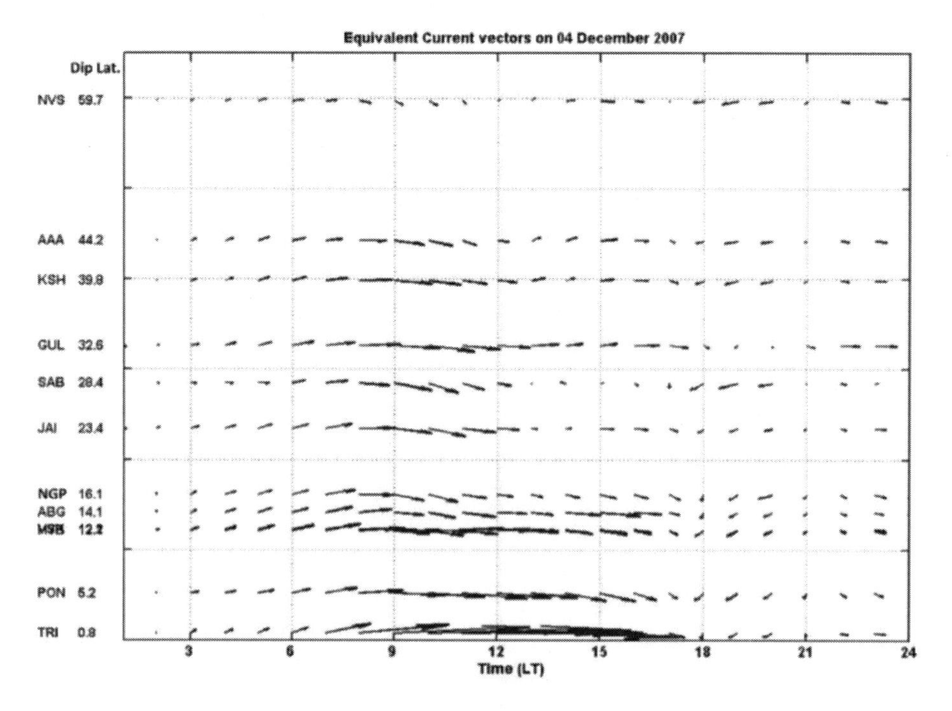

Figure 2.3.16. Equivalent current vectors on 4 December 2007 – No Sq loop formation. According to Vichare et al. (2017). Permission from COSPAR.

2.3.27. The Influence of IMF Clock Angle on the Cross Section of the Tail Bow Shock

Wang et al. (2016) study effects of the IMF orientation on the terrestrial tail bow shock location and shape by using global MHD magnetosphere model and empirical bow shock models. It is shown that the tail bow shock cross section is well approximated by an ellipse with the direction of the major axis roughly perpendicular to the IMF clock angle direction. With the increasing IMF clock angle, the eccentricity of the bow shock cross section increases for northward IMF but decreases for southward IMF.

2.3.28. Magnetopause Reconnection Layer Bounded by Switch-Off Shocks: Part 2. Pressure Anisotropy

The jump conditions are analyzed by Sonnerup et al. (2016b) in detail for two slow shocks bounding a reconnection plasma jet, observed on 3 August 2008 by the spacecraft THEMIS D on the dayside, low-latitude magnetopause. Both shocks are near the switch-off limit. They have been previously examined by Sonnerup et al. (2016a), on the basis of the simplest MHD version of the jump conditions. In Sonnerup et al. (2016b), those conditions now include the pressure anisotropy, normal heat fluxes, and a finite normal magnetic field component, the effects of all of which are found to be small. Sonnerup et al. (2016b) also present and discuss the, mostly field-aligned, measured total heat fluxes, which are found to be substantial and directed away from the reconnection site. It is shown that the double-adiabatic (Chew-Goldberger-Low) invariants are far from invariant. Their combination indicates a large entropy increase across the shock on the magnetospheric side with a much smaller increase across the shock on the magnetosheath side. The detailed cause of the entropy changes remains unclear but appears to involve irreversible transfer of energy between thermal motion parallel and perpendicular to the magnetic field. The new results confirm the previously found presence of heavy ions and the values of the effective ion mass on both sides of the event. They also confirm the need for an ion pressure correction in the shock on the magnetospheric side.

2.3.29. The Plasmasphere Electron Content Paradox

As outlined Krall and Huba (2016), measurements show that plasmasphere refilling rates decrease with increasing solar activity, while paradoxically, the vertical integration of the plasmasphere electron density (pTEC) increases with increasing solar activity. Using the Naval Research Laboratory SAMI2 (Sami2 is Another Model of the Ionosphere) and SAMI3 (Sami3 is Also a Model of the Ionosphere) codes, Krall and Huba (2016) simulate plasmasphere refilling following a model storm, reproducing this observed phenomenon. In doing so, they find that the refilling rate and resulting pTEC values are sensitive to the oxygen profile in the thermosphere and exosphere: the supply of H^+ in the topside ionosphere is limited by the local O^+ density, through $H + O^+ \rightarrow H^+ + O$ charge exchange. At solar minimum, the O^+ supply simply increases with the O density in the exosphere. At solar maximum, Krall and Huba (2016) find that O-O^+ collisions limit the O^+ density in the topside ionosphere such that it decreases with increasing O density. The paradox occurs because the pTEC metric gives electrons in the topside ionosphere more weight than electrons in the plasmasphere.

2.3.30. Where Does the Plasmasphere Begin? Revisit to Topside Ionospheric Profiles in Comparison with Plasmaspheric TEC from Jason-1

According to Lee et al. (2016), topside ionospheric profiles have been measured by Alouette 1 and ISIS 1/2 in the periods of 1962–1972 and 1972–1979, respectively. The profiles cover from the orbital altitude of 1000 km to the F2 peak and show large variations over local time, latitude, and seasons. Lee et al. (2016) analyze these variations in comparison with plasmaspheric total electron contents (pTECs) that were measured by Jason-1 satellite from the altitude of 1336 km to 20,200 km (GPS orbit). The scale heights of the profiles are generally smaller in the daytime than nighttime but show large day-to-day variations, implying that the ionospheric profiles at 1000 km are changing dynamically, rather than being in diffusive equilibrium. Lee et al. (2016) also derived transition heights between O^+ and H^+, which show a clear minimum at dawn for low-latitude profiles due to decreasing O^+ density at night. To compare with pTEC, they compute topside ionospheric total electron content (tiTEC) by integrating over 800–1336 km using the slope of the profiles. The tiTEC varies in a clear diurnal pattern from ~0.3 to ~1 and ~3 total electron content unit (TECU, 1 TECU = 10^{16} el.m^{-2}) for low and high solar activity, respectively, whereas Jason-1 pTEC values are distributed over 2–6 TECU and 4–8 TECU for low and high solar activity, respectively, with no apparent diurnal modulation. Latitudinal variations of tiTEC show distinctive hemispheric asymmetry while that of Jason-1 pTEC is closely symmetric about the magnetic equator. The local time and latitudinal variations of tiTEC basically resemble those of the ionosphere but are characteristically different from those of Jason-1 pTEC. Based on the difference between tiTEC and pTEC variations, Lee et al. (2016) propose that the region above ~1300 km should be considered as the plasmasphere. Lower altitudes for the base of "plasmaspheric TEC," as used in some studies, would cause contamination of ionospheric influence.

2.3.31. Reverse Flow Events and Small-Scale Effects in the Cusp Ionosphere

Spicher et al. (2016) report in situ measurements of plasma irregularities associated with a reverse flow event (RFE) in the cusp F region ionosphere. The Investigation of Cusp Irregularities 3 (ICI-3) sounding rocket, while flying through a RFE, encountered several regions with density irregularities down to meter scales. Spicher et al. (2016) address in detail the region with the most intense small-scale fluctuations in both the density and in the AC electric field, which were observed on the equatorward edge of a flow shear, and coincided with a double-humped jet of fast flow. Due to its long-wavelength and low-frequency character, the Kelvin-Helmholtz instability (KHI) alone cannot be the source of the observed irregularities. Using ICI-3 data as inputs, Spicher et al. (2016) perform a numerical stability analysis of the inhomogeneous energy-density-driven instability (IEDDI) and demonstrate that it can excite electrostatic ion cyclotron waves in a wide range of wave numbers and frequencies for the electric field configuration observed in that region, which can give rise to the observed small-scale turbulence. According to Spicher et al. (2016), the IEDDI can seed as a secondary process on steepened vortices created by a primary KHI. Such an interplay between macroprocesses and microprocesses could be an important mechanism for ion heating in relation to RFEs.

2.3.32. Forces Driving Fast Flow Channels, Dipolarizations, and Turbulence in the Magnetotail

As outlined El-Alaoui et al. (2016), fast flow channels are a major contributor to earthward transport in the magnetotail. Fast flow channels originate at reconnection sites in the magnetotail and terminate with dipolarizations, earthward moving regions of enhanced magnetic field. El-Alaoui et al. (2016) have used a global magnetohydrodynamic simulation of magnetotail dynamics during a substorm on 7 February 2009 to investigate the stresses in the plasma sheet. In particular, we present an analysis of forces in and around flow channels and dipolarizations. Earthward of the dipolarization magnetic and thermal pressure forces are nearly in equilibrium. Tailward of the dipolarization the pattern of stresses is complex, but several general features are evident. In the earthward flow channel the magnetic tension force dominates. In the dipolarization a tailward magnetic force is partially balanced by the earthward pressure force. The stresses in azimuth (y) are variable. In some places they cause the flow stream to meander, while in others they cause the stream to become wider or narrower. A major consequence of fast earthward flow channels is that they drive large-scale vortices that initiate turbulence.

2.3.33. Generalized Magnetotail Equilibria: Effects of the Dipole Field, Thin Current Sheets, and Magnetic Flux Accumulation

Generalizations of the class of quasi-1-D solutions of the 2-D Grad-Shafranov equation, first considered by Schindler (1972), are investigated by Sitnov and Merkin (2016). It is shown that the effect of the dipole field, treated as a perturbation, can be included into the original 1972 class solution by modification of the boundary conditions. As outlined Sitnov and Merkin (2016), some of the solutions imply the formation of singularly thin current sheets. Equilibrium solutions for such sheets resolving their singular current structure on the scales comparable to the thermal ion gyroradius can be obtained assuming anisotropic and nongyrotropic plasma distributions. It is shown that one class of such equilibria with the dipole-like boundary perturbation describes bifurcation of the near-Earth current sheet. Another class of weakly anisotropic equilibria with thin current sheets embedded into a thicker plasma sheet helps explain the formation of thin current sheets in a relatively distant tail, where such sheets can provide ion Landau dissipation for spontaneous magnetic reconnection. The free energy for spontaneous reconnection can be provided due to accumulation of the magnetic flux at the tailward end of the closed field line region. The corresponding hump in the normal magnetic field profile B_z $(x, z = 0)$ creates a nonzero gradient along the tail. The resulting gradient of the equatorial magnetic field pressure is shown to be balanced by the pressure gradient and the magnetic tension force due to the higher-order correction of the latter in the asymptotic expansion of the tail equilibrium in the ratio of the characteristic tail current sheet variations across and along the tail.

2.3.34. Magnetotail Magnetic Flux Monitoring Based on Simultaneous Solar Wind and Magnetotail Observations

As outlined Shukhtina et al. (2016), the magnetotail magnetic flux (MTF) is an important global variable to describe the magnetospheric state and dynamics. Existing methods of MTF estimation on the basis of the polar cap area, inferred from observations of global auroras and field-aligned currents, do not allow benchmarking due to the absence of a gauge for comparison; besides, they rarely allow a systematic nearly real time MTF monitoring. Shukhtina et al. (2016) describe three modifications (F_0, F_1, and F_2) of the method to calculate the MTF, based on simultaneous spacecraft observations in the magnetotail and in the solar wind, suitable for real-time MTF monitoring. The MTF dependence on the solar wind parameters and the observed tail lobe magnetic field is derived from the pressure balance conditions. An essential part of this study is the calibration of the approximate method against global 3-D MHD simulations and the empirical T14 magnetospheric field model. The calibration procedure provides all variables required to evaluate F_0, F_1, and F_2 quantities and, at the same time, computes the reference MTF value through any tail cross section. It allowed for Shukhtina et al. (2016) to extend the method to be used in the near tail, investigate its errors, and define the applicability domain. The method was applied to Cluster and THEMIS measurements and compared with methods of polar cap area calculation based on IMAGE and AMPERE observations. Shukhtina et al. (2016) also discuss possible applications and some recent results based on the proposed method.

2.3.35. Stability of Magnetotail Equilibria with a Tailward Bz Gradient

Analytical properties of two-dimensional magnetotail equilibria with a region of magnetic flux accumulation (a "Bz hump") are considered by Merkin and Sitnov (2016) with respect to tearing and interchange stability. Since the former mode does not initially involve a change in topology, Merkin and Sitnov (2016) denote it the Magnetic Flux Release Instability (MFRI). The region earthward of the Bz peak is found most unstable to both modes due to the presence of a tailward Bz gradient there. Stability is studied as a function of proximity of the Bz accumulation region to Earth. Interchange is found more difficult to stabilize closer to Earth as the level of the background pressure required for stabilization is higher there. At the same time, the kinetic MFR mode has the largest potential for destabilization farther in the tail as long as the current sheet is sufficiently thin to allow Landau dissipation, while the tearing (MFRI) Cd parameter is sufficiently large. An MHD analog of the kinetic MFR mode is allowed still farther from Earth, and the regions of kinetic and MHD instability may overlap dependent on the parameters of the equilibrium. Merkin and Sitnov (2016) provide analytical results describing the MFR stability as a function of a number of key parameters, including the level of current sheet stretching and its length, as well as the amplitude and the scale size of the Bz gradient.

2.3.36. Magnetospheric Response and Reconfiguration Times Following IMF B_y Reversals

As note Tenfjord et al. (2017), the interaction between the IMF and the geomagnetic field at the dayside magnetopause leads to transfer of momentum and energy which changes the magnetospheric configuration, but only after a certain time. Tenfjord et al. (2017) quantify this time, to advance our understanding of the causes for the delayed response of the magnetosphere. They study the response and reconfiguration time of the inner magnetosphere to IMF B_y reversals. A superposed epoch analysis of magnetic field measurements from four Geostationary Operational Environmental Satellite spacecraft at different local times both for negative to positive IMF B_y reversals and for positive to negative reversals is presented. The magnetospheric response time at geosynchronous orbit to the sudden change of IMF B_y is less than 15 (~10) min from the bow shock (magnetopause) arrival time, while the reconfiguration time is less than 46 (~41) min. These results are consistent with a B_y component induced on closed magnetic field lines due to the asymmetric loading of flux following asymmetric dayside reconnection when IMF $B_y \neq 0$. Obtained results also confirm the earlier studies by Tenfjord et al. (2015) that nightside reconnection is not required for generating a B_y component on closed field lines.

2.3.37. The Influence of Kinetic Effect on the MHD Scalings of a Thin Current Sheet

In Sasunov et al.(2017), assuming that magnetic moment μ is conserved on periodic segments of charged-particle trajectories, it has been shown analytically that the half-thickness of a stable thin current sheet (TCS) is practically independent on VT/Vbulk, where VT and Vbulk are thermal and bulk velocities of the TCS-forming particles, respectively. This fact has been confirmed by the PIC modeling of the TCS without any special assumptions on the particle trajectory type. Such coincidence of analytical and numeric results indicates that an equilibrium TCS indeed is formed by counterstreaming proton flows with μ = constant. The weak dependence of the TCS half-thickness on V_T/V_{bulk} also results in a reverse relation between the background plasma number density n_0 and V_{bulk}, so that the increase of V_{bulk} results in a decrease of n_0.

2.3.38. IMF Dependence of Energetic Oxygen and Hydrogen Ion Distributions in the Near-Earth Magnetosphere

As underlind Luo et al. (2017), energetic ion distributions in the near-Earth plasma sheet can provide important information for understanding the entry of ions into the magnetosphere and their transportation, acceleration, and losses in the near-Earth region. In the study of Luo et al. (2017), 11 years of energetic proton and oxygen observations (> ~274 keV) from Cluster/Research with Adaptive Particle Imaging Detectors were used to statistically study the energetic ion distributions in the near-Earth region. The dawn-dusk asymmetries of the distributions in three different regions (dayside magnetosphere, near-Earth nightside plasma sheet, and tail plasma sheet) are examined in Northern and Southern Hemispheres. The results show that the energetic ion distributions are influenced by the dawn-dusk interplanetary magnetic field (IMF) direction. The enhancement of ion intensity largely correlates with the location of the magnetic reconnection at the magnetopause. The results of Luo et al. (2017) imply that substorm-related acceleration processes in the magnetotail are not the only source of energetic ions in the dayside and the near-Earth magnetosphere. Energetic ions delivered through reconnection at the magnetopause significantly affect the energetic ion population in the magnetosphere. Luo et al. (2017) also believe that the influence of the dawn-dusk IMF direction should not be neglected in models of the particle population in the magnetosphere.

2.3.39. Evolution of the magnetic field structure outside the magnetopause under radial IMF conditions

Pi et al. (2017) use the THEMIS data to investigate the magnetic field structure just outside the magnetopause and its time evolution for radial IMF events. When the magnetic field drapes around the magnetopause in the magnetosheath region, an asymmetric magnetic field orientation in different hemispheres is expected. The two-case study of Pi et al. (2017) reveals some conflicts with the predicted draped field configuration in the Southern Hemisphere. The magnetosheath B_z component had a different sign depending on the upstream IMF B_x component's polarity at the beginning of the radial IMF intervals. With time, the observed B_z became northward in both cases with increasing positive values through the events. The increasing value of the B_z component may be explained by two possible mechanisms: by a change of the upstream IMF and by a reconnection between magnetosheath and magnetospheric field lines. The study of Pi et al. (2017) shows that both mechanisms contributed to the observed changes. Thus, there was a correlation between the change of the upstream IMF conditions and an increase in the magnetosheath northward magnetic field component. The observed formation of the boundary layer near the magnetopause proves that the reconnection process was ongoing at least for a part of the time. Pi et al. (2017) suggest two possible reconnection scenarios: one near subsolar point and another tailward of the one cusp due to lobe reconnection. The asymmetry of reconnection locations causes rearrangement of the magnetic field structure near the magnetopause and turns the observed magnetosheath B_z component even further into positive values.

2.4. Interplanetary Electric Field, Reconnection, Relativistic Electrons, and Electrodynamics of Magnetosphere-Ionosphere System

2.4.1. The Step Response Function Relating the Interplanetary Electric Field to the Dayside Magnetospheric Reconnection Potential

Blanchard and Baker (2010) present a statistical analysis of the response of the magnetic reconnection rate between the IMF and the magnetosphere to southward turnings of the IMF. The magnetic reconnection rate is calculated from SuperDARN measurements. The polar cap boundary is identified as the offset circle (3° toward midnight from the magnetic pole) that best separates high spectral width backscatter, indicative of open magnetic field lines, from low spectral width backscatter. The electric potential on the boundary is determined from the best fit of the line–of–sight F region plasma velocity to an eighth- order spherical harmonic function of ionospheric electrical potential. The reconnection rate is determined from the electric potential on the boundary and the expansion rate of the polar cap. Solar wind data (velocity and magnetic field) are obtained from Wind spacecraft measurements and propagated to the magnetopause. The relationship between interplanetary electric field and the magnetic reconnection rate is analyzed by calculating linear response functions. Blanchard and Baker (2010) determine that the convection term in the reconnection rate measurement exhibits a unimodal response to the interplanetary electric field, whereas the boundary motion term in the reconnection rate measurement exhibits a bimodal response, with a positive mode (increase in polar cap expansion rate) followed by a negative mode. The response of the dayside reconnection rate to a step change in the interplanetary electric field exhibits a transient response of one hour duration followed by a steady state response at a level that is half of the peak response.

2.4.2. The Effect of Different Solar Wind Parameters upon Significant Relativistic Electron Flux Dropouts in the Magnetosphere

Superposed epoch analyses were performed by Gao et al. (2015) on 193 significant relativistic electron flux dropout events, in order to study the roles of different solar wind parameters in driving the depletion of relativistic electrons, using ~16 years of data from the POES and GOES missions, and the OMNIWEB solar wind database. It's find that the solar wind dynamic pressure and IMF B_Z play key roles in causing the relativistic electron flux dropouts, but also that either large solar wind dynamic pressure or strong southward IMF B_Z by itself is capable of producing the significant depletion of relativistic electrons. The relativistic electron flux dropouts occur not only when the magnetopause is compressed closer to the Earth but also when the magnetopause is located very far (> ~10 R_E). Importantly, obtained results show that in addition to the large solar wind dynamic pressure, which pushes the magnetopause inward strongly and causes the electrons to escape from the magnetosphere, relativistic electrons can also be scattered into the loss cone and precipitate into the Earth's atmosphere during periods of strong southward IMF B_Z, which preferentially provides a source of free energy for electromagnetic ion cyclotron (EMIC) wave excitation. This is supported by the fact that the strongest electron precipitation into the atmosphere is found in the dusk sector, where EMIC waves are typically observed in the high-density plasmasphere or plume and cause efficient electron precipitation down to ~1 MeV.

2.4.3. Electrodynamics of Magnetosphere-Ionosphere Coupling and Feedback on Magnetospheric Field Line Resonances

Lu et al. (2007) present a new dynamic model that describes coupling between standing inertial or ion-acoustic-gyroradius-scale shear Alfvén waves, compressional modes, and auroral density disturbances. The model is applied to the excitation of field line resonances (FLRs) in dipolar and stretched geomagnetic fields in Earth’s magnetosphere. Magnetosphere-ionosphere coupling is included by accounting for the closure of magnetospheric field-aligned currents (FACs) through Pedersen currents in the ionosphere. A second new aspect is that the height-integrated Pedersen conductivity is treated as a dynamic parameter by electrodynamically coupling the two-dimensional finite element wave code ‘Topo’ to the ionospheric ionization model ‘Global Airglow Model’ (GLOW). Lu et al. (2007) demonstrate that field line stretching brings the equatorial plasma β above unity, where the reduced MHD formulism for low-frequency plasma breaks down. This new magnetosphere-ionosphere model makes it possible to investigate magnetosphere-ionosphere coupling (through field aligned currents closing by ionospheric cross-field currents) with feedback effects due to auroral electron precipitations. As outlined Lu et al. (2007), this is a significant extension of Prakash et al. (2003), mainly in that: (1) rather than using an envelop approximation, they treat the full nonlinear MHD equations, i.e., include the full compressional modes and their coupling with FLRs; (2) they use a more self-consistent, interactive ionospheric conductivity model considering the effects of field-aligned potential drops and the magnetic mirror force on electron precipitation. Lu et al. (2007) noted that this new magnetosphere-ionosphere coupling model (coupled Topo MHD code and GLOW model) makes it possible to investigate also the effect of ionospheric conductivity on the physics of magnetosphere-ionosphere coupling and diurnal, seasonal and solar cycle variations due to the feedback. It is shown that the auroral electron precipitation induced Pedersen conductivity enhancement can lead to strong feedback effects on magnetospheric FLR wave amplitudes and density perturbation. It has been shown that the nonlinearity from SAW ponderomotive forces can steepen the local Alfvén speed gradient and then significantly affect the dynamic evolution of a FLR (Lu et al., 2005a). The ponderomotive force pushes plasma from high latitudes to the equatorial magnetosphere, generating the ionospheric (equatorial) cavities (bumps) (Lu et al., 2005b).

As an application of developed model, Lu et al. (2007) study a specific FLR event observed on 31 January 1997, when the NASA FAST satellite was over the Canadian Auroral Network for the OPEN Program Unified Study (CANOPUS) Gillam station. Using geomagnetic fields computed from the T96 magnetic field model (Tsyganenko, 1996), Lu et al. (2007) show that auroral electron precipitation produces strong Pedersen conductivity enhancements that control the final amplitude and width of the excited FLR, along with the amplitude of associated density fluctuations. The predictions of the model are generally consistent with observations of this event. Lu et al. (2007) summarize main obtained results as following:

(i) Precipitation energies required to initiate the feedback effects using GLOW or Robinson et al. (1987) models are higher than the envelop calculation using Reiff formula (Reiff, 1984).

(ii) The ionospheric feedback depends on the competition between precipitation energy and wave damping. Although the higher wave damping may reduce the fieldaligned current, the stretching of magnetic field can bring about a larger parallel current and magnetic perturbation. It is shown that the precipitation energies required to initiate the feedback effect in a stretched case are lower than in the case of a dipolar field. This conclusion is different from that of Prakash et al. (2003), where the precipitation energies to initiate the feedback effect in a dipolar field is lower.

(iii) Nonlinear effects can produce strongly localized FLRs and density perturbation is strongly enhanced when auroral electron precipitation is included. The distribution of the density perturbation in full MHD calculations is different from the case with a reduced MHD calculation, mainly in that there exist significant movements of plasma across stretched field lines in full MHD while the density perturbation is mainly along the field lines in reduced MHD computation. This implies that in high-β situations, where reduced MHD breaks down, new behavior can occur owing to high plasma pressure effects.

(iv) Finally, it was applied the new interactive magnetosphere-ionosphere model to a specific case for 31 January 1997 FLR event, using the magnetic fields from T96 and parameters approximating this event. Lu et al. (2007) find that the modeled spatial scale of FLR arcs is around 35 km, quite comparable to the observed width in optical data for this event, and the amplitudes of wave fields ($E_{\perp} \sim 20$ mV/m, $j_{\parallel} \sim 2$ μA/m^2, by $b_Y \sim 50$ nT) are all reasonable in agreement with observations.

(v) Lu et al. (2007) mentioned that their work is limited to small current amplitude systems. For large currents (>10 μA/m^2), ionospheric electrons can be heated by resonant standing shear Alfvén waves through Joule dissipation, which may produce significant ionization and feedback on the FLR amplitude and structure (Lu et al., 2005a,b). Lu et al. (2007) plan to incorporate this effect into the interactive magnetosphere-ionosphere coupling model to investigate the feedback of ionospheric conductivity on the physics of magnetosphere-ionosphere coupling and diurnal, seasonal and solar cycle variations due to this feedback.

2.4.4. Reconnection Guide Field and Quadrupolar Structure Observed by MMS on 16 October 2015 at 1307 UT

Denton et al. (2016) estimate the guide field near the X point, B_{M0}, for a magnetopause crossing by the Magnetospheric Multiscale (MMS) spacecraft at 1307 UT on 16 October 2015 that showed features of electron-scale reconnection. This component of the magnetic field is normal to the reconnection plane L-N containing the reconnection magnetic field, B_L, and the direction e_N normal to the current sheet. The B_M field component appears to approximately have quadrupolar structure close to the X point. Using several different methods to estimate values of the guide field near the X point, some of which use an assumed quadrupolar symmetry, we find values ranging between −3.1 nT and −1.2 nT, with a nominal value of about −2.5 nT. The rough consistency of these values is evidence that the quadrupolar structure exists.

2.4.5. Ion Larmor Radius Effects Near a Reconnection X Line at the Magnetopause: THEMIS Observations and Simulation Comparison

Phan et al. (2016) report a THEMIS-D spacecraft crossing of a magnetopause reconnection exhaust ~9 ion skin depths (d_i) downstream of an X line. The crossing was characterized by ion jetting at speeds substantially below the predicted reconnection outflow speed. In the magnetospheric inflow region THEMIS detected (a) penetration of magnetosheath ions and the resulting flows perpendicular to the reconnection plane, (b) ion outflow extending into the magnetosphere, and (c) enhanced electron parallel temperature. Comparison with a simulation suggests that these signatures are associated with the gyration of magnetosheath ions onto magnetospheric field lines due to the shift of the flow stagnation point toward the low-density magnetosphere. Observations of Phan et al. (2016) indicate that these effects, ~2–3 d_i in width, extend at least 9 d_i downstream of the X line. The detection of these signatures could indicate large-scale proximity of the X line but do not imply that the spacecraft was upstream of the electron diffusion region.

2.4.6. Magnetic Reconnection at the Dayside Magnetopause: Advances with MMS

As outlined Burch and Phan (2016), magnetic reconnection is known to be an important process for coupling solar wind mass and momentum into the Earth's magnetosphere. Reconnection is initiated in an electron-scale dissipation/diffusion region around an X line, but its consequences are large scale. While past experimental efforts have advanced our understanding of ion-scale physics and the consequences of magnetic reconnection, much higher spatial and temporal resolutions are needed to understand the electron-scale processes that cause reconnection. The Magnetospheric Multiscale (MMS) mission was implemented to probe the electron scale of reconnection. This article reports on results from the first scan of the dayside magnetopause with MMS. Specifically, Burch and Phan (2016) introduce a new event involving the radial traversal of guide-field reconnection to illustrate features of reconnection physics on the electron scale.

2.4.7. Peculiarities of the Formation of a Thin Current Sheet in the Earth's Magnetosphere

Domrin et al.(2016) investigate the process of the self-consistent formation of a thin current sheet with a thickness close to the ion Larmor gyroradius in the presence of decreasing magnetic field's normal component Bn. This behavior is typical of the current sheet of the Earth's magnetospheric tail during geomagnetic substorms. It has been shown that, in a numerical model of the current sheet, based on the particle-in-cell method, the appearance of self-consistent electric field component Ey in the current sheet vicinity can lead to its significant thinning and, eventually, to the formation of a multiscale configuration with a thin current sheet (TCS) in the central region supported by transient particles. The structure of the resulting equilibrium is determined by the initial parameters of the model and by the particle dynamics during the sheet thinning. Under certain conditions, the particle drift in the crossed electric and magnetic fields leads to a significant portion of ions becoming trapped near the neutral sheet and, in this way, to the formation of a wider configuration with an embedded thin current sheet. The population of trapped particles produces diamagnetic negative currents that manifest in the form of negative wings at the periphery of the sheet. Correspondingly, in the direction perpendicular to the sheet, a nonmonotonic coordinate dependence of the magnetic field appears. The mechanisms of the evolution of the current sheet in the Earth's magnetotail and the formation of a multiscale structure are discussed.

2.4.8. Three-Dimensional Development of Front Region of Plasma Jets Generated by Magnetic Reconnection

A three-dimensional fully kinetic particle-in-cell simulation of antiparallel magnetic reconnection is performed by Nakamura et al. (2016) to investigate the three-dimensional development of reconnection jet fronts treating three instabilities: the lower hybrid drift instability (LHDI), the ballooning/interchange instability (BICI), and the ion-ion kink instability. Sufficiently large system size and high ion-to-electron mass ratio of the simulation allow us to see the coupling among the three instabilities in the fully kinetic regime for the first time. As the jet fronts develop, the LHDI and BICI become dominant over the ion-ion kink instability. The rapid growth of the LHDI enhances the BICI growth and the resulting formation of finger-like structures. The small-scale front structures produced by these instabilities are similar to recent high-resolution field observations of the dipolarization fronts in the near-Earth magnetotail using THEMIS and Cluster spacecraft and pose important questions for a future full high-resolution observation by the Magnetospheric Multiscale (MMS) mission.

2.4.9. On the Occurrence of Magnetic Reconnection Equatorward of the Cusps at the Earth's Magnetopause during Northward IMF Conditions

As note Trattner et al. (2017), magnetic reconnection changes the topology of magnetic field lines. This process is most readily observable with in situ instrumentation at the Earth's magnetopause as it creates open magnetic field lines to allow energy and momentum flux to flow from the solar wind to the magnetosphere. Most models use the direction of the IMF to determine the location of these magnetopause entry points, known as reconnection lines. Dayside locations of magnetic reconnection equatorward of the cusps are generally found during sustained intervals of southward IMF, while high-latitude region regions poleward of the cusps are observed for northward IMF conditions. In this study Trattner et al. (2017) discuss Double Star magnetopause crossings and a conjunction with a Polar cusp crossing during northward IMF conditions with a dominant IMF B_Y component. During all seven dayside magnetopause crossings, Double Star detected switching ion beams, a known signature for the presence of reconnection lines. In addition, Polar observed a cusp ion-energy dispersion profile typical for a dayside equatorial reconnection line. Using the cutoff velocities for the precipitating and mirrored ion beams in the cusp, the distance to the reconnection site is calculated, and this distance is traced back to the magnetopause, to the vicinity of the Double Star satellite locations. Analysis of Trattner et al. (2017) shows that, for this case, the predicted line of maximum magnetic shear also coincides with that dayside reconnection location.

2.4.10. Oxygen Acceleration in Magnetotail Reconnection

Motivated by the observed high concentration of oxygen ions in the magnetotail during enhanced geomagnetic activity, Liang et al. (2017) investigated the oxygen acceleration in magnetotail reconnection by using 2.5-D implicit particle-in-cell simulations. It is found that lobe oxygen ions can enter the downstream outflow region, i.e., the outflow region downstream of the dipolarization fronts (DFs) or the reconnection jet fronts. Without entering the reconnection exhaust, they are accelerated by the Hall electric field. They can populate the downstream outflow region before the DFs arrive there. This acceleration is in addition to acceleration in the exhaust by the Hall and reconnection electric fields. Oxygen ions in the preexisting current sheet are reflected by the propagating DF creating a reflected beam with a hook shape in phase space. This feature can be applied to deduce a history of the DF speed. However, it is difficult to observe for protons because their typical thermal velocity in the plasma sheet is comparable those of the DF and the reflection speed. The oxygen ions from the lobes and the preexisting current sheet form multiple beams in the distribution function in front of the DF. By comparing oxygen concentrations of 50%, 5%, and 0% with the same current sheet thickness, Liang et al. (2017) found that the DF thickness is proportional to the oxygen concentration in the preexisting current sheet. All the simulation results can be used to compare with the observations from the Magnetospheric Multiscale mission.

2.5. LEOPARD: A Grid-Based Dispersion Relation Solver for Arbitrary Gyrotropic Distributions

As note Astfalk and Jenko (2017), particle velocity distributions measured in collisionless space plasmas often show strong deviations from idealized model distributions. Despite this observational evidence, linear wave analysis in space plasma environments such as the solar wind or Earth's magnetosphere is still mainly carried out using dispersion relation solvers based on Maxwellians or other parametric models. To enable a more realistic analysis, Astfalk and Jenko (2017) present the new grid-based kinetic dispersion relation solver LEOPARD (Linear Electromagnetic Oscillations in Plasmas with Arbitrary Rotationally-symmetric Distributions) which no longer requires prescribed model distributions but allows for arbitrary gyrotropic distribution functions. Astfalk and Jenko (2017) discuss the underlying numerical scheme of the code and and it was shown a few exemplary benchmarks. Furthermore, Astfalk and Jenko (2017) demonstrate a first application of LEOPARD to ion distribution data obtained from hybrid simulations. In particular, it is shown that in the saturation stage of the parallel fire hose instability, the deformation of the initial bi-Maxwellian distribution invalidates the use of standard dispersion relation solvers. A linear solver based on bi-Maxwellians predicts further growth even after saturation, while LEOPARD correctly indicates vanishing growth rates. Astfalk and Jenko (2017) also discuss how this complies with former studies on the validity of quasilinear theory for the resonant fire hose.

2.6. Inner Magnetosphere Coupling: Recent Advances

As outlined Usanova and Shprits (2017), the dynamics of the inner magnetosphere is strongly governed by the interactions between different plasma populations that are coupled through large-scale electric and magnetic fields, currents, and wave-particle interactions. Inner magnetospheric plasma undergoes self-consistent interactions with global electric and magnetic fields. Waves excited in the inner magnetosphere from unstable particle distributions can provide energy exchange between different particle populations in the inner magnetosphere and affect the ring current and radiation belt dynamics. The ionosphere serves as an energy sink and feeds the magnetosphere back through the cold plasma outflow. The precipitating inner magnetospheric particles influence the ionosphere and upper atmospheric chemistry and affect climate. Satellite measurements and theoretical studies have advanced the understanding of the dynamics of various plasma populations in the inner magnetosphere. However, the knowledge of the coupling processes among the plasmasphere, ring current, radiation belts, global magnetic and electric fields, and plasma waves generated within these systems is still incomplete. This special issue incorporates extended papers presented at the Inner Magnetosphere Coupling III conference held 23–27 March 2015 in Los Angeles, California, USA, and includes modeling and observational contributions addressing interactions within different plasma populations in the inner magnetosphere (plasmasphere, ring current, and radiation belts), coupling between fields and plasma populations, as well as effects of the inner magnetosphere on the ionosphere and atmosphere.

2.7. A Statistical Study on Hot Flow Anomaly Current Sheets

As note Zhao et al. (2017), hot flow anomalies (HFAs) are phenomena frequently observed near Earth's bow shock and form when the interplanetary discontinuities interact with Earth's bow shock. Zhao et al. (2017) perform a statistical study to determine what kinds of discontinuities are more efficient to generate HFAs. Zhao et al. (2017) use strict criteria to identify classic HFAs, excluding similar foreshock phenomena such as spontaneous hot flow anomalies (SHFAs) and foreshock bubbles. Obtained results show that magnetic field on at least one side of the interplanetary discontinuities has to be connected to the bow shock in order to form HFAs. Discontinuities with large shear angles are more efficient to form HFAs. The thickness of current sheets and the thickness of HFAs are strongly correlated and current sheets with thickness from 1000 km to 3162 km are more efficient to form HFAs. Of the HFAs, 74% have the electric field pointing toward the current sheet on the leading side and 72% have the electric field pointing toward the current sheet on the trailing side. In addition, the variations of plasma parameters and the magnetic field of HFAs with E inward on both sides are more dramatic than those with E inward on only one side. An HFA is more likely to form when the reflected flow from the bow shock is along the discontinuity.

2.8. Structure of Current and Plasma in Current Sheets Depending on the Conditions of Sheet Formation

According to Frank et al. (2017), the structure of current sheets created under laboratory conditions is characterized by a large variety, which depends substantially on the conditions under which the sheet is formed. Frank et al. (2017) present the results of an experimental study of the structure and evolution of current sheets that were formed in magnetic configurations with a singular line of the X type. It has been shown that the change in the transverse magnetic field gradient, the strength of the longitudinal magnetic field, and the mass of the ions in plasma makes it possible to significantly vary the main parameters of the current sheets. This offers the challenges of using laboratory experimental results for analyzing and simulating the cosmophysical processes.

2.9. Temperature of the Plasmasphere from Van Allen Probes HOPE

Genestreti et al.(2017) introduce two novel techniques for estimating temperatures of very low energy space plasmas using, primarily, in situ data from an electrostatic analyzer mounted on a charged and moving spacecraft. The techniques are used to estimate proton temperatures during intervals where the bulk of the ion plasma is well below the energy bandpass of the analyzer. Both techniques assume that the plasma may be described by a one-dimensional $\mathbf{E} \times \mathbf{B}$ drifting Maxwellian and that the potential field and motion of the spacecraft may be accounted for in the simplest possible manner, i.e., by a linear shift of coordinates. The first technique involves the application of a constrained theoretical fit to a measured distribution function. The second technique involves the comparison of total and partial-energy number densities. Both techniques are applied to Van Allen Probes Helium, Oxygen, Proton, and Electron (HOPE) observations of the proton component of the plasmasphere during two orbits on 15 January 2013. Genestreti et al. (2017) find that the temperatures calculated from these two order-of-magnitude-type techniques are in good agreement with typical ranges of the plasmaspheric temperature calculated using retarding potential analyzer-based measurements—generally between 0.2 and 2 eV (2000–20,000 K). Genestreti et al. (2017) also find that the temperature is correlated with L shell and hot plasma density and is negatively correlated with the cold plasma density. It is posit that the latter of these three relationships may be indicative of collisional or wave-driven heating of the plasmasphere in the ring current overlap region. Genestreti et al. (2017) note that these techniques may be easily applied to similar data sets or used for a variety of purposes.

2.10. Timescales for the Penetration of IMF By into the Earth's Magnetotail

According to Browett et al. (2017), previous studies have shown that there is a correlation between the B_Y component of the IMF and the B_Y component observed in the magnetotail lobe and in the plasma sheet. However, studies of the effect of IMF B_Y on several magnetospheric processes have indicated that the B_Y component in the tail should depend more strongly on the recent history of the IMF B_Y rather than on the simultaneous measurements of the IMF. Estimates of this timescale vary from ~25 min to ~4 h. Browett et al. (2017) present a statistical study of how promptly the IMF B_Y component is transferred into the neutral sheet, based on Cluster observations of the neutral sheet from 2001 to 2008, and solar wind data from the OMNI database. Five thousand nine hundred eighty-two neutral sheet crossings during this interval were identified, and starting with the correlation between instantaneous measurements of the IMF and the magnetotail (recently reported by Cao et al. (2014)), Browett et al. (2017) vary the time delay applied to the solar wind data. Obtained results suggest a bimodal distribution with peaks at ~1.5 h and ~3 h. The relative strength of each peak appears to be well controlled by the sign of the IMF B_Z component with peaks being observed at 1 h of lag time for southward IMF and up to 5 h for northward IMF conditions, and the magnitude of the solar wind velocity with peaks at 2 h of lag time for fast solar wind and 4 h for slow solar wind conditions.

2.11. Suprathermal Electron Acceleration in the Near-Earth Flow Rebounce Region

As outlined Liu et al. (2017), flux pileup regions (FPRs) are traditionally referred to the strong-Bz bundles behind dipolarization fronts (DFs) in the Earth's magnetotail and can appear both inside earthward and tailward bursty bulk flows. It has been widely reported that suprathermal electrons (40–200 keV) can be efficiently accelerated inside earthward FPRs, leaving the electron acceleration inside tailward FPRs as an open question. Liu et al. (2017) focus on the electron acceleration inside a tailward FPR that is formed due to the flow rebounce in the near-Earth region ($X_{GSM} \approx -12\ R_E$) and compare it quantitatively with the acceleration inside an earthward FPR. By examining the Cluster data in 2008, Liu et al. (2017) sequentially observe an earthward FPR and a tailward FPR in the near-Earth region, with the earthward one belonging to decaying type and the tailward one belonging to growing type. Inside the earthward FPR, Fermi acceleration and betatron cooling of suprathermal electrons are found, while inside the tailward FPR, Fermi and betatron acceleration occur. Whistler-mode waves are observed inside the tailward FPR; their generation process may still be at the early stage. Liu et al. (2017) notice that the suprathermal electron fluxes inside the tailward FPR are about twice as large as those inside the earthward FPR, suggesting that the acceleration of suprathermal electrons is more efficient in the flow rebounce region. These acceleration processes have been successfully reproduced using an analytical model; they emphasize the role of flow rebounce in accelerating suprathermal electrons and further reveal how the MHD-scale flow modulates the kinetic-scale electron dynamics in the near-Earth magnetotail.

2.12. A Statistical Study of Single Crest Phenomenon in the Equatorial Ionospheric Anomaly Region Using Swarm A Satellite

As outlined Fathy and Ghamry (2017), though the Equatorial Ionospheric Anomaly (EIA) is represented by two crests within ±15° latitude, a single crest is also observed in the entire ionosphere. Few studies have addressed single crest phenomena. A statistical study of 2237 single crest phenomenon from the in situ electron density measurements of Swarm A satellite was investigated during December 2013–December 2015. The analysis of Fathy and Ghamry (2017) focused on local time, seasonal, and both geographic and geomagnetic latitudinal variations. Obtained results are as following: 1 – The maximum number of events peaks mainly in the dayside region around 0800–1200 LT and these occur mainly within the magnetic equator; 2 – The maximum amplitude of the single crests take place most prominently during equinoxes; 3 – The majority of single crests occur in the northern hemisphere; 4 – The seasonal distribution of the events shows that the summer events are located further from the magnetic equator in the northern hemisphere and shift their locations into the southern hemisphere in winter, while spring events are centered along the magnetic equator; 5 – Dayside single crest events appear close to the magnetic equator and more centered on the equator in winter season; 6 – Dawn, night and dusk side events reverse their location from northern hemisphere in summer to southern hemisphere in winter. Figures 2.12.1 – 2.12.5 are illustrate some obtained results.

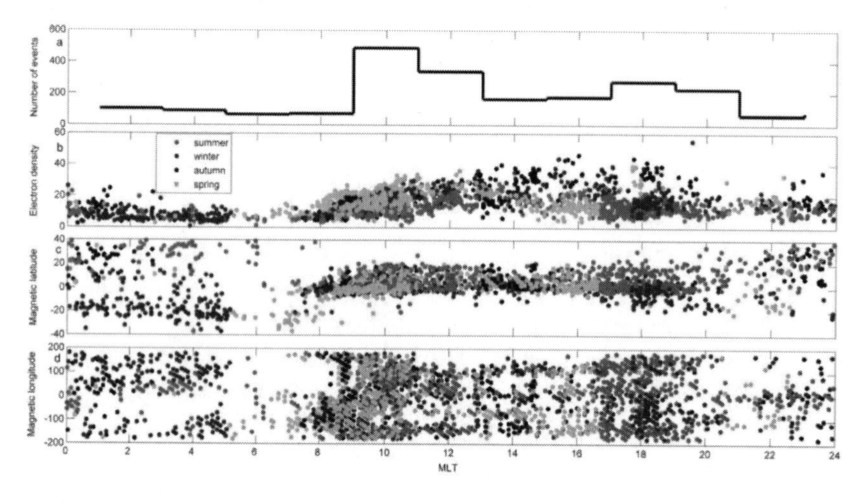

Figure 2.12.1. Local time distribution of the single crests versus (**a**) number of events each 2 h (**b**) the amplitude (**c**) the latitudinal distribution (**d**) the longitudinal distribution. According to Fathy and Ghamry (2017). Permission from COSPAR.

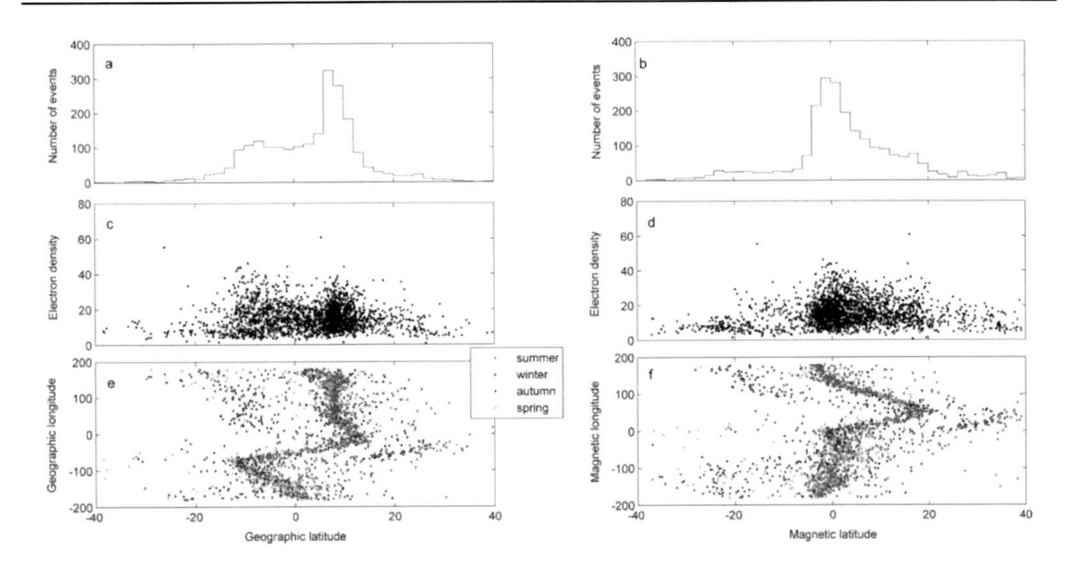

Figure 2.12.2. Distribution of the single crests with respect to the geographic and magnetic latitudes: (**a** and **b**) show histogram of a number of events versus geographic and magnetic latitudes, respectively, every two degrees. (**c** and **d**) illustrate the geographic and magnetic latitudes, respectively, versus the amplitude of the crests. Longitudinal geographic and magnetic distribution of crests with respect to the geographic and magnetic latitude is given in (**e** and **f**), respectively. According to Fathy and Ghamry (2017). Permission from COSPAR.

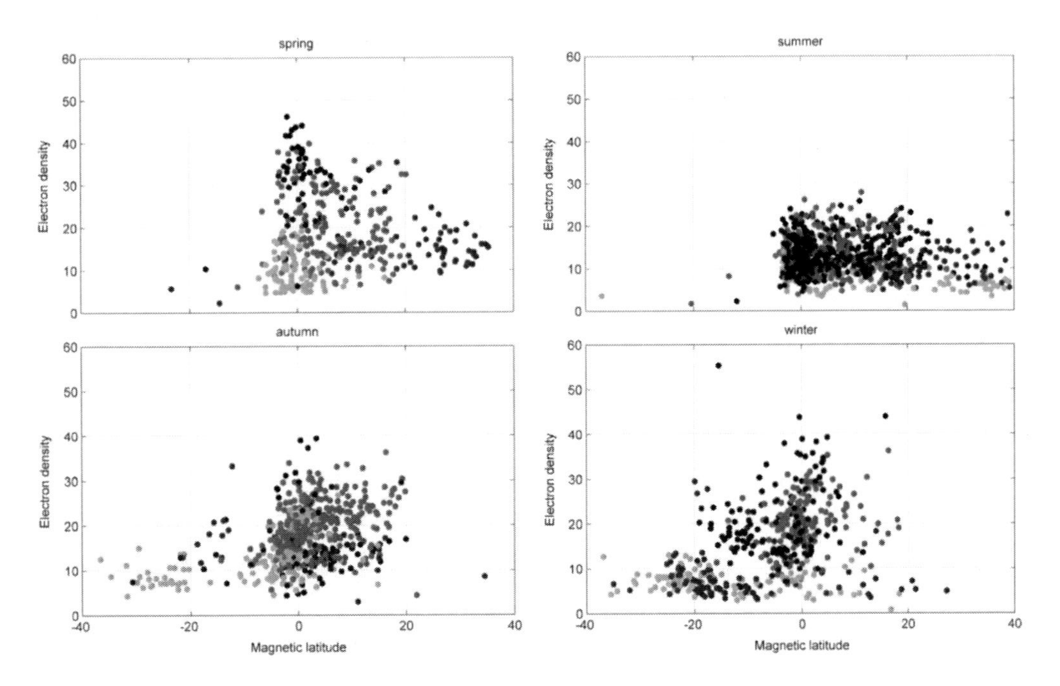

Figure 2.12.3. The single crests amplitude distribution in the four seasons (spring, summer, autumn and winter), with respect different local times. Red, blue, green, and black corresponding to day, night, dawn, and dusk, respectively. According to Fathy and Ghamry (2017). Permission from COSPAR.

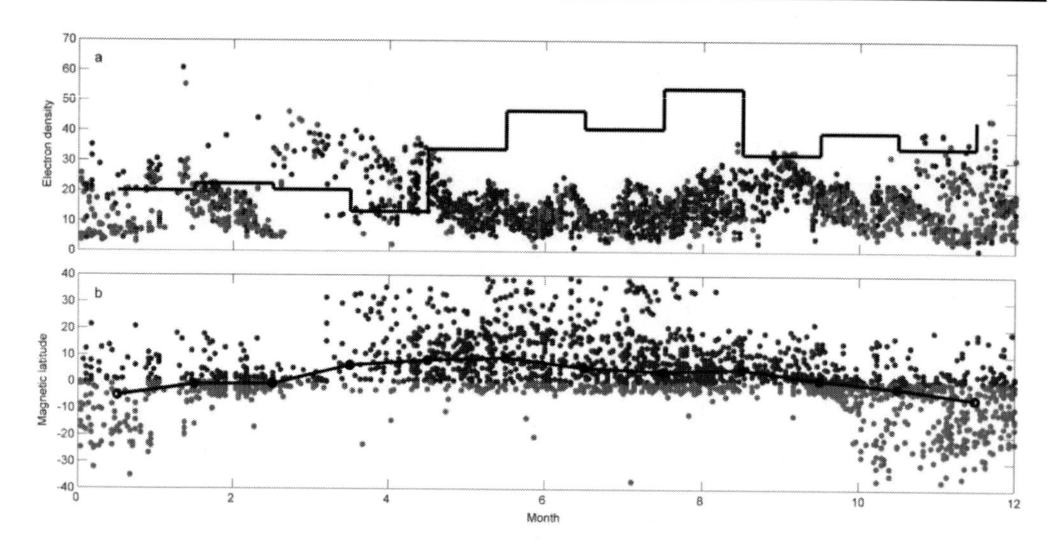

Figure 2.12.4. Seasonal variation of the single crests versus (**a**) amplitude of the crest (**b**) the magnetic latitude. Solid line indicate the total number each month. Blue and red dots corresponding to northern and southern hemisphere, respectively. According to Fathy and Ghamry (2017). Permission from COSPAR.

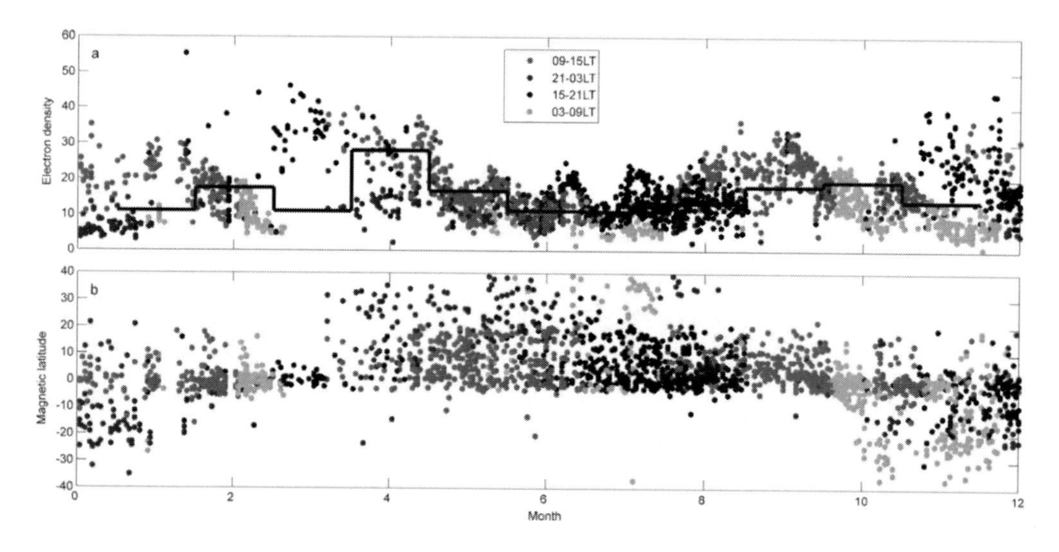

Figure 2.12.5. The same as Figure 2.12.4 but for local time dependence. Red color for 09:00–15:00 LT, black color for 15:00–21:00 LT, blue color for 21:00–03:00 LT and green color for 03:00–09:00 LT. According to Fathy and Ghamry (2017). Permission from COSPAR.

2.13. A New Methodology for the Development of High-Latitude Ionospheric Climatologies and Empirical Models

As notes Chisham (2017), many empirical models and climatologies of high-latitude ionospheric processes, such as convection, have been developed over the last 40 years. One common feature in the development of these models is that measurements from different times are combined and averaged on fixed coordinate grids. Chisham (2017) outlined that this methodology ignores the reality that high-latitude ionospheric features are organized relative to the location of the ionospheric footprint of the boundary between open and closed geomagnetic field lines (OCB). This boundary is in continual motion, and the polar cap that it encloses is continually expanding and contracting in response to changes in the rates of magnetic reconnection at the Earth's magnetopause and in the magnetotail. As a consequence, models that are developed by combining and averaging data in fixed coordinate grids heavily smooth the variations that occur near the boundary location. Chisham (2017) propose that the development of future models should consider the location of the OCB in order to more accurately model the variations in this region. It is present a methodology which involves identifying the OCB from spacecraft auroral images and then organizing measurements in a grid where the bins are placed relative to the OCB location. Chisham (2017) demonstrate the plausibility of this methodology using ionospheric vorticity measurements made by the Super Dual Auroral Radar Network radars and OCB measurements from the IMAGE spacecraft FUV auroral imagers. This demonstration shows that this new methodology results in sharpening and clarifying features of climatological maps near the OCB location. Chisham (2017) discuss the potential impact of this methodology on space weather applications.

2.14. The Latitudinal Structure of Nighttime Ionospheric TEC and Its Empirical Orthogonal Functions Model over North American Sector

Le et al. (2017) collected total electron content (TEC) data in the longitudinal sector of 60°W–90°W during 1999–2015 to investigate the latitudinal variation of nighttime middle- and high-latitude ionosphere. The midlatitude trough is one of the important features of the nighttime ionosphere. The statistical analysis provides unprecedented detail of the local time, seasonal, solar activity, and geomagnetic activity variations of the total electron content in the latitude range of 40°N–75°N, focusing on the variation of midlatitude trough. The results show that the trough minimum position has significant local time, seasonal, and geomagnetic activity dependences but slight solar activity dependence. In addition, an empirical model of the TEC in the middle to high latitudes was constructed by empirical orthogonal function analysis. The empirical model of Le et al. (2017) can reconstruct latitudinal profiles of TEC and well reproduce the dependence of the midlatitude trough on local time, seasonal, solar cycle, and geomagnetic activity. In addition, Le et al. (2017) also analyzed the geomagnetic activity dependence of TEC at different latitudes and different local time sectors.

2.15. Periodicity in the Occurrence of Equatorial Plasma Bubbles Derived from C/NOFS Observations in 2008–2012

As note Choi et al. (2017), the quasi-periodic occurrence of equatorial plasma bubbles is understood in terms of seeding mechanisms in the bottomside F region. However, no quantitative investigation has been conducted to identify how often quasi-periodic bubbles occur. The study of Choi et al. (2017) investigates the wave property in the bubble occurrence (or spacing between bubbles) using the measurements of the plasma density in 2008–2012 by the Planar Langmuir Probe on board the Communication/Navigation Outage Forecasting System (C/NOFS) satellite. The wave property is investigated using the Lomb-Scargle periodograms derived from 664 segments of series of bubbles. In the majority of segments, the spacing between bubbles is represented by the combination of several wave components. Periodic bubbles whose property is represented by a few pronounced wave components are rare events. These results indicate that the spacing between bubbles is generally irregular. The manner of bubble occurrence does not show any notable variation with longitude and season. Choi et al. (2017) conclude that because a consistent wave property does not exist in the occurrence of bubbles and the appearance of bubbles in the topside is affected by many factors, the manner of bubble occurrence in satellite observations does not provide a precise diagnostic of seeding mechanisms.

2.16. On the Contribution of Thermal Excitation to the Total 630.0 nm Emissions in the Northern Cusp Ionosphere

As note Kwagala et al. (2017), direct impact excitation by precipitating electrons is believed to be the main source of 630.0 nm emissions in the cusp ionosphere. However, the paper of Kwagala et al. (2017) investigates a different source, 630.0 emissions caused by thermally excited atomic oxygen O(1D) when high electron temperature prevail in the cusp. On 22 January 2012 and 14 January 2013, the European Incoherent Scatter Scientific Association (EISCAT) radar on Svalbard measured electron temperature enhancements exceeding 3000 K near magnetic noon in the cusp ionosphere over Svalbard. The electron temperature enhancements corresponded to electron density enhancements exceeding 10^{11} m^{-3} accompanied by intense 630.0 nm emissions in a field of view common to both the EISCAT Svalbard radar and a meridian scanning photometer. This offered an excellent opportunity to investigate the role of thermally excited O(1D) 630.0 nm emissions in the cusp ionosphere. The thermal component was derived from the EISCAT Radar measurements and compared with optical data. For both events the calculated thermal component had a correlation coefficient greater than 0.8 to the total observed 630.0 nm intensity which contains both thermal and particle impact components. Despite fairly constant solar wind, the calculated thermal component intensity fluctuated possibly due to dayside transients in the aurora.

2.17. Estimating Some Parameters of the Equatorial Ionosphere Electrodynamics from Ionosonde Data in West Africa

As informed Grodji et al. (2017), during the International Equatorial Electrojet Year (IEEY), an IPS-42 ionosonde located at Korhogo (9.33°N, 5.42°W, −1.88° dip-lat) and a meridian chain of 10 magnetic stations were setup in West Africa (5°West longitude). Some characteristic parameters of the equatorial electrojet were estimated by Grodji et al. (2017) on the basis of the IPS-42 ionosonde data at Korhogo during the years 1993 and 1994. The study consisted of determining the zonal electric field through an estimate of the plasma vertical drift velocity. The daytime plasma vertical drift velocity was estimated from the time rates of change of the F-layer virtual height variations and a correction term that takes into account the ionization production and recombination effects. This method resulted in an improved vertical drift velocity, which was found to be comparable to the results of previous studies. The estimated vertical drift velocity was used in a semi-empirical approach which involved the IRI-2012 model for the Pedersen and Hall conductivities and the IGRF-10 model for the geomagnetic main field intensity. As outlined Grodji et al. (2017), the zonal and polarization electric fields on one hand, and the eastward Pedersen, Hall and the equatorial electrojet current densities on the other hand, were estimated. Furthermore the integrated peak current density at the EEJ center was estimated from ionosonde observations and compared with that inferred from magnetometer data. The integrated EEJ peak current densities obtained from both experiments were found to be in the same order and their seasonal variations exhibit the same trends as well. Some obtained results are illustrated by Figures 2.17.1 – 2.17. and Table 2.17.1.

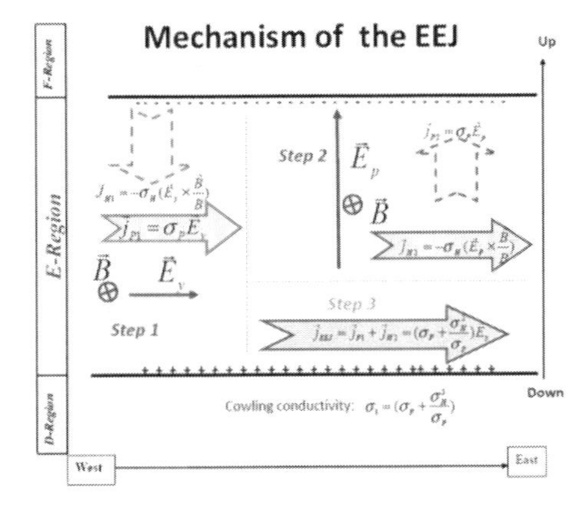

Figure 2.17.1. Mechanism of the equatorial electrojet current flow in the E-region is described in three steps. Step 1: zonal electric $\mathbf{E}_y$ (red arrow), magnetic field **B** (green cross-circle), downward Hall current density (dashed wide arrow) and the eastward Pedersen current (solid wide arrow). Step 2: vertical polarization electric $\mathbf{E}_\mathbf{p}$ (blue arrow), magnetic field **B** (green cross-circle), upward Pedersen current density (dashed wide blue arrow) and the eastward Hall current density (solid wide blue arrow). Step 3: The total EEJ current density (wide pink arrow). According to Grodji et al. (2017). Permission from COSPAR.

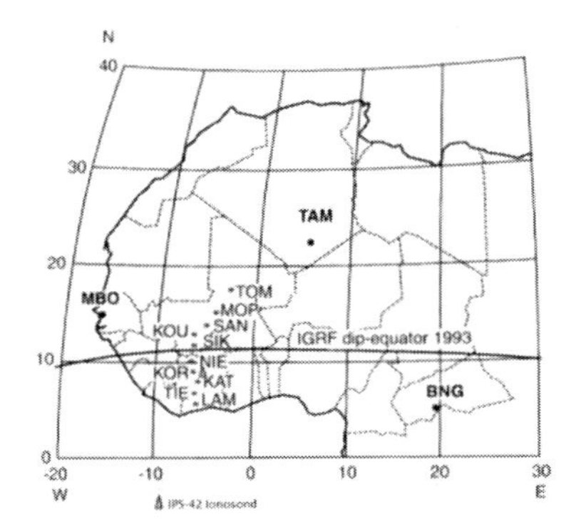

Figure 2.17.2. The 5°W meridian chain magnetic stations (asterisk) and the ionosonde (triangle) installed in West Africa during the International Equatorial Elec-trojet Year (IEEY). According to Grodji et al. (2017). Permission from COSPAR.

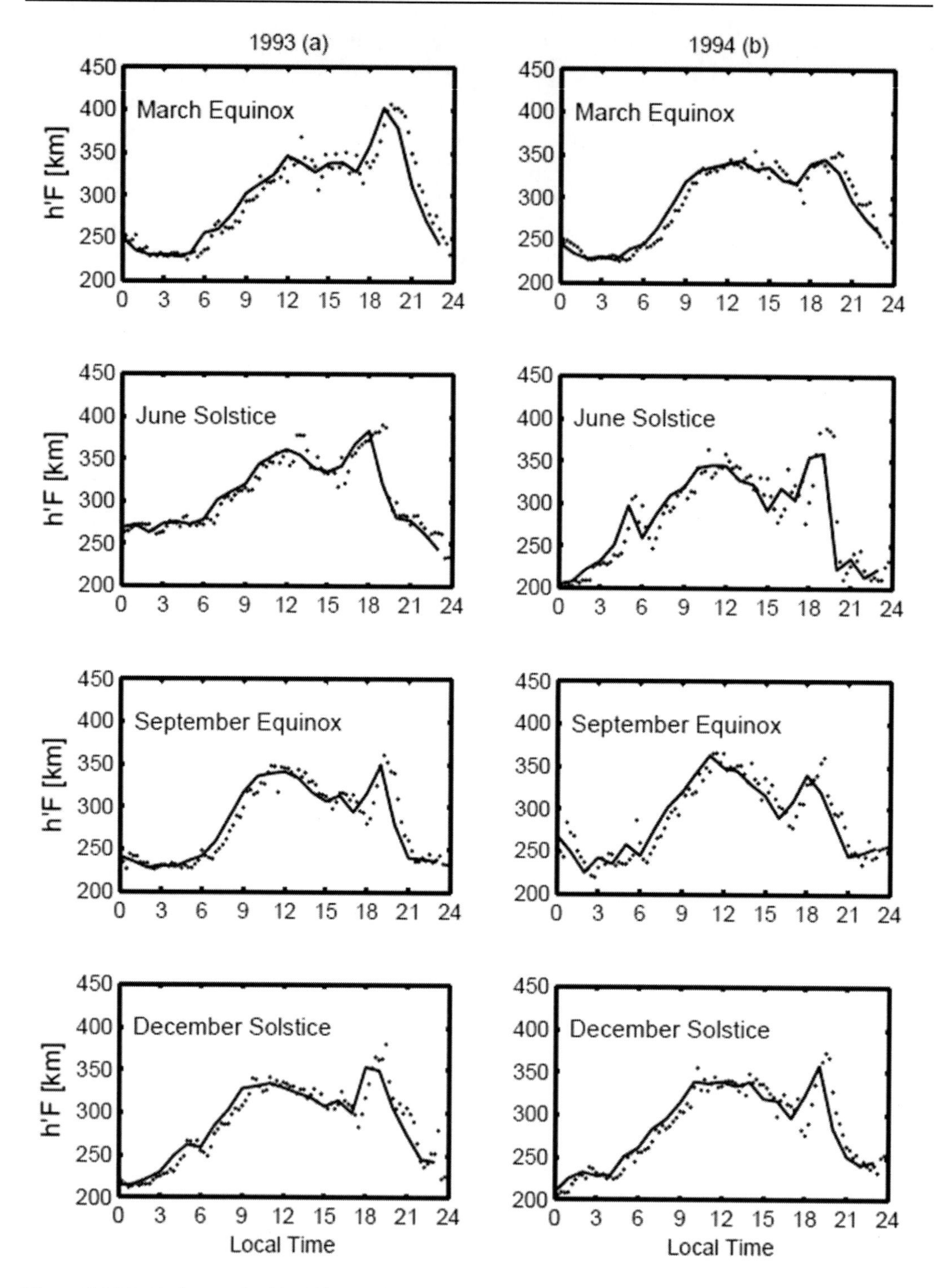

Figure 2.17.3. Daily trend of the F layer virtual heights h′F from the ISP-42 ionosonde data in 1993 (a) and 1994 (b). The dots indicate the seasonal average of the F layer virtual height at every 15 min and the solid lines represent the hourly averages. According to Grodji et al. (2017). Permission from COSPAR.

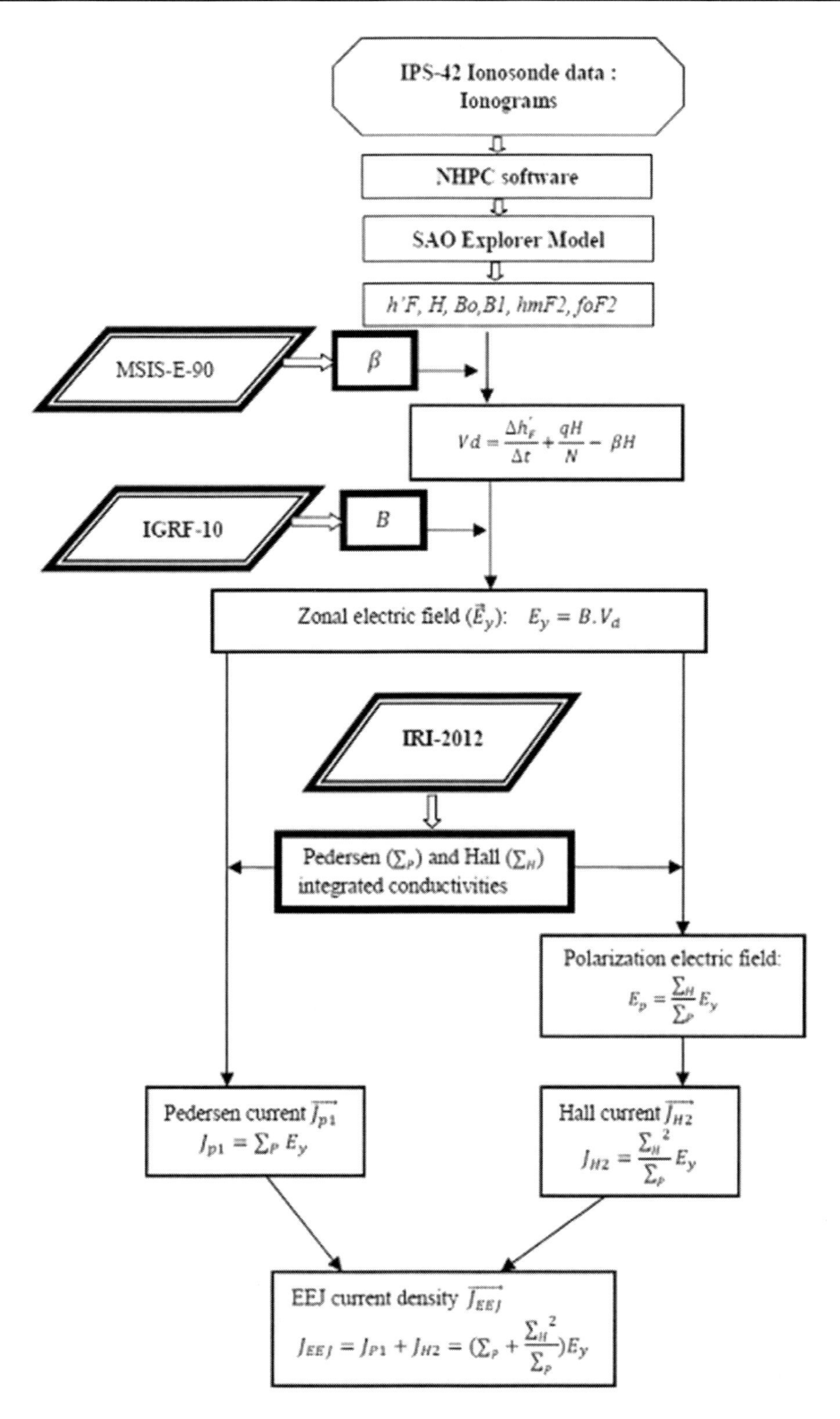

Figure 2.17.4. Diagrams of the method used to process ionosonde data for the estimation of the electric fields and current densities of the equatorial ionosphere in West Africa. According to Grodji et al. (2017). Permission from COSPAR.

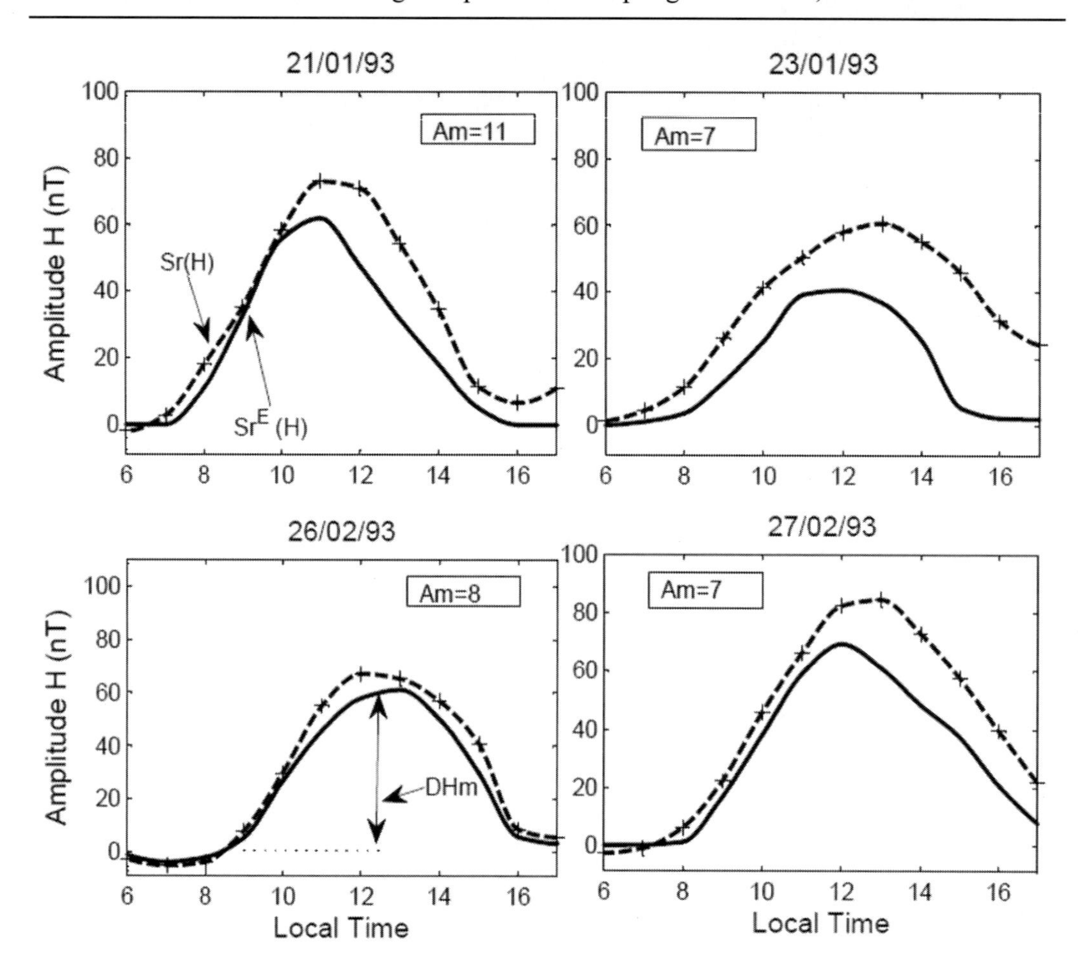

Figure 2.17.5. Daily regular variation of the horizontal component (H) of the geomagnetic field. The dashed lines represent the total daily regular variation $S_r(H)$, and the solid lines represent the equatorial electrojet contribution $S_r^E(H)$ in the daily regular variation $S_r(H)$. According to Grodji et al. (2017). Permission from COSPAR.

Table 2.17.1. Seasonal averages of the plasma vertical drift velocity (V_d), the zonal (E_y) and polarization (E_p) electric fields, the eastward Pedersen (J_P) and Hall (J_H) current densities, the net eastward current density (J_{EEJ}) and the peak current density (I_{0i}) at the center of the equatorial electrojet around noontime in 1993 and 1994. According to Grodji et al. (2017). Permission from COSPAR

	1993				1994			
	M. eq.	J. sol.	S. eq.	D. sol.	M. eq.	J. sol.	S. eq.	D. sol.
V_d (m/s)	15.61	15.01	15.30	13.81	15.03	14.01	15.01	13.56
E_y (mV/m)	0.49	0.47	0.48	0.44	0.48	0.47	0.48	0.43
E_p (mV/m)	9.51	8.41	8.98	7.61	9.22	7.93	8.81	7.63
J_{P1} (10^{-3} A/km^2)	15.8	16.3	15.3	13.0	13.6	13.6	14.0	11.7
J_{H2} (A/km^2)	5.80	5.02	5.19	3.90	5.06	4.35	4.75	3.63
J_{EEJ} (A/km^2)	5.82	5.04	5.21	3.92	5.07	4.36	4.76	3.64
I_{oi} (A/km)	290.80	252.03	260.55	196.12	253.93	218.43	238.68	182.14

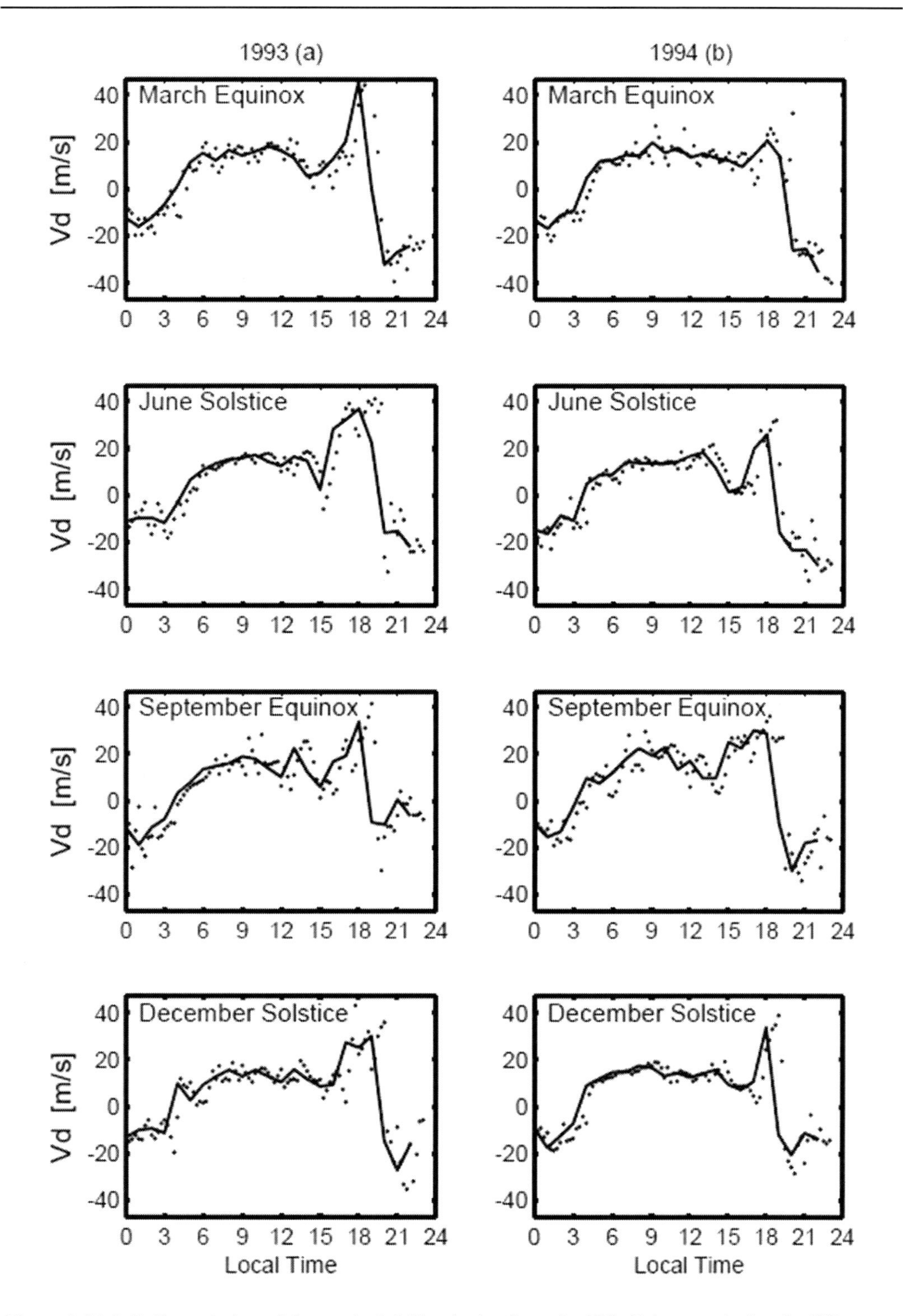

Figure 2.17.6. Daily variation of the vertical drift velocity from the ISP-42 ionosonde data in different seasons of 1993 (a) and 1994 (b). The dots indicate the seasonal averages of the vertical drift velocity at every 15 min and the solid lines represent the hourly averages. According to Grodji et al. (2017). Permission from COSPAR.

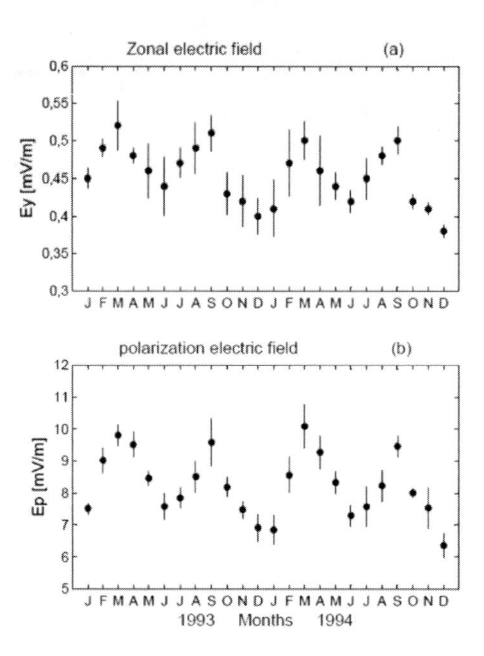

Figure 2.17.7. Seasonal variations of the Zonal electric field E_y (a) and Polarization electric field E_p (b). The dots represent the monthly averages of E_y respectively E_p from January 1993 to December 1994, the error bars indicate the standard deviations. According to Grodji et al. (2017). Permission from COSPAR.

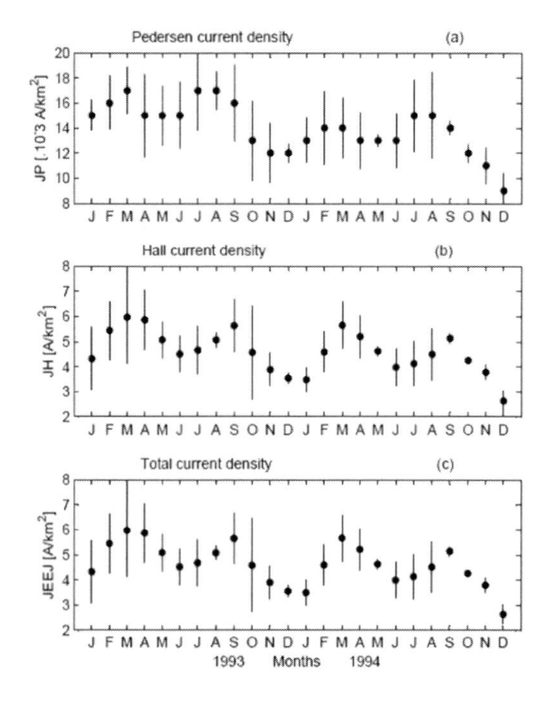

Figure 2.17.8. Variation of the eastward Pedersen (a), Hall (b) and total electrojet (c) current densities obtain from the IPS-42 ionosonde data during the two years (1993 and 1994). According to Grodji et al. (2017). Permission from COSPAR.

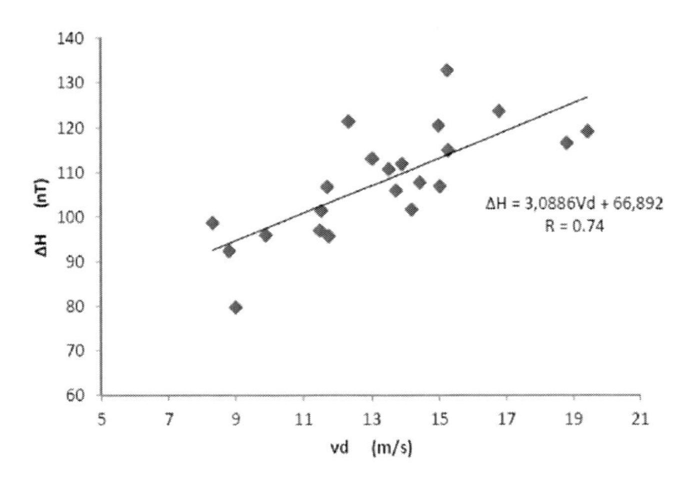

Figure 2.17.9. Correlation between the EEJ magnetic effect DH and the vertical drift velocity V_d. According to Grodji et al. (2017). Permission from COSPAR.

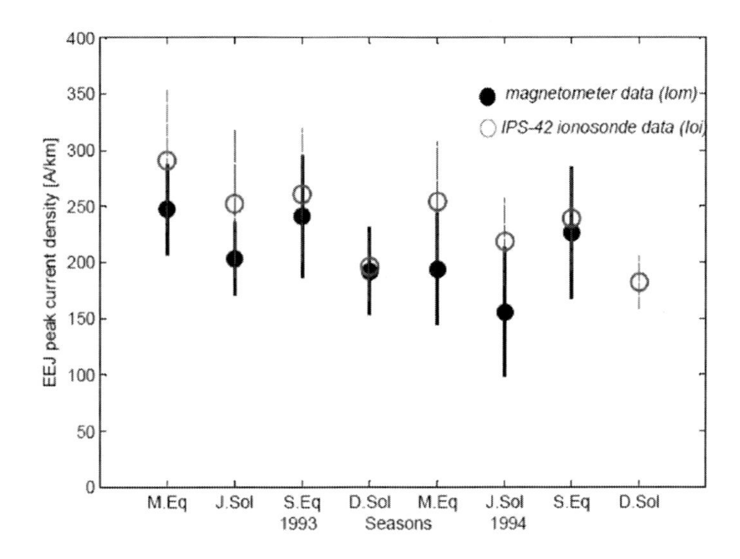

Figure 2.17.10. Comparison of the peak current density (I_o) of the equatorial electrojet obtained from the IPS-42 ionosonde data (red circles I_{oi}) and magnetometer data (black dots I_{om}). According to Grodji et al. (2017). Permission from COSPAR.

2.18. A Model-Assisted Radio Occultation Data Inversion Method Based on Data Ingestion Into NeQuick

As outlined Shaikh et al. (2017), the Inverse Abel transform is the most common method to invert radio occultation (RO) data in the ionosphere and it is based on the assumption of the spherical symmetry for the electron density distribution in the vicinity of an occultation event. It is understood that this 'spherical symmetry hypothesis' could fail, above all, in the presence of strong horizontal electron density gradients. As a consequence, in some cases wrong electron density profiles could be obtained. In Shaikh et al. (2017), in order to incorporate the knowledge of horizontal gradients, it is suggested an inversion technique based on the adaption of the empirical ionospheric model, NeQuick2, to RO-derived TEC. The method relies on the minimization of a cost function involving experimental and model-derived TEC data to determine NeQuick2 input parameters (effective local ionization parameters) at specific locations and times. These parameters are then used to obtain the electron density profile along the tangent point (TP) positions associated with the relevant RO event using NeQuick2. The main focus of Shaikh et al. (2017) research has been laid on the mitigation of spherical symmetry effects from RO data inversion without using external data such as data from global ionospheric maps (GIM). By using RO data from Constellation Observing System for Meteorology Ionosphere and Climate (FORMOSAT-3/COSMIC) mission and manually scaled peak density data from a network of ionosondes along Asian and American longitudinal sectors, Shaikh et al. (2017) have obtained a global improvement of 5% with 7% in Asian longitudinal sector, in the retrieval of peak electron density (NmF2) with model-assisted inversion as compared to the Abel inversion. Mean errors of NmF2 in Asian longitudinal sector are calculated to be much higher compared to American sector. Main obtained results are illustrated by Figures 2.18.1 – 2.18.4 and by Table 2.18.1.

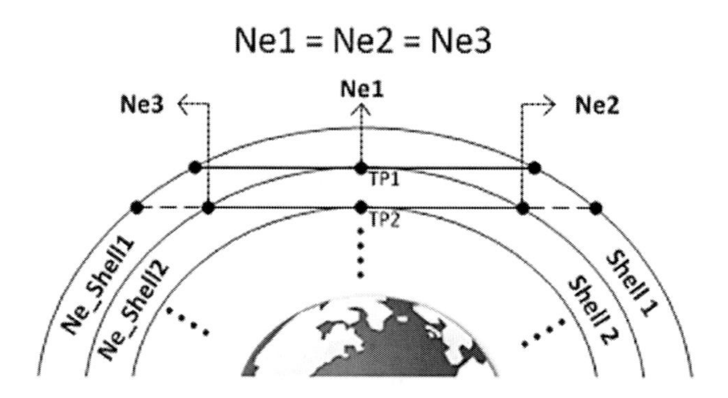

Figure 2.18.1. Illustration of the idea of spherical symmetry hypothesis used in standard Abel inversion. According to Shaikh et al. (2017). Permission from COSPAR.

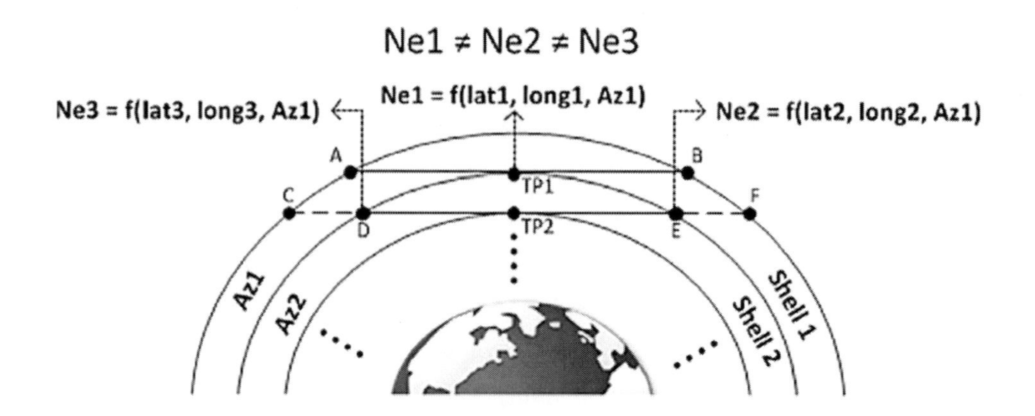

Figure 2.18.2. Illustration of the idea of how the concept of Az (effective F10.7) has been used to apply Model-assisted Inversion. According to Shaikh et al. (2017). Permission from COSPAR.

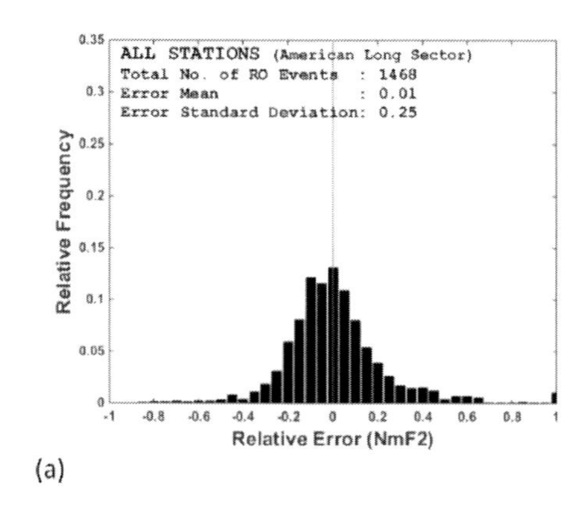

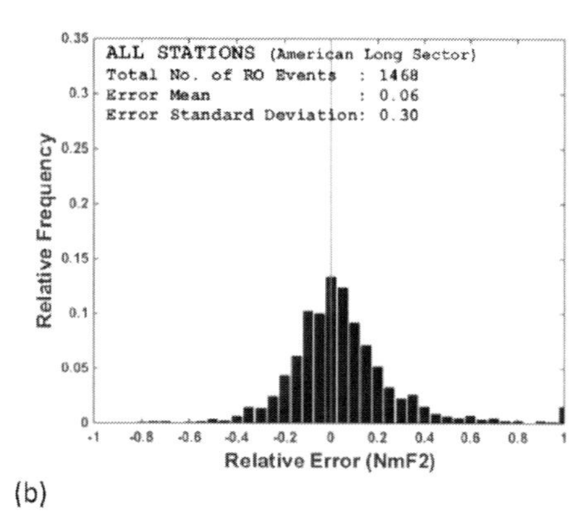

Figure 2.18.3. Relative NmF2 error for stations in American longitudinal sector: (a) Model-assisted inversion (black); (b) Standard Abel inversion (blue). According to Shaikh et al. (2017). Permission from COSPAR.

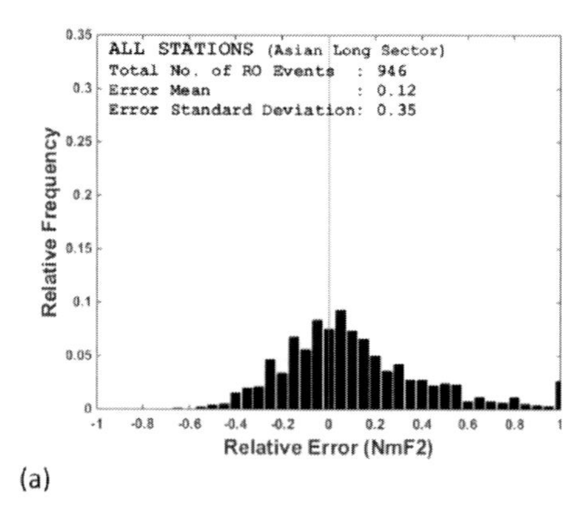

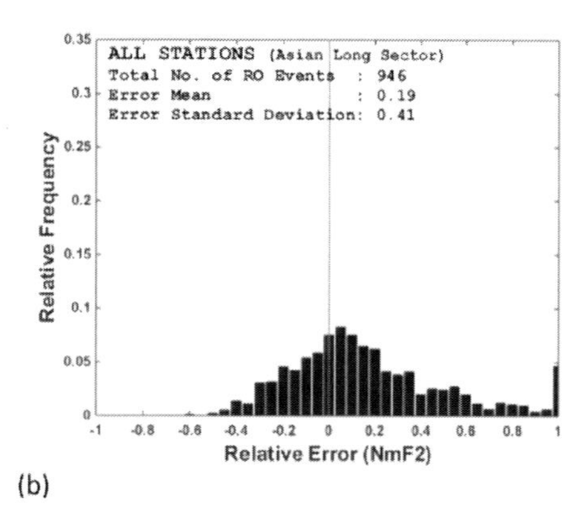

Figure 2.18.4. Relative NmF2 error for stations in Asian longitudinal sector: (a) Model-assisted inversion (black); (b) Standard Abel inversion (blue). According to Shaikh et al. (2017). Permission from COSPAR.

Table 2.18.1. Summary of sector-wise results (considering all stations in one longitudinal sector). According to Shaikh et al. (2017). Permission from COSPAR

Result Type	Relative Error Mean	
	Model-assisted Inversion	Std. Abel Inversion
Including all factors		
American Sector	0.01	0.06
Asian Sector	0.12	0.19
Day time		
American Sector	0.02	0.07
Asian Sector	0.16	0.20
Night time		
American Sector	−0.01	0.02
Asian Sector	0.07	0.13
Medium-high solar activity		
American Sector	0.05	0.09
Asian Sector	0.22	0.27

2.19. Long-Term Changes in Space Weather Effects on the Earth's Ionosphere

As outlined Tsagouri et al. (2017), certain limitations that have been identified in existing ionospheric prediction capabilities indicate that the deeper understanding and the accurate formulation of the ionospheric response to external forcing remain always high priority tasks for the research community. In this respect, the paper of attempts Tsagouri et al. (2017) an investigation of the long-term behavior of the ionospheric disturbances from the solar minimum between the solar cycles 23 and 24 up to the solar maximum of solar cycle 24. The analysis is based on observations of the foF2 critical frequency and the hmF2 peak electron density height obtained in the European region, records of the Dst and AE indices, as well as measurements of energetic particle fluxes from NOAA/POES satellites fleet. The discussion of the ionospheric behavior in a wide range of geophysical conditions within the same solar cycle facilitates the determination of general trends in the ionospheric response to different faces of space weather driving. According to the evidence, the disturbances in the peak electron density reflect mainly the impact of geoeffective solar wind structures on the Earth's ionosphere. The intensity of the disturbances may be significant (greater than 20% with respect to normal conditions) in all cases, but the ionospheric response tends to have different characteristics between solar minimum and solar maximum conditions. In particular, in contrast to the situation in solar maximum, in solar minimum years the solar wind impact on the Earth's ionosphere is mainly built on the occurrence of ionization increases, which appear more frequent and intense than ionization depletions. The ionization enhancements are apparent in all local time sectors, but they peak in the afternoon hours, while a significant part of them seems not related with an F2 layer uplifting. Taking into account the main interplanetary drivers of the disturbances in each case, i.e. high speed streams (HSSs) and corotating interaction regions (CIRs) in solar minimum and coronal mass ejections (CME) in solar maximum, Tsagouri et al. (2017) argue that the identified tendency may be considered as evidence of the ionospheric response to different solar wind drivers. Some obtained results are illustrated by Figures 2.19.1 – 2.19.3 and Table 2.19.1. For obtaining more precisely results, the work of Tsagouri et al. (2017) takes substantial advantage from the confidence score (CS), introduced by Galkin et al. (2014) as automatic quality metric that was assigned to observations.

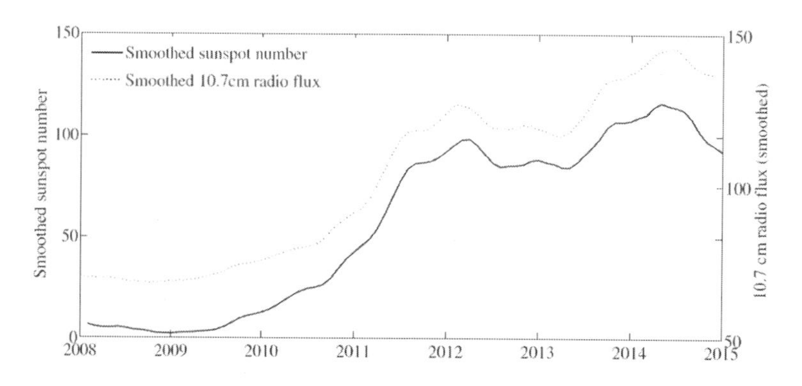

Figure 2.19.1. The smoothed monthly solar sunspot number and 10.7 cm radio flux estimates for the interval 2008–2014. According to Tsagouri et al. (2017). Permission from COSPAR.

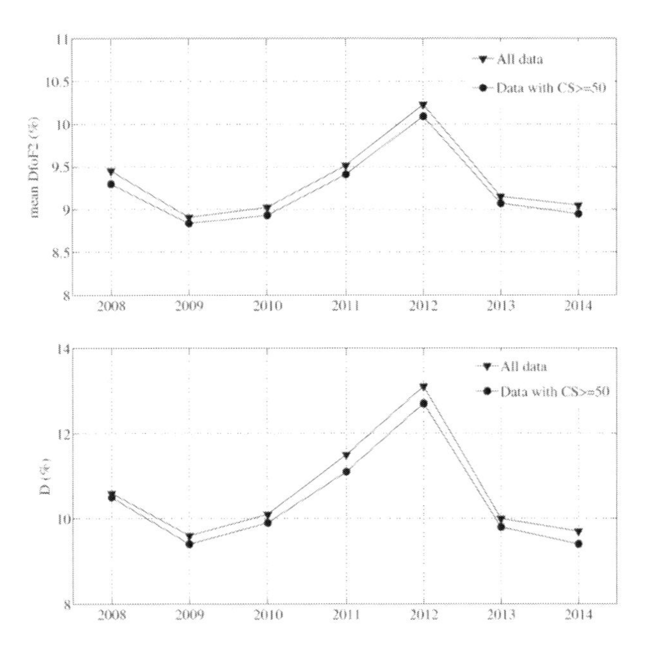

Figure 2.19.2. The annual mean of |DfoF2| (at the top), and the relative occurrence D of significant ionospheric disturbances (at the bottom) within a year as two indicators of the ionospheric disturbance level for each year of the analysis. According to Tsagouri et al. (2017). Permission from COSPAR.

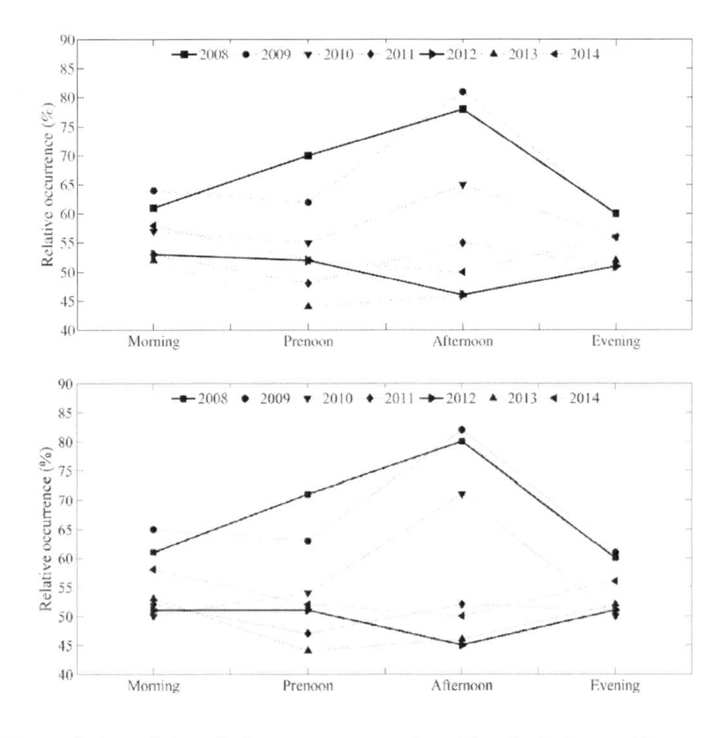

Figure 2.19.3. The variation of the relative occurrence of positive deviations with respect to the total disturbances versus the local time sector for each year of the analysis when all available data (top panel) and data marked with CSP50 (bottom panel) are analyzed. According to Tsagouri et al. (2017). Permission from COSPAR.

Table 2.19.1. The percentage of the ionospheric disturbances (DfoF2 greater than 20%) that are accompanied by an increase in hmF2 in each local time sector during low (2008) and high (2012) solar activity years. The values in the parentheses represent the results in case the analysis includes only the observations marked with CS ≥ 50. According to Tsagouri et al. (2017). Permission from COSPAR.

Year	All	Morning	Pre-noon	Afternoon	Evening
2008	51% (51%)	45% (45%)	55% (55%)	63% (63%)	46% (47%)
2012	57% (57%)	51% (50%)	58% (58%)	84% (85%)	48% (48%)

2.20. A Two-Dimensional Global Simulation Study of Inductive-Dynamic Magnetosphere-Ionosphere Coupling

Tu and Song (2016) present the numerical methods and results of a global two-dimensional multifluid-collisional-Hall MHD simulation model of the ionosphere-thermosphere system, an extension of their one-dimensional three-fluid MHD model. The model solves, self-consistently, Maxwell's equations, continuity, momentum, and energy equations for multiple ion and neutral species incorporating photochemistry, collisions among the electron, ion and neutral species, and various heating sources in the energy equations. The inductive-dynamic approach (solving self-consistently Faraday's law and retaining inertia terms in the plasma momentum equations) used in the model retains all possible MHD waves, thus providing faithful physical explanation (not merely description) of the magnetosphere-ionosphere/thermosphere (M-IT) coupling. In the present study, Tu and Song (2016) simulate the dawn-dusk cross-polar cap dynamic responses of the ionosphere to imposed magnetospheric convection. It is shown that the convection velocity at the top boundary launches velocity, magnetic, and electric perturbations propagating with the Alfvén speed toward the bottom of the ionosphere. Within the system, the waves experience reflection, penetration, and rereflection because of the inhomogeneity of the plasma conditions. The reflection of the Alfvén waves may cause overshoot (stronger than the imposed magnetospheric convection) of the plasma velocity in some regions. The simulation demonstrates dynamic propagation of the field-aligned currents and ionospheric electric field carried by the Alfvén waves, as well as formation of closure horizontal currents (Pedersen currents in the *E* region), indicating that in the dynamic stage the M-I coupling is via the Alfvén waves instead of field-aligned currents or electric field mapping as described in convectional M-I coupling models.

2.21. Postmidnight Ionospheric Troughs in Summer at High Latitudes

Voiculescu et al. (2016) identify possible mechanisms for the formation of postmidnight ionospheric troughs during summer, in sunlit plasma. Four events were identified in measurements of European Incoherent Scatter and ESR radars during CP3 experiments, when the ionosphere was scanned in a meridional plan. The spatial and temporal variation of plasma density, ion, and electron temperatures were analyzed for each of the four events. Super Dual Auroral Radar Network plasma velocity measurements were added, when these were available. For all high-latitude troughs the ion temperatures are high at density minima (within the trough), at places where the convection plasma velocity is eastward and high. There is no significant change in electron temperature inside the trough, regardless of its temporal evolution. Voiculescu et al. (2016) find that troughs in sunlit plasma form in two steps: the trough starts to form when energetic electron precipitation leads to faster recombination in the *F* region, and it deepens when entering a region with high eastward flow, producing frictional heating and further depleting the plasma. The high-latitude plasma convection plays an important role in formation and evolution of troughs in the postmidnight sector in sunlit plasma. During one event a second trough is identified at midlatitudes, with different characteristics, which is most likely produced by a rapid subauroral ion drift in the premidnight sector.

2.22. Relationship between Ionospheric F2-Layer Critical Frequency, F10.7, and F10.7P around African EIA Trough

As outlined Ikubanni and Adeniyi (2017), improved ionospheric modeling requires a better understanding of the relationship between ionospheric parameters and their influencing solar and geomagnetic sources. Published reports of the validation of the International Reference Ionosphere (IRI) for quiet-time revealed either underestimation or overestimation at a greater magnitude during high solar fluxes, especially at low latitude. With daily foF2 data from Ouagadougou (geor. 12.4°N, 1.5°W) covering a solar cycle, Ikubanni and Adeniyi (2017) have presented preliminary results from the analysis of solar dependence of six different classifications of the data: (i) daily values, (ii) monthly mean, (iii) daily quiet values (with Ap $\leqslant$ 20), (iv) monthly-quiet-mean values, (v) monthly median, and (vi) monthly-quiet-median values. All six classifications show good nonlinear relationship with both F10.7 and F10.7P, however, the differences between the dependence of classes (i) and (iii) of foF2 on the two solar indices is more substantial than those of classes (ii), (iv), (v), and (vi). Of all the six classes, the monthly averages are best related to both solar activity indices. Further analysis shows that magnetic disturbances are non-influential in the variations of the monthly mean of both solar activity indices; this makes both good indices for quiet-time modeling. Likewise, F10.7 and F10.7P are indistinguishable for long-term modeling around the African EIA trough region. While monthly median values may be best for mid-latitude region, either the mean/median values could be used for low-latitude region. However, as note Ikubanni and Adeniyi (2017), it could be worthwhile to examine the distribution of the data from the station under consideration. Some obtained results are illustrated by Figures 2.22.1 – 2.22.3.

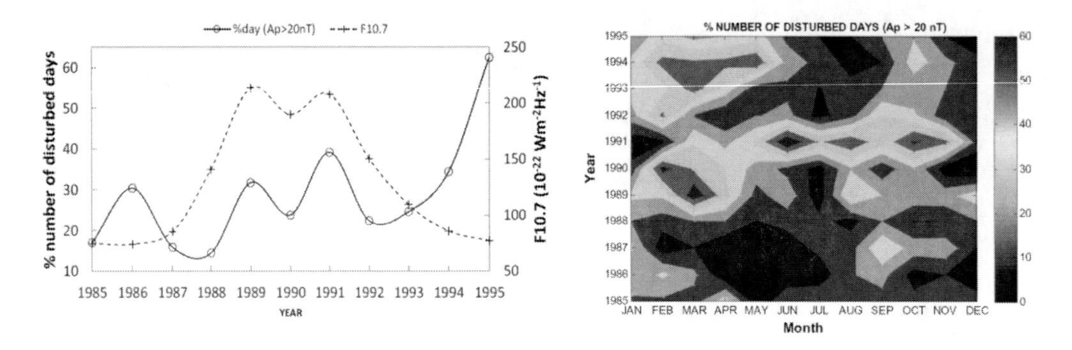

Figure 2.22.1. Left panel – percentage of disturbed days in each year and the F10.7 yearly mean; Right panel – percentage of disturbed days in each month of each year. According to Ikubanni and Adeniyi (2017). Permission from COSPAR.

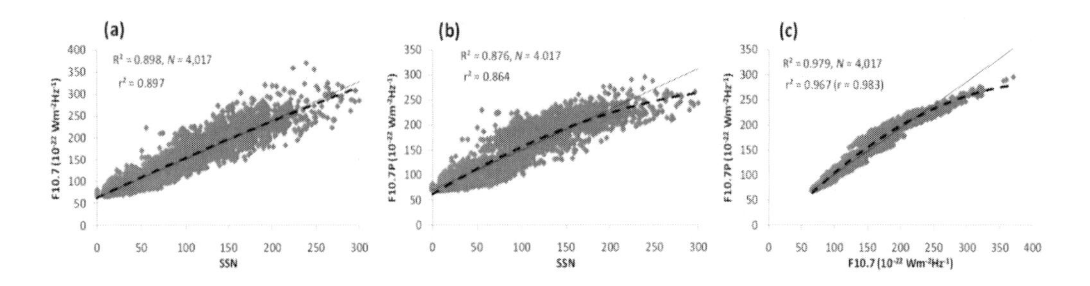

Figure 2.22.2. Regression plots describing the relationship between three solar activity proxies during solar cycle 22 (solid line – linear fit; broken line – quadratic fit; N – number of data points). [(a) F10.7 vs. SSN, (b) F10.7P vs. SSN, (c) F10.7P vs. F10.7]. According to Ikubanni and Adeniyi (2017). Permission from COSPAR.

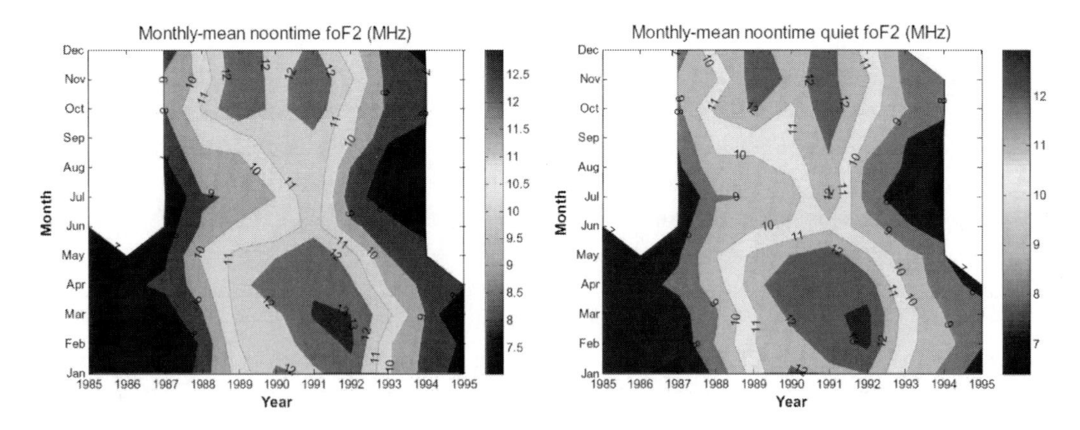

Figure 2.22.3. Contour plot of the monthly-mean noontime foF2 (Left panel) and monthly-mean noontime-quiet foF2 (Right panel) during solar cycle 22 [1985– 1995]. According to Ikubanni and Adeniyi (2017). Permission from COSPAR.

2.23. Variation of TEC and Related Parameters over the Indian EIA Region from Ground and Space Based GPS Observations during the Low Solar Activity Period of May 2007–April 2008

The annual variations of ionospheric Total Electron Content (TEC), F-region peak ionization (N_mF_2) and the ionospheric slab thickness (τ) over the Indian region during the low solar activity period of May 2007–April 2008 have been studied by Chakravarty et al. (2017). For this purpose the ground based TEC data obtained from GAGAN measurements and the space based data from GPS radio occultation technique using CHAMP have been utilised. The results of these independent measurements are combined to derive additional parameters such as the equivalent slab thickness of the total and the bottom-side ionospheric regions (τ_T and τ_B). The one year hourly average values of all these parameters over the ionospheric anomaly latitude region (10–26°N) are presented in Chakravarty et al. (2017) along with the statistical error estimates. It is expected that these results are potentially suited to be used as base level values during geomagnetically quiet and undisturbed solar conditions. Some main obtained results are illustrated by Figures 2.23.1 – 2.23.6.

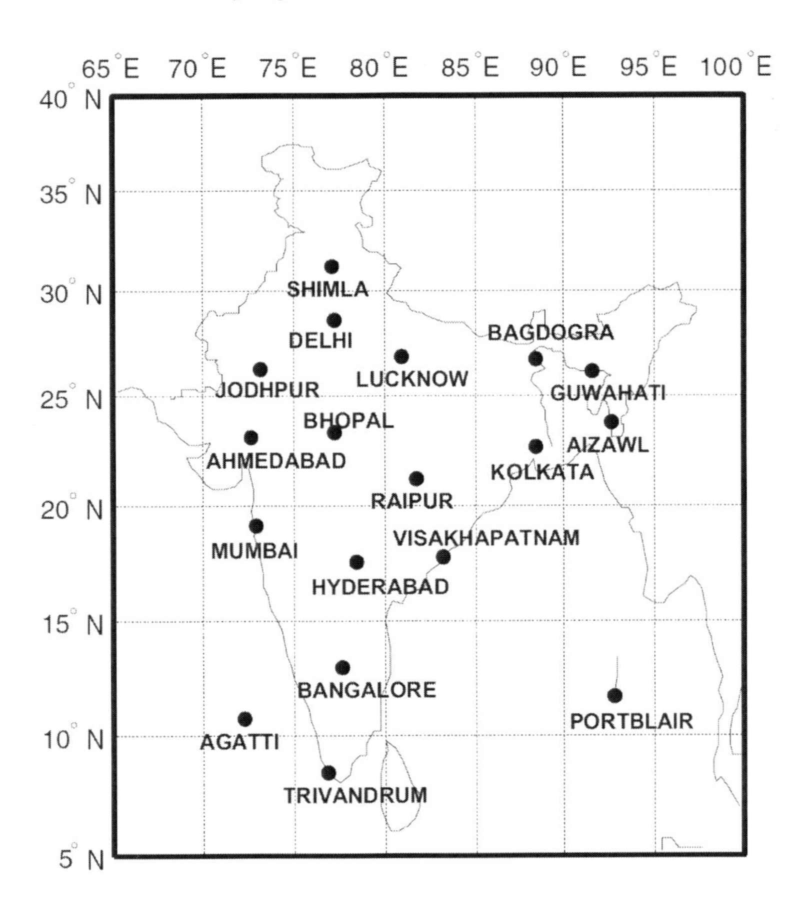

Figure 2.23.1. Distribution of GPS receivers of the GAGAN ground station network. According to Chakravarty et al. (2017). Permission from COSPAR.

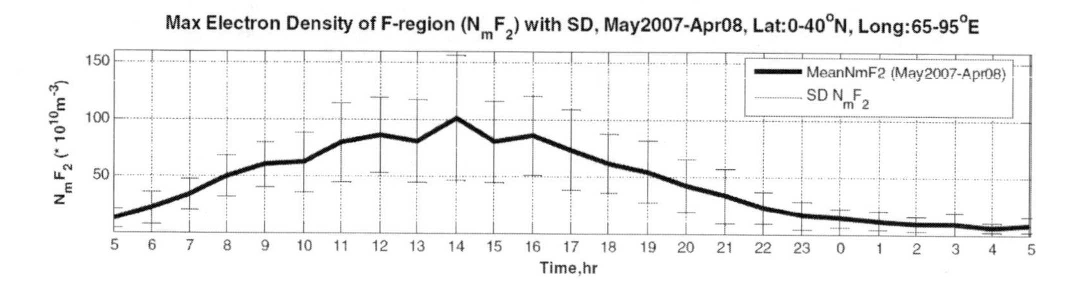

Figure 2.23.2. Diurnal variation of the electron density NmF2 during May 2007–April 2008, covering geographic latitudes 0–40°N and longitude 65–95°E from IRO-CHAMP observations. According to Chakravarty et al. (2017). Permission from COSPAR.

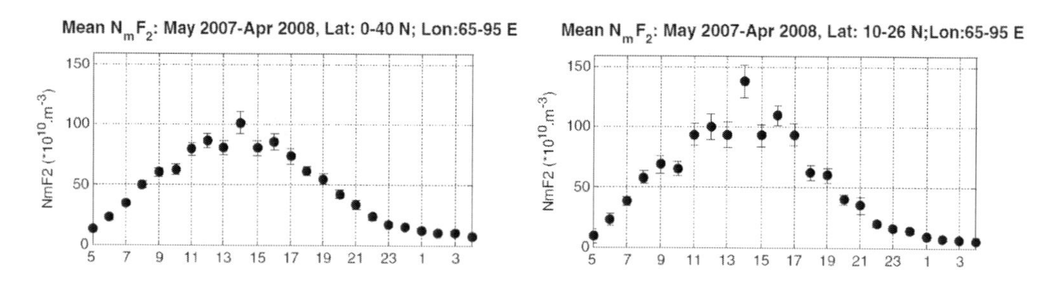

Figure 2.23.3. Diurnal variation of the peak electron density NmF2 during May 2007–April 2008, covering Latitudes 0–40°N and 10–26°N for the same longitudes (65–95°E). The vertical bars are the estimates of the standard errors. According to Chakravarty et al. (2017). Permission from COSPAR.

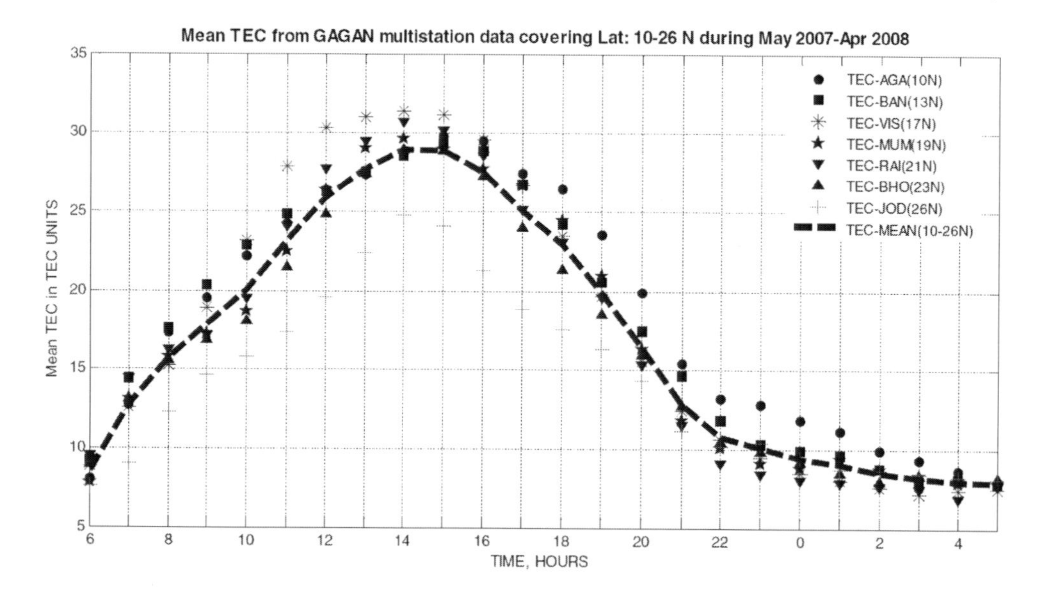

Figure 2.23.4. Hourly mean TEC values during May 2007–April 2008 for individual stations covering 10–26°N in latitude and the annual mean combined for all stations. According to Chakravarty et al. (2017). Permission from COSPAR.

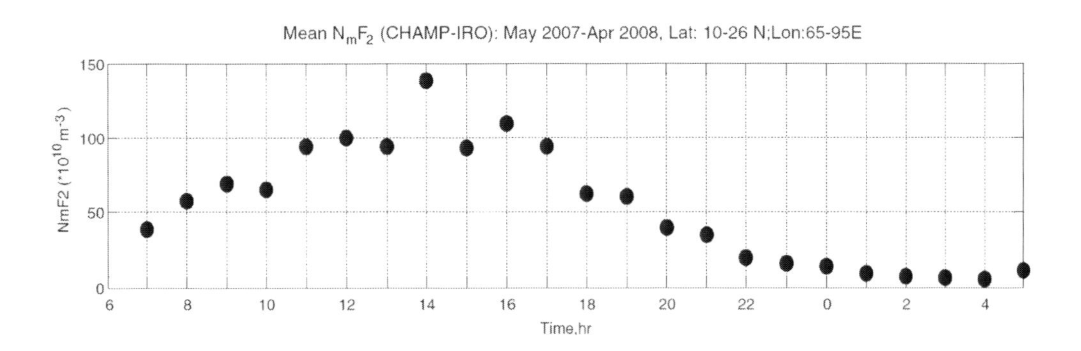

Figure 2.23.5. Hourly mean NmF2 (CHAMP) values of the period May 2007–April 2008 over the Indian EIA region. According to Chakravarty et al. (2017). Permission from COSPAR.

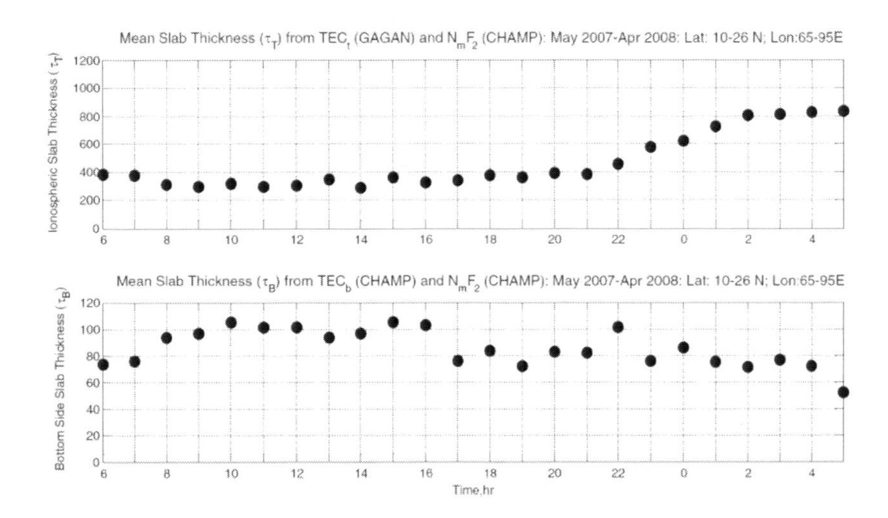

Figure 2.23.6. Mean values of ionospheric slab thickness (both total and bottom side) using TEC from GAGAN and NmF2/TECb from CHAMP observations during May 2007–April 2008 over the Indian EIA region. According to Chakravarty et al. (2017). Permission from COSPAR.

2.24. Dual E × B Flow Responses in the Dayside Ionosphere to a Sudden IMF B_Y Rotation

Eriksson et al. (2017) report for the first time a dual transition state in the dayside ionosphere following a sudden rotation of the IMF in the upstream magnetosheath from IMF $B_y < 0$ to $B_y > 0$ during $B_z < 0$. **E** × **B** drifts respond with different time delays in the dayside auroral zone and high-latitude polar cap with an initial 11 min transition state of oppositely directed **E** × **B** drifts coexisting at these locations. This is followed by a 6–8 min rotation of lower latitude **E** × **B** flow from dusk to dawn. Eriksson et al. (2017) propose that this sequence of events is consistent with two separate X lines coexisting on the subsolar and lobe magnetopause. Time delays are proposed for merged flux of the draped preceding IMF to exit the subsolar region before the new IMF may be processed along a newly reconfigured component reconnection X line. Finally, a strong direct correlation is observed between magnetosheath plasma density and auroral zone **E** × **B** speeds. As outlined Eriksson et al. (2017), it is often assumed that the ionosphere responds very fast to changes of the solar wind magnetic field in the dayside sector of the auroral zone. Eriksson et al. (2017) show, for the first time, that this is not the case for very fast (5 s) rotations of the east-west component of the upstream solar wind magnetic field. It is very likely related to a processing delay of magnetic field reconnection at the Earth's outermost magnetic field boundary with the solar wind.

2.25. Longitudinal Variations of Topside Ionospheric and Plasmaspheric TEC

The upward looking ionospheric total electron content (TEC) from the MetOp-A and TSX satellites during 2008–2015 has been used by Zhong et al. (2017) to systematically study the longitudinal variations of the topside ionosphere and plasmasphere. The results of this study are summarized by Zhong et al. (2017) as follows: **(1)** There are significant longitudinal variations in the topside ionosphere and plasmasphere at low latitudes. The TEC maximum during the June solstice over the Western and Central Pacific Ocean corresponds to a TEC minimum at the same location during the December solstice, but the opposite behavior occurs over South America and the Atlantic Ocean. **(2)** During the solstices, the relative longitudinal variations in the geomagnetic equatorial region do not have a strong dependence on local time and solar activity. **(3)** The TEC in the winter hemisphere decreases with increasing solar activity, especially at higher altitudes and at night. The topside TEC depletion with solar activity depends on longitude. **(4)** The solstice-like longitudinal pattern lasts much longer than the equinox-like patterns, with the June solstice pattern lasting the longest. Furthermore, the equinox-like longitudinal patterns occur in March when expected, whereas they extend from the autumnal equinox until the end of October. **(5)** The longitudinal variations of upward looking TEC are different from the corresponding longitudinal variations of electron densities around the F_2 peak and orbital altitudes. This indicates that the topside ionosphere structure is strongly influenced by the physical processes in the topside region, rather than being a pure reflection of the ionospheric F_2 peak structure.

2.26. Simulations of the Ionospheric Annual Asymmetry: Sun-Earth Distance Effect

As outlined Dang et al. (2017), it has been suggested that the difference of the Sun-Earth distance between the December and June solstices has a great impact on the ionospheric annual asymmetry. In the study of Dang et al. (2017), the physical mechanisms of the Sun-Earth distance effects on the ionospheric annual asymmetry are investigated using Thermosphere-Ionosphere Electrodynamics General Circulation Model simulations. As conclude Dang et al. (2017), the main findings are the following: **(1)** The Sun-Earth distance affects the ionospheric annual asymmetry mainly through photochemical processes. **(2)**

During the daytime, this photochemical process results from the combined effect of ionization rate of atomic oxygen and the recombination with neutral species; the solar irradiation variation between December and June directly leads to about 6% December-June electron density difference via ionization of atomic oxygen, whereas the recombination with neutral composition contributes to 12% – 15% December-June electron density difference. **(3)** In the plasma fountain-prominent region (between 20° and 40° magnetic latitude), ambipolar diffusion can also be modulated by the Sun-Earth distance effect and contribute to the ionospheric annual asymmetry. **(4)** During the nighttime, the Sun-Earth distance effect impacts the annual asymmetry by changing thermospheric composition and ionospheric diffusion.

References

Abraham A., G. Renuka, and L. Cherian, 2010. "Interplanetary magnetic field–geomagnetic field coupling and vertical variance index", *J. Geophys. Res.*, **115**, A01207, doi:10.1029/2008JA013890.

Akasofu S.-I., 1981. "Energy coupling between the solar wind and the magnetosphere", *Space Sci. Rev.*, **28**, No. 2, 121-190.

Anad F., C. Amory-Mazaudier, M. Hamoudi, S. Bourouis, A. Abtout, E. Yizengaw, 2016. "Sq solar variation at Medea Observatory (Algeria), from 2008 to 2011", *Advances in Space Research*, **58**, 1682–1695.

Anderson D., A. Anghel, J. Chau, and O. Veliz, 2004. "Daytime vertical E x B drift velocities inferred from ground-based magnetometer observations at low latitudes", *Space Weather*, **2**, S11001, doi:10.1029/2004SW000095.

Anderson D., A. Anghel, K. Yumoto, M. Ishitsuka, and E. Kudeki, 2002. "Estimating daytime vertical ExB drift velocities in the equatorial F region using ground-based magnetometer observations", *Geophys. Res. Lett.*, **29**, No. 12, 1596, doi:10.1029/2001GL014562.

Andreeva V.A. and N.A. Tsyganenko, 2016. "Reconstructing the magnetosphere from data using radial basis functions", *J. Geophys. Res. Space Physics*, **121**, 2249–2263, doi:10.1002/2015JA022242.

Astfalk P. and F. Jenko, 2017. "LEOPARD: A grid-based dispersion relation solver for arbitrary gyrotropic distributions", *J. Geophys. Res. Space Physics*, **122**, 89–101, doi:10.1002/2016JA023522.

Axford W.I. and C.O. Hines, 1961. "A Unifying Theory of High-Latitude Geophyscial Phenomena and Geomagnetic Storms", *Can. J. Phys.*, **39**, No. 10, 1433-1464.

Bandić M., G. Verbanac, M.B. Moldwin, V. Pierrard, and G. Piredda, 2016. "MLT dependence in the relationship between plasmapause, solar wind, and geomagnetic activity based on CRRES: 1990–1991", *J. Geophys. Res. Space Physics*, **121**, 4397–4408, doi:10.1002/2015JA022278.

Baumjohann W. and G. Paschmann, 1989. "Determination of the polytropic index in the plasma sheet", *Geophys. Res. Lett.*, **16**, doi:10.1029/GL016i004p00295, 295–298.

Baumjohann W., 1993. "The near-Earth plasma sheet: An AMPTE/IRM perspective", *Space Sci. Rev.*, **64**, doi:10.1007/BF00819660, 141–163.

Baumjohann W., G. Paschmann, and C.A. Cattell, 1989. "Average plasma properties in the central plasma sheet", *J. Geophys. Res.*, **94**, No. A5, doi:10.1029/JA094iA06p06597, 6597– 6606.

Berchem J., R.L. Richard, C.P. Escoubet, S. Wing, and F. Pitout, 2016. "Asymmetrical response of dayside ion precipitation to a large rotation of the IMF", *J. Geophys. Res. Space Physics*, **121**, No. 1, doi:10.1002/2015JA021969, 263–273.

Birn J., J. Raeder, Y.L. Wang, R.A. Wolf, and M. Hesse, 2004. "On the propagation of bubbles in the geomagnetic tail", *Ann. Geophys.*, **22**, 1773–1786.

Birn J., M. Hesse, and K. Schindler, 2006a, "On the role of entropy conservation and entropy loss governing substorm phases", *Int. Conf. Substorms-8*, 19–24.

Birn J., M. Hesse, and K. Schindler, 2006b, "Entropy conservation in simulations of magnetic reconnection", *Phys. Plasmas*, **13**, 092117-1–092117-6, doi:10.1063/1.2349440.

Blanchard G.T. and K.B. Baker, 2010. "Analysis of the step response function relating the interplanetary electric field to the dayside magnetospheric reconnection potential", *J. Geophys. Res.*, **115**, A05211, doi:10.1029/2009JA014681.

Borg A.L., Øieroset M., Phan T.D., Mozer F.S., Pedersen A., Mouikis C., McFadden J.P., Twitty C., Balogh A., and Rème H., 2005. "Cluster encounter of a magnetic reconnection diffusion region in the near-Earth magnetotail on September 19, 2003", *Geophys. Res. Lett.*, **32**, L19105, doi:10.1029/2005GL023794.

Borovsky J.E., 2008. "The rudiments of a theory of solar wind/magnetosphere coupling derived from first principles", *J. Geophys. Res.*, **113**, A08228, doi:10.1029/2007JA012646.

Borovsky J.E., M.F. Thomsen, R.C. Elphic, T.E. Cayton, and D.J. McComas, 1998. "The transport of plasma sheet material from the distant tail to geosynchronous orbit", *J. Geophys. Res.*, **103**, No. A9, doi:10.1029/97JA03144, 20297–20331.

Browett S.D., R.C. Fear, A. Grocott, and S.E. Milan, 2017. "Timescales for the penetration of IMF By into the Earth's magnetotail", *J. Geophys. Res. Space Physics*, **122**, 579–593, doi:10.1002/2016JA023198.

Burch J.L. and T.D. Phan, 2016. "Magnetic reconnection at the dayside magnetopause: Advances with MMS", *Geophys. Res. Lett.*, **43**, 8327–8338, doi:10.1002 /2016GL069787.

Burch J.L., D.G. Mitchell, B.R. Sandel, P.C. Brandt, and M. Wüest, 2001b. "Global dynamics of the plasmasphere and ring current during magnetic storms", *Geophys. Res. Lett.*, 28, No. 6, 1159-1162.

Burch J.L., S.B. Mende, D.G. Mitchell, T.E. Moore, C.J. Pollack, B.W. Reinisch, B.R. Sandel, S.A. Fuselier, D.L. Gallagher, J.L. Green, J.D. Perez, and P.H. Reiff, 2001a. "Views of Earth's magnetosphere with the IMAGE satellite", *Science*, **291**, No. 5504, 619-624.

Burlaga L.F., J.F. Lemaire, and J.M. Turner, 1977. "Interplanetary current sheets at 1 AU", *J. Geophys. Res.*, **82**, doi:10.1029/JA082i022p03191, 3191–3200.

Cao J., A. Duan, M. Dunlop, X. Wei, and C. Cai, 2014. "Dependence of IMF By penetration into the neutral sheet on IMF Bz and geomagnetic activity", *J. Geophys. Res. Space Physics*, **119**, 5279–5285, doi:10.1002/2014JA019827.

Carpenter D.L. and J. Lemaire, 1997. "Erosion and recovery of the plasmasphere in the plasmapause region", *Space Sci. Rev.*, **80**, No. 1-2, 153-179.

Cassak P.A. and Shay M.A., 2007. “Scaling of asymmetric reconnection: General theory and collisional simulations”, *Phys. Plasmas*, **14**, 102114, doi:10.1063/1.2795630.

Chakravarty S.C., K. Nagaraja, and N. Jakowski, 2017, “Variation of TEC and related parameters over the Indian EIA region from ground and space based GPS observations during the low solar activity period of May 2007–April 2008”, *Advances in Space Research*, **59**, 1223–1233.

Chané E., J. Raeder, J. Saur, F.M. Neubauer, K.M. Maynard, and S. Poedts, 2015. “Simulations of the Earth’s magnetosphere embedded in sub-Alfvénic solar wind on 24 and 25 May 2002”, *J. Geophys. Res. Space Physics*, **120**, No. 10, doi:10.1002/2015JA021515, 8517–8528.

Chappell C.R., K.K. Harris, and G.W. Sharp, 1970. “A study of the influence of magnetic activity on the location of the plasmapause as measured by OGO 5”, *J. Geophys. Res.*, **75**, No. 1, 50-56.

Chisham G., 2017. “A new methodology for the development of high-latitude ionospheric climatologies and empirical models”, *J. Geophys. Res. Space Physics*, **122**, 932–947, doi:10.1002/2016JA023235.

Choi J.-M., H. Kil, Y.-S. Kwak, J. Park, W.K. Lee, and Y.H. Kim, 2017. “Periodicity in the occurrence of equatorial plasma bubbles derived from the C/NOFS observations in 2008–2012”, *J. Geophys. Res. Space Physics*, **122**, 1137–1145, doi:10.1002/2016JA023528.

Colella P. and P.R. Woodward, 1984. “The piecewise parabolic method (PPM) for gas-dynamical simulations”, *J. Comput. Phys.*, **54**, 174– 201.

Cooling B.M.A., C.J. Owen, and S.J. Schwartz, 2001. “Role of the magnetosheath flow in determining the motion of open flux tubes”, *J. Geophys. Res.*, **106**, No. A9, 18763-18776.

Cowley S.W.H., 1976. “Comments on the merging of nonantiparallel magnetic fields”, *J. Geophys. Res.*, **81**, No. 19, 3455-3458.

Cowley S.W.H. and M. Lockwood, 1992. “Excitation and decay of solar wind-driven flows in the magnetosphere-ionosphere system”, *Ann. Geophys., Atmos. Hydrospheres Space Sci.* (France), **10**, No. 1-2, 103-115.

Crooker N.U., 1979. "Dayside merging and cusp geometry", *J. Geophys. Res.*, **84**, No. A3, 951-959.

Crooker N.U., E.W. Cliver, and B.T. Tsurutani, 1992. “The semiannual variation of great geomagnetic storms and the post-shock Russel-McPherron effect preceding coronal mass ejecta”, *Geophys. Res. Lett.*, **19**, No. 5, 429-432.

Dang T., W. Wang, A. Burns, X. Dou, W. Wan, and J. Lei, 2017. “Simulations of the ionospheric annual asymmetry: Sun-Earth distance effect”, *J. Geophys. Res. Space Physics*, **122**, 6727–6736, doi:10.1002/2017JA024188.

Denton R.E., B.U.Ö. Sonnerup, H. Hasegawa, T.D. Phan, C.T. Russell, R.J. Strangeway, B.L. Giles, and R.B. Torbert, 2016. “Reconnection guide field and quadrupolar structure observed by MMS on 16 October 2015 at 1307 UT”, *J. Geophys. Res. Space Physics*, **121**, 9880–9887, doi:10.1002/2016JA023323.

Domrin V.I., H.V. Malova, A.V. Artemyev, and A.P. Kropotkin, 2016. “Peculiarities of the Formation of a Thin Current Sheet in the Earth’s Magnetosphere”, *Cosmic Research*, **54**, No. 6, 423–437. Original Russian Text published in *Kosmicheskie Issledovaniya*, **54**, No. 6, 463–478.

Dorelli J., A. Bhattacharjee, and J. Raeder, 2007. "Separator reconnection at Earth's dayside magnetopause under northward interplanetary magnetic field conditions", *J. Geophys. Res.*, **112**, A02202, doi:10.1029/2006JA011877.

Dryer M., 1973. "Bow shock and its interaction with interplanetary shocks", *Radio Sci.*, **8**, No. 11, 893-901.

Dryer M., D.L. Merritt, and P.M. Aronson, 1967. "Interaction of a plasma cloud with the Earth's magnetosphere", *J. Geophys. Res.*, **72**, No. 11, 2955-2962.

Dungey J.W., 1961. "Interplanetary magnetic field and the auroral zones", *Phys. Rev. Lett.*, **6**, No. 2, 47-48.

Dungey J.W., 1963. "The structure of the exosphere, or, adventures in velocity" in *Geophysics, The Earth's Environment*, Proceedings of the 1962 Les Houches Summer School, edited by C. DeWitt, J. Hiebolt, and A. Lebeau, Gordon and Breach, New York, 505-549.

El-Alaoui M., R.L. Richard, Y. Nishimura, and R.J. Walker, 2016. "Forces driving fast flow channels, dipolarizations, and turbulence in the magnetotail", *J. Geophys. Res. Space Physics*, **121**, 11,063–11,076, doi:10.1002/2016JA023139.

Eriksson S., S.R. Elkington, T.D. Phan, S.M. Petrinec, H. Reme, M.W. Dunlop, M. Wiltberger, A. Balogh, R.E. Ergun, and M. Andre, 2004. "Global control of merging by the interplanetary magnetic field: Cluster observations of dawnside flank magnetopause reconnection", *J. Geophys. Res.*, **109**, No. A12, A12203, doi:10.1029/2003JA010346, 1-21.

Eriksson S., M. Maimaiti, J. B. H. Baker, K. J. Trattner, D. J. Knipp, and F. D. Wilder, 2017. "Dual E × B flow responses in the dayside ionosphere to a sudden IMF By rotation", *Geophys. Res. Lett.*, **44**, 6525–6533, doi:10.1002/2017GL073374.

Fathy A. and E. Ghamry, 2017. "A statistical study of single crest phenomenon in the equatorial ionospheric anomaly region using Swarm A satellite", *Advances in Space Research*, **59**, 1539–1547.

Fedder J.A., C.M. Mobarry, and J.G. Lyon, 1991 "Reconnection voltage as a function of IMF clock angle", *Geophys. Res. Lett.*, **18**, No. 6, 1047–1050.

Fedder J.A., J.G. Lyon, S.P. Slinker, and C.M. Mobarry, 1995a. "Topological structure of the magnetotail as a function of interplanetary magnetic field direction", *J. Geophys. Res.*, **100**, No. A3, 3613–3621.

Fedder J.A., S.P. Slinker, J.G. Lyon, and R.D. Elphinstone, 1995b. "Global numerical simulation of the growth phase and the expansion onset for a substorm observed by Viking", *J. Geophys. Res.*, **100**, A10, 19083-19093.

Fejer B.G. and L. Scherliess, 1995. "Time dependent response of equatorial ionospheric electric fields to magnetospheric disturbances", *Geophys. Res. Lett.*, **22**, doi:10.1029/95GL00390, 851–854.

Fejer B.G. and L. Scherliess, 1998. "Mid- and low-latitude prompt penetration ionospheric zonal plasma drifts", *Geophys. Res. Lett.*, **25**, doi:10.1029/98GL02325, 3071–3074.

Frank A.G., G.V. Ostrovskaya, E.V. Yushkov, A.V. Artemyev, and S.N. Satunin, 2017. "Structure of Current and Plasma in Current Sheets Depending on the Conditions of Sheet Formation", *Cosmic Research*, **55**, No. 1, 46–56.

Fu X.R., Q.M. Lu, and S. Wang, 2006. "The process of electron acceleration during collisionless magnetic reconnection", *Phys. Plasmas*, **13**, 012309, doi:10.1063 /1.2164808.

Fujii R., T. Iijima, T.A. Potemra, and M. Sugiura, 1981. "Seasonal dependence of large-scale Birkeland currents", *Geophys. Res. Lett.*, **8**, No. 10, 1103-1106.

Fujimoto M., T. Terasawa, T. Mukai, Y. Saito, T. Yamamoto, and S. Kokubun, 1998. "Plasma entry from the flanks of the near-Earth magnetotail: Geotail observations", *J. Geophys. Res.*, **103**, No. A3, 4391-4408.

Galkin I.A., B.W. Reinisch, X. Huang, and G.M. Khmyrov, 2014. "Confidence Score of ARTIST Autoscaling", *INAG Bulletin*, No. 73, Available at <http://www.ursi.org/files/CommissionWebsites/INAG/web-73/confidence_score.pdf> (March 1, 2014).

Gao X., W. Li, J. Bortnik, R.M. Thorne, Q. Lu, Q. Ma, X. Tao, and S. Wang, 2015. "The effect of different solar wind parameters upon significant relativistic electron flux dropouts in the magnetosphere", *J. Geophys. Res. Space Physics*, **120**, No. 6, doi:10.1002/2015JA021182, 4324–4337.

Garner T.W., R.A. Wolf, R.W. Spiro, M.F. Thomsen, and H. Korth, 2003. "Pressure balance inconsistency exhibited in a statistical model of magnetospheric plasma", *J. Geophys. Res.*, **108**, No. A8, 1331, doi:10.1029/2003JA009877, SMP12-1-14.

Genestreti K.J., J. Goldstein, G.D. Corley, W. Farner, L.M. Kistler, B.A. Larsen, C.G. Mouikis, C. Ramnarace, R.M. Skoug, and N.E. Turner, 2017. "Temperature of the plasmasphere from Van Allen Probes HOPE", *J. Geophys. Res. Space Physics*, **122**, 310–323, doi:10.1002/2016JA023047.

Goldstein J., B.R. Sandel, W.T. Forrester, and P.H. Reiff, 2003a. "IMF driven plasmasphere erosion of 10 July 2000", *Geophys. Res. Lett.*, **30**, No. 3, 1146, doi:10.1029 /2002GL016478, 46-1-4.

Goldstein J., B.R. Sandel, P.H. Reiff, and M.R. Hairston, 2003b. "Control of plasmaspheric dynamics by both convection and sub-auroral polarization stream", *Geophys. Res. Lett.*, **30**, No. 24, 2243, doi:10.1029/2003GL018390, SSC6-1-5.

Goldstein J., B.R. Sandel, M.F. Thomsen, M. Spasojević, and P.H. Reiff, 2004. "Simultaneous remote-sensing and in situ observations of plasmaspheric drainage plumes", *J. Geophys. Res.*, **109**, No. A3, A03202, doi:10.1029/2003JA010281, 1-11.

Goldstein J., B.R. Sandel, W.T. Forrester, M.F. Thomsen, and M.R. Hairston, 2005. "Global plasmasphere evolution 22-23 April 2001", *J. Geophys. Res.*, **110**, No. A12, A12218, doi:10.1029/2005JA011282, 1-15.

Gosling J.T., R.M. Skoug, D.J. McComas, and C.W. Smith, 2005. "Direct evidence for magnetic reconnection in the solar wind near 1 AU", *J. Geophys. Res.*, **110**, A01107, doi:10.1029/2004JA010809.

Grib S.A., 1971. "On the interaction of the shock waves with the magnetosphere of the Earth during geomagnetic storm with sudden commencement", in *Program and Abstracts for the XV IUGG General Assembly*, Nauka, Moscow, p. 452.

Grib S.A., 1972. "Interaction of the solar-wind shock wave with the Earth's magnetosphere", *Dokl. Akad. Nauk Belorussian SSR.*, **16**, No. 6, 493-496. In Russian.

Grib S.A., 1982. "Interaction of non-perpendicular/parallel solar wind shock waves with the Earth's magnetosphere", *Space Sci. Rev.*, **32**, No.1-2, 43-48.

Grib S.A., B.E. Briunelli, M. Dryer, and W.-W. Shen, 1979. "Interaction of interplanetary shock waves with the bow shock-magnetopause system", *J. Geophys. Res.*, **84**, No. A10, 5907-5922.

Grodji F.O., V. Doumbia, K. Boka, C. Amory-Mazaudier, Y. Cohen, and R. Fleury, 2017. "Estimating some parameters of the equatorial ionosphere electrodynamics from ionosonde data in West Africa", *Advances in Space Research*, **59**, 311–325.

Hairston M.R., K.A. Drake, and R. Skoug, 2005. "Saturation of the ionospheric polar cap potential during the October-November 2003 superstorms", *J. Geophys. Res.*, **110**, A09S26, doi:10.1029/2004JA010864.

Hashimoto K.K., T. Kikuchi, and Y. Ebihara, 2002. "Response of the magnetospheric convection to sudden interplanetary magnetic field changes as deduced from the evolution of partial ring currents", *J. Geophys. Res.*, **107**, No. A11, 1337, doi:10.1029/2001JA009228, SMP1-1-14.

Hu Y.-Q., X.C. Guo., G.-Q. Li, C. Wang, and Z.-H. Huang, 2005. "Oscillation of quasi-steady Earth's magnetosphere", *Chin. Phys. Lett.*, **22**, No. 10, 2723–2726.

Hu Y.Q., X.C. Guo, and C. Wang, 2007. "On the ionospheric and reconnection potentials of the Earth: Results from global MHD simulations", *J. Geophys. Res.*, **112**, A07215, doi:10.1029/2006JA012145.

Hu Y.Q., Z. Peng, C. Wang, and J.R. Kan, 2009. "Magnetic merging line and reconnection voltage versus IMF clock angle: Results from global MHD simulations", *J. Geophys. Res.*, **114**, A08220, doi:10.1029/2009JA014118.

Huang C.Y. and L.A. Frank, 1994. "A statistical survey of the central plasma sheet", *J. Geophys. Res.*, **99**, No. A1, doi:10.1029/93JA01894, 83–95.

Iijima T., T.A. Potemra, L.J. Zanetti, and P.F. Bythrow, 1984. "Largescale Birkeland currents in the dayside polar region during strongly northward IMF: A new Birkeland current system", *J. Geophys. Res.*, **89**, No. A9, 7441-7452 (1984).

Ikubanni S.O. and J.O. Adeniyi, 2017, "Relationship between ionospheric F2-layer critical frequency, F10.7, and F10.7P around African EIA trough", *Advances in Space Research*, **59**, 1014–1022.

Ivanov K.G., 1964. "Interaction between advancing shock waves and strong discontinuities in space in the Earth's neighborhood", *Geomagn. Aeron.*, **4**, No. 4, 626-629. In Russian: *Geomagnetism and Aeronomy*, **4**, No. 4, 803-806.

Janhunen P., 1996. "GUMICS-3: A global ionosphere-magnetosphere coupling simulation with high ionospheric resolution", in *ESA Symposium Proceedings on "Environmental Modelling for Space-based applications"*, ESTEC, Noordwijk, NL, edited by W.Burke and T.-D. Guyenne, ESA SP-392, ESA Publ. Div., Noordwijk, 233-239.

Johnson C.Y., J.M. Young, and J.C. Holmes, 1971. "Magnetoglow: A new geophysical resource", *Science*, **171**, No. 3969, 379-381.

Johnson J.R. and S. Wing, 2009. "Northward interplanetary magnetic field plasma sheet entropies", *J. Geophys. Res.*, **114**, A00D08, doi:10.1029/2008JA014017.

Kan J.R. and L.C. Lee, 1979. "Energy coupling function and solar wind magnetosphere dynamo", *Geophys. Res. Lett.*, **6**, No. 7, 577– 580.

Kaufmann R.L., W.R. Paterson, and L.A. Frank, 2004. "Pressure, volume, density relationships in the plasma sheet", *J. Geophys. Res.*, **109**, No. A8, A08204, doi:10.1029/2003JA010317, 1-12.

Kikuchi T., H. Lühr, T. Kitamura, O. Saka, and K. Schlegel, 1996. "Direct penetration of the polar electric field to the equator during a DP2 event as detected by the auroral and equatorial magnetometer chains and the EISCAT radar", *J. Geophys. Res.*, **101**, No. A8, 17161-17173.

Kistler L.M., E. Mobius, W. Baumjohann, G. Paschmann, and D.C. Hamilton, 1992. "Pressure changes in the plasma sheet during substorm injections", *J. Geophys. Res.*, **97**, No. A3, doi:10.1029/91JA02802, 2973–2983.

Kokubun S., T. Yamamoto, M.H. Acuna, K. Hayashi, K. Shiokawa, and H. Kawano, 1994. "The GEOTAIL magnetic field experiment", *J. Geomagn. Geoelectr.*, **46**, No. 1, 7-21.

Koval A., J. Šafránková, Z. Němeček, L. Prech, A.A. Samsonov, and J.D. Richardson, 2005. "Deformation of interplanetary shock fronts in the magnetosheath", *Geophys. Res. Lett.*, **32**, No. 15, L15101, doi:10.1029/2005GL023009, 1-4.

Koval A., J. Šafránková, Z. Němeček, and L. Přech, 2006a. "Propagation of interplanetary shocks through the solar wind and magnetosheath", *Adv. Space Res.*, **38**, No. 3, 552-558.

Koval A., J. Šafránková, Z. Němeček, A.A. Samsonov, L. Přech, J.D. Richardson, and M. Hayosh, 2006b. "Interplanetary shock in the magnetosheath: Comparison of experimental data with MHD modeling", *Geophys. Res. Lett.*, **33**, L11102, doi:10.1029 /2006GL025707, 1-5.

Kozelova T.V. and B.V. Kozelov, 2015. "Particle injections observed at the morning sector as a response to IMF turning", *Adv. Space Res.*, **56**, No. 10, 2106–2116.

Krall J. and J.D. Huba, 2016. "The plasmasphere electron content paradox", *J. Geophys. Res. Space Physics*, **121**, 8924–8935, doi:10.1002/2016JA023008.

Krauss-Varban D., and N. Omidi, 1995. "Large-scale hybrid simulations of the magnetotail during reconnection", *Geophys. Res. Lett.*, **22**, doi:10.1029/95GL03414, 3271–3274.

Kwagala N.K., K. Oksavik, D.A. Lorentzen, and M.G. Johnsen, 2017. "On the contribution of thermal excitation to the total 630.0 nm emissions in the northern cusp ionosphere", *J. Geophys. Res. Space Physics*, **122**, 1234–1245, doi:10.1002/2016JA023366.

Laitinen T.V., M. Palmroth, T.I. Pulkkinen, P. Janhunen, and H.E.J. Koskinen, 2007. "Continuous reconnection line and pressure-dependent energy conversion on the magnetopause in a global MHD model", *J. Geophys. Res.*, **112**, No. A11, A11201, doi:10.1029/2007JA012352, 1-13.

Lavraud B., M.F. Thomsen, J.E. Borovsky, M.H. Denton, and T.I. Pulkkinen, 2006b. "Magnetosphere preconditioning under northward IMF: Evidence from the study of coronal mass ejection and corotating interaction region geoeffectiveness", *J. Geophys. Res.*, **111**, A09208, doi:10.1029/2005JA011566, 1-10.

Le H., N. Yang, L. Liu, Y. Chen, and H. Zhang, 2017. "The latitudinal structure of nighttime ionospheric TEC and its empirical orthogonal functions model over North American sector", *J. Geophys. Res. Space Physics*, **122**, 963–977, doi:10.1002/2016JA023361.

Lee H.-B., Y.H. Kim, E. Kim, J. Hong, and Y.-S. Kwak, 2016 "Where does the plasmasphere begin? Revisit to topside ionospheric profiles in comparison with plasmaspheric TEC from Jason-1", *J. Geophys. Res. Space Physics*, **121**, 10,091–10,102, doi:10.1002/2016JA022747.

Lemaire J., 1974. "The 'Roche-limit' of ionospheric plasma and the formation of the plasmapause", *Planet. Space Sci.*, **22**, No. 5, 757-766.

Lepping R.P., M.H. Acuna, L.E. Burlaga, W.M. Farrell, J.A. Slavin, K.H. Schatten, F. Mariani, N.F. Ness, F.M. Neubauer, Y.C. Whang, J.B. Byrnes, R.S. Kennon, P.V. Panetta, J. Scheifele, and E.M. Worley, 1995."The WIND magnetic field investigation", *Space Sci. Rev.*, **71**, No. 1-4, 207-229.

Liang H., G. Lapenta, R.J. Walker, D. Schriver, M. El-Alaoui, and J. Berchem, 2017. "Oxygen acceleration in magnetotail reconnection", *J. Geophys. Res. Space Physics*, **122**, 618–639, doi:10.1002/2016JA023060.

Lin Y., 1997. "Generation of anomalous flows near the bow shock by the interaction of interplanetary discontinuities", *J. Geophys. Res.*, **102**, doi:10.1029/97JA01989, 24265–24281.

Lin Y., 2001. "Global hybrid simulation of the magnetopause reconnection layer and associated field-aligned currents", *J. Geophys. Res.*, **106**, doi:10.1029/2000JA000184, 25451–25465.

Lin Y. and D.W. Swift, 1996. "A two-dimensional hybrid simulation of the magnetotail reconnection layer", *J. Geophys. Res.*, **101**, doi:10.1029/96JA01457, 19859–19870.

Lin Y. and X.Y. Wang, 2005. "Three-dimensional global hybrid simulation of dayside dynamics associated with the quasi-parallel bow shock", *J. Geophys. Res.*, **110**, A12216, doi:10.1029/2005JA011243.

Lin R.L., X.X. Zhang, S.Q. Liu, Y.L. Wang, and J.C. Gong, 2010. "A three-dimensional asymmetric magnetopause model", *J. Geophys. Res. Space Physics*, **115**, A04207, doi: 10.1029/2009JA014235.

Liu C.M., H.S. Fu, Y. Xu, T.Y. Wang, J.B. Cao, X.G. Sun, and Z.H. Yao, 2017. "Suprathermal electron acceleration in the near-Earth flow rebounce region", *J. Geophys. Res. Space Physics*, **122**, 594–604, doi:10.1002/2016JA023437.

Lockwood M., S.W.H. Cowley, and M.P. Freeman, 1990. "The excitation of plasma convection in the high-latitude ionosphere", *J. Geophys. Res.*, **95**, No. A6, 7961-7972.

Lopez R.E., 2016. "The integrated dayside merging rate is controlled primarily by the solar wind", *J. Geophys. Res. Space Physics*, **121**, 4435–4445, doi:10.1002/2016JA022556.

Lottermoser R.F., M. Scholer, and A.P. Matthews, 1998. "Ion kinetic effects in magnetic reconnection: Hybrid simulations", *J. Geophys. Res.*, **103**, doi:10.1029/97JA01872, 4547–4559.

Lu G., T.E. Holzer, D. Lummerzheim, J.M. Ruohoniemi, P. Stauning, O. Troshichev, P.T. Newell, M. Brittnacher, and G. Parks, 2002. "Ionospheric response to the interplanetary magnetic field southward turning: Fast onset and slow reconfiguration", *J. Geophys. Res.*, **107**, No. A8, 1153, doi:10.1029/2001JA000324, SIA2-1-9.

Lu J.Y., R. Rankin, R. Marchand, and V.T. Tikhonchuk, 2005a. "Nonlinear electron heating by resonant shear Alfvén waves in the ionosphere", *Geophys. Res. Lett.*, **32**, No. 1, L01106, doi:10.1029/2004GL021830.

Lu J.Y., R. Rankin, R. Marchand, and V.T. Tikhonchuk, 2005b. "Reply to comment by J.-P. St.-Maurice on "Nonlinear electron heating by resonant shear Alfvén waves in the ionosphere"", *Geophys. Res. Lett.*, **32**, No. 13, L13103, doi:10.1029/2005GL023149, 1-3.

Lu J.Y., R. Rankin, R. Marchand, I.J. Rae, W. Wang, S.C. Solomon, and J. Lei, 2007. "Electrodynamics of magnetosphere-ionosphere coupling and feedback on magnetospheric field line resonances", *J. Geophys. Res.*, **112**, A10219, doi:10.1029/2006JA012195, 1-11.

Luhmann J.G., R.J. Walker, C.T. Russell, N.U. Crooker, J.R. Spreiter, and S.S. Stahara, 1984. "Patterns of potential magnetic field merging sites on the dayside magnetopause", *J. Geophys. Res.*, **89**, No. A3, 1739-1742.

Luo H., E.A. Kronberg, K. Nykyri, K.J. Trattner, P.W. Daly, G.X. Chen, A.M. Du, and Y.S. Ge, 2017. "IMF dependence of energetic oxygen and hydrogen ion distributions in the near-Earth magnetosphere", *J. Geophys. Res. Space Physics*, **122**, 5168–5180, doi:10.1002/2016JA023471.

Lyatsky W., S. Lyatskaya, and A. Tan, 2007. "A coupling function for solar wind effect on geomagnetic activity", *Geophys. Res. Lett.*, **34**, No. 2, L02107, doi:10.1029/2006GL027666.

Lyons L.R. and T.W. Speiser, 1982. "Evidence for current sheet acceleration in the geomagnetic tail", *J. Geophys. Res.*, **87**, No. A4, 2276-2286.

Lyons L.R., G.T. Blanchard, J.C. Samson, R.P. Lepping, T.Yamamoto, and T. Moretto, 1997. "Coordinated observations demonstrating external substorm triggering", *J. Geophys. Res.*, **102**, doi:10.1029/97JA02639, 27039–27051.

Maynard N.C., B.U.Ã. Sonnerup, G.L. Siscoe, D.R. Weimer, K.D. Siebert, G.M. Erickson, W.W. White, J.A. Schoendorf, D.M. Ober, G.R. Wilson, and M.A. Heinemann, 2002. "Predictions of magnetosheath merging between IMF field lines of opposite polarity", *J. Geophys. Res.*, **107**, No. A12, 1456, doi:10.1029/2002JA009289.

McComas D.J., S.J. Bame, P. Barker, W.C. Feldman, J.L. Phillips, and P. Riley, 1998. "Solar wind electron proton alpha monitor (SWEPAM) for the Advanced Composition Explorer", *Space Sci. Rev.*, **86**, No. 1-4, 563-612.

Moore T.E., M.-C. Fok, and M.O. Chandler, 2002. "The dayside reconnection X line", *J. Geophys. Res.*, **107,** No. A10, 1332, doi:10.1029/2002JA009381, SMP26-1-7.

Mukai T., S. Machida, Y. Saito, M. Hirahara, T. Terasawa, N. Kaya, T. Obara, M. Ejiri, and A. Nishida, 1994. "The low energy particle (LEP) experiment onboard the GEOTAIL satellite", *J. Geomagn. Geoelectr.*, **46**, No. 8, 669-692.

Murakami G., M. Hirai, and I. Yoshikawa, 2007. "The plasmapause response to the southward turning of the IMF derived from sequential EUV images", *J. Geophys. Res.*, **112**, A06217, doi:10.1029/2006JA012174, 1-7.

Nagata D., S. Machida, S. Ohtani, Y. Saito, and T. Mukai, 2007. "Solar wind control of plasma number density in the near-Earth plasma sheet", *J. Geophys. Res.*, **112**, A09204, doi:10.1029/2007JA012284, 1-9.

Nakamura T.K.M., R. Nakamura, W. Baumjohann, T. Umeda, and I. Shinohara, 2016. "Three-dimensional development of front region of plasma jets generated by magnetic reconnection", *Geophys. Res. Lett.*, **43**, 8356–8364, doi:10.1002/2016GL070215.

Němeček Z., J. Šafrankova, O. Kruparova, L. Přech, K. Jelínek, Š. Dušík, J. Šimůnek, K. Grygorov, and J.-H. Shue, 2015. "Analysis of temperature versus density plots and their relation to the LLBL formation under southward and northward IMF orientations", *J. Geophys. Res. Space Physics,* **120**, No. 5, doi:10.1002/2014JA020308, 3475-3488.

Newell P.T., T. Sotirelis, K. Liou, C.-I. Meng, and F.J. Rich, 2006. "Cusp latitude and the optimal solar wind coupling function", *J. Geophys. Res.*, **111**, A09207, doi:10.1029/2006JA011731, 1-11.

Newell P.T., T. Sotirelis, K. Lion, C.-I. Meng, and F.J. Rich, 2007, "A nearly universal solar wind-magnetosphere coupling function inferred from 10 magnetospheric state variables", *J. Geophys. Res.*, **112**, A01206, doi:10.1029/2006JA012015.

Nishida A., 1966. "Formation of plasmapause, or magnetospheric plasma knee, by the combined action of magnetospheric convection and plasma escape from the tail, *J. Geophys. Res.*, **71**, 5669-5679.

Øieroset M., H. Lühr, J. Moen, T. Moretto, and P.E. Sandholt, 1996. "Dynamical auroral morphology in relation to ionospheric plasma convection and geomagnetic activity:

Signatures of magnetopause X line dynamics and flux transfer events", *J. Geophys. Res.*, **101**, doi:10.1029/96JA00613, 13275–13292.

Øieroset M., T.-D. Phan, M. Fujimoto, and R.P. Lepping, 2001. "In situ detection of collisionless reconnection in the Earth's magnetotail", Nature, **412**, doi:10.1038/35086520, 414–417.

Omidi N. and D.G. Sibeck, 2007. "Flux transfer events in the cusp", *Geophys. Res. Lett.*, **34**, L04106, doi:10.1029/2006GL028698.

Omidi N., T. Phan, and D.G. Sibeck, 2009. "Hybrid simulations of magnetic reconnection initiated in the magnetosheath", *J. Geophys. Res.*, **114**, A02222, doi:10.1029/2008JA013647.

Osmane A., A.P. Dimmock, R. Naderpour, T.I. Pulkkinen, and K. Nykyri, 2015. "The impact of solar wind ULF Bz fluctuations on geomagnetic activity for viscous timescales during strongly northward and southward IMF", *J. Geophys. Res. Space Physics*, **120**, No. 11, doi:10.1002/2015JA021505, 9307–9322.

Palmroth M., T.I. Pulkkinen, P. Janhunen, and C.-C. Wu, 2003. "Stormtime energy transfer in global MHD simulation", *J. Geophys. Res.*, **108**, No. A1, 1048, doi:10.1029/2002JA009446, SMP24-1-12.

Pang Y., Y. Lin, X.H. Deng, X.Y. Wang, and B. Tan, 2010. "Three-dimensional hybrid simulation of magnetosheath reconnection under northward and southward interplanetary magnetic field", *J. Geophys. Res.*, **115**, A03203, doi:10.1029/2009JA014415.

Papitashvili V.O., O.A. Troshichev, D.S. Faermark, and A.N. Zaitzev, 1981. "Linear dependence of the intensity of geomagnetic variations in the polar region on the magnitudes of the southern and northern components of the interplanetary magnetic field", *Geomagn. Aeron.,* **21**, No. 4, 565-566.

Paschmann G., 2008. "Recent in-situ observations of magnetic reconnection in near-Earth space", *Geophys. Res. Lett.*, **35**, L19109, doi:10.1029/2008GL035297.

Perreault P. and S.-I. Akasofu, 1978. "A study of geomagnetic storms", *Geophys. J. R. Astron. Soc. (UK).*, **54**, No. 3, 547-583.

Phan T.-D., J.T. Gosling, M.S. Davis, R.M. Skoug, M. Øieroset, R.P. Lin, R.P. Lepping, D.J. McComas, C.W. Smith, H. Reme and A. Balogh, 2006a. "A magnetic reconnection X-line extending more than 390 Earth radii in the solar wind", *Nature*, **439**, No. 7073, 175–178, doi:10.1038/nature04393.

Phan T.D., H. Hasegawa, M. Fujimoto, M. Oieroset, T. Mukai, R.P. Lin, and W.R. Paterson, 2006b. "Simultaneous Geotail and Wind observations of reconnection at the subsolar and tail flank magnetopause", *Geophys. Res. Lett.*, **33**, L09104, doi:10.1029/2006GL025756.

Phan T.D., Paschmann G., Twitty C., Mozer F.S., Gosling J.T., Eastwood J.P., Øieroset M., Rème H., and Lucek E.A., 2007. "Evidence for magnetic reconnection initiated in the magnetosheath", *Geophys. Res. Lett.*, **34**, L14104, doi:10.1029/2007GL030343.

Phan T.D., Shay M.A., Haggerty C.C., et al. 2016. "Ion Larmor radius effects near a reconnection X line at the magnetopause: THEMIS observations and simulation comparison", *Geophys. Res. Lett.*, **43**, 8844–8852, doi:10.1002/2016GL070224.

Pi G., J.-H. Shue, K. Grygorov, H.-M. Li, Z. Němeček, J. Šafránková, Y.-H. Yang, and K. Wang, 2017. "Evolution of the magnetic field structure outside the magnetopause under radial IMF conditions", *J. Geophys. Res. Space Physics*, **122**, 4051–4063, doi:10.1002/2015JA021809.

Pierrard V. and J. Cabrera, 2005. "Comparisons between EUV/IMAGE observations and numerical simulations of the plasmapause formation", *Ann. Geophys.*, **23**, No. 7, 2635-2646.

Pierrard V. and J. Cabrera, 2006. "Dynamical simulations of plasmapause deformations", *Space Sci. Rev.*, **122**, No. 1-4, 119-126.

Prakash M., R. Rankin, and V.T. Tikhonchuk, 2003. "Precipitation and nonlinear effects in geomagnetic field line resonances", *J. Geophys. Res.*, **108**, No. A4, 8014, doi:10.1029/2002JA009383, COA 15-1.

Pritchett P.L., 2006. "Relativistic electron production during guide field magnetic reconnection", *J. Geophys. Res.*, **111**, A10212, doi:10.1029/2006JA011793.

Pu Z.Y., X.G. Zhang, X.G. Wang, J. Wang, X.-Z. Zhou, M.W. Dunlop, L. Xie, C.J. Xiao, Q.G. Zong, S.Y. Fu, Z.X. Liu, C. Carr, Z.W. Ma, C. Shen, E. Lucek, H. Rème, and P. Escoubet, 2007. "Global view of dayside magnetic reconnection with the dusk-dawn IMF orientation: A statistical study for Double Star and Cluster data", *Geophys. Res. Lett.*, **34**, L20101, doi:10.1029/2007GL030336, 1-6.

Pudovkin M.I., B.P. Besser, and S.A. Zaitseva, 1998. "Magnetopause stand-off distance in dependence on the magnetosheath and solar wind parameters", *Ann. Geophys. (Germany)*, **16**, No. 4, 388-396.

Pulkkinen T.I., M. Palmroth, and R.L. McPherron, 2007. "What drives magnetospheric activity under northward IMF conditions?", *Geophys. Res. Lett.*, **34**, L18104, doi:10.1029/2007GL030619, 1-4.

Pulkkinen T.I., A.P. Dimmock, A. Osmane, and K. Nykyri, 2015. "Solar wind energy input to themagnetosheath and at the magnetopause", *Geophys. Res. Lett.*, 42, No. 12, doi:10.1002/2015GL064226, 4723-4730.

Pushkar E.A., A.A. Barmin, and S.A. Grib, 1991. "Investigation in the MHD approximation of the incidence of the solar-wind shock wave on the near-Earth bow shock", *Geomagnetism and Aeronomy*, **31**, No. 3, 410-412.

Raeder J., 2003. "Global geospace modeling: Tutorial and review", in *Space Plasma Simulation*, edited by J. Büchner, C. T. Dum, and M. Scholer, Lecture Notes Phys., **615**, Springer, Berlin, 212-246.

Reiff P.H., 1984. "Models of auroral-zone conductance", in *Magnetospheric Currents*, Chapman Conference, Irvington, VA, Selected Papers, Geophys. Monogr. Ser., **28**, edited by Thomas A. Potemra, AGU, Washington, D.C., 180-191.

Retin A., Sundkvist D., Vaivads A., Mozer F.S., André M., and Owen C.J., 2007. "In situ evidence of magnetic reconnection in turbulent plasma", *Nat. Phys.*, **3**, No. 4, 235–238.

Ridley A.J., 2005. "A new formulation for the ionospheric cross polar cap potential including saturation effects", *Ann. Geophys.*, **23**, No. 11, 3533-3547.

Ridley A.J., 2007. "Alfvén wings at Earth's magnetosphere under strong interplanetary magnetic fields", *Ann. Geophys.*, **25**, No. 2, 533-542.

Ridley A.J., G. Lu, C.R. Clauer, and V.O. Papitashvili, 1998. "A statistical study of the ionospheric convection response to changing interplanetary magnetic field conditions using the assimilative mapping of ionospheric electrodynamics technique", *J. Geophys. Res.*, **103,** No. A3, 4023-4039.

Robinson R.M., R.R. Vondrak, K. Miller, T. Dabbs, and D. Hardy, 1987. "On calculating ionospheric conductances from the flux and energy of precipitating electrons", *J. Geophys. Res.*, **92**, No. A3, 2565-2569.

Rong Z.J., A.T.Y. Lui, W.X. Wan, Y.Y. Yang, C. Shen, A.A. Petrukovich, Y.C. Zhang, T.L. Zhang, and Y. Wei, 2015. "Time delay of interplanetary magnetic field penetration into Earth's magnetotail", *J. Geophys. Res. Space Physics,* **120**, No. 5, doi:10.1002/2014JA020452, 3406-3414.

Ruohoniemi J.M. and R.A. Greenwald, 1998. "Response of high-latitude convection to a sudden southward IMF turning", *Geophys. Res. Lett.*, **25**, No. 15, 2913-2916.

Russell C.T., J.T. Gosling, R.D. Zwickl, and E.J. Smith, 1983. "Multiple spacecraft observations of interplanetary shocks: ISEE three-dimensional plasma measurements", *J. Geophys. Res.*, **88**, No. A12, 9941-9947.

Russell C.T., Y.L. Wang, J. Raeder, R.L. Tokar, C.W. Smith, K.W. Ogilvie, A.J. Lazarus, R.P. Lepping, A. Szabo, H. Kawano, T. Mukai, S. Savin, Y.I. Yermolaev, X.-Y. Zhou, and B.T. Tsurutani, 2000. "The interplanetary shock of September 24, 1998: Arrival at Earth", *J. Geophys. Res.*, **105**, No. A11, 25143-25154.

Ŝafránková J., Z. Němeček, L. Přech, G. Zastenker, K.I. Paularena, N. Nikolaeva, M. Nozdrachev, A. Skalsky, and T. Mukai, 1998. "The January 10–11, 1997 magnetic cloud: Multipoint measurements", *Geophys. Res. Lett.*, **25**, No. 14, 2549-2552.

Ŝafránková J., Z. Němeček, L. Přech, A.A. Samsonov, A. Koval, and K. Andréeová, 2007. "Modification of interplanetary shocks near the bow shock and through the magnetosheath", *J. Geophys. Res.*, **112**, A08212, doi:10.1029/2007JA012503, 1-9.

Samsonov A.A. and D. Hubert, 2004. "Steady state slow shock inside the Earth's magnetosheath: To be or not to be? Part 2. Numerical three-dimensional MHD modeling", *J. Geophys. Res.*, **109**, No. A1, A01218, doi:10.1029/2003JA010006, 1-12.

Samsonov A.A. and M.I. Pudovkin, 2000. "Application of the bounded anisotropy model for the dayside magnetosheath", *J. Geophys. Res.*, **105**, No. A6, 12859-12867.

Samsonov A.A., Z. Němeček, and J. Šafránková, 2006. "Numerical MHD modeling of propagation of interplanetary shock through the magnetosheath", *J. Geophys. Res.*, **111**, No. A8, A08210, doi:10.1029/2005JA011537, 1-9.

Sandel B.R., A.L. Broadfoot, C.C. Curtis, R.A. King, T.C. Stone, R.H. Hill, J. Chen, O.H.W. Siegmund, R. Raffanti, D. Allred, S. Turley, and D.I. Gallagher, 2000. "The Extreme Ultraviolet Imager investigation for the IMAGE mission", *Space Sci. Rev.*, **91**, No. 1-2, 197-242.

Sandel B.R., R.A. King, W.T. Forrester, D.L. Gallagher, A.L. Broadfoot, and C.C. Curtis, 2001. "Initial results from the IMAGE extreme ultraviolet imager", *Geophys. Res. Lett.*, **28**, No. 8, 1439-1442.

Sasunov Y.L., M.L. Khodachenko, I.I. Alexeev, E.S. Belenkaya, O.V. Mingalev, and M.N. Melnik, 2017. "The influence of kinetic effect on the MHD scalings of a thin current sheet", *J. Geophys. Res. Space Physics*, **122**, 493–500, doi:10.1002/2016JA023162.

Scherliess L. and B.G. Fejer, 1999. "Radar and satellite global equatorial F region vertical drift model", *J. Geophys. Res.*, **104,** doi:10.1029/1999JA900025, 6829– 6842.

Schindler K., 1972. "A self consistent theory of the tail of the magnetosphere", in *Earth's Magnetospheric Processes*, edited by B. M. McCormac, pp. 200–209, D. Reidel, Dordrecht, Holland.

Shaikh M.M., B. Nava, and A. Kashcheyev, 2017. "A model-assisted radio occultation data inversion method based on data ingestion into NeQuick", *Advances in Space Research*, **59**, 326–336.

Shen W.-W. and M. Dryer, 1972. "Magnetohydrodynamic theory for the interaction of an interplanetary double-shock ensemble with the Earth's bow shock", *J. Geophys. Res.*, **77**, No. 25, 4627-4644.

Shepherd S.G., 2007, "Polar cap potential saturation: Observations, theory, and modeling", *J. Atmos. Sol. Terr. Phys.*, **69, No. 3**, 234–248.

Shepherd S., R. Greenwald, and J. Ruohoniemi, 2002. "Cross polar cap potentials measured with Super Dual Auroral Radar Network during quasi-steady solar wind and interplanetary magnetic field conditions", *J. Geophys. Res.*, **107**, No. A7, 1094, doi:10.1029/2001JA000152.

Shue J.-H., P. Song, C.T. Russell, J.T. Steinberg, J.K. Chao, G. Zastenker, O.L. Vaisberg, S. Kokubun, H.J. Siager, T.R. Detman, and H. Kawano, 1998. "Magnetopause location under extreme solar wind conditions", *J. Geophys. Res.*, **103**, No. A8, 17691-17700.

Shukhtina M.A., E.I. Gordeev, V.A. Sergeev, N.A. Tsyganenko, L.B.N. Clausen, and S.E. Milan, 2016. "Magnetotail magnetic flux monitoring based on simultaneous solar wind and magnetotail observations", *J. Geophys. Res. Space Physics*, **121**, 8821–8839, doi:10.1002/2016JA022911.

Siscoe G.L. and Huang T.S., 1985. "Polar cap inflation and deflation", *J. Geophys. Res.*, **90**, No. A1, 543-547.

Siscoe G.L., G.M. Erickson, B.U.Ö Sonnerup, N.C. Maynard, K.D. Siebert, D.R. Weimer, and W.W. White, 2001. "Global role of $E_{||}$ in magnetopause reconnection: An explicit demonstration", *J. Geophys. Res.*, **106, No.** A7, 13015–13022.

Siscoe G.L., N.U. Crooker, and K.D. Siebert, 2002a. "Transpolar potential saturation: Roles of region 1 current system and solar wind ram pressure", *J. Geophys. Res.*, **107**, No. A10, 1321, doi:10.1029/2001JA009176.

Siscoe G.L., G.M. Erickson, B.U.O. Sonnerup, N.C. Maynard, J. A. Schoendorf, K. D. Siebert, D. R. Weimer, W. W. White, and G. R. Wilson, 2002b. "Hill model of transpolar potential saturation: Comparisons with MHD simulations", *J. Geophys. Res.*, **107**, No. A6, 1075, doi:10.1029/2001JA000109.

Sitnov M.I. and V.G. Merkin, 2016. "Generalized magnetotail equilibria: Effects of the dipole field, thin current sheets, and magnetic flux accumulation", *J. Geophys. Res. Space Physics*, **121**, 7664–7683, doi:10.1002/2016JA023001.

Slivka K.Y., V.S. Semenov, N.V. Erkaev, N.P. Dmitrieva, I.V. Kubyshkin, and H. Lammer, 2015. "Peculiarities of magnetic barrier formation for southward and northward directions of the IMF", *J. Geophys. Res. Space Physics*, **120**, No. 11, doi:10.1002/2015JA021250, 9471–9483.

Smith C.W., J. L'Heureux, N.F. Ness, M.H. Acuna, L.F. Burlaga, and J. Scheiffele, 1998. "The ACE magnetic fields experiment", *Space Sci. Rev.*, **86**, No. 1-4, 613-632.

Sonnerup B.U.O., 1974. "Magnetopause reconnection rate", *J. Geophys. Res.*, **79**, No. 10, 1546-1549.

Sonnerup B.U.O., G. Paschmann, I. Papamastorkis, N. Sckopke, G. Haerendel, S.J. Bame, J.R. Asbridge, J.T. Gosling, and C.T. Russell, 1981. "Evidence for magnetic field reconnection at the Earth's magnetopause", *J. Geophys. Res.*, **86**, No. A12, 10049-10067.

Sonnerup B.U.Ö., G. Paschmann, and T.-D. Phan, 1995. "Fluid aspects of reconnection at the magnetopause: In situ observations", in *Physics of the Magnetopause*, Geophys. Monogr. Ser, **90**, edited by P. Song, B.U.Ö. Sonnerup and M.F. Thomsen, AGU, Washington, D.C., 1-167.

Sonnerup B., G. Paschmann, S. Haaland, T. Phan, and S. Eriksson, 2016a. "Reconnection layer bounded by switch-off shocks: Dayside magnetopause crossing by THEMIS D", *J. Geophys. Res. Space Physics*, **121**, 3310–3332, doi:10.1002/2016JA022362.

Sonnerup B., S. Haaland, G. Paschmann, T. Phan, and S. Eriksson, 2016b. "Magnetopause reconnection layer bounded by switch-off shocks: Part 2. Pressure anisotropy", *J. Geophys. Res. Space Physics*, **121**, 9940–9955, doi:10.1002/2016JA023250.

Spasojević M., J. Goldstein, D.L. Carpenter, U.S. Inan, B.R. Sandel, M.B. Moldwin, and B.W. Reinisch. 2003. "Global response of the plasmasphere to a geomagnetic disturbance", *J. Geophys. Res.*, **108**, No. A9, 1340, doi:10.1029/2003JA009987, SMP4-1-14.

Spence H.E. and M.G. Kivelson, 1993. "Contributions of the low-latitude boundary layer to the finite width magnetotail convection model", *J. Geophys. Res.*, **98**, No. A9, 15487-15496.

Spence H.E., M.G. Kivelson, R.J. Walker, and D J. McComas, 1989. "Magnetospheric plasma pressures in the midnight meridian: Observations from 2.5 to 35 R_E", *J. Geophys. Res.*, **94**, No. A5, doi:10.1029/JA094iA05p05264, 5264–5272.

Spicher A., A.A. Ilyasov, W.J. Miloch, A.A. Chernyshov, L.B.N. Clausen, J.I. Moen, T. Abe, and Y. Saito, 2016. "Reverse flow events and small-scale effects in the cusp ionosphere", *J. Geophys. Res. Space Physics*, **121**, 10,466–10,480, doi:10.1002/2016JA022999.

Spreiter J.R. and S.S. Stahara, 1992. "Computer modeling of solar wind interaction with Venus and Mars", in *Venus and Mars: Atmospheres, Ionospheres and Solar Wind Interactions*, Geophys. Monogr. Ser., **66**, edited by J.G. Luhmann, M. Tatrallyay, and R.O. Pepin, AGU, Washington, D.C., 345-383.

Spreiter J.R. and S.S. Stahara, 1994. "Gasdynamic and magnetohydrodynamic modeling of the magnetosheath: A tutorial", *Adv. Space Res.*, **14**, No. 7, 5-19.

Svalgaard L. and E.W. Cliver, 2007. "Interhourly variability index of geomagnetic activity and its use in deriving the long-term variation of solar wind speed", *J. Geophys. Res.*, **112**, A10111, doi:10.1029/2007JA012437.

Swift D.W., 1996. "Use of a hybrid code for a global-scale plasma simulation", *J. Comput. Phys.*, **126**, doi:10.1006/jcph.1996.0124, 109–121.

Szabo A., 2005. "Interplanetary discontinuities and shocks in the Earth's magnetosheath", in *Multiscale Processes in the Earth's Magnetosphere: From Interball to Cluster*, NATO Sci. Ser., vol. 178, edited by J.-A. Sauvaud and Z. Němeček, doi:10.1007/1-4020-2768-0-4, Springer, New York, 57–71.

Szabo A., R.P. Lepping, J. Merka, C.W. Smith, and R.M. Skoug, 2001. "The evolution of interplanetary shocks driven by magnetic clouds", in *Solar Encounter: Proceedings of the First Solar Orbiter Workshop*, edited by B. Battrick et al., Eur. Space Agency, Spec. Publ., ESA-493, 383-387.

Tenfjord P., N. Østgaard, K. Snekvik, K.M. Laundal, J.P. Reistad, S. Haaland, and S.E. Milan, 2015. "How the IMF By induces a By component in the closed magnetosphere and how it leads to asymmetric currents and convection patterns in the two hemispheres", *J. Geophys. Res. Space Physics*, **120**, 9368–9384, doi:10.1002/2015JA021579.

Tenfjord P., N. Østgaard, R. Strangeway, S. Haaland, K. Snekvik, K.M. Laundal, J.P. Reistad, and S.E. Milan, 2017. "Magnetospheric response and reconfiguration times following IMF By reversals", *J. Geophys. Res. Space Physics*, **122**, 417–431, doi:10.1002/2016JA023018.

Terasawa T., M. Fujimoto, T. Mukai, I. Shinohara, Y. Saito, T. Yamamoto, S. Machida, S. Kokubun, A.J. Lazarus, J.T. Steinberg, and R.P. Lepping, 1997. "Solar wind control of density and temperature in the near-Earth plasma sheet: WIND/GEOTAIL collaboration", *Geophys. Res. Lett.*, **24**, No. 8, 935-938.

Thomsen M.F., 2004. "Why Kp is such a good measure of magnetospheric convection", *Space Weather*, **2**, No. 11, S11004, doi:10.1029/2004SW000089.

Thomsen M.F., J.E. Borovsky, and R.M. Skoug, 2003. "Delivery of cold, dense plasma sheet material into the near-Earth region", *J. Geophys. Res.*, **108**, No. A4, 1151, doi:10.1029/2002JA009544.

Trattner K.J., S.A. Fuselier, and S.M. Petrinec, 2004. "Location of the reconnection line for northward interplanetary magnetic field", *J. Geophys. Res.*, **109**, No. A3, A03219, doi:10.1029/2003JA009975, 1-10.

Trattner K.J., S. Thresher, L. Trenchi, S.A. Fuselier, S.M. Petrinec, W.K. Peterson, and M.F. Marcucci, 2017. "On the occurrence of magnetic reconnection equatorward of the cusps at the Earth's magnetopause during northward IMF conditions", *J. Geophys. Res. Space Physics*, **122**, 605–617, doi:10.1002/2016JA023398.

Tsagouri I., I. Galkin, and T. Asikainen, 2017. "Long-term changes in space weather effects on the Earth's ionosphere", *Advances in Space Research*, **59**, 351–365.

Tsurutani B.T. and W.D. Gonzalez, 1997. "The Interplanetary Causes of Magnetic Storms: A Review", in *Magnetic Storms*, eds B.T. Tsurutani, W.D. Gonzalez, Y. Kamide, and J.K. Arballo, Geophys. Monogr. Ser., AGU, Washington, D.C., 77-89.

Tsyganenko N.A., 1996. "Effects of the solar wind conditions on the global magnetospheric configuration as deduced from data-based field models", *Proc. of 3-rd Intern. Conf. on Substorms* (ICS-3), Versailles, France, 12–17 May 1996, ESA SP-389, Eur. Space Agency, Paris, 181–185.

Tsyganenko N.A. and V.A. Andreeva, 2015. "A forecasting model of the magnetosphere driven by an optimal solar wind coupling function", *J. Geophys. Res. Space Physics*, **120**, No. 10, doi:10.1002/2015JA021641, 8401–8425.

Tsyganenko N.A. and V.A. Andreeva, 2016 "An empirical RBF model of the magnetosphere parameterized by interplanetary and ground-based drivers", *J. Geophys. Res. Space Physics*, **121**, 10,786–10,802, doi:10.1002/2016JA023217.

Tsyganenko N.A. and T. Mukai, 2003. "Tail plasma sheet models derived from Geotail particle data", *J. Geophys. Res.*, 108, No. A3, 1136, doi:10.1029/2002JA009707, SMP23-1-15.

Tu J. and P. Song, 2016, "A two-dimensional global simulation study of inductive-dynamic magnetosphere-ionosphere coupling", *J. Geophys. Res. Space Physics*, **121**, 11,861–11,881, doi:10.1002/2016JA023393.

Usadi A., R.A. Wolf, M. Heinemann, and W. Horton, 1996, "Does chaos alter the ensemble averaged drift equation?", *J. Geophys. Res.*, **101**, No. A7, doi:10.1029/96JA00522, 15491–15514.

Usanova M.E. and Y.Y. Shprits, 2017. "Inner magnetosphere coupling: Recent advances", *J. Geophys. Res. Space Physics*, **122**, 102–104, doi:10.1002/2016JA023614.

Vichare G., R. Rawat, M. Jadhav, A.K. Sinha, 2017. "Seasonal variation of the Sq focus position during 2006–2010", *Advances in Space Research*, **59**, 542–556.

Voiculescu M., T. Nygren, A.T. Aikio, H. Vanhamaki, and V. Pierrard, 2016, "Postmidnight ionospheric troughs in summer at high latitudes", *J. Geophys. Res. Space Physics*, **121**, 12,171–12,185, doi:10.1002/2016JA023360.

Wang C.-P., L.R. Lyons, J.M. Weygand, T. Nagai, and R.W. McEntire, 2006. "Equatorial distributions of the plasma sheet ions, their electric and magnetic drifts, and magnetic fields under different interplanetary magnetic field *Bz* conditions", *J. Geophys. Res.*, **111**, No. A4, A04215, doi:10.1029/2005JA011545, 1-11.

Wang C.-P., L.R. Lyons, T. Nagai, J.M. Weygand, and R.W. McEntire, 2007. "Sources, transport, and distributions of plasma sheet ions and electrons and dependences on interplanetary parameters under northward interplanetary magnetic field", *J. Geophys. Res.*, **112**, A10224, doi:10.1029/2007JA012522, 1-12.

Wang M., J.Y. Lu, K. Kabin, H.Z. Yuan, X. Ma, Z.-Q. Liu, Y.F. Yang, J.S. Zhao, and G. Li, 2016. "The influence of IMF clock angle on the cross section of the tail bow shock", *J. Geophys. Res. Space Physics*, **121**, 11,077–11,085, doi:10.1002/2016JA022830.

Watanabe M., K. Kabin, G. J. Sofko, R. Rankin, T. I. Gombosi, A. J. Ridley, and C. R. Clauer, 2005. "Internal reconnection for northward interplanetary magnetic field", *J. Geophys. Res.*, **110**, A06210, doi:10.1029/2004JA010832.

Weimer D.R., D.M. Ober, N.C. Maynard, M.R. Collier, D.J. McComas, N.F. Ness, C.W. Smith, and J. Watermann, 2003. "Predicting interplanetary magnetic field (IMF) propagation delay times using the minimum variance technique", *J. Geophys. Res.*, **108**, No.A1, 1026, doi:10.1029/2002JA009405, SMP16-1-12.

Wilder F.D., C.R. Clauer, and J.B.H. Baker, 2008. "Reverse convection potential saturation during northward IMF", *Geophys. Res. Lett.*, **35**, No. 12, L12103, doi:10.1029/2008GL034040.

Wilder F.D., C.R. Clauer, and J.B.H. Baker, 2009. "Reverse convection potential saturation during northward IMF under various driving conditions", *J. Geophys. Res.*, **114**, A08209, doi:10.1029/2009JA014266.

Williams D.J., 1997. "Considerations of source, transport, acceleration/heating and loss processes responsible for geomagnetic tail particle populations", *Space Sci. Rev.*, **80**, No. 1-2, 369-389.

Williams D.J., E.C. Roelof, and D.G. Mitchell, 1992. "Global magnetospheric imaging", *Rev. Geophys.*, **30**, No. 3, 183-208.

Williams D.J., R.W. McEntire, C. Schlemm, II, A.T.Y. Lui, G. Gloeckler, S.P. Christon, and F. Gliem, 1994. "Geotail energetic particles and ion composition instrument", *J. Geomagn. Geoelectr.*, **46**, No. 1, 39-57.

Wing S. and J.R. Johnson, 2009. "Substorm entropies", *J. Geophys. Res.*, **114**, A00D07, doi:10.1029/2008JA013989.

Wing S. and P.T. Newell, 2002. "2D plasma sheet ion density and temperature profiles for northward and southward IMF", *Geophys. Res. Lett.*, **29**, No. 9, 1307, doi:10.1029/2001GL013950, 21-1-4.

Wing S., J.R. Johnson, P.T. Newell, and C.-I. Meng, 2005. "Dawn-dusk asymmetries, ion spectra, and sources in the northward interplanetary magnetic field plasma sheet", *J. Geophys. Res.*, **110**, No. A8, A08205, doi:10.1029/2005JA011086, 1-17.

Wing S., J.W. Gjerloev, J.R. Johnson, and R.A. Hoffman, 2007. "Substorm plasma sheet ion pressure profiles", *Geophys. Res. Lett.*, **34**, No. 16, L16110, doi:10.1029/2007GL030453.

Wolf R.A., 1983. "The quasi-static (slow-flow) region of the magnetosphere", in *Solar-Terrestrial Physics*, edited by R.L. Carovillano and J.M. Forbes, D. Reidel, Hingham, Mass., 303–368.

Wu C.-C., 2003a. "Shock wave interaction with the magnetopause", *Space Sci. Rev.*, **107**, No. 1-2, 219-226.

Yan M. and L.C. Lee, 1996. "Interaction of interplanetary shocks and rotational discontinuities with the Earth's bow shock", *J. Geophys. Res.*, **101**, No. A3, 4835-4848.

Yeh T., 1976. "Day side reconnection between a dipolar geomagnetic field and a uniform interplanetary field", *J. Geophys. Res.*, **81**, No. 13, 2140–2144.

Yoshikawa I., M. Nakamura, M. Hirahara, Y. Takizawa, K. Yamashita, H. Kunieda, T. Yamazaki, K. Misaki, and A. Yamaguchi, 1997. "Observation of He II emission from the plasmasphere by a newly developed EUV telescope on board sounding rocket S-520-19", *J. Geophys. Res.*, **102**, No. A9, 19897-19902.

Yoshikawa I., A. Yamazaki, K. Shiomi, M. Nakamura, K. Yamashita, and Y. Takizawa, 2000a. "Evolution of the outer plasmasphere during low geomagnetic activity observed by the EUV scanner onboard Planet-B", *J. Geophys. Res.*, **105**, No. A12, 27777-27789.

Yoshikawa I., A. Yamazaki, K. Shiomi, K. Yamashita, Y. Takizawa, and M. Nakamura, 2000b. "Photometric measurement of cold helium ions in the magnetotail by an EUV scanner onboard Planet-B: Evidence of the existence of cold plasmas in the near-Earth plasma sheet", *Geophys. Res. Lett.*, **27**, No. 21, 3657-3570.

Yoshikawa I., A. Yamazaki, K. Shiomi, K. Yamashita, Y. Takizawa, and M. Nakamura, 2001. "Interpretation of the He II (304 Å) EUV image of the inner magnotosphere by using empirical models", *J. Geophys. Res.*, **106**, No. A11, 25745-25758.

Zhang L.Q., L. Dai, W. Baumjohann, A.T.Y. Lui, C. Wang, H. Rème, and M.W. Dunlop, 2016a. "Temporal evolutions of the solar wind conditions at 1 AU prior to the near-Earth X lines in the tail: Superposed epoch analysis", *J. Geophys. Res. Space Physics*, **121**, 7488–7496, doi:10.1002/2016JA022687.

Zhang L.Q., W. Baumjohann, C. Wang, L. Dai, and B.B. Tang, 2016b. "Bursty bulk flows at different magnetospheric activity levels: Dependence on IMF conditions", *J. Geophys. Res. Space Physics*, **121**, 8773–8789, doi:10.1002/2016JA022397

Zhao L.L., H. Zhang, and Q.-G. Zong, 2017. "A statistical study on hot flow anomaly current sheets", *J. Geophys. Res. Space Physics*, **122**, 235–248, doi:10.1002/2016JA023319.

Zhong J., J. Lei, W. Wang, A.G. Burns, X. Yue, and X. Dou, 2017. "Longitudinal variations of topside ionospheric and plasmaspheric TEC", *J. Geophys. Res. Space Physics*, **122**, 6737–6760, doi:10.1002/2017JA024191.

Zhuang H.C., C.T. Russell, E.J. Smith, and J.T. Gosling, 1981. "Three dimensional interaction of interplanetary shock waves with the bow shock and magnetopause: A comparison of theory with ISEE observations", *J. Geophys. Res.*, **86**, No. A7, 5590-5600.

Chapter 3

EFFICIENCY OF SOLAR WIND-MAGNETOSPHERE COUPLING FOR SW PRESSURE VARIATIONS

3.1. SOLAR WIND–MAGNETOSPHERE COUPLING EFFICIENCY FOR SOLAR WIND PRESSURE IMPULSES

Palmroth et al. (2007) carry out the first statistical study on the role of solar wind dynamic pressure impulses in the solar wind–magnetosphere coupling efficiency. They identify pressure impulses from the solar wind, and compute the coupling efficiency using four different definitions, utilizing both magnetometer data as well as Super Dual Auroral Radar Network (SuperDARN) observations. They also investigate the coupling energy as parameterized by the local AE index obtained from the IMAGE magnetometer array. It was noted that the effect of solar wind dynamic pressure on magnetospheric energetics has gained interest during recent years. Hubert et al. (2006) presented evidence that solar wind dynamic pressure impulses increase the reconnection rate in the tail, which implies an increased production of closed flux in the nightside. This leads to shrinking of the polar cap towards the magnetic pole as a response to the pressure impulse, a phenomenon that has been observed with ultraviolet cameras onboard polar-orbiting spacecraft (e.g., Boudouridis et al., 2005). In addition to the well known effect of the equatorial magnetic field increase, dynamic pressure has also been suggested to increase ionospheric Joule heating through strengthening the Chapman-Ferraro currents at the magnetopause, which then connect to Region 1 currents and the ionosphere (Palmroth et al., 2004b). Solar wind dynamic pressure impulses have also been suggested to trigger substorms and pseudo-breakups in several case studies (e.g., Zhou and Tsurutani, 2001). As described Palmroth et al. (2007), the coupling efficiency dependence on dynamic pressure has been studied by computing the ratio of the polar cap potential to the solar wind-imposed magnetospheric potential. Boyle et al. (1997) concluded that coupling efficiency does not depend on the instantaneous values of the dynamic pressure. Boudouridis et al. (2005) studied a set of events during which there was a dynamic pressure impulse, and concluded that the coupling efficiency increases after a pressure impulse during southward IMF. Both Boyle et al. (1997) and Boudouridis et al. (2005) used DMSP measurements that can determine the polar cap potential with a high accuracy, but an instantaneous change in the potential is detected only if the spacecraft constellation is in a suitable location as it traverses the polar cap. Furthermore, the coupling efficiency in these two investigations is somewhat different, as Boyle et al. (1997) measure the available magnetospheric potential over a fixed magnetospheric size, while Boudouridis et al. (2005) compute this size from the dynamic pressure and IMF using the Shue et al. (1998) formulation.

The solar wind observations in Palmroth et al. (2007) are recorded by the ACE spacecraft. The solar wind density and velocity are determined by the Solar Wind Electron Proton Alpha Monitor (SWEPAM) instrument (McComas et al., 1998), while the IMF observations are recorded by the magnetic field instrument MAG (Smith et al., 1998). Using data from the SWEPAM instrument, Palmroth et al. (2007) looked for stepwise changes in the solar wind dynamic pressure during the period 1998–2002. They required that the pressure step was larger than 1 nPa compared to a 30-min average prior to the pulse, and that the pressure fluctuations during the 30 minutes before and after the pulse were small. The second criterion was tested by comparing periods having a larger than 1 nPa change in the dynamic pressure with a Heaviside step function. A total of 236 one-hour periods for which the correlation coefficient with a Heaviside function was more than 0.95 were found. These events were divided into two groups according to whether the total magnetic field increases (fast type discontinuities, FTD) or decreases (slow type discontinuities, STD) at the time of the dynamic pressure impulse. A total of 105 events follow the FTD characteristics, while 131 events have an STD type of evolution. The naming convention follows the physical properties of fast and slow interplanetary shocks, for which the total magnetic field increases and decreases, respectively.

Palmroth et al. (2007) use several different data sets to infer the coupling efficiency between the solar wind and the magnetosphere. The polar cap potential is determined from the SuperDARN radar observations with two-minute temporal cadence by fitting spherical harmonic functions to the data, after which the difference of the minimum and maximum potential is obtained (Ruohoniemi and Baker, 1998). Together with the polar cap potential, it was used the northern polar cap index (PCN), which is a single magnetometer recording near the magnetic pole (Troshichev et al., 2006). Furthermore, Palmroth et al. (2007) use the recordings of the meridional IMAGE magnetometer array (ranging from Svalbard to Southern Estonia) to characterize the energy deposition in the ionosphere by computing a local AE index (IMAGE E, or IE). The IE has been shown to correspond to the AE index during 16-04 UT when IMAGE is in the night sector (Kauristie et al., 1997), but since pressure pulse responses are not necessarily limited to the night sector, IE should characterize the disturbances also at other local times. In Palmroth et al. (2007) were used four definitions Eff_1 – Eff_4 for the coupling efficiency:

$$\mathrm{Eff}_1 = \mathrm{PC}/\Phi_{\mathrm{MAG}} = \mathrm{PC}/EL_{\mathrm{MAG}}, \tag{3.1.1}$$

where

$$E = V_{SW}\sqrt{B_Y^2 + B_Z^2}\ , \tag{3.1.2}$$

and V_{SW} is the solar wind velocity, **B** is the IMF, and L_{MAG} is the magnetospheric size (according to Shue et al., 1998);

$$\mathrm{Eff}_2 = (\mathrm{PC}/L_{\mathrm{REC}})/E, \tag{3.1.3}$$

where L_{REC} is the distance between ionospheric convection cells;

$$\mathrm{Eff}_3 = \mathrm{PCN}/\Phi_{MAG} \text{ and } \mathrm{Eff}_4 = \mathrm{PCN}/E, \tag{3.1.4}$$

where PCN is the polar cap index.

According to Eq. 3.1.1, Eff_1 is the ratio of the polar cap potential to the available magnetospheric potential, where the latter is a product of the IEF and the size of the magnetosphere measured at the terminators using the Shue et al. (1998) formulation. This definition is also used by Boudouridis et al. (2005). As the IEF ideally maps down to the ionosphere along equipotential field lines, the coupling efficiency may also be defined from the ratio between the reconnection-induced electric field in the ionosphere and the IEF. This definition is valid both for low-latitude reconnection as well as during lobe reconnection, during which the ionospheric footprint of the reconnection line can be found in between the reverse convection cells (Chisham et al., 2004). Hence, for Eff_2 (see Eq. 3.1.3), Palmroth et al. (2007) first approximate the reconnection-induced electric field in the ionosphere as the ratio of the polar cap potential to the distance between the centers of the convection cells (L_{REC}) that is assumed to be proportional to the length of the reconnection line in the ionosphere. This is then divided by the IEF. Eff_3 is defined as the ratio of the PCN index and the available magnetospheric potential, while Eff_4 is the ratio of the PCN index with the IEF in the YZ_{GSM} plane (see Eq. 3.1.4). The use of Eff_3 and Eff_4 is justified, because the PCN correlates well with the polar cap electric field measurements from DMSP (Troshichev et al., 2000). To explore the different definitions of the coupling efficiency, Palmroth et al. (2007) looked for events having large amount of radar echoes from both cells at the dayside; for these events the SuperDARN polar cap potential is most accurate. Six events in 2000-2001 were selected for closer inspection (2 during northward, 2 during southward, and 2 during near-zero IMF B_Z).

Palmroth et al. (2007) came to following conclusions:

(i) It was investigated the solar wind–magnetosphere coupling efficiency in response to solar wind dynamic pressure impulses.

(ii) Due to the selection criteria, obtained results apply to steady MHD discontinuities.

(iii) Pressure impulses are internally different, and they can be characterized based on simultaneous changes in the IMF magnitude.

(iv) The evolution of the IMF is crucial when computing the coupling efficiency, as it affects the evolution of the IEF.

(v) Hence, if the superposed epoch analysis is carried out for the entire data set without grouping according to the impulse structure, the coupling efficiency is constant in time.

(vi) However, for increasing IMF (the FTD events), the coupling efficiency is observed to decrease, while for decreasing IMF cases (the STD events) the coupling efficiency increases.

(vii) This intuitively clear feature is consistent in all four definitions of coupling efficiency used here, and present even if the events are grouped according to the IMF B_Z direction prior to the impulse.

(viii) It is the property of the pressure impulse itself that determines the magnetospheric and ionospheric response.

(ix) Although the coupling efficiency decreases for stronger driver (the FTD events), the FTD events show a stronger response in coupling energy (this is seen in the superposed epoch analysis of the local AE index obtained from the IMAGE magnetometer array).

(x) The decreasing coupling efficiency for the FTD events is interpreted as energy transmitted through the system, while during the weaker driver (the STD events) the energy is more easily extracted from the solar wind; this is consistent with a result of Pulkkinen et al. (2007a), who find that the ionospheric activity, ring current intensification, and geostationary field stretching during stormtime activations are relatively larger for weaker IEF

(xi) While described study concerns the response to pressure impulses, it was also find that a weaker driver leads to more effective energy coupling.

3.2. Association of Reconnection Enhancements by Solar Wind Dynamic Pressure Increases

3.2.1. Dayside Reconnection Enhancement Resulting from a Solar Wind Dynamic Pressure Increase

As noted Boudouridis et al. (2007), it has recently been established that, in addition to changes in the IMF, sudden solar wind dynamic pressure enhancements also have profound effects on the magnetosphere-ionosphere system (e.g., Boudouridis et al., 2003; Liou, 2006). These effects include widening of the auroral oval and closing of the polar cap (Lyons, 2000; Zesta et al., 2000a; Boudouridis et al., 2003, 2004a, 2005; Hubert et al., 2006), intensification of the aurora (Zhou and Tsurutani, 1999; Lyons, 2000; Zesta et al., 2000b; Boudouridis et al., 2003, 2004a; Zhou et al., 2003; Meurant et al., 2004; Kozlovsky et al., 2005), intensification of ionospheric and magnetospheric current systems (e.g., Zesta et al., 2000a; Shue and Kamide, 2001, and references therein), geosynchronous and inner magnetosphere energetic particle enhancements (Li et al., 1998, 2003; Boudouridis et al., 2003; Lee et al., 2004), modification of the magnetic field at geosynchronous with dipolarization-like response on the nightside during strongly southward IMF (Lee and Lyons, 2004), increase in ionospheric Joule heating (Palmroth et al., 2004a,b), and enhancements of ionospheric convection, the cross-polarcap potential, the coupling efficiency between the solar wind and the magnetosphere (Boudouridis et al., 2004a, 2004b, 2005; Ober et al., 2006), and the ring current asymmetry (Shi et al., 2005, 2006).

Boudouridis et al. (2007) outlined that one of the most intriguing effects of solar wind dynamic pressure fronts is the poleward expansion of the auroral oval and the closing of the polar cap observed over a wide range of Magnetic Local Times (MLTs), often with the exception of the near-noon region. The amount of closing depends on the prior background IMF B_Z conditions which can be viewed as 'preconditioning' of the magnetosphere (Boudouridis et al., 2003). Boudouridis et al. (2004a) have shown that the closing is greater under strong southward IMF orientation as compared to near-zero IMF B_Z conditions prior to the pressure-front impact. Perhaps, it also depends on the pressure-front characteristics, namely total (ΔP_{SW}) or relative ($\Delta P_{SW}/P_{SW}$) strength of the pressure increase, but the results are not yet conclusive (Boudouridis et al., 2005). The polar cap shrinking can range from a few degrees of magnetic latitude up to about 10° in some cases over certain MLT ranges.

Such a dramatic closing of the polar cap implies that an enhancement of magnetotail reconnection is induced by the pressure front (Boudouridis et al., 2003, 2004a; Milan et al., 2004; Hubert et al., 2006). As noted Boudouridis et al. (2007), the increased solar wind pressure compresses the magnetosphere leading to a burst of increased tail reconnection. In a case study of two consecutive shock impacts, Hubert et al. (2006) estimated the tail reconnection potential at 132 and 114 kV after each shock, respectively, as compared to undisturbed values of ~30 and ~20 kV, respectively. Milan et al. (2004) studied the response of the magnetotail to a sudden increase in solar wind dynamic pressure. They found that the compression led to a reduction in the open magnetic flux in the Northern Hemisphere from 0.5 to 0.2 GWb, down to 2.5% of the total hemispheric flux from a nominal value of 7–8%, corresponding to a tail reconnection potential of 150 kV. Similar observations of polar cap closing have also been made at Saturn, and the same interpretation of pressure-induced magnetic flux closure in the tail has been given to account for Saturn's auroral dynamics, with the additional complication of a rapidly rotating ionosphere (e.g., Badman et al., 2005; Cowley et al., 2005).

As mentioned above, the dayside poleward oval boundary is usually the only part of the auroral oval boundary that does not move poleward in response to an abrupt increase in solar wind pressure. This may reflect the dayside part of the oval being the least affected by the changes in the reconnection rate in the tail discussed above. Of all seven pressure enhancement cases studied by Boudouridis et al. (2003, 2004a, 2005) using Defense Meteorological Satellite Program (DMSP) precipitating particle measurements and/or POLAR Ultra-Violet Imager (UVI) images, only two showed some poleward expansion of the oval and closing of the polar cap near noon. Boudouridis et al. (2007) noted that noon closure would normally imply a dayside reconnection reduction, which is directly driven by IMF changes. For one of those two events, a simultaneous northward turning was observed, which could account for, at least, part of the near-noon oval closure. However, the second event had no changes in IMF. The explanation given by Boudouridis et al. (2004a) for the near-noon polar cap closing of this event was that a simultaneous increase in ionospheric convection resulting from the increase in solar wind pressure had sufficient magnitude to transport newly closed field lines from the flanks to the dayside in short timescales to produce the poleward expansion of the oval there. The sudden pressure enhancement for this event occurred after a long period of strongly southward IMF ($B_z < -12$ nT) which, according to Boudouridis et al. (2004a), is a favorable condition for a strong and widespread in MLT closing of the polar cap. For the rest of the events studied by Boudouridis et al. (2003, 2004a, 2005), dayside reconnection changes may have a bigger effect on the dayside poleward oval boundary motion than nightside reconnection changes. However, how exactly does the dayside oval respond to the impact of a solar wind pressure front? Is there an increase or decrease in the dayside reconnection rate? MHD simulations of the solar wind density effect on the energy transfer from the solar wind to the magnetosphere performed by Lopez et al. (2004) have shown that the dayside reconnection rate increases with increasing density and thus pressure, and the effect is more pronounced when the background IMF is strongly southward. Ober et al. (2006) conducted MHD simulations of the pressure pulse event on 10 January 1997 previously studied by Boudouridis et al. (2004b). They managed to reproduce the observed increase of the transpolar potential and attributed the variation entirely to an enhancement of dayside merging (even though Boudouridis et al. (2004b) attributed it mostly to nightside reconnection). Kozlovsky et al. (2005) studied the response of the dayside auroral oval to a sudden impulse in solar wind dynamic pressure using global ultraviolet auroral images from the Imager for Magnetopause-to-Aurora Global Exploration (IMAGE) satellite. They found that the middle of the dayside auroral oval moves poleward by about 3° in magnetic latitude, but this motion is due to redistribution of the luminosity within the oval and is not associated with a corresponding motion of the dayside poleward oval boundary. The analytical formulation of the magnetopause reconnection potential given by Siscoe et al. (2002) includes only a factor of $P_{SW}^{-1/6}$ for the solar wind pressure dependence arising from Chapman-Ferraro scaling and therefore predicts a weak decrease in dayside merging with pressure. Many solar wind-magnetosphere coupling functions have been introduced in the past: (see Newell et al. (2007) for a list), but only a few contain the solar wind dynamic pressure explicitly. While the solar wind velocity appears in these relations by itself or via the solar wind electric field, changes in solar wind density dominate the dynamic pressure enhancements in the previous studies of Boudouridis et al. (2003, 2004a, 2004b, 2005) including the events studied in Boudouridis et al. (2007).

As outlined Boudouridis et al. (2007), it is thus unclear from previous results, especially observationally, how the dayside reconnection rate reacts to a sudden enhancement in solar wind pressure, P_{SW}. Boudouridis et al. (2007) investigate the effect of solar wind dynamic pressure fronts on the dayside reconnection rate by looking at dayside ionospheric convection changes using Super Dual Auroral Radar Network (SuperDARN) observations during three solar wind pressure-front events. These events are three of the seven cases studied by Boudouridis et al. (2003, 2004a, 2005) for which good dayside SuperDARN coverage exists. In all cases, we observe a significant increase in the horizontal velocities coinciding with the time of the pressure-front impact. The ionospheric flow enhancements are concentrated mostly near the expected location of the cusp in accordance with the prevailing direction of the IMF B_Y component. Furthermore, for the two cases with southward IMF conditions throughout the pressure jump, the variations in the average flow magnitude follow the respective variations in solar wind pressure very closely, pointing to a definite association of the dayside convection, and by extension of the dayside reconnection rate, with the solar wind pressure.

According to Boudouridis et al. (2007), SuperDARN is a network of high frequency (HF) coherent scatter radars that can be used to monitor the plasma convection over the high-latitude ionosphere by receiving backscatter from small-scale plasma density irregularities in the ionospheric F region (Greenwald et al., 1995; Ruohoniemi and Baker, 1998). Boudouridis et al. (2007) used plasma flow vectors measured by nine Northern Hemisphere stations. Each radar scans at 16 different azimuthal settings separated by 3.3°. One full scan is completed every 2 min. The horizontal line-of-sight velocity data from each station are averaged to an altitude-adjusted corrected geomagnetic (AACGM) latitude-longitude grid (Baker and Wing, 1989). The grid cells are 1° in latitude (~111 km projected on the surface of the Earth) and as close as possible to 111 km in longitude. This means that there are fewer grid cells in each latitude as we move toward the pole but ensures that the grid elements have more or less the same size in both latitude and longitude (see Ruohoniemi and Baker (1998) for details and an illustration). Boudouridis et al. (2007) used series of SuperDARN flow snapshots to evaluate changes in dayside ionospheric convection after a sudden increase in solar wind dynamic pressure. They present three case studies of dayside convection enhancements after a sudden increase in solar wind dynamic pressure: 6 November 2000, 19 February 1999, and 18 February 1999. The first two pressure fronts occur under steady southward IMF, while the last one exhibits a rotation of the IMF B_Z component from very southward to very northward. Boudouridis et al. (2005) studied this event and showed solar wind and IMF data from the Advanced Composition Explorer (ACE) spacecraft located at (219, 15.7, –23) R_E in the undisturbed solar wind (not shown in the current paper). Before the pressure-front impact, the solar wind pressure was stable at ~4 nPa. At impact, the pressure rises quickly (within 10 min) to ~25 nPa for about 5 min and then drops down to ~10–15 nPa, still higher than before the front impact, for about 35 min. At this point, the pressure exhibits another peak, rising up to almost 20 nPa for ~20 min. During this whole period, IMF B_Z remained steady southward in the –(8–12) nT range, except for a brief (~5 min) excursion into positive territory at the time of the first pressure peak. Boudouridis et al. (2005) used DMSP data in their study of this event, which yield low time resolution of the observed convection response (once in every DMSP orbit, ~101 min), and thus they were only interested in the large-scale effect of the pressure enhancement, considering only low versus high pressure domains before and after the pressure front, respectively. Boudouridis et al. (2007) used 2-min resolution SuperDARN scans to monitor the dayside convection changes and study the details of the pressure variation effect focusing on the two individual pressure peaks. Since SuperDARN has high temporal resolution, Boudouridis et al. (2007) need more precise timings of the front arrival; and they show solar wind and IMF data from the Geotail spacecraft which observed the pressure front within the dawn magnetosheath at $X_{GSE} \approx -8$ R_E. The two main peaks of the pressure variation are evident in the Geotail data, at ~18:06 and ~18:48 UT, though with larger magnitude than seen by ACE due to the solar wind compression in the magnetosheath. The same holds for the IMF component magnitudes. Boudouridis et al. (2007) examine the response of the dayside convection to each pressure peak separately.

The second solar wind pressure-front event occurred on 19 February 1999. ACE, located at Geocentric Solar Ecliptic (GSE) (243, –0.7, 24.2) R_E, encountered the pressure front at around 11:47 UT. Using concurrent IMP8 spacecraft data inside the Earth's dawnside magnetosheath, Boudouridis et al. (2005) estimated a magnetopause impact time of ~12:30–12:35 UT. A more accurate estimate can result by propagating the ACE IMF data according to the technique of Weimer et al. (2003). This yields an arrival time of the solar wind pressure front at (17, 0, 0) R_E of ~12:32 UT. The solar wind pressure was originally at a low value, ~2–3 nPa, until ~12:25 UT. After an initial slow increase for ~5 min, it exhibits a sharp increase at 12:32 UT. After this point, it remained at ~6–10 nPa for about 2 hours. The IMF B_z component was steady southward at ~ –(6–7) nT before the pressure-front impact and began a slow decline in magnitude after, eventually turning northward at ~14:05 UT.

According to Boudouridis et al. (2007), unlike the previous two events for which the pressure jump occurred under steady southward IMF conditions, in the 18 February 1999 event, there is a simultaneous northward turning of the IMF. IMF B_Z data from three solar wind monitors (ACE, WIND, and IMP-8), presented by Boudouridis et al. (2003), show a remarkable agreement in the recording of this northward turning and the accompanying pressure front. ACE detected the changes in pressure and IMF while located at GSE (243.2, 0.6, 24.2) R_E. The pressure front reached ACE at ~09:51 UT. The same front was detected by IMP-8 located at –30 R_E downtail at ~1038 UT. On the basis of these measurements,

Boudouridis et al. (2003) estimated a magnetopause impact time at ~1030 UT. This is confirmed by the solar wind propagation technique of Weimer et al. (2003), which puts the front arrival at (17, 0, 0) R_E at 1028–1030 UT. Boudouridis et al. (2007) noted that it is not immediately clear what is the reason for the second flow peak as it occurs right after the return of the solar wind pressure to more moderate values (~10 nPa) and with no significant changes observed in the IMF B_Z and B_Y components. It seems to be the result of simultaneous flow enhancements in two polar regions, the 06–10 MLT range, where the original postfront enhancement was observed, and a new near-noon (11–12 MLT) patch. The initial response of the convection closely follows the increase in solar wind pressure with only ~4 min time difference.

Boudouridis et al. (2007) summarize obtained results as following:

(i) It was found the changes in the dayside ionospheric convection observed by the SuperDARN network and their association with changes in the solar wind dynamic pressure.

(ii) More specifically, it was shown that abrupt increases in the solar wind pressure lead to corresponding strong enhancements in the magnitude of the dayside flow speeds.

(iii) It was presented three case studies, two under steady southward IMF and one with a simultaneous northward turning of the IMF.

(iv) In the first southward IMF event, a solar wind pressure front on 6 November 2000 consists of two characteristic peaks in its pressure structure. Each of these peaks leads to a respective increase in the ionospheric flow speeds near the expected location of the cusp. The variation in convection flows matches closely the variation in dynamic pressure with only ~4 min time difference between the maximum flow response and the pressure peak in each case.

(v) In the second southward IMF event, a broad region of high pressure on 19 February 1999 leads to elevated convection around noon MLT for the duration of the high-pressure regime, again with a 4-min time difference to reach the maximum convection response.

(vi) In the last event, a high pressure front and a simultaneous northward turning on 18 February 1999 result in increased flows in the morning-to-noon sector despite the northward turning of the IMF. These persist while the IMF is northward, and the flow pattern is consistent with that expected under these IMF conditions. The time for maximum ionospheric response is again about 4 min after the arrival of the pressure front.

(vii) The 4-min time to reach the maximum response reflects the time it takes the high-latitude ionosphere to respond fully to changes in the solar wind and/or the IMF. Ruohoniemi and Greenwald (1998) studied the response of the high-latitude convection to a sudden southward turning of the IMF. They found that the convection starts to respond almost immediately with characteristic timescales of 2–4 min for a full-pattern reconfiguration. Murr and Hughes (2001) studied the reconfiguration timescales of ionospheric convection for several events with a southward turning of the IMF using ground magnetometers. They found that the onset of changes in convection occurs at all local times within 1–2 min, and that full reconfiguration near noon occurs in ~5 min, which they attributed to the time needed for several bounces back and forth to the ionosphere of Alfvén waves launched at the magnetopause after the southward turning of the IMF. Described here results are consistent with these values, suggesting that a nearly immediate ionospheric response that maximizes within 4 min also applies to the response of the high-latitude convection to sudden increases in solar wind dynamic pressure.

(viii) In all of considered cases, the convection changes observed are near the location where the cusp is expected to be according to the existing IMF conditions (Crooker, 1979). In the first event, the highest variation of ionospheric flows occurs in the 09–11 MLT sector and above ~70° MLAT for both solar wind pressure peaks.

(ix) For southward IMF with positive IMF B_Y, this is the predicted position of the cusp for the Northern Hemisphere. In the second event, the IMF is again southward, but the IMF B_Y exhibits small-scale variations, from ~–4 nT to ~+4 nT, before the pressure-front impact. This result leads to anticipate a cusp location near noon MLT, which is where the increased flows are observed after the front arrival.

(x) In the last case, IMF B_Z is strongly negative, and IMF B_Y is strongly positive before the pressure front, which shifts the cusp much toward the dawn. That is where the largest increases in ionospheric flows are observed after the increase in pressure.

(xi) When eventually the convection reconfigures ~15 min later to accommodate the simultaneous IMF northward turning, enhanced equatorward and dawnward flows appear near noon, consistent with the new IMF configuration.

(xii) The above observations suggest that solar wind pressure enhancements increase the rate of dayside reconnection, which then manifests itself as enhanced dayside ionospheric convection. The enhanced convection seems to follow closely the variations of the solar wind pressure, at least for southward IMF conditions, implying a direct correlation between the solar wind dynamic pressure and the dayside reconnection rate.

(xiii) For considered three cases, the solar wind velocity slightly decreases after the pressure-front impact, either gradually (events 1 and 2) or abruptly (event 3). So at least for these cases, the above correlation is between the solar wind density and the dayside reconnection rate.

(xiv) This of course does not exclude a possible role of the solar wind velocity in enhancing dayside reconnection through an increase in pressure. However, this case would be more difficult to evaluate because of a simultaneous increase in the solar wind electric field and the difficulty in separating the two effects.

(xv) Transient phenomena, such as Traveling Convection Vortices (TCVs), cannot be responsible for the stationary and sustained convection enhancements observed in considered events. Case and statistical studies alike (Murr et al., 2002; Zesta et al., 2002) have shown that these structures propagate away from noon at speeds of 3–11 km/s. Even at the low end of this range, they will move far away from noon during the 20-min span of the Super-DARN observations.

(xvi) Boudouridis et al. (2004a) suggested that solar wind pressure fronts induce enhanced magnetotail reconnection on the basis of the nightside closing of the polar cap. Therefore solar wind pressure fronts have a considerable effect on the global electrodynamics of the magnetosphere. Enhanced dayside and nightside reconnection compete for determining the size and shape of the polar cap and may possibly both contribute to changes of the cross-polar-cap potential (Cowley and Lockwood, 1992; Boudouridis et al., 2005; Milan et al., 2007).

3.2.2. Nightside Flow Enhancement Associated with Solar Wind Dynamic Pressure Driven Reconnection

As outlined Boudouridis et al. (2008), over the past few years the prominent role of solar wind dynamic pressure in enhancing dayside and nightside reconnection and driving enhanced ionospheric convection has been documented by both ground and spaceborn instruments. SuperDARN observations show that solar wind pressure fronts induce significantly enhanced ionospheric convection in the dayside ionosphere. In parallel, Defense Meteorological Satellite Program (DMSP) precipitating particle measurements and POLAR Ultra-Violet Imager (UVI) images have demonstrated that sudden solar wind pressure increases also significantly reduce the size of the polar cap, especially on the nightside, suggesting an enhancement of magnetotail reconnection. MHD models of the interaction of the magnetosphere with solar wind pressure fronts have reproduced the enhancement of dayside reconnection but have failed so far to account for the observed closing of the polar cap on the nightside and the suggested magnetotail reconnection increase. It was used SuperDARN measurements of ionospheric convection within the nightside polar ionosphere, including near the magnetic separatrix, to evaluate the strength of the observed nightside reconnection enhancement after an abrupt increase in solar wind dynamic pressure on 6 November 2000 and compare it with similar observations on the dayside. Boudouridis et al. (2008) show that an enhancement of nightside convection occurs after a sudden increase in solar wind pressure, delayed by about 40 min compared with the observed dayside convection enhancement. The nightside enhanced flows are observed crossing the open-closed boundary determined by POLAR UVI data, indicating an enhancement of tail reconnection that is possibly due to the pressure increase and is, in addition to the tail reconnection, associated with the more immediate closing of the polar cap.

3.2.3. The Relation between Transpolar Potential and Reconnection Rates during Sudden Enhancement of Solar Wind Dynamic Pressure: OpenGGCM-CTIM Results

Connor et al. (2014) investigate how solar wind energy is deposited into the magnetosphere-ionosphere system during sudden enhancements of solar wind dynamic pressure (Psw), using the coupled Open Geospace General Circulation Model–Coupled Thermosphere Ionosphere Model (OpenGGCM-CTIM) 3-D global magnetosphere-ionosphere-thermosphere model. Connor et al. (2014) simulate three unique events of solar wind pressure enhancements that occurred during negative, near-zero, and positive IMF B_Z. Then, it was examined the behavior of the dayside and nightside reconnection rates and quantify their respective contributions to cross polar cap potential (CPCP), a proxy of ionospheric plasma convection strength. The modeled CPCP increases after a Psw enhancement in all three cases, which agrees well with observations from the Defense Meteorological Satellite Program spacecraft and predictions from the assimilative mapping of ionospheric electrodynamics technique. In the OpenGGCM-CTIM model, dayside reconnection increases within 9–13 min of the pressure impact, while nightside reconnection intensifies about 13–25 min after the pressure increase. As the strong Psw compresses the dayside magnetosheath and, subsequently, the magnetotail, their magnetic fields intensify and activate stronger antiparallel reconnection on the dayside magnetopause first and near the central plasma sheet second. For southward IMF, dayside reconnection contributes to the CPCP enhancement 2–4 times more than nightside reconnection. For northward IMF, the dayside contribution weakens, and nightside reconnection contributes more to the CPCP enhancement. It was found that high-latitude magnetopause reconnection during northward IMF produces sunward ionospheric plasma convection, which decreases the typical dawn-to-dusk ionosphere electric field. This results in a weaker dayside reconnection contribution to the CPCP during northward IMF.

3.3. Coupling of SW Dynamic Pressure Events with Magnetosphere at High Latitudes

3.3.1. Global Auroral Response to a Solar Wind Pressure Pulse

According to Brittnacher et al. (2000), a global intensification of the aurora was observed by the Ultraviolet Imager on the NASA Polar spacecraft in conjunction with the arrival of the sheath from a solar coronal mass ejection. The aurora was first observed to brighten on the dayside and then the intensification progressed rapidly toward the nightside. During this time the IMP-8 spacecraft in the solar wind recorded a 35-minute period of increased solar wind dynamic pressure. A small substorm (or, possibly pseudobreakup) occurred within a minute of the arrival of the auroral intensification on the nightside in conjunction with a second peak in the dynamic pressure. Brittnacher et al. (2000) propose that the intensification of the aurora can be explained on the basis of the compression of the magnetopause and the generation of hydromagnetic waves by the rapid increase in the solar wind dynamic pressure. It is also evident that the substorm was triggered by waves, generated by a second rise in the dynamic pressure, that propagated to flux tubes connected to the premidnight auroral region. Obtained results are illustrated by Figures 3.3.1 – 3.3.2.

3.3.2. Coupling of Magnetic Activity in the Polar Caps with Solar Wind Dynamic Pressure and Geoeffective Electric Field

Response of the polar cap (PC) magnetic activity to sudden changes in solar wind dynamic pressure is analyzed by Troshichev et al. (2007) with use of the northern and southern polar cap indices for 1998–2002 and corresponding solar wind parameters: geoeffective interplanetary electric field (Em) and the solar wind dynamic pressure (P_{SW}), reduced to the magnetopause. As noted Troshichev et al. (2007), the method of the polar cap (PC) index derivation was formulated about 20 years ago (Troshichev and Andrezen, 1985; Troshichev et al., 1988), and from that time the main principles of the method remain invariable. The PC index is by definition proportional to the intensity of the polar cap magnetic disturbance and calibrated with respect to the geoeffective interplanetary electric field (Kan and Lee, 1979):

$$\mathrm{Em} = V_{SW}B_T\sin^2(\theta/2) = V_{SW}\left(B_Z^2 + B_Y^2\right)^{1/2}\sin^2(\theta/2) \qquad (3.3.1)$$

fixed in the solar wind near the magnetosphere. The following concepts were adopted as principal guidelines while elaborating the unified technique for calculation of the PC index in the northern (PCN) and southern (PCS) polar caps (Troshichev et al., 2006): (1) The PC index value should be representative, as closely as possible, of the geoeffective interplanetary electric field Em, and (2) the values of the PC index in the northern and southern polar caps should be similar under conditions of insignificant values of the IMF B_Y component. As outlined Troshichev et al. (2007), the first concept deserves a special comment. Right from the start, the PC index was intended to characterize the geoeffective interplanetary electric field proceeding from ground measurements of the polar cap geomagnetic variations. These variations are generated by the ionospheric electric currents, which are determined, in turn, by the ionospheric conductance and the electric field applied to the ionosphere. Under ordinary conditions the polar cap ionospheric conductance is regulated by the intensity of the solar UV light. Hence its daily and seasonal variations can easily be taken into account, except during extraordinary enhancements of the ionospheric conductance in response to bombardment of the polar caps by energetic particles related to solar flares or substorms. The electric field in the near-pole region is determined by the cross-polar cap voltage, which under ordinary conditions, is mainly controlled by the interplanetary electric field but can also be affected by other agents. The solar wind dynamic pressure pulses were suggested as one of the main contributors to the PC index, other than the interplanetary electric field (see, e.g., Lukianova, 2003; Lee et al., 2004; Liou et al., 2004).

As noted Troshichev et al. (2007), it is well known that the dayside magnetopause shifts earthward and outward under conditions of the solar wind pressure increase. Sawtooth oscillations in the magnetic field and energetic particle flux directly driven by solar wind dynamic pressure enhancements were observed at geosynchronous orbit (Erlandson et al., 1991; Lee et al., 2004). The compressional waves propagated down to the polar ionosphere, where they produced transient ionospheric flows at latitudes equatorward of the convection reversal boundary (Sibeck and Croley, 1991), an increase of the cusp and low-latitude boundary layer (LLBL) area projection (Newell and Meng, 1994), an occurrence of the aurora near the local noon and propagation of the waves along the auroral oval (Zhou and Tsurutani, 1999; Zhou et al., 2003; Meurant et al., 2004), an increase of the auroral oval width, and intensification of auroral particle precipitation (Boudouridis et al., 2003). As for influence of the solar wind pressure pulses on substorm development the results turned out to be contradictory. It was noted that shock-induced aurora, starting in the noon sector, may eventually trigger the development of the auroral substorm on the nightside (Meurant et al., 2004) and the latitudes of the substorm onset are in clear anticorrelation with the solar wind dynamic pressure (Gérard et al., 2004). However, Liou et al. (2003, 2004) state that shock compression is not likely to trigger substorms but enhances magnetospheric currents and auroral particle precipitation, contributing equally with IMF B_Z to the westward auroral electrojet. It is significant that all the above results concern the phenomena in the auroral oval and in the auroral magnetosphere, respectively. There were only a few observations that indicated that polar cap magnetic activity responds to the solar wind pressure pulses as well. According to Erlandson et al. (1991), compressions and rarefactions were recorded by the Viking satellite in the magnetosphere above the polar cap, and Motoba et al. (2003) indicate that geomagnetic field fluctuations, coherent with pressure variations, appeared on a global scale at stations located from polar to equatorial regions. Influence of pressure pulses on the PC index was found by Lukianova and Troshichev (2002). After a more detailed examination of the relationship between the interplanetary electric field, the solar wind pressure pulses, and the PC and AE indices, Lukianova (2003) came to the conclusion that the PC index directly responds to an enhancement in the solar wind dynamic pressure P_{SW} and that the influence of P_{SW} pulses can be as large as influence of southward IMF. The subsequent studies (Lee et al., 2004; Huang, 2005) suggested a key role of the interplanetary electric field in the relationship between the PC index and P_{SW}. Lee et al. (2004) indicated that even a modest dynamic pressure enhancement can result in significant changes in the magnetosphere if the IMF stays strongly southward for a long interval. Huang (2005) concluded that the PC index responds to pressure pulses only under conditions of the southward IMF, and the rate of response depends on the southward IMF magnitude. All these conclusions were based on case study results. The superposed epoch analysis on the PC index for 43 pressure impulses was presented by Liou et al. (2004). Liou et al. (2004) concluded that the shock-induced magnetic disturbances at high latitudes are associated with DP-2 current enhancements but did not take into account the effect of southward IMF in these enhancements.

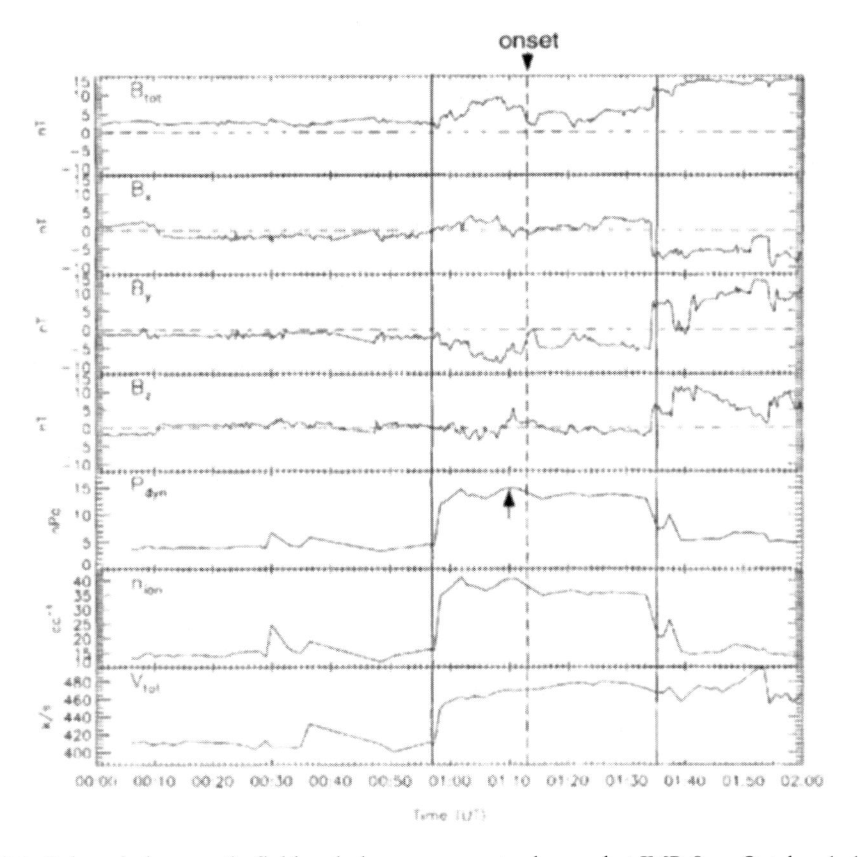

Figure 3.3.1. Solar wind magnetic field and plasma moments observed at IMP-8 on October 1, 1997. The solid lines mark the approximate beginning and end of the pressure pulse. The arrow indicates the second rise in the dynamic pressure that results in auroral intensifications that progress toward the nightside and a substorm. The dashed line marks the onset of the small substorm at 0113 UT. According to Brittnacher et al. (2000). Permission from COSPAR.

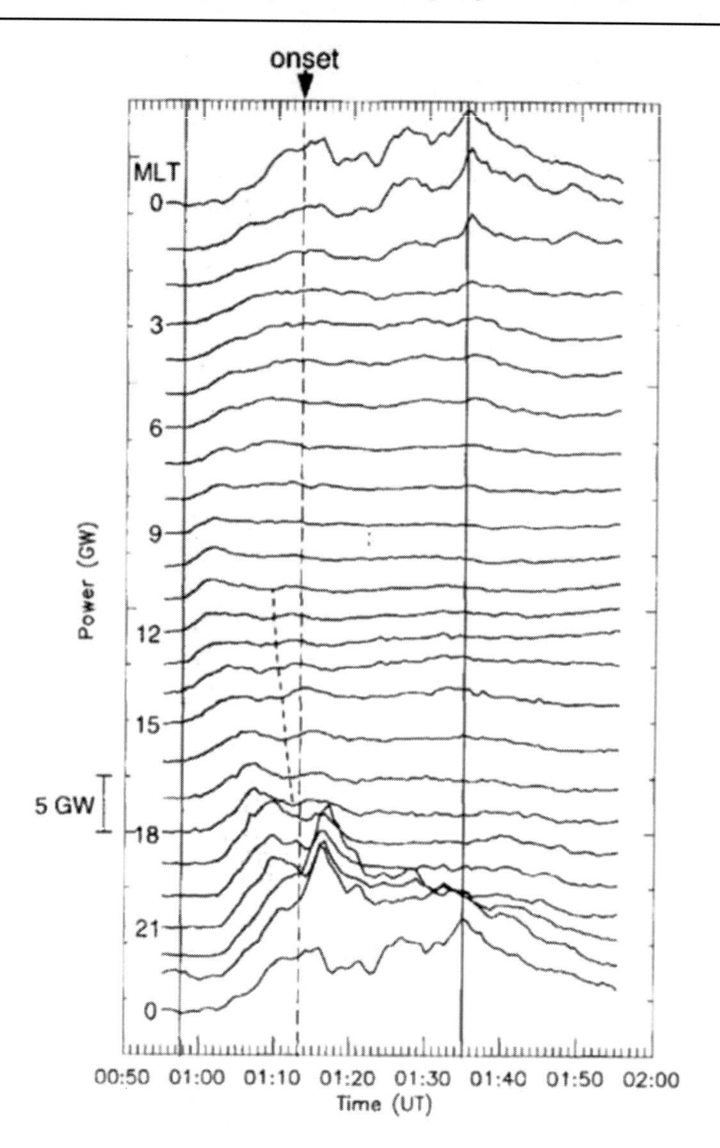

Figure 3.3.2. Stackplot of auroral power for one hour MLT segments of the aurora. The solid lines mark the approximate beginning and end of the pressure pulse. The dashed line marks the onset of the substorm at 0113 UT. The gray area highlights the initial ranping up of auroral power following the arrival of the pressure pulse. The short-dashed line between the 11 and 18 MLT line plots shows the leading edge of the auroral intensification due to the second peak in the dynamic pressure as it propagates along the duskside toward the premidnight region. According to Brittnacher et al. (2000). Permission from COSPAR.

According to Troshichev et al. (2007), the unified PC index is determined as a function of polar cap magnetic activity calibrated according to the geoeffective interplanetary electric field Em (Troshichev et al., 2006). The calculation procedure has been regulated such that the close relationship between the unified PC indices and Em was ensured. In this approach the discrepancy between the calculated PC indices and the 'ideal' PC index, equal in magnitude to the Em (and measured in mV/m), could be considered as a result of the influence of factors, other than Em, on the PC index value. Hence, if the solar wind dynamic pressure influence on the PC index takes place, it can be estimated quantitatively by this means. In the analysis were used the 5-min unified PCN/PCS indices for 1998–2002. The corresponding solar wind parameters measured on board ACE satellite were estimated at the magnetopause to calculate the geoeffective interplanetary electric field Em according to Eq. 3.3.1 and the solar wind dynamic pressure as

$$P_{SW} = nm(V_{SW})^2, \qquad (3.3.2)$$

where n is proton density and m is mass. The time of the signal passage from the ACE location to magnetopause was taken into account with allowance for the real solar wind speed for each particular event. As a result, in the analysis there are compared the behavior of PC index with appropriate changes of the solar wind characteristics Em and P_{SW} in the vicinity of the magnetopause. The impact time of the pressure shock could be precisely estimated with sudden brightenings of the dayside aurora and sudden magnetic impulses (Liou et al., 2003); however, such accuracy seems to be excessive for statistical treatment of the 5-min averages of the PC index. The additional delay time $\tau_D \sim 20$ min typical of the Em signal passage from the magnetopause to the polar cap and its transformation into the magnetic activity (Troshichev and Andrezen, 1985) is ignored in the analysis.

As outlined Troshichev et al. (2007), only sudden pressure pulses, starting against the background of a steady quiet pressure level during the preceding 6 hours, were taken for the analysis. At first, the initial and final levels of the pressure jump are determined, the pressure level being calculated as the mean for 6 hours before and 3 hours after the sudden pressure jump. Then the difference between the initial and final levels is fixed as a pressure gradient δP_{SW}. Thereafter, the moment of maximum derivative $\delta P_{SW}/\delta t$ is identified as a pressure pulse onset. This specific moment is taken as a key ('zero') date for the epoch superposition method, with the characteristics Em, PC, and real P_{SW} being related to the key date. When there are examined the average pressure gradient growth rate, Troshichev et al. (2007) calculated the value $\delta P_{SW}/\delta t$, where δt is the time of the given pressure gradient growth. This time for the sudden pressure jumps is usually limited to 5–10 min. Excluded from this study are the extremely large pressures rising slowly over a period of some hours. We shall classify the PC indices for summer and winter conditions (PC summer and PC winter), instead of PCN and PCS indices, since the results of Janzhura et al. (2007) revealed a regular discrepancy in the response of the summer and winter PC indices to Em influence in spite of the general consistency of the unified PCN and PCS indices.

As noted Troshichev et al. (2007), the sharp increase of the plasma dynamic pressure usually takes place on fronts of the interplanetary shocks related to the coronal mass ejections. The interplanetary shocks are accompanied by strong oscillations of the IMF and appropriate variations of the geoeffective electric field. It was investigate 62 pressure pulses with $\delta P_{SW} > 4$ nPa, taking place in 1998–2002. Troshichev et al. (2007) begin with the relationship between averaged P_{SW}, Em, and PC quantities under varying restrictions imposed on the interplanetary electric field: $1 > Em > 0$ mV/m, $3 > Em > 1$ mV/m, and $Em > 3$mV/m for practically arbitrary values of the pressure jump (pressure gradient $\delta P_{SW} > 2$ nPa and derivative $\delta P_{SW}/\delta t > 0.04$ nPa/min). The average electric field Em demonstrates only an insignificant and short-lived rise in connection with the pressure jump (zero moment), which can be related to the subsequent insignificant rises (~0.4 mV/m) in both PC summer and PC winter indices, occurring with a 20-min delay. After that the electric field reduces below pre-jump level, and both PC indices diminish as well. While increasing the average Em to 2.5 mV/m, the PC indices increase to 2.5–3 mV/m. Growth of summer and winter PC indices start within a few minutes after the pressure jump, almost simultaneously with the Em increase, but maximal magnetic activity in the polar caps is reached about 15–30 min after the maximum in Em. When the interplanetary electric field increases over 3 mV/m, the average PC indices enhance up to 3.5–5 mV/m. Again, both PC indices start to grow at 'zero moment' but reach their maximal values 15–30 min after the maximum of Em. Troshichev et al. (2007) summarize main obtained results as following: (i) The geoeffective interplanetary electric field Em determines the behavior of the PC index, the value of which is directly influenced by the Em increase and reaches maximum with a delay time of about 15–30 min relative to the maximum Em. (ii) The solar wind pressure growth rate (i.e., jump power $\delta P_{SW}/\delta t$) appears to be the second most important factor for the PC index increase: the value $\delta P_{SW} = 1$ nPa being approximately equivalent to action of $\delta Em = 0.33$ mV/m and PC indices starting to grow within a few minutes after the pressure jump. (iii) The pressure growth rate seems to be the sole factor for the PC index increase under conditions of northward IMF B_Z and for the PC index decrease after a negative dynamic pressure drop. (iv) The summer PC index increases more quickly and reaches greater values than the winter PC index under the influence of the pressure jump.

3.3.3. Geomagnetic Response to Solar Wind Dynamic Pressure Impulse Events at High-Latitude Conjugate Points

As note Kim et al. (2013), it is commonly assumed that geomagnetic activity is symmetrical between interhemispheric conjugate locations. However, in many cases, such an assumption proved to be wrong. Especially in high-latitude regions where the magnetosphere and the ionosphere are coupled in a more complex and dynamic fashion, asymmetrical features in geomagnetic phenomena are often observed. Kim et al. (2013) present investigations of geomagnetic responses to sudden change in solar wind pressure to examine interhemispheric conjugate behavior of magnetic field variations, which have rarely been made mainly due to the difficulty of facilitating conjugate-point measurements. In this study, using magnetometer data from three conjugate stations in Greenland and Antarctica, solar wind pressure impulse events (>5 nPa in <16 min) and their geomagnetic responses, typically seen as magnetic impulse events, have been examined. Obtained results suggest that asymmetry in ground response patterns between the conjugate locations often shows little correlation with IMF orientation, season, and ionospheric conductivity, indicating that much more complex mechanism might be involved in creating interhemispheric conjugate behavior.

3.3.4. Day-Night Coupling by a Localized Flow Channel Visualized by Polar Cap Patch Propagation

Nishimura et al. (2014) present unique coordinated observations of the dayside auroral oval, polar cap, and nightside auroral oval by three all-sky imagers, two SuperDARN radars, and Defense Meteorological Satellite Program (DMSP)-17. This data set revealed that a dayside poleward moving auroral form (PMAF) evolved into a polar cap airglow patch that propagated across the polar cap and then nightside poleward boundary intensifications (PBIs). SuperDARN observations detected fast antisunward flows associated with the PMAF, and the DMSP satellite, whose conjunction occurred within a few minutes after the PMAF initiation, measured enhanced low-latitude boundary layer precipitation and enhanced plasma density with a strong antisunward flow burst. The polar cap patch was spatially and temporally coincident with a localized antisunward flow channel. The propagation across the polar cap and the subsequent PBIs suggests that the flow channel originated from dayside reconnection and then reached the nightside open-closed boundary, triggering localized nightside reconnection and flow bursts within the plasma sheet.

3.4. Response to Solar Wind Dynamic Pressure Pulses on the Geosynchronous Orbit

3.4.1. Response of the Magnetic Field in the Geosynchronous Orbit to Solar Wind Dynamic Pressure Pulses

As noted Wang et al. (2007), the solar wind dynamic pressure P_{SW} is one of the main factors that affect magnetosphere. The change of P_{SW} could be in the form of an increase, an decrease, or a pulse (roughly resembles a step function, where the solar wind dynamic pressure increases/decreases followed by a decrease/increase within a limited time). Changes in P_{SW} produce various responses in the magnetosphere-ionosphere system, including a change in the geosynchronous magnetic field. Early study by Ogilvie et al. (1968) and Burlaga and Ogilvie (1969) showed that the pulses in the geomagnetic field were cased by density enhancements rather than shock pairs. Rufenach et al. (1992) showed that, for quiet conditions, the average value of the magnetospheric magnetic field measured by the GOES satellites at geosynchronous orbit generally increases when the hourly-averaged solar wind dynamic pressures increases. Borodkova et al. (1995) and Sibeck et al. (1996) found a direct correspondence between dayside magnetospheric magnetic field changes and step function decreases and increases in the solar wind P_{SW}. Using hourly-averaged data, Wing and Sibeck (1997) studied the correlation of the geosynchronous magnetic field with the IMF B_Z component and the solar wind P_{SW}, and showed that increases in the solar wind P_{SW} and the geosynchronous magnetic field are correlated near noon but are anticorrelated on the nightside. Lee and Lyons (2004) found that the dayside geosynchronous response to a P_{SW} enhancement when the IMF is southward is mostly compressional, while on the nightside the response is similar to dipolarization. Borodkova et al. (2005) demonstrated that sharp increases (decreases) in the solar wind P_{SW} always result in increases (decreases) in the geosynchronous magnetic field strength with the maximum amplitude near noon. Further study by Borodkova et al. (2006a,b) indicated that the amplitude of magnetic field response in geosynchronous orbit strongly depend on the location of the observer relative to noon meridian, the value of pressure before disturbance and the change in the amplitude of the pressure. Sanny et al. (2002) showed that the variability of the geosynchronous magnetic field strength near local noon was strongly affected by changes in the P_{SW} but independent of the

IMF B_Z. Generally speaking, disturbances in the magnetospheric magnetic field must be associated with corresponding changes in the large-scale magnetospheric current systems, i.e., the Chapman-Ferraro, ring, cross-tail and Birkeland currents. The amplitudes of the responses of each current system to solar wind changes should be different and vary with local time. A fast response of the geosynchronous magnetic field to a large and sharp solar wind P_{SW} change is mainly the manifestation of Chapman-Ferraro currents. As outlined Wang et al. (2007), solar wind dynamic pressure pulses are relatively small structures in the solar wind, and it is of interest to study their geoeffectiveness. In Wang et al. (2007) there are investigate correlations of the P_{SW} pulses and their geosynchronous magnetic field responses. Since the magnetospheric response to sudden changes in P_{SW} strongly depends on whether the magnetosphere is quiet or disturbed (Wing and Sibeck, 1997; Lukianova, 2003), Wang et al. (2007) focus on events observed under quiet magnetospheric conditions, i.e., $Dst > -50$ nT. In an attempt to understand the observations, Wang et al. (2007) finally used a newly developed PPMLR-MHD code, which models the global behavior of the solar wind-magnetosphere-ionosphere system, to model the effects of pressure pulses on the magnetosphere and compare these model results with the observations.

While the ACE spacecraft is providing continuous solar wind data at the Earth-Sun L1 point since its launch in 1998, the WIND spacecraft also observes solar wind near Earth. Since the solar wind structures may undergo evolution from L1 point to the region near Earth, it is preferable to compare the observations from both spacecraft. The instruments onboard the WIND spacecraft provide solar wind data starting in 1994. Wang et al. (2007) used data from times when the WIND spacecraft is located more than 30 r_E upstream of Earth. The solar wind data from WIND with 90-s time resolution were obtained from ftp://space.mit.edu/pub/plasma/wind and data from ACE with 64-s time resolution were obtained from http://www.srl.caltech.edu for the period 1998–2005. Wang et al. (2007) used geosynchronous observations with 1-min resolution from GOES-8, 9, 10, 11 and 12 magnetospheric magnetic field observations (http://goes.ngdc.noaa.gov) which cover the time range from 1998 to 2005. They select large the solar wind dynamic pressure P_{SW} pulses in the solar wind according to the following criteria: P_{SW} should have a increase more than 1 nPa in less than 10 min in the start and then have the similar large and sharp decrease in the end; the whole duration should be less than 70 min and the variation within the structure should be less than 60%. Then Wang et al. (2007) associate these pulses with disturbances in the geosynchronous magnetic field. Wang et al. (2007) first used an auto-search computer program to search for potential dynamic pulses in the ACE and WIND data sets, and then visually inspected each case. They identified more than 1000 P_{SW} pulses in the solar wind data between 1998 and 2005. However, only 111 of these pulses were observed in geomagnetically quiet times with $Dst > -50$ nT with geosynchronous magnetic field responses. Most of these solar dynamic pressure pulses (~74%) were observed both by ACE and WIND, while almost all of the geosynchronous magnetic field responses were observed by two or three GOES spacecraft at different local times. Wang et al. (2007) perform a statistical survey on these 111 cases and their geosynchronous magnetic field responses. For 93.6% of the events, the relative change of solar wind dynamic pressure $\delta P_{SW}/P_{SW}$ ranges from 0 to 6. There are classified these cases into four categories based on the values of $\delta P_{SW}/P_{SW}$ (0–1, 1–3, 3–4, 4–6), so in each category the value of $\delta P_{SW}/P_{SW}$ does not change significantly.

In order to understand the above described observations, Wang et al. (2007) used the recently developed PPMLR-MHD code (Hu et al., 2007), which models the global behavior of the solar wind-magnetosphere- ionosphere system, to conduct numerical experiments. This PPMLR-MHD code has been extended to its parallel version and used to study the interaction of the interplanetary shocks with the magnetosphere (Wang et al., 2005, 2006; Huang et al., 2007). Wang et al. (2007) performed a statistical survey of solar wind dynamic pressure P_{SW} pulses and geosynchronous magnetic fields observed between 1998 and 2005. In geomagnetic quiet times, with $Dst > -50$ nT, they found 111 solar wind dynamic pressure pulses which are clearly associated with geosynchronous magnetic field responses. These responses are often observed by two or three GOES spacecraft at different local times.

The main points of this study are summarized as follows:

(i) The magnitudes of the responses of the geosynchronous magnetic field δB_Z to solar wind dynamic pressure pulses δP_{SW} peak near the noon meridian, which is in consistent with previous work focusing on the responses to the sharp and large solar wind dynamic pressure changes. However, the relative magnitudes of the responses of the geosynchronous magnetic field $\delta B_Z/\langle B_Z\rangle$ depend only weakly on local time.

(ii) The magnitudes of δB_Z are proportional to $\langle B_Z\rangle$ under the condition of roughly constant solar wind dynamic pressure. When the solar wind $\delta P_{SW}/P_{SW}$ increases, the rate of geosynchronous magnetic field variation increases correspondingly. The direction of the interplanetary magnetic field does not affect the response significantly.

(iii) The amplitude of δB_Z and the amplitude of $\langle B_Z\rangle$ are correlated with a correlation coefficient of ~0.7. These results imply that the magnitude of the geosynchronous magnetic field response could be determined by $\langle B_Z\rangle$.

(iv) The global MHD simulation results have reproduced the main characters of the observations.

3.4.2. Large and Sharp Changes of Solar Wind Dynamic Pressure and Disturbances of the Magnetospheric Magnetic Field at Geosynchronous Orbit Caused by These Variations

The large and sharp changes of solar wind dynamic pressure, found from the INTERBALL-1 satellite and WIND spacecraft data, are compared by Borodkova et al. (2006a) with simultaneous magnetic field disturbances in the magnetosphere measured by geosynchronous GOES-8, GOES-9, and GOES-10 satellites. For this purpose, about 200 events in the solar wind, associated with sharp changes of the dynamic pressure, were selected from the INTERBALL-1 satellite data obtained during 1996–1999. The large and sharp changes of the solar wind dynamic pressure were shown to result in rapid variations of the magnetic field strength in the outer magnetosphere, the increase (drop) of the solar wind dynamic pressure always lead to an increase (drop) of the geosynchronous magnetic field magnitude. The value of the geomagnetic field variation strongly depends on the local time of the observation point, reaching a maximum value near the noon meridian. It is shown that the direction of the B_Z component of the IMF has virtually no effect on the geomagnetic field variation because of a sharp jump of pressure. The time shift between an event in the solar wind and its response in the magnetosphere at a geosynchronous orbit essentially depends on the inclination of the front of a solar wind disturbance to the Sun-Earth line.

3.4.3. Geosynchronous Magnetic Field Response to the Large and Fast Solar Wind Dynamic Pressure Change

Borodkova et al. (2008) present a comparison of large and sharp solar wind dynamic pressure changes, observed by several spacecraft, with fast disturbances in the magnetospheric magnetic field measured by the GOES-8, 9 and 10 geosynchronous satellites. Almost 400 solar wind pressure changes in the period 1996 2003 were selected for this study. Using the large statistics Borodkova et al. (2008) confirmed that increases (decreases) in the dynamic pressure always results in increases (decreases) in the magnitude of geosynchronous B_Z component. The amplitude of the geosynchronous B_Z response strongly depends on the location of observer relative to the noon meridian, from the value of solar wind pressure before the disturbance arriving and firstly from the amplitude of the pressure change. It is shown that 3D MHD numerical simulation of geosynchronous B_Z response performed for different pressure changes inside 9-15 LT range is consistent well with the experimental results. Obtained results are illustrated by Figures 3.4.1 – 3.4.5.

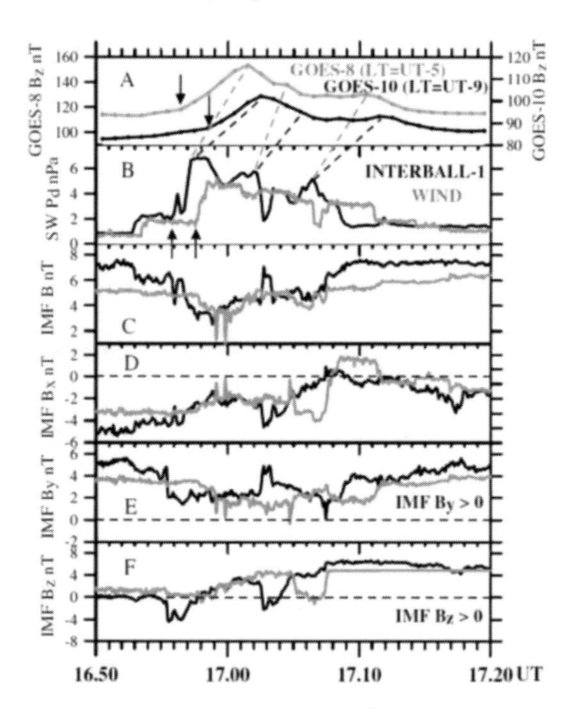

Figure 3.4.1. Example of geosynchronous dayside magnetic field response to a solar wind pressure change on June 6, 2000. Panels, from top to bottom, show simultaneous measurements of geosynchronous B_z-component at GOES-8 and GOES-10, solar wind pressure P_d, magnetic field module B, B_x, B_y, B_z at INTERBALL-1 (black color) and WIND (grey color). Arrows mark the start of the solar wind pressure P_d and geosynchronous B_z changes. Dashed lines indicate peak-to-peak correspondence between solar wind pressure behavior and geosynchronous B_z response. According to Borodkova et al. (2008). Permission from COSPAR.

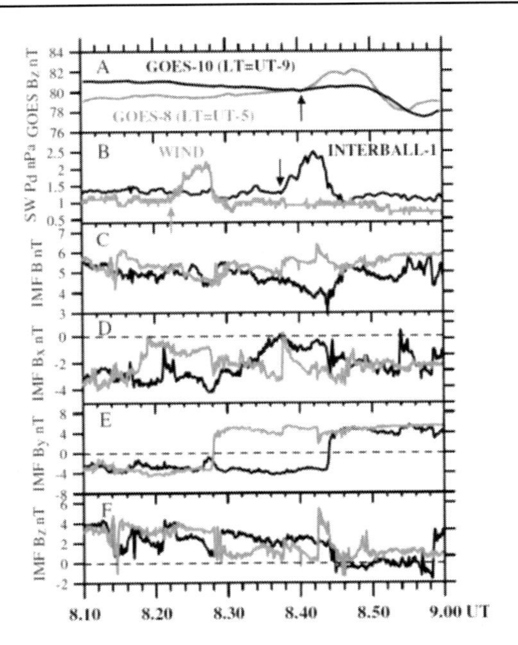

Figure 3.4.2. Example of geosynchronous nightside magnetic field response to solar wind pressure pulse on April 25, 2000. Designations are the same as for Figure 3.4.1. According to Borodkova et al. (2008). Permission from COSPAR.

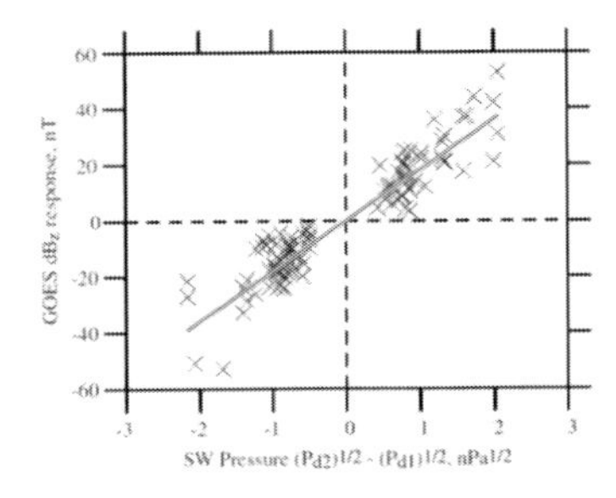

Figure 3.4.3. The relationship between the amplitude of geosynchronous B_z response and difference of the square roots of the solar wind pressure value across the discontinuity. The solid line is the least squares fitting of all data. According to Borodkova et al. (2008). Permission from COSPAR.

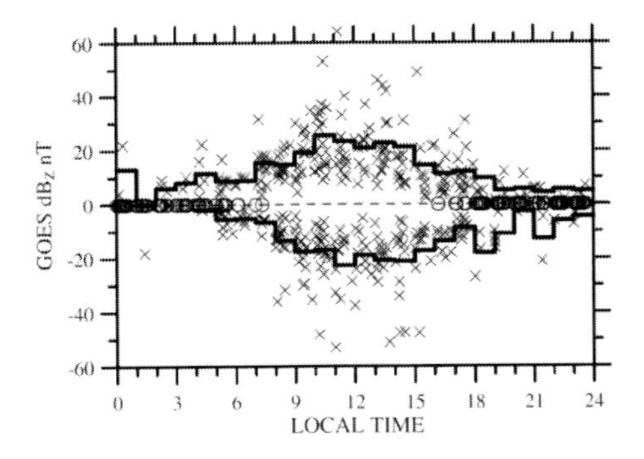

Figure 3.4.4. The dependence of geosynchronous B_z response to solar wind pressure change on local time. Solid lines indicate the histogram of averaged geosynchronous dB_z responses in each 1-h local time bin for pressure increase and decrease events. Circles along x-axis denote solar wind pressure change events without geosynchronous B_z response. According to Borodkova et al. (2008). Permission from COSPAR.

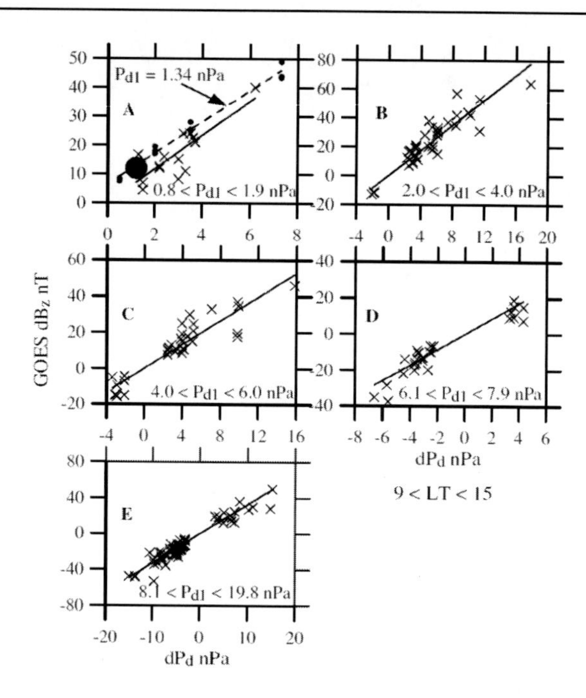

Figure 3.4.5. Relationship between the amplitude of geosynchronous B_z response dB_z and solar wind pressure change dP_d (crosses) for events in the 9–15 LT range. Each panel presents data related to different narrow ranges of upstream pressure P_{dl} values indicated on the panel. Solid lines represent best fits to the data. Black points and dashed line on the panel A are results of numerical simulation. Small circle marks event with the same initial pressure P_{dl} = 1.34 nPa as was taken for numerical simulation. According to Borodkova et al. (2008). Permission from COSPAR.

3.4.4. Effect of Large and Sharp Changes of Solar Wind Dynamic Pressure on the Earth's Magnetosphere: Analysis of Several Events

The magnetosphere and ionosphere response to arrival of large changes of the solar wind dynamic pressure with sharp fronts to the Earth is considered by Borodkova (2010). It is shown that, under an effect of an impulse of solar wind pressure, the magnetic field at a geosynchronous orbit changes: it grows with increasing solar wind pressure and decreases, when the solar wind pressure drops. Energetic particle fluxes also change: on the dayside of the magnetosphere the fluxes grow with arrival of an impulse of solar wind dynamic pressure, and on the nightside the response of energetic particle fluxes depends on the IMF direction. Under the condition of negative B_Z-component of the IMF on the nightside of the magnetosphere, injections of energetic electron fluxes can be observed. It is shown, that large and fast increase of solar wind pressure, accompanied by a weakly negative B_Z -component of the IMF, can result in particless' precipitation on the dayside of the auroral oval, and in the development of a pseudobreakup or substorm on the nightside of the oval. The auroral oval dynamics shows that after passage of an impulse of solar wind dynamic pressure the auroral activity weakens. In other words, the impulse of solar wind pressure in the presence of weakly negative IMF can not only cause the pseudobreakup/substorm development, but control this development as well.

3.4.5. ULF Waves Excited by Negative/Positive Solar Wind Dynamic Pressure Impulses at Geosynchronous Orbit

As outlined Zhang et al. (2010), when a solar wind dynamic pressure impulse impinges on the magnetosphere, ultra-low-frequency (ULF) waves can be excited in the magnetosphere and the solar wind energy can be transported from interplanetary space into the inner magnetosphere. Zhang et al. (2010) have systematically studied ULF waves excited at geosynchronous orbit by both positive and negative solar wind dynamic pressure pulses. It is identified 270 ULF events excited by positive solar wind dynamic pressure pulses and 254 ULF events excited by negative pulses from 1 January 2001 to 31 March 2009. It is found that the poloidal and toroidal waves excited by positive and negative pressure pulses oscillate in a similar manner of phase near 06:00 local time (LT) and 18:00 LT, but in antiphase near 12:00 LT and 0:00 LT. Furthermore, it is shown that excited ULF oscillations are in general stronger around local noon than those in the dawn and dusk flanks. It is demonstrated that disturbances induced by negative impulses are weaker than those by positive ones, and the poloidal wave amplitudes are stronger than the toroidal wave amplitudes both in positive and negative events. The potential impact of these excited waves on energetic electrons at geosynchronous orbit has also been discussed.

3.4.6. Sudden Impulses at Geosynchronous Orbit and at Ground

As outlined Villante and Piersanti (2011), sudden impulses of the magnetospheric and geomagnetic field are caused by sudden increases in the dynamic pressure of the solar wind, generally associated with the Earth's arrival of interplanetary shock waves or discontinuities. Villante and Piersanti (2011) compared the observed field jumps, at geosynchronous orbit and at ground, with those predicted by a theoretical model (T04, Tsyganenko and Sitnov, 2005) for transitions between two steady states representations of the magnetosphere under different solar wind conditions. The geosynchronous response (mostly along the northward component, B_Z) in the dayside hemisphere is basically consistent with the magnetic field jump expected for changes of the magnetopause current alone. A similar conclusion holds for the positive changes observed in the nightside hemisphere; by contrast, the competing contributions of several current systems (from the magnetopause, cross-tail current, ring current, field aligned currents) determine in this time sector a large variety of small amplitude and negative responses that cannot be univocally interpreted. A different situation emerges in ground observations, at low latitudes. Here the observed responses along the northward component, H, in the dawn sector are consistent with changes of the magnetopause current; by contrast, in the entire postnoon hemisphere, they are greater than expected. The H variations are also typically accompanied by a significant change of the eastward component, D. These aspects reveal additional ionospheric contributions which determine a negative change of the D component through the entire day and a remarkable enhancement (≈50%) of the H component in the postnoon sector. The comparison between observations and model representations in several cases allows to discriminate between the roles of the current systems which compete in determining the field variations. Occasionally, SI manifestations are accompanied by trains of ULF fluctuations ($f \approx$ 1-100 mHz); Villante and Piersanti (2011) present a case in which the experimental observations suggest evidence for a cavity/waveguide oscillation at f ≈ 3.2 mHz.

3.4.7. Geostationary Magnetic Field Response to Solar Wind Pressure Variations: Time Delay and Local Time Variation

The relationship between solar wind dynamic pressure changes and geosynchronous magnetic field response is studied by Jackel et al. (2012) using 15 years of OMNI2 and GOES data at 1-minute resolution. Significant magnetospheric response to solar wind-forcing is found to be most frequent near noon (30% of all intervals), and virtually absent on the night-side. The strongest response occurs when IMF B_Z is strongly northward and the effect of reducing IMF B_Z is most pronounced in the dusk sector. Approximately 25% of dayside B_Z variance for related intervals can be attributed to direct response from solar wind dynamic pressure forcing. Time lag between changes in the solar wind at the bow shock nose and similar fluctuations in the magnetosphere at geosynchronous orbit (6.6 R_E) is typically 2 to 4 minutes, with responses occurring first in the post-noon sector and approximately 2 minutes later near dawn. The OMNI2 HRO time-shifting algorithm appears to be quite effective, with a slight (2 minute) systematic increase in lag and no increased scatter for the most distant upstream solar wind satellite location.

3.4.8. Analysis of Trends between Solar Wind Velocity and Energetic Electron Fluxes at Geostationary Orbit Using the Reverse Arrangement Test

As outlined Aryan et al. (2013), a correlation between solar wind velocity (V_{SW}) and energetic electron fluxes (EEF) at the geosynchronous orbit was first identified more than 30 years ago. However, recent studies have shown that the relation between V_{SW} and EEF is considerably more complex than was previously suggested. The application of process identification technique to the evolution of electron fluxes in the range 1.8 − 3.5 MeV has also revealed peculiarities in the relation between V_{SW} and EEF at the geosynchronous orbit. It has been revealed that for a constant solar wind density, EEF increase with V_{SW} until a saturation velocity is reached. Beyond the saturation velocity, an increase in V_{SW} is statistically not accompanied with EEF enhancement. The present study is devoted to the investigation of saturation velocity and its dependency upon solar wind density using the reverse arrangement test. In general, the results indicate that saturation velocity increases as solar wind density decreases. This implies that solar wind density plays an important role in defining the relationship between V_{SW} and EEF at the geosynchronous orbit.

3.4.9. Joint Responses of Geosynchronous Magnetic Field and Relativistic Electrons to External Changes in Solar Wind Dynamic Pressure and Interplanetary Magnetic Field

Li et al. (2013) studied statistically the joint responses of magnetic field and relativistic (>0.5 MeV) electrons at geosynchronous orbit to 201 interplanetary perturbations during

6 years from 2003 (solar maximum) to 2008 (solar minimum). The statistical results indicate that during geomagnetically quiet times ($H_{SYM} > -30$ nT, and $AE < 200$ nT), ~47.3% changes in the geosynchronous magnetic field and relativistic electron fluxes are caused by the combined actions of the enhancement of solar wind dynamic pressure (P_d) and the southward turning of IMF ($\Delta P_d > 0.4$ nPa and IMF $B_Z < 0$ nT), and only ~18.4% changes are due to single dynamic pressure increase ($\Delta Pd > 0.4$ nPa, but $B_Z > 0$ nT), and ~34.3% changes are due to single southward turning of IMF (IMF $B_Z < 0$ nT, but $|\Delta Pd| < 0.4$ nPa). Although the responses of magnetic field and relativistic electrons to the southward turning of IMF are weaker than their responses to the dynamic pressure increase, the southward turning of IMF can cause significant dawn-dusk asymmetric perturbations that the magnetic field and relativistic electron fluxes increase on the dawnside (LT ~ 00:00–12:00) but decrease on the duskside (LT ~ 13:00–23:00) during the quiet times. Furthermore, the variation of relativistic electron fluxes is adiabatically controlled by the magnitude and elevation angle changes of magnetic field during the single IMF southward turnings. However, the variation of relativistic electron fluxes is independent of the change in magnetic field in some magnetospheric compression regions during the solar wind dynamic pressure enhancements (including the single pressure increases and the combined external perturbations), indicating that nonadiabatic dynamic processes of relativistic electrons occur there.

3.4.10. The Analysis of Electron Fluxes at Geosynchronous Orbit Employing a NARMAX Approach

According to Boynton et al. (2013), the methodology based on the Error Reduction Ratio (ERR) determines the causal relationship between the input and output for a wide class of nonlinear systems. In the present study, ERR is used to identify the most important solar wind parameters, which control the fluxes of energetic electrons at geosynchronous orbit. The results show that for lower energies, the fluxes are indeed controlled by the solar wind velocity, as was assumed before. For the lowest energy range studied here (24.1 keV), the solar wind velocity of the current day is the most important control parameter for the current day's electron flux. As the energy increases, the solar wind velocity of the previous day becomes the most important factor. For the higher energy electrons (around 1 MeV), the solar wind velocity registered 2 days in the past is the most important controlling parameter. Such a dependence can, perhaps, be explained by either local acceleration processes due to the interaction with plasma waves or by radial diffusion if lower energy electrons possess higher mobility. However, in the case of even higher energies (2.0 MeV), the solar wind density replaces the velocity as the key control parameter. Such a dependence could be a result of solar wind density influence on the dynamics of various waves and pulsations that affect acceleration and loss of relativistic electrons. The study also shows that statistically the variations of daily high energy electron fluxes show little dependence on the daily averaged B_Z, daily time duration of the southward IMF, and daily integral inline image (where B_S is the southward component of IMF).

3.5. Low Latitude Geomagnetic Response to Solar Wind Pressure Variations

3.5.1. Some Aspects of the Low Latitude Geomagnetic Response under Different Solar Wind Conditions

Villante et al. (2003) review some aspects of low latitudes ($L \leq 2$) geomagnetic field variations associated with magnetospheric pulsations as well as with continuous and impulsive variations of the solar wind pressure.

3.5.2. On the Nature of Response of Dayside Equatorial Geomagnetic H-Field to Sudden Magnetospheric Compressions

As noted Sastri et al. (2006), a step-like increase in the solar wind ram pressure causes a sudden compression of the magnetosphere which manifests as a sudden commencement (SC) in the geomagnetic field. SC manifests in two basic forms in the geomagnetic H-field of the dayside dip equatorial region, namely, as a positive impulse termed SC(+), and as a positive impulse preceded by a sharp negative impulse termed, SC(−, +) or SC*. Unlike at high latitudes, the factor(s) that determine this bi-modal response of the equatorial geomagnetic H-field to sudden magnetospheric compressions are not known. As an exploratory step in identifying the causative factor(s), Sastri et al. (2006) have performed a statistical study of the characteristics of the daytime (06-18 IST) SCs recorded at the equatorial station, Kodaikanal (10.25N, 77.5E, dipole latitude 0.6N) and the low latitude station, Alibag (18.6N, 72.9E, dipole latitude 9.5N) over the period 1957–2002. A total of 304 SCs have been analyzed out of which 99 (32.6 per cent) were of SC* type. We find that the average value of the amplitude of the positive main impulse (MI) in H-component of SC* is higher than that of the conventional SC(+). This difference in the MI amplitude is statistically significant and persists even when the analysis is restricted to the interval of SC* occurrence, namely, 0730–1730 IST. Such a behavior is not seen in the amplitude of the MI at Alibag, away from the influence of the equatorial electrojet. It is found for the first time that a linear relationship exists between the amplitudes of the preliminary reverse impulse (PRI) and MI in SC* events. The study by Sastri et al. (2006) of the H-field response at dayside dip equatorial stations of the Circum-pan Pacific Magnetometer Network (CPMN) to sudden magnetospheric compressions under different IMF orientation (northward and southward), showed that the impact of an impulse in solar wind dynamic pressure under southward IMF is associated with SC* in both the events considered. On the other hand, under northward IMF the dynamic pressure increase consistently led to SC(+) in the two events studied. The statistical and case study results imply that, when compared to SC(+), the contribution of ionospheric currents of polar origin gain prominence when SC* is excited at dayside dip equator, and the orientation of IMF B_Z may be one of the (if not the only one) deterministic factors underlying the bi-modal response of the dayside equatorial H-field to a sudden increase in solar wind dynamic pressure.

3.6. Global Geomagnetic and Auroral Response to Solar Wind Dynamic Pressure Variations

3.6.1. Global Geomagnetic Response to a Sharp Compression of the Magnetosphere and IMF Variations on October 29, 2003

Solovyev et al. (2004) study the response of the ionosphere and magnetosphere to a sudden commencement (SC) on October 29, 2003, at 06:11 UT. It is shown that the geomagnetic response had the form of two successive stages. In the first ~5 min after the SC, a strong intensification of a two-vortex current system of the DP2 type was observed in latitudes ~ 67°-65°, with variations of Delta H -4000 nT (+700 nT). At the same time, energetic electrons were injected without dispersion to geosynchronous orbits simultaneously in the sectors ~ 16, ~ 04, and ~ 07 MLT. In the subsequent 5-15 min, a new intensification of the western electrojet took place in all time sectors at latitudes ~70°. Around midnight, this electrojet was extended in the poleward direction up to the polar cap latitudes ($\Phi' \approx$ 75°-83°). It had an unusually high velocity of extension (up to ~ 5.0 km/s) and was accompanied by typical dispersionless substorm injections, but only at meridians ~ 04 and 07 MLT. From comparing the development of electrojets with the data of satellite observations in the solar wind and magnetosphere, Solovyev et al. (2004) suggest that ~ 3-5 min after the SC onset a dipolization of the magnetic field at the geosynchronous orbit occurred. It was connected with the decay of the current flowing across the magnetotail. The subsequent extension of the region of current decay into the tail up to 150 R_E proceeded with a velocity of ≥ 1000 km/s, which exceeds the known velocities of such an extension by a factor of ~ 5.

3.6.2. Global Geomagnetic and Auroral Response to Variations in the Solar Wind Dynamic Pressure on April 1, 1997

The geomagnetic and auroral response to the variations in the solar wind dynamic pressure (P_d) are investigated by Solovyev et al. (2006) in the periods of positive values of the IMF B_Z component. It is shown that the growth of P_d results in the intensification of luminosity along the auroral oval and in the poleward expansion of the poleward boundary of luminosity (PBL) in the nightside part of the oval by ~7° in latitude at a velocity of ~0.5 km/s and is accompanied by an enhancement of the DP2-type current system. A decrease in P_d, accompanied by an abrupt reversal of the IMF B_Y polarity from positive to negative, results in an enhancement of the westward electrojet and in a poleward shift of PBL and electrojet center. The conclusion of Solovyev et al. (2006) has been made that the available three types of auroral response to P_d variations differ in the azimuthal velocity of the luminosity region or particle precipitation along the auroral oval: V_1 ~ 30–40 km/s, V_2 ~ 10 km/s, and V_3 ~ 1 km/s.

3.6.3. Response of Dayside Auroras to Abrupt Increases in the Solar Wind Dynamic Pressure at Positive and Negative Polarity of the IMF B_Z Component

The optical observations on Heiss Island ($\Phi' = 75.0°$) have been used by Vorobjev et al. (2009) to study the characteristics of auroras in the near-noon MLT sector after abrupt increases in the solar wind dynamic pressure at negative and positive polarity of the IMF B_Z component. It has been found out that the 427.8 and 557.7 nm emission intensities considerably increased at $B_Z < 0$ both equatorward of the dayside red luminosity band and within this band. The value of the emission intensities at a red luminosity maximum (I6300/I5577 ~ 0.5) indicates that energetic electron precipitation is of the magnetospheric origin. At $B_Z > 0$, fluxes of harder ($E > 1$ keV) precipitating electrons were superimposed on the soft spectrum of precipitating particles in the equatorial part of the red luminosity band. This red band part was hypothetically caused by the low-latitude boundary layer (LLBL) on closed lines of the geomagnetic field, the estimated thickness of which is ~3 R_E. The 557.7 nm emission intensity increased during 3-5 min after SC/SI and was accompanied by the displacement of the red band equatorward boundary toward lower latitudes. The displacement value was ~150-200 km when the dynamic pressure abruptly increased by a factor of 3-5. After SC/SI, the 630.0 nm emission intensity continued increasing during 16-18 min. It is assumed that the time of an increase in the red line intensity corresponds to the time of saturation of the magnetospheric boundary layers with magnetosheath particles after an abrupt increase in their density.

3.6.4. Simulations of Observed Auroral Brightening Caused by Solar Wind Dynamic Pressure Enhancements under Different IMF Conditions

In Peng et al. (2011) solar wind dynamic pressure (P_{dyn}) enhancements have been observed to cause large-scale auroral brightening. The mechanism for this kind of auroral brightening is still a topic of current space research. Using the global piecewise parabolic method Lagrangian remap (PPMLR)-MHD simulation model, Peng et al. (2011) investigate three auroral brightening events caused by dynamic pressure enhancement under different IMF conditions: (1) $B_Z < 0$ and $B_y > 0$ on 11 August 2000, (2) $B_Z < 0$ and $B_Y < 0$ on 8 May 2001, and (3) $B_Z \geq 0$ on 21 January 2005. Peng et al. (2011) show that the auroral location depends on the IMF conditions. Under southward IMF conditions, when B_Y is negative, the duskside aurora is located more equatorward at around 70° magnetic latitude (MLAT) for all magnetic local times; when B_Y is positive, the duskside aurora can even reach beyond 80° MLAT. A smaller and more localized response is seen when the IMF B_Z is nearly zero or northward, as shown in previous studies. The simulation results of Peng et al. (2011) are consistent with these observations, indicating that the observed aurora activities could be caused by solar wind dynamic pressure enhancements. The simulation results suggest that the enhancement of P_{dyn} can increase the ionospheric transpolar potential and the corresponding field-aligned currents, leading to the observed auroral brightening.

3.7. Magnetosphere - Solar Wind Dynamic Pressure Coupling during Magnetic Storms

3.7.1. Modeling Magnetospheric Current Response to Solar Wind Dynamic Pressure Enhancements during Magnetic Storms: Methodology and Results of the 25 September 1998 Peak Main Phase Case

Shi et al. (2008a) present a methodology for using the modular Tsyganenko storm magnetic field model (TS04) as a tool to investigate the response of magnetospheric currents to solar wind dynamic pressure enhancements during magnetic storms. It was demonstrated the technique by examining the contribution of each model current to the observed dawn-dusk asymmetric ground H perturbation during a peak storm main phase event. Shi et al. (2008a) add present pressure terms to the parameterizations of several individual model currents and fit them to the observed low-latitude and midlatitude H perturbations. It was found that the asymmetric H perturbation for this case is primarily due to a net field-aligned current (FAC) system, with upward FACs peaking in the evening and downward FACs peaking in the morning, and the equatorial portion of the partial ring current (PRC). This net FAC system includes the region 1 and region 2 FACs and the closure FAC of the PRC, with the largest contribution from the PRC closure FACs. The model results show that in this main phase case, the PRC played a more important role in causing the dawn-dusk asymmetric H perturbation than did the other currents owing to its strength and asymmetry. The model results do not show significant contributions from the symmetric ring current and tail current, but the magnetopause current gives a significant positive perturbation at all MLTs.

3.7.2. Modeling Magnetospheric Current Response to Solar Wind Dynamic Pressure Enhancements during Magnetic Storms: Application to Different Storm Phases

Two patterns of ground H perturbation as response to pressure enhancements during different phases of storms are found by Shi et al. (2008b). One is polarity asymmetry under southward IMF B_Z, e.g., 25 September 1998, 29 May 2003, and 10 January 1997 events; the other is positive H perturbation everywhere under northward IMF B_Z, e.g., 7 November 2000 event. The polarity asymmetry can be further divided into two categories: dawn-dusk asymmetry with negative H perturbation at dusk extending through noon to late morning (25 September 1998 event) and positive H perturbation elsewhere, and day-night asymmetry with a smaller region of negative H perturbation extending less far toward dusk (29 May 2003 and 10 January 1997 events). The net field-aligned current (FAC) system from the R1, R2, and the partial ring current (PRC) closure FAC, the equatorial portion of the PRC, and Chapman-Ferraro (CF) current are responsible for the polarity asymmetry. The stronger effect of the R1 is primarily responsible for the negative H perturbation at noon and positive at midnight. The PRC strength at the onset of pressure enhancements determines how far the negative H perturbation at noon extends toward dusk and which pattern is most likely. A stronger (weaker) PRC extends negative H perturbation more (less) toward dusk and thus increases the likelihood of dawn-dusk (day-night) asymmetry. Positive H perturbation everywhere results from dominant contribution of the CF and lack of significant preexisting PRC and R1 and R2.

3.8. Magnetospheric Vortices Associated with Solar Wind Pressure Enhancements

3.8.1. Plasma Flow Vortices in the Tail Plasma Sheet Associated with Solar Wind Pressure Enhancement

According to Tian et al. (2010), a series of earthward-moving (~140 km/s) plasma flow vortices with anticlockwise (when viewed from above the ecliptic plane) rotation was detected in the dawnside tail plasma sheet between 12:55 and 14:00 UT on 6 July 2003. These flow vortices were observed under the condition of northward IMF with an enhanced solar wind dynamic pressure. Analyzing the plasma and magnetic field data from the Cluster spacecraft and using the Grad-Shafranov streamline reconstruction technique, Tian et al. (2010) show that the vortex-like plasma structures have a very similar shape: a V_X component dominant in the dawnside, while a distinct V_Y component appears in the duskside, and each structure has a size of about $1.8R_E \times 0.68R_E$, approximately in the xy plane of GSM coordinates. It is found that the vortices contain both magnetosphere-originated hot (N ~ 0.1 cm^{-3}, E > 3 keV) and magnetosheath-originated denser and colder (N > 0.2 cm^{-3}, E < 1 keV) populations on the closed field lines. The vortices involve fast earthward flows ($V_X > 200$ km/s) of mainly sheath-originated plasmas. Tian et al. (2010) suggest that these observed plasma flow vortices are generated inside the magnetotail during the prolonged and intensified compression of the magnetosphere by the enhanced solar wind dynamic pressure.

3.8.2. Solar Wind Pressure Pulse-Driven Magnetospheric Vortices and Their Global Consequences

Shi et al. (2014) report the in situ observation of a plasma vortex induced by a solar wind dynamic pressure enhancement in the nightside plasma sheet using multipoint measurements from THEMIS satellites. The vortex has a scale of 5–10 R_E and propagates several R_E downtail, expanding while propagating. The features of the vortex are consistent with the prediction of the Sibeck (1990) model, and the vortex can penetrate deep (~8 R_E) in the dawn-dusk direction and couple to field line oscillations. Global MHD simulations are carried out, and it is found that the simulation and observations are consistent with each other. Data from THEMIS ground magnetometer stations indicate a poleward propagating vortex in the ionosphere, with a rotational sense consistent with the existence of the vortex observed in the magnetotail.

3.9. Geomagnetic Pulsations and ULF Associated with Solar Wind Pressure Changes

3.9.1. Bursts of Geomagnetic Pulsations in the Frequency Range 0.2–5 Hz Excited by Large Changes of the Solar Wind Pressure

Parkhomov et al. (2010) present the results of studying the magnetospheres's response to sharp changes of the solar wind flow (pressure) based on observations of variations of the ions flux of the solar wind onboard the Inreball-1 satellite and of geomagnetic pulsations (the data of two mid-latitude observatories and one auroral observatory are used). It is demonstrated that, when changes of flow runs into the magnetosphere, in some cases short (duration ~ < 5 min) bursts of geomagnetic pulsations are excited in the frequency range Δf~ 0.2-5 Hz. The bursts of two types are observed: noise bursts without frequency changes and wide-band ones with changing frequency during the burst. A comparison is made of various properties of these bursts generated by pressure changes at constant velocity of the solar wind and by pressure changes on the fronts of interplanetary shock waves at different directions of the vertical component of the IMF.

3.9.2. Multiple Responses of Magnetotail to the Enhancement and Fluctuation of Solar Wind Dynamic Pressure and the Southward Turning of IMF

According to Li et al. (2011), during the interval from 06:15 to 07:30 UT on 24 August 2005, the Chinese Tan-Ce 1 (TC1) satellite observed the multiple responses of the near-Earth magnetotail to the combined changes in solar wind dynamic pressure and IMF. The magnetotail was highly compressed by a strong interplanetary shock because of the dynamic pressure enhancement (~15 nPa), and the large shrinkage of magnetotail made a northern lobe and plasma mantle move inward to the position of the inbound TC1 that was initially in the plasma sheet. Meanwhile, the dynamic pressure fluctuations (~0.5–3 nPa) behind the shock drove the quasi-periodic oscillations of the magnetopause, lobe-mantle boundary, and geomagnetic field at the same frequencies: one dominant frequency was around 3 mHz and the other was around 5 mHz. The quasi-periodic oscillations of the lobe-mantle boundary caused the alternate entries of TC1 into the northern lobe and the plasma mantle. In contrast to a single squeezed or deformed magnetotail by a solar wind discontinuity moving tailward, the compressed and oscillating magnetotail can better indicate the dynamic evolution of magnetotail when solar wind dynamic pressure increases and fluctuates remarkably, and the near-Earth magnetotail is quite sensitive even to some small changes in the solar wind dynamic pressure when it is highly compressed. Furthermore, it is found that a considerable amount of oxygen ions (O^+) appeared in the lobe after the southward turning of IMF.

3.9.3. Specific Features of Daytime Long-Period Pulsations Observed during the Solar Wind Impulse against a Background of the Substorm of August 1, 1998

Long-period geomagnetic pulsations in the (1.7–6.7) mHz frequency range at 18.25–18.48 UT on August 1, 1998, caused by several successive sudden changes in the solar wind (SW) dynamic pressure, are studied by Klibanova et al. (2014) against a background of substorm intensification. The data of the ground stations, which were near local noon (the CANOPUS Canadian network) and on the nightside (the auroral stations in Yakutia and at the IMAGE network), and the INTERBALL-1, ACE, WIND, and GOES 8, extra magnetospheric satellites are used. The effect of the SW plasma and IMF parameters, SW inhomogeneity front inclination, and geomagnetic activity on the pulsation propagation and polarization direction and amplitude is discussed. The properties of pulsations, recorded before the substorm, correspond to the pulsation excitation by the inhomogeneity front incident on the magnetopause during the magnetically quiet period: pulsations propagate from the contact point onto the nightside when the amplitude increases and the polarization sense of rotation is opposite on the dawn and dusk sides. Substorm intensification results in the propagation direction reversal and in a more complex behavior of the pulsation amplitude and polarization on the dayside.

3.9.4. Solar Wind Driving of Magnetospheric ULF Waves: Field Line Resonances Driven by Dynamic Pressure Fluctuations

As noted Claudepierre et al. (2010), several observational studies suggest that solar wind dynamic pressure fluctuations can drive magnetospheric ultralow-frequency (ULF) waves on the dayside. To investigate this causal relationship, Claudepierre et al. (2010) present results from Lyon-Fedder-Mobarry (LFM) global, three-dimensional MHD simulations of the solar wind–magnetosphere interaction. These simulations are driven with synthetic solar wind input conditions where idealized ULF dynamic pressure fluctuations are embedded in the upstream solar wind. In three of the simulations, a monochromatic, sinusoidal ULF oscillation is introduced into the solar wind dynamic pressure time series. In the fourth simulation, a continuum of ULF fluctuations over the 0–50 mHz frequency band is introduced into the solar wind dynamic pressure time series. In this numerical experiment, the idealized solar wind input conditions allow to study only the effect of a fluctuating solar wind dynamic pressure, while holding all of the other solar wind driving parameters constant. It is shown that the monochromatic solar wind dynamic pressure fluctuations drive toroidal mode field line resonances (FLRs) on the dayside at locations where the upstream driving frequency matches a local field line eigenfrequency. In addition, it is shown that the continuum of upstream solar wind dynamic pressure fluctuations drives a continuous spectrum of toroidal mode FLRs on the dayside. The characteristics of the simulated FLRs agree well with FLR observations, including a phase reversal radially across a peak in wave power, a change in the sense of polarization across the noon meridian, and a net flux of energy into the ionosphere.

3.10. Influence of SW Dynamic Pressure Fluctuations on Neutral Atoms, Secondary Rarefaction Waves, Cavity Modes, and IEF Penetration into Geomagnetosphere

3.10.1. Energetic Neutral Atom Response to Solar Wind Dynamic Pressure Enhancements

Lee at al. (2007a) have investigated the response of the ring current to solar wind dynamic pressure (P_{dyn}) enhancement impacts on the magnetosphere by using energetic neutral atom (ENA) images obtained by the High-Energy Neutral Atom (HENA) imager on board the IMAGE spacecraft. As noted Lee at al. (2007a), abrupt enhancements of P_{dyn} when impacting the magnetosphere lead to disturbances of the magnetosphere-ionosphere system in various ways. These include ground magnetic disturbances (e.g., Russell et al., 1994, and references therein), a geosynchronous magnetic field response (e.g., Wing et al., 2002; Lee and Lyons, 2004, and references therein), a geosynchronous energetic particle disturbance (e.g., Lee et al., 2005), a response throughout the magnetotail (e.g., Nakai et al., 1991; Kawano et al., 1992; Fairfield and Jones, 1996; Collier et al., 1998; Ostapenko and Maltsev, 1998; Kim et al., 2004), an auroral disturbance (e.g., Lyons et al., 2000, 2005; Zesta et al., 2000b; Chua et al., 2001; Boudouridis et al., 2003), and a polar cap convection change (e.g., Lukianova, 2003; Boudouridis et al., 2003). Some of the typical compression effects are as following. First, an obvious effect by an enhanced P_{SW} when impacting the magnetosphere is an increase of the ground geomagnetic H component at low latitudes due to enhanced magnetopause current density (e.g., Russell et al., 1994, and references therein). Also, the magnetospheric magnetic field should be compressed, and geosynchronous spacecraft observations can easily confirm the magnetic compression effect (e.g., Wing et al., 2002; Lee and Lyons, 2004, and references therein). Auroral brightening takes place in regions other than near-midnight where substorm onset brightening usually occurs, and it sometimes appears to be relatively global, covering both the dayside and nightside (e.g., Lyons et al., 2005; Chua et al., 2001).

As outlined Lee at al. (2007a), another compression feature is that the charged particles trapped within the magnetosphere are accelerated adiabatically. Local measurements of energetic charged particles, for example, at geosynchronous orbit, often reveal particle flux increases as a result of the compression (e.g., Li et al., 2003). However, they sometimes show flux decreases or little change as well. Lee et al. (2005) suggested the spatial profiles of the background adiabatic particle distributions determine whether or not locally measured fluxes show an increase. Another important effect related to the P_{dyn} enhancement is substorm triggering under certain conditions. The P_{dyn} trigger has been discussed by many researchers (e.g., Heppner, 1955; Schieldge and Siscoe, 1970; Kawasaki et al., 1971; Burch, 1972a; Kokubun et al., 1977; Zhou and Tsurutani, 2001; Liou et al., 2003; Lyons et al., 2005; Lee et al., 2005). Using the Polar UV image data, Zhou and Tsurutani (2001) tested substorm triggering possibility by P_{SW} enhancement events associated with an interplanetary shock and reported that substorm triggering was observed for ~44% of the studied events, for which the IMF B_Z was 'strongly southward' for >1.5 hours. However it should be noted that most of their events actually correspond to weakly southward IMF conditions (B_Z ~0 to –4 nT) prior to the shock. In addition, Liou et al. (2003) pointed out that what Zhou and Tsurutani (2001) meant by substorm triggering was not necessarily the substorm onset auroral breakup. By considering whether or not there was a substorm auroral breakup for a larger number of events, Liou et al. (2003) concluded that the probability of substorm triggering by a shock compression was very low for their studied events. However, Lyons et al. (2005) and Lee et al. (2005) suggested that substorm triggering by a P_{SW} enhancement is possible if the accompanied IMF is strongly southward whereas for northward and weakly southward IMF conditions, P_{SW} enhancements result in only typical compressive disturbances without a substorm onset. In fact, Lyons et al. (2005) and Lee et al. (2005) reported several examples of substorm triggering by a P_{SW} enhancement following ≥ 1-hour period of IMF $B_Z \leq -8$ nT. However, for most of the events studied by Liou et al. (2003), the accompanied IMF was not strongly southward, having led to low probability of triggering. Although there has yet to be a realistic determination of the precise magnitude and length of the P_{SW} enhancement required for triggering a substorm as a function of the preceding IMF B_Z and P_{SW}, it seems clear that prolonged and/or strong southward IMF is generally a favorable condition for a P_{SW} enhancement to trigger a substorm.

As described Lee at al. (2007a), when a P_{SW} enhancement triggers a substorm, the direct compression effect still exists. The substorm effect is dominant in the pre-midnight sector and the direct compressional effect is seen elsewhere. Lyons et al. (2005) and Lee et al. (2005) referred to this as a two-mode type response. For example, the magnetopause current enhancement leads to increases in the dayside ground H, whereas quasi-simultaneously the ground H component shows larger increases on the nightside owing to the wedge current formation of the triggered substorm (Lyons et al., 2005; Lee et al., 2005). Also, auroral brightening consists of substorm breakup near the Harang reversal region and additional brightening elsewhere, which is the direct results of compression (Lyons et al., 2005). The geosynchronous measurements of energetic charged particles by Los Alamos National Laboratory (LANL) spacecraft show particle flux enhancements due to substorm injections as well as compressive flux changes (Lee et al., 2005).

Lee at al. (2007a) outlined that they are interested in examining global responses of energetic particles to a P_{dyn} enhancement. Previous information on such global responses has only been obtained from studies using local measurements. The energetic neutral atom (ENA) images, which are used in Lee at al. (2007a), is a useful tool for studying the global aspect. ENAs are produced by charge-exchange interactions between energetic trapped ions and cold ambient neutral atoms in the geocorona. These ENAs travel freely without being affected by electromagnetic fields in the magnetosphere, and they can be used to remotely measure the magnetospheric ions. This powerful tool for space plasma diagnostics has been used substantially for ring current study (e.g., Roelof, 1987; Henderson et al., 1997; Brandt et al., 2002a; Ohtani et al., 2006), for substorm or storm-substorm relation research (e.g., Jorgensen et al., 2000; Reeves and Henderson, 2001; Brandt et al., 2002a - e; Reeves et al., 2004; Ohtani et al., 2005; Pollock et al., 2003), for oxygen studies (e.g., Lui et al., 2005b; Mitchell et al., 2003; Nosé et al., 2005), and for other issues (e.g., Perez et al., 2004; Vallat et al., 2004). Several spacecraft have made ENA observations, such as ISEE 1 (Roelof, 1987), Astrid (Barabash, 1995; Brandt et al., 1999), Geotail (Lui et al., 1996), POLAR (Henderson et al., 1997), Cassini (Mitchell et al., 1998) and IMAGE (Mitchell et al., 2000).

Lee at al. (2007a) examine the global response of the ENA for six P_{SW} enhancement events. The P_{SW} increase for all events is abrupt and clearly noticeable. Specifically, the relative enhancement of P_{SW} is ≥100%. Also the P_{dyn} enhancement impact for all events is well separated or clearly distinguished from any earlier substorm occurrence. Most importantly, the IMAGE spacecraft at the time of a P_{SW} event has to be at a proper position that is not too close to the Earth, preferably near apogee, and with a good look direction. This is a critical requirement for a reliable study on ENA responses to P_{SW} enhancements, and all of investigated six events meet this requirement to a reasonable degree. Also, all but one of these events are relatively free from the solar contamination problem that often limits the selection of suitable events. Four P_{SW} events are under a northward or weakly southward IMF condition for which the magnetospheric response involves no substorm triggering, and two P_{SW} events are under a strongly southward IMF condition where a substorm is triggered by the P_{SW} enhancement impact. Lyons et al. (2005) suggested that there is a possible interplay effect between simultaneous changes of IMF and P_{SW} particularly when the accompanied IMF B_Z is strongly southward. In fact, it is quite often the case that both IMF and P_{SW} change together. The latter two events under strongly southward IMF conditions were selected as they meet the condition to a reasonable degree of not having a significant IMF change, in particular in IMF B_Z, around the time of the P_{SW} enhancement to avoid the possible interplay effect (it did not require this condition when the IMF is northward).

As noted Lee at al. (2007a), the IMAGE satellite, launched in March 2000 (Burch, M2000), is the first satellite equipped with instruments that can globally image ENA with high resolution. The HENA instrument is one of the neutral atom imaging instruments on the satellite, and detects charge exchange neutrals of high energies, 10 keV to ~200 keV, with 2 min time resolution (Mitchell et al., 2000). HENA can also image the ENA emissions separately in hydrogen and oxygen, but only after August 2001; before then, the instrument could measure the oxygen ENA flux only with the lowest-energy (<10 keV/nucleon) hydrogen channel and only for disturbed periods (Mitchell et al., 2003). Also, while the effective angular resolution of the instrument is several degrees for hydrogen (~8° or better above the energy of ~50 keV/nucleon), it is tens of degrees for oxygen (Mitchell et al., 2003). In addition to the IMAGE HENA ENA data, Lee at al. (2007a) use geosynchronous magnetic and charged-particle flux data, ground geomagnetic H-component data, and auroral images in order to check for the compression effect and to determine whether or not a substorm occurred.

Lee at al. (2007a) summarized main obtained results as following:

(i) It was distinguished between pure compression events (without substorm effects) and events where a substorm is triggered by the P_{SW} enhancement.

(ii) The pure compression events, all of which were for northward IMF conditions, result in weak-to-modest ENA emission enhancement: the total ENA emission rate relative to averages over 1 hour prior to the pressure impact increases by ~25% to ~40% in both hydrogen and oxygen channels for P_{SW} enhancements of ~100% to ~450%. This pure compression-induced ENA emission must be primarily due to adiabatically energized charged particles by the pressure compression. It should also be noted that the resultant ENA emission enhancement should be more than what is expected only from energization because the compression pushes the ions toward the Earth and the neutral atom density increases sharply (Ohtani et al., 2005; Brandt et al., 2002c). This implies that the ENA emission enhancement is not proportional to the ion flux enhancement. Future work to compare the ENA and ion flux enhancements would help to clarify this issue. The enhanced ENA emission drops as the P_{SW} decreases, implying that the ENA responses were directly caused by the adiabatic compression and decompression processes. It was also seen that the pure compression events lead to overall global, quasi-simultaneous in MLT, increases of ENA in contrast to pure substorm-induced ENA enhancements that are initially localized in the nightside region followed by the azimuthal drift effect.

(iii) For the events where the P_{SW} enhancement triggered a substorm, there are presented two events under strongly southward IMF conditions, both of which are during major storms with Dst_{min} less than –100 nT. Both events indicate significantly larger ENA emission than that for the pure compression events. For the 11 August 2000 event where the P_{SW} enhancement is ~100% under IMF B_Z ~ –11 nT, the total ENA emission rate relative to averages over 1 hour prior to the pressure impact increased by ~170% for 60–198 keV hydrogen and ~130% for <160 keV oxygen. They are still significant, ~110% for 60–198 keV hydrogen and 94% for <160 keV oxygen, even when the near-Earth region where the low-altitude contribution is dominant is excluded from the estimation. For the 14 October 2000 event, a quantitative estimation of the total ENA emission rate was limited owing to solar contamination, but the significant increase of the ENA emission was clear. The significant ENA emission for both events is of course mainly due to ions generated by the triggered substorm. Also, its appearance is not as global in MLT as for the pure compression events because of the predominant substorm effect that starts near midnight and spreads soon in MLT, but the global response to the compression exists as well.

(iv) The selection of six events consisted of four pure compression events during primarily northward IMF conditions and two P_{SW}-triggered substorms during strongly southward IMF conditions. This selection gives sharp contrast between the two types of events; however there are many intermediate and more complex types of events. For example, it is one additional event (the pure substorm event at 08:54 UT on 29 September 2001) where the preceding IMF condition is neither simply northward nor strongly southward, where no obvious pressure trigger existed and where the ENA emission rate increase was intermediate. Clearly, a statistical analysis based on a larger number of events than included here is desired to check the generality of the present result. From a statistical viewpoint, it would be useful to determine whether or not ENA enhancement is generally larger when a substorm is triggered than when there is only the compressional effect. It is also interesting to note the suggestion by Jorgensen et al. (2000) that the ENA emission is a new type of substorm signature.

(v) The selection of events is limited by several conditions. One of the stringent conditions is the requirement of no significant IMF changes at the time of the P_{SW} enhancement when the accompanied IMF is southward. This requirement was set because such a simultaneous IMF turning could interact with the P_{SW} enhancement (Lyons et al., 2005), which can make it difficult to clarify the P_{SW} effect alone. Quite often, both P_{SW} and IMF in reality change together, and it would be interesting to examine the possible interplay effect using the ENA images along with other data.

(vi) Extraction of the ion distribution from the observed ENA data is a challenging task in general, and there have been such attempts by some researchers (e.g., DeMajistre et al., 2004; Brandt et al., 2002c; Perez et al., 2001; Roelof and Skinner, 2000). It would be interesting to examine the extracted ion distributions in response to the dynamic pressure impact and compare them with local measurements of ions in the inner magnetosphere. Also, use of both the LANL SOPA and MPA data would be interesting for comparison with the ions extracted from the IMAGE HENA ENA data. Furthermore, several researchers have suggested that the O^+ ions during substorm expansion are energized more significantly than protons owing to their non-adiabatic behavior in response to the substorm dipolarization (e.g., Fok et al., 2006; Jones et al., 2006; Nosé et al., 2000; Delcourt et al., 1990). The non-adiabatic behavior is expected because the oxygen gyro-period is comparable to the timescale of dipolarization. It would be interesting to check whether or not a similar non-adiabatic response of O^+ ions can occur for the pure compression events where the solar wind dynamic pressure increases over a sufficiently short timescale to break down the first adiabatic invariant and/or the second adiabatic invariant. This would, however, require a detailed numerical analysis based on a realistic magnetospheric model.

3.10.2. Effect of the Interplanetary Secondary Rarefaction Waves on the Geomagnetic Field

As noted Grib and Belakhovsky (2009), the arrival of the interplanetary shock wave to the magnetosphere causes compression of this region and SSC, usually warning about the beginning of a geomagnetic storm, or a sudden impulse (SI) event. The interplanetary shock wave, which is a fast MHD shock, is characterized by an abrupt increase in the velocity, solar wind density, temperature, and magnetic field value. A SSC event observed on the Earth is the superposition of the low disturbances (DL) and polar disturbances (DP) fields. The DL field is caused by the magnetopause currents, and information about a change in the field is transmitted deep into the magnetosphere via the fast magnetosonic wave. The response to a change in the DL field is most pronounced at low and middle latitudes. The DP field is decomposed into two parts (DP_{PI} and DP_{MI}), corresponding to the preliminary and main impulses:

$$D_{SSC} = DL + DP_{PI} + DP_{MI}. \tag{3.10.1}$$

The PI impulse originates in the polar region and is related to an enhancement of the ionospheric and field-aligned currents. The MI impulse is related to an enhancement of the magnetopause currents during its compression (Araki, 1994). The SSC amplitude depends on the observation point latitude and local time. For middle and low latitudes, the SSC amplitude is maximal at the equator and decreases with increasing latitude owing to weakening of the effect of the magnetopause currents. The SSC amplitude at low latitudes (Φ = 15°–30°) weakly depends on MLT and is maximal near noon and minimal near midnight (Russell et al., 1992). The SSC impulse appears several minutes earlier at high latitudes than at low ones. Grib and Belakhovsky (2009) outlined that when a head-on collisions of the interplanetary shock wave with the bow shock was considered (Grib and Martynov, 1977; Grib et al., 1979), it was for the first time indicated that the interplanetary shock wave is refracted into the magnetosheath in the form of a weaker fast shock. The solution of the Riemann problem by the trial calculation method was used in this case. The same method was subsequently used to consider the reflection of a refracted shock from the magnetopause, represented in the form of a tangential discontinuity, and it was indicated that a return fast rarefaction wave is reflected from this discontinuity. The reflection of this wave from the bow shock rear was subsequently considered and the origination of the secondary direct fast rarefaction wave, changing the magnetosheath parameters and affecting the magnetopause, was indicated. Set of following equations schematically describes the mechanism by which the secondary rarefaction wave originates:

$$S_{2\rightarrow}S_{1bow} \rightarrow S_{3bow}TS_{4\rightarrow},\ S_{4\rightarrow}T_m \rightarrow R_{\leftarrow}T_mS_{5\rightarrow},\ S'_{bow}R_{\leftarrow} \rightarrow S''_{bow}TR'_{\rightarrow} \quad (3.10.2)$$

The interaction of the S_2 interplanetary shock wave with the S_{1bow} bow shock results in the origination of two (direct S_4 and return S_{3bow}) shocks and the tangential discontinuity (T) between them. A return shock wave is a weakened detached bow shock S_{3bow}. The direction of an arrow in Eq. 3.10.2 indicates the wave direction: toward the Sun (to the left) and toward the Earth (to the right). A direct shock wave (S_4) subsequently moves through the magnetosheath without distortions in a first approximation. It is assumed that the Bernoulli's MHD equation holds true at zero streamline in the magnetic transition layer. A fast rarefaction wave (R), moving in the magnetosheath toward the bow shock, originates when the shock wave interacts with the magnetopause (Tm). One more rarefaction wave (R) originates when the rarefaction wave is reflected from the bow shock back (or rear). This secondary rarefaction wave moves toward the magnetopause and causes this region to move back toward the Sun; this wave should also gradually decrease the magnetic field value after an abrupt increase during SSC. When oblique interaction between the fast shock wave and the bow shock was considered (Gribe, 1982), it was obtained that this interaction results in the origination of the direct fast shock wave, direct slow rarefaction wave, contact discontinuity, slow return shock wave, and fast return shock wave (distorted bow shock). The situation considered in Grib (1982) does not contradict Eq. 3.10.2 but completes it. Based on the ISEE satellite data, Zhuang et al. (1981) confirmed the presence of the MHD rarefaction wave reflected from the magnetopause. The observations in the magnetosheath are complex owing to turbulence and vortices; therefore, it is still difficult to obtain the evidence of the secondary MHD rarefaction wave from these observations. Samsonov et al. (2006, 2007) performed the MHD modeling of the interaction of the interplanetary shock wave with the bow shock and the magnetopause when a modified fast shock wave crosses the magnetosheath. It was obtained that the interaction between the fast shock wave and the magnetopause results in the origination of a shock or a rarefaction wave. However, the interaction even between the shock wave reflected from the magnetopause and the bow shock rear results in the origination of a rarefaction wave, which coincides with the result of Grib et al. (1979). As noted Grib and Belakhovsky (2009), based on the MHD modeling, Ŝafránková et al. (2007) indicated that the bow shock moves first antisunward and then sunward during the interaction between the interplanetary shock wave and bow shock. The bow shock backward motion possibly results from the interaction between the interplanetary shock wave and magnetopause and the origination of the secondary rarefaction wave in the magnetosheath. The aim of Grib and Belakhovsky (2009) work is to reveal the effect of the secondary rarefaction wave, originating in the magnetospheric magnetosheath, on the geomagnetic field behavior during SSC.

Grib and Belakhovsky (2009), based on the WIND and GOES satellite data on the solar wind and IMF parameters and the data on the surface magnetic field, show that the secondary

MHD rarefaction wave can affect the geomagnetic field during SSC event. As a result, the secondary rarefaction wave originates in the magnetosheath when the shock wave interacts with the Earth's magnetosphere. They considered three events, when the interplanetary shock wave approached the Earth's magnetosphere: on January 17 and May 6, 2005, and July 9, 2006. Considered events indicate that the magnetic field at the ground stations and in geostationary orbit decreased earlier than the solar wind dynamic pressure. This effect can be explained by the origination of the **secondary rarefaction wave** in the magnetosheath and by the effect of this wave on the geomagnetic field behavior.

3.10.3. Magnetospheric Cavity Modes Driven by Solar Wind Dynamic Pressure Fluctuations

As outlined Claudepierre et al. (2009), several observational studies suggest that some dayside magnetospheric ultra-low frequency (ULF) pulsations may be directly driven by ULF fluctuations in the solar wind dynamic pressure. For example, Kepko and Spence (2003) examine six events where discrete ULF fluctuations are observed in the solar wind dynamic pressure. Kepko and Spence (2003) show a one-to-one correspondence between these solar wind dynamic pressure fluctuations and discrete spectral peaks in dayside GOES magnetic field data. They argue that the dayside magnetospheric ULF pulsations are directly driven by the corresponding solar wind dynamic pressure fluctuations. Other observational studies (Sibeck et al., 1989; Korotova and Sibeck, 1995; Matsuoka et al., 1995; Han et al., 2007) also suggest that solar wind dynamic pressure fluctuations can directly drive dayside magnetic field ULF pulsations. Very recent work (Viall et al., 2009) concludes that approximately half of the variations observed in magnetospheric ULF waves are likely directly driven by solar wind dynamic pressure fluctuations. Claudepierre et al. (2009) investigate, through the use of global MHD simulations, the magnetospheric response to ULF solar wind dynamic pressure (henceforth, p_{dyn}) fluctuations. Here, 'ULF' refers to frequencies in the 0.5 to 50 mHz range (Pc3-Pc5 bands according to classification of Jacobs et al., 1964), though they make no distinction between continuous and irregular magnetospheric pulsations.

Claudepierre et al. (2009) present results from Lyon-Fedder-Mobarry (LFM) global, three-dimensional MHD simulations of the solar wind-magnetosphere interaction. They use these simulations to investigate the role that solar wind dynamic pressure fluctuations play in the generation of magnetospheric ULF pulsations. The simulations are driven with idealized solar wind input conditions. In four of the simulations, Claudepierre et al. (2009) introduce monochromatic ULF fluctuations in the upstream solar wind dynamic pressure. In the fifth simulation, they introduce a continuum of ULF frequencies in the upstream solar wind dynamic pressure fluctuations. In this numerical experiment, the idealized nature of the solar wind driving conditions allows to study the magnetospheric response to only a fluctuating upstream dynamic pressure, while holding all other solar wind driving parameters constant. The simulation results suggest that ULF fluctuations in the solar wind dynamic pressure can drive magnetospheric ULF pulsations in the electric and magnetic fields on the dayside. Moreover, the simulation results suggest that when the driving frequency of the solar wind dynamic pressure fluctuations matches one of the natural frequencies of the magnetosphere, magnetospheric cavity modes can be energized.

3.10.4. Effects of Continuous Solar Wind Pressure Variations on the Long-Lasting Penetration of the IEF during Southward IMF

As noted Yuan and Deng (2007), it is believed that the interplanetary electric field (IEF) can penetrate into the mid- or low-latitude ionosphere. For the penetration of the IEF, the relationship between the solar wind pressure and the penetrated electric field in the equatorial ionosphere is yet an unsolved topic. With observations of the IEF, solar wind pressure obtained from the Wind satellite, and the geomagnetic field on four magnetometers from high latitudes to magnetic equator, Yuan and Deng (2007) present an event of a long-lasting penetration of the IEF into the dayside equatorial ionosphere during the main phase of a magnetic storm on April 17, 1999. In the event, variations of the geomagnetic field both at high latitudes and at magnetic equator show coherence with those of the solar wind pressure, indicating that a series of enhancements of solar wind pressures can drive those coherent increases of the penetrated electric field in dayside equatorial ionosphere during stable and southward IMF. Therefore, not only the dawn-dusk component of the IEF, but also the solar wind pressure can strongly control the long-lasting penetration of the magnetospheric electric field into low-latitude or equatorial ionosphere during southward IMF. In addition, result of Yuan and Deng (2007) suggests that a combination of equatorial and low-latitude ground magnetograms may be used to monitor the solar wind dynamic pressure variation not only in the northward IMF, but also in the southward IMF. Some obtained results are illustrated by Figures 3.10.1 – 3.10.3.

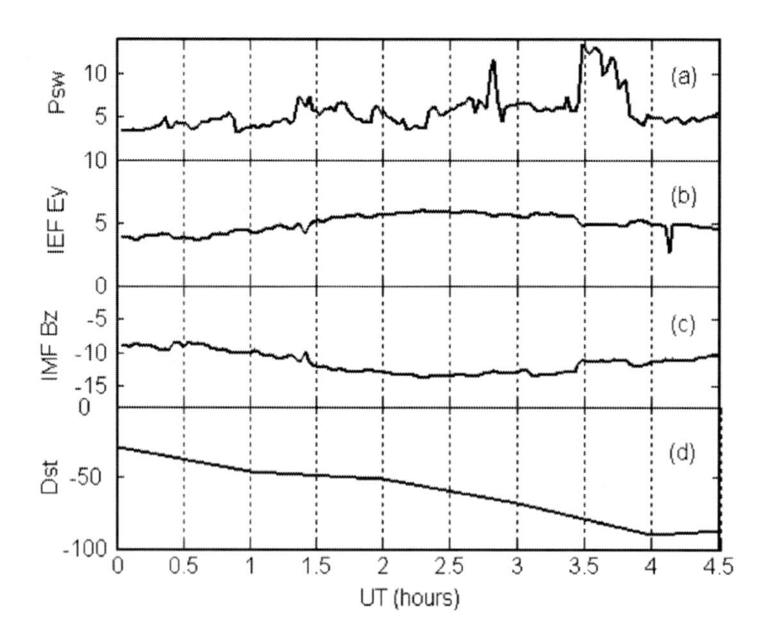

Figure 3.10.1. (a) Solar wind dynamic pressure (nPa), (b) interplanetary electric field (IEF) E_y component (mV/m), (c) B_z component (nT) of IMF measured by the Wind spacecraft, and (d) *Dst* index (nT) between 0000 and 0430 UT on April, 17, 1999. The spacecraft data are displayed in GSM coordinates. According to Yuan and Deng (2007). Permission from COSPAR.

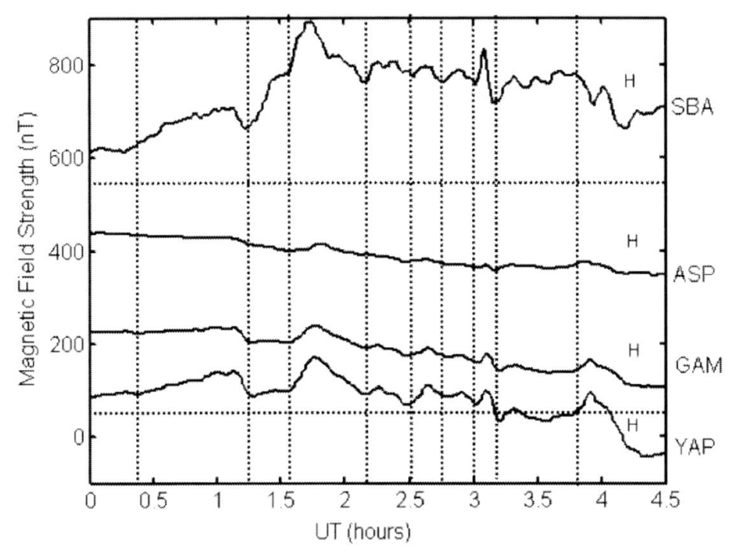

Figure 3.10.2. H component of magnetic field at high-latitude Scott Base (SBA), midlatitude Alice Springs (ASP), low-latitude Guam (GAM), and equatorial latitude Yap (YAP) magnetometer stations in the daytime. Two horizontal dashed lines denote the quiet nighttime magnetic level for SBA and YAP stations. Nine vertical dashed lines denote the negative peaks for the H component for the YAP station. According to Yuan and Deng (2007). Permission from COSPAR.

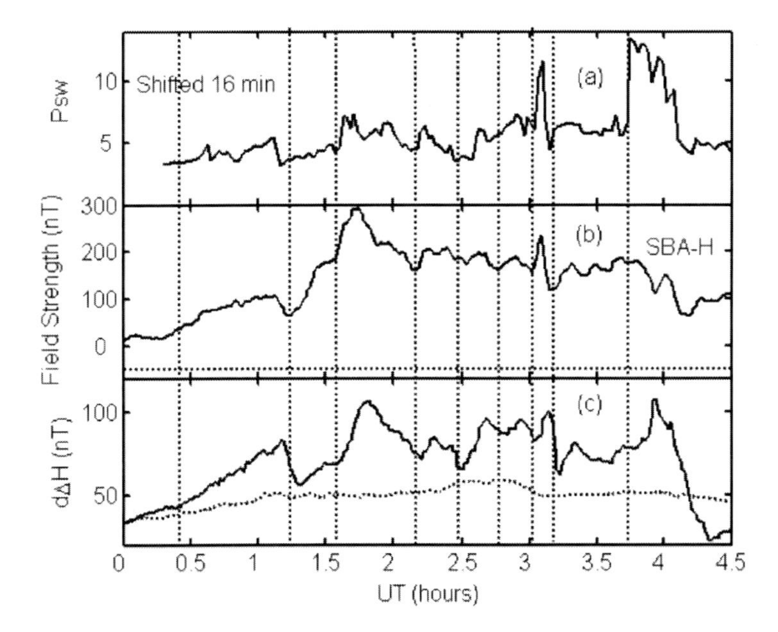

Figure 3.10.3. (a) Solar wind dynamic pressure (nPa), (b) H component of magnetic field at high-latitude (SBA) magnetometer station, (c) difference in the H component (DH) between YAP and GAM stations between 0000 and 0430 UT on April, 17, 1999. Nine vertical dashed lines denote the negative peaks for the H component of the SBA station. The dotted curves in panels (b and c) present the H component of SBA station, (DH) between YAP and GAM stations on the quiet day, i.e., April 15, 1999, respectively. According to Yuan and Deng (2007). Permission from COSPAR.

3.11. Ionospheric Convection Variation, IMF Modulation and Cavity Mode, Asymmetry of Magnetosheath Flows and Magnetopause Shape, Nightside Magnetospheric Current Circuit, Sunward Magnetosheath Flows and Magnetopause Motion in Response to Sudden Increases in the Solar Wind Dynamic Pressure

3.11.1. Global-Scale Observations of Ionospheric Convection Variation in Response to Sudden Increases in the Solar Wind Dynamic Pressure

Gillies et al. (2012) have used a superposed epoch analysis to study 205 sudden commencement (SC) events detected with ground-based magnetometers between the years 2000 and 2007. The strength of the SC events was clearly correlated to the magnitude of the jump in the solar wind dynamic pressure, regardless of whether or not the SC events were followed by a magnetic storm. Data from the SuperDARN demonstrated that both the ionospheric plasma drift speed and the number of echoes increased in the noon sector in response to the increase in solar wind dynamic pressure. In contrast, the number of SuperDARN echoes in the midnight sector decreased as the solar wind dynamic pressure increased, even though the average drift speed in the midnight sector also increased. Gillies et al. (2012) also uncovered that the ionosphere and ring current evolve differently in response to the pressure pulses. The SYM-H index, which represents changes in both the magnetopause and ring currents, responded immediately and either rapidly returned to pre-SC values or progressed into the main phase of a geomagnetic storm. In contrast, the ionospheric convection data were affected for a much longer time. The implication is that the ring current reacts to a sudden compression of the magnetosphere on a time scale of 10 min, while the convection pattern itself is affected for as long as the increase in solar wind dynamic pressure is sustained, or until a geomagnetic storm was triggered, as is the case in the sudden storm commencement (SSC) subset of events.

3.11.2. Interaction of Solar Wind Pressure Pulses with the Magnetosphere: IMF Modulation and Cavity Mode

According to Liu et al. (2012), energy transfer through the magnetopause involves an interplay of two processes. On one hand, microphysics of reconnection determines how easily the magnetopause can be opened. On the other hand, the global state of the solar wind and magnetosphere determines how much energy is available for transfer and whether there exist "resonant" interactions whereby the transfer is particularly efficient. In the case of solar wind pressure pulses, empirical evidence has suggested that the solar wind-magnetosphere interaction can become unusually intense, leading to large-scale global auroral response and occasionally geomagnetic storms. In this study, for the first time, magnetic reconnection and global magnetospheric oscillation known as the cavity mode are integrated to give a comprehensive description of energy transfer through the dayside magnetopause. Using a heuristic model in which the inflow into the magnetopause is proportional to the magnitude of pressure pulse and an IMF proxy, Liu et al. (2012) derive the fractions of energy converted to reconnection and field-line resonance per unit incident compressional energy in a pressure pulse. It is found that the magnitude of energy transfer is modulated by the IMF proxy, whereas the frequency spectrum of the transfer is modulated by the cavity mode. Under extreme conditions, reconnection can transfer almost 100% of incident compressional energy at the maximum absorption bands. Even under the typical value reconnection rate (~0.1), approximately 30% of the incident energy can be absorbed in these bands. The frequency response of reconnection transfer has pulse-like peaks in the > 3 mHz range and rather insensitive to the solar wind and wave parameters. In contrast, the frequency response of the shear-Alfvénic transfer centers in the 1–4 mHz range and has a more broadband shape that is significantly influenced by the solar wind density.

3.11.3. Asymmetry of Magnetosheath Flows and Magnetopause Shape during Low Alfvén Mach Number Solar Wind

Lavraud et al. (2013) report statistical, observational results that pertain to changes in the magnetosheath flow distribution and magnetopause shape as a function of solar wind M_A and IMF clock angle orientation. It was used Cluster 1 data in the magnetosheath during the period 2001–2010, using an appropriate spatial superposition procedure, to produce magnetosheath flow distributions as a function of location in the magnetosheath relative to the IMF and other parameters. The results demonstrate that enhanced flows in the magnetosheath are expected at locations quasi-perpendicular to the IMF direction in the plane perpendicular to the Sun-Earth line; in other words, for the special case of a northward IMF, enhanced flows are observed on the dawn and dusk flanks of the magnetosphere, while much lower flows are observed above the poles. The largest flows are adjacent to the magnetopause. Using appropriate magnetopause crossing lists (for both high and low M_A), we also investigate the changes in magnetopause shape as a function of solar wind M_A and IMF orientation. Comparing observed magnetopause crossings with predicted positions from an axisymmetric semi-empirical model, we statistically show that the magnetopause is generally circular during high M_A, while is it elongated (albeit with moderate statistical significance) along the direction of the IMF during low M_A. These findings are consistent with enhanced magnetic forces that prevail in the magnetosheath during low M_A. The component of the magnetic forces parallel to the magnetopause produces the enhanced flows along and adjacent to the magnetopause, while the component normal to the magnetopause exerts an asymmetric pressure on the magnetopause that deforms it into an elongated shape.

3.11.4. Nightside Magnetospheric Current Circuit: Time Constants of the Solar Wind-Magnetosphere Coupling

The study of Ohtani and Uozumi (2014) addresses the characteristics of the nightside magnetospheric current system using the analogy of an electric circuit. The modeled circuit consists of the generator (V: solar wind), inductor (L: tail lobes), capacitor (C: plasma sheet convection), and resistor (R: particle energization). The electric circuit has three time constants:

$$\tau_{CR} = CR,\ \tau_{LC} = \sqrt{LC},\ \text{and}\ \tau_{L/R} = L/R. \qquad (3.11.1)$$

Here τ_{CR} is of the order of the ion gyroperiod in the plasma sheet, τ_{LC} is a global timescale ($2\pi\ \tau_{LC}$ is several tens of minutes), and $\tau_{L/R}$ is even longer (several hours). Despite uncertainty in the estimate of each circuit element, $\tau_{CR} \ll \tau_{LC} \ll \tau_{L/R}$ holds generally for the magnetosphere, which characterizes the electric circuit as overdamped. The following implications are obtained: (1) During the substorm growth phase the cross-tail current increases continuously even if interplanetary magnetic field (IMF) B_Z does not change after southward turning; (2) the magnetotail current weakens following northward turnings if the change of IMF B_Z is comparable to the preceding southward IMF B_Z; otherwise it may strengthen continuously if more gradually; (3) during the early main phase of magnetospheric storms the enhancement of the lobe magnetic energy is far more prominent than the enhancements of the kinematic and kinetic energies of the plasma sheet plasma; (4) The efficiency of the solar wind-magnetosphere coupling changes on a timescale of several hours ($\tau_{L/R}$) through the change of the tail flaring, and so does the cross polar-cap potential; and (5) the magnetospheric current system does not resonate to an oscillatory external driver, and therefore, the periodicity of some magnetotail phenomena reflects that of their triggers.

3.11.5. The Role of Pressure Gradients in Driving Sunward Magnetosheath Flows and Magnetopause Motion

As outlined Archer et al. (2014), while pressure balance can predict how far the magnetopause will move in response to an upstream pressure change, it cannot determine how fast the transient response will be. Using THEMIS data, Archer et al. (2014) present multipoint observations revealing, for the first time, strong (thermal + magnetic) pressure gradients in the magnetosheath due to a foreshock transient, most likely a hot flow anomaly, which decreased the total pressure upstream of the bow shock. By converting the spacecraft time series into a spatial picture, Archer et al. (2014) quantitatively show that these pressure gradients caused the observed acceleration of the plasma, resulting in fast sunward magnetosheath flows ahead of a localized outward distortion of the magnetopause. The acceleration of the magnetosheath plasma was fast enough to keep the peak of the magnetopause bulge at approximately the equilibrium position, i.e., in pressure balance. Therefore, it was shown that pressure gradients in the magnetosheath due to transient changes in the total pressure upstream can directly drive anomalous flows and in turn are important in transmitting information from the bow shock to the magnetopause.

3.11.6. A Numerical Study of the Interhemispheric Asymmetry of the Equatorial Ionization Anomaly in Solstice at Solar Minimum

In the study of Dang et al. (2016), the mechanisms of interhemispheric asymmetry of the equatorial ionization anomaly (EIA) in June solstice at solar minimum were investigated through a series of simulations using the Thermosphere Ionosphere Electrodynamics General Circulation Model. Obtained results indicate that the trans-equatorial neutral wind is the main cause of the interhemispheric asymmetry. The trans-equatorial wind transports plasma from the summer hemisphere to winter hemisphere, leading to an enhancement in the winter EIA crest. Meanwhile, the ion production and loss are also important factors in producing the EIA asymmetry through photochemical processes. The longitudinal variations of the EIA interhemispheric asymmetry are also explored through imposing each term of ion continuity equation longitudinally independent. The term analysis results suggest that the neutral wind variation dominates the longitudinal patterns of the interhemispheric asymmetry, while the **E** × **B** drift and photochemical process contribute less.

3.12. Response of the Night Aurora to a Negative Sudden Impulse

Data from the meridian scanning photometers of the NORSTAR network and all-sky cameras of the THEMIS network were used by Belakhovsky and Vorobjev (2016) for a detailed study of the response of night auroras to the sharp decrease of the solar wind dynamic pressure on September 28, 2009. The decrease in dynamic pressure was accompanied by a corresponding depression of the magnetic field in the SYM-H index and the origin of a negative sudden impulse (SI–) with a duration of 5–8 min and amplitude of 150–200 nT in the horizontal component of the magnetic field at stations of the night sector of the auroral zone. The magnetic impulse was preceded by a long calm magnetic period, although the IMF Bz-component was negative for ~1.5 hour before the SI–. The commencement of the SI–, which was determined by variations in the magnetic field at ~0650 UT, was accompanied by a sharp increase in the intensity of discrete forms of polar auroras in the midnight sector of the auroral zone and their fast propagation to the pole. Approximately 6–8 min after the SI–, the auroral intensity in the emissions, which were excited by the fluxes of precipitated electrons and protons, quickly began to decrease in the night sector. Analysis of the optical observations showed the two-stage character of the response of the night auroras to the SI– in the considered event: first, fast movement of the discrete aurora forms to the pole with a significant increase in their intensity, and a further fast decrease in auroral intensity with a delay of ~6–8 min relative to the SI–. The possible reasons for such aurora behavior are discussed by Belakhovsky and Vorobjev (2016).

3.13. IMF B_Y Effects on Ground Magnetometer Response to Increased Solar Wind Dynamic Pressure Derived from Global MHD Simulations

As note Ozturk et al. (2017), during sudden solar wind dynamic pressure enhancements, the magnetosphere undergoes rapid compression resulting in a reconfiguration of the global current systems, most notably the field-aligned currents (FACs). Ground-based magnetometers are traditionally used to study such compression events. However, as outlined

Ozturk et al. (2017), factors affecting the polarity and magnitude of the ground-based magnetic perturbations are still not well understood. In particular, interplanetary magnetic field (IMF) B_y is known to create significant asymmetries in the FAC patterns. Ozturk et al. (2017) use the University of Michigan Block Adaptive Tree Roe Upwind Scheme (BATS'R'US) MHD code to investigate the effects of IMF B_y on the global variations of ground magnetic perturbations during solar wind dynamic pressure enhancements. Using virtual magnetometers in three idealized simulations with varying IMF B_y, Ozturk et al. (2017) find asymmetries in the peak amplitude and magnetic local time of the ground magnetic perturbations during the preliminary impulse (PI) and the main impulse (MI) phases. These asymmetries are especially evident at high-latitude ground magnetometer responses where the peak amplitudes differ by 50 nT at different locations. Ozturk et al. (2017) show that the FACs related with the PI are due to magnetopause deformation, and the FACs related with the MI are generated by vortical flows within the magnetosphere, consistent with other simulation results. The perturbation FACs due to pressure enhancements and their magnetospheric sources do not differ much under different IMF B_y polarities. However, as outlined Ozturk et al. (2017), the conductance profile affected by the superposition of the preexisting FACs and the perturbation FACs including their closure currents is responsible for the magnitude and location asymmetries in the ground magnetic perturbations.

REFERENCES

Araki T., 1994. "A Physical Model of the Geomagnetic Sudden Commencement," *Geophys. Monogr.* Am. Geophys. Union, **81**, 183–200.

Archer M.O., D.L. Turner, J.P. Eastwood, T.S. Horbury, and S.J. Schwartz, 2014. "The role of pressure gradients in driving sunward magnetosheath flows and magnetopause motion", *J. Geophys. Res. Space Physics*, **119**, No. 10, doi:10.1002/2014JA020342, 8117-8125.

Aryan H., R.J. Boynton, and S.N. Walker, 2013. "Analysis of trends between solar wind velocity and energetic electron fluxes at geostationary orbit using the reverse arrangement test", *J. Geophys. Res. Space Physics*, **118**, No. 2, doi:10.1029/2012JA 018216, 636-641.

Badman S.V., E.J. Bunce, J.T. Clarke, S.W.H. Cowley, J.-C. Gérard, D. Grodent, and S.E. Milan, 2005. "Open flux estimates in Saturn's magnetosphere during the January 2004 Cassini-HST campaign, and implications for reconnection rates", *J. Geophys. Res.*, **110**, No. A11, A11216, doi:10.1029/2005JA011240, 1-16.

Baker K.B. and S. Wing, 1989. "A new magnetic coordinate system for conjugate studies at high latitudes", *J. Geophys. Res.*, **94**, No. A7, 9139-9143.

Barabash S., 1995. "Satellite Observations of the Plasma-Neutral Coupling Near Mars and the Earth", *PhD thesis*, Swedish Institute of Space Physics, Kiruna, Sweden, 1-77.

Belakhovsky V.B. and V.G. Vorobjev, 2016. "Response of the Night Aurora to a Negative Sudden Impulse", *Geomagnetism and Aeronomy*, **56**, No. 6, 682–693. Original Russian Text in *Geomagnetizm i Aeronomiya*, **56**, No. 6, 733–744.

Borodkova N.L., 2010. “Effect of large and sharp changes of solar wind dynamic pressure on the earths’s magnetosphere: analysis of several events”, *Cosmic Res.*, **48**, No. 1, 41-55. Russian Text: *Kosmicheskie Issledovaniya*, **48**, No. 1, 43-57.

Borodkova N.L., G.N. Zastenker, and D.G. Sibeck, 1995. “A case and statistical study of transient magnetic field events at geosynchronous orbit and their solar wind origin”, *J. Geophys. Res.*, **100**, No. A4, 5643-5656.

Borodkova N.L., G.N. Zastenker, M. Riazantseva, and J.D. Richardson, 2005. “Large and sharp solar wind dynamic pressure variations as a source of geomagnetic field disturbances at the geosynchronous orbit”, *Planet. Space Sci.*, **53**, No. 1-3, 25-32.

Borodkova N.L., G.N. Zastenker, M.O. Ryazantseva, and J. Richardson, 2006a. “Large and sharp changes of solar wind dynamic pressure and disturbances of the magnetospheric magnetic field at geosynchronous orbit caused by these variations”, *Cosmic Res.*, **44**, No. 1, 1-8. Russian Text: *Kosmicheskie Issledovaniya*, **44**, No. 1, 3-11.

Borodkova N.L., J.B. Liu, Z.H. Huang, G.N. Zastenker, C. Wang, and P.E. Eiges, 2006b. “Effect of change in large and fast solar wind dynamic pressure on geosynchronous magnetic field”, *Chin. Phys.*, **15**, No. 10, 2458-2464.

Borodkova N.L., J.B. Liu, Z.H. Huang, and G.N. Zastenker, 2008. “Geosynchronous magnetic field response to the large and fast solar wind dynamic pressure change”, *Adv.Space Res.*, **41**, No. 8, 1220-1225.

Boudouridis A., E. Zesta, L.R. Lyons, P.C. Anderson, and D. Lummerzheim, 2003. "Effect of solar wind pressure pulses on the size and strength of the auroral oval", *J. Geophys. Res.*, **108**, No. A4, 8012, doi:10.1029/2002JA009373, COA 13-1-16.

Boudouridis A., E. Zesta, L.R. Lyons, P.C. Anderson, and D. Lummerzheim, 2004a. "Magnetospheric reconnection driven by solar wind pressure fronts", *Ann. Geophys.*, **22**, No. 4, 1367-1378.

Boudouridis A., E. Zesta, L.R. Lyons, and P.C. Anderson, 2004b. "Evaluation of the Hill-Siscoe transpolar potential saturation model during a solar wind dynamic pressure pulse", *Geophys. Res. Lett.*, **31**, No. 23, L23802, doi:10.1029/2004GL021252, 1-4.

Boudouridis A., E. Zesta, L.R. Lyons, P.C. Anderson, and D. Lummerzheim, 2005. “Enhanced solar wind geoeffectiveness after a sudden increase in dynamic pressure during southward IMF orientation”, *J. Geophys. Res.*, **110**, No. A5, A05214, doi:10.1029/2004JA010704, 1-15.

Boudouridis A., L.R. Lyons, E. Zesta, and J.M. Ruohoniemi, 2007. "Dayside reconnection enhancement resulting from a solar wind dynamic pressure increase", *J. Geophys. Res.*, **112**, A06201, doi:10.1029/2006JA012141, 1-12.

Boudouridis A., L.R. Lyons, E. Zesta, J.M. Ruohoniemi, and D. Lummerzheim, 2008. “Nightside flow enhancement associated with solar wind dynamic pressure driven reconnection”, *J. Geophys. Res.*, **113**, No. A12, A12211, 1-11, doi:10.1029/2008JA013489.

Boyle C.B., P.H. Reiff, and M.R. Hairston, 1997. “Empirical polar cap potentials”, *J. Geophys. Res.*, **102**, No. A1, 111-126.

Boynton R.J., M.A. Balikhin, S.A. Billings, G.D. Reeves, N. Ganushkina, M. Gedalin, O.A. Amariutei, J.E. Borovsky, and S.N. Walker, 2013. “The analysis of electron fluxes at geosynchronous orbit employing a NARMAX approach”, *J. Geophys. Res. Space Physics*, **118**, No. 4, doi:10.1002/jgra.50192, 1500-1513.

Brandt P.C., S. Barabash, O. Norberg, R. Lundin, E.C. Roelof, and C.J. Chase, 1999. "Energetic neutral atom imaging at low altitudes from the Swedish microsatellite Astrid: Images and spectral analysis", *J. Geophys. Res*., 104, No. A2, 2367-2379.

Brandt P.C., S. Ohtani, D.G. Mitchell, M.-C. Fok, E.C. Roelof, and R. Demajistre, 2002a. "Global ENA observations of the storm mainphase ring current: Implications for skewed electric fields in the inner magnetosphere", *Geophys. Res. Lett*., **29**, No. 20, 1954, DOI 10.1029/2002GL015160, 15-1-3.

Brandt P.C., S. Ohtani, D.G. Mitchell, R. Demajistre, and E.C. Roelof, 2002b. "ENA observations of a global substorm growth phase dropout in the nightside magnetosphere", *Geophys. Res. Lett*., **29**, No. 20, 1962, doi:10.1029/2002GL015057, 23-1-3.

Brandt P.C., R. Demajistre, E.C. Roelof, S. Ohtani, D.G. Mitchell, and S. Mende, 2002c. "IMAGE/high-energy energetic neutral atom: Global energetic neutral atom imaging of the plasma sheet and ring current during substorms", *J. Geophys. Res*., **107**, No. A12, 1454, doi:10.1029/2002JA009307, SMP21-1-13.

Brandt P.C., D.G. Mitchell, Y. Ebihara, B.R. Sandel, E.C. Roelof, J.L. Burch, and R. Demajistre, 2002d. "Global IMAGE/HENA observations of the ring current: Examples of rapid response to IMF and ring current-plasmasphere interaction", *J. Geophys. Res*., **107**, No.11, 1359, doi:10.1029/2001JA000084, SMP12-1-12.

Brandt P.C., S. Ohtani, D.G. Mitchell, M.-C. Fok, E.C. Roelof, and R. Demajistre, 2002e. "Global ENA observations of the storm mainphase ring current: Implications for skewed electric fields in the inner magnetosphere", *Geophys. Res. Lett*., **29**, No. 20, 1954, doi:10.1029/2002GL015160, 15-1-3.

Brittnacher M, M Wilber, M Fillingim, D Chua, G Parks, J Spann, and G Germany, 2000. "Global auroral response to a solar wind pressure pulse", *Adv.Space Res*., **25**, No. 7-8, 1377-1385.

Burch J.L, 1972a. "Preconditions for the triggering of polar magnetic substorms by storm sudden commencements", *J. Geophys. Res*., **77**, No. 28, 5629-5632.

Burlaga L.F., and K.W. Ogilvie, 1969. "Causes of sudden commencements and sudden impulses", *J. Geophys. Res*., **74**, No. 11, 2815-2825.

Chisham G., M.P. Freeman, I.J. Coleman, M. Pinnock, M.R. Hairston, M. Lester, and G. Sofko, 2004. "Measuring the dayside reconnection rate during an interval of due northward interplanetary magnetic field", *Ann. Geophys*., **22**, No. 12, 4243-4258 (2004).

Chua D., G. Parks, M. Brittnacher, W. Peria, G. Germany, J. Spann, and C. Carlson, 2001. "Energy characteristics of auroral electron precipitation: A comparison of substorms and pressure pulse related auroral activity", *J. Geophys. Res*., **106**, No. A4, 5945-5956.

Claudepierre S.G., M. Wiltberger, S.R. Elkington, W. Lotko, and M.K. Hudson, 2009. "Magnetospheric cavity modes driven by solar wind dynamic pressure fluctuations", *Geophys. Res. Lett*., **36**, L13101, doi:10.1029/2009GL039045.

Claudepierre S.G., M.K. Hudson, W. Lotko, J.G. Lyon, and R.E. Denton, 2010. "Solar wind driving of magnetospheric ULF waves: Field line resonances driven by dynamic pressure fluctuations", *J. Geophys. Res*., **115**, No. A11, A11202, doi:10.1029/2010JA015399, 1-17.

Collier M.R., J.A. Slavin, R.P. Lepping, K. Ogilvie, A. Szabo, H. Laakso, and S. Taguchi, 1998. "Multispacecraft observations of sudden impulses in the magnetotail caused by solar wind pressure discontinuities: WIND and IMP 8", *J. Geophys. Res*., **103**, No. A8, 17293-17305.

Connor H.K., E. Zesta, D.M. Ober, and J. Raeder, 2014. “The relation between transpolar potential and reconnection rates during sudden enhancement of solar wind dynamic pressure: Open GGCM-CTIM results”, *J. Geophys. Res. Space Physics*, **119**, No. 5, doi:10.1002/2013JA019728, 3411-3429.

Cowley S.W.H. and M. Lockwood, 1992. “Excitation and decay of solar wind-driven flows in the magnetosphere-ionosphere system”, *Ann. Geophys., Atmos. Hydrospheres Space Sci.* (France), **10**, No. 1-2, 103-115.

Cowley S.W.H., S.V. Badman, E.J. Bunce, J.T. Clarke, J.-C. Gérard, D. Grodent, C.M. Jackman, S.E. Milan, and T.K. Yeoman, 2005. “Reconnection in a rotation-dominated magnetosphere and its relation to Saturn’s auroral dynamics”, *J. Geophys. Res.*, **110**, No. A2, A02201, doi:10.1029/2004JA010796, 1-19.

Crooker N.U., 1979. "Dayside merging and cusp geometry", *J. Geophys. Res.*, **84**, No. A3, 951-959.

Dang T., X. Luan, J. Lei, X. Dou, and W. Wan, 2016. “A numerical study of the interhemispheric asymmetry of the equatorial ionization anomaly in solstice at solar minimum”, *J. Geophys. Res. Space Physics*, **121**, 9099–9110, doi:10.1002/2016JA023012.

Delcourt D., J.A. Sauvaud, and A. Pedersen, 1990. "Dynamics of single particle orbits during substorm expansion", *J. Geophys. Res.*, **95**, No. A12, 20853-20856.

DeMajistre R., E.C. Roelof, P.C. Brandt, and D.G. Mitchell, 2004. "Retrieval of global magnetospheric ion distributions from high energy neutral atom measurements made by the IMAGE/HENA instrument", *J. Geophys. Res.*, **109**, No. A4, A04214, doi:10.1029/2003JA010322, 1-11.

Erlandson R.E., D.G. Siebeck, R.E. Lopez, L.J. Zanetti, and T.A. Potemra, 1991. “Observations of solar wind pressure initiated fast mode waves at geostationary orbit and in the polar cap”, *J. Atmos. Terr. Phys. (UK)*, **53**, No, 3-4, 231–239.

Fairfield D.H. and J. Jones, 1996. "Variability of the tail lobe field strength", *J. Geophys. Res.*, **101**, No. A4, 7785-7791.

Fok M.-C., T.E. Moore, P.C. Brandt, D.C. Delcourt, S.P. Slinker, and J.A. Fedder, 2006. "Impulsive enhancements of oxygen ions during substorms", *J. Geophys. Res.*, **111**, No. A10, A10222, doi:10.1029/2006JA011839, 1-14.

Gérard J.C., J. Gustin, A. Saglam, J.T. Clarke, and J.T. Trauger, 2004. “Characteristics of Saturn’s FUV aurora observed with the Space Telescope Imaging Spectrograph”, *J. Geophys. Res.*, **109**, No. A9, A09207, doi:10.1029/2004JA010513, 1-10.

Gillies D.M., J.-P. St.-Maurice, K.A. McWilliams, and S. Milan, 2012. “Global-scale observations of ionospheric convection variation in response to sudden increases in the solar wind dynamic pressure”, *J. Geophys. Res.*, **117**, No. A4, A04209, doi:10.1029/2011JA017255, 1-18.

Greenwald R.A.et al., 1995. "DARN/Super DARN: A global view of the dynamics of high-latitude convection", *Space Sci. Rev.*, **71**, No. 1-4, 761-796.

Grib S.A., 1982. “Interaction of non-perpendicular/parallel solar wind shock waves with the Earth’s magnetosphere”, *Space Sci. Rev.* **32**, No.1, 43–48.

Grib S.A. and V.B. Belakhovsky, 2009. “Effect of the Interplanetary Secondary Rarefaction Waves on the Geomagnetic Field”, *Geomagnetism and Aeronomy,* **49**, No. 6, 733–740.

Grib S.A. and Martynov M.V., 1977.“Shock Wave Formation in the Magnetic Transition Layer before the Earth’s Magnetosphere”, *Geomagn. Aeron.* **17**, No. 2, 252–258.

Grib S.A., B.E. Briunelli, M. Dryer, and W.-W. Shen, 1979. "Interaction of interplanetary shock waves with the bow shock-magnetopause system", *J. Geophys. Res.*, **84**, No. A10, 5907-5922.

Han D.-S., H.-G. Yang, Z.-T. Chen, T. Araki, M.W. Dunlop, M. Nosé, T. Iyemori, Q. Li, Y.-F. Gao, and K. Yumoto, 2007. "Coupling of perturbations in the solar wind density to global Pi3 pulsations: A case study", *J. Geophys. Res.*, **112**, A05217, doi:10.1029/2006JA011675.

Henderson M.G., G.D. Reeves, H.E. Spence, R.B. Sheldon, A.M. Jorgensen, J.B. Blake, and J.F. Fennell, 1997. "First energetic neutral atom images from Polar", *Geophys. Res. Lett.*, **24**, No. 10, 1167-1170.

Heppner J.P., 1955. "Note on the occurrence of world-wide SSC's during the onset of negative bays at College", *J. Geophys. Res.*, **60**, No. 1, 29-32.

Hu Y.Q., X.C. Guo, and C. Wang, 2007. "On the ionospheric and reconnection potentials of the earth: Results from global MHD simulations", *J. Geophys. Res.*, **112**, A07215, doi:10.1029/2006JA012145, 1-8.

Huang C.-S., Reeves G.D., Le G., and Yumoto K., 2005. "Are sawtooth oscillations of energetic plasma particle fluxes caused by periodic substorms or driven by solar wind pressure enhancements?", *J. Geophys. Res.*, **110**, No. A7, A07207, doi:10.1029/2005JA011018, 1-19.

Huang C.-S., S. Sazykin, J. Chao, N. Maruyama, and M.C. Kelley, 2007. "Penetration electric fields: Efficiency and characteristic time scale", *J. Atmos. Sol. Terr. Phys.*, **69**, No. 10-11, doi: 10.1016/j.jastp.2006.08.016, 1135-1146.

Hubert B., M. Palmroth, T.V. Laitinen, P. Janhunen, S.E. Milan, A. Grocott, S.W.H. Cowley, T. Pulkkinen, and J.-C. Gérard, 2006. "Compression of the Earth's magnetotail by interplanetary shocks directly drives transient magnetic flux closure", *Geophys. Res. Lett.*, **33**, No. 10, L10105, doi:10.1029/2006GL026008, 1-4.

Jackel B.J., B. McKiernan, and H.J. Singer, 2012. "Geostationary magnetic field response to solar wind pressure variations: Time delay and local time variation", *J. Geophys. Res.*, **117**, No. A5, A05203, doi:10.1029/2011JA017210, 1-11.

Jacobs J.A., Y. Kato, S. Matsushita, and V.A. Troitskaya, 1964. "Classification of geomagnetic micropulsations", *J. Geophys. Res.*, **69**, No. 1, 180–181.

Janzhura A., O. Troshichev, and P. Stauning, 2007. "Unified PC indices: Relation to the isolated magnetic substorms", *J. Geophys. Res.*, **112**, No. A9, A09207, doi:10.1029/2006JA012132.

Jones S.T., M.-C. Fok, and P.C. Brandt, 2006. "Modeling global O^+ substorm injection using analytic magnetic field model", *J. Geophys. Res.*, **111**, No. A11, A11S07, doi:10.1029/2006JA011607, 1-8.

Jorgensen A.M., L. Kepko, M.G. Henderson, H.E. Spence, G.D. Reeves, J.B. Sigwarth, and L.A. Frank, 2000. "Association of energetic neutral atom bursts and magnetospheric substorms", *J. Geophys. Res.*, **105**, No. A8, 18753-18763.

Kan J.R. and L.C. Lee, 1979. "Energy coupling function and solar wind-magnetosphere dynamo", *Geophys. Res. Lett.*, **6**, No. 7, 577-580.

Kauristie K., T.I. Pulkkinen, R.J. Pellinen, and H.J. Opgenoorth, 1997. "What can we tell about global auroral-electrojet activity from a single meridional magnetometer chain?", *Ann. Geophys.*, **14**, No. 11, 1177-1185

Kawano H., T. Yamamoto, S. Kokubun, and R.P. Lepping, 1992. "Rotational polarities of sudden impulses in the magnetotail lobe", *J. Geophys. Res.*, **97**, No. A11, 17177-17182.

Kawasaki K., S.-I. Akasofu, F. Yasuhara, and C.-I. Meng, 1971. "Storm sudden commencements and polar magnetic substorms", *J. Geophys. Res.*, **76**, No. 28, 6781-6789.

Kepko L., and H.E. Spence, 2003. "Observations of discrete, global magnetospheric oscillations directly driven by solar wind density variations", *J. Geophys. Res.*, **108**, No. A6, 1257, doi:10.1029/2002JA009676.

Kim K.-H., Cattell C.A., Lee D.-H., Balogh A., Lucek E., Andre M., Khotyaintsev Y., and Rème H., 2004. "Cluster observations in the magnetotail during sudden and quasiperiodic solar wind variations", *J. Geophys. Res.*, **109**, No. A4, A04219, doi:10.1029/2003JA010328, 1-10.

Kim H., X. Cai, C.R. Clauer, B.S.R. Kunduri, J. Matzka, C. Stolle, and D.R. Weimer, 2013. "Geomagnetic response to solar wind dynamic pressure impulse events at high-latitude conjugate points", *J. Geophys. Res. Space Physics*, **118**, No. 10, doi:10.1002/jgra.50555, 6055-6071.

Klibanova Yu.Yu., V.V. Mishin, and B. Tsegmed, 2014. "Specific features of daytime long-period pulsations observed during the solar wind impulse against a background of the substorm of August 1, 1998", *Cosmic Res.*, **52**, No. 6, 421-429. Russian text: *Kosmicheskie Issledovaniya*, **52**, No. 6, 459-467.

Kokubun S., R.L. McPherron, and C.T. Russell, 1997. "Triggering of substorms by solar wind discontinuities", *J. Geophys. Res.*, **82**, No. 1, 74-86.

Korotova G.I. and D.G. Sibeck, 1995. "A case study of transient event motion in the magnetosphere and in the ionosphere", *J. Geophys. Res.*, **100**, No. A1, 35–46.

Kozlovsky A., V. Safargaleev, N. Østgaard, T. Turunen, A. Koustov, J. Jussila, and A. Roldugin, 2005. "On the motion of dayside auroras caused by a solar wind pressure pulse", *Ann. Geophys.*, **23**, No. 2, 509-521.

Lavraud B., E. Larroque, E. Budnik, V. Génot, J.E. Borovsky, M.W. Dunlop, C. Foullon, H. Hasegawa, C. Jacquey, K. Nykyri, A. Ruffenach, M.G.G.T. Taylor, I. Dandouras, and H. Rème, 2013. "Asymmetry of magnetosheath flows and magnetopause shape during low Alfve´n Mach number solar wind", *J. Geophys. Res. Space Physics*, **118**, No. 3, doi:10.1002/jgra.50145, 1089-1100.

Lee D.-Y. and L.R. Lyons, 2004. "Geosynchronous magnetic field response to solar wind dynamic pressure pulse", *J. Geophys. Res.*, **109**, No. A4, A04201, doi:10.1029/2003JA010076, 1-16.

Lee D.-Y., L.R. Lyons, and K. Yumoto, 2004. "Sawtooth oscillations directly driven by solar wind dynamic pressure enhancements", *J. Geophys. Res.*, **109**, No. A4, A04202, doi:10.1029/2003JA010246, 1-16.

Lee D.-Y., L.R. Lyons, and G.D. Reeves, 2005 "Comparison of geosynchronous energetic particle flux responses to solar wind dynamic pressure enhancements and substorms", *J. Geophys. Res.*, **110**, No. A9, A09213, doi:10.1029/2005JA011091, 1-20.

Lee D.-Y., S. Ohtani, P.C. Brandt, and L.R. Lyons, 2007a. "Energetic neutral atom response to solar wind dynamic pressure enhancements", *J. Geophys. Res.*, **112**, A09210, doi:10.1029/2007JA012399, 1-19.

Lee D.-Y., L.R. Lyons, J.M. Weygand, and C.-P. Wang, 2007b. "Reasons why some solar wind changes do not trigger substorms", *J. Geophys. Res.*, **112**, A06240, doi:10.1029/2007JA012249, 1-17.

Li X., D.N. Baker, M. Temerin, T. Cayton, G.D. Reeves, T. Araki, H. Singer, D. Larson, R.P. Lin, and S.G. Kanekal, 1998. "Energetic electron injections into the inner magnetosphere during the Jan. 10–11, 1997 magnetic storm", *Geophys. Res. Lett.*, **25**, No. 14, 2561-2564.

Li X., D.N. Baker, S. Elkington, M. Temerin, G.D. Reeves, R.D. Belian, J.B. Blake, H.J. Singer, W. Peria, and G. Parks, 2003. "Energetic particle injections in the inner magnetosphere as a response to an interplanetary shock", *J. Atmos. Sol. Terr. Phys.*, **65**, No. 2, 233-244.

Li L.Y., J.B. Cao, G.C. Zhou, T.L. Zhang, D. Zhang, I. Dandouras, H. Rème, and C.M. Carr, 2011. "Multiple responses of magnetotail to the enhancement and fluctuation of solar wind dynamic pressure and the southward turning of interplanetary magnetic field", *J. Geophys. Res.*, **116**, No. A12, A12223, doi:10.1029/2011JA016816, 1-11.

Li L.Y., J.B. Cao, J.Y. Yang, and Y.X. Dong, 2013. "Joint responses of geosynchronous magnetic field and relativistic electrons to external changes in solar wind dynamic pressure and interplanetary magnetic field", *J. Geophys. Res. Space Physics*, **118**, No. 4, doi:10.1002/jgra.50201, 1472-1482.

Liou K., P.T. Newell, C.-I. Meng, C.-C. Wu, and R.P. Lepping, 2003. "Investigation of external triggering of substorms with Polar ultraviolet imager observations", *J. Geophys. Res.*, **108**, No. A10, 1364, doi:10.1029/2003JA009984, SIA2-1-14.

Liou K., P.T. Newell, C.-I. Meng, C.-C. Wu, and R.P. Lepping, 2004. "On the relationship between shock-induced polar magnetic bays and solar wind parameters", *J. Geophys. Res.*, **109**, No. A6, A06306, doi:10.1029/2004JA010400, 1-8.

Liou K., P.T. Newell, T. Sotirelis, and C.-I. Meng, 2006. "Global auroral response to negative pressure impulses", *Geophys. Res. Lett.*, **33**, No. 11, L11103, doi:10.1029 /2006GL025933, 1-5.

Liu W.W., 2012. "Interaction of solar wind pressure pulses with the magnetosphere: IMF modulation and cavity mode", *J. Geophys. Res.*, **117**, No. A8, A08234, doi:10.1029/2012JA017904, 1-13.

Lopez R.E., M. Wiltberger, S. Hernandez, and J.G. Lyon, 2004. "Solar wind density control of energy transfer to the magnetosphere", *Geophys. Res. Lett.,* **31**, No. 8, L08804, doi:10.1029/2003GL018780, 1-4.

Lui A.T.Y., D.J. Williams, E.C. Roelof, R.W. McEntire, and D.G. Mitchell, 1996. "First composition measurements of energetic neutral atoms", *Geophys. Res. Lett.*, **23**, No. 19, 2641-2644.

Lui A.T.Y., C. Jacquey, G.S. Lakhina, R. Lundin, T. Nagai, T.-D. Phan, Z.Y. Pu, M. Roth, Y. Song, R.A. Treumann, M. Yamauchi, and L.M. Zelenyi, 2005a. "Critical issues on magnetic reconnection in space plasmas", *Space Sci. Rev.*, **116**, No. 3-4, 497-521.

Lui A.T.Y., P.C. Brandt, and D.G. Mitchell, 2005b. "Observations of energetic neutral oxygen by IMAGE/HENA and Geotail/EPIC", *Geophys. Res. Lett.*, **32**, No. 13, L13104, doi:10.1029/2005GL022851, 1-4.

Lukianova R., 2003. "Magnetospheric response to the solar wind dynamic pressure inferred from polar cap index", *J. Geophys. Res.*, **108**, No. A12, 1428, doi:10.1029/2002JA 009790, SMP11-1-16.

Lukianova R. and O. Troshichev, 2002. “Magnetospheric response to the solar wind dynamic pressure inferred from the polar cap index”, in *Proceedings of the Sixth International Conference on Substorms, (ICS-6)*, edited by R.M. Winglee, Univ. of Wash. Press, Seattle, 99-104.

Lyons L.R., 2000. "Geomagnetic disturbances: Characteristics of distinction between types, and relations to interplanetary conditions", *J. Atmos. Sol. Terr. Phys.*, **62**, No. 12, 1087-1114.

Lyons L.R., E. Zesta, J.C. Samson, and G.D. Reeves, 2000. "Auroral disturbances during the January 10, 1997 magnetic storm", *Geophys. Res. Lett.*, **27**, No. 20, 3237-3240.

Lyons L.R., D.-Y. Lee, C.-P. Wang, and S. Mende, 2005. “Global auroral responses to abrupt solar wind changes: Dynamic pressure, substorm, and null events”, *J. Geophys. Res.*, **110**, No. A8, A08208, doi:10.1029/2005JA011089, 1-15.

Matsuoka H., K. Takahashi, K. Yumoto, B.J. Anderson, and D.G. Sibeck, 1995. “Observation and modeling of compressional Pi 3 magnetic pulsations”, *J. Geophys. Res.*, **100**, No. A7, 12103–12115.

McComas D.J., S.J. Bame, P. Barker, W.C. Feldman, J.L. Phillips, and P. Riley, 1998. “Solar wind electron proton alpha monitor (SWEPAM) for the Advanced Composition Explorer”, *Space Sci. Rev.*, **86**, No. 1-4, 563-612.

Meurant M., J.-C. Gérard, C. Blockx, B. Hubert, and V. Coumans, 2004. "Propagation of electron and proton shock-induced aurora and the role of the interplanetary magnetic field and solar wind", *J. Geophys. Res.*, **109**, No. A10, A10210, doi:10.1029/2004JA010453, 1-19.

Milan S.E., S.W.H. Cowley, M. Lester, D.M. Wright, J.A. Slavin, M. Fillingim, C.W. Carlson, and H.J. Singer, 2004. “Response of the magnetotail to changes in open flux content of the magnetosphere”, *J. Geophys. Res.*, **109**, No. A4, A04220, doi:10.1029/2003JA010350, 1-16.

Milan S.E., G. Provan, and B. Hubert, 2007. “Magnetic flux transport in the Dungey cycle: A survey of dayside and nightside reconnection rates”, *J. Geophys. Res.*, **112**, A01209, doi:10.1029/2006JA011642, 1-13.

Mitchell D.G., S.M. Krimigis, A.F. Cheng, S.E. Jaskulek, E.P. Keath, B.H. Mauk, R.W. McEntire, E.C. Roelof, C.E. Schlemm, B.E. Tossman, and D.J. Williams, 1998. "The Imaging Neutral Camera for the Cassini Mission to Saturn and Titan", in *Measurement Technique in Space Plasmas: Fluids*, Geophys. Monogr. Ser., **103**, edited by R.F. Pfaff, J.E. Borovsky, and D.T. Young, AGU, Washington, D.C., 281-288.

Mitchell D.G., S.E. Jaskulek, C.E. Schlemm, E.P. Keath, R.E. Thompson, B.E. Tossman, J.D. Boldt, J.R. Hayes, G.B. Andrews, N. Paschalidis, D.C. Hamtiton, R.A. Lundgren, E.O. Tums, P. Wilson IV, H.D. Voss, D. Prentice, K.C. Hsieh, C.C. Curtis, and F.R. Powell, 2000. “High energy neutral atom (HENA) imager for the IMAGE mission”, *Space Sci. Rev.*, **91**, No. 1-2, 67-112.

Mitchell D.G., P.C:Son Brandt, E.C. Roelof, D.C. Hamilton, K.C. Retterer, and S. Mende, 2003."Global imaging of O^+ from IMAGE/HENA", *Space Sci. Rev.*, **109**, No. 1, 63-75.

Motoba T., T. Kikuchi, T. Okuzawa, and K. Yumoto, 2003. “Dynamical response of the magnetosphere-ionosphere system to a solar wind dynamic pressure oscillation”, *J. Geophys. Res.*, **108**, No. A5, 1206, doi:10.1029/2002JA009696.

Murr D.L. and W.J. Hughes, 2001. "Reconfiguration timescales of ionospheric convection", *Geophys. Res. Lett.*, **28**, No. 11, 2145-2148.

Murr D.L., W.J. Hughes, A.S. Rodger, E. Zesta, H.U. Frey, and A.T. Weatherwax, 2002. "Conjugate observations of traveling convection vortices: The field-aligned current system", *J. Geophys. Res.*, **107**, No. A10, 1306, doi:10.1029/2002JA009456, SIA14-1-13.

Nakai H., Y. Kamide, and C.T. Russell, 1991. "Influences of solar wind parameters and geomagnetic activity on the tail lobe magnetic field: A statistical study", *J. Geophys. Res.*, **96**, No. A4, 5511-5523.

Newell P.T. and I. Meng, 1994. "Ionospheric projections of magnetospheric regions under low and high solar wind pressure conditions", *J. Geophys. Res.*, **99**, No. A1, 273-286.

Newell P.T., T. Sotirelis, K. Liou, C.-I. Meng, and F.J. Rich, 2007. "A nearly universal solar wind-magnetosphere coupling function inferred from 10 magnetospheric state variables", *J. Geophys. Res.*, **112**, A01206, doi:10.1029/2006JA012015, 1-16.

Nishimura Y., L.R. Lyons, Y. Zou, K. Oksavik, J.I. Moen, L.B. Clausen, E.F. Donovan, V. Angelopoulos, K. Shiokawa, J.M. Ruohoniemi, N. Nishitani, K.A. McWilliams, and M. Lester, 2014. "Day-night coupling by a localized flow channel visualized by polar cap patch propagation", *Geophys. Res. Lett.*, **41**, No. 11, doi:10.1002/2014GL060301, 3701-3709.

Nosé M., A.T.Y. Lui, S. Ohtani, B.H. Mauk, R.W. McEntire, D.J. Williams, T. Mukai, and K. Yumoto, 2000. "Acceleration of oxygen ions of ionospheric origin in the near-Earth magnetotail during substorms", *J. Geophys. Res.*, **105**, No. A4, 7669-7677.

Nosé M., S. Taguchi, K. Hosokawa, S.P. Christon, R.W. McEntire, T.E. Moore, and M.R. Collier, 2005. "Overwhelming O^+ contribution to the plasma sheet energy density during the October 2003 superstorm: Geotail/EPIC and IMAGE/LENA observations", *J. Geophys. Res.*, **110**, No. A9, A09S24, doi:10.1029/2004JA010930, 1-8.

Ober D.M., G.R. Wilson, N.C. Maynard, W.J. Burke, and K.D. Siebert, 2006. "MHD simulation of the transpolar potential after a solar-wind density pulse", *Geophys. Res. Lett.*, **33**, No. 4, L04106, doi:10.1029/2005GL024655, 1-4.

Ogilvie K.W., L.F. Burlaga, and T.D. Wilkerson, 1968. "Plasma observations on Explorer 34", *J. Geophys. Res.*, **73**, No. 21, 6809-6824.

Ohtani S. and T. Uozumi, 2014. "Nightside magnetospheric current circuit: Time constants of the solar wind-magnetosphere coupling", *J. Geophys. Res. Space Physics*, **119**, No. 5, doi:10.1002/2013JA019680, 3558-3572.

Ohtani S., P.C. Brandt, D.G. Mitchell, H. Singer, M. Nosé, G.D. Reeves, and S.B. Mende, 2005 "Storm-substorm relationship: Variations of the hydrogen and oxygen energetic neutral atom intensities during storm-time substorms", *J. Geophys. Res.*, **110**, No. A7, A07219, doi:10.1029/2004JA010954, 1-14.

Ohtani S., P.C. Brandt, H. Singer, D.G. Mitchell, and E.C. Roelof, 2006. "Statistical Characteristics of the Hydrogen and Oxygen ENA Intensities of the Storm-time Ring Current", *J. Geophys. Res.*, **111**, A06209, doi:10.1029/2005JA011201, 1-9.

Ostapenko A.A. and Y.P. Maltsev, 1998. "Three-dimensional magnetospheric response to variations in the solar wind dynamic pressure", *Geophys. Res. Lett.*, **25**, No. 3, 261-263.

Ozturk D.S., S. Zou, and J.A. Slavin, 2017. "IMF B_y effects on ground magnetometer response to increased solar wind dynamic pressure derived from global MHD simulations", *J. Geophys. Res. Space Physics*, **122**, 5028–5042, doi:10.1002/2017JA023903.

Palmroth M., P. Janhunen, T.I. Pulkkinen, and H.E.J. Koskinen, 2004a. "Ionospheric energy input as a function of solar wind parameters: Global MHD simulation results", *Ann. Geophys.*, **22**, No. 2, 549-566.

Palmroth M., T.I. Pulkkinen, P. Janhunen, D.J. McComas, C.W. Smith, and H.E.J. Koskinen, 2004b. "Role of solar wind dynamic pressure in driving ionospheric Joule heating", *J. Geophys. Res.*, **109**, No. A11, A11302, doi:10.1029/2004JA010529, 1-7.

Palmroth M., N. Partamies, J. Polvi, T.I. Pulkkinen, D.J. McComas, R.J. Barnes, P. Stauning, C.W. Smith, H.J. Singer, and R. Vainio, 2007. "Solar wind-magnetosphere coupling efficiency for solar wind pressure impulses", *Geophys. Res. Lett.*, **34**, No. 11, L11101, doi:10.1029/2006GL029059, 1-5.

Parkhomov V.A., G.N. Zastenker, M.O. Riazantseva, B. Tsegmed, and T.A. Popova, 2010. "Bursts of geomagnetic pulsations in the frequency range 0.2-5 Hz excited by large changes of the solar wind pressure", *Cosmic Res.*, **48**, No. 1, 86-100. Russian Text: *Kosmicheskie Issledovaniya*, **48**, No. 1, 87-101.

Peng Z., C. Wang, Y.Q. Hu, J.R. Kan, and Y.F. Yang, 2011. "Simulations of observed auroral brightening caused by solar wind dynamic pressure enhancements under different interplanetary magnetic field conditions", *J. Geophys. Res.*, **116**, No. A6, A06217, doi:10.1029/2010JA016318, 1-11.

Perez J.D., G. Kozlowski, P.C. Brandt, D.G. Mitchell, J.-M. Jahn, C.J. Pollock, and X.X. Zhang, 2001. "Initial ion equatorial pitch angle distributions from energetic neutral atom images obtained by IMAGE", *Geophys. Res. Lett.*, **28**, No. 6, 1155-1158.

Perez J.D., X.-X. Zhang, P.C. Brandt, D.G. Mitchell, J.-M. Jahn, and C.J. Pollock, 2004. "Dynamics of ring current ions as obtained from IMAGE HENA and MENA ENA images", *J. Geophys. Res.*, **109**, A05208, doi:10.1029/2003JA010164.

Pollock C.J., P.C: Son-Brandt, J.L. Burch, M.G. Henderson, J.-M. Jahn, D.J. McComas, S.B. Mende, D.G. Mitchell, G.D. Reeves, E.E. Scime, R.M. Skoug, M. Thomsen, and P. Valek, 2003. "The role and contributions of energetic neutral atom (ENA) imaging in magnetospheric substorm research", *Space Sci. Rev.*, **109**, No. 1, 155-182.

Pulkkinen T.I., N. Partamies, R.L. McPherron, M. Henderson, G.D. Reeves, M.F. Thomsen, and H.J. Singer, 2007a. "Comparative statistical analysis of storm time activations and sawtooth events", *J. Geophys. Res.*, **112**, A01205, doi:10.1029/2006JA012024.

Reeves G.D. and M.G. Henderson, 2001. "The storm-substorm relationship: Ion injections in geosynchronous measurements and composite energetic neutral atom images, *J. Geophys. Res.*, **106**, No. A4, 5833-5844.

Reeves G., M.C. Henderson, R.M. Skoug, M.F. Thomsen, J.E. Borovsky, H.O. Funsten, P. C. Son Brandt, D.J. Mitchell, J.-M. Jahn, C.J. Pollock, D.J. McComas, and S.B. Mende, 2004. "Image, polar, and geosynchronous observations of substorm and ring current ion injection", in *Disturbances in Geospace: The Storm-Substorm Relationship*, Geophys. Monogr. Ser., **142**, edited by A.S. Sharma, Y. Kamide, and G.S. Lakhina, AGU, Washington, D.C., 91-101.

Roelof E.C., 1987. "Energetic neutral atom image of a storm-time ring current", *Geophys. Res. Lett.*, **14**, No. 6, 652-655.

Roelof E.C. and A.J. Skinner, 2000. "Extraction of ion distributions from magnetospheric ENA and EUV images", *Space Sci. Rev.*, **91**, No. 1-2, 437-459.

Rufenach C.L., R.L. McPherron, and J. Schaper, 1992. "The quit geomagnetic field at geosynchronous orbit and its dependence on solar wind dynamic pressure", *J. Geophys. Res.*, **97**, No. A1, 25-42.

Ruohoniemi J.M. and K.B. Baker, 1998. "Large-scale imaging of highlatitude convection with Super Dual Auroral Radar Network HF radar observations", *J. Geophys. Res.*, **103**, No. A9, 20797-20811.

Ruohoniemi J.M. and R.A. Greenwald, 1998. "Response of high-latitude convection to a sudden southward IMF turning", *Geophys. Res. Lett.*, **25**, No. 15, 2913-2916.

Russell C.T., Ginskey M., Petrinec S., and Le G., 1992. "The Effect of Solar Wind Dynamic Pressure Changes on Low and Mid-Latitude Magnetic Records", *Geophys. Res. Lett.* **19**, No. 12, 1227–1230.

Russell C.T., M. Ginskey, and S.M. Petrinec, 1994. "Sudden impulses at low-latitude stations: Steady state response for northward interplanetary magnetic field, *J. Geophys. Res.*, **99**, No. A1, 253-261.

Ŝafránková J., Z. Němeček, L. Přech, A.A. Samsonov, A. Koval, and K. Andréeová, 2007. "Modification of interplanetary shocks near the bow shock and through the magnetosheath", *J. Geophys. Res.*, **112**, A08212, doi:10.1029/2007JA012503, 1-9.

Samsonov A.A., Z. Němeček, and J. Šafránková, 2006. "Numerical MHD modeling of propagation of interplanetary shock through the magnetosheath", *J. Geophys. Res.*, **111**, No. A8, A08210, doi:10.1029/2005JA011537, 1-9.

Samsonov A.A., D.G. Sibeck, and J. Imber, 2007. "MHD simulation for the interaction of an interplanetary shock with the Earth's magnetosphere", *J. Geophys. Res.*, **112**, A12220, doi:10.1029/2007JA012627, 1-9.

Sanny J., J.A. Tapia, D.G. Sibeck, and M.B. Moldwin, 2002. "Quiet time variability of the geosynchronous magnetic field and its response to the solar wind", *J. Geophys. Res.*, **107**, No. A12, 1443, doi:10.1029/2002JA009448, SMP16-1-10.

Sastri J.H., K. Yumoto, J.V.S.V. Rao, and R. Subbiah, 2006. "On the nature of response of dayside equatorial geomagnetic H-field to sudden magnetospheric compressions", *J. Atmos. Solar-Terr. Phys.*, **68**, No.14, 1642-1652.

Schieldge J.P. and G.L. Siscoe, 1970. "A correlation of the occurrence of simultaneous sudden magnetospheric compression and geomagnetic bay onsets with selected geophysical indices", *J. Atmos. Terr. Phys.*, **32**, No. 11, 1819-1830.

Shi Y., E. Zesta, L.R. Lyons, A. Boudouridis, K. Yumoto, and K. Kitamura, 2005. "Effect of solar wind pressure enhancements on storm time ring current asymmetry", *J. Geophys. Res.*, **110**, No. A10, A10205, doi:10.1029/2005JA011019, 1-19.

Shi Y., E. Zesta, L.R. Lyons, K. Yumoto, and K. Kitamura, 2006. "Statistical study of effect of solar wind dynamic pressure enhancements on dawn-to-dusk ring current asymmetry", *J. Geophys. Res.*, **111**, No. A10, A10216, doi:10.1029/2005JA011532, 1-11.

Shi Y., E. Zesta, and L.R. Lyons, 2008a. "Modeling magnetospheric current response to solar wind dynamic pressure enhancements during magnetic storms: 1. Methodology and results of the 25 September 1998 peak main phase case", *J. Geophys. Res.*, **113**, No. A10, A10218, doi:10.1029/2008JA013111, 1-13.

Shi Y., E. Zesta, and L.R. Lyons, 2008b. "Modeling magnetospheric current response to solar wind dynamic pressure enhancements during magnetic storms: 2. Application to different storm phases", *J. Geophys. Res.*, **113**, No. A10, A10219, doi:10.1029/2008JA013420, 1-17.

Shi Q.Q., M.D. Hartinger, V. Angelopoulos, A.M. Tian, S.Y. Fu, Q.-G. Zong, J. M. Weygand, J. Raeder, Z.Y. Pu, X.Z. Zhou, M.W. Dunlop, W.L. Liu, H. Zhang, Z.H. Yao, and X.C. Shen, 2014. "Solar wind pressure pulse-driven magnetospheric vortices and their global consequences", *J. Geophys. Res. Space Physics*, **119**, No. 6, doi:10.1002/2013JA019551, 4274-4280.

Shue J.-H. and Y. Kamide, 2001. "Effects of solar wind density on auroral electrojets", *Geophys. Res. Lett.*, **28**, No. 11, 2181-2184.

Shue J.-H., P. Song, C.T. Russell, J.T. Steinberg, J.K. Chao, G. Zastenker, O.L. Vaisberg, S. Kokubun, H.J. Siager, T.R. Detman, and H. Kawano, 1998. "Magnetopause location under extreme solar wind conditions", *J. Geophys. Res.*, **103**, No. A8, 17691-17700.

Sibeck D.G., 1990. "A model for the transient magnetospheric response to sudden solar wind dynamic pressure variations", *J. Geophys. Res.*, **95**, No. A4, doi:10.1029/JA095iA04 p03755, 3755–3771.

Sibeck D.G. and D.J. Croley, Jr., 1991. "Solar wind dynamic pressure variations and possible ground signatures of flux transfer events", *J. Geophys. Res.*, **96**, No. A2, 1669-1683.

Sibeck D.G., W. Baumjohann, R.C. Elphic, D.H. Fairfield, J.F. Fennell, W.B. Gail, L.J. Lanzerotti, R.E. Lopez, H. Luehr, A.T.Y. Lui, C.G. Maclennan, R.W. McEntire, T.A. Potemra, T.J. Rosenberg, and K. Takahashi, 1989. "The magnetospheric response to 8-minute period strong-amplitude upstream pressure variations", *J. Geophys. Res.*, **94**, No. A3, doi:10.1029/JA094iA03p02505, 2505-2519.

Sibeck D.G., N.L. Borodkova, and G.N. Zastenker, 1996. "Solar wind variations as a source of short-term magnetic field disturbances in the dayside magnetosphere", *Cosmic Res.*, **34**, No. 3, 228-242.

Siscoe G.L., G.M. Erickson, B.U.Ö. Sonnerup, N.C. Maynard, J.A. Schoendorf, K.D. Siebert, D.R. Weimer, W.W. White, and G.R. Wilson, 2002. "Hill model of transpolar potential saturation: Comparisons with MHD simulations", *J. Geophys. Res.*, **107**, No. A6, 1075, doi:10.1029/2001JA000109, SMP8-1-8.

Smith C.W., J. L'Heureux, N.F. Ness, M.H. Acuna, L.F. Burlaga, and J. Scheiffele, 1998. "The ACE magnetic fields experiment", *Space Sci. Rev.*, **86**, No. 1-4, 613-632.

Solovyev S.I., A.V. Moiseyev, V.A. Mullayarov, A. Du, M. Engebretson, and L. Newitt, 2004. "Global geomagnetic response to a sharp compression of the magnetosphere and IMF variations on October 29, 2003", *Cosmic Res.*, **42**, No. 6, 597-606. Russian Text: *Kosmicheskie Issledovaniya*, **42**, No. 6, 622-631.

Solovyev S.I., A.V. Moiseev, E.S. Barkova, K. Yumoto, and M. Engebretson, 2006. "Global geomagnetic and auroral response to variations in the solar wind dynamic pressure on April 1, 1997", *Geomagnetism and Aeronomy*, **46**, No. 1, 41-51. Russian Text: *Geomagnetizm i Aeronomiya*, **46**, No. 1, 44-54.

Tian A.M., Q.G. Zong, Y.F. Wang, Q.Q. Shi, S.Y. Fu, and Z.Y. Pu, 2010. "A series of plasma flow vortices in the tail plasma sheet associated with solar wind pressure enhancement", *J. Geophys. Res.*, **115**, No. A9, A09204, doi:10.1029/2009JA014989, 1-15.

Troshichev O.A. and V.G. Andrezen, 1985. "The relationship between interplanetary quantities and magnetic activity in the southern polar cap", *Planet. Space Sci.*, **33**, No. 4, 415-419.

Troshichev O.A., V.G. Andrezen, S. Vennerstrøm, and E. Friis-Christensen, 1988. "Magnetic activity in the polar cap-A new index", *Planet. Space Sci.*, **36**, No. 11, 1095-1102.

Troshichev O.A., R. Yu. Lukianova, V.O. Papitashvili, F.J. Rich, and O. Rasmussen, 2000. "Polar cap index (PC) as a proxy for ionospheric electric field in the near-pole region", *Geophys. Res. Lett.*, **27**, No. 23, 3809-3812.

Troshichev O., A. Janzhura, and P. Stauning, 2006. "Unified PCN and PCS indices: Method of calculation, physical sense, and dependence on the IMF azimuthal and northward components", *J. Geophys. Res.*, **111**, No. A5, A05208, doi:10.1029/2005JA011402, 1-10.

Troshichev O.A., A.S. Janzhura, and P. Stauning, 2007. "Magnetic activity in the polar caps: Relation to sudden changes in the solar wind dynamic pressure", *J. Geophys. Res.*, **112**, A11202, doi:10.1029/2007JA012369, 1-10.

Tsyganenko N.A. and I.M. Sitnov, 2005. "Modeling the dynamics of the inner magnetosphere during strong geomagnetic storms", *J. Geophys. Res.*, **110**, No. A3, A03208, doi:10.1029/2004JA010798, 1-16.

Vallat C. et al., 2004. "First comparisons of local ion measurements in the inner magnetosphere with energetic neutral atom magnetospheric image inversions: Cluster-CIS and IMAGE-HENA observations", *J. Geophys. Res.*, **109**, No. A4, A04213, doi:10.1029/2003JA010224, 1-10.

Viall N.M., L. Kepko, and H.E. Spence, 2009. "Relative occurrence rates and connection of discrete frequency oscillations in the solar wind density and dayside magnetosphere", *J. Geophys. Res.*, **114**, A01201, doi:10.1029/2008JA013334.

Villante U. and M. Piersanti, 2011. "Sudden impulses at geosynchronous orbit and at ground", *J. Atmos. Solar-Terr. Phys.*, **73**, No.1, 61-76.

Villante U., P. Francia, M. Vellante, and P. Di Giuseppe, 2003. "Some aspects of the low latitude geomagnetic response under different solar wind conditions", *Space Sci. Rev.*, **107**, No. 1-2, 207-217.

Vorobjev V.G., V.L. Zverev, and O.I. Yagodkina, 2009. "Response of Dayside Auroras to Abrupt Increases in the Solar Wind Dynamic Pressure at Positive and Negative Polarity of the IMF Bz Component", *Geomagnetism and Aeronomy*, **49**, No. 6, 712-721. Russian Text: *Geomagnetizm i Aeronomiya*, **49**, No. 6, 746-756.

Wang C., Z.H. Huang, Y.Q. Hu, and X.C. Guo, 2005 "3D global simulation of the interaction of the interplanetary shocks with the magnetosphere", In *The physics of collisionless shocks*, 4th Annual IGPP Intern. Astrophysics Conf., edited by Li G., G.P. Zank, C.T. Russell, AIP Conf. Proc. (USA), **781**, 320-324.

Wang C., C.X. Li, Z.H. Huang, and J.D. Richardson, 2006. "The effect of interplanetary shock strengths and orientations on storm sudden commencement rise times", *Geophys. Res. Lett.*, **33**, L14104, doi:10.1029/2006GL025966, 1-3.

Wang C., J.B. Liu, Z.H. Huang, and J.D. Richardson, 2007. "Response of the magnetic field in the geosynchronous orbit to solar wind dynamic pressure pulses", *J. Geophys. Res.*, **112**, A12210, doi:10.1029/2007JA012664, 1-8.

Weimer D.R.et al., 2003. "Predicting interplanetary magnetic field (IMF) propagation delay times using the minimum variance technique", *J. Geophys. Res.*, **108**, No.A1, 1026, doi:10.1029/2002JA009405, SMP16-1-12.

Wing S. and D.G. Sibeck, 1997. "Effects of interplanetary magnetic field z component and the solar wind dynamic pressure on the geosynchronous magnetic field", *J. Geophys. Res.*, **102**, No. A4, 7207-7216.

Wing S., D.G. Sibeck, M. Wiltberger, and H. Singer, 2002. "Geosynchronous magnetic field temporal response to solar wind and IMF variations",. *J. Geophys Res.*, **107**, No.A8, 1222, doi:10.1029/2001JA009156, SMP32-1-10.

Yuan Z. and X. Deng, 2007. "Effects of continuous solar wind pressure variations on the long-lasting penetration of the interplanetary electric field during southward interplanetary magnetic field", *Adv. Space Res.*, **39**, No. 8, 1342-1346.

Zesta E.et al., 2000a. "The effect of the January 10, 1997 pressure pulse on the magnetosphere–ionosphere current system", in *Magnetospheric Current Systems*, edited by S. Ohtani, R. Fujii, M. Hesse, and R.L. Lysak, Geophys. Monogr. Ser., **118**, AGU, Washington, D. C., 217-226.

Zesta E., L.R. Lyons, and E. Donovan, 2000b. "The auroral signature of Earthward flow bursts observed in the magnetotail", *Geophys. Res. Lett.*, **27**, No. 20, 3241-3244.

Zesta E., W.J. Hughes, and M.J. Engebretson, 2002. "A statistical study of traveling convection vortices using the Magnetometer Array for Cusp and Cleft Studies", *J. Geophys. Res.*, **107**, No. A10, 1317, doi:10.1029/1999JA000386, SIA18-1-21 (2002).

Zhang X.Y. et al., 2010. "ULF waves excited by negative/positive solar wind dynamic pressure impulses at geosynchronous orbit", *J. Geophys. Res.*, **115**, No. A10, A10221, doi:10.1029/2009JA015016, 1-14.

Zhou X. and B.T. Tsurutani, 1999. "Rapid intensification and propagation of the dayside aurora: Large scale interplanetary pressure pulses (fast shocks)", *Geophys. Res. Lett.*, **26**, No. 8, 1097-1100.

Zhou X. and B.T. Tsurutani, 2001. "Interplanetary shock triggering of nightside geomagnetic activity: Substorms, pseudobreakups, and quiescent events", *J. Geophys. Res.*, **106**, No. A9, 18957-18968.

Zhou X.-Y.et al., 2003. "Shock aurora: FAST and DMSP observations", *J. Geophys. Res.*, **108**, No. A4, 8019, doi:10.1029/2002JA009701.

Zhuang H.C., C. T. Russel, E. J. Smith, and J. T. Gosling, 1981."Three-Dimensional Interaction of Interplanetary Shock Waves with the Bow Shock and Magnetopause: A Comparison of Theory with ISEE Observations", *J. Geophys. Res.*, **86**, No. A7, 5590-5600.

Chapter 4

COUPLING SOLAR WIND AND IMF WITH HIGH LATITUDE MAGNETOSPHERIC AND AURORAL ACTIVITY

4.1. COUPLING OF SOLAR WIND AND IMF WITH AURORAL AND SUBSTORMS ACTIVITY

4.1.1. IMF B_Y Control of Dayside Auroras

The effect of the IMF B_Y component on the dayside auroral oval from Viking UV measurements for March–November 1986 is studied by Trondsen et al. (1999). Observations of dayside auroras from Viking UV images for large positive (15 cases) and negative (22 cases) IMF B_Y ($|B_Y|$>4 nT), suggest that: (1) the intensity of dayside auroras tends to increase for negative IMF By and to decrease for positive By, so that negative IMF B_Y conditions seem preferable for observations of dayside auroras; (2) for negative IMF B_Y, the auroral oval tends to be narrow and continuous throughout the noon meridian without any noon gap or any strong undulation in the auroral distribution. For positive IMF B_Y, a sharp decrease and spreading of auroral activity is frequently observed in the post-noon sector, a strong undulation in the poleward boundary of the auroral oval around noon, and the formation of auroral forms poleward of the oval; and (3) the observed features of dayside auroras are in reasonable agreement with the expected distribution of upward field-aligned currents associated with the IMF B_Y in the noon sector.

4.1.2. Radial Plasma Pressure Gradients in the High Latitude Magnetosphere as Sources of Instabilities Leading to the Substorm Onset

As noted Stepanova et al. (2004), various types of instabilities have been proposed as sources of inner magnetosphere substorm expansion phase onset. Distribution of plasma pressure in the plasma sheet, beta parameter, volume of magnetic flux tube are the key parameters for understanding the processes leading the instability development. The radial distribution of plasma pressure was recovered by Stepanova et al. (2004) using the low-altitude polar-orbiting Aureol-3 precipitating particle data. Corresponding variations in the magnetic pressure and in the volume of magnetic flux tube were estimated using Tsyganenko 96 magnetic field model. It is shown that development of an instability responsible for the substorm expansion phase onset takes place in the regions of low beta or at the boundary of the regions of high and low beta. A possible role of the flute-like (interchange-like) instabilities is investigated. It is shown that the obtained profiles of radial plasma pressure gradients apparently are not sharp enough to develop the instability related to radial plasma pressure gradients. The role of the instability related to the existence of the azimuthal plasma pressure gradients is discussed. Such an instability can develop when the density of the field-aligned current reaches a definite threshold value. The instability develops faster in the region of upward field-aligned current, where the existing field-aligned potential drop leads to the magnetosphere-ionosphere decoupling. Obtained results are illustrated by Figure 4.1.1 – 4.1.5 and by Table 4.1.1.

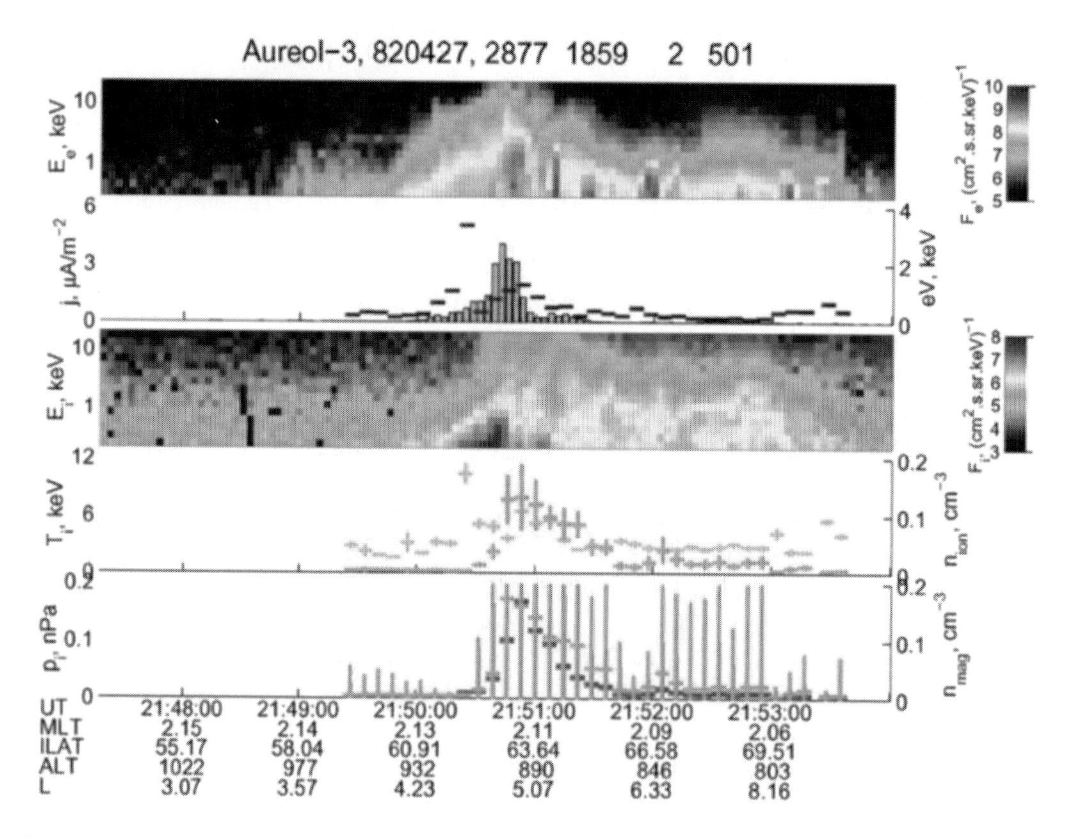

Figure 4.1.1. The April 27, 1982 event observed by the Aureol-3: (a) electron spectrogram, (b) field-aligned current (histrogram, left axis), and field-aligned potential drop (bars, right axis), (c) ion spectrogram, (d) ion temperature (gray bars, left axis), and ion concentration obtained from the ion fluxes (black bars, right axis), (e) magnetospheric pressure (gray bars, left axis), and ion concentration in the magnetosphere (black bars, right axis). According to Stepanova et al. (2004). Permission from COSPAR.

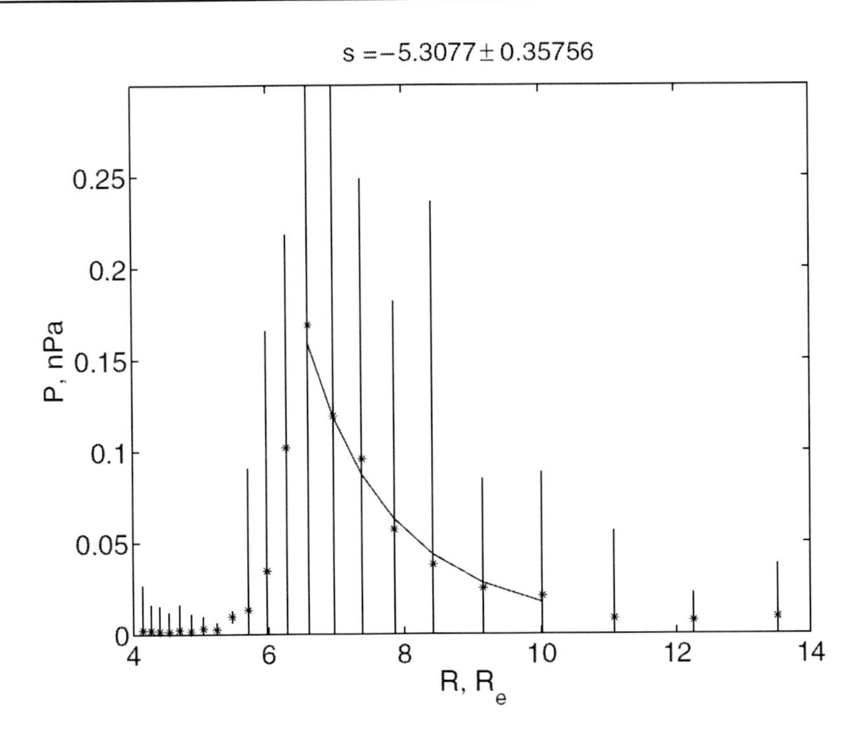

Figure 4.1.2. Radial profile of plasma pressure obtained for the April 27, 1982 event. According to Stepanova et al. (2004). Permission from COSPAR.

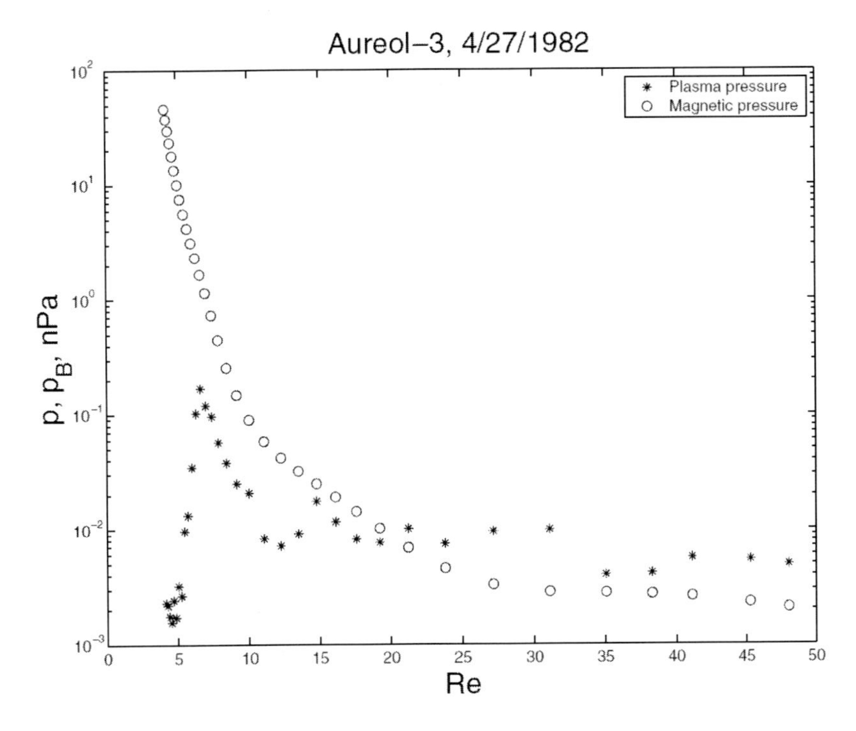

Figure 4.1.3. Plasma and magnetic pressure radial profiles for the April 27, 1982 event. Magnetic pressure was calculated using Tsyganenko 96 magnetic field model. According to Stepanova et al. (2004). Permission from COSPAR.

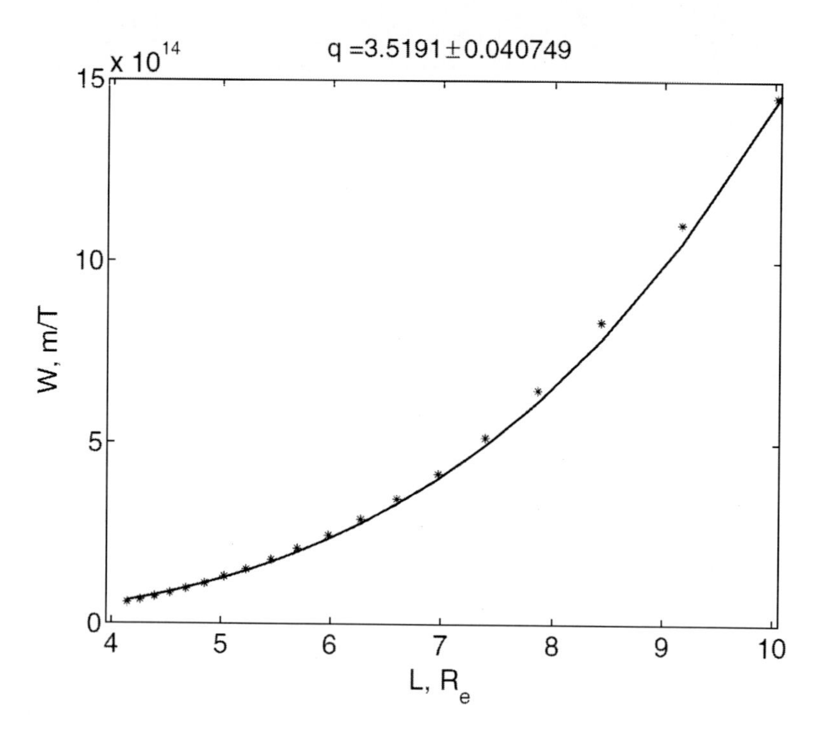

Figure 4.1.4. Variation of the volume of the magnetic flux tube, calculated using Tsyganenko 96 magnetic field model for the April 27, 1982 event. According to Stepanova et al. (2004). Permission from COSPAR.

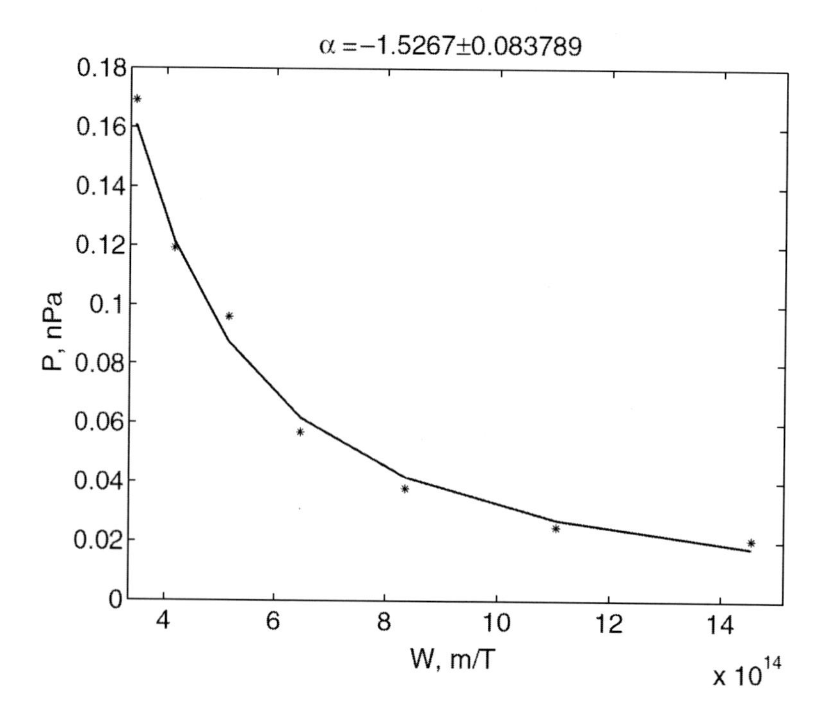

Figure 4.1.5. Variation of the pressure with the volume of the magnetic flux tube for the April 27, 1982 event. According to Stepanova et al. (2004). Permission from COSPAR.

Table 4.1.1. Principal characteristics of nearly onset events including solar wind parameters used in Tsyganenko 96 model and results of fitting of $p \propto R^{-s}$, $W \propto R^{q}$, and $p \propto W^{-\alpha}$. According to Stepanova et al. (2004). Permission from COSPAR

	Day	n (cm^{-3})	V (km/s)	B_y (nT)	B_z (nT)	D_{st}	s	q	α
A	March 4, 1982	5.5	530	2.7	0.2	−41	−5.9 ± 0.3	2.38 ± 0.03	−2.7 ± 0.1
B	May 3, 1982	3.4	627	2.3	3.6	−16	−3.5 ± 0.5	3.50 ± 0.06	−0.67 ± 0.09
C	December 14, 1982						No ion data		
D	November 5, 1981	15.5	378	4.9	2.1	−7	−7.0 ± 1.0	3.17 ± 0.06	−2.4 ± 0.2
E	April 27, 1982	11.3	469	−4.5	−2.0	−19	−5.3 ± 0.4	3.52 ± 0.04	−1.53 ± 0.08
F	June 15, 1982	7.3	451	5.5	−2.4	−4	−10 ± 2	3.42 ± 0.01	−3.0 ± 0.5
G	November 3, 1981	21.3	364	1.6	−2.6	+1	−6.0 ± 0.9	3.68 ± 0.07	−1.3 ± 0.2

4.1.3. "Compression Aurora": Particle Precipitation Driven by Long-Duration High Solar Wind RAM Pressure

Liou et al. (2007) present results from an observational study on global auroral response to large solar wind dynamic pressure, using global auroral images acquired by the ultraviolet imager on board the Polar spacecraft. They noted that it is generally accepted that the state of the magnetosphere is controlled by the solar wind and its embedded IMF through two distinct physical processes: magnetic merging/reconnection (Dungey, 1961) and viscous forcing (Axford and Hines, 1961). Magnetic merging is a principal mechanism for transferring solar wind mass, momentum, and energy to the magnetosphere, whereas viscous interactions in the sheared flow on the flanks of the magnetopause are considered less effective. Studies, particularly during the last decade, have shown that auroras, auroral electrojets, and cross-polar cap potential drops are enhanced immediately after an impingement of solar wind dynamic pressure enhancements, and the effect is especially pronounced for southward IMF. Thus solar wind dynamic pressure is an important source of magnetic activity and enhanced convection; however, the underlying physical process involved is still poorly understood. In the past, studies of the effects of solar wind dynamic pressure on the aurora have been predominantly conducted for interplanetary shocks and large-pressure pulses (e.g., Vorobev, 1974; Craven et al., 1986; Spann et al., 1998; Zhou and Tsurutani, 1999; Chua et al., 2001; Liou et al., 2002; Zhang et al., 2003). Shock-induced auroras are transient and therefore are considered associated with the steep pressure perturbations in the shock/pressure pulse front. The arrival of a shock at the front of the magnetosphere triggers sudden brightening of the aurora on either side or both sides of noon (Zhou and Tsurutani, 1999). The initial brightening is usually localized near noon and is followed by a quick antisunward expansion reaching the night sector in ~20 min. On the other hand, a sharp pressure drop can cause reductions of the dawn/dusk aurora within ~10 min of a negative impulse (Liou et al., 2006). The effect of magnetospheric compression by interplanetary shocks/pressure pulses can produce short-lived (~5–6 min) patchy auroras in the midday subauroral zone from particle sources deep in the inner magnetosphere (Liou et al., 2002; Zhang et al., 2003). In the night sector, delayed and enhanced auroral activity can also occur (Chua et al., 2001; Zhou and Tsurutani, 2001; Liou et al., 2003; Boudouridis et al., 2003), widening the auroral zone and reducing the polar cap size (Chua et al., 2001; Boudouridis et al., 2003, 2004a). As outlined Liou et al. (2007), precipitation of unstructured central plasma sheet (CPS) electrons has been identified as the major source of the enhancement of aurora (Zhou et al., 2003a; Liou et al., 2006).

Mechanisms causing such an enhanced CPS electron precipitation remain unclear. Pitch angle diffusion by electrostatic electron cyclotron harmonic (ECH) (e.g., Kennel et al., 1970; Lyons, 1974) and whistler mode waves (e.g., Johnstone et al., 1993) is generally considered responsible for diffuse auroral particle precipitation. When the magnetosphere is compressed by shocks, perpendicular heating owing to conservation of the first adiabatic invariant can increase the temperature anisotropy of the particles and trigger loss cone instability; pre-existing, trapped CPS particles can be scattered into the loss cone by wave-particle interactions (Zhou and Tsurutani, 1999; Tsurutani et al., 2001). On the other hand, Liou et al. (2002, 2006) argued that reductions in the mirror ratio caused by compression of the magnetosphere can be a major alternative. The study of Liou et al. (2007) is intended to resolve some of the outstanding questions regarding the underlying physical processes that lead to the enhancement of auroral particle precipitation during high solar wind dynamic pressure. As all previous studies have focused on the auroral transient effects associated with pressure perturbations, Liou et al. (2007) will take a slightly different approach in this study by considering the effect of large and 'steady' solar wind dynamic pressure, for example, downstream of a shock and/or a long-duration (>1 h) pressure pulse. Liou et al. (2007) argue that auroral responses to such solar wind structures are prompt and lasting because of reductions in the mirror ratio.

In Liou et al. (2007), auroral images acquired from the ultraviolet imager (UVI) (Torr et al., 1995) on board the Polar spacecraft are used to provide global auroral displays. They studied a number of events of large long-duration (>1 h) pressure pulses, two of which are presented. *The first event* is one of the well-studied geomagnetic storm events associated with a coronal mass ejection on 10 January 1997 (e.g., Fox et al., 1998). Auroral morphology during the prestorm period has been associated with the so-called 'two-cell' aurora (Shue et al., 2002, 2006). *The second event*, occurring on 26 September 1999, is selected because concurrent observations from Wind, Defense Meteorological Satellite Program (DMSP), and Geotail exist, allowing for a detail study of source mechanisms. As noted Liou et al. (2007), ultimately the enhancement of aurora is associated with an increase of precipitating particle energy flux. Studying the magnetospheric source regions of auroral particles provides an important clue for the mechanism of precipitation. To explore possible causes of the increase of precipitating particles, Liou et al. (2007) study the characteristics of precipitating particles from DMSP satellites. The 26 September 1999 event provides an excellent opportunity for such a study because data are available for a comparison of particle characteristics inside and outside of the pressure pulse at nearly identical local times for both dawn and dusk sectors.

Liou et al. (2007) discuss and summarize main obtained results as following:

(i) There are shown explicitly two events that demonstrate that enhancements of auroras in the dawn and dusk sectors associated with increases in the solar wind dynamic pressure can persist, especially in the dawn sector, during the entire magnetospheric compression. It were also examined a few other events (such as 1 May 1997 and 1 October 1997) and found similar effects, though the degree of auroral enhancements varies from one event to the other. In the dusk sector the compression effect is not as obvious as that in the dawn sector. This is probably due to the fact that earthward and eastward drifting CPS electrons rarely reach the post-noon sector. Very often that identification of the compression aurora is hindered by the dayside expansion of auroral activity initiated in the night sector. This is to say events similar to the 26 September 1999 with a clear two-cell structure for the entire compression period are rare. This fact has prevented from performing statistical analysis of the compression-induced aurora. Nonetheless, this work presents the first evidence of persistent auroral activity in the dawn/dusk sector during continuing magnetospheric compression.

(ii) Note that enhanced auroral emissions in the dawn and dusk (two-cell) have been reported to occur during the growth phase of substorm (Shue et al., 2002). Differences in terms of auroral morphology between the two-cell aurora and the compression-induced aurora are not obvious. Two-cell auroras are convection-driven because they occur during southward IMF conditions, whereas the aurora phenomenon we report here is compression-driven. The two examples shown by Shue et al. (2002) are also associated with large dynamic pressure.

(iii) Previous statistical work relating the aurora to solar wind parameters has found the highest association with a proxy for the merging rate (Liou et al., 1998; Newell et al., 2007). These studies found the best correlation involved a time delay of 40–60 min for the pre-midnight aurora (Liou et al., 1998) and/or integration over 2–3 h for the total auroral power (Newell et al., 2007). It has shown that the pressure effect can be both prompt and lasting. The promptness with which auroral luminosity responds to pressure changes contrasts sharply with the response to merging rate changes. The former depends essentially on immediate conditions; the later effect occurs only over an extended time integration (although larger in magnitude).

(iv) A key feature associated with magnetospheric compression is the widening of the dawn/dusk main oval, as demonstrated by measurements made from DMSP and UVI for the 26 September 1999 event. This event suggests that the overall area of the polar cap changes with solar wind dynamic pressure and is generally consistent with the finding of Chua et al. (2001) and Boudouridis et al. (2003). Boudouridis et al. (2003) found a significant increase of the auroral zone width and a decrease of the polar cap size when the solar wind dynamic pressure increases under steady southward IMF conditions, while there is little change when IMF B_z was nearly zero for more than 1 h prior to the pressure pulse. They proposed that preconditioning of the magnetosphere by a southward IMF that moves the poleward boundary of the auroral oval equatorward may be necessary for a pressure enhancement to cause a broadening of the oval. The 26 September 1999 event shows that significant broadening of the oval width and a decrease in polar cap size can also occur after a long duration (~2 h) of northward IMF and the broadening can persist throughout the high-pressure period.

(v) Observations of precipitating particles made by DMSP series flights for the 26 September 1999 event indicate that the enhancement of the dawn/dusk auroras was associated mainly with CPS electron precipitations. Previous studies have also shown a CPS electron dominant precipitation associated with transient dayside auroral enhancements/reductions during sudden compression/decompression (Liou et al., 2002, 2006; Zhou et al., 2003b). Diffuse auroras are generally believed to result from pitch angle diffusion by electrostatic ECH (e.g., Kennel et al., 1970; Lyons, 1974) and whistler mode waves (e.g., Johnstone et al., 1993). In this scenario, when the magnetosphere is compressed by shocks, perpendicular heating owing to conservation of the first adiabatic invariant can increase the temperature anisotropy of the particles and trigger loss cone instability ($T_\perp/T_\parallel > 1$); preexisting trapped CPS particles can be scattered into the loss cone by wave-particle interactions and produce aurora (Zhou and Tsurutani, 1999; Tsurutani et al., 2001). Enhancements of plasma waves during magnetospheric compression have been reported by Anderson and Hamilton (1993) as electromagnetic ion cyclotron waves, by Lauben et al. (1998) as very low frequency chorus emission in the magnetosphere, and recently by Zhou et al. (2003a) as broadband electromagnetic waves in the ionosphere.

(vi) Wave observations by Geotail during the fortuitous event of 26 September 1999 did not provide evidence to support the proposed ECH and whistler mode wave mechanism. Therefore particle scattering by these waves can be ruled out at least from this event. It is worthwhile to note that adiabatic heating should be limited to periods of increasing dynamic pressure, which is not consistent with described observations. Furthermore, adiabatic compression causes $V_\perp$ to increase and consequently moves particles away from the loss cone. Observations from FAST reported by Zhou et al. (2003a) also suggest that enhanced diffuse precipitation is not associated with plasma waves and field-aligned currents in the equatorward part of the oval.

(vii) Alternatively, reductions in the mirror ratio caused by compression of the magnetosphere can be the major cause of enhanced aurora as proposed by Liou et al. (2002, 2006). This reduction is because magnetospheric compression causes the Earth's magnetic field to increase most at the equator and reduces mirror ratio ($R_m = B_m/B_{eq}$). The equatorial magnetic field strength before and during the compression can be estimated by pressure balance between ram and magnetic pressures at the magnetopause. The increase in ram pressure from 7 to 19 nPa would cause an increase of ~60% in the equatorial magnetic field strength. This is consistent with the magnetic field strength increase at Geotail (from ~38 nT before the compression (~19:05 UT) to ~60 nT at ~19:56 UT). The increase of magnetic field strength would reduce the mirror ratio about 1.6 times; the equatorial loss cone pitch angle during compression correspondingly increases on 60%. The energy flux at the dawnside increased from ~5.5–6.5 Gw/h to about 10–11 Gw/h (~75% increase). So at least for this particular event the main part of the increase, if not all, can be attributed to this mechanism.

(viii) Furthermore, this simple mechanism requires no pitch angle diffusion and can hold as long as the high solar wind dynamic pressure persists. Because it takes seconds to empty the flux tube for keV electrons, this mechanism also requires a non-interrupted plasma supply. This is generally true because diffuse auroras exist continuously. Liou et al. (2007) believe that pitch angle scattering loss by wave-particle interactions may play a role; however, the relative role the two processes play in the compression-triggered diffuse aurora may vary from case to case. Further studies will help resolve this issue.

(ix) As a final remark, past studies have established a prompt auroral response to sudden changes in the solar wind dynamic pressure. Although the involved underlying physical process is still poorly understood, these studies suggest that dynamic pressure variations are an important source of geomagnetic activity in addition to magnetic merging (Dungey, 1961) and viscosity forcing (Axford and Hines, 1961). One of the important implications of this result is the continuing loss of plasma sheet particles through particle precipitation into the magnetosphere during high solar wind dynamic pressure. The magnetospheric plasma loss can be significant for long-duration high solar wind dynamic pressure, which occurs frequently during large solar disturbances such as corona mass ejections and recurrent corotating interaction regions. During such circumstances, anomalous enhancements in ionospheric conductivity and joule heating can occur and affect the electrodynamics of the magnetosphere-ionosphere system and the vertical profile of the thermospheric neutral composition. Therefore the solar wind dynamic pressure should be included in empirical functions for gauging energy coupling between the solar wind and the magnetosphere, should be considered and included in any theoretical and model work.

4.1.4. Response of Dayside Auroras to Abrupt Increases in the Solar Wind Dynamic Pressure at Positive and Negative Polarity of the IMF Bz

According to Vorobjev et al. (2009), the optical observations on Heiss Island ($\Phi = 75.0°$) have been used to study the characteristics of auroras in the near-noon MLT sector after abrupt increases in the solar wind dynamic pressure at negative and positive polarity of the IMF B_z component. They considered in detail the behavior of auroras in the near-noon sector of the magnetic local time (MLT) during abrupt changes in the solar wind dynamic pressure on December 18, 1985, and November 28, 1989, when the IMF B_z component was negative and positive, respectively. It has been found out that the 427.8 and 557.7 nm emission intensities considerably increased at $B_z < 0$ both equatorward of the dayside red luminosity band and within this band. The value of the emission intensities at a red luminosity maximum (I6300/I5577 $\approx$ 0.5) indicates that energetic electron precipitation is of the magnetospheric origin. At $B_z > 0$, fluxes of harder ($E > 1$ keV) precipitating electrons were superimposed on the soft spectrum of precipitating particles in the equatorial part of the red luminosity band. This red band part was hypothetically caused by the low-latitude boundary layer (LLBL) on closed lines of the geomagnetic field, the estimated thickness of which is ~3 r_E. The 557.7 nm emission intensity increased during 3–5 min after SC/SI and was accompanied by the displacement of the red band equatorward boundary toward lower latitudes. The displacement value was ~150–200 km when the dynamic pressure abruptly increased by a factor of 3–5. After SC/SI, the 630.0 nm emission intensity continued increasing during 16–18 min. It is assumed that the time of an increase in the red line intensity corresponds to the time of saturation of the magnetospheric boundary layers with magnetosheath particles after an abrupt increase in their density.

4.1.5. Auroral Electrojets Variations Caused by Recurrent High-Speed Solar Wind Streams during the Extreme Solar Minimum of 2008

The IMAGE network magnetic measurements are used by Guo et al. (2012) to investigate the response of the auroral electrojets to the recurrent high-speed solar wind streams (HSSs) during the extreme solar minimum period of 2008. Guo et al. (2012) first compare the global AU/AL indices with the corresponding IU/IL indices determined from the IMAGE magnetometer chain and find that the local IMAGE chain can better monitor the activity in MLT sectors 1230–2230 for IU and 2230–0630 for IL during 2008. In the optimal MLT sectors, the eastward and westward electrojets and their central latitude reveal clear 9-day periodic variations associated with the recurrent HSSs. For the 9-day perturbations, both the eastward and westward electrojet currents are better correlated with parallel electric field (E_{PAR}) and electron hemispheric power (HPe) than with other forcing parameters. Interestingly, the eastward electrojet shows good correlations (CC > 0.6) with EPAR and HPe only in part of its optimal MLT-sector, roughly 1200–1800, while the westward electrojet shows good correlations (CC < −0.6) with E_{PAR} and HPe in its whole optimal MLT sector. The poor correlations between the eastward electrojet and E_{PAR} and HPe in the MLT sector 1800–2200 might be attributed to the impact of other magnetosphere-ionosphere coupling processes. The sensitivities of the eastward and westward electrojet currents to E_{PAR} are close to 0.06 MA/(mV/m) and −0.12 MA/(mV/m), respectively, and the sensitivities of their central latitudes to E_{PAR} are close to −2.83 Deg/(mV/m) and −2.14 Deg/(mV/m), respectively. The observed auroral electrojet response to the recurrent solar wind forcing provides new opportunities to study the physical processes governing the eastward and westward auroral electrojets.

4.1.6. Influence of IMF and Solar Wind on Auroral Brightness in Different Regions

By integrating and averaging the auroral brightness from Polar Ultraviolet Imager auroral images, which have the whole auroral ovals, and combining the observation data of IMF and solar wind from NASA Operating Missions as a Node on the Internet (OMNI), Yang et al. (2013) investigate the influence of IMF and solar wind on auroral activities, and analyze the separate roles of the solar wind dynamic pressure, density, and velocity on aurora, respectively. Yang et al. (2013) statistically analyze the relations between the interplanetary conditions and the auroral brightness in dawnside, dayside, duskside, and nightside. It is found that the three components of the IMF have different effects on the auroral brightness in the different regions. Different from the nightside auroral brightness, the dawnside, dayside, and duskside auroral brightness are affected by the IMF B_X, and B_Y components more significantly. The IMF B_X and B_Y components have different effects on these three regional auroral brightness under the opposite polarities of the IMF B_Z. As expected, the nightside aurora is mainly affected by the IMF B_Z, and under southward IMF, the larger the $|B_Z|$, the brighter the nightside aurora. The IMF B_X and B_Y components have no visible effects. On the other hand, it is also found that the aurora is not intensified singly with the increase of the solar wind dynamic pressure: when only the dynamic pressure is high, but the solar wind velocity is not very fast, the aurora will not necessarily be intensified significantly. These results can be used to qualitatively predict the auroral activities in different regions for various interplanetary conditions.

4.1.7. Fine-Scale Transient Arcs Seen in a Shock Aurora

Motoba et al. (2014) report, for the first time, fine-scale transient arcs that emerged successively within the initial 1–2 min evolutionary interval of a postnoon shock aurora on 14 July 2012. Data were acquired from ~2 Hz temporal resolution imaging of dayside aurora with a white light all-sky camera (ASC) at South Pole Station (magnetic latitude = −74.3°, magnetic local time = UT −3.5 h). Just after 1809:50 UT at which the initial response to an interplanetary (IP) shock was detected in the postnoon geosynchronous magnetic field, the ASC observed three successive transient arcs of which the locations shifted equatorward with an abrupt jump by ~0.2° in latitude. All of the transient arcs occurred in a closed field line region, ~1.0°–1.5° equatorward of the polar cap or open/closed field line boundary inferred from the intensity ratio of I630.0/I557.7 but just poleward of the shock-induced proton and diffuse-type electron aurorae. Each of the transient arcs had a latitudinal width of ~0.1° and a short lifetime of ~20–30 s. Although the obvious mechanism has still remained unclear, possible interpretations of the fine-scale transient arc features are discussed in terms of a local process of each of the magnetospheric origin (mode conversion) and ionospheric origin (feedback interaction) that may be induced by IP shock.

4.1.8. Hemispheric Asymmetry of the Structure of Dayside Auroral Oval

According to Hu et al. (2014), a comprehensive analysis of long-term and multispectral auroral observations made in the Arctic and Antarctica demonstrates that the dayside auroral ovals in two hemispheres are both presented in a two-peak structure, namely, the prenoon 09:00 magnetic local time (MLT) and postnoon 15:00 MLT peaks. The two-peak structures of dayside ovals, however, are asymmetric in the two hemispheres; i.e., the postnoon average auroral intensity is more than the prenoon one in the Northern Hemisphere but less in the Southern Hemisphere. The hemispheric asymmetry cannot be accounted for by the effect of the IMF B_Y component and the seasonal difference of ionospheric conductivities in the two hemispheres, which were used to interpret satellite-observed real-time auroral intensity asymmetries in the two hemispheres in previous studies. Hu et al. (2014) suggest that the hemispheric asymmetry is the combined effect of the prenoon-postnoon variations of the magnetosheath density and local ionospheric conductivity.

4.1.9. Relative Brightness of the O^+(2D-2P) Doublets in Low Energy Aurora

The ratio of the emission line doublets from O^+ at 732.0 nm (I732) and 733.0 nm (I733) has been measured by Whiter et al. (2014) in auroral conditions of low energy electron precipitation from Svalbard (78.20° north, 15.83° east) during two winters between 2003 and 2006. The value obtained for R = I732/I733 for the 2003-2004 season is 1.38 ± 0.02. This result is slightly higher than theoretical values, and values obtained in airglow, but is lower than earlier measurements in similar auroral conditions. Most of the data from 2003-2004 were from 06 UT to 13 UT, which straddles the time when Svalbard is under the magnetospheric cusp region. The value obtained for the 2005-2006 season is R = 1.45 ± 0.08. The data from this season have lower spectral resolution, and contain much more scatter than those from the earlier season. Higher ratio values mostly occur between 14 UT and 17 UT, when Svalbard is outside the cusp, and when more energetic precipitation may dominate. One of the motivations of the work is the need for accurate modelling of the emission doublet at 732.0 nm. It is one of the emissions measured by the Auroral Structure and Kinetics (ASK) instrument, which is also located at Svalbard. Accurate determination of R = I732/I733 provides a powerful method for separating the density of the upper 2P states, information which is needed for ionospheric modelling of emissions. The work of Whiter et al. (2014) is especially relevant for studies of plasma flows in the ionosphere utilising the long lifetime of the O^+ emission. Although the present work is a statistical study, the work shows that it is necessary to determine whether there are significant variations in the ratio resulting from auroral energy deposition, large electric fields, and changes in composition.

4.1.10. An Optimum Solar Wind Coupling Function for the AL Index

McPherron et al. (2015) define a coupling function as a product of solar wind factors that partially linearizes the relation between it and a magnetic index. It is consider functions that are a product of factors of solar wind speed V, density N, transverse magnetic field $B_\perp$, and interplanetary magnetic field (IMF) clock angle θ_c each raised to a different power. The index is the auroral lower (AL index) which monitors the strength of the westward electrojet. Solar wind data 1995–2014 provide hour averages of the factors needed to calculate optimum exponents. Nonlinear inversion determines both the exponents and linear prediction filters of short data segments. The averages of all exponents are taken as optimum exponents and for V, N, $B_\perp$, and $\sin(\theta_c/2)$ are [1.92, 0.10, 0.79, 3.67] with errors in the second decimal. Hourly values from 1966 to 2014 are used next to calculate the optimum function and the functions VBs, epsilon, and universal coupling function. A year-long window is advanced by 27 days calculating linear prediction filters for the four functions. The mention above functions predict 43.7, 61.2, 65.6, and 68.3% of AL variance, respectively. The optimum function is 2.74% better than universal coupling function with a confidence interval 2.60–2.86%. Coupling strength defined as the sum of filter weights (nT/mV/m) is virtually identical for all functions and varies systematically with the solar cycle being strongest (188 nT/mV/m) at solar minimum and weakest (104 nT/mV/m) at solar maximum. Saturation of the polar cap potential approaching solar maximum may explain the variation.

4.1.11. Electrodynamics and Energy Characteristics of Aurora at High Resolution by Optical Methods

As outlined Dahlgren et al. (2016), technological advances leading to improved sensitivity of optical detectors have revealed that aurora contains a richness of dynamic and thin filamentary structures, but the source of the structured emissions is not fully understood. In addition, high-resolution radar data have indicated that thin auroral arcs can be correlated with highly varying and large electric fields, but the detailed picture of the electrodynamics of auroral filaments is yet incomplete. The Auroral Structure and Kinetics (ASK) instrument is a state-of-the-art ground-based instrument designed to investigate these smallest auroral features at very high spatial and temporal resolution, by using three electron multiplying CCDs in parallel for three different narrow spectral regions. ASK is specifically designed to utilize a new optical technique to determine the ionospheric electric fields. By imaging the long-lived O^+ line at 732 nm, the plasma flow in the region can be traced, and since the plasma motion is controlled by the electric field, the field strength and direction can be estimated at unprecedented resolution. The method is a powerful tool to investigate the detailed electrodynamics and current systems around the thin auroral filaments. The two other ASK cameras provide information on the precipitation by imaging prompt emissions, and the emission brightness ratio of the two emissions, together with ion chemistry modeling, is used to give information on the energy and energy flux of the precipitating electrons. Dahlgren et al. (2016) discuss these measuring techniques and give a few examples of how they are used to reveal the nature and source of fine-scale structuring in the aurora.

4.1.12. Relationship between Auroral Oval Poleward Boundary Intensifications and Magnetic Field Variations in the Solar Wind

Kornilov et al. (2016) note that bright auroral arcs, as a rule, evolve near the poleward boundary of the auroral oval at the growth phase of a substorm, a phenomenon that is known to occur near the poleward edge of the auroral oval. The closeness of these arcs to the projection of the magnetic separatrix on the night side suggests that their generation is related to magnetic reconnection in the magnetospheric tail in a particular way. In Kornilov et al. (2016) this suggestion is confirmed by the fact that integral brightness of the auroral oval at the poleward edge correlates with magnetic field structures in the solar wind that are observed by ACE and Wind satellites at distances of 50– 300 R_E upstream and are shifted towards the magnetospheric tail with time delays of ~ 10–80 min, consistent with measurements of the solar wind velocity. About 50 examples of this correlation have been found. The possible physical mechanisms of the effect observed are discussed.

4.1.13. Investigation of Triggering of Poleward Moving Auroral Forms Using Satellite-Imager Coordinated Observations

As note Wang et al. (2016a), poleward moving auroral forms (PMAFs) are thought to be an ionospheric signature of dayside magnetic reconnection. While PMAFs are more likely to occur when the IMF is southward, how often PMAFs are triggered by changes in solar wind parameters is still an open question. To address this issue, Wang et al. (2016a) used one of the Automatic Geophysical Observatories all-sky imagers in Antarctica and the THEMIS B and C satellites, which can give solar wind measurements much closer to the subsolar bow shock than by Wind or ACE, to examine if PMAFs occurred in association with IMF orientation changes. Wang et al. (2016a) identified 60 PMAFs in conjunction with THEMIS B and C during 2008, 2009, and 2011 and 70% of events show reduction of Bz before PMAF onset indicating that IMF southward turning plays an important role in triggering a majority of PMAFs. In contrast, the magnitude of the IMF Bz reduction in OMNI data was smaller and the reduction occurred in a slightly smaller percentage of events (40–60%). This suggests that solar wind structures that missed the L1 point or evolution of solar wind between the L1 point and THEMIS may be important for identifying IMF changes responsible for transient dayside reconnection. Additionally, 17 PMAFs that did not have substantial IMF southward turnings are correlated well with foreshock events, indicating that foreshock phenomena may also play a role in triggering PMAFs.

4.1.14. Ion Gyroradius Effects on Particle Trapping in Kinetic Alfvén Waves along Auroral Field Lines

In the study of Damiano et al. (2016), a 2-D self-consistent hybrid gyrofluid-kinetic electron model is used to investigate Alfvén wave propagation along dipolar magnetic field lines for a range of ion to electron temperature ratios. The focus of the investigation is on understanding the role of these effects on electron trapping in kinetic Alfvén waves sourced in the plasma sheet and the role of this trapping in contributing to the overall electron energization at the ionosphere. This work also builds on authors previous effort (Damiano et al., 2015) by considering a similar system in the limit of fixed initial parallel current, rather than fixed initial perpendicular electric field. It is found that the effects of particle trapping are strongest in the cold ion limit and the kinetic Alfvén wave is able to carry trapped electrons a large distance along the field line yielding a relatively large net energization of the trapped electron population as the phase speed of the wave is increased. However, as the ion temperature is increased, the ability of the kinetic Alfvén wave to carry and energize trapped electrons is reduced by more significant wave energy dispersion perpendicular to the ambient magnetic field which reduces the amplitude of the wave. This reduction of wave amplitude in turn reduces both the parallel current and the extent of the high-energy tails evident in the energized electron populations at the ionospheric boundary (which may serve to explain the limited extent of the broadband electron energization seen in observations). Even in the cold ion limit, trapping effects in kinetic Alfvén waves lead to only modest electron energization for the parameters considered (on the order of tens of eV) and the primary energization of electrons to keV levels coincides with the arrival of the wave at the ionospheric boundary.

4.1.15. Is Diffuse Aurora Driven from above or below?

According to Khazanov et al. (2017), in the diffuse aurora, magnetospheric electrons, initially precipitated from the inner plasma sheet via wave-particle interaction processes, degrade in the atmosphere toward lower energies, and produce secondary electrons via impact ionization of the neutral atmosphere. These initially precipitating electrons of magnetospheric origin can also be additionally reflected back into the magnetosphere, leading to a series of multiple reflections by the two magnetically conjugate atmospheres that can greatly impact the initially precipitating flux at the upper ionospheric boundary (700–800 km). The resultant population of secondary and primary electrons cascades toward lower energies and escape back to the magnetosphere. Escaping upward electrons traveling from the ionosphere can be trapped in the magnetosphere, as they travel inside the loss cone, via Coulomb collisions with the cold plasma, or by interactions with various plasma waves. Even though this scenario is intuitively transparent, this magnetosphere-ionosphere coupling element is not considered in any of the existing diffuse aurora research. Nevertheless, as we demonstrate in this letter, this process has the potential to dramatically affect the formation of electron precipitated fluxes in the regions of diffuse auroras.

4.2. Coupling of Solar Wind and IMF with Processes in Polar Cap

4.2.1. Generation of High-Density Plasma in the Polar Cap Observed by the Akebono Satellite

As note Abe et al. (2005), the polar cap ionosphere is known as a comparatively low-density (< 10^3 cm^{-3} above 3000 km) plasma region and less active than the cusp or auroral region in terms of particle precipitation. Because of the low density, very limited information on the thermal plasma in the high-altitude polar cap is available in the literature. On very rare occasions, the Akebono satellite encounters regions of unusually high-density plasma above 4000 km altitude, in which both the electron temperature (< 3000° K) and parallel ion drift velocity (< 1 km/s) are distinctively low. The ion drift meter observations on the DMSP satellite show that the plasma convection in the polar cap is predominantly directed from the dayside to nightside on such occasions, suggesting that anti-sunward convection is a necessary condition. The low electron temperature and ion velocity accompanied by the high plasma density may suggest that a small-amplitude ambipolar electric field is also another necessary condition for increasing the plasma density. Obtained results are illustrated by Figures 4.2.1 – 4.2.3.

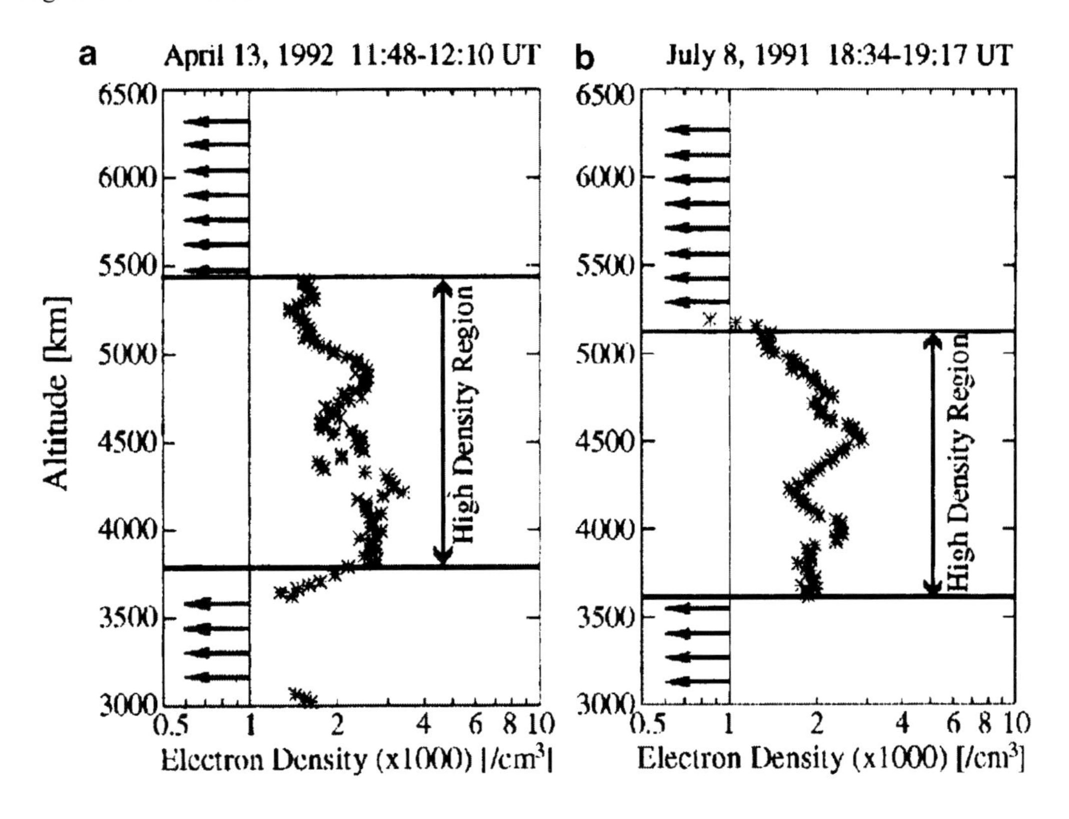

Figure 4.2.1. Altitude profiles of measured electron density on (a) April 13, 1992 and (b) July 8, 1991. The arrows indicate density below 10^3 cm^{-3}. According to Abe et al. (2005). Permission from COSPAR.

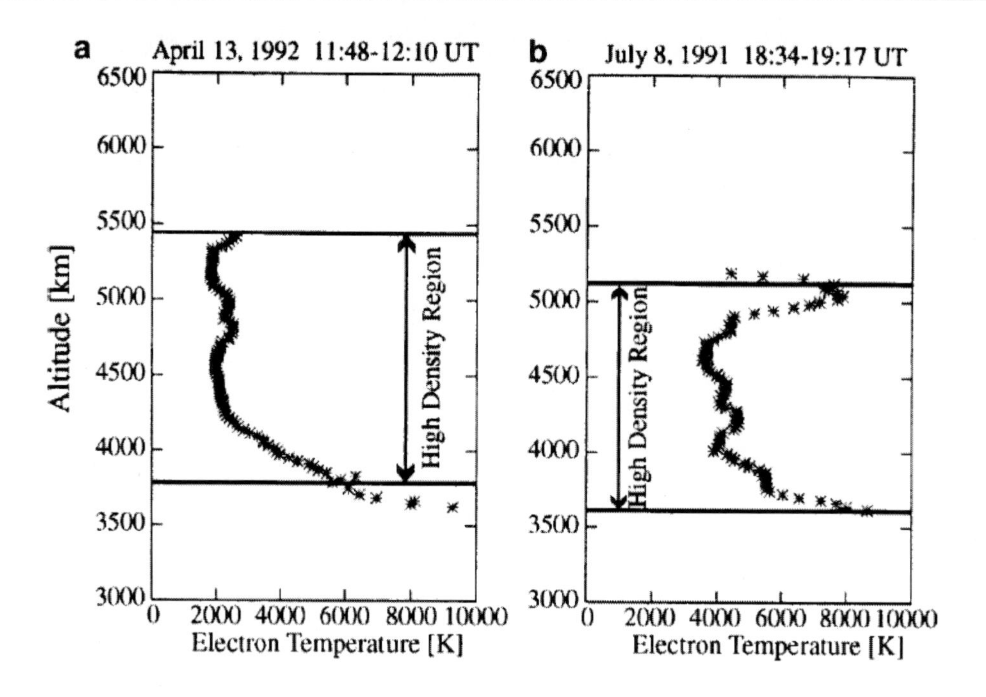

Figure 4.2.2. Altitude profiles of measured electron temperature on (a) April 13, 1992 and (b) July 8, 1991. According to Abe et al. (2005). Permission from COSPAR.

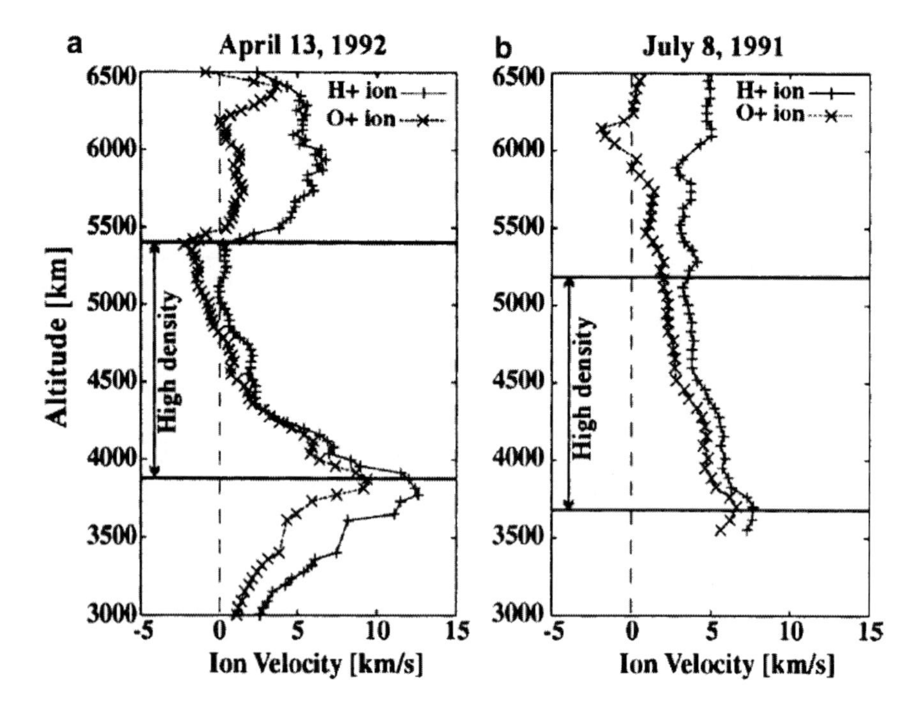

Figure 4.2.3. Altitude profiles of parallel ion drift velocity on (a) April 13, 1992 and (b) July 8, 1991. According to Abe et al. (2005). Permission from COSPAR.

4.2.2. PC-Index Fluctuations and Intermittency of the Magnetospheric Dynamics

One minute resolution Polar Cap (PC) index was used by Stepanova et al. (2005) for the analysis of magnetospheric dynamics. The 1995-2000 time series analysis revealed that the power spectrum of the PC-index fluctuations is a power law in a wide range of frequencies. However, the obtained exponents differ for low and high frequency regions. The probability distribution functions of the PC-index fluctuations show a strong non-gaussian shape, depending on the time of increment. This indicates that the PC-index exhibits intermittency, previously detected in solar wind and auroral electrojet index fluctuations. The PC-index probability distribution functions were fitted by the functional form proposed by Castaing et al. (1990) to describe intermittency phenomena in ordinary turbulent fluid flows. The agreement between the fitting parameters obtained for the PC index and those reported before for solar wind magnetic field fluctuations is within 30%; which is noticeably less than the difference between the same parameters of solar wind and the AE-index fluctuations. This fact indicates that the PC index reflects the solar wind influence on the high-latitude magnetosphere, especially during the summer. Obtained results are illustrated by Figures 4.2.4 – 4.2.6.

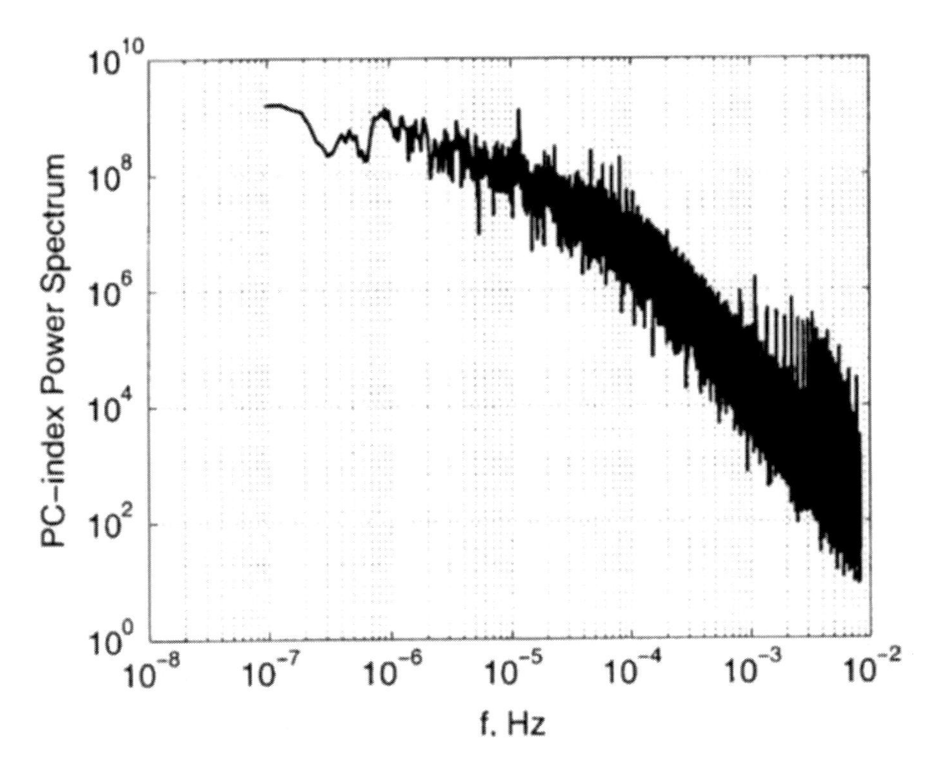

Figure 4.2.4. Power spectrum of the PC-index fluctuations obtained using a four year record of one-minute-resolution PC-index. According to Stepanova et al. (2005). Permission from COSPAR.

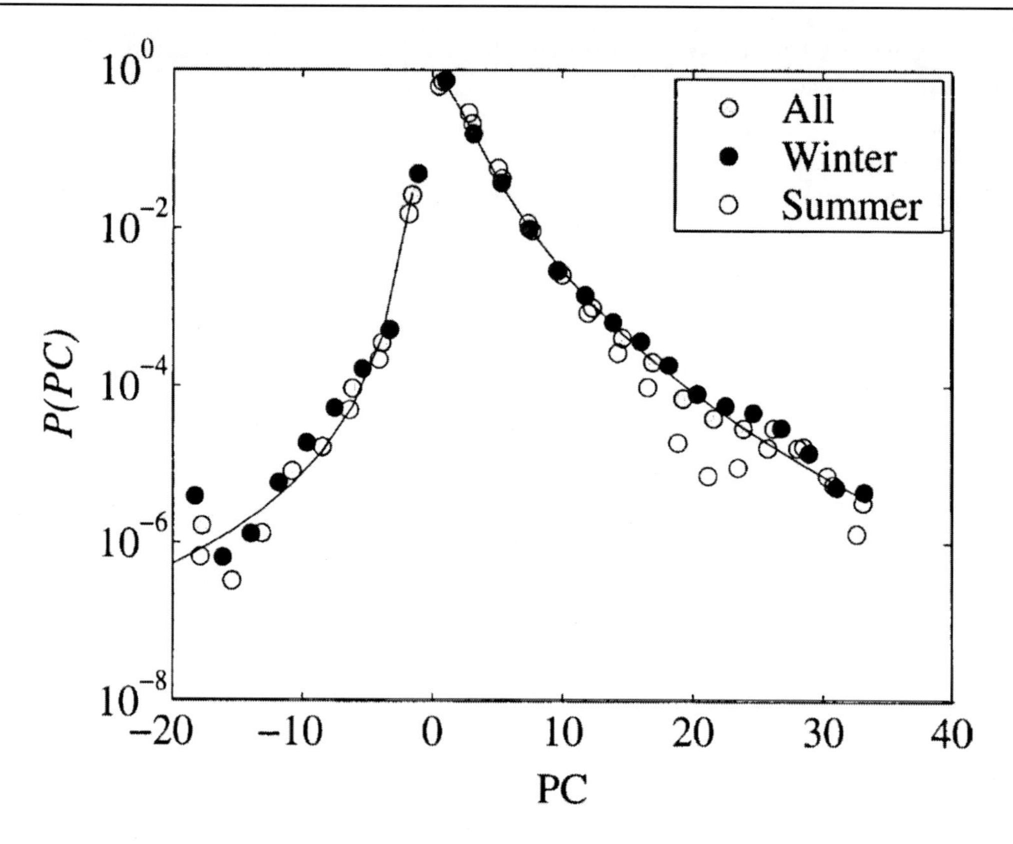

Figure 4.2.5. Probability density function of the PC-index *P*(PC). Black solid lines refer to the best fit of *P*(PC) by two log-normal distributions for positive and negative branches of the PC-index. According to Stepanova et al. (2005). Permission from COSPAR.

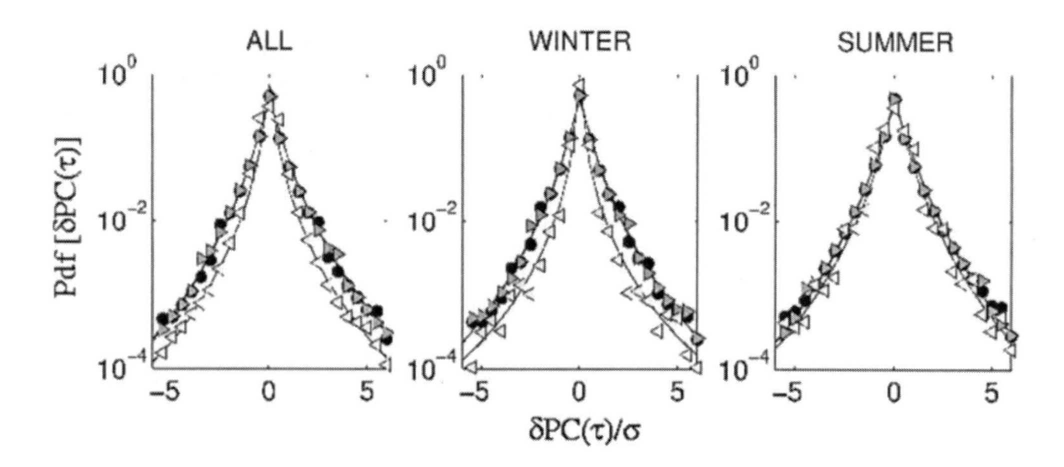

Figure 4.2.6. Probability distribution function Pdf [δPC(τ)] for time scale $\tau = 8$ min for total (o), positive (▷), and negative (◁) PC-index datasets; σ is a standard deviation of δPC(τ). Solid lines correspond to best fits of data. The statistical error bars are smaller than the sign sizes. According to Stepanova et al. (2005). Permission from COSPAR.

4.2.3. Enhanced High-Altitude Polar Cap Plasma and Magnetic Field Values in Response to the Interplanetary Magnetic Cloud

According to Osherovich et al. (2007) the magnetospheric electron number density and the magnetic field strength near 8 R_E over the polar cap increased dramatically after the arrival of an interplanetary magnetic cloud on 31 March 2001. These parameters were determined with high accuracy from the plasma resonances stimulated by the Radio Plasma Imager (RPI) on Imager for Magnetopause-to-Aurora Global Exploration (IMAGE) near apogee during both quiet (30 March 2001) and disturbed (31 March 2001) days. The quiet day and disturbed day values were each compared with magnetospheric magnetic field and electron density models; good agreement was found with the former but not the latter. The magnetospheric response was also expressed in terms of the ratio of the electron plasma frequency

$$f_{pe}[\text{kHz}] = \{80.6\, N_e\, [\text{cm}^{-3}]\}^{1/2} \tag{4.2.1}$$

to the electron cyclotron frequency

$$f_{ce}\, [\text{kHz}] \approx 0.028\, |\mathbf{B}\, [\text{nT}]|, \tag{4.2.2}$$

which is proportional to the ratio of the electron gyroradius to the Debye radius. Simultaneous Wind measurements of the solar wind magnetic field strength, speed, and plasma density were used in Osherovich et al. (2007) to calculate the solar wind quasi-invariant QI:

$$\text{QI} = (B^2/8\pi)/(\rho v^2/2) = (1/M_A)^2. \tag{4.2.3}$$

This index is equivalent to the ratio of the solar wind magnetic pressure to the solar wind ram pressure or to the inverse of the magnetic Mach number squared. These non-dimensional quantities, QI and f_{pe}/f_{ce}, have fundamental meanings in the solar wind MHD regime and in the relation between electric and magnetic forces on electrons in the magnetosphere, respectively. During the large 31 March 2001 storm, IMAGE was at the right place at the right time so as to enable comparisons between RPI f_{pe}/f_{ce} and Wind QI determinations. Both QI and f_{pe}/f_{ce} formed maxima during 6-hour observing intervals during this storm that were found to be highly correlated (87%) with a magnetospheric time lag of about 3 hours for f_{pe}/f_{ce}

As Osherovich et al. (2007) outlined, that in the case of planetary magnetospheres the similarity of the spectrum of sounder-stimulated plasma resonances in the Earth's ionosphere and magnetosphere (Benson et al., 2003) and in Jupiter's Io plasma torus (Osherovich et al., 1993; Stone et al., 1992), and observations indicating that this spectrum is much more sensitive to variations in f_{pe}/f_{ce} than to variations in the individual parameters f_{pe} or f_{ce}, or even T_e, (Benson et al., 2001), suggests that this plasma parameter f_{pe}/f_{ce} may be considered to have a quasi-invariant nature. Even though it may vary from less than 0.2 in auroral kilometric radiation (AKR) source regions (Benson and Calvert, 1979) to more than 10 in the low-altitude equatorial ionosphere (Benson, 1972, 1974), there are extended regions in Earth's ionosphere and magnetosphere and Jupiter's Io plasma torus where this parameter is in the range from 1 to 8 in spite of much larger variations in f_{pe}, f_{ce}, and T_e between these three plasmas (Osherovich et al., 2005). Combining the present results, comparing solar wind QI and magnetospheric polar cap f_{pe}/f_{ce} variations, with these earlier findings suggests the value of creating three-dimensional f_{pe}/f_{ce} maps for magnetically quiet and disturbed times in addition to such maps for **B** and N_e.

4.2.4. Polar Cap Potential Saturation: An Energy Conservation Perspective

As noted Liu (2007), there is mounting evidence that the polar cap potential drop saturates when the solar wind electric field tends to large values (Russell et al., 2001; Shepherd et al., 2002; Hairston et al., 2003, 2005), and explanation of this phenomenon has turned into a significant theoretical preoccupation (Siscoe et al., 2002; Raeder and Lu, 2005; Ridley, 2005; MacDougall and Jayanchandran, 2006). Most theories concern themselves with limiting factors on the transfer into the magnetosphere of the solar wind electric field E and are therefore characterized as first order. In Liu (2007) it was proposed a second-order theory that limits the term $|E|^2$. The motivating consideration is that the energy driving magnetosphere-ionosphere processes must come from the solar wind in the long run. Therefore, the Perrault-Akasofu parameter (PAP) (Perreault and Akasofu, 1978) or similar measures of solar wind input sets a natural upper limit on the energy budget of the magnetosphere-ionosphere system. In particular, the ionospheric Joule heating, which is proportional to $|E|^2$, should generally fall under such a limit. Liu (2007) shows that this simple consideration based on energy conservation can lead to situations where a significant polar cap potential saturation is predicted.

Liu (2007) turn the attention to the poleward boundary, which is also associated with the boundary of the polar cap. Without affecting the argument and physics, Liu (2007) assumes that the poleward boundary is a circle with radius a. The open flux through the polar cap is $\Psi = \pi a^2 B_0$, where B_0 is the magnetic field in the polar cap. The potential drop across the polar cap is $V_{pc} = 2E_{pc}a$ (for a first-order potential distribution $\Phi \propto \sin\phi$, where ϕ is magnetic longitude). Combing the two relations one obtains

$$V_{pc} = \frac{2v_x}{\pi a}\Psi, \tag{4.2.4}$$

where $v_x = E_{pc}/B_0$ is the anti-sunward convection speed in the polar cap. Since v_x and a are expected to vary in tandem, the ratio v_x/a should be a slower-varying parameter in comparison to Ψ. In Liu (2007) it was holded v_x/a to be a constant. Effectively, this assumption makes the polar cap potential and open flux linear functions of each other. As Liu (2007) outlined, Troshichev et al. (1996) studied the correlation between the polar cap size and polar cap index (PCI), as well as that between polar cap size with the polar cap potential and PCI, using Akebono electric field and particle data. By eliminating the PCI from these two relations, a direct correlation between the polar cap potential and open flux is obtained. Ridley and Kihn (2004) performed a study similar to Troshichev et al. (1996) using, however, AMIE as the method for polar cap potential and polar cap size evaluation. Whereas some discrepancies in detail were noted, the results of Ridley and Kihn (2004) showed similar trends to Troshichev et al. (1996) in terms of potential polar cap area relationship. Liu (2007), by using fitting formulas to results of Troshichev et al. (1996) and Ridley and Kihn (2004) obtained that they can be further fitted to a linear formula

$$V_{pc} = C_1 A + C_0 \text{ kV} \tag{4.2.5}$$

where A is the area of the polar cap, and C_0 and C_1 are fitting constants. Liu (2007) simplified the empirical Eq. 4.2.4 to

$$V_{pc} \propto a^2, \tag{4.2.6}$$

which implies that the polar cap will expand to a diameter 52° for V_{pc} = 160 kV. Liu (2007) emphasized that the qualitative behaviors essential for the theory are (1) both oval boundaries move equatorward as V_{pc} increases, but (2) the poleward boundary moves faster than the equatorward boundary. Liu (2007) considered in details also conductivity model, saturation in the contained dipole convection model and in the presence of a distributed region of currents, comparison theory with observations, and came to following conclusions: **(i)** When the electric field in the solar wind increases, the magnetosphere becomes more active. **(ii)** It was assert that the following are among the geospace responses to the escalation: (1) the auroral oval moves equatorward but with less closed magnetic flux involved in magnetospheric convection and (2) The Pedersen conductivity in the oval increases. **(iii)** It was working from the assertion that in a long-term average, the ionospheric Joule heating is less than the solar wind energy input function to the magnetosphere. **(iv)** If to make the reasonable assumption that substorm energy release feeds mostly auroral intensification, ring current injection, and plasmoid ejection, with relatively little going to ionospheric Joule heating, the above proposition applies approximately on an instantaneous basis as well. **(v)** It was shown that polar cap potential saturation emerges as a natural solution of energy conservation equation in the above context. **(vi)** It was shown that the saturation is robust qualitatively, as it is a result of dimensional analysis, as well as quantitatively, as it is relatively insensitive to the conductivity and convection models used.

4.2.5. On the Problem of the Polar Cap Area Saturation

As outlined Merkin and Goodrich (2007), the phenomenon of the transpolar potential saturation has received a lot of interest over the past few years from experimentalists, modelers, as well as theorists. There now appears to be enough observational evidence of the saturation effect despite the relative rarity of events driven strongly enough to cause it (see Shepherd (2007) for a review). Global MHD models confirm the non-linear nature of the relation between Φ_{PC} and the solar wind convective electric field. A number of explanations based on these simulations as well as theoretical considerations have been suggested (Siscoe et al., 2004). The importance of this effect follows from the notion that Φ_{PC} is a measure of the fraction of the solar wind potential that is applied to the magnetosphere. Of equal importance for understanding the state of the magnetosphere on a global scale is the open magnetic flux threading the ionospheric polar caps. This quantity is to very good accuracy proportional to A_{PC}, since the dipole magnetic field (the approximation used in global MHD models) does not vary significantly across the polar cap. Φ_{PC} is sometimes regarded as a measure of the open magnetic flux. However, by virtue of Faraday's law, a change in the total closed magnetic flux content in the magnetosphere is due to the inductive electric field. Such a change, resulting, for instance, from a southward rotation of the IMF B_Z component and accompanied by an expansion of the polar cap on the dayside, is observed in the ionosphere on time scales of a few minutes to ~15 minutes depending on the magnetic local time (MLT), and the corresponding ionospheric convection signatures are not directly mapped into the magnetosphere (Lockwood et al., 1990). Not only it is the inductive electric field that is responsible for changing the open (closed) magnetic flux, but during such transient periods magnetic field lines cannot be considered equipotential (Hesse et al., 1997), and therefore the instantaneous value of Φ_{PC}, defined as the difference between the maximum and the minimum electrostatic potential in the ionosphere, is in no direct relation to the rate of change of the open magnetic flux. However, in a hypothetical steady state magnetosphere-ionosphere configuration the two quantities can be related indirectly. Indeed, both the transpolar potential and the open magnetic flux (or A_{PC}), being global indicators of the state of the solar windmagnetosphere- ionosphere system, can be functions of the system geometry. For instance, in the oversimplified example of a circular polar cap (certainly a bad approximation as viewed in global MHD simulations), the two are related via the polar cap radius, provided the ionospheric electric field is known. Using the Lyon-Fedder-Mobarry (LFM) global MHD model, Merkin and Goodrich (2007) investigate the behavior of A_{PC} under the influence of the southward IMF of varying strength and compare it to the behavior of Φ_{PC}. In particular, we address the question of whether A_{PC} and the open magnetic flux saturate as the solar wind convective electric field increases, and consider how this behavior is affected by changing other parameters in the system, e.g., the solar wind dynamic pressure and the ionospheric conductance. Kabin et al. (2004) used the BATS-R-US global MHD model to investigate the geometry and the position of the open-closed field line boundary under a wide range of idealized solar wind and IMF conditions. That study, while exploring a broad range of input parameters, in particular the IMF strength and orientation, did not investigate specifically the saturation of A_{PC}.

In Merkin and Goodrich (2007) the low resolution LFM simulation code was used with a grid of 53 × 24 × 32, or 40704, cells in total, corresponding to the radial, polar and azimuthal directions in the LFM sense (the polar axis is aligned with the Solar Magnetic X-axis) (Lyon et al., 2004). The inflow solar wind boundary conditions were fixed during a given run and varied between the different runs as follows: the solar wind velocity $V_X = -500$ km/s, $V_Y = V_Z = 0$ km/s; IMF $B_Z = -2.5$, 5, 10, 15, 20, 30, 40, 50 nT, $B_X = B_Y = 0$ nT; and the solar wind number density $n = 10$ cm^{-3}. Merkin and Goodrich (2007) repeated this series of runs with two different values of the ionospheric Pedersen conductance SP, which was set to 5 and 10 S uniformly over the entire polar cap, while the Hall conductance was set to 0. In addition, for SP = 5 S Merkin and Goodrich (2007) repeated the series of simulation runs with a lower solar wind number density $n = 5$ cm^{-3}. Thus it became able to probe the influence of the ionospheric conductance and the solar wind dynamic pressure on A_{PC} and the open magnetic flux. Merkin and Goodrich (2007) summarized the obtained main results and discussed the perspective of future work as following: **(i)** There are shown results of idealized global MHD simulations and provided their phenomenological interpretation. **(ii)** A lot remains to be done; in particular, a more rigorous analysis confirming discussed conjecture that it is the change in the geometry of the dayside magnetopause that results in the polar cap area saturation should be possible. **(iii)** Effects of the ram pressure and the ionospheric conductance have to be examined in detail. **(iv)** It still remains to be considered how the described results would be affected by including a more realistic model of the ionospheric conductance, i.e., the day-night gradient, the Hall conductance, and the auroral oval contributions.

4.2.6. Extreme Polar Cap Density Enhancements along Magnetic Field Lines during an Intense Geomagnetic Storm: Ionosphere as Matter Source for Polar Magnetosphere

As noted Tu et al. (2007), the polar ionosphere is regarded as a major source of plasma from the ionosphere to the magnetosphere through the pervasive polar wind (primarily hydrogen ions H^+) and intermittent heavy ions (primarily oxygen ions O^+) outflows at least during geomagnetically active times (e.g., Moore and Delcourt, 1995; Chappell et al., 2000) Although still controversial, Chappell et al. (1987) concluded that the ionosphere alone was able to supply the entire plasma content of the magnetosphere under all geomagnetic conditions by estimating volumes and ion residence times for the plasmasphere, plasma trough, plasma sheet, and magnetotail lobes. Thermal plasma of ionospheric origin has been observed, at one time or another, nearly everywhere in the magnetosphere from in situ observations. Therefore the study of extreme plasma density enhancements in the polar cap is of considerable interest. Plasma densities in the polar cap ionosphere and magnetosphere are highly variable and structured because of complex chemical and dynamic processes. For example, localized electron density (N_e) enhancements (polar patches) can form at ionospheric F-layer altitudes when the midlatitude ionosphere plasma is transported across the dayside cusp into the polar cap by time varying plasma convection (e.g., Whitteker et al. 1976; Rodger et al. 1994; Sojka et al. 2006). The O^+ density trough (with density as low as 0.01 cm^{-3}) at about 1 r_E in altitude is observed on the nightside polar cap when there is a lack of source O^+ in the ionospheric F-layer (Zeng et al. 2004). The plasma density at polar magnetosphere altitudes can be increased by the enhanced cleft ion fountain or the increased ion outflows locally in the polar cap ionosphere (Horwitz and Moore, 1997; Schunk and Sojka, 1997; Schunk, 2000; Tu et al., 2004).

As outlined Tu et al. (2007), in the dayside cusp/auroral region the heating/acceleration processes cause ionospheric ion outflows along the magnetic field lines and outflowing ions are meanwhile transported to the polar magnetosphere by the anti-sunward convection. The heating/acceleration processes and anti-sunward convection together were termed as cleft ion fountain, which can supply plasma to the polar magnetosphere (Horwitz, 1984; Lockwood et al. 1985; Tsunoda et al., 1989; André et al., 1990; Horwitz and Moore, 1997; Moore et al., 1999; Dubouloz et al., 2001; Tu et al., 2005a). The cleft ion fountain can be enhanced by the increased energy input (through increased Poynting flux and soft-electron precipitation flux) to the polar ionosphere at times of geomagnetic disturbances. This is because the increased energy input increases both the effects of the heating/acceleration processes and the ionospheric ion (primarily O^+) density in the heating/acceleration region that provides an augmented source population for the ion outflows. The cleft ion fountain may be even more strongly increased during major magnetic storms when there is a large flux of midlatitude ionospheric plasma transporting to the dayside cusp/auroral region. During the major magnetic storms, the dayside midlatitude ionospheric density can be enhanced by the uplift of the low-latitude F-layer plasma and the diffusion of the lifted plasma along magnetic field lines to higher latitudes (e.g., Warren, 1969; Evans, 1973; Tanaka, 1979; Foster, 1993; Buonsanto, 1999; Tsurutani et al., 2004; Huang et al. 2005). Rapid poleward convection then carries a large flux of midlatitude ionospheric plasma through the dayside cusp into the polar cap as the feet of the flux tubes move from the middle latitudes to the higher latitudes (e.g., Pryse et al., 2004; Foster et al., 2005). Meanwhile, the plasmaspheric plasma in the outer region of the plasmasphere (close to the plasmapause) is stripped away and transported to the dayside magnetopause by subauroral polarization stream-associated sunward convection, which is seen as a plasmaspheric drainage plume from extreme ultraviolet images taken by the IMAGE satellite (e.g., Burch et al., 2001). The plasmaspheric drainage plume and the midlatitude ionospheric plasma transported to high latitudes have recently been observed to occur along common flux tubes extending from the ionosphere to the plasmasphere (Garcia et al., 2003; Foster et al., 2002, 2004). A recent estimate made by Foster et al. (2004) using combined observations with a network of GPS total electron content (TEC) receivers, the Millstone Hill incoherent scatter radar, and DMSP and IMAGE satellites indicates a rate of plasma transported toward the cusp up to 10^{26} ions/s (primarily O^+) during a large geomagnetic disturbance. The transported plasma includes contribution from both the ionosphere and the plasmasphere. From the opinion of Tu et al. (2007), a major contribution to the transported plasma is believed to be from the ionosphere because of the much higher ionospheric density. Such a large rate of ionospheric plasma transporting to high latitudes provides a strongly increased source population for the potentially large flux of plasma outflow to the magnetosphere from the dayside cusp/auroral region (Foster et al., 2005), i.e., strongly enhanced cleft ion fountain. It is expected that the plasma density in the polar cap magnetosphere be raised through the strongly enhanced cleft ion fountain. Assuming that this mechanism generates density enhancements, questions arise such as how strong and up to what altitude these density enhancements can be observed. The answers to these questions are important because they may provide clues to the strength and altitude distribution of the heating/acceleration processes in the dayside cusp/auroral region.

According to Tu et al. (2007), the RPI instrument alternated between making passive radio wave measurements and radio sounding measurements. Each of these modes of operation is analyzed and displayed differently. The design and measurement characteristics of the IMAGE RPI have been described in detail by Reinisch et al. (2000) and the inversion technique used to derive field-aligned N_e profiles has been discussed in previous studies (Reinisch et al., 2001, 2004; Huang et al., 2004). In brief, the RPI, in active sounding modes, transmitted coded signals stepping through frequencies from 3 kHz to 3 MHz and detected the echoes. The received echoes are plotted in a 'plasmagram' with the analysis software known as BinBrowser (Galkin et al., 2004). A plasmagram is a color-coded display of signal amplitude as a function of frequency and echo delay time (represented as virtual range: one-half of the echo delay time multiplied by the speed of light in free space). Multifrequency echoes can form distinct traces in a plasmagram. It has been found that when discrete traces are observed they correspond to reflected signals that propagate along the magnetic field line threading the satellite (Reinisch et al., 2001; Fung et al., 2003; Fung and Green, 2005).

Applying a new inversion algorithm (Huang et al., 2004), one can derive the N_e distribution along a field line almost instantaneously (in ≤ 1 min). The transmitted signals also stimulate local (at satellite locations) plasma resonances that can be used to determine the in situ electron cyclotron harmonic frequencies and plasma frequency (Benson et al., 2003).

Tu et al. (2007) investigate two data groups: group 1 represents the measurements made on the nightside and group 2 on the dayside. It is noted that group 2 measurements were made during a decrease of the Dst index with Dst ~ –260 nT in the storm recovery phase. This Dst decrease was driven by a period of persistently negative IMF B_Z. The solar wind transition time from the ACE location to the magnetopause was taken into account. It is seen that the nightside measurements were carried out around IMF B_Z reorientation from northward to southward and IMF B_Y was generally negative. The dayside measurements were made in a period of persistently southward IMF with IMF B_Z = –20 nT. The IMF B_Y during the period of dayside measurements was positive. Measurements were made within the period of a TEC plume extending to above 65° (up to 70°) magnetic latitude, but additional measurement was obtained after the TEC plume had retreated to latitudes below 60°.

Tu et al. (2007) present observations of the field-aligned N_e profiles measured by the IMAGE RPI in the polar cap. As mentioned previously, the sounding measurement programs of the RPI that captured traces during the 31 March 2001 storm used long transmitter pulses (51.2 ms). Such long transmitter pulses prevented some short-duration (low virtual range on the plasmagrams) resonances from being detected, making it more difficult to determine f_{ce} and f_{pe} at the satellite. This difficulty can be overcome, however, by first identifying some of the stronger resonances at the nf_{ce} harmonics, with the aid of a model for the magnetic field B, so as to determining f_{ce} at the satellite. Using this f_{ce} value, the ambient f_{pe} value is obtained from a self-consistent identification of the sounder-stimulated resonances at f_{pe} and at the upper hybrid frequency f_{uh} according to

$$f_{uh}^2 = f_{pe}^2 + f_{ce}^2, \tag{4.2.7}$$

as well as the projection of the X-mode trace to its cutoff value f_x at the satellite given by

$$f_x = (f_{ce}/2)\left[1 + \left(1 + 4f_{pe}^2/f_{ce}^2\right)\right]^{1/2}. \tag{4.2.8}$$

As outlined Tu et al. (2007), an additional aid in the self-consistent identification of the above features is provided by two series of resonances often observed between the nf_{ce} harmonics. One series, known as the Qn resonances, are observed at frequencies greater than f_{uh} and are due to sounder-stimulated electrostatic Bernstein-mode waves with group velocity comparable to the satellite velocity (Muldrew, 1972). The other, known as the Dn resonances, are observed at frequencies less than f_{pe} and have been explained in terms of sounder-stimulated electromagnetic cylindrical plasma oscillations as described in the review by Osherovich et al. (2005). The frequencies of the resonances in both of these series are dependent mainly on the ratio f_{pe}/f_{ce}, so they provide an extra degree of confidence in the identification of the proper features as f_{pe} and f_{uh} on RPI plasmagrams, particularly when f_x is present, as described by Benson et al. (2003). The accuracy of the ambient f_{ce} and f_{pe} determinations is dependent on the frequency resolution of the RPI sounding mode in use.

This resolution is typically a few percent for the modes used for long-range sounding. Thus the ambient f_{ce} and f_{pe} can be determined to this accuracy. The accuracy of the ambient N_e is less but is typically better than 10%. The remote N_e profile along the magnetic field direction is obtained by inverting the X-mode virtual range/frequency trace; confidence in this approach is obtained by using the resulting N_e profile to correctly calculate the virtual range/frequency traces of other independent modes of propagation when they are observed (Huang et al., 2004). Tests of the accuracy of such N_e profiles have been performed in the topside ionosphere. For example, conjunction comparisons between Langmuir-probe measurements on the low-altitude Dynamics Explorer-2 satellite and topside-sounder measurements on the higher-altitude International Satellites for Ionospheric Studies (ISIS1 and ISIS2) demonstrated the good accuracy of the topside N_e profiles at all altitudes from the satellite down to the F-region peak density (where the agreement was typically about 30%) (Hoegy and Benson, 1988).

Tu et al. (2007) discussed and summarized obtained main results as following: **(i)** There are presented the IMAGE RPI measurements of the field-aligned electron density distributions from an IMAGE polar cap pass from nightside to dayside during the 31 March 2001 magnetic storm. **(ii)** The RPI observed density enhancements on both the nightside and dayside in the polar cap. The nightside measurements were made at altitudes below about $5R_E$ while the observed dayside density profiles extended to $\sim 7R_E$ altitude. **(iii)** Density observations at such high altitudes and even higher altitudes have been made previously using in situ measurements (e.g., Marklund et al., 1990; Laakso et al., 2002; Osherovich et al., 2001, 2007; Benson et al., 2003, 2006). **(iv)** Instantaneously measured field-aligned density profiles in the polar cap have also been reported previously for the altitudes below about 4 R_E (Nsumei et al., 2003; Tu et al., 2005b). **(v)** However, the instantaneously measured fieldaligned density profiles up to 7 R_E altitude are, for the first time, presented in Tu et al. (2007). It is shown that the N_e profiles were strongly enhanced at least up to 7 R_E altitude during the storm on 31 March 2001. The density at 7 R_E reached about 10 cm^{-3}, an unusually high value at that altitude because it is even higher than the average density observed previously at 4 R_E altitude (~ 6 cm^{-3}). **(vi)** Compared to the average density profiles obtained from IMAGE RPI measurements in the work of Nsumei et al. (2003), the measured density profiles in Tu et al. (2007) show large density enhancements at all altitudes. **(vii)** During the 31 March 2001 magnetic storm, a strong ionospheric plume, extending from midlatitude ionosphere across the dayside auroral/cusp to high latitudes, existed from 18:00 to 20:35 UT. **(viii)** The plume continuously provided an enhanced source population for about 2.5 hours for the potential outflow of the large flux of ionospheric plasma along magnetic field lines. **(ix)** The high-latitude convection as shown by the DMSP-F15 observations was anti-sunward in the polar cap and remained anti-sunward from 14:15 to 22:00 UT as revealed by Digisonde F-region electron drift measurements in the central polar cap. **(x)** The cleft ion fountain operates when the polar cap convection is anti-sunward. During geomagnetic disturbances, the cleft ion fountain effect may be strengthened by increased energy deposit to the high-latitude ionosphere and thus result in enhanced polar cap densities at high altitudes (e.g., Tu et al., 2004). **(xi)** The ionospheric plasma plume may cause even more greatly enhanced source population for the cleft ion fountain effect. This is perhaps the cause of the strongest density enhancements shown by the density profiles 1–4 observed on the dayside. **(xii)** During considered period strong ionospheric ion outflows have been observed by the Cluster satellites within the main phase of the storm from 08:00 to 09:30 UT (Korth et al., 2004). Within the period of the second decrease of the Dst driven by strongly negative IMF B_Z which lasted for more than 5 hours, ionospheric ion outflows are also expected to occur. **(xiii)** The discussed observations suggest that the most strongly enhanced density extending to high altitudes in the polar cap is likely caused by the greatly enhanced cleft ion fountain that supplies plasma into the polar cap magnetosphere with the enhanced source population in the cusp/auroral region. **(xiv)** The TEC plume intruding to the polar cap also includes a contribution from the plasmaspheric plasma, which is stripped away from the outer edge of the plasmasphere. **(xv)** The stripped plasmaspheric plasma is transported to the dayside magnetopause and may directly account for part of the density enhancements in the polar magnetosphere. Nevertheless, the major contribution to the TEC plume is from the ionosphere (Foster et al., 2004; Huang et al., 2005), which provided greatly strengthened source population for the cleft ion fountain. Therefore the cleft ion fountain that transports outflowing ionospheric plasma to the polar magnetosphere may be the primary source of the density enhancements, particularly for the dayside density enhancements. The described observations, however, do not allow to examine in detail the possible link between the plasma plume and the strongest density enhancements (for example, the time delay between the arrival of the TEC plume to the cusp/auroral region and the polar magnetospheric density enhancements), as this requires information about the global high-latitude convection pattern and history (which may be obtained by the SuperDARN). **(xvi)** In addition, one needs information about the spatial and temporal variations of the heating/acceleration processes in the dayside cusp/aurora region in order to evaluate the plasma transport. This information is not yet available. The described observations can provide constrains to the simulations that explore the spatial and temporal variations of those heating/acceleration processes. **(xvii)** It is necessary to note that the topside ionosphere density in the polar cap was increased during the period when the TEC plume was observed to intrude across the dayside auroral/cusp to the polar cap. This increased ionospheric ion (primarily O^+) density can also induce increased ion outflows locally in the polar cap and thus may contribute to the density enhancements at high altitudes. **(xviii)** However, the DMSP-F15 satellite observed downward O^+ ion vertical velocity in the polar cap for the period from 18:51 to 19:03 UT. Therefore the outflows of O^+ from the polar cap may be ruled out for the dayside measurements. Tu et al. (2007) do not have enough information about the H^+ outflows (polar wind), so they cannot assess the relative importance of the enhanced cleft ion fountain effects and the locally enhanced polar wind in causing the observed polar magnetosphere density enhancements. **(xix)** Nevertheless, it is reasonable to suggest that the greatly enhanced cleft ion fountain is responsible for the density enhancements, at least for the strongest enhancements on the dayside because of the considerably weaker density enhancement when the plasma plume subsided to below 60° MLAT. When the plasma plume subsided, significantly less (even absent) flux of ionospheric plasma is transported to the dayside auroral/cusp for the ion outflows, resulting in a weaker cleft ion fountain and thus causing weaker density enhancement at high altitudes in the polar cap. **(xx)** Other evidence that supports this conjecture is that the outflow flux of ionospheric ions is increased primarily due to an increase in the O^+ outflow flux (e.g., Moore et al., 1999). Without locally observed O^+ outflow in the polar cap, the cleft ion fountain is thus the most likely mechanism for the dayside density enhancements observed in the polar magnetosphere. **(xxi)** The described IMAGE RPI observations display extreme electron density enhancements along magnetic field lines in the polar cap extending to 7 r_E altitude. The enhanced cleft ion fountain effects are believed to be responsible for the observed density enhancements although some of the nightside density enhancements may be caused by the increase of the ion outflows locally in the polar cap. **(xxii)** The strongest density enhancements observed on the dayside are also likely associated with the plasma plume transporting primarily ionospheric plasma from lower latitudes to the dayside auroral/cusp region that further strengthened the cleft ion fountain. **(xxiii)** The study of Tu et al. (2007) and the works of Osherovich et al. (2001, 2007) and Benson et al. (2003, 2006) show that during magnetic storms the plasma density is globally redistributed not only in the ionosphere and plasmasphere, as is already well known, but also in the polar magnetosphere.

4.2.7. Temporal Evolution of the Transpolar Potential after a Sharp Enhancement in Solar Wind Dynamic Pressure

As noted Boudouridis et al. (2008), the significant effect of sudden enhancements in solar wind dynamic pressure to many aspects of magnetospheric dynamics has been amply demonstrated in the past few years (Boudouridis et al., 2003, 2007; Liou, 2006, and references therein). One of the results of a sudden increase in dynamic pressure is the enhancement of ionospheric convection as seen by several low-altitude Defense Meteorological Satellite Program (DMSP) spacecraft (Boudouridis et al., 2004a,b, 2005) and Super Dual Auroral Radar Network (SuperDARN) observations (Boudouridis et al., 2007). For cases of southward IMF before and after the increase in pressure, Boudouridis et al. (2005) has shown that the solar wind/magnetosphere coupling efficiency increases after the pressure enhancement. They defined the coupling efficiency as the ratio of the transpolar potential (measured by DMSP spacecraft) to the potential in the undisturbed solar wind across the width of the magnetosphere (calculated from solar wind parameters). Their result suggests that the abrupt increase in dynamic pressure contributes to the ionospheric convection enhancement independently from any concurrent changes in the IMF. A recent statistical study by Palmroth et al. (2007) shows that the coupling efficiency, as defined by Boudouridis et al. (2005), increases on the average after pressure-front events with a concurrent decrease in the total IMF magnitude (slow type discontinuities), while it decreases after pressure-front events with a simultaneous increase in the total IMF magnitude (fast type discontinuities). The three cases studied by Boudouridis et al. (2005) exhibit either steady or decreasing total field magnitude and therefore are consistent with the results of Palmroth et al. (2007).

Boudouridis et al. (2008) outlined that one of the most striking effects of solar wind dynamic pressure fronts is the poleward expansion of the auroral oval and the closing of the polar cap observed over a wide range of Magnetic Local Times (MLTs), often with the exception of the near-noon region (Boudouridis et al., 2003, 2004a, 2005). This can range from a few degrees magnetic latitude (MLAT) up to 10° in some cases over certain MLT ranges. The dramatic shrinking of the polar cap suggests an enhancement of magnetotail reconnection induced by the pressure front (Boudouridis et al., 2003, 2004a; Milan et al., 2004; Hubert et al., 2006). Hubert et al. (2006) estimated the tail reconnection potential for two consecutive pressure front impacts. They found it to increase from initial values of ~30 kV and ~20 kV to 132 kV and 114 kV, respectively. Milan et al. (2004), studying a similar case, observed a reduction of the open magnetic flux in the Northern Hemisphere from 0.5

GWb to 0.2 GWb, down to 2.5% of the total hemispheric flux from a nominal value of 7–8%. They report a corresponding tail reconnection potential of 150 kV. Most recently Boudouridis et al. (2007) have investigated changes in dayside reconnection after impacts of solar wind dynamic pressure fronts by looking at dayside ionospheric convection changes using SuperDARN observations. They observed a significant increase in ionospheric velocities coinciding with the time of the pressure front impact. The flow enhancements were concentrated mostly near the expected location of the cusp in accordance with the prevailing direction of the IMF B_Y component. Furthermore, for the two cases with southward IMF conditions throughout the pressure jump, the variations in the average flow magnitude follow the respective variations in solar wind pressure very closely, pointing to a definite association of the dayside convection with solar wind pressure. Considering that the dayside poleward boundary of plasma sheet precipitation near the location of the observed convection enhancement did not move significantly during these two events (Boudouridis et al., 2005), these results suggest a close connection between the dayside reconnection rate and the solar wind pressure. Observations therefore suggest that solar wind pressure fronts have a considerable effect on the global electrodynamics of the magnetosphere. Enhanced dayside and nightside reconnection compete in determining the size and shape of the polar cap, and may both contribute to changes in ionospheric convection and the transpolar potential.

Boudouridis et al. (2008) report on DMSP observations and Assimilative Mapping of Ionospheric Electrodynamics (AMIE) runs of the transpolar (or cross-polar-cap) potential for a long-lasting step-like increase in solar wind pressure on 30 April 1998. They seek to evaluate the temporal evolution of the transpolar potential for several hours after the increase in pressure and while the pressure remains high, and examine any possible connection to changes in dayside and nightside reconnection rates. Boudouridis et al. (2004a) presented DMSP precipitating particle observations for the 30 April 1998 pressure-front event to demonstrate the nightside closing of the polar cap by about 5° M_{LAT}. Boudouridis et al. (2004a) also showed solar wind pressure and IMF B_Z measurements taken by three solar wind monitors, ACE, IMP8, and WIND. The AMIE technique (Richmond and Kamide, 1988) utilizes a large number of observations from various sources (ground magnetometers, DMSP satellites, and radars) to determine the high-latitude convection pattern by means of a weighted, least squares fit of coefficients. It yields a number of desired ionospheric electrodynamics quantities including transpolar potential, hemispheric power (HP, a measure of auroral precipitation power), Joule heating (JH), AE and Dst indices (e.g., Ridley et al., 1998; Kihn et al., 2006). For event 30 April 1998, AMIE was run with 1-min resolution using only ground magnetometers (with the number of stations varying from 116 to 119).

Boudouridis et al. (2008) discuss and summarize main obtained results as following: **(i)** There are presented observations associated with a sudden step increase in solar wind dynamic pressure. The considered case on 30 April 1998 is unusual in that the dynamic pressure remained elevated and nearly constant under relatively steady IMF conditions for ~4 hours, allowing to evaluate the internal long-term response of the magnetosphere-ionosphere system following the pressure enhancement. **(ii)** There is examined the temporal evolution of the transpolar potential measured with two different methods, low-altitude DMSP spacecraft measurements and the AMIE technique. Both methods show the potential more than doubling, reaching a maximum about an hour after the pressure enhancement. Despite the solar wind and IMF conditions then remaining relatively steady, in the next 2.5 hours the potential slowly decreased from its peak value but still ended up at a value higher than its pre-front level. **(iii)** Boudouridis et al. (2007) studied the response of dayside ionospheric convection after an abrupt step increase in solar wind dynamic pressure on 19 February 1999 using SuperDARN convection data. They showed that an enhancement of dayside reconnection is initiated by the pressure increase, and that the enhanced dayside reconnection rate remains high while the solar wind pressure is high. Such an increase of dayside reconnection leads to an increased transpolar potential. Since the enhanced reconnection correlates well with the pressure variation in the case presented by Boudouridis et al. (2007), it was expected that if the transpolar potential increase was due solely to the enhancement of dayside reconnection, then the potential would remain high while the solar wind pressure stays high for the 30 April 1998 pressure front too. This is not the case for this event. **(iv)** Ober et al. (2006) suggested a transient inductive response of the dayside reconnection rate to an increase in pressure. In their scenario the transpolar potential behavior is well characterized by an L-R circuit equation derived from integrating Faraday's law around the Region 1 current loop. The potential first rises quickly after the increase in pressure, and then returns slowly to previous levels in the course of ~15 min. This response was verified for the 30 April 1998 event also using the AMIE technique (Ober et al., 2007). The 15 min response time is much shorter than the 3–4 hours found here, though it could account for the 15 min sharp spike observed by AMIE at ~09:30 UT. It is therefore clear that enhanced dayside reconnection alone cannot fully account for the transpolar potential evolution seen after the step increase in solar wind dynamic pressure for this event. **(v)** Boudouridis et al. (2004a), based on DMSP observations of the closing of the polar cap after a pressure enhancement, suggested that a solar wind pressure front can also induce enhanced magnetotail reconnection. The enhanced tail reconnection might have a limited lifetime, as evidenced by the fact that the nightside closing of the polar cap does not continue indefinitely after a pressure increase. The initial fast poleward motion of the polar cap boundary on the nightside, and the subsequent decrease of its poleward speed have been observed by DMSP spacecraft for the 30 April 1998 event. Boudouridis et al. (2004a) have shown that the nightside polar cap boundary moved poleward by ~5° to ~77° MLAT, as observed by DMSP F11 at 09:46–10:00 UT. DMSP F13, passing over the same region at 10:12–10:28 UT observed the same boundary above 80° MLAT. Subsequent F11 and F13 orbits from 11:26 UT to 13:49 UT showed the polar cap boundary located at the vicinity of 80° MLAT with no further poleward motion. Thus after the magnetosphere adjusts to a new compressed state, the increased reconnection rate in the tail might slowly fade away. **(vi)** In addition to reducing the size of the polar cap, it has been suggested (e.g., Cowley and Lockwood, 1992; Boudouridis et al., 2005) that magnetotail reconnection may contribute to ionospheric convection and the transpolar potential. If true then magnetotail reconnection might contribute to the enhanced ionospheric convection observed immediately after an increase in solar wind dynamic pressure (Boudouridis et al., 2004a, 2005; Hubert et al., 2006). If the observed convection enhancement after a solar wind pressure increase is related in part to the enhanced tail reconnection, it would also exhibit a transient behavior, despite the fact that solar wind and IMF conditions remain relatively steady. If instead it were related solely to external forcing, i.e., high solar wind pressure through an enhancement of dayside reconnection, it should remain steadily high throughout the high-pressure environment as observed by Boudouridis et al. (2007) locally near the dayside cusp. **(vii)** The considered case study supports the transient nature interpretation and thus suggests a combined contribution to the transpolar potential from both dayside and nightside reconnection after a sharp enhancement in solar wind dynamic pressure. The enhanced dayside reconnection provides a steady contribution varying proportionally to the enhanced pressure. The nightside reconnection on the other hand provides a transient component that manifests as an initial increase of the transpolar potential which then fades away in a few hours when the tail reconnection rate returns to lower values. They both might be modulated by other factors, such as the background IMF conditions or the pressure-front characteristics, and hence their timescales and/or magnitudes may vary from case to case. There is also evidence that there remains a residual effect on the transpolar potential during the high-pressure regime that can possibly be associated with dayside reconnection, nightside reconnection, or both. **(viii)** Described observations also suggest that models which parameterize the state of the magnetosphere as a function of instantaneous solar wind and IMF conditions may lack an important aspect of magnetospheric dynamics, as they do not take into account long term effects (a few hours in duration) like the one discussed here. **(ix)** Another important issue is how dayside and nightside reconnection combine to produce the observed transpolar potential after a sharp increase in solar wind pressure. Does one dominate at any single time or both contribute at all times? In other words, is the transpolar potential controlled by the maximum of the two reconnection rates or is it always a function of both processes? A possible way to test the relative importance of dayside/nightside reconnection would be to use SuperDARN convection data during events with simultaneous dayside/nightside coverage, and concurrent polar cap boundary determinations from auroral images. Additional events are needed to establish the consistency of described results and the connection between solar wind dynamic pressure fronts, enhanced dayside/nightside reconnection, and the temporal evolution of the transpolar potential.

4.2.8. Statistical Study of the Effect of ULF Fluctuations in the IMF on the Cross Polar Cap Potential Drop for Northward IMF

As outlined Kim et al. (2011), recent studies showed that, regardless of the orientation of the IMF, ULF wave activity in the solar wind can substantially enhance the convection in the high latitude ionosphere, suggesting that ULF fluctuations may also be an important contributor to the coupling of the solar wind to the magnetosphere-ionosphere system. Kim et al. (2011) conduct a statistical study to understand the effect of ULF power in the IMF on the cross polar cap potential, primarily focusing on northward IMF. They have analyzed the AMIE calculations of the polar cap potential, a IMF ULF index that is defined as the logarithm of Pc5 ULF power in IMF, and solar wind velocity and dynamic pressure for 249 days in 2003. It was found that, separated from the effects of solar wind speed and dynamic pressure, the average cross polar cap potentials show a roughly linear dependence on the ULF index, with a partial correlation coefficient of 0.19. Highly structured convection flow patterns with a number of localized vortices are often observed under fluctuating northward IMF. For such a convection configuration, it is hard to estimate properly the cross polar cap potential drop, as the enhanced flows around the vortices that may be associated with IMF fluctuations do not necessarily yield a large potential drop. Thus, despite the relatively small correlation coefficient, the found linear trend gives support to the significant role of IMF ULF fluctuations on the coupling of the solar wind to the magnetosphere-ionosphere system.

4.2.9. The Nonlinear Response of the Polar Cap Potential under Southward IMF: A Statistical View

Wilder et al. (2011) report the results of an investigation into the effect of solar wind properties on the saturation of the cross polar cap potential (CPCP) during periods of strongly southward IMF. Wilder et al. (2011) use propagated solar wind data to search for periods between 1998 and 2007 when the interplanetary electric field (IEF) is stable for more than 50 min and placed further conditions on the availability of SuperDARN and DMSP velocity data. CPCP values are calculated from these data sets and various fits of the polar cap potential to the IEF are compared. It is found that the trend is nonlinear, with a square root function fitting better than a straight line, and that the CPCP does not appear to exhibit asymptotic behavior. The nonlinearity of the CPCP is then correlated with various interplanetary parameters to test the various models of polar cap potential saturation. It is also found that the deviation of the CPCP from a linear fit has statistically significant correlation with solar wind Alfvènic Mach number and no significant correlation with solar wind dynamic pressure.

4.2.10. A Relationship between the Auroral Absorption and the Magnetic Activity in the Polar Cap

As outlined Frank-Kamenetsky and Troshichev (2012), the auroral absorption is an absorption of the cosmic radio noise; it depends on the electron concentration in the ionospheric D-layer, which is strongly affected by the precipitating flux of high-energy (E > 30 keV) electrons. The data from Canadian chain of riometers were used by Frank-Kamenetsky and Troshichev (2012) to study the absorption as a function of magnetic local time (MLT) and geomagnetic latitude for different levels of geomagnetic activity, characterized by the Polar Cap (PC) index (Troshichev et al., 2006). The 1 min and hourly values of absorption were divided into two sets depending on the level of activity and examined as a function of MLT and geomagnetic latitude for quiet conditions (<PC> = 0.56 mV/m) and a higher activity (<PC> = 1.87 mV/m). It is shown that the auroral absorption reaches distinct maxima at latitudes of 66°-68° (equatorward zone) and 71°-72° (poleward zone), both of them being dependent on level of activity determined by the PC-index. The maximum correlation between the absorption and PC index is observed in the after-midnight MLT sector. The results show that the PC index is a good indicator of the solar wind energy entry into the magnetosphere and can be useful for monitoring the state of low ionosphere in the auroral zone.

4.2.11. Response of the Polar Magnetic Field Intensity to the Exceptionally High Solar Wind Streams in 2003

According to Lukianova et al. (2012), the exceptionally high solar wind stream activity in 2003 caused a record intensity in the auroral electrojet currents, leading to a major reduction of the horizontal field at auroral latitudes and to a notable strengthening of the vertical geomagnetic field in the polar cap. This strengthening is clearly visible in the observatory annual values as a significant deflection in the corresponding secular variation. A similar but weaker deflection also occurs during the strongest high speed stream years of the earlier solar cycles, e.g., in 1983 and 1994. It is also found that, in addition to the disturbed times, the westward electrojet was often enhanced even during the most quiet times of the strongest high speed stream years. The quiet time level was more disturbed in 2003 than in other high speed stream years, when an exceptionally clear signal was seen in the polar cap Z intensity even in the annual mean curve in this year. Lukianova et al. (2012) exclude other current systems like the ring current or the DPY current as possible explanations.

4.2.12. Simulation of the Polar Cap Potential during Periods with Northward IMF

Bhattarai et al. (2012) examine the response of the ionospheric cross-polar cap potential to steady, purely northward IMF using the Lyon-Fedder-Mobarry global MHD simulation of the Earth's magnetosphere. The simulation produces the typical, high-latitude “reversed cell” convection that is associated with northward IMF, along with a two cell convection pattern at lower latitude that is interpret as being driven by the viscous interaction. The behavior of the potential can be divided into two basic regions: the viscous dominated region and the reconnection dominated region. The viscous dominated region is characterized by decreasing viscous potential with increasing northward IMF. The reconnection dominated region may be further subdivided into a linear region, where reconnection potential increases with increasing magnitude of northward IMF, and the saturation region, where the value of the reconnection potential is relatively insensitive to the magnitude of the northward IMF. The saturation of the cross-polar cap potential for northward IMF has recently been documented using observations and is established as a feature of a global MHD simulation as well. The region at which the response of the potential transitions from the linear region to the saturation region is also the region in parameter space at which the magnetosheath transitions from being dominated by the plasma pressure to being dominated by the magnetic energy density. This result is supportive of the recent magnetosheath force balance model for the modulation of the reconnection potential. Within that framework, and including the described understanding of the viscous potential, Bhattarai et al. (2012) present a conceptual model for understanding the full variation of the polar cap potential for northward IMF, including the simulated dependencies of the potential on solar wind speed and ionospheric conductivity.

4.2.13. Day-Night Coupling by a Localized Flow Channel Visualized by Polar Cap Patch Propagation

Nishimura et al. (2014) present unique coordinated observations of the dayside auroral oval, polar cap, and nightside auroral oval by three all-sky imagers, two SuperDARN radars, and DMSP-17. This data set revealed that a dayside poleward moving auroral form (PMAF) evolved into a polar cap airglow patch that propagated across the polar cap and then nightside poleward boundary intensifications (PBIs). SuperDARN observations detected fast antisunward flows associated with the PMAF, and the DMSP satellite, whose conjunction occurred within a few minutes after the PMAF initiation, measured enhanced low-latitude boundary layer precipitation and enhanced plasma density with a strong antisunward flow burst. The polar cap patch was spatially and temporally coincident with a localized antisunward flow channel. The propagation across the polar cap and the subsequent PBIs suggests that the flow channel originated from dayside reconnection and then reached the nightside open-closed boundary, triggering localized nightside reconnection and flow bursts within the plasma sheet.

4.2.14. Polar Cap Response to the Solar Wind Density Jump under Constant Southward IMF

As note Belenkaya et al. (2014), sharp changes of the solar wind parameters determining the dynamic pressure jump lead to strong magnetosphere-ionosphere disturbances. The effect on the Earth's ionospheric high latitudes of the solar wind dynamic pressure pulse caused only by the increase of the interplanetary plasma density under southward constant IMF is considered by Belenkaya et al. (2014). It is investigate reaction of the cross-polar cap potential on the increase of AL index and/or jump of the solar wind density. It is found that for the case of 10 January 1997 the main contribution to the polar cap potential drop increase gave the growth of AL index relative to the input of the solar wind density jump. It is also study the influence of the solar wind density increase on the cross-polar cap potential for the quiet magnetospheric conditions. It occurred that the polar cap potential difference decreases with the great increase of the interplanetary plasma density. For the disturbed magnetosphere the main role in the polar cap potential drop increase plays increase of AL. Thus, Belenkaya et al. (2014) found the change of the cross-polar cap potential due to the AL index variations and/or the solar wind density drop even in a case when the interplanetary electric field is constant.

4.2.15. Transpolar Arc Observation after Solar Wind Entry into the High-Latitude Magnetosphere

As note Mailyan et al. (2015), recently, Cluster observations have revealed the presence of new regions of solar wind plasma entry at the high-latitude magnetospheric lobes tailward of the cusp region, mostly during periods of northward IMF. Observations from the Global Ultraviolet Imager (GUVI) experiment on board the TIMED spacecraft and Wideband

Imaging Camera imager on board the IMAGE satellite are used by Mailyan et al. (2015) to investigate a possible link between solar wind entry and the formation of transpolar arcs in the polar cap. Mailyan et al. (2015) focus on a case when transpolar arc formation was observed twice right after the two solar wind entry events were detected by the Cluster spacecraft. In addition, GUVI and IMAGE observations show a simultaneous occurrence of auroral activity at low and high latitudes after the second entry event, possibly indicating a two-part structure of the continuous band of the transpolar arc.

4.2.16. Investigation of a Rare Event Where the Polar Ionospheric Reverse Convection Potential Does Not Saturate during a Period of Extreme Northward IMF Solar Wind Driving

As outlined Clauer et al. (2016), a variety of statistical studies have shown that the ionospheric polar potential (produced by solar wind-magnetosphere-ionosphere coupling) is linear for weak to moderate solar wind driving but becomes nonlinear during periods of very strong driving. It has been shown that this applies to the two-cell convection potential that develops during southward IMF and also to the reverse convection cells that develop during northward IMF. This has been described as polar potential saturation, and it appears to begin when the driving solar wind electric field becomes greater than 3 mV/m. Utilizing measurements from the Resolute Incoherent Scatter Radar (RISR-N), Clauer et al. (2016) examine ionospheric data near local noon within the reverse convection cells that developed during a period of very strong northward IMF on 12 September 2014. During this period Clauer et al. (2016) measure the electric field within the throat of the reverse convection cells to be near 150 mV/m at a time when the IMF is nearly 28 nT northward. This is far in excess of the 30–40 mV/m expected for polar potential saturation of the reverse convection cells. In fact, the development of the electric field responds linearly to the IMF B_Z component throughout this period of extreme driving. The conditions in the solar wind show the solar wind velocity near 600 km/s, number density near 20 ions/cm^3, and the Alfvén velocity about 75 km/s giving an Alfvén Mach number of 8. A search of several years of solar wind data shows that these values occur together 0.035% of the time. These conditions imply a high plasma β in the magnetosheath. Clauer et al. (2016) believe that condition of high β along with high mass density and a strong merging electric field in the magnetosheath are the significant parameters that produce the linear driving of the ionospheric electric field during this unusual period of extreme solar wind conditions. A discussion of current theories to account for cross-polar cap potential saturation is given by Clauer et al. (2016) with the conclusion that theories that utilize magnetosheath parameters as they affect the reconnection rate appear to be the most relevant to the cross-polar cap potential saturation solution.

4.2.17. New Evidence of Dayside Plasma Transportation over the Polar Cap to the Prevailing Dawn Sector in the Polar Upper Atmosphere for Solar-Maximum Winter

As note Yang et al. (2016), it is well known that owing to the transport of high-density sunlit plasma from dayside to nightside primarily by convection, polar cap tongue of ionization (TOI), polar cap patches, and blobs are common features in the polar ionosphere. The steep density gradients at the edges of these structures lead to severe problems in applications involving radio waves traversing the ionosphere. To better understand the evolution of TOI/patches/blobs, it is essential to examine how the transported sunlit plasma is distributed – outlined Yang et al. (2016). Through averaging the hourly total electron content in solar-maximum winter, Yang et al. (2016) present complete distribution of polar ionospheric plasma and find that the dayside plasma can be transported through cusp, over polar cap, and eventually to the prevailing dawnside, showing asymmetric distribution around magnetic midnight. The negative IMF By or Bz component is favored for the plasma transportation from dayside to the prevailing dawn sector. This provides direct evidence for the plasma source of the dawnside high-density plasma structure. The same corotating convection direction as convection at auroral dawnside is responsible for the prevailing dawn sector transportation. According to opinion of Yang et al. (2016), this finding is significant for forecasting TOI/patches/blobs in conducting space weather in the polar ionosphere.

4.2.18. A Polar Cap Absorption Model Optimization Based on the Vertical Ionograms Analysis

As outlined Zaalov and Moskaleva (2016), space weather events significantly affect the high frequency (HF) radio wave propagation. The now-casting and forecasting of HF radio wave absorption is important for the HF communication industries. The paper of Zaalov and Moskaleva (2016) assimilates vertical sounding data into an absorption model to improve its performance as a now-casting tool. The approach is a modification of the algorithm elaborated by Sauer and Wilkinson (2008), which is based on the riometer data. The optimization is focused on accounting for short timescale variation of the absorption. It should be noted that the expression of the frequency dependence of absorption induced by the energetic particle precipitation employed in Sauer and Wilkinson (2008) model is based on the riometer data at frequencies of 20, 30, and 50 MHz. The approach suggested by Zaalov and Moskaleva (2016) provides an opportunity for expanding the frequency dependence of the absorption for frequencies below 10 MHz. The simulation of the vertical ionograms in the polar cap region uses a computational model designed to overcome the high frequency wave propagation problem in high latitude of the Earth. HF radio wave absorption induced by solar UV illumination, X-ray flares and energetic particles precipitation is taken into consideration in Zaalov and Moskaleva (2016) model. The absorption caused by the energetic particle precipitation is emphasized, because the study is focused on HF wave propagation in polar cap region. A comparison of observed and simulated vertical ionograms enables the coefficients, which relate absorption (day-time and night-time) to integral proton flux to be refined. The values of these coefficients determined from evaluation of the data recorded by any reliable ionosonde are valid for absorption calculation in high latitude region. Some obtained results are illustrated by Figures 4.2.7 – 4.2.10.

4.2.19. Polar cap patch transportation beyond the classic scenario.

Zhang et al. (2016) report the continuous monitoring of a polar cap patch, encompassing its creation, and a subsequent evolution that differs from the classic behavior. The patch was formed from the storm-enhanced density plume, by segmentation associated with a subauroral polarization stream generated by a substorm. Its initial antisunward motion was halted due to a rapidly changing of IMF conditions from strong southward to strong eastward with weaker northward components, and the patch subsequently very slowly evolved behind the duskside of a lobe reverse convection cell in afternoon sectors, associated with high-latitude lobe reconnection, much of it fading rapidly due to an enhancement of the ionization recombination rate. As outlined Zhang et al. (2016), this differs from the classic scenario where polar cap patches are transported across the polar cap along the streamlines of twin-cell convection pattern from day to night. This observation provides by new important insights into patch formation and control by the IMF, which has to be taken into account in F region transport models and space weather forecasts.

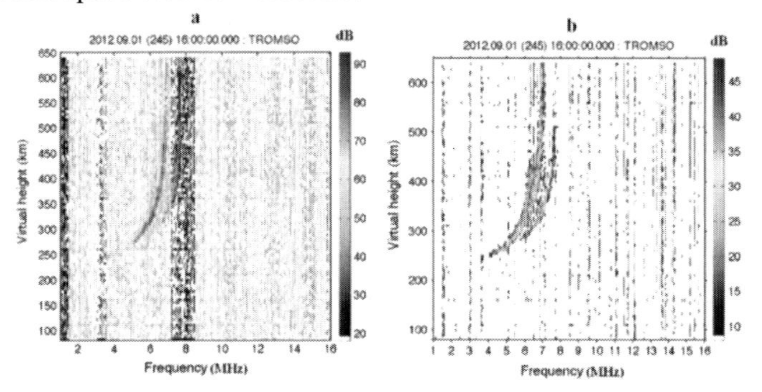

Figure 4.2.7. Vertical sounding data obtained in Tromsø on September 1 2012 at 16:00 UT. Panel (a): ionogram imported from SAO Explorer site [http://umlcar.uml.edu/SAO-X/SAO-X.html]. Panel (b): refined ionogram accounting effects of the AGC operation. According to Zaalov and Moskaleva (2016). Permission from COSPAR.

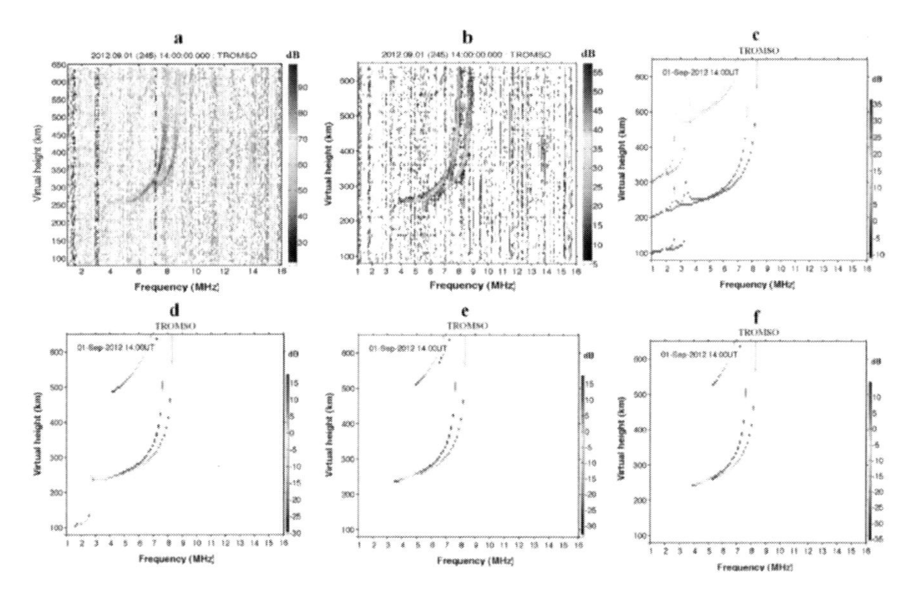

Figure 4.2.8. Ionogram evolution corresponding to Tromsø ionosonde on September 1 2012 at 14:00 UT. Panel (a): original ionogram downloaded from [http://umlcar.uml.edu/SAO-X/SAO-X.html]. Panel (b): refined ionogram processed to eliminate the AGC influence. Panel (c): ionogram simulated without accounting for absorption. Panel (d): simulated ionogram, where absorption was calculated in accordance with the D-RAP model. Panel (e): ionogram simulated with modified frequency dependence of the absorption ($m_d = 0.115$ and $m_n = 0.02$, i.e. correspond to D-RAP model). Panel (f): ionogram simulated with modified frequency dependence of the absorption and optimized coefficients ($m_d = 0.150$ and $m_n = 0.02$). According to Zaalov and Moskaleva (2016). Permission from COSPAR.

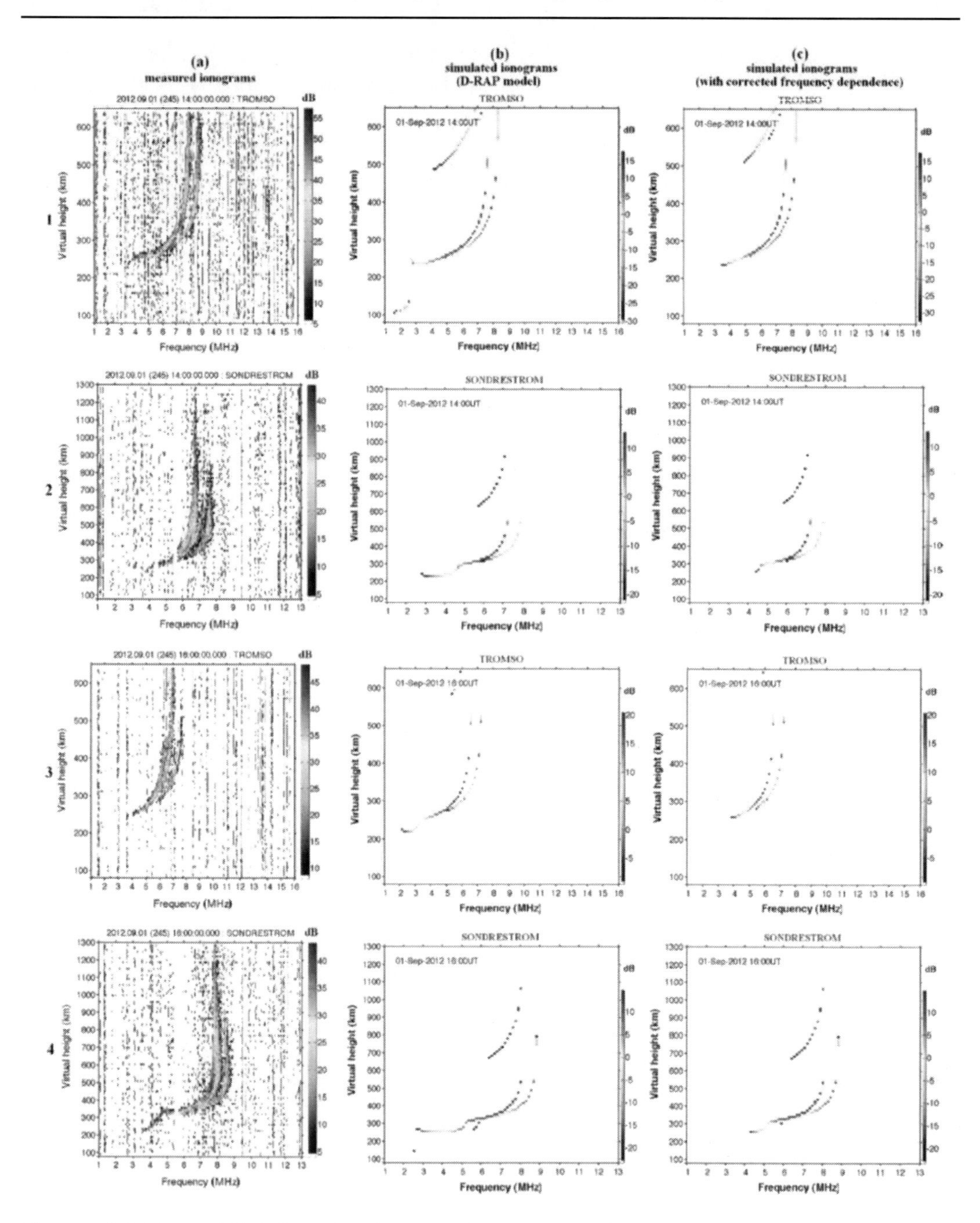

Figure 4.2.9. The modified absorption frequency dependence. Column (a): ionosonde data. Column (b): ionograms simulated in accordance with frequency dependence based on D-RAP model. Column (c): ionograms simulated in accordance with modified frequency dependence. Row 1: Tromsø ionogram on September 1 2012 at 14:00 UT. Row 2: Sondrestrom ionogram on September 1 2012 at 14:00 UT. Row 3: Tromsø ionogram on September 1 2012 at 16:00 UT. Row 4: Sondrestrom ionogram on September 1 2012 at 16:00 UT. The coefficients $m_d = 0.115$ and $m_n = 0.02$ correspond to the D-RAP model. According to Zaalov and Moskaleva (2016). Permission from COSPAR.

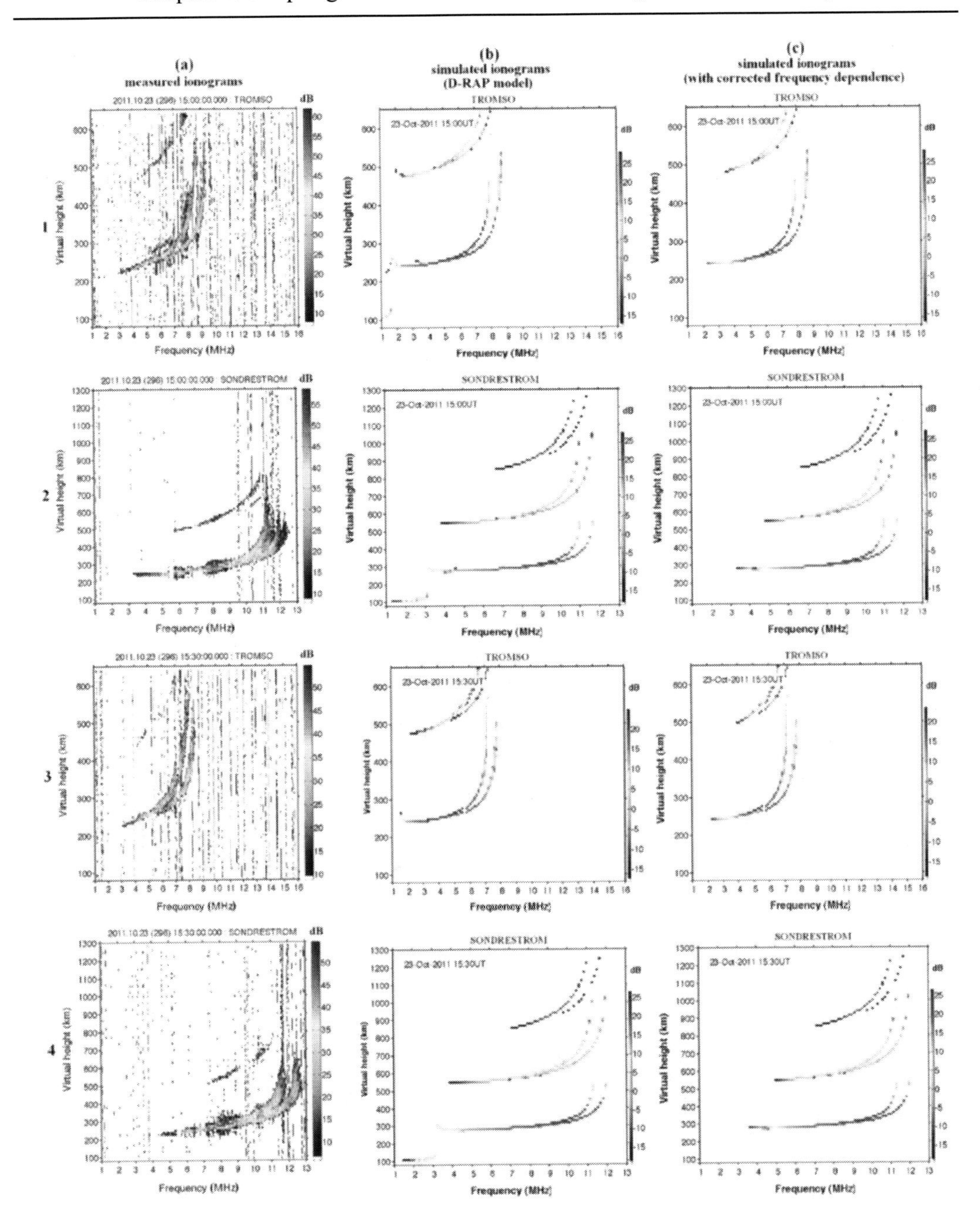

Figure 4.2.10. The modified absorption frequency dependence. Column (a): ionosonde data. Column (b): ionograms simulated in accordance with frequency dependence based on D-RAP model. Column (c): ionograms simulated in accordance with modified frequency dependence. Row 1: Tromsø ionogram on October 23 2011 at 15:00 UT. Row 2: Sondrestrom ionogram on October 23 2011 at 15:00 UT. Row 3: Tromsø ionogram on October 23 2011 at 15:30 UT. Row 4: Sondrestrom ionogram on October 23 2011 at 15:30 UT. The coefficients $m_d = 0.115$ and $m_n = 0.02$ correspond to the D-RAP model. According to Zaalov and Moskaleva (2016). Permission from COSPAR.

4.3. Broadband Electrostatic Waves in the Auroral Region

Grabbe and Menietti (2006) note that intense broadband electrostatic waves (BEN) in the plasma sheet boundary layer (PSBL) of the magnetotail, originally discovered in the 1970s (Gurnett et al., 1976) were shown to be correlated with the occurrence of ion beams in the PSBL from ISEE-1 observations (Eastman et al., 1984, 1985). Ion beam models were developed and widely investigated as capable of producing the broad spectrum that is observed (Grabbe and Eastman, 1984; Grabbe, 1985a, 1985b, 1987; Akimoto and Omidi, 1986; Schriver and Ashour-Abdalla, 1987, 1989, 1990; Nishikawa et al., 1987, 1988). A variation on the Grabbe-Eastman (GE) model was proposed and debated (Dusenbery and Lyons, 1985; Dusenbery, 1986, 1987, 1988; Omidi and Akimoto, 1988). The consistency of ion beam models with signatures in the wave data on ISEE-1 were examined by Grabbe (1989). All of these works supported the view that multiple streaming instabilities or sources were necessary to account for the origin of BEN. Separate models involving trapped particle modes were also proposed for narrow-band electrostatic noise observed in the more distant magnetotail (Coroniti and Ashour-Abdalla, 1989; Coroniti et al., 1993). Spiky pulses a few milliseconds in length on BEN in the PSBL were reported as observed by the waveform capture receiver on Geotail in 1994, and a new model for BEN was proposed involving Bernstein-Greene-Kruskal (BGK, or trapped particle) modes, as the observations provided the first clear evidence for waves showing nonlinear effects in BEN (Kojima et al., 1994; Matsumoto et al., 1994). Several models and simulations were analyzed centered around these trapped particle modes (Omura et al., 1996; Krasovsky et al., 1997) and compared with subsequent observations of similar spiky turbulence in data from Geotail (Matsumoto et al., 1998, 1999; Omura et al., 1999). These BGK models examined generally ignored the influence of the magnetic field. However, as Robinson (1988) pointed out, there are rather stringent conditions to make an unmagnetized model good for a weakly magnetized plasma. Simulation of these models with the inclusion of the magnetic field in particle trapping was analyzed by Miyake et al. (1998a). This study showed that the magnitude of the magnetic field critically affects the formation of BGK-type electrostatic solitary waves and can prevent their formation.

BGK models were reexamined by Grabbe (1998, 2000a, 2000b) with the magnetic field included and other more realistic parameters used, considering not only the BGK modes proposed from the Geotail data, but also a magnetized kinetic model for the generation of part or all of the wave spectrum. BEN wave data from ISEE-1 and ISEE-3 observed in the mid-magnetotail and distant magnetotail, which show evidence of large angles of propagation with respect to the magnetic field for frequencies near and below f_{ce} (Grabbe and Eastman, 1984; Coroniti et al., 1990), were discussed as evidence of non-BGK modes. An alternative model was proposed for the mid-magnetotail and distant magnetotail, which has both nonlinear BGK-type waves producing the highest frequencies that are closely field aligned, and standard beam instabilities driving the bulk of the broadband spectrum of frequencies below that, which are propagating obliquely to the magnetic field. The full magnetized kinetic plasma model with instabilities driven by electron and ion beams for realistic kappa distributions was examined for parameters typical of the mid-magnetotail (typically at $r \sim 15$ R_E), and found to produce substantial growth rates, up to about 10–20% of the plasma frequency f_{pe}. The ISEE-1 and ISEE-3 data discussed support this model with generation of these waves by beam instabilities, but not by trapped particle modes, so these studies imply instability processes for standard linear modes play an important role for a large part of the BEN observed, up to approximately the electron cyclotron frequency f_{ce}. However, trapping effects can be important at frequencies well above f_{ce}., where the trapped particle models examined in simulation studies (developed for magnetotail plasmas closer to the Earth) are approximately valid. These highest frequencies are observed only in a narrower angular range with respect to the magnetic field for distance beyond about 10 r_E (lower magnetic field), and are consistent with BGK model predictions. This model predicts that, at further radial distances out into the magnetotail, BGK-type solitary waves should only be present in the source region, but not in BEN that has propagated well outside the source region.

Franz et al. (1998) described Polar observation of solitary wave structure in the high-altitude polar magnetosphere. Cattell et al. (1999) described observations on Polar of solitary waves in the high-altitude cusp. The structures exhibited bipolar electric fields. Observations from Polar of BEN obtained pole-ward of and within the near-Earth extension of the plasma sheet boundary layer (PSBL) were analyzed by Grabbe and Menietti (2002). The wave data examined for BEN pole-ward of the PSBL (in the plasma mantle) exhibited essentially little or no evidence of solitary waves. However, the wave data for crossing into the plasma sheet boundary layer source region shows both a very turbulent region immediately at the crossing and the appearance of solitary waves a very short time later. The low-frequency portion of the observed electric field wave data fits the theory of standard beam instabilities, but the higher-frequency portion running up to about f_{pe} exhibits nonlinear characteristics, with the magnetic field apparently playing an important role in that nonlinearity. Grabbe (2002, 2003) made a theoretical analysis of nonlinear electrostatic waves using coupled plasma equations for a magnetized plasma, yielding a BGK-like equation generalized for trapped particle distributions that produce these structures in the guiding center approximation for a magnetized plasma. This was extended by Grabbe (2005), who included finding and analyzing the solution of this magnetized BGK-like equation. In these studies a conditional requirement was derived for electron trapping, and the amount of trapping shows a marked decrease as the angle of the electric field relative to the background magnetic field increases, generally ceasing at a critical finite angle. The criterion for trapping found by Grabbe (2002, 2003) was

$$\cos\theta > \sqrt{1+\left(v_b/v_{D\max}\right)^2+\left(v_{tw}/v_{D\max}\right)^2}-\left(v_{tw}/v_{D\max}\right), \tag{4.3.1}$$

where θ is the angle with respect to the magnetic field, $v_{D\max} = E/cB$ (the maximum **E × B** drift velocity), $v_b = |\mathbf{v}_b|/c$, $v_{tw} = \Delta w \omega_{ce}/2c$, with E and Δw the BGK mode electric field and width, $\mathbf{v}_b$ is the electron beam velocity, and $\omega_{ce} = Be/m_e$ is the cyclotron frequency. The results were applied to broadband electrostatic waves (BEN) in the magnetotail, and the analysis showed upper limits existed on the angular range with respects to the background magnetic field at which these BGK solitary waves can exist. It was found that trapping can occur over larger angles in the near-Earth case because of the large magnetic field there, accounting for oblique solitary waves. However, trapped particle modes are confined to virtual alignment (within about 15°–20°) with the magnetic field at distances out at 10 R_E and above in the magnetotail, and cannot exist outside those small-angle ranges. This strongly supports the conclusion made by Grabbe (2000a, 2000b) that in cases further out into the magnetotail that solitary waves are highly field aligned. Cattell et al. (2005) reported observations on Cluster of electron holes near the outer edge of the plasma sheet when there were 'narrow' (in pitch angle) electron beams present but not when the beams were broad in pitch angle. This is expected from the theoretical predictions of Grabbe (2002, 2003) because farther out in magnetotail, electron holes can only exist propagating along **B**, formed from beams with energies focused along **B**. They also reported that the velocity and scale sizes of e holes are consistent with Drake et al. (2003) model for reconnection. The purpose of the Grabbe and Menietti (2006) study is to make comparisons of some BEN data showing solitary waves from Polar with predictions of the theory on BGK modes with a magnetic field present, as well as other aspects of theory for the generation of those waves. These observations are confined to the auroral region and magnetotail close to the Earth, and the angular limitations discussed by Grabbe (2002, 2003) may not be significant here. Thus the focus will be on comparing the electric fields of the observed waves with the electric field predictions from BGK theory.

Polar is the first satellite to have 3 orthogonal electric antennas (E_u, E_v, and E_z), 3 triaxial magnetic search coils, and a magnetic loop antenna, as well as an advanced plasma wave instrument (Gurnett et al., 1995). This combination can potentially provide the polarization and direction of arrival of a signal without any prior assumptions. The Plasma Wave Instrument (PWI) on the Polar spacecraft is designed to provide measurements of plasma waves in the Earth's polar regions over the frequency range from 0.1 Hz to 800 kHz. Five receiver systems are used to process the data: a wideband receiver, a high-frequency waveform receiver (HFWR), a low-frequency waveform receiver (LFWR), two multi-channel analyzers, and a pair of sweep frequency receivers (SFR). The SFR has a frequency range from 26 Hz to 808 kHz in 5 frequency bands. The frequency resolution is about 3% at the higher frequencies. In the log mode a full frequency spectrum can be obtained every 33 s. Grabbe and Menietti (2006) selected Polar auroral region passes for the northern hemisphere for presentation. The data were observed over distances ranging from $r = 6.58\ R_E$ to $r = 8.76\ R_E$ out, where the plasma frequency $f_{pe} \sim f_{ce}$ (the electron cyclotron frequency). Grabbe and Menietti (2006) came to the following conclusions: **(i)** The auroral data, when compared with the theoretically predicted electric field required for the existence of BGK modes, shows that condition is not satisfied for any of the solitary wave cases that were examined. These apparently nonlinear forms of BEN are distinct from the more common forms of BEN which appear consistent with standard beam instabilities. Thus these solitary waves may well originate from a different nonlinear process than electron trapping. **(ii)** The observations do not rule out the possibility that BEN of sufficiently large electric fields near the source region which is confined at sufficiently narrow angles with respect to the magnetic field, can satisfy the criteria for BGK waves. However, well outside these source regions the trapping process is suppressed, but the solitary waveforms may still exist because the forms have not yet broken up. BEN much more commonly takes on its standard apparently linear form, so the trapped electron BGK mode forms of BEN would be a much rarer form of BEN confined to near the source region. **(iii)** BGK solitary waves could still exist out to larger angles with respect to the magnetic field direction at these distances out under 10 R_E because $f_{ce} > f_p$ (equivalently $\lambda_{De} > r_{ce}$), provided that source regions are present where the trapping criterion is satisfied. That is because, to lowest order the wave is experiencing the 'infinite magnetic field' approximation in all directions, except at very large angles with respect to the magnetic field. However, at distance >10 R_E we have $f_{ce} < f_p$, or equivalently $\lambda_{De} < r_{ce}$. Thus the larger electron cyclotron radius is preventing the one-dimensional solitary wave from forming except at angles very close to the magnetic field direction, where the cyclotron effects are no longer significant. The minimum electric field requirement still must be satisfied for those BGK waves to exist, so they would typically be confined to a very narrow angle with respect to the magnetic field right in the source region (the plasma sheet boundary layer in those regions). **(iv)** However, the solitary waves are likely produced by a process other than particle trapping, such as by the nonlinear stages of beam instabilities for a magnetized plasma (Goldman et al., 1999; Singh et al., 2001; Oppenheim et al., 1999, 2001; Lu et al., 2005). Those processes should be further explored for realistic models so comparisons between the theory and experiment can adequately supply strong and clear evidence for it. Future developments on models for the nonlinear waves and comparisons with the BEN data in potential source regions will help answer the question as to whether such alternative processes are the probable source of these solitary waves.

4.4. Subauroral Ion Drifts and Magnetosphere-Ionosphere-Thermosphere Coupling

4.4.1. Polar Thermospheric Response to Solar Magnetic Cloud/Coronal Mass Ejection Interactions with the Magnetosphere

As note Sivjee (1999), the Solar Magnetic Cloud (SMC)/Coronal Mass Ejection (CME) event of January1997 triggered auroral displays in all sectors of the auroral oval as well as in the polar cap region. Near infrared emissions from these auroras were recorded simultaneously in the night sector over Sondrestromfjord (Sonde), Greenland, in the day sector over Longyearbyen, Svalbard and in the polar cap region over Eureka, Canada. The spectral distributions of these emissions indicate precipitation of electrons with average energy (E_{AV}) of (500 ± 100) eV, dissipating most of their energy around (180± 20) km height (h_{max}) in the thermosphere. These findings are consistent with the concurrent auroral ionization profiles recorded by the Incoherent Scatter Radar soundings at Sonde. In contrast, most of the night time auroras, not related to SMC/CME events, are excited by electrons with E_{AV} > a few keV and peak in the lower thermosphere with h_{max} around 110 km. Similarly, normal dayside cusp auroras and polar cap drizzle excited emissions emanate from the upper thermosphere above 200 km altitude. SMC/CME related auroras were also observed in October 1995 at Sonde, and in May 1996 as well as in May 1997 at the South Pole Station in Antarctica. Spectral characteristics, and hence E_{AV} and h_{max}, of all these other SMC/CME related auroras, are similar to those of the January 1997event. These observations suggest that during a significant part of the period when SMC/CME plasmas and fields interact with the magnetosphere, relatively low energy electrons precipitate in the thermosphere. Such SMC/CME triggered auroras interact with the middle thermosphere constituents in the 160-200 km height region. The latter region is inaccessible for remote sensing its composition and thermodynamics in normal auroras, which generally peak at lower heights; the SMC/CME events provide the opportunity for such investigations.

4.4.2. Cluster and DMSP Satellite Observations of Subauroral Ion Drifts: Magnetosphere-Ionosphere Coupling along an Entire Field Line

As noted Puhl-Quinn et al. (2007), polarization jets (Galperin et al., 1974) or subauroral ion drifts (SAID) (Spiro et al., 1979) are latitudinally narrow, enhanced streams of westward convection ($\langle V_{west} \rangle \geq 1$ km/s) equatorward of the electron plasma sheet. They are driven by meridional electric fields intensified in a low-conductive nightside subauroral ionosphere conjugate to the ring current/plasmasphere overlap region (RCPO). SAID are considered a subset of the subauroral polarization streams (SAPS) that include plasma flow events with both broad and narrow extents in latitude (Foster and Burke, 2002). Understanding the SAPS/SAID phenomenon is important for studies of how the structure, dynamics, and chemistry of the midlatitude ionosphere-thermosphere system are coupled to the structure and dynamics of the RCPO region. The coupling occurs via electric fields and currents. Energy is transferred from the magnetospheric RCPO region into the ionosphere and thermosphere, causing, in some cases, extreme, geomagnetic activity-related variability. The ability to understand and model this phenomenon requires both ionospheric and magnetospheric observations. Puhl-Quinn et al. (2007) outlined that while SAPS/SAID characteristics are well documented at ionospheric altitudes (e.g., Spiro et al., 1979; Karlsson et al., 1998; Foster and Vo, 2002; Figueiredo et al., 2004), investigations at higher altitudes are much more scarce. Anderson et al. (2001) found over 110 examples of SAID at altitudes ~9000 km from the Akebono satellite. Several of them had ionospheric counterparts at altitudes ~840 km from DMSP, which were nearly identical when electrostatically mapped to a common altitude. This indicated no significant field-aligned potential drops between 840 and 9000 km within the SAID channel. The near-equatorial, magnetospheric characteristics of SAPS/SAID are even less documented. This is largely due to the relatively long magnetospheric satellite orbital period combined with the extreme variability of both plasmapause position and geomagnetic conditions. In other words, being 'in the right place at the right time' is a rarely satisfied condition. Maynard et al. (1980) presented the only two examples of SAID found near the plasmapause at $L = 4$ in a cursory survey of 2 years worth of the ISEE-1 electric field data set. Other studies have focused more on SAPS rather than SAID observations: Measurements from the Active Magnetospheric Particle Tracer Explorers IRM satellite showed the presence of enhanced irregular meridional (and outward) electric fields with $\langle E_Y \rangle \leq 3$ mV/m in RCPO near dusk (LaBelle et al., 1988). Mapped to the ionosphere, these fields would produce broad irregular SAPS with $\langle V_{west} \rangle \leq 1$ km/s; during the magnetic storm of 5 June 1991, the Combined Release and Radiation Effects Satellite observed several structured SAPS events in the RCPO alike those measured by DMSP in the topside ionosphere (Burke et al., 2000; Mishin and Burke, 2005). Puhl-Quinn et al. (2007) report on magnetically conjugate Cluster and the DMSP satellite observations of subauroral ion drifts (SAID) during moderate geomagnetic activity levels on 8 April 2004, when the field-aligned separation of DMSP and Cluster (≈28,000 km) is the largest separation ever analyzed with respect to the SAID phenomenon. Nonetheless, it can be seen coherent, subauroral magnetosphere-ionosphere coupling along an entire field line in the post-dusk sector. The four Cluster satellites crossed SAID electric field channels with meridional magnitude E_M of 25 mV/m in situ and latitudinal extent $\Delta\Lambda \approx 0.5°$ in the southern and northern hemispheres near 07:00 and 07:30 UT, respectively. Cluster was near perigee ($r \approx 4\ R_E$) and within 5°

(15°) of the magnetic equator for the southern (northern) crossing. The SAID were located near the plasmapause–within the ring current-plasmasphere overlap region. Downward field-aligned current signatures were observed across both SAID crossings. The most magnetically and temporally conjugate SAID field from DMSP F16A at 07:12 UT was practically identical in latitudinal size to that mapped from Cluster. Since the DMSP ion drift meter saturated at 3000 m/s (or ~114 mV/m) and the electrostatically mapped value for EM from Cluster exceeded 300 mV/m, a magnitude comparison of EM was not possible. Although the conjugate measurements show similar large-scale SAID features, the differences in substructure highlight the physical and chemical diversity of the conjugate regions. Puhl-Quinn et al. (2007) summarized obtained results as following: **(i)** There are presented multispacecraft (4 Cluster and DMSP satellite) observations of a strong SAID event on 8 April 2004. **(ii)** Taken as an ensemble of 12 separate SAID crossings, the observations show a temporal stability of at least 1.5 hours, a center location at 58° ILAT and 21:53 MLT, and an azimuthal extent of 45°. **(iii)** The highest degree of magnetic and temporal conjugacy was between DMSP F16A and Cluster in the Southern Hemisphere (Δ_{UT} = 12 min., Δ_{ILAT} ~ 0°, Δ_{MLT} ~10°). They were separated by a field-aligned distance of ≈28,000 km or 4.4 r_E (this is the largest separation ever analyzed with respect to the SAID phenomenon). **(iv)** Large-scale features found to be common to both the ionospheric and magnetospheric data sets include a latitudinal width of 0.5° and an ionospheric downward FAC strength of ~0.32 mA/m^2. **(v)** Comparison of the E_M peak magnitude was not possible due to the saturation of the ion drift meter. Smaller-scale features differ between the two data sets. **(vi)** DMSP shows the plasmapause at the poleward edge of the channel, whereas Cluster shows the plasmapause across the entire channel. **(vii)** DMSP shows a deep density trough on the equatorward side of the channel, whereas there is no indication of this in the Cluster data set. **(viii)** The density irregularities shown by Cluster are much smaller than the projection of the ionospheric trough; these differences highlight the complexity of magnetosphere-ionosphere coupling at smaller scales. **(ix)** The SAID generation mechanism was not discussed but rather will be saved for a future study of multiple DMSP-Cluster conjugate events. The observations support the notion of active, strong magnetosphere-ionosphere coupling along the entire field lines in the dusk-side ring current/plasmasphere overlap region during moderate geomagnetic activity. More measurements of this type are necessary to parameterize the strength of this coupling and possibly to validate future SAPS/SAID models of the inner magnetosphere (e.g., Goldstein et al., 2005a,b).

4.4.3. Mapping High-Latitude Ionospheric Electrodynamics with SuperDARN and AMPERE

An assimilative procedure for mapping high-latitude ionospheric electrodynamics is developed by Cousins et al. (2015a) for use with plasma drift observations from the Super Dural Auroral Radar Network (SuperDARN) and magnetic perturbation observations from the Active Magnetosphere and Planetary Electrodynamics Response Experiment (AMPERE). This procedure incorporates the observations and their errors, as well as two background models and their error covariances (estimated through empirical orthogonal function analysis) to infer complete distributions of electrostatic potential and vector magnetic potential in the high-latitude ionosphere. The assimilative technique also enables objective error analysis of the results. Various methods of specifying height-integrated ionospheric conductivity, which is required by the procedure, are implemented and evaluated quantitatively. The benefits of using both SuperDARN and AMPERE data to solve for both electrostatic and vector magnetic potentials, rather than using the data sets independently or solving for just electrostatic potential, are demonstrated. Specifically, solving for vector magnetic potential improves the specification of field-aligned currents (FACs), and using both data sets together improves the specification of features in regions lacking one type of data (SuperDARN or AMPERE). Additionally, using the data sets together results in a better correspondence between large-scale features in the electrostatic potential distribution and those in the FAC distribution, as compared to using SuperDARN data alone to infer electrostatic potential and AMPERE data alone to infer FACs. Finally, the estimated uncertainty in the results decreases by typically ~20% when both data sets rather than just one are included.

4.4.4. Optimal Interpolation Analysis of High-Latitude Ionospheric Hall and Pedersen Conductivities: Application to Assimilative Ionospheric Electrodynamics Reconstruction

McGranaghan et al. (2016) have developed a new optimal interpolation (OI) technique to estimate complete high-latitude ionospheric conductance distributions from Defense Meteorological Satellite Program particle data. The technique combines particle precipitation-based calculations of ionospheric conductances and their errors with a background model and its error covariance (modeled with empirical orthogonal functions) to infer complete distributions of the high-latitude ionospheric conductances. McGranaghan et al. (2016) demonstrate this technique for the 26 November through 2 December 2011 period and analyze a moderate geomagnetic storm event on 30 November 2011. Quantitatively and qualitatively, this new technique provides better ionospheric conductance specification than past statistical models, especially during heightened geomagnetic activity. It's provide initial evidence that auroral images from the Defense Meteorological Satellite Program Special Sensor Ultraviolet Spectrographic Imager instrument can be used to further improve the OI conductance maps. The developed OI conductance patterns allow assimilative mapping of ionospheric electrodynamics reconstructions driven separately by radar and satellite magnetometer observations to be in closer agreement than when other, commonly used, conductance models are applied. This work (1) supports better use of the diverse observations available for high-latitude ionospheric electrodynamics specification and (2) supports the Cousins et al. (2015b) assertion that more accurate models of the ionospheric conductance are needed to robustly assimilate ground- and space-based observations of ionospheric electrodynamics(see Subsection 4.4.3). McGranaghan et al. (2016) find that the OI conductance distributions better capture the dynamics and locations of discrete electron precipitation that modulate the coupling of the magnetosphere-ionosphere-thermosphere system.

4.4.5. Plasma and Convection Reversal Boundary Motions in the High-Latitude Ionosphere

Chen et al. (2016) present a statistical study of the high-latitude ionospheric plasma motion at the convection reversal boundary (CRB) and its dependence on the location of the CRB and the IMF orientation by using the Defense Meteorological Satellite Program (DMSP) F13 and F15 measurements over the period from 2000 to 2007. During periods of stable southward IMF, Chen et al. (2016) find a smaller variability in plasma drifts across the CRB over a 4 h segment in magnetic local time (MLT) around dawn and dusk compared to that for variable IMF. Across these segments, the plasma motion at the CRB is directed poleward at local times closer to local noon and equatorward at local times closer to midnight on both the dawn and dusk sides with a total potential drop ~10 kV, suggesting that the CRB behaves much like an adiaroic line. For variable IMF with no stability constraint, Chen et al. (2016) see a relatively narrow distribution of plasma drifts across the CRB only in the 6–7h and 17–18 h MLT and equatorward/poleward motions of the CRB when the CRB is located at the highest/lowest latitudes. The smaller local time extent of the adiaroic line for variable IMF (~1 h) may be associated with rotation of the dayside merging gap in local time or local contractions and expansions of the polar cap boundary.

4.4.6. A Comparison of the Relative Effect of the Earth's Quasi-DC and AC Electric Field on Gradient Drift Waves in Large-Scale Plasma Structures in the Polar Regions

As outlined Burston et al. (2016), radio signals traversing polar cap plasma patches and other large-scale plasma structures in polar regions are prone to scintillation. This implies that irregularities in electron concentration often form within such structures. The current standard theory of the formation of such irregularities is that the primary gradient drift instability drives a cascade from larger to smaller wavelengths that manifest as variations in electron concentration. The electric field can be described as the sum of a quasi-DC and an AC component. While the effect of the quasi-DC component has been extensively investigated in theory and by modeling, the contribution of the AC component has been largely neglected. The paper of Burston et al. (2016) investigates the relative contributions of both components, using data from the Dynamics Explorer 2 satellite. It concludes that the contribution of the AC electric field to irregularity growth cannot be neglected. This has consequences for the understanding of large-scale plasma structures in polar regions (and any associated radio scintillation) as the AC electric field component varies in all directions. Hence, its effect is not limited to the trailing edge of such structures, as it is for the quasi-DC component. This raises the need for new experimental and modeling investigations of these phenomena.

4.5. Correlations between the Coronal Hole Area, Solar Wind Velocity, and Local Magnetic Indices in the Canadian Region during the Decline Phase of Cycle 23

As noted Shugai et al. (2009), large coronal holes (CHs), which stretch to the solar equator and exist for many solar rotations, are usually observed during the solar cycle decline phase. Such CHs cause recurrent high-speed solar wind streams, which exist for a long time and result in pronounced (especially at high latitudes) but not very strong recurrent geomagnetic disturbances that are registered during several or even many days (up to the entire solar rotation). For a long time, these disturbances have been known as recurrent storms in the literature (Bartels, 1934; Chapman and Bartels, M1962; Akasofu and Chapman, M1972). Such solar wind streams are characterized by the presence of Alfvén waves propagating from the Sun with medium and large amplitudes. These waves are 'a remnant' of the energy flux, which was mainly expended in increasing the solar wind plasma velocity in the corona and inner Heliosphere. Studying the related complex of phenomena on the Sun, in the Heliosphere, and on the Earth has attracted attention until recently (Tsurutani et al., 2006). Stronger sporadic disturbances can originate under the action coronal mass ejections, the duration of which is usually shorter than that of streams from CHs. The statistical studies of the relations between CH parameters and geomagnetic activity are rather extensive; nevertheless, most works were devoted to mid-latitude geomagnetic indices (see, e.g., Thomson, 1995; Watari, 1989). At the same time numerous observations indicated that high-speed solar wind streams dramatically affect the state of the geomagnetic field at high latitudes.

Shugai et al. (2009) used the hourly indices of geomagnetic activity, presented by the Canadian network of geomagnetic observatories for the polar, auroral, and subauroral zones of geomagnetic activity. The daily indices were subsequently calculated as daily average values of the hourly indices. The data from Resolute Bay (74.7° N, 265.1° E), Fort Churchill (58.8° N, 265.9° E), and Ottawa (45.4° N, 284.5° E) observatories were used to calculate the values of the polar (Pol), auroral (Aur), and subauroral (Sub) magnetic indices, respectively. In Shugai et al. (2009) geomagnetic disturbances in the Canadian region are compared with their solar and heliospheric sources during the decline phase of solar activity (2003-2007), when recurrent solar wind streams from low-latitude coronal holes were clearly defined. A linear correlation analysis has been performed using the following data: the daily and hourly indices of geomagnetic activity, solar wind velocity, and coronal hole area. The obtained correlation coefficients were rather low between the coronal hole areas and geomagnetic activity (0.17–0.48), intermediate between the coronal hole areas and the solar wind velocity (0.40–0.65), and rather high between the solar wind velocity and geomagnetic activity (0.50–0.70). It has been indicated by Shugai et al. (2009) that the correlation coefficient values can be considerably increased (by tens of percent in the first case and about twice in the second case) if variations in the studied parameters related to changes in the ionosphere (different illumination during a year) and variations in the heliolatitudinal shift of the coordinate system between the Earth, the Sun, and a spacecraft are more accurately taken into account.

4.6. Coupling of Energetic Particles Precipitation at High Latitudes with Solar Wind and IMF

4.6.1. Energetic Electron Bursts at High Magnetic Latitudes: Correlation with Magnetospheric Activity

As outlined Lugo-Solis et al. (2005), there is considerable interest on the origin of high energy electron bursts recently observed at high magnetic latitudes by dosimeters on the GPS satellites. The energy of these electron bursts can be as high as 500 keV. These bursts are seen in regions that map onto open field lines and where normally observed fluxes are at background levels. In this study Lugo-Solis et al. (2005) examine the distribution of these events, and correlate them with magnetic indices and solar wind magnetic field data. Obtained results indicate that many of these electron bursts are likely to be substorm-associated. Main results are illustrated by Figures 4.6.1 – 4.6.5.

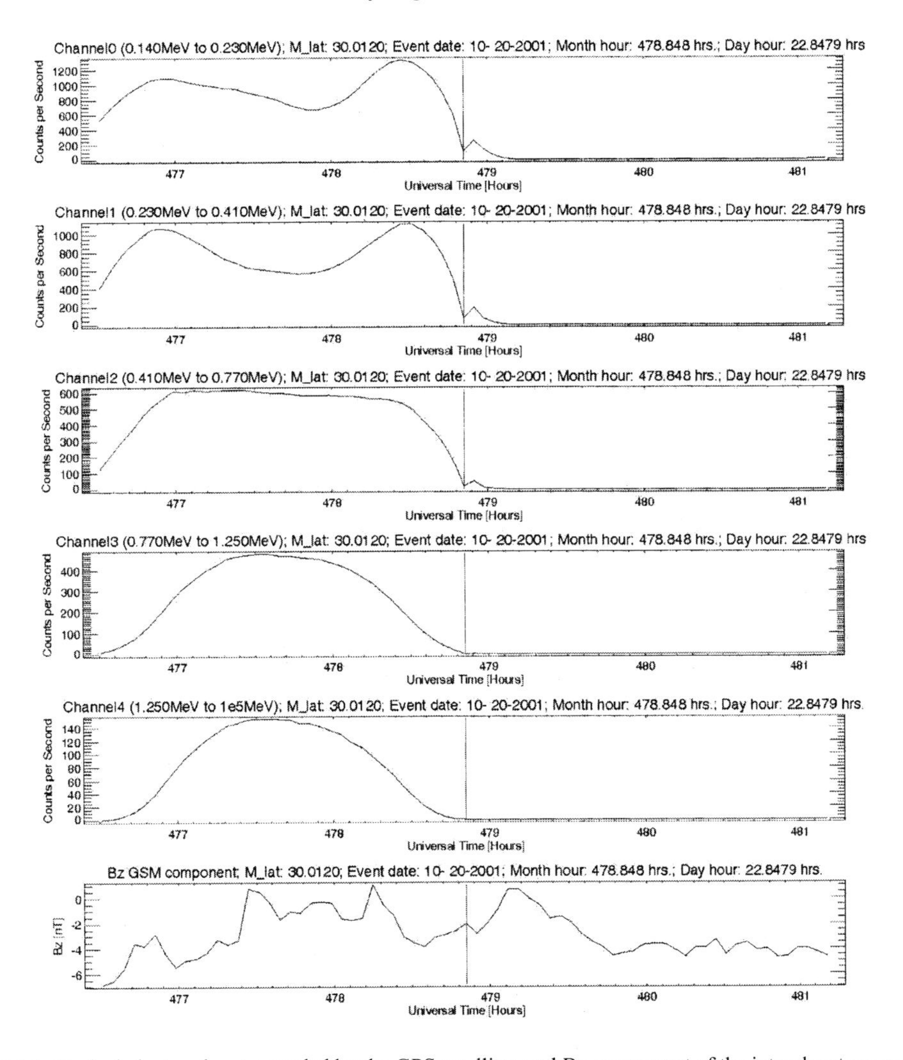

Figure 4.6.1. Typical electron burst recorded by the GPS satellites and Bz component of the interplanetary magnetic field (IMF). According to Lugo-Solis et al. (2005). Permission from COSPAR.

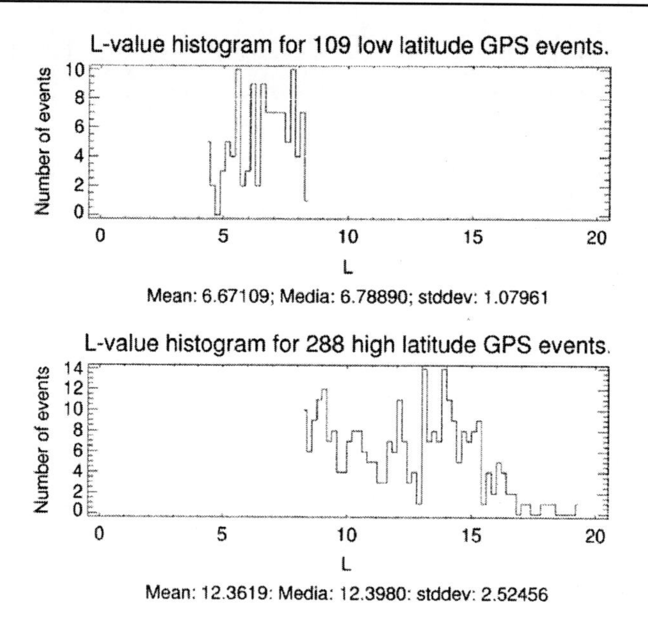

Figure 4.6.2. L-Value histograms for low (<45°) and high magnetic latitude (>45°) GPS events. According to Lugo-Solis et al. (2005). Permission from COSPAR.

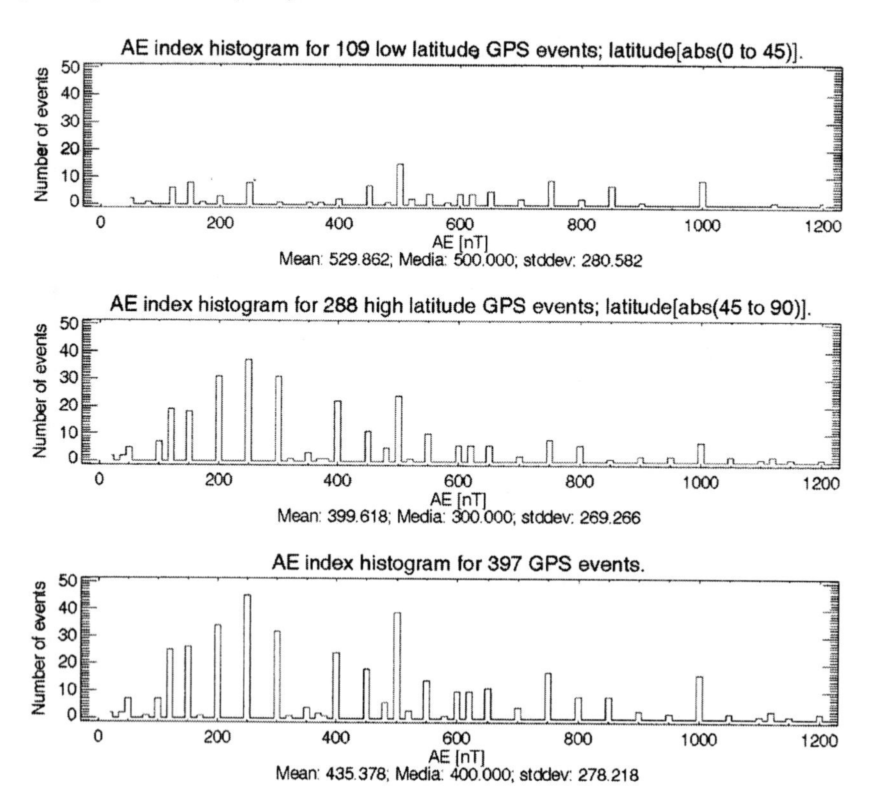

Figure 4.6.3. AE histograms for both low and high latitude events. According to Lugo-Solis et al. (2005). Permission from COSPAR.

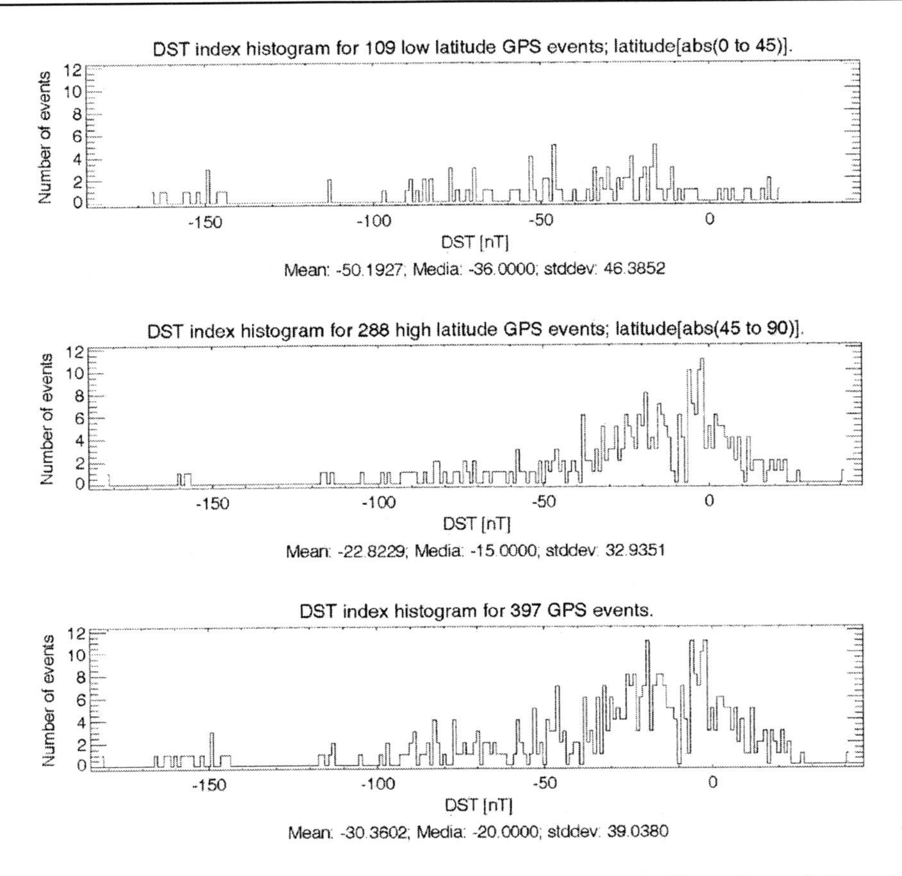

Figure 4.6.4. *Dst* histograms for both low and high latitude events. According to Lugo-Solis et al. (2005). Permission from COSPAR.

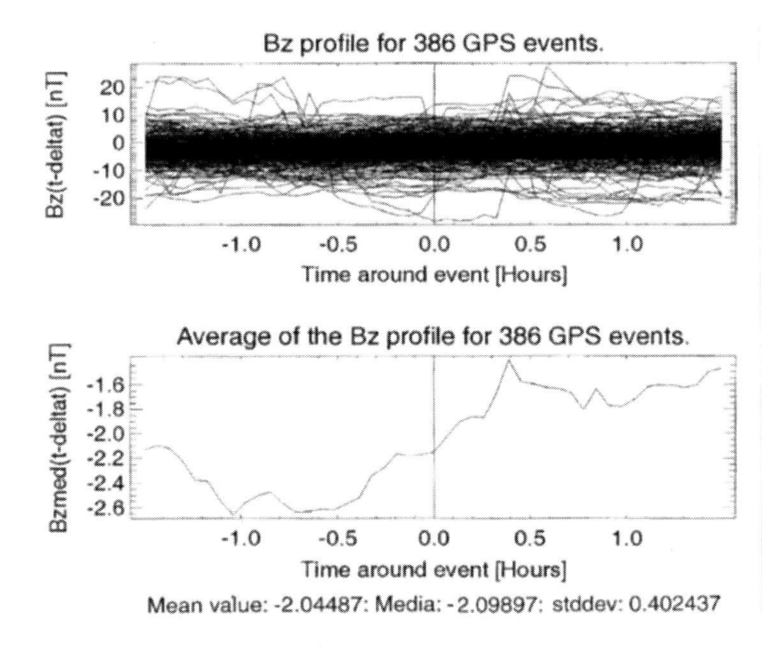

Figure 4.6.5. Superposed epoch plot for 386 GPS events and the associated solar wind magnetic profile. According to Lugo-Solis et al. (2005). Permission from COSPAR.

4.6.2. Seasonal and Hemispheric Variations of the Total Auroral Precipitation Energy Flux from TIMED/GUVI

The auroral hemispheric power (HP) has been calculated by Luan et al. (2010) from the averaged energy flux derived from Far-ultraviolet emission observations made by the global ultraviolet imager (GUVI) instrument on board the Thermosphere Ionosphere Mesosphere Energetics and Dynamics (TIMED) satellite during 2002–2007. This HP was used to study how variations in seasonal and hemispheric asymmetries changed with changing geomagnetic activity. Obtained results showed that there were persistent seasonal and hemispheric differences in quiet conditions. There were HP differences of about 1–3 GW between the summer and winter seasons in each hemisphere and also between the two hemispheres during each solstice period for low geomagnetic activity ($Kp \leq 3$). The summer-winter asymmetry was 4%–35% when HP was low. These summer-winter differences became negligible when geomagnetic activity was moderate to active. Similarly, there were also HP differences of about 2 GW between local summers of the two hemispheres during geomagnetically quiet conditions ($Kp < 3$) but not during higher Kp conditions. The hemispheric asymmetries between the two summer solstices were about 10%–20% during quiet conditions, whereas there was no apparent hemispheric asymmetry between the two winter solstices under all Kp 1–5 conditions. Solar illumination effects were probably the primary cause of the seasonal and hemispheric variations of the auroral hemispheric power for these geomagnetically quiet conditions. During moderate and active conditions the conductivities were driven more by the production of ionization due to precipitation, so the precipitation was more symmetric.

4.6.3. Seasonal and UT Variations of the Position of the Auroral Precipitation and Polar Cap Boundaries

The data of the DMSP F7 spacecraft are used by Vorobjev and Yagodkina (2010) for studying the influence of the geomagnetic dipole tilt angle on the latitudinal position of auroral precipitation boundaries in the nighttime (2100–2400 MLT) and daytime (0900–1200 MLT) sectors. It is shown that, in the nighttime sector, the high-latitude zone of soft diffuse precipitation (SDP) and the boundary of the polar cap (PC) at all levels of geomagnetic activity are located at higher and lower latitudes relative to the equinox period in winter and summer, respectively. The position of boundaries of the diffuse auroral precipitation zone (DAZ) located equatorward from the auroral oval does not depend on the season. In the daytime sector, the inverse picture is observed: the SDP precipitation zone takes the most low-latitude and high-latitude positions in the winter and summer periods, respectively. The total value of the displacements from winter to summer of both the nighttime and daytime boundaries of the PC is ~2.5°. A diurnal wave in the latitudinal position of the nighttime precipitation boundaries is detected. The wave is most pronounced in the periods of the winter and fall seasons, is much weaker in the spring period, and is almost absent in summer. The diurnal variations of the position of the boundaries are quasi-sinusoidal oscillations with the latitude maximum and minimum at 0300–0500 and 1700–2100 UT, respectively. The total value of the diurnal displacement of the boundaries is ~2.5° of latitude. The results obtained show that, undergoing seasonal and diurnal variations, the polar cap is shifted as a whole in the direction opposite to the changes in the tilt angle of the geomagnetic dipole. The seasonal displacements of the polar cap and its diurnal variations in the winter period occur without any substantial changes in its area.

4.6.4. Energetic Electron Precipitation Events Recorded in the Earth's Polar Atmosphere

As outlined Makhmutov et al. (2011), there are three main populations of ionizing particles in the Earth's atmosphere, namely, galactic and solar cosmic rays and energetic electrons precipitating from the magnetosphere during interplanetary and geomagnetic perturbations. All these particle populations were observed in the long term balloon cosmic ray measurements in the atmosphere at several geomagnetic locations including northern and southern polar regions and mid latitudes. The measurements are carried out by the Lebedev Physical Institute Russian Academy of Science from 1957 till now. Makhmutov et al. (2011) focus on energetic electron precipitation events observed in the Earth's polar atmosphere in 1958-2010. There are discuss: (1) the origin of these events and their relation to the geomagnetic disturbances, and (2) the relationship between the solar activity level and the occurrence rate of electron precipitation events in the polar atmosphere. Obtained results are illustrated in Figures 4.6.6 – 4.6.7.

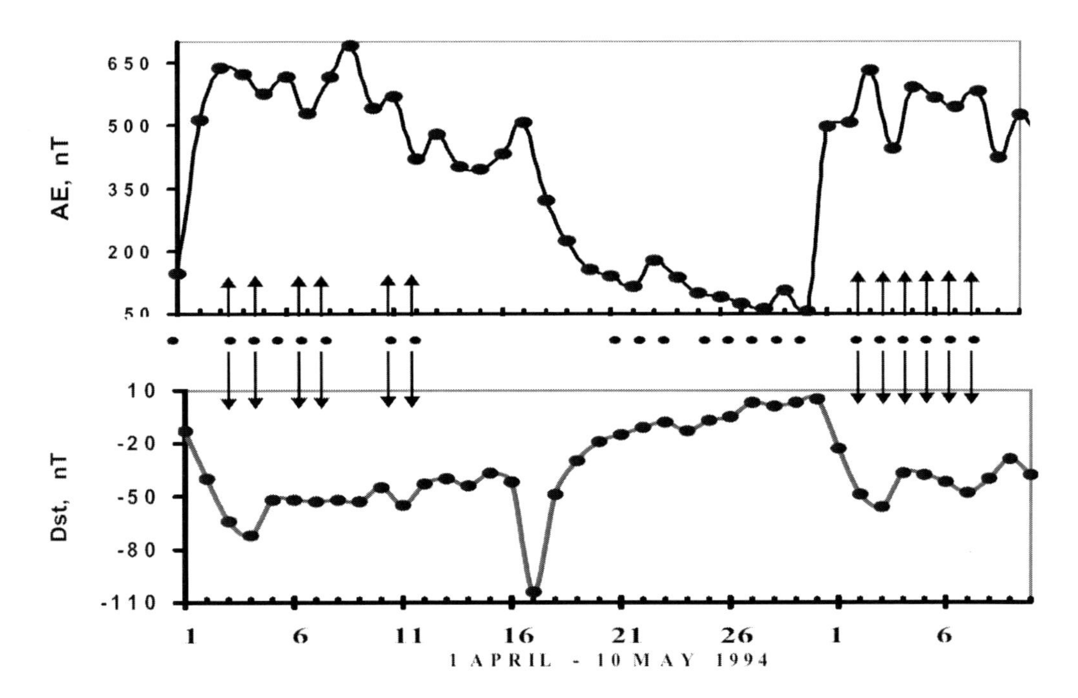

Figure 4.6.6. Geomagnetic auroral electrojet index (AE; upper panel)) and equatorial *Dst* index (bottom) time variations during April,1 – May, 10 1994. Dates of balloon flights at Murmansk region are noted by circles in middle panel, arrows indicate the days of EPE observation in the atmosphere. According to Makhmutov et al. (2011). Permission from Editors of ICRC Proceedings.

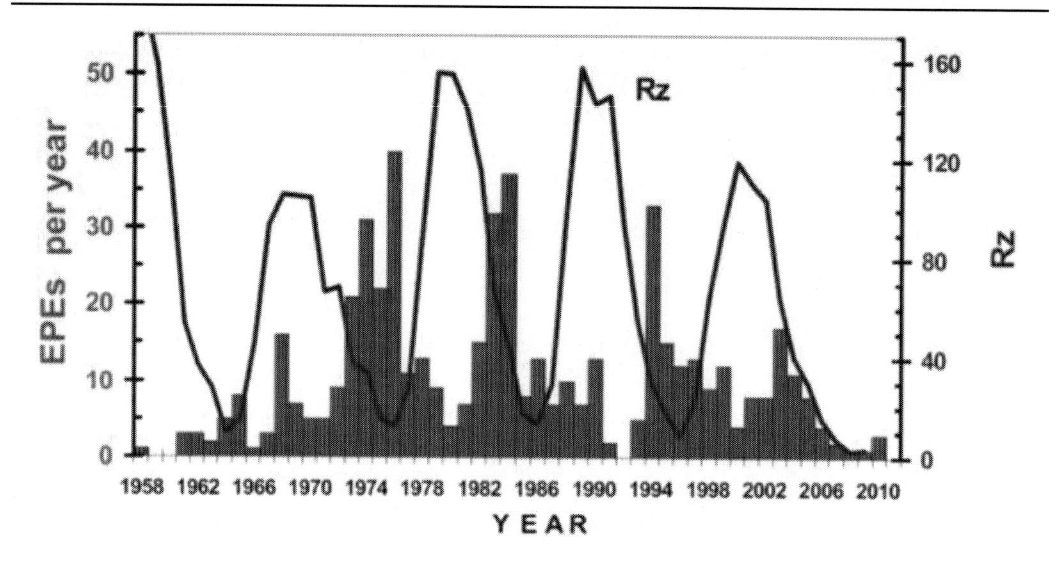

Figure 4.6.7. Yearly number of energetic electron precipitation events (EPEs) observed at Northern polar atmosphere (Murmansk region; histogram) and yearly sunspot number Rz (thick line; right Y-axis). According to Makhmutov et al. (2011). Permission from Editors of ICRC Proceedings.

4.6.5. APES: Acute Precipitating Electron Spectrometer—A High Time Resolution Monodirectional Magnetic Deflection Electron Spectrometer

Michell et al. (2016) present a description of the Acute Precipitating Electron Spectrometer (APES) that was designed and built for the Ground-to-Rocket Electron Electrodynamics Correlative Experiment (GREECE) auroral sounding rocket mission. The purpose was to measure the precipitating electron spectrum with high time resolution, on the order of milliseconds. The trade-off made in order to achieve high time resolution was to limit the aperture to only one look direction. The energy selection was done by using a permanent magnet to separate the incoming electrons, such that the different energies would fall onto different regions of the microchannel plate and therefore be detected by different anodes. A rectangular microchannel plate (MCP) was used (15 mm × 100 mm), and there was a total of 50 discrete anodes under the MCP, each one 15 mm × 1.5 mm, with a 0.5 mm spacing between anodes. The target energy range of APES was 200 eV to 30 keV.

4.6.6. A New Ionospheric Electron Precipitation Module Coupled with RAM-SCB within the Geospace General Circulation Model

As outlined Yu et al. (2016), electron precipitation down to the atmosphere due to wave-particle scattering in the magnetosphere contributes significantly to the auroral ionospheric conductivity. In order to obtain the auroral conductivity in global MHD models that are incapable of capturing kinetic physics in the magnetosphere, MHD parameters are often used to estimate electron precipitation flux for the conductivity calculation. Such an MHD approach, however, lacks self-consistency in representing the magnetosphere-ionosphere coupling processes. Yu et al. (2016) improve the coupling processes in global models with a more physical method. It is calculate the physics-based electron precipitation from the ring current and map it to the ionospheric altitude for solving the ionospheric electrodynamics. In particular, Yu et al. (2016) use the BATS-R-US (Block Adaptive Tree Scheme-Roe type-Upstream) MHD model coupled with the kinetic ring current model RAM-SCB (Ring current-Atmosphere interaction Model with Self-Consistent Magnetic field (B)) that solves pitch angle-dependent electron distribution functions, to study the global circulation dynamics during the 25–26 January 2013 storm event. Since the electron precipitation loss is mostly governed by wave-particle resonant scattering in the magnetosphere, Yu et al. (2016) further investigate two loss methods of specifying electron precipitation loss associated with wave-particle interactions: (1) using pitch angle diffusion coefficients $D_{\alpha\alpha}(E, \alpha)$ determined from the quasi-linear theory, with wave spectral and plasma density obtained from statistical observations (named as "diffusion coefficient method") and (2) using electron lifetimes $\tau(E)$ independent on pitch angles inferred from the above diffusion coefficients (named as "lifetime method"). Yu et al. (2016) found that both loss methods demonstrate similar temporal evolution of the trapped ring current electrons, indicating that the impact of using different kinds of loss rates is small on the trapped electron population. However, for the precipitated electrons, the lifetime method hardly captures any precipitation in the large L shell (i.e., $4 < L < 6.5$) region, while the diffusion coefficient method produces much better agreement with NOAA/POES measurements, including the spatial distribution and temporal evolution of electron precipitation in the region from the premidnight through the dawn to the dayside. Further comparisons of the precipitation energy flux to DMSP observations indicates that the new physics-based precipitation approach using diffusion coefficients for the ring current electron loss can explain the diffuse electron precipitation in the dawn sector, such as the enhanced precipitation flux at auroral latitudes and flux drop near the subauroral latitudes, but the traditional MHD approach largely overestimates the precipitation flux at lower latitudes.

4.6.7. Localized Polar Cap Precipitation in Association with Nonstorm Time Airglow Patches

As outlined Zou et al. (2017), although airglow patches are traditionally regarded as high-density ionospheric plasma unrelated to local precipitation, past observations were limited to disturbed conditions. Recent nonstorm time observations show patches to be associated with ionospheric flow channels and localized field-aligned currents. Zou et al. (2017) examine whether nonstorm time patches are related also to polar cap precipitation using Fast Auroral Snapshot-imager conjunctions. Zou et al.(2017) have identified localized precipitation that is enhanced within patches in comparison to the weak polar rain outside patches. The precipitation consists of structured or diffuse soft electron fluxes. While the latter resembles polar rain only with higher fluxes, the former consists of discrete fluxes enhanced by 1–2 orders of magnitude from several to several hundred eV. Therefore, patches should be regarded as part of a localized magnetosphere-ionosphere coupling system along open magnetic field lines and their transpolar evolution as a reflection of mesoscale magnetotail lobe processes. The precipitation expected to be a minor contributor to patch ionization.

4.6.8. The effect of ring current electron scattering rates on magnetosphere-ionosphere coupling

The simulation study of Perlongo et al. (2017) investigated the electrodynamic impact of varying descriptions of the diffuse aurora on the magnetosphere-ionosphere (M-I) system. Pitch angle diffusion caused by waves in the inner magnetosphere is the primary source term for the diffuse aurora, especially during storm time. The magnetic local time (MLT) and storm-dependent electrodynamic impacts of the diffuse aurora were analyzed using a comparison between a new self-consistent version of the Hot Electron Ion Drift Integrator with varying electron scattering rates and real geomagnetic storm events. The results were compared with Dst and hemispheric power indices, as well as auroral electron flux and cross-track plasma velocity observations. It was found that changing the maximum lifetime of electrons in the ring current by 2–6 h can alter electric fields in the nightside ionosphere by up to 26%. The lifetime also strongly influenced the location of the aurora, but the model generally produced aurora equatorward of observations.

4.7. Geomagnetic Pulsations at Polar Latitudes and Its Coupling with Solar Wind and IMF

4.7.1. Relation between Auroral and Geomagnetic Pulsations at Polar Latitudes According to the Observations on Spitsbergen

Geomagnetic pulsations of the Pc4–5 type at the Barentsburg Observatory on Spitsbergen for December 2007 to January 2008 are compared by Roldugin and Roldugin (2010) with the auroral intensity variations based on the photometric records at the same observatory. In all cases, auroral pulsations similar in shape are also observed simultaneously with geomagnetic pulsations. In the morning and daytime hours, the pulsation radiance maxima fall on the positive half-periods in the H component at the observation point; in the evening and nighttime hours, they fall on the negative half-periods.

4.7.2. Relation between Sudden Increases in the Solar Wind Dynamic Pressure, Auroral Proton Flashes, and Geomagnetic Pulsations in the Pc1 Range

The interrelation between sudden increases in the solar wind dynamic pressure, auroral proton flashes on the dayside equatorward of the oval, and geomagnetic pulsations in the Pc1 range is considered by Popova et al. (2010) on the basis of simultaneous observations of the solar wind plasma parameters, proton auroras on the IMAGE satellite, and geomagnetic pulsations at the Lovozero Observatory. It is indicated that proton luminosity flashes were observed in 70% of cases equatorward of the auroral oval during sudden changes in the solar wind pressure. In this case, flashes of proton auroras were observed in 85% of cases during sudden changes in the pressure, which were related to interplanetary shocks. Increases in pressure during tangential discontinuities were accompanied by flashes of proton auroras only in 45% of cases. When the ground station was conjugate to the region occupied by a proton aurora flash, the appearance or intensification of existent pulsations in the Pc1 range was observed in 96% of cases. When the ground station was not conjugate to the region of a proton luminosity flash, the response in geomagnetic pulsations was observed in 32% of events. When a sudden change in the solar wind pressure was not accompanied by a proton luminosity flash, the response in pulsations in the Pc1 range was hardly observed.

4.7.3. A Comparison between Large-Scale Irregularities and Scintillations in the Polar Ionosphere

In Wang et al. (2016b) a comparison tool has been developed by mapping the global GPS total electron content (TEC) and large coverage of ionospheric scintillations together on the geomagnetic latitude/magnetic local time coordinates. Using this tool, a comparison between large-scale ionospheric irregularities and scintillations is pursued during a geomagnetic storm. Irregularities, such as storm enhanced density, middle-latitude trough, and polar cap patches, are clearly identified from the TEC maps. At the edges of these irregularities, clear scintillations appeared but their behaviors were different. Phase scintillations (σ_φ) were almost always larger than amplitude scintillations (S_4) at the edges of these irregularities, associated with bursty flows or flow reversals with large density gradients. An unexpected scintillation feature appeared inside the modeled auroral oval where S_4 were much larger than σ_φ, most likely caused by particle precipitations around the exiting polar cap patches.

4.7.4. Observations and Modeling of Ionospheric Scintillations at South Pole during Six X-Class Solar Flares in 2013

Using two B-spline basis functions of degree 4 and the ionospheric scintillation data from a Global Positioning Satellite System (GPS) scintillation receiver at South Pole, Priyadarshi et al. (2016) reproduced ionospheric scintillation indices for the periods of the six X-class solar flares in 2013. These reproduced indices have filled the data gaps, and they are serving as a smooth replica of the real observations. In either event, these modeled scintillation indices are minimizing the geometrical effects between GPS satellite and the receiver. Six X-class solar flares have been studied during the summer and winter months, using the produced scintillation indices based on the observations from the GPS receiver at South Pole and the in situ plasma measurement from the associated passing of Defense Meteorological Satellite Program. Obtained results show that the solar flare peak suppresses the scintillation level and builds time-independent scintillation patterns; however, after a certain time from the solar flare peak, complicated scintillation patterns develop at high-latitude ionosphere and spread toward the polar cap boundary region. Substantial consistency has been found between moderate proton fluxes and scintillation enhancement.

4.7.5. Scintillation and Irregularities from the Nightside Part of a Sun-Aligned Polar Cap Arc

Van der Meeren et al. (2016) study the presence of irregularities and scintillation in relation to the nightside part of a long-lived, Sun-aligned transpolar arc on 15 January 2015. The arc was observed in DMSP UV and particle data and lasted at least 3 h between 1700 and 2000 UT. The arc was more intense than the main oval during this time. From all-sky imagers on Svalbard Van der Meeren et al. (2016) were able to study the evolution of the arc, which drifted slowly westward toward the dusk cell. The intensity of the arc as observed from ground was 10–17 kR in 557.7 nm and 2–3.5 kR in 630.0 nm, i.e., significant emissions in both green and red emission lines. Van der Meeren et al. (2016) have used high-resolution raw data from global navigation satellite systems (GNSS) receivers and backscatter from Super Dual Auroral Radar Network (SuperDARN) radars to study irregularities and scintillation in relation to the polar cap arc. Even though the literature has suggested that polar cap arcs are potential sources for irregularities, obtained results indicate only very weak irregularities. This may be due to the background density in the northward IMF polar cap being too low for significant irregularities to be created.

4.8. Universal Time Variations of the Auroral Hemispheric Power and Their Interhemispheric Asymmetry from TIMED/GUVI Observations

The paper of Luan et al. (2016) quantitatively analyzes the auroral hemispheric power (HP) and its interhemispheric asymmetry as a function of universal time (UT) for geomagnetically quiet conditions (Kp = 1–3) from Thermosphere Ionosphere Mesosphere Energetics and Dynamics/Global Ultraviolet Imager (TIMED/GUVI) imaging observations. The HP variation with UT can be approximately characterized as two cases: One is for similar HP variations in the equinoxes in the Northern Hemisphere and for the June solstices of both hemispheres, and the other is for similar HP patterns in the equinoxes in the Southern Hemisphere and for the December solstices of both hemispheres. In the equinoxes, the HP variations are interhemispherically asymmetric due to different occurrence time of the HP peak. In the solstices, the HP is generally interhemispherically symmetric in its diurnal variations, but there is interhemispheric asymmetry in the magnitudes of the maximum HP. For geomagnetically quiet conditions (Kp = 2), in the equinoxes relative interhemispheric differences are typically between 0 and 20%, with respect to the averaged HP from the two hemispheres, while during the solstices, the maximum relative interhemispheric asymmetry can be as large as 30% in December, but it is only ~15% in June. These two cases are mainly associated with variations of auroral precipitation power in the night side sector (21:00–03:00 magnetic local time, MLT), which are primarily controlled by solar illumination conditions in both hemispheres and are also attributed to the difference in the geographical area of the auroral oval in the two hemispheres. Furthermore, as outlined Luan et al. (2016), the general interhemispheric symmetry of the HP variations in solstices suggests that auroral acceleration is not only controlled locally by solar illumination conditions, which has been well known previously, but also might be affected by processes in the precipitation source region.

4.9. Morphology of High-Latitude Plasma Density Perturbations as Deduced from the Total Electron Content Measurements Onboard the Swarm Constellation

Park et al. (2017) investigate the climatology of high-latitude total electron content (TEC) variations as observed by the dual-frequency Global Navigation Satellite Systems (GNSS) receivers onboard the Swarm satellite constellation. The distribution of TEC perturbations as a function of geographic/magnetic coordinates and seasons reasonably agrees with that of the Challenging Minisatellite Payload observations published earlier. Categorizing the high-latitude TEC perturbations according to line-of-sight directions between Swarm and GNSS satellites, Park et al. (2017) can deduce their morphology with respect to the geomagnetic field lines. In the Northern Hemisphere, the perturbation shapes are mostly aligned with the L shell surface, and this anisotropy is strongest in the nightside auroral (substorm) and subauroral regions and weakest in the central polar cap. The results are consistent with the well-known two-cell plasma convection pattern of the high-latitude ionosphere, which is approximately aligned with L shells at auroral regions and crossing different L shells for a significant part of the polar cap. In the Southern Hemisphere, the perturbation structures exhibit noticeable misalignment to the local L shells. Here the direction toward the Sun has an additional influence on the plasma structure, which Park et al. (2017) attribute to photoionization effects. The larger offset between geographic and geomagnetic poles in the south than in the north is responsible for the hemispheric difference.

4.10. Inverse Electron Energy Dispersion from Moving Auroral Forms

As outlined Cameron and Knudsen (2016), numerous published examples of energy-dispersed bursts show electron energies reaching as high as several keV and decaying to lower energies over a fraction of 1 s. This signature has been interpreted by some authors as due to impulsive acceleration to a broad range of energies in a localized region and by others as the result of impulsive, dispersive Alfvén waves, in which case the acceleration takes place over an extended distance along magnetic field lines. According to Cameron and Knudsen (2016), a survey by the Suprathermal (0–350 eV) Electron Imager on the Enhanced Polar Outflow Probe (ePOP) in the topside ionosphere has produced examples of high-to-low ("regular") energy dispersion, but also a smaller number of examples exhibiting low-to-high ("inverse") dispersion, which has not been reported before. Motivated by a recent report of regular electron dispersion produced by auroral rays moving faster than the $\mathbf{E} \times \mathbf{B}$ drift speed, Cameron and Knudsen (2016) investigate a heuristic model of electron acceleration within a region of uniform electric field parallel to B which extends a distance *La* along magnetic field lines. It is shown that in addition to a broad range of energies, this model produces inverse dispersion when the detector is less than *La* beneath the bottom of the acceleration region and regular dispersion for detector distances larger than *La*. This simple model is meant to inform future efforts to construct a more physical model of suprathermal electron acceleration within moving auroral forms and suggests that inverse dispersion indicates relative proximity to an altitude-extended acceleration region.

4.11. Properties of the F2-Layer Critical Frequency Median in the Nocturnal Subauroral Ionosphere during Low and Moderate Solar Activity

According to Deminov et al. (2016), based on an analysis of data from the European ionospheric stations at subauroral latitudes, it has been found that the main ionospheric trough (MIT) is not characteristic for the monthly median of the F2-layer critical frequency (foF2), at least for low and moderate solar activity. In order to explain this effect, the properties of foF2 in the nocturnal subauroral ionosphere have been additionally studied for low geomagnetic activity, when the MIT localization is known quite reliably. It has been found that at low and moderate solar activity during night hours in winter, the foF2 data from ionospheric stations are often absent in the MIT area. For this reason, a model of the foF2 monthly median, which was constructed from the remaining data of these stations, contains no MIT or a very weakly pronounced MIT.

4.12. Non-Parametric Data Analysis of Low-Latitude Auroras and Naked-Eye Sunspots in the Medieval Epoch

Bekli et al. (2017) have studied solar activity by analyzing naked-eye sunspot observations and aurorae borealis observed at latitudes below 45°. They focused on the medieval epoch by considering the non-telescopic observations of sunspots from AD 974 to 1278 and aurorae borealis from AD 965 to 1273 that are reported in several Far East historical sources, primarily in China and Korea. After setting selection rules, Bekli et al. (2017) analyzed the distribution of these individual events following the months of the Gregorian calendar. In December, an unusual peak is observed with data recorded in both China and Japan, but not within Korean data. In extreme conditions, where the collection of events is reduced and discontinuous in some temporal intervals, Bekli et al. (2017) used the non-parametric kernel method. It is opted for the plugin approach of Sheather and Jones (1991) instead of cross-validation techniques to estimate the probability density functions (PDF) of the events. Bekli et al. (2017) obtained optimized bandwidths of 13.29 years for sunspots and 9.06 years for auroras, and 95% confidence intervals. The PDF curves exhibit multiple peaks occurring at quasi-periodic times with a very high positive.

4.13. Extreme Plasma Convection and Frictional Heating of the Ionosphere: ISR Observations

Extremely elevated ion temperatures observed with the Poker Flat and Resolute Bay incoherent scatter radars in the E and F regions of the polar ionosphere are discussed by

Akbari at al. (2017). Their observations include one-dimensional, line-of-sight ion temperatures ($T_{i,1d}$) that at times can rise to up to~8000 ∘ K. While examining the accuracy of the derived temperatures is difficult due to the several potential sources of error and uncertainty, Akbari at al. (2017) find that at altitudes of~130–250 km the line-of-sight ion temperatures obtained at relatively small aspect angles (for instance, at 22.5° away from the magnetic field direction) are below the expected values for the average or three-dimensional ion temperatures (T_i). It is find that this difference matches well with the theoretical expectation from anisotropic ion velocity distributions that emerge from polarization elastic scattering collisions between NO^+ and N_2 . These results are therefore the clearest detection of anisotropic ion velocity distributions with the Poker Flat Incoherent Scatter Radar and Resolute Bay Incoherent Scatter Radar facilities. Moreover, at higher altitudes (>~300 km), where the resonant charge exchange collisions prevail, and for a very high ion temperature event (T_i ~8000°K), Akbari at al. (2017) observe that the incoherent scatter radar (ISR) spectra obtained at large aspect angles (e.g., 55°) look like those expected from non-Maxwellian plasma. At similar altitudes, the measured one-dimensional temperatures at an aspect angle of 22.5° are found to exceed the expected values from the frictional heating and resonant charge exchange collisions by about 1000°K. According to opinion of Akbari at al. (2017), a portion of this offset is thought to reflect the role of Coulomb collisions.

4.14. Effects of Geomagnetic Disturbances in Daytime Variations of the Atmospheric Electric Field in Polar Regions

Kleimenova et al. (2017) have studied the influence of daytime polar substorms (geomagnetic bays under the IMF $B_z > 0$) on variations of the vertical gradient of the atmospheric electric field potential (E_z) observed at the Polish Hornsund Station (Svalbard, Norway). Only the observations of E_z under "fair weather" conditions were used, i.e. in the absence of strong wind, precipitations, low cloud cover, etc. Kleimenova et al. (2017) studied more than 20 events of daytime polar substorms registered by the Scandinavian chain of IMAGE magnetometers in 2010–2014 during the "fair weather" periods at the Hornsund Station. Analysis of the observations showed that E_z significantly deviates from the its background variations during daytime, as a rule, when the Hornsund Station is in the region of projection of the daytime auroral oval, the position of which was determined from OVATION data. It was shown that the development of a daytime polar substorm leads to fluctuating enhance of E_z values. It was found by Kleimenova et al. (2017) that E_z surges are accompanied by intensification of field-aligned electric currents outflowing from the ionosphere, which were calculated from the data of low-orbit communication satellites of the AMPERE project.

4.15. Photoelectrons in the Quiet Polar Wind

The study of Glocer et al. (2017) presents a newly coupled model capable of treating the superthermal electron population in the global polar wind solution. The model combines the hydrodynamic Polar Wind Outflow Model (PWOM) with the kinetic SuperThermal Electron Transport (STET) code. The resulting PWOM-STET coupled model is described and then used to investigate the role of photoelectrons in the polar wind. Glocer et al. (2017) present polar wind results along single stationary field lines under dayside and nightside conditions, as well as the global solution reconstructed from nearly 1000 moving field lines. The model results show significant day-night asymmetries in the polar wind solution owing to the higher ionization and photoelectron fluxes on the dayside compared to the nightside. Field line motion is found by Glocer et al. (2017) to modify this dependence and create global structure by transporting field lines through different conditions of illumination and through the localized effects of Joule heating.

4.16. Latitudinal and MLT Dependence of the Seasonal Variation of Geomagnetic Field Around Auroral Zone

Seasonal variation of geomagnetic field around auroral zone is analyzed by Zhu et al. (2017) in dependence of geomagnetic latitude, magnetic local time (MLT) and geomagnetic conditions. The study of Zhu et al. (2017) uses horizontal component (H) of geomagnetic field obtained from 6 observatories located in geomagnetic latitudes 57.8°N–73.8°N along 115°E longitudinal line. The results indicate that seasonal variations of geomagnetic field around auroral zone are different combinations of annual and semiannual variations at different latitudinal ranges. According to Zhu et al. (2017), both annual and semiannual variations show distinct MLT dependency: **(1)** At dayside auroral latitudes (around 72°N geomagnetic latitude), geomagnetic field shows distinct annual variation under both quiet and disturbed conditions. Furthermore, the annual component is mainly contributed by data of dusk sector. **(2)** At nightside auroral latitudes (around 65°N), geomagnetic field shows semiannual dominated seasonal variation. Under quiet conditions the annual component is comparable to the semiannual component, while under disturbed conditions, the semiannual component is twice as much as the annual component. Under quiet conditions, the semiannual component is mainly contributed by 1300–1400 MLT, while the annual component has two peaks: one is around 1100–1300 MLT and the other is around 2000–2200 MLT. Under disturbed conditions, the semiannual component is mainly contributed by data around midnight, while the annual component is mainly contributed by dusk sector. **(3)** At subauroral latitudes (around 60°N), annual variation is comparable to semiannual variation under both quiet and disturbed conditions. Both annual and semiannual components show similar MLT dependencies as that of nightside auroral latitudes.

References

Abe T., Y. Ichikawa, and A.W. Yau, 2005. "Generation of high-density plasma in the polar cap observed by the Akebono satellite", *Adv. Space Res.*, **36**, No. 10, 1872-1877.

Akbari H., L.V. Goodwin, J. Swoboda, J.-P. St.-Maurice, and J.L. Semeter, 2017. "Extreme plasma convection and frictional heating of the ionosphere: ISR observations", *J. Geophys. Res. Space Physics*, **122**, 7581–7598, doi:10.1002/2017JA023916.

Akimoto K. and N. Omidi, 1986. "The generation of broadband electrostatic noise by an ion beam in the magnetotail", *Geophys. Res. Lett.*, **13**, No. 2, 97-100.

Anderson B.J. and D.C. Hamilton, 1993. "Electromagnetic ion cyclotron waves stimulated by modest magnetospheric compressions", *J. Geophys. Res.*, **98**, No. A7, 11369-11382.

André M., G.B. Crew, W.K. Peterson, A.M. Persoon, C.J. Pollock, and M.J. Engebretson, 1990. "Ion heating by broadband low-frequency waves in the cusp/cleft", *J. Geophys. Res.*, **95**, No. A12, 20809-20823.

Axford W.I. and C.O. Hines, 1961. "A Unifying Theory of High-Latitude Geophyscial Phenomena and Geomagnetic Storms", *Can. J. Phys.*, **39**, No. 10, 1433-1464.

Bartels J., 1934. "Twenty-Seven Day Recurrences in Terrestrial-Magnetic and Solar Activity," *J. Terr. Magn.*, **39**, No. 201-202, 1923–1933.

Bekli M.R., N. Zougab, A. Belabbas, and I. Chadou, 2017. "Non-parametric Data Analysis of Low-latitude Auroras and Naked-eye Sunspots in the Medieval Epoch", *Solar Phys.*, **292**, 52, DOI: 10.1007/s11207-017-1084-5.

Belenkaya E.S., V.V. Kalegaev, and M.S. Blokhina, 2014. "Polar cap response to the solar wind density jump under constant southward IMF", *Geomagnetism and Aeronomy*, **54**, No. 6, 702-711.

Benson R.F., 1972. "Ionospheric plasma resonances: Time durations vs. latitude, altitude, and fN/fH", *Planet. Space Sci.*, **20**, No. 5, 683-706.

Benson R.F., 1974. "Stimulation of the Harris instability in the ionosphere", *Phys. Fluids*, **17**, No. 5, 1032-1037.

Benson R.F. and W. Calvert, 1979. "ISIS 1 observations at the source of auroral kilometric radiation", *Geophys. Res. Lett.*, **6**, No. 6, 479-482.

Benson R.F., V.A. Osherovich, J. Fainberg, A.-F. Vinas, and D.R. Ruppert, 2001. "An interpretation of banded magnetospheric radio emissions", *J. Geophys. Res.*, **106**, No. A7, 13179-13190.

Benson R.F., V.A. Osherovich, J. Fainberg, and B.W. Reinisch, 2003. "Classification of IMAGE/RPI-stimulated plasma resonances for the accurate determination of magnetospheric electron-density and magnetic field values", *J. Geophys. Res.*, **108**, No. A5, 1207, doi:10.1029/2002JA009589, SMP 16-1.

Benson R.F., V.A. Osherovich, J. Fainberg, J.L. Green, S.A. Boardsen, L.N. Garcia, N. Tsyganenko, and B.W. Reinisch, 2006. "RPI/IMAGE plasma and magnetic signatures of solar-wind/magnetosphere interactions, *Eos Trans. AGU*, **87**(36), Jt. Assem. Suppl., Abstract SM42A-05.

Bhattarai S.K., R.E. Lopez, R. Bruntz, J.G. Lyon, and M. Wiltberger, 2012. "Simulation of the polar cap potential during periods with northward interplanetary magnetic field", *J. Geophys. Res. Space Physics*, **117**, No. A4, A04219, doi:10.1029/2011JA017143, 1-12.

Boudouridis A., E. Zesta, L.R. Lyons, P.C. Anderson, and D. Lummerzheim, 2003. "Effect of solar wind pressure pulses on the size and strength of the auroral oval", *J. Geophys. Res.*, **108**, No. A4, 8012, doi:10.1029/2002JA009373, COA 13-1-16.

Boudouridis A., E. Zesta, L.R. Lyons, P.C. Anderson, and D. Lummerzheim, 2004a. "Magnetospheric reconnection driven by solar wind pressure fronts", *Ann. Geophys.*, **22**, No. 4, 1367-1378.

Boudouridis A., E. Zesta, L.R. Lyons, and P.C. Anderson, 2004b. "Evaluation of the Hill-Siscoe transpolar potential saturation model during a solar wind dynamic pressure pulse", *Geophys. Res. Lett.*, **31**, No. 23, L23802, doi:10.1029/2004GL021252, 1-4.

Boudouridis A., E. Zesta, L.R. Lyons, P.C. Anderson, and D. Lummerzheim, 2005. "Enhanced solar wind geoeffectiveness after a sudden increase in dynamic pressure during southward IMF orientation", *J. Geophys. Res.*, **110**, No. A5, A05214, doi:10.1029/2004JA010704, 1-15.

Boudouridis A., L.R. Lyons, E. Zesta, and J.M. Ruohoniemi, 2007. "Dayside reconnection enhancement resulting from a solar wind dynamic pressure increase", *J. Geophys. Res.*, **112**, A06201, doi:10.1029/2006JA012141, 1-12.

Boudouridis A., E. Zesta, L.R. Lyons, P.C. Anderson, and A.J. Ridley, 2008. "Temporal evolution of the transpolar potential after a sharp enhancement in solar wind dynamic pressure", *Geophys. Res. Lett.*, **35**, L02101, doi:10.1029/2007GL031766, 1-5.

Buonsanto M.J., 1999. "Ionospheric storms – A review", *Space Sci. Rev.*, **88**, No. 3-4, 563-601.

Burch J.L., S.B. Mende, D.G. Mitchell, T.E. Moore, C.J. Pollack, B.W. Reinisch, B.R. Sandel, S.A. Fuselier, D.L. Gallagher, J.L. Green, J.D. Perez, and P.H. Reiff, 2001. "Views of Earth's magnetosphere with the IMAGE satellite", *Science*, **291**, No. 5504, 619-624.

Burke W., A.G. Rubin, N.C. Maynard, L.C. Gentile, P.J. Sultan, F.J. Rich, O. de La Beaujardiere, Huang C.Y., II, and G.R. Wilson, 2000. "Ionospheric disturbances observed by DMSP at middle to low latitudes during the magnetic storm of June 4–6, 1991", *J. Geophys. Res.*, **105**, No. A8, 18391-18405.

Cameron T. and D. Knudsen, 2016, "Inverse electron energy dispersion from moving auroral forms", *J. Geophys. Res. Space Physics*, **121**, 11,896–11,911, doi:10.1002/2016JA023045.

Castaing B., Y. Gagne, and E.J. Hopfinger, 1990. "Velocity probability distribution functions of high Reynolds number turbulence", *Physica D*, **46**, No. 2, 177-200.

Cattell C.A., Dombeck J., Wygant J.R., Hudson M.K., Mozer F.S., Temerin M.A., Peterson W.K., Kletzing C.A., Russell C.T., and Pfaff R.F., 1999. "Comparisons of Polar satellite observations of solitary wave velocities in the plasma sheet boundary and the high altitude cusp to those in the auroral zone", *Geophys. Res. Lett.*, **26**, No. 3, 425-428.

Cattell C.A., J. Dombeck, J. Wygant, J.F. Drake, M. Swisdak, M.L. Goldstein, W. Keith, A. Fazakerley, M. André, E. Lucek, and A. Balogh, 2005. "Cluster observations of electron holes in association with magnetotail reconnection and comparison to simulations", *J. Geophys. Res.*, **110**, No. 1, A01211, doi:10.1029/2004JA010519, 1-16.

Chappell C.R., T.E. Moore, and J.H. Waite Jr., 1987. "The ionosphere as a fully adequate source of plasma for the Earth's magnetosphere", *J. Geophys. Res.*, **92**, No. A6, 5896-5910.

Chappell C.R., B.L. Giles, T.E. Moore, D.C. Delcourt, P.D. Craven, and M.O. Chandler, 2000. "The adequacy of the ionospheric source in supplying magnetospheric plasma", *J. Atmos. Sol-Terr. Phys.*, **62**, No. 6, 421-436.

Chen Y.-J., R.A. Heelis, and J.A. Cumnock, 2016. "Plasma and convection reversal boundary motions in the high-latitude ionosphere", *J. Geophys. Res. Space Physics*, **121**, 5752–5763, doi:10.1002/2016JA022796.

Chua D., G. Parks, M. Brittnacher, W. Peria, G. Germany, J. Spann, and C. Carlson, 2001. "Energy characteristics of auroral electron precipitation: A comparison of substorms and pressure pulse related auroral activity", *J. Geophys. Res.*, **106**, No. A4, 5945-5956.

Clauer C.R., Z. Xu, M. Maimaiti, J.M. Ruohoneimi, W. Scales, M.D. Hartinger, M. J. Nicolls, S. Kaeppler, F.D. Wilder, and R.E. Lopez, 2016. "Investigation of a rare event where the polar ionospheric reverse convection potential does not saturate during a period of extreme northward IMF solar wind driving", *J. Geophys. Res. Space Physics*, **121**, 5422–5435, doi:10.1002/2016JA022557.

Coroniti F.V. and M. Ashour-Abdalla, 1989. "Electron velocity space hole modes and narrowband electrostatic noise in the distant tail", *Geophys. Res. Lett.*, **16**, No. 16, 747-750.

Coroniti F.V., E.W. Greenstadt, B.T. Tsuritani, E.J. Smith, R.D. Zwickel, and J.T. Gosling, 1990. "Plasma waves in the distant geomagnetic tail: ISEE 3", *J. Geophys. Res.*, **95**, No. A12, 20977-20995.

Coroniti F.V., M. Ashour-Abdalla, and R.L. Richard, 1993. "Electron velocity space hole modes", *J. Geophys. Res.*, **98**, No. 7, 11349-11358.

Cousins E.D.P., T. Matsuo, and A.D. Richmond, 2015a. "Mapping high-latitude ionospheric electrodynamics with SuperDARN and AMPERE", *J. Geophys. Res. Space Physics*, **120**, No. 7, doi:10.1002/2014JA020463, 5854–5870.

Cousins E.D.P., T. Matsuo, A.D. Richmond, and B.J. Anderson, 2015b. "Dominant modes of variability in large-scale Birkeland currents", *J. Geophys. Res. Space Physics*, **120**, 6722–6735, doi:10.1002/2014JA020462.

Cowley S.W.H. and M. Lockwood, 1992. "Excitation and decay of solar wind-driven flows in the magnetosphere-ionosphere system", *Ann. Geophys., Atmos. Hydrospheres Space Sci.* (France), **10**, No. 1-2, 103-115.

Craven J.D., L.A. Frank, C.T. Russell, E.J. Smith, and R.P. Lepping, 1986. "Global auroral responses to magnetospheric compressions by shocks in the solar wind -Two case studies", in *Solar Wind Magnetosphere Coupling*, edited by Y. Kamide and J. A. Slavin, Terra Scientific Publ. Co., Tokyo, 367-380.

Dahlgren H., B.S. Lanchester, N. Ivchenko, and D.K. Whiter, 2016. "Electrodynamics and energy characteristics of aurora at high resolution by optical methods", *J. Geophys. Res. Space Physics*, **121**, 5966–5974, doi:10.1002/2016JA022446.

Damiano P.A., J.R. Johnson, and C.C. Chaston, 2015. "Ion temperature effects on magnetotail Alfvén wave propagation and electron energization", *J. Geophys. Res. Space Physics*, **120**, 5623–5632, doi:10.1002/2015JA021074.

Damiano P.A., J.R. Johnson, and C.C. Chaston, 2016. "Ion gyroradius effects on particle trapping in kinetic Alfvén waves along auroral field lines", *J. Geophys. Res. Space Physics*, **121**, 10,831–10,844, doi:10.1002/2016JA022566.

Deminov M.G., R.G. Deminov, and V.N. Shubin, 2016. "Properties of the F2-layer Critical Frequency Median in the Nocturnal Subauroral Ionosphere during Low and Moderate Solar Activity", *Geomagnetism and Aeronomy*, **56**, No. 6, 750–756. Original Russian Text in *Geomagnetizm i Aeronomiya*, **56**, No. 6, 789–795.

Drake J.F., Swisdak M.S., Cattell C., Shay M.A., Rogers B., and Zeiler A., 2003. "Formation of electron holes and particle energization during magnetic reconnection", *Science*, **299**, No. 5608, doi:10.1126/science.1080333, 873-877.

Dubouloz N., M. Bouhram, C. Senior, D. Delcourt, M. Malingre, and J.-A. Sauvaud, 2001. "Spatial structure of the cusp/cleft ion fountain: A case study using a magnetic conjugacy between Interball AP and a pair of Super DARN radars", *J. Geophys. Res.*, **106**, No. A1, 261-274.

Dungey J.W., 1961. "Interplanetary ma, 1961.gnetic field and the auroral zones", *Phys. Rev. Lett.*, **6**, No. 2, 47-48.

Dusenbery P., 1986. "Generation of broadband noise in the magnetotail by the beam acoustic instability", *J. Geophys. Res.*, **91**, No. A11, 12005-12016.

Dusenbery P.B., 1987. "Convective growth of broadband turbulence in the plasma sheet boundary layer", *J. Geophys. Res.*, **92**, No. A3, 2560-2564.

Dusenbery P., 1988. "Reply", *J. Geophys. Res.*, **93**, No. A12, 14729-14731.

Dusenbery P. and L.R. Lyons, 1985. "The generation of electrostatic noise in the plasma sheet boundary layer", *J. Geophys. Res.*, **90**, No. 11, 10935-10943.

Eastman T.E., L.A. Frank, W.K. Peterson, and W. Lennartson, 1984. "The plasma sheet boundary layer", *J. Geophys. Res.*, **89**, No. A3, 1553-1572.

Eastman T.E., L.A. Frank, W. Peterson, and C. Huang, 1985. "The boundary layers as the primary transport regions of the Earth's magnetotail", *J. Geophys. Res.*, **90**, No. A10, 9541-9560.

Evans J.V., 1973. "The causes of storm-time increases of the F-layer at mid-latitudes", *J. Atmos. Sol. Terr. Phys.*, **35**, No. 4, 593-616.

Figueiredo S., T. Karlsson, and G.T. Marklund, 2004. "Investigation of subauroral ion drifts and related field-aligned currents and ionospheric Pedersen conductivity distribution", *Ann. Geophys.*, **22**, No. 3, 923-934.

Foster J.C., 1993. "Storm time plasma transport at middle and high latitudes", *J. Geophys. Res.*, **98**, No. A2, 1675-1689.

Foster J.C. and W.J. Burke, 2002. "SAPS: A new categorization for subauroral electric fields", *Eos Trans. AGU*, **83**, No. 36, 393-394.

Foster J.C. and Vo H.B., 2002. "Average characteristics and activity dependence of the subauroral polarization stream", *J. Geophys. Res.*, **107**, A12, 1475, doi:10.1029/2002JA009409, SIA16-1-10.

Foster J.C., P.J. Erickson, A.J. Coster, J. Goldstein, and F.J. Rich, 2002. "Ionospheric signatures of plasmaspheric tails", *Geophys. Res. Lett.*, **29**, No. 13, 1623, doi:10.1029/2002GL015067, 1-1-4.

Foster J.C., A.J. Coster, P.J. Erickson, F.J. Rich, and B.R. Sandel, 2004. "Stormtime observations of the flux of plasmaspheric ions to the dayside cusp/magnetopause", *Geophys. Res. Lett.*, **31**, No. 8, L08809, doi:10.1029/2004GL020082, 1-4.

Foster J.C., A.J. Coster, P.J. Erickson, J.M. Holt, F.D. Lind, W. Rideout, M. McCready, A. van Eyken, R.J. Barnes, R.A. Greenwald, and F.J. Rich, 2005. "Multiradar observations of the polar tongue of ionization", *J. Geophys. Res.*, **110**, No. A9, A09S31, doi:10.1029/2004JA010928, 1-12.

Fox N.J., M. Peredo, and B.J. Thompson, 1998. "Cradle to grave tracking of the January 6-11, 1997 Sun-Earth connection event", *Geophys. Res. Lett.*, **25**, No. 14, 2461-2464.

Frank-Kamenetsky A. and O. Troshichev, 2012. "A relationship between the auroral absorption and the magnetic activity in the polar cap", *J. Atmos. Sol. Terr. Phys.*, **77**, 40-45.

Franz J.R., P.M. Kintner, and J.S. Pickett, 1998. "Polar observations of coherent electric field structure", *Geophys. Res. Lett.*, **25**, No. 8, 1277-1280.

Fung S.F. and J.L. Green, 2005. "Modeling of field-aligned radio echoes in the plasmasphere", *J. Geophys. Res.*, **110**, No. A1, A01210, doi:10.1029/2004JA010658, 1-9.

Fung S.F., R.F. Benson, D.L. Carpenter, J.L. Green, V. Jayanti, I.A. Galkin, and B.W. Reinisch, 2003. "Guided echoes in the magnetosphere: Observations by radio plasma imager on IMAGE", *Geophys. Res. Lett.*, **30**, No. 11, 1589, doi:10.1029/2002GL016531, 43-1-4.

Galkin I., B.W. Reinisch, G. Grinstein, G. Khmyrov, A. Kozlov, X. Huang, and S.F. Fung, 2004. "Automated exploration of the radio plasma imager data", *J. Geophys. Res.*, **109**, No. A12, A12210, doi:10.1029/2004JA010439, 1-12.

Galperin Y., Y. Ponomarev, and A. Zosimova, 1974. "Plasma convection in the polar ionosphere", *Ann. Geophys.*, **30**, No. 1, 1-7.

Garcia L.N., S.F. Fung, J.L. Green, S.A. Boardsen, B.R. Sandel, and B.W. Reinisch, 2003. "Observations of the latitudinal structure of plasmaspheric convection plumes by IMAGE-RPI and EUV", *J. Geophys. Res.*, **108**, No. A8, 1321, doi:10.1029/2002JA 009496, SMP6-1-12.

Glocer A., G. Khazanov, and M. Liemohn, 2017. "Photoelectrons in the quiet polar wind", *J. Geophys. Res. Space Physics*, **122**, 6708–6726, doi:10.1002/2017JA024177.

Goldman M.V., M. M. Oppenheim, and D.L. Newman, 1999. "Nonlinear two-stream instabilities as an explanation for auroral bipolar wave structures", *Geophys. Res. Lett.*, **26**, No. 13, 1821-1824.

Goldstein J., J.L. Burch, and B.R. Sandel, 2005a. "Magnetospheric model of subauroral polarization stream", *J. Geophys. Res.*, **110**, No. A9, A09222, doi:10.1029/2005JA 011135, 1-10.

Goldstein J., J.L. Burch, B.R. Sandel, S.B. Mende, P.C. son Brandt, and M.R. Hairston, 2005b. "Coupled response of the inner magnetosphere and ionosphere on 17 April 2002", *J. Geophys. Res.*, **110**, No. A3, A03205, doi:10.1029/2004JA010712, 1-22.

Grabbe C.L., 1985a. "New results on the generation of broadband electrostatic waves in the magnetotail", *Geophys. Res. Lett.*, **12**, No. 8, 483-486.

Grabbe C.L., 1985b. "Generation of broadband electrostatic noise in the magnetotail", in *Advances in Space Plasma Physics*, edited by B. Buti, World Sci., Hackensack, N.J., 356-381.

Grabbe C.L., 1987. "Numerical study of the spectrum of broadband electrostatic noise in the magnetotail", *J. Geophys. Res.*, **92**, No. A2, 1185-1192.

Grabbe C.L., 1989. "Wave propagation effects of ion beam instabilities on broadband electrostatic noise in the magnetotail", *J. Geophys. Res.*, **94**, No. A12, 17299-17304.

Grabbe C.L., 1998. "Broadband electrostatic waves in the plasma sheet for kappa particle distributions", *report*, Univ. of Iowa, Iowa City.

Grabbe C.L., 2000a. "Generation of broadband electrostatic waves in Earth's magnetotail", *Phys. Rev. Lett.*, **84**, No. 16, 3614-3617.

Grabbe C.L., 2000b. "Origins of broadband electrostatic waves in the magnetotail", in *Recent Research Developments in Plasmas*, edited by Pandalai, Transworld Res. Network, Trivandrum, India, **1**, 89-100.

Grabbe C.L., 2002. "Solitary wave structure in a magnetized plasma and the source region of BEN", *Geophys. Res. Lett.*, **29**, No. 16, 1804, doi:10.1029/2002GL015265, 51-1-4.

Grabbe C.L., 2003. "Correction to "Solitary wave structure in magnetized plasmas and the source region of BEN" by C.L. Grabbe", *Geophys. Res. Lett.*, **30**, No. 15, 1797, doi:10.1029/2003GL017894, SSC2-1-1.

Grabbe C.L., 2005. "Trapped-electron solitary wave structures in a magnetized plasma", *Phys. Plasmas*, **12**, No. 7, doi:10.1063/1.1978888, 072311-1-6.

Grabbe C.L. and T.E. Eastman, 1984. "Generation of broadband electrostatic noise by ion beam instabilities in the magnetotail", *J. Geophys. Res.*, **89**, No. A6, 3865-3872.

Grabbe C.L. and Menietti J.D., 2002. "Electrostatic wave variety and the origins of BEN", *Planet. Space Sci.*, **50**, No. 3, 335-341.

Grabbe C.L. and Menietti J.D., 2006. "Broadband electrostatic wave observations in the auroral region on Polar and comparisons with theory", *J. Geophys. Res.*, **111**, No 10, A10226, doi:10.1029/2006JA011602, 1-12.

Guo J., X. Feng, T.I. Pulkkinen, E.I. Tanskanen, W. Xu, J. Lei, and B.A. Emery, 2012. "Auroral electrojets variations caused by recurrent high-speed solar wind streams during the extreme solar minimum of 2008", *J. Geophys. Res. Space Physics*, **117**, No. A4, A04307, doi:10.1029/2011JA017458, 1-9.

Gurnett D.A., L.A. Frank, and D. Lepping, 1976. "Plasma waves in the distant magnetotail", *J. Geophys. Res.*, **81**, No. 34, 6059-6071.

Gurnett D.A., A.M. Persoon, R.F. Randall, D.L. Odem, S.L. Remington, T.F. Averkamp, M.M. Debower, G.B. Hospodarsky, R.L. Huff, D.L. Kirchner, M.A. Mitchell, B.T. Pham, J.R. Phillips, W.J. Schintler, P. Sheyko, and D.R. Tomash, 1995. "The Polar plasma wave instrument", *Space Sci. Rev.*, **71**, No. 1-4, 597-622.

Hairston M.R., T.W. Hill, and R.A. Heelis, 2003. "Observed saturation of the ionospheric polar cap potential during the 31 March 2001 storm", *Geophys. Res. Lett.*, **30**, No. 6, 1325, doi:10.1029/2002GL015894, 58-1-4.

Hairston M.R., K.A. Drake, and R. Skoug, 2005. "Saturation of the ionospheric polar cap potential during the October–November 2003 superstorms", *J. Geophys. Res.*, **110**, No. A9, A09S26, doi:10.1029/2004JA010864, 1-12.

Hesse M., J. Birn, and R.A. Hoffman, 1997. "On the mapping of ionospheric convection into the magnetosphere", *J. Geophys. Res.*, **102**, No. A5, 9543-9551.

Hoegy W.R. and R.F. Benson, 1988. "DE/ISIS conjunction comparisons of high-latitude electron density features", *J. Geophys. Res.*, **93**, No. A6, 5947-5954.

Horwitz J.L., 1984. "Features of ion trajectories in the polar magnetosphere", *Geophys. Res. Lett.*, **11**, No. 11, 1111-1114.

Horwitz J.L. and T.E. Moore, 1997. "Four contemporary issues concerning ionospheric plasma flow to the magnetosphere", *Space Sci. Rev.*, **80**, No. 1-2, 49-76.

Hu Z.-J., Y. Ebihara, H.-G. Yang, H.-Q. Hu, B.-C. Zhang, B. Ni, R. Shi, and T.S. Trondsen, 2014. "Hemispheric asymmetry of the structure of dayside auroral oval", *Geophys. Res. Lett.*, **41**, No. 24, doi:10.1002/2014GL062345, 8696-8703.

Huang X., B.W. Reinisch, P. Song, P. Nsumei, J.L. Green, and D.L. Gallagher, 2004. "Developing an empirical density model of the plasmasphere using IMAGE/RPI observations", *Adv. Space Res.*, **33**, No. 6, 829-832.

Huang C.S., J.C. Foster, L.P. Goncharenko, P.J. Erickson, W. Rideout, and A.J. Coster, 2005. "A strong positive phase of ionospheric storms observed by the Millstone Hill incoherent scatter radar and global GPS network", *J. Geophys. Res.*, **110**, No. A6, A06303, doi:10.1029/2004JA010865.

Hubert B., M. Palmroth, T.V. Laitinen, P. Janhunen, S.E. Milan, A. Grocott, S.W.H. Cowley, T. Pulkkinen, and J.-C. Gérard, 2006. "Compression of the Earth's magnetotail by interplanetary shocks directly drives transient magnetic flux closure", *Geophys. Res. Lett.*, **33**, No. 10, L10105, doi:10.1029/2006GL026008, 1-4.

Johnstone A.D., D.M. Walton, R. Liu, and D.A. Hardy, 1993. "Pitch angle diffusion of low-energy electrons by whistler mode waves", *J. Geophys. Res.*, 98, No. A4, 5959-5967.

Kabin K., R. Rankin, G. Rostoker, R. Marchand, I.J. Rae, A.J. Ridley, T.I. Gombosi, C.R. Clauer, and D.L. DeZeeuw, 2004. "Open-closed field line boundary position: A parametric study using an MHD model", *J. Geophys. Res.*, **109**, No. A5, A05222, doi:10.1029/2003JA010168.

Karlsson T., G. Marklund, L. Blomberg, and A. Mälkki, 1998. "Subauroral electric fields observed by Freja satellite: A statistical study", *J. Geophys. Res.*, **103**, No. A3, 4327-4341.

Kennel C.F., F.L. Scarf, R.W. Fredricks, J.H. McGhee, and F.V. Coroniti, 1970. "VLF electric field observations in the magnetosphere", *J. Geophys. Res.*, **75**, No. 31, 6136-6152.

Khazanov G.V., D.G. Sibeck, and E. Zesta, 2017. "Is diffuse aurora driven from above or below?", *Geophys. Res. Lett.*, **44**, 641–647, doi:10.1002/2016GL072063.

Kihn E.A., R. Redmon, A.J. Ridley, and M.R. Hairston, 2006. "A statistical comparison of the AMIE derived and DMSP-SSIES observed high-latitude ionospheric electric field", *J. Geophys. Res.*, **111**, No. A8, A08303, doi:10.1029/2005JA011310, 1-11.

Kim H.-J., L. Lyons, A. Boudouridis, V. Pilipenko, A.J. Ridley, and J.M. Weygand, 2011. "Statistical study of the effect of ULF fluctuations in the IMF on the cross polar cap potential drop for northward IMF", *J. Geophys. Res. Space Physics*, **116**, No. A10, A10311, doi:10.1029/2011JA016931, 1-5.

Kleimenova N.G., M. Kubicki, A. Odzimek, L.M. Malysheva, and L.I. Gromova, 2017. "Effects of Geomagnetic Disturbances in Daytime Variations of the Atmospheric Electric Field in Polar Regions", *Geomagnetism and Aeronomy*, **57**, No. 3, 266–273. Original Russian Text published in *Geomagnetizm i Aeronomiya*, **57**, No. 3, 290–297.

Kojima H., H. Matsumoto, T. Miyatake, I. Nagano, A. Fujita, L.A. Frank, T. Mukai, W.R. Paterson, Y. Saito, S. Machida, and R.R. Anderson, 1994. "Relation between electrostatic solitary waves and hot plasma flow in the plasma sheet boundary layer: GEOTAIL observations", *Geophys. Res. Lett.*, **21**, No. 25, 2919-2922.

Kornilov I.A., T.A. Kornilova, and I.V. Golovchanskaya, 2016. "Relationship between Auroral Oval Poleward Boundary Intensifications and Magnetic Field Variations in the Solar Wind", *Geomagnetism and Aeronomy*, **56**, No. 3, 268–275. Original Russian Text published in *Geomagnetizm i Aeronomiya*, **56**, No. 3, 288–295.

Korth A., M. Franz, Q.-G. Zong, T.A. Fritz, J.-A. Sauvaud, H. Reme, I. Dandouras, R. Friedel, C.G. Mouikis, L.M. Kistler, E. Mobius, M.F. Marcucci, M. Wilber, G. Parks, A. Keiling, R. Lundin, and P.W. Daly, 2004. "Ion injections at auroral latitude during the March 31, 2001 magnetic storm observed by Cluster", *Geophys. Res. Lett.*, **31**, No. 20, L20806, doi:10.1029/2004GL020356, 1-4.

Krasovsky V.I., H. Matsumoto, and Y. Omura, 1997. "Bernstein-Greene-Kruskal analysis of electrostatic solitary waves observed with Geotail", *J. Geophys. Res.*, **102**, No. A10, 22131-22140.

Laakso H., R. Pfaff, and P. Janhunen, 2002. "Polar observations of electron density distribution in the Earth's magnetosphere: 2. Density profiles", *Ann. Geophys.*, **20**, No. 11, 1725-1735.

LaBelle J., R. Treumann, W. Baumjohann, G. Haerendel, N. Sckopke, G. Paschmann, and H. Lühr, 1988. "The duskside plasmapause/ring current interface: Convection and plasma wave observations", *J. Geophys. Res.*, **93**, No. A4, 2573-2590.

Lauben D.S., U.S. Inan, T.F. Bell, D.L. Kirchner, G.B. Hospodarsky, and J.S. Pickett, 1998. "VLF chorus emissions observed by POLAR during the January 10, 1997, magnetic cloud", *Geophys. Res. Lett.*, **25**, No. 15, 2995-2998.

Liou K., 2006. "Global auroral response to interplanetary media with emphasis on solar wind dynamic pressure enhancements", in *Recurrent Magnetic Storms: Corotating Solar Wind Streams*, edited by B. Tsurutani et al., Geophys. Monogr. Ser., **167**, AGU, Washington, D.C., 197-212.

Liou K., P.T. Newell, C.-I. Meng, M. Brittnacher, and G. Parks, 1998. "Characteristics of the solar wind controlled auroral emissions ", *J. Geophys. Res.*, **103**, No. A8, 17543-17557.

Liou K., C.-C. Wu, R.P. Lepping, P.T. Newell, and C.-I. Meng, 2002. "Midday sub-auroral patches (MSPs) associated with interplanetary shocks", *Geophys. Res. Lett.*, **29**, No. 16, 1771, doi:10.1029/2001GL014182, 18-1-4.

Liou K., P.T. Newell, C.-I. Meng, C.-C. Wu, and R.P. Lepping, 2003. "Investigation of external triggering of substorms with Polar ultraviolet imager observations", *J. Geophys. Res.*, **108**, No. A10, 1364, doi:10.1029/2003JA009984, SIA2-1-14.

Liou K., P.T. Newell, T. Sotirelis, and C.-I. Meng, 2006. "Global auroral response to negative pressure impulses", *Geophys. Res. Lett.*, **33**, No. 11, L11103, doi:10.1029/2006 GL025933, 1-5.

Liou K., P.T. Newell, J.-H. Shue, C.-I. Meng, Y. Miyashita, H. Kojima, and H. Matsumoto, 2007. "Compression aurora': Particle precipitation driven by long-duration high solar wind ram pressure", *J. Geophys. Res.*, **112**, A11216, doi:10.1029/2007JA012443, 1-10.

Liu W.W., 2007. "Polar cap potential saturation: An energy conservation perspective", *J. Geophys. Res.*, **112**, A07210, doi:10.1029/2007JA012392, 1-8.

Lockwood M., M.O. Chandler, J.L. Horwitz, J.H. Waite Jr., T.E. Moore, and C.R. Chappell, 1985. "The cleft ion fountain", *J. Geophys. Res.*, **90**, No. A10, 9736-9748.

Lockwood M., S.W.H. Cowley, and M.P. Freeman, 1990. "The excitation of plasma convection in the high-latitude ionosphere", *J. Geophys. Res.*, **95**, No. A6, 7961-7972.

Lu Q.M., D.Y. Wang, and S. Wang, 2005. "Generation mechanism of electrostatic structures in the Earth's auroral zone", *J. Geophys. Res.*, **110**, No. A3, A03223, doi:10.1029/2004JA010739, 1-8.

Luan X., W. Wang, A. Burns, S. Solomon, Y. Zhang, and L.J. Paxton, 2010. "Seasonal and hemispheric variations of the total auroral precipitation energy flux from TIMED/GUVI", *J. Geophys. Res. Space Physics*, **115**, No. A11, A11304, doi:10.1029/2009JA015063, 1-15.

Luan X., W. Wang, A. Burns, and X. Dou, 2016 "Universal time variations of the auroral hemispheric power and their interhemispheric asymmetry from TIMED/GUVI observations", *J. Geophys. Res. Space Physics*, **121**, 10,258–10,268, doi:10.1002/2016JA022730.

Lugo-Solis A., R. Lopez, J.C. Ingraham, and R. Friedel, 2005. “Energetic electron bursts at high magnetic latitudes: Correlation with magnetospheric activity”, Adv. Space Res., **36**, No. 10, 1840-1844.

Lukianova, R,. K. Mursula, and A. Kozlovsky, 2012. “Response of the polar magnetic field intensity to the exceptionally high solar wind streams in 2003”, *Geophys. Res. Lett.*, **39**, No. 4, L04101, doi:10.1029/2011GL050420, 1-5.

Lyon J.G., J.A. Fedder, and C.M. Mobarry, 2004. “The Lyon–Fedder–Mobarry (LFM) global MHD magnetospheric simulation code”, *J. Atmos. Sol. Terr. Phys.*, **66**, No. 15-16, 13333-13350.

Lyons L.R., 1974. “Electron diffusion driven by magnetospheric electrostatic waves”, *J. Geophys. Res.*, **79**, No. 4, 575-580.

MacDougall J.W. and P.T. Jayanchandran, 2006. “Polar cap voltage saturation”, *J. Geophys. Res.*, **111**, No. A12, A12306, doi:10.1029/2006JA011741, 1-5.

Makhmutov V.S., G.A. Bazilevskaya, Y.I. Stozhkov, A.K. Svirzhevskaya, and N.S. Svirzhevsky, 2011. “Energetic electron precipitation events recorded in the Earth's polar atmosphere”, *32nd Inter. Cosmic Ray Conf.*, **6**, Beijing, China, 17-20.

Marklund G.T., L.G. Blomberg, C.-G. Falthammar, R.E. Erlandson, and T.A. Potemra, 1990. “Signatures of the high-altitude polar cusp and dayside auroral regions as seen by the Viking electric field experiment”, *J. Geophys. Res.*, **95**, No. A5, 5767-5780.

Matsumoto H., H. Kojima, T. Miyatake, Y. Omura, M. Okada, I. Nagano, and M. Tsutsui, 1994. “Electrostatic solitary waves (ESW) in the magnetotail: BEN wave forms observed by GEOTAIL”, *Geophys. Res. Lett.*, **21**, No. 25, 2915-2918.

Matsumoto H., H. Kojima, Y. Omura, and I. Nagano, 1998. “Plasma waves in Geospace: GEOTAIL observations”, in *New Perspectives on the Earth’s Magnetotail*, Geophys. Monogr. Ser., vol. 105, edited by A. Nishida, D. N. Baker, and S.W.H. Cowley, AGU, Washington, D.C., 259-319.

Matsumoto H., L.A. Frank, Y. Omura, H. Kojima, W.R. Paterson, M. Tsuitsui, R.R. Anderson, S. Horiyama, S. Kokubun, and T. Yamamoto, 1999. “Generation mechanism of electrostatic waves based on GEOTAIL plasma wave observations”, *Geophys. Res. Lett.*, **26**, No. 3, 421-424.

Mailyan B., Q. Q. Shi, A. Kullen, R. Maggiolo, Y. Zhang, R.C. Fear, Q.-G. Zong, S.Y. Fu, X.C. Gou, X. Cao, Z.H. Yao, W.J. Sun, Y. Wei, and Z.Y. Pu, 2015. “Transpolar arc observation after solar wind entry into the high-latitude magnetosphere”, *J. Geophys. Res. Space Physics,* **120**, No. 5, 10.1002/2014JA020912, 3525-3534.

Maynard N.C., T.L. Aggson, and J.P. Heppner, 1980. “Magnetospheric observation of large subauroral electric fields”, *Geophys. Res. Lett.*, **7**, No. 11, 881-884.

McGranaghan R., D.J. Knipp, T. Matsuo, and E. Cousins, 2016. “Optimal interpolation analysis of high-latitude ionospheric Hall and Pedersen conductivities: Application to assimilative ionospheric electrodynamics reconstruction”, *J. Geophys. Res. Space Physics*, **121**, 4898–4923, doi:10.1002/2016JA022486.

McPherron R.L., T.-S. Hsu, and X. Chu, 2015. “An optimum solar wind coupling function for the AL index”, *J. Geophys. Res. Space Physics,* **120**, No. 4, doi:10.1002/2014JA020619, 2494-2515.

Merkin V.G. and C.C. Goodrich, 2007. "Does the polar cap area saturate?", *Geophys. Res. Lett.*, **34**, L09107, doi:10.1029/2007GL029357, 1-5.

Michell R.G., M. Samara, G. Grubbs II, K. Ogasawara, G. Miller, J.A. Trevino, J. Webster, and J. Stange, 2016. "APES: Acute Precipitating Electron Spectrometer—A high time resolution monodirectional magnetic deflection electron spectrometer", *J. Geophys. Res. Space Physics*, **121**, 5959–5965, doi:10.1002/2016JA022637.

Milan S.E., S.W.H. Cowley, M. Lester, D.M. Wright, J.A. Slavin, M. Fillingim, C.W. Carlson, and H.J. Singer, 2004. "Response of the magnetotail to changes in open flux content of the magnetosphere", *J. Geophys. Res.*, **109**, No. A4, A04220, doi:10.1029/2003JA010350, 1-16.

Mishin E. and W. Burke, 2005. "Stormtime coupling of the ring current, plasmasphere, and topside ionosphere: Electromagnetic and plasma disturbances", *J. Geophys. Res.*, **110**, No. A7,A07209, doi:10.1029/2005JA011021, 1-13.

Miyake T., Y. Omura, H. Matsumoto, and H. Kojima, 1998. "Twodimensional computer simulations of electrostatic solitary waves observed by Geotail spacecraft", *J. Geophys. Res.*, **103**, No. A6, 11841-11850.

Moore T.E. and D.C. Delcourt, 1995. "The geopause", *Rev. Geophys., (USA)*, **33**, No. 2, 175–209.

Moore T.E., R. Lundin, D. Alcayde, M. André, S.B. Ganguli, M. Temerin, and A. Yau, 1999. "Source processes in the high-latitude ionosphere", in *Magnetospheric Plasma Sources and Losses*, edited by B. Hultqvist, M. Øieroset, G. Paschmann, and R. Treumann, Springer, New York, 7-84.

Motoba T., Y. Ebihara, A. Kadokura, and A.T. Weatherwax, 2014. "Fine-scale transient arcs seen in a shock aurora transient arcs seen in a shock aurora", *J. Geophys. Res. Space Physics*, **119**, No. 8, doi:10.1002/2014JA020229, 6249-6255.

Muldrew D.B., 1972. "Electrostatic resonances associated with the maximum frequencies of cyclotron-harmonic waves", *J. Geophys. Res.*, **77**, No. 10, 1794-1801.

Newell P.T., T. Sotirelis, K. Liou, C.-I. Meng, and F.J. Rich, 2007. "A nearly universal solar wind-magnetosphere coupling function inferred from 10 magnetospheric state variables", *J. Geophys. Res.*, **112**, A01206, doi:10.1029/2006JA012015, 1-16.

Nishikawa K.-I., L.A. Frank, T.E. Eastman, and C.Y. Huang, 1987. "Simulation of electrostatic turbulence in the plasma sheet boundary layer with electron currents and ion beams", in *Magnetotail Physics*, edited by A. Lui, Johns Hopkins Univ. Press, Baltimore, Md., 313-320.

Nishikawa K.-I., L.A. Frank, T.E. Eastman, and C.Y. Huang, 1988. "Simulation of electrostatic turbulence in the plasma sheet boundary layer with electron currents and bean-shaped ion beams", *J. Geophys. Res.*, **93**, No. A6, 5929-5935.

Nishimura Y., L.R. Lyons, Y. Zou, K. Oksavik, J.I. Moen, L.B. Clausen, E.F. Donovan, V. Angelopoulos, K. Shiokawa, J.M. Ruohoniemi, N. Nishitani, K.A. McWilliams, and M. Lester, 2014. "Day-night coupling by a localized flow channel visualized by polar cap patch propagation", *Geophys. Res. Lett.*, **41**, No. 11, doi:10.1002/2014GL060301, 3701-3709.

Nsumei P.A., X. Huang, B.W. Reinisch, P. Song, V.M. Vasyliūnas, J.L. Green, S.F. Fung, R.F. Benson, and D.L. Gallagher, 2003. "Electron density distribution over the northern polar region deduced from IMAGE/radio plasma imager sounding", *J. Geophys. Res.*, **108**, No. A2, 1078, doi:10.1029/2002JA009616, SMP7-1-10.

Ober D.M., G.R. Wilson, N.C. Maynard, W.J. Burke, and K.D. Siebert, 2006. "MHD simulation of the transpolar potential after a solar-wind density pulse", *Geophys. Res. Lett.*, **33**, No. 4, L04106, doi:10.1029/2005GL024655, 1-4.

Ober D.M., G.R. Wilson, W.J. Burke, N.C. Maynard, and K.D. Siebert, 2007. "Magnetohydrodynamic simulations of transient transpolar potential responses to solar wind density changes", *J. Geophys. Res.*, **112**, No. A10, A10212, doi:10.1029/2006 JA012169.

Omidi N. and K. Akimoto, 1988. "Comment on 'Generation of broadband noise in the magnetotail by the beam acoustic instability", *J. Geophys. Res.*, **93**, No. A12, 14725-14728.

Omura Y., H. Matsumoto, T. Miyake, and H. Kojima, 1996. "Electron beam instabilities as generation mechanism of electrostatic solitary waves in the magnetotail", *J. Geophys. Res.*, **101**, No. A2, 2685-2697.

Omura Y., H. Kojima, N. Miki, T. Mukai, H. Matsumoto, and R. Anderson, 1999. "Electrostatic solitary waves carried by diffused electron beams observed by the Geotail spacecraft", *J. Geophys. Res.*, **104**, No. A7, 14627-14637.

Oppenheim M.M., D.L. Newman, and M.V. Goldman, 1999. "Evolution of electron phase-space holes in a 2D magnetized plasma", *Phys. Rev. Lett. (USA).*, **83**, No. 12, 2344-2347.

Oppenheim M.M., G. Veltoulis, D.L. Newman, and M.V. Goldman, 2001. "Evolution of electron phase-space holes in 3D", *Geophys. Res. Lett.*, **28**, No. 9, 1891-1894.

Osherovich V.A., R.F. Benson, J. Fainberg, R.G. Stone, and R.J. MacDowall, 1993. "Sounder stimulated Dn Resonances in Jupiter's Io plasma torus", *J. Geophys. Res.*, **98**, No. E10, 18751-18756.

Osherovich V.A., R.F. Benson, J. Fainberg, and B.W. Reinisch, 2001. "Increased magnetospheric f_{pe}/f_{ce} values in response to magnetic-cloud enhancements of the solar-wind quasi-invariant on March 31 2001", *Eos Trans. AGU*, **82**(47), Fall Meet. Suppl., Abstract SM31B-0758.

Osherovich V.A., R.F. Benson, and J. Fainberg, 2005. "Electromagnetic bounded states and challenges of plasma spectroscopy", *IEEE Trans. Plasma Sci., (USA)*, **33**, No. 2, Part 2, 599-608.

Osherovich V.A., R.F. Benson, J. Fainberg, J.L. Green, L. Garcia, S. Boardsen, N. Tsyganenko, and B.W. Reinisch, 2007. "Enhanced high-altitude polar cap plasma and magnetic field values in response to the interplanetary magnetic cloud that caused the great storm of 31 March 2001: A case study for a new magnetospheric index", *J. Geophys. Res.*, **112**, A06247, doi:10.1029/2006JA012105, 1-12.

Palmroth M., N. Partamies, J. Polvi, T.I. Pulkkinen, D.J. McComas, R.J. Barnes, P. Stauning, C.W. Smith, H.J. Singer, and R. Vainio, 2007. "Solar wind-magnetosphere coupling efficiency for solar wind pressure impulses", *Geophys. Res. Lett.*, **34**, No. 11, L11101, doi:10.1029/2006GL029059, 1-5.

Park J., H. Luehr, G. Kervalishvili, J. Rauberg, C. Stolle, Y.-S. Kwak, and W.K. Lee, 2017. "Morphology of high-latitude plasma density perturbations as deduced from the total electron content measurements onboard the Swarm constellation", *J. Geophys. Res. Space Physics*, **122**, 1338–1359, doi:10.1002/2016JA023086.

Perlongo N.J., A.J. Ridley, M.W. Liemohn, and R.M. Katus, 2017. "The effect of ring current electron scattering rates on magnetosphere-ionosphere coupling", *J. Geophys. Res. Space Physics*, **122**, 4168–4189, doi:10.1002/2016JA023679.

Perreault P. and S.-I. Akasofu, 1978. “A study of geomagnetic storms”, *Geophys. J. R. Astron. Soc. (UK).*, **54**, No. 3, 547-583.

Popova T.A., A.G. Yahnin, T.A. Yahnina, H. Frey, 2010. “Relation between sudden increases in the solar wind dynamic pressure, auroral proton flashes, and geomagnetic pulsations in the Pc1 range”, *Geomagnetism and Aeronomy*, **50**, No. 5, 568-575. Russian Text: *Geomagnetizm i Aeronomiya*, **50**, No. 5, 595-602.

Priyadarshi S., Q.-H. Zhang, Y.-Z. Ma, Y. Wang, and Z.-Y. Xing, 2016. “Observations and modeling of ionospheric scintillations at South Pole during six X-class solar flares in 2013”, *J. Geophys. Res. Space Physics*, **121**, 5737–5751, doi:10.1002/2016JA022833.

Pryse S.E., R.W. Sims, J. Moen, L. Kersley, D. Lorentzen, and W.F. Denig, 2004. “Evidence for solar production as a source of polar cap plasma”, *Ann. Geophys.*, **22**, No. 4, 1093-1102.

Puhl-Quinn P.A., H. Matsui, E. Mishin, C. Mouikis, L. Kistler, Y. Khotyaintsev, P.M.E. Décréau, and E. Lucek, 2007. "Cluster and DMSP observations of SAID electric fields", *J. Geophys. Res.*, **112**, A05219, doi:10.1029/2006JA012065, 1-10.

Raeder J. and G. Lu, 2005. “Polar cap saturation during large geomagnetic storms”, *Adv. Space Res.*, **36**, No. 10, 1804-1808.

Reinisch B.W., D.M. Haines, K. Bibl, G. Cheney, J.A. Galkin, X. Huang, S.H. Myers, G.S. Sales, R. Benson, S. Fung, J.L. Green, S. Boardsen, W.L. Taylor, J.-L. Bougeret, R. Manning, N. Meyer-Vernet, M. Moncuquet, D.L. Carpenter, D.L. Gallagher, and P. Reiff, 2000. “The Radio Plasma Imager investigation on the IMAGE spacecraft”, *Space Sci. Rev.*, **91**, No. 1-2, 319-359.

Reinisch B.W., X. Huang, P. Song, G.S. Sales, S.F. Fung, J.L. Green, D.L. Gallagher, and V.M. Vasyliūnas, 2001. “Plasma density distribution along the magnetospheric field: RPI observations from IMAGE”, *Geophys. Res. Lett.*, **28**, No. 24, 4521-4524.

Reinisch B.W., X. Huang, P. Song, J.L. Green, S.F. Fung, V.M. Vasyliūnas, D.L. Gallagher, and B.R. Sandel, 2004. “Plasmaspheric mass loss and refilling as a result of a magnetic storm”, *J. Geophys. Res.*, **109**, No. A1, A01202, doi:10.1029/2003JA009948, 1-11.

Richmond A.D. and Y. Kamide, 1988. “Mapping electrodynamic features of the high-latitude ionosphere from local observations: Technique”, *J. Geophys. Res.*, **93**, No. A6, 5741-5759.

Ridley A.J., 2005. “A new formulation for the ionospheric cross polar cap potential including saturation effects”, *Ann. Geophys.*, **23**, No. 11, 3533-3547.

Ridley A.J. and E.A. Kihn, 2004. “Polar cap index comparisons with AMIE cross polar cap potential, electric field, and polar cap area”, *Geophys. Res. Lett.*, **31**, No. 7, L07801, doi:10.1029/2003GL019113, 1-5.

Ridley A.J., G. Lu, C.R. Clauer, and V.O. Papitashvili, 1998. “A statistical study of the ionospheric convection response to changing interplanetary magnetic field conditions using the assimilative mapping of ionospheric electrodynamics technique”, *J. Geophys. Res.*, **103,** No. A3, 4023-4039.

Robinson P.A., 1988. “Conditions for the validity of unmagnetized-plasma theory in describing weakly magnetized plasmas”, *Phys. Fluids*, **31**, No. 3, 525-534.

Rodger A.S., M. Pinnock, J.R. Dudeney, K.B. Baker, and R.A. Greenwald, 1994. “A new mechanism for polar patch formation”, *J. Geophys. Res.*, **99**, No. A4, 6425-6436.

Roldugin V.C. and A.V. Roldugin, 2010. “Relation between auroral and geomagnetic pulsations at polar latitudes according to the observations on spitsbergen”, *Geomagnetism and Aeronomy*, **50**, No. 5, 606-615. Russian Text: *Geomagnetizm i Aeronomiya*, **50**, No. 5, 634-643.

Russell C.T., J.G. Luhmann, and G. Lu, 2001. "Nonlinear response of the polar cap ionosphere to large values of the interplanetary electric field", *J. Geophys. Res.*, **106**, No. A9, 18495-18504.

Sauer H.H. and D.C. Wilkinson, 2008. "Global mapping of ionospheric HF/VHF radio wave absorption due to solar energetic protons". *Space Weather*, **6**, S12002. http://dx.doi.org/10.1029/2008SW000399.

Schriver D. and M. Ashour-Abdalla, 1987. "Generation of high frequency broadband electrostatic noise: The role of cold electrons", *J. Geophys. Res.*, **92**, No. A6, 5807-5819.

Schriver D. and M. Ashour-Abdalla, 1989. "Broadband electrostatic noise due to field-aligned currents", *Geophys. Res. Lett.*, **16**, No. 8, 899-902.

Schriver D. and M. Ashour-Abdalla, 1990. "Cold plasma heating in the plasma sheet boundary layer: Theory and simulation", *J. Geophys. Res.*, **95**, No. A4, 3987-4005.

Schunk R.W. and J.J. Sojka, 1997. "Global ionosphere-polar wind system during changing magnetic activity", *J. Geophys. Res.*, **102**, No. A6, 11625-11651.

Sheather S.J. and M.C. Jones, 1991. "A reliable data-based bandwidth selection method for kernel density estimation", *J. Roy. Stat. Soc.*, **B 53**, 683.

Shepherd S.G., 2007. "Polar cap potential saturation: Observations, theory, and modeling", *J. Atmos. Sol. Terr. Phys.*, **69**, No. 3, 234-248.

Shepherd S.G., R.A. Greenwald, and J.M. Ruohoniemi, 2002. "Cross polar cap potentials measured with Super Dual Auroral Radar Network during quasi-steady solar wind and interplanetary magnetic field conditions", *J. Geophys. Res.*, **107**, No. A7, 1094, doi:10.1029/2001JA000152, SMP5-1-11.

Shue J.-H., P.T. Newell, K. Liou, C.-I. Meng, Y. Kamide, and R.P. Lepping, 2002. "Two-component auroras", *Geophys. Res. Lett.*, **29**, No. 10, 1379, doi:10.1029/2002GL014657, 17-1-4.

Shue J.-H., P.T. Newell, K. Liou, C.-I. Meng, M.R. Hairston, and F.J. Rich, 2006. "Ionospheric characteristics of the dusk-side branch of the two-cell aurora", *Ann. Geophys.*, **24**, No. 1, 203-214.

Shugai Yu.S., I.S. Veselovsky, and L.D. Trichtchenko, 2009. "Studying Correlations between the Coronal Hole Area, Solar Wind Velocity, and Local Magnetic Indices in the Canadian Region during the Decline Phase of Cycle 23", *Geomagnetism and Aeronomy*, **49**, No. 4, 415–424.

Singh S.V., R.V. Reddy, and G.S. Lakhina, 2001. "Broadband electrostatic noise due to nonlinear electron-acoustic waves", *Adv. Space Res.*, **28**, No. 11, 1643-1648.

Siscoe G.L., G.M. Erickson, B.U.Ö. Sonnerup, N.C. Maynard, J.A. Schoendorf, K.D. Siebert, D.R. Weimer, W.W. White, and G.R. Wilson, 2002. "Hill model of transpolar potential saturation: Comparisons with MHD simulations", *J. Geophys. Res.*, **107**, No. A6, 1075, doi:10.1029/2001JA000109, SMP8-1-8.

Siscoe G., J. Raeder, and A.J. Ridley, 2004. "Transpolar potential saturation models compared", *J. Geophys. Res.*, **109**, No. A9, A09203, doi:10.1029/2003JA010318, 1-10.

Sivjee G.G., 1999. "Polar thermospheric response to solar magneticcloud/coronal mass ejection interactions with themagnetosphere", *J. Atmos. Sol. Terr. Phys.*, **61**, No. 3-4, 207-215.

Sojka J.J., C. Smithtro, and R.W. Schunk, 2006. "Recent developments in ionosphere-thermosphere modeling with an emphasis on solar-variability", *Adv. Space Res.*, **37**, No. 2, 369-379.

Spann J.F., M. Brittnacher, R. Elsen, G.A. Germany, and G.K. Parks, 1998. "Initial response and complex polar cap structures of the aurora in response to the January 10, 1997 magnetic cloud", *Geophys. Res. Lett.*, **25**, No. 14, 2577-2580.

Spiro R., R. Heelis, and W. Hanson, 1979. "Rapid subauroral ion drifts observed by Atmospheric Explorer C", *Geophys. Res. Lett.*, **6**, No. 8, 657-660.

Stepanova M.V., E.E. Antonova, J.M. Bosqued, and R. Kovrazhkin, 2004. "Radial plasma pressure gradients in the high latitude magnetosphere as sources of instabilities leading to the substorm onset", Adv. Space Res., **33**, No. 5, 761-768.

Stepanova M., E. Antonova, and O. Troshichev, 2005. "PC-index fluctuations and intermittency of the magnetospheric dynamics", Adv. Space Res., **36**, No. 12, 2423-2427.

Stone R.G., B.M. Pedersen, C.C. Harvey, P. Canu, N. Cornilleau-Wehrlin, M.D. Desch, C. de Villedary, J. Fainberg, W.M. Farrell, K. Goetz, R.A. Hess, S. Hoang, M.L. Kaiser, P.J. Kellogg, A. Lecacheux, N. Lin, R.J. MacDowall, R. Manning, C.A. Meetre, N. Meyer-Vernet, M. Moncuquet, V. Osherovich, M.J. Reiner, A. Tekle, J. Thiessen, and P. Zarka, 1992. "Ulysses radio and plasma wave observations in the Jupiter environment", *Science*, **257**, No. 5076, 1524-1531.

Tanaka T., 1979. "The worldwide distribution of positive ionospheric storms", *J. Atmos. Terr. Phys.*, **41**, No. 2, 103-110.

Thomson A.W.P., 1995. "A Statistical Relationship between Coronal Hole Central Meridian Passage and Geomagnetic Activity", *J. Geomagn. Geoelectr.* **47**, No. 12, 1263–1275.

Torr M.R., D.G. Torr, M. Zukic, R.B. Johnson, J. Ajello, P. Banks, K. Clark, K. Cole, C. Keffer, G. Parks, B. Tsurutani, and J. Spann, 1995. "A far ultraviolet imager for the International Solar-Terrestrial Physics mission", *Space Sci. Rev.*, 71, No. 1-4, 329-383.

Trondsen T.S., W. Lyatsky, L.L. Cogger, and J.S. Murphree, 1999. "Interplanetary magnetic field By control of dayside auroras", *J. Atmos. Sol. Terr. Phys.*, **61**, No. 11, 829-840.

Troshichev O., H. Hayakawa, A. Matsuoka, T. Mukai, and K. Tsuruda, 1996. "Cross polar cap diameter and voltage as a function of PC index and interplanetary quantities", *J. Geophys. Res.*, **101**, No. A6, 13429-13435.

Troshichev O., A. Janzhura, and P. Stauning, 2006. "Unified PCN and PCS indices: Method of calculation, physical sense, and dependence on the IMF azimuthal and northward components", *J. Geophys. Res.*, **111**, No. A5, A05208, doi:10.1029/2005JA011402, 1-10.

Tsunoda R.T., R.C. Livingston, J.F. Vickrey, R.A. Heelis, W.B. Hanson, F.J. Rich, and P.F. Bythrow, 1989. "Dayside observations of thermal-ion upwelling at 800-km altitude: An ionospheric signature of the cleft ion fountain", *J. Geophys. Res.*, **94**, No. A11, 15277-15290.

Tsurutani B.T., X.-Y. Zhou, J.K. Arballo, W.D. Gonzalez, G.S. Lakhina, V. Vasyliunas, J.S. Pickett, T. Araki, H. Yang, G. Rostoker, T.J. Hughes, R.P. Lepping, and D. Berdichevsky, 2001. "Auroral zone dayside precipitation during magnetic storm initial phases", *J. Atmos. Sol. Terr. Phys.*, **63**, No. 5, 513-522.

Tsurutani B., A. Mannucci, B. Iijima, Ali Abdu Mangalathayil, J.H.A. Sobral, W. Gonzalez, F. Guarnien, T. Tsuda, A. Saito, K. Yumoto, B. Fejer, T.J. Fuller-Rowell, J. Kozyra, J.C. Foster, A. Coster, and V.M. Vasyliunas, 2004. "Global dayside ionospheric uplift and enhancement associated with interplanetary electric fields", *J. Geophys. Res.*, **109**, No. A8, A08302, doi:10.1029/2003JA010342, 1-16.

Tsurutani B.T. et al., 2006. "Corotating Solar Wind Streams and Recurrent Geomagnetic Activity: A Review", *J. Geophys. Res.* 111, A07S06.

Tu J.-N., J.L. Horwitz, P.A. Nsumei, P. Song, X.-Q. Huang, and B.W. Reinisch, 2004. "Simulation of polar cap field-aligned electron density profiles measured with the IMAGE radio plasma imager", *J. Geophys. Res.*, **109**, No. A7, A07206, doi:10.1029/2003JA010310, 1-9.

Tu J., J.L. Horwitz, and T.E. Moore, 2005a. "Simulating the cleft ion fountain at polar perigee altitudes", *J. Atmos. Sol. Terr. Phys.*, **67**, No. 5, 465-477.

Tu J, P. Song, B.W. Reinisch, X. Huang, J.L. Green, H. U. Frey, and P.H. Reiff, 2005b. "Electron density images of the middle- and high latitude magnetosphere in response to the solar wind", *J. Geophys. Res.*, **110**, No. A12, A12210, doi:10.1029/2005JA011328, 1-12.

Tu J.-N., M. Dhar, P. Song, B.W. Reinisch, J.L. Green, R.F. Benson, and A.J. Coster, 2007. "Extreme polar cap density enhancements along magnetic field lines during an intense geomagnetic storm", *J. Geophys. Res.*, **112**, A05201, doi:10.1029/2006JA012034, 1-15.

Van der Meeren C., K. Oksavik, D.A. Lorentzen, L.J. Paxton, and L.B.N. Clausen, 2016. "Scintillation and irregularities from the nightside part of a Sun-aligned polar cap arc", *J. Geophys. Res. Space Physics*, **121**, 5723–5736, doi:10.1002/2016JA022708.

Vorobjev V.G., 1974. "SC-associated effects in auroras", *Geomagn. Aeron. (USA)*, **14**, No. 1, 72-74.

Vorobjev V.G. and O.I. Yagodkina, 2010. "Seasonal and UT Variations of the Position of the Auroral Precipitation and Polar Cap Boundaries", *Geomagnetism and Aeronomy*, **50**, No. 5, 597-605. Russian Text: *Geomagnetizm i Aeronomiya*, **50**, No. 5, 625-633.

Vorobjev V.G., V.L. Zverev, and O.I. Yagodkina, 2009. "Response of Dayside Auroras to Abrupt Increases in the Solar Wind Dynamic Pressure at Positive and Negative Polarity of the IMF B_z Component", *Geomagnetism and Aeronomy*, **49**, No. 6, 712–721.

Wang Y., Q.-H. Zhang, P.T. Jayachandran, M. Lockwood, S.-R. Zhang, J. Moen, Z.-Y. Xing, Y.-Z. Ma, and M. Lester, 2016a. "A comparison between large-scale irregularities and scintillations in the polar ionosphere", *Geophys. Res. Lett.*, **43**, 4790–4798, doi:10.1002/2016GL069230.

Wang B., Y. Nishimura, Y. Zou, L.R. Lyons, V. Angelopoulos, H. Frey, and S. Mende, 2016b. "Investigation of triggering of poleward moving auroral forms using satellite-imager coordinated observations", *J. Geophys. Res. Space Physics*, **121**, 10,929–10,941, doi:10.1002/2016JA023128.

Warren E.S., 1969. "The topside ionosphere during geomagnetic storms", *Proc. IEEE*, **57**, No. 5, 1029-1036.

Watari S.-I., 1989. "The Latitudinal Distribution of Coronal Holes and Geomagnetic Storms Due to Coronal Holes," in Proceedings of the Solar–Terrestrial Prediction Workshop, Leura, Australia, 1989.

Whiter D.K., B.S. Lanchester, B. Gustavsson, N. I. B. Jallo, O. Jokiaho, N. Ivchenko, and H. Dahlgren, 2014. "Relative brightness of the O+(2D-2P) doublets in low energy aurora", *Astrophys. J.*, 797, No. 1, 64, doi:10.1088/0004-637X/797/1/64, 1-9.

Whitteker J.H., L.H. Brace, E.J. Maier, J.R. Burrows, W.H. Dodson, and J.D. Winningham, 1976. "A snapshot of the polar ionosphere", *Planet. Space Sci.*, **24**, No. 1, 25-28.

Wilder F.D., C.R. Clauer, J.B.H. Baker, E.P. Cousins, and M.R. Hairston, 2011. “The nonlinear response of the polar cap potential under southward IMF: A statistical view”, *J. Geophys. Res. Space Physics*, **116**, No. A12, A12229, doi:10.1029/2011JA016924, 1-13.

Yang Y.F., J.Y. Lu, J.-S. Wang, Z. Peng, and L. Zhou, 2013. “Influence of interplanetary magnetic field and solar wind on auroral brightness in different regions”, *J. Geophys. Res. Space Physics*, **118**, No. 1, doi:10.1029/2012JA017727, 209-217.

Yang S.-G., B.-C. Zhang, H.-X. Fang, Y. Kamide, C.-Y. Li, J.-M. Liu, S.-R. Zhang, R.-Y. Liu, Q.-H. Zhang, and H.-Q. Hu, 2016. “New evidence of dayside plasma transportation over the polar cap to the prevailing dawn sector in the polar upper atmosphere for solar-maximum winter”, *J. Geophys. Res. Space Physics*, **121**, 5626–5638, doi:10.1002/2015JA022171.

Yu Y., V.K. Jordanova, A.J. Ridley, J.M. Albert, R.B. Horne, and C.A. Jeffery, 2016. “A new ionospheric electron precipitation module coupled with RAM-SCB within the geospace general circulation model”, *J. Geophys. Res. Space Physics*, **121**, 8554–8575, doi:10.1002/2016JA022585.

Zaalov N.Y. and E.V. Moskaleva, 2016. “A polar cap absorption model optimization based on the vertical ionograms analysis”, *Advances in Space Research*, **58**, 1763–1777.

Zeng W., J.L. Horwitz, P.D. Craven, F.J. Rich, and T.E. Moore, 2004. “The O^+ density trough at 5000 km altitude in the polar cap”, *J. Geophys. Res.*, **109**, No. A3, A03220, doi:10.1029/2003JA010210, 1-13.

Zhang Y., L.J. Paxton, T.J. Immel, H.U. Frey, and S.B. Mende, 2003. “Sudden solar wind dynamic pressure enhancements and dayside detached auroras: IMAGE and DMSP observations”, *J. Geophys. Res.*, **108**, No. A4, 8001, doi:10.1029/2002JA009355.

Zhang Q.-H. et al., 2016. “Polar cap patch transportation beyond the classic scenario”, *J. Geophys. Res. Space Physics*, **121**, 9063–9074, doi:10.1002/2016JA022443

Zhou X. and B.T. Tsurutani, 1999. "Rapid intensification and propagation of the dayside aurora: Large scale interplanetary pressure pulses (fast shocks)", *Geophys. Res. Lett.*, **26**, No. 8, 1097-1100.

Zhou X. and B.T. Tsurutani, 2001. “Interplanetary shock triggering of nightside geomagnetic activity: Substorms, pseudobreakups, and quiescent events”, *J. Geophys. Res.*, **106**, No. A9, 18957-18968.

Zhou X.-Y., R.J. Strangeway, P.C. Anderson, D.G. Sibeck, B.T. Tsurutani, G. Haerendel, H.U. Frey, and J.K. Arballo, 2003a. "Shock aurora: FAST and DMSP observations", *J. Geophys. Res.*, **108**, No. A4, 8019, doi:10.1029/2002JA009701.

Zhou G.C., C.L. Cai, H. Reme, A. Balogh, J.B. Cao, and D.J. Wang, 2003b. “Quasi-collisionless magnetic reconnection events in the near- Earth magnetotail”, *Chin. J. Space Sci.*, **23**, No. 1, 25-33.

Zhu J., A. Du, J. Ou, and W. Xu, 2017. “Latitudinal and MLT dependence of the seasonal variation of geomagnetic field around auroral zone”, *Advances in Space Research*, **60**, 667–676.

Zou Y., Y. Nishimura, L.R. Lyons, and K. Shiokawa, 2017. “Localized polar cap precipitation in association with nonstorm time airglow patches”, *Geophys. Res. Lett.*, **44**, 609–617, doi:10.1002/2016GL071168.

Chapter 5

Coupling of Interplanetary Shock Waves, Coronal Mass Ejections, Corotating Interaction Regions, and Other Solar Wind Discontinuities with Geomagnetosphere

5.1. Coupling of Interplanetary Shock Waves (ISWs) with Geomagnetosphere

5.1.1. Response of the Dawn-Side, High-Altitude, High-Latitude Magnetosphere to the Arrival of an Interplanetary Shockwave

Wüest et al. (2002) investigate in detail the magnetospheric response observed with instruments on the Polar spacecraft during the arrival of a shock wave on January 10, 1997 at 0105 UT. Magnetic field increases have been observed in the solar wind, the magnetosheath, at the Polar location and on the ground. Particle density increases have been observed in the solar wind, the magnetosheath, but not at the position of the Polar spacecraft. Due to the compression of the magnetosphere Wüest et al. (2002) also would have expected a particle density increase, but in fact, a small particle density decrease was observed with low and medium energy particle instruments. The effect is attributed to an electric field pulse that convects downflowing particles onto another trajectory. Polar was located at that time at 8.4 R_E, 6.1 MLT and 61.1 MLAT. Obtained results are illustrated by Figures 5.1.1 and 5.1.2.

5.1.2. Direct Interplanetary Shock Triggering of Substorms: WIND and POLAR

WIND solar wind data and POLAR UV imaging data are used by Tsurutani and Zhou (2003) to study magnetospheric responses to interplanetary shocks. It is found that the solar wind preconditions determine the different auroral responses. A ~ 1.5 hr "precondition" (upstream of the interplanetary shocks) gives good empirical results. If the upstream IMF B_Z is strongly southward preceding interplanetary shocks, a substorm expansion phase results

(44% of all events). If the IMF B_Z is ~ 0, pseudobreakups occur. If B_Z is northward, there is no obvious resultant nightside geomagnetic activity. Some results are illustrated by Figures 5.1.1 – 5.1.3 and Table 5.1.1.

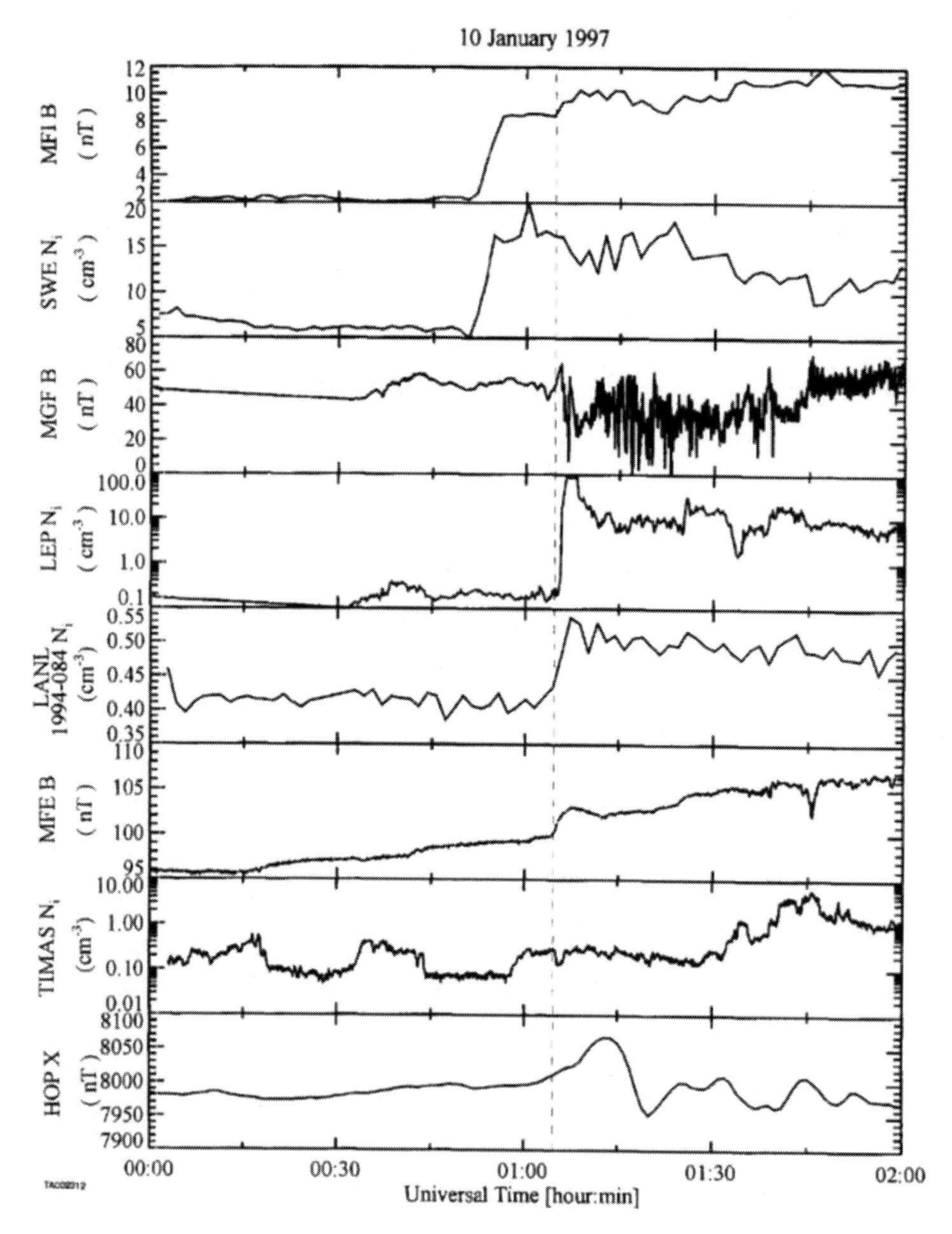

Figure 5.1.1. Magnetic field strength and number densities as function of Universal Time for January 10, 1997. The top panel shows Wind MFI total magnetic field. The second panel shows number density derived from the WIND SWE instrument. The third panel shows Geotail MGF total magnetic field. The fourth panel shows Geotail LEP ion densities. The next panel displays ion density as observed by the LANL 1994-084 geosynchronous satellite. The next two panels show Polar MFE total magnetic field and TIMAS ion densities. The last panel shows the X component of the magnetic field measured at the HOP station of the IMAGE magnetic network. The dashed line is at 01:04:40 UT. According to Wüest et al. (2002). Permission from COSPAR.

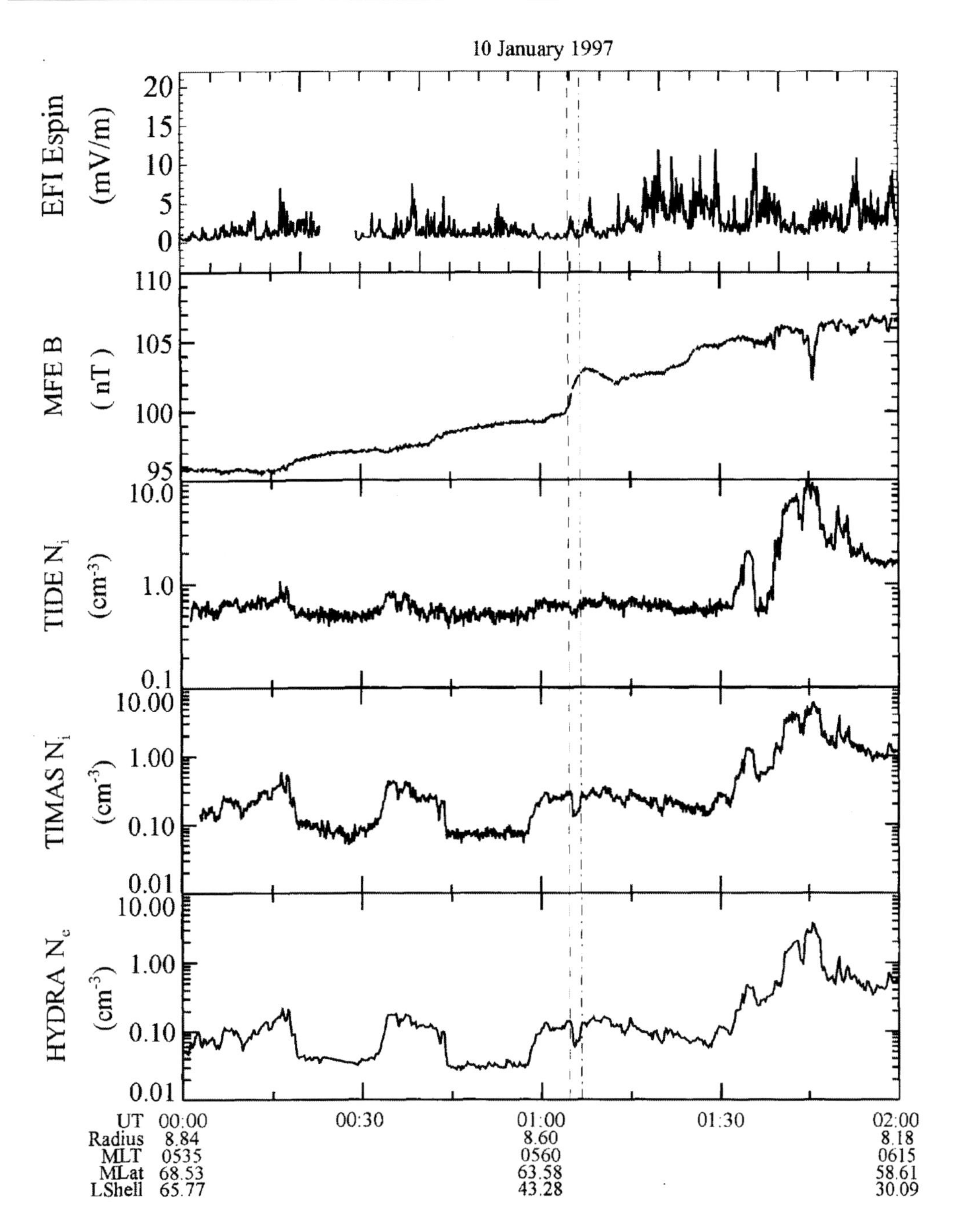

Figure 5.1.2. Electric and magnetic field strength, ion and electron densities for the time period 0000-0200 UT on January 10, 1997. The top panel shows EFI electric field key parameter data in the spacecraft spin plane. The next panel shows MFE total magnetic field strength. The next two panels show TIDE and TIMAS ion densities. The last panel shows HYDRA electron density. The first, dashed vertical line is at 01:04:40 UT. The second, later vertical line is drawn two minutes later. According to Wüest et al. (2002). Permission from COSPAR.

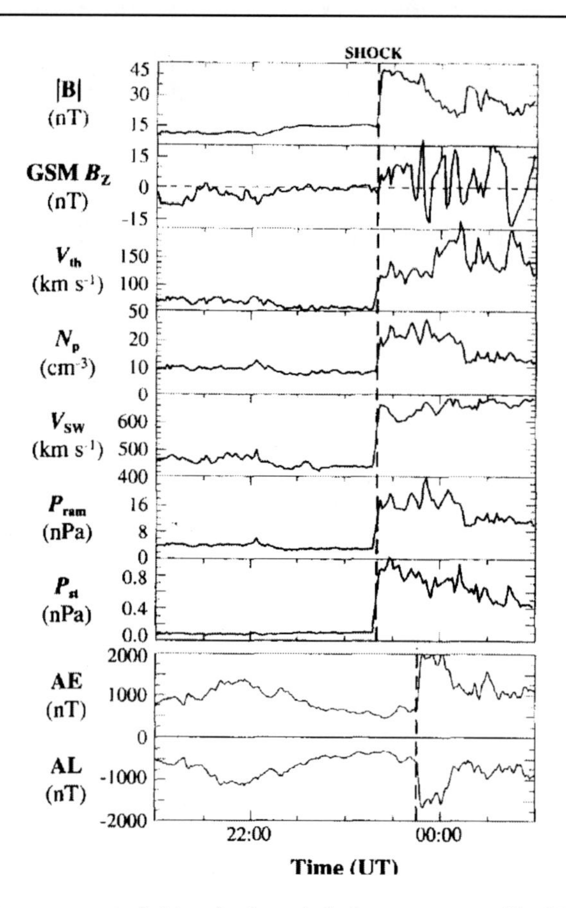

Figure 5.1.3. Interplanetary magnetic field and solar wind plasma measured by WIND at (183, 13, -9) R_E GSM during the September 24, 1998 event. The vertical dashed line in the upper solar wind parameter panels indicates the interplanetary shock at 2320 UT. The vertical dashed line in the bottom AE and AL panels shows the interplanetary shock arrival time at the magnetopause, at ~ 2345 UT. According to Tsurutani and Zhou (2003). Permission from COSPAR.

Table 5.1.1. Preconditions of 18 interplanetary (IP) shock-associated events. According to Tsurutani and Zhou (2003). Permission from COSPAR

Type	Date	IMF |B| (nT)	IMF B_S (nT)	IMF B_N (nT)	IMF B_z (nT)	V_{SW} (km s^{-1})	N_p (cm^{-3})	P_{ram} (nPa)	P_{st} (nPa×10^{-3})	AL_{ps} (nT)	AL_{ss} Peak (nT)
SS	09/18/97	6.7	6.4	0.0	-6.4	331	12.2	3.5	20.3	-302	-491
	10/01/97	3.5	1.5	0.9	-0.05	429	12.6	4.1	11.4	-68	-185
	10/10/97	8.9	1.6	1.8	-0.03	414	11.8	4.5	34.6	-43	-190
	11/22/97	6.4	2.3	1.3	-0.7	349	12.6	3.4	32.7	-37	-363
	04/07/98	7.1	1.2	0.0	-1.2	293	10.4	2.0	23.0	-35	-263
	05/03/98	3.2	1.4	0.0	-1.4	430	4.7	2.0	4.6	-32	-136
	05/29/98	11.3	4.2	1.9	-1.7	518	8.8	5.3	90.4	-165	-955
	09/24/98	12.3	2.7	0.7	-1.8	446	8.6	3.9	94.8	-498	-1670
PB	01/10/97	2.4	0.1	0.4	0.3	375	7.7	2.4	5.5	-5	-24
	08/09/97	3.9	0.5	0.4	0.3	335	15.0	3.8	10.8	0	-35
	10/23/97	5.0	0.3	1.4	1.3	300	7.5	1.5	14.9	-8	-9
	11/01/97	6.0	1.6	1.1	0.1	341	30.7	8.0	21.9	-42	-22
	12/10/97	6.2	1.1	2.4	2.0	286	11.0	2.0	19.6	-5	-4
	09/08/98	9.0	1.5	1.0	0.2	328	4.7	1.1	33.0	-61	-111
	10/02/98	6.6	0.4	1.3	0.6	511	3.7	2.2	25.9	-22	-299
QE	06/13/98	4.6	0.0	1.4	1.4	315	3.4	0.8	9.0	3	-15
	06/25/98	11.1	0.0	10.0	10.0	400	14.1	5.1	51.4	9	-9
	08/10/98	4.9	0.0	0.7	0.7	400	4.8	1.7	11.0	-89	-123

5.1.3. Energetic Particle Injections in the Inner Magnetosphere as a Response to an Interplanetary Shock

As outlined Li et al. (2003), the response of the magnetosphere to interplanetary shocks or pressure pulses can result in sudden injections of energetic particles into the inner magnetosphere. On August 26, 1998, an interplanetary shock caused two injections of energetic particles in close succession: one directly from the dayside and the other indirectly from the nightside associated with a sudden magnetic field enhancement induced by the shock's effect on the magnetotail. The latter injection was different from a typical substorm injection in that the nightside magnetic field at geosynchronous orbit enhanced almost simultaneously over a wide range of local times within after the arrival of the shock. Available observations and simulations of Li et al. (2003) show that like the dayside, the nightside magnetosphere can also inject energetic particles into the inner magnetosphere from a wide local time region in response to a shock impact. The nightside particle injection was due to changes in magnetic and electric fields over a large region of space and thus shows that the magnetic and electric fields in the magnetotail can respond globally to the shock impact.

5.1.4. Some Aspects of the Interaction of Interplanetary Shocks with the Earth's Magnetosphere: An Estimate of the Propagation Time through the Magnetosheath

Villante et al. (2004) analyzed a number of forward shocks detected by Wind in the interplanetary medium to determine the orientation and speed of the shock fronts. Assuming a planar shock geometry and a constant propagation speed (both conformed by a comparison with available Geotail observations), Villante et al. (2004) determined when the shock would hit the bow shock (BS) and evaluated the propagation time of associated disturbances between the BS and ground. It was found delay times were ~ 5 min: for almost radially propagating structures, this delay time would imply magnetosheath speeds of shock associated disturbances of ~ 1/3 – 1/4 of the external shock speed.

5.1.5. Quasi-Parallel Shock Structure and Processes

As outlined in their extended review Burgess et al. (2005), when the IMF is oriented such that the angle between the upstream magnetic field and the nominal bow shock normal is small ($\theta_{Bn} < 45°$), a much more complex shock is observed than in the quasi-perpendicular case. Historically, this has made interpreting single spacecraft data more difficult, so that for a long time the quasi-parallel shock remained relatively poorly understood. The difficulties arise, as we now understand, because the supercritical quasi-parallel shock is a spatially extended and inhomogeneous transition, with smaller length scale features cyclically reforming within it. Under the quasi-parallel magnetic geometry ions are able to escape into the region upstream of the shock, where they give rise to and interact with the waves which populate the foreshock (e.g., Le and Russell, 1992), as discussed in Eastwood et al. (2005). Of particular importance is the association of energetic ions (10–300 keV) with the foreshock and quasi-parallel shock. The role of the quasi-parallel shock as the site of particle acceleration has been fundamental in the development and testing of theories of particle acceleration. The understanding gained has direct implications for other solar system and astrophysical shocks. Before the launch of Cluster a picture of the quasi-parallel shock had been developed from single and dual spacecraft observations, together with results from numerical simulations. Observations of the magnetic field strength changes within the shock showed that they could not all be explained simply by the in-and-out motion of a single shock surface, which would produce nested signatures. A nested signature arises when the boundary crossings observed at one spacecraft are contained, in time, within those observed by a second spacecraft. Instead there were coherent, short scale, magnetic pulsations embedded within the overall transition. These pulsations had sunward directed velocities, but were convected anti-sunwards in the solar wind plasma flow (e.g., Thomsen et al., 1988). A standard model was developed in which the quasi-parallel shock transition was viewed as being composed of a patchwork of magnetic field enhancements, which grew from the interaction between upstream waves propagating sunward in the plasma frame and a gradient in the supra-thermal particle pressure (e.g., Giacalone et al., 1993; Dubouloz and Scholer, 1995). These field enhancements were characterized by a region where the magnetic field magnitude was a factor of 4 or so greater than the background field. They were also somewhat separated from surrounding fluctuations, such that they appeared as discrete structures. These magnetic field enhancements were termed SLAMS (short large amplitude magnetic structures) (Schwartz and Burgess, 1991) and were proposed to be the 'building blocks' of the shock. Intrinsic to this picture were the concepts of a spatially extended and patchy shock transition, since the SLAMS collectively caused the thermalisation of the plasma, and temporal or cyclic evolution: new waves growing and steepened as they were convected back toward the shock, replacing those SLAMS which had passed downstream (Burgess, 1989; Schwartz and Burgess, 1991). These pulsations have formed the major focus for Cluster work related to the quasi-parallel shock. At the time Cluster was launched there were many questions remaining unanswered about the nature of the quasi-parallel shock. Statistical studies of dual spacecraft observations suggested that SLAMS-like pulsations had a shorter correlation length than ULF waves: 1000 km (Greenstadt et al., 1982) compared with 0.5RE (Le and Russell, 1990), but without multi-spacecraft observations it was not possible to determine their overall size or shape perpendicular to the plasma flow direction, over what scale they were coherent, or whether they had internal structure. Their effect on the plasma was not well understood, and although the downstream region showed evidence for variations in the ion reflection properties of the shock (Thomsen et al., 1990), the relative contributions of a spatially extended and temporally varying shock were not well established. Information on SLAMS growth rates was limited by having only two point measurements with an intrinsic temporal/spatial variation ambiguity, and it was not known on what timescales they developed as they approached the shock surface. Simulation results suggested that they had a rapid growth rate, of the order of seconds or less (Giacalone et al., 1994), and that they might be refracted in a direction parallel to the shock as they were convected anti-sunward (Dubouloz and Scholer, 1995). More recently, predictions were also made of the evolution of the shape of the SLAMS structure with time from a ULF wave, to a symmetric magnetic field enhancement and finally to a steepened, asymmetric shape (e.g., Tsubouchi and Lemb`ege, 2004). Simulations also predicted that steepened SLAMS should reflect a portion of the incoming solar wind flow, since their leading edges behave locally as quasi-perpendicular shocks.

5.1.6. Quasi-Perpendicular Shock Structure and Processes

As outlined in their extended review Bale et al. (2005), in the two decades prior to the launch of Cluster, collisionless shocks at which the magnetic field in the unshocked plasma is nearly perpendicular to the shock normal ('quasi-perpendicular shocks') received considerable attention. This is due, in part, to their relatively clean, laminar appearance in the time series data. The tendency of the magnetic field to bind particles together owing to their (perpendicular) gyromotion gives rise to this appearance, which facilitated deeper studies into the collisionless processes responsible for the overall thermalization of the principle plasma populations as well as the acceleration of an energetic non-thermal component. Despite the considerable effort, key questions remained unanswered or remained open to interpretation. Single, and at best dual, spacecraft studies were unable to place quantitative limits on the important spatial scales, nor assess the role of non-stationary aspects in the overall shock transition. By taking advantage of the sharp, quasi-perpendicular shock transitions, Cluster investigations have been able to address the shock orientation and motion via now-standard four spacecraft techniques. As a consequence, Cluster has been able to probe the internal shock scales (and hence physics). Additionally, the multispacecraft strategy has enabled definitive studies of where energetic particles do, and don't, come from. This review summarizes many of these achievements.

5.1.7. Propagation of Interplanetary Shocks into the Earth's Magnetosphere

The paper of Andréeová and Přech (2007) is devoted to the study of propagation of disturbances caused by interplanetary shocks (IPS) through the Earth's magnetosphere. Using simultaneous observations of various fast forward shocks by different satellites in the solar wind, magnetosheath and magnetosphere from 1995 till 2002, Andréeová and Přech (2007) traced the interplanetary shocks into the Earth's magnetosphere and calculated the velocity of their propagation into the Earth's magnetosphere, and analyzed fronts of the disturbances. From the onset of disturbances at different satellites in the magnetosphere Andréeová and Přech (2007) obtained speed values ranging from 500 to 1300 km/s in the direction along the IP shock normal, that is in a general agreement with results of previous numerical MHD simulations. The paper discusses in detail a sequence of two events on November 9th, 2002. For the two cases Andréeová and Přech (2007) estimated the propagation speed of the IP shock caused disturbance between the dayside and nightside magnetosphere to be ~590 km/s and ~714–741 km/s, respectively. This increase partially attributed to higher Alfven speed in the outer magnetosphere due to the compression of the magnetosphere as a consequence of the first event, and partially to the faster and stronger driving interplanetary shock. High-time resolution GOES magnetic field data revealed a complex structure of the compressional wave fronts at the dayside geosynchronous orbit during these events, with initial very steep parts (~10 s). Main results are illustrated with Figures 5.1.4 - 5.1.7.

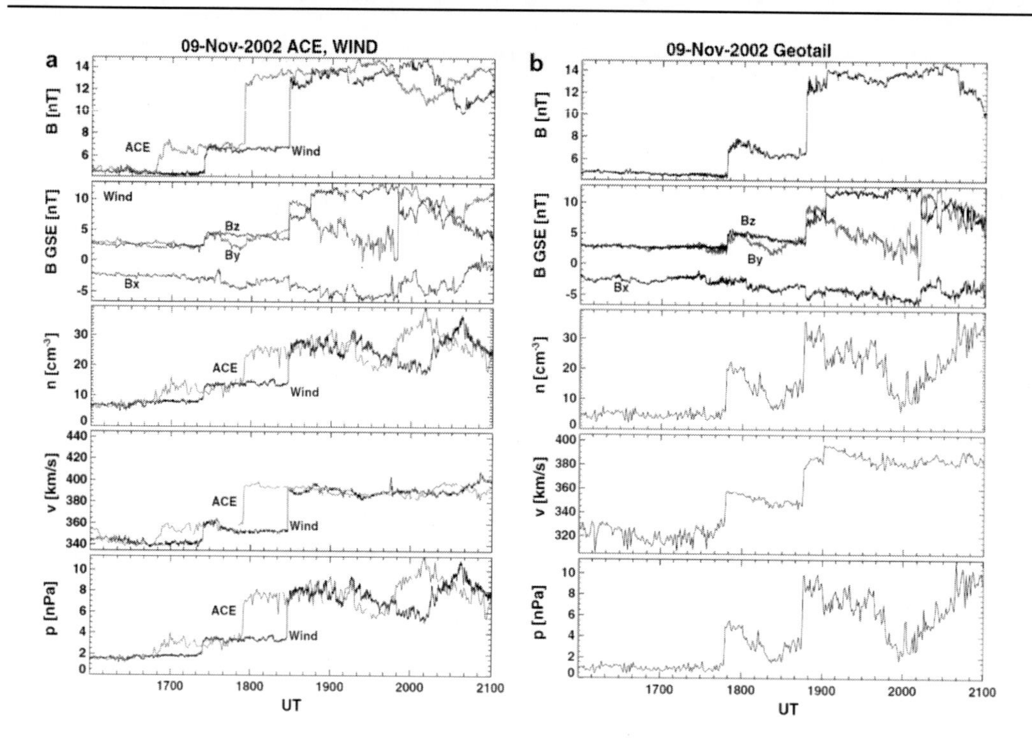

Figure 5.1.4. November 9th, 2002 event. (a) Magnetic field magnitude and components (Wind only), proton density and speed, and the solar wind dynamic pressure from the satellites ACE and Wind located far in the solar wind. (b) Observations of the same solar wind parameters from the Geotail satellite close to the bow shock. The bottom three panels show the CPI instrument data. According to Andréeová and Přech (2007). Permission from COSPAR.

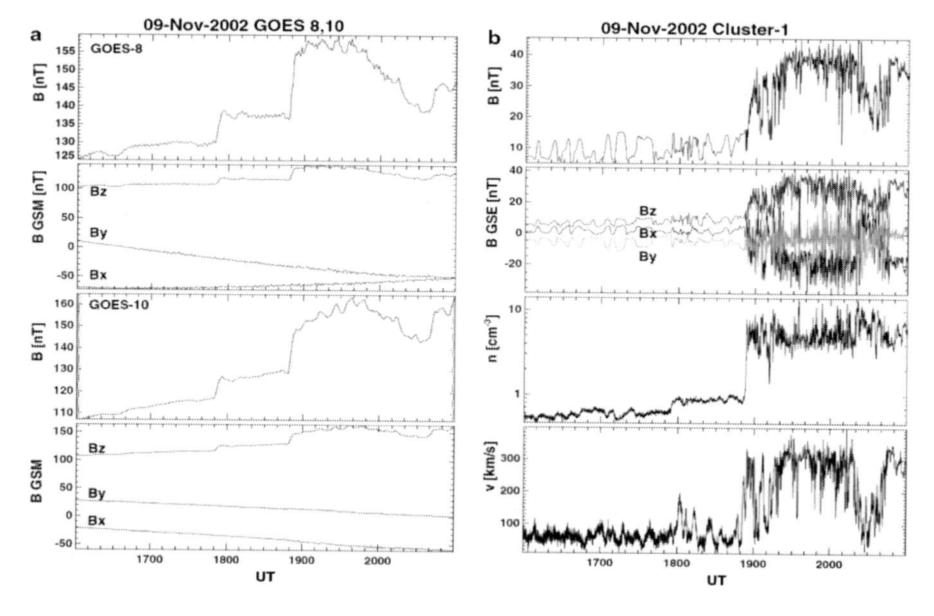

Figure 5.1.5. November 9th, 2002 event. (a) Observations of the magnetic fields from the geosynchronous satellites GOES-8 and GOES-10 in the GSM coordinates. (b) Cluster SC1 primary parameters. The top two panels show the magnetic field, next two panels plot the ion density and speed. According to Andréeová and Přech (2007). Permission from COSPAR.

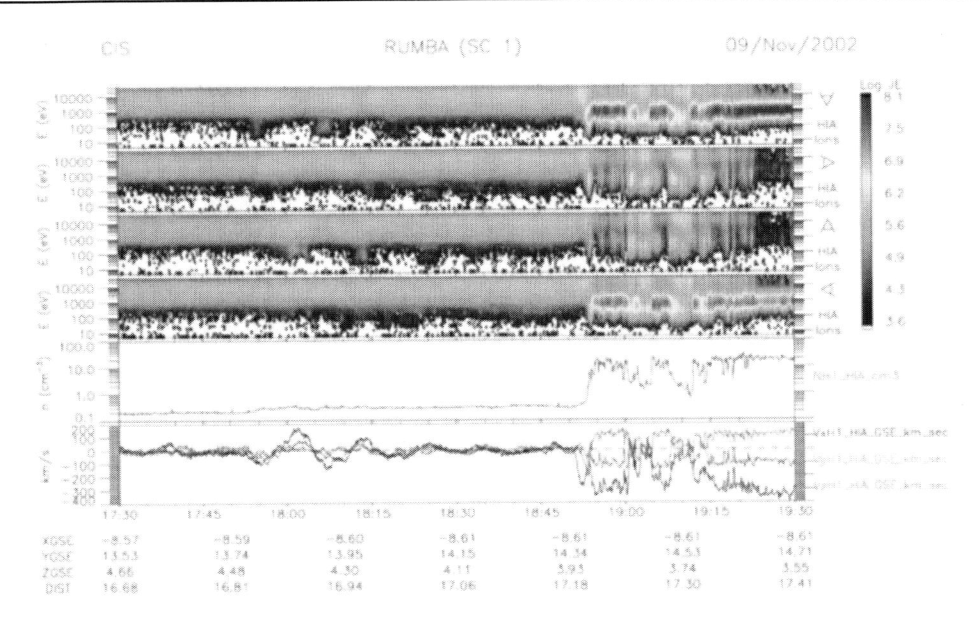

Figure 5.1.6. November 9th, 2002 event: Data plot of the CIS experiment onboard Cluster SC1. The top 4 panels show spectrograms from the HIA sensor, for ions arriving in the 90_ · 180_ sectors with a field-of-view pointing in the Sun, dusk, tail, and dawn directions, respectively (in particle energy flux: keV/ (cm2 s sr keV)). The next two panels show the ion density and ion bulk velocity. According to Andréeová and Přech (2007). Permission from COSPAR.

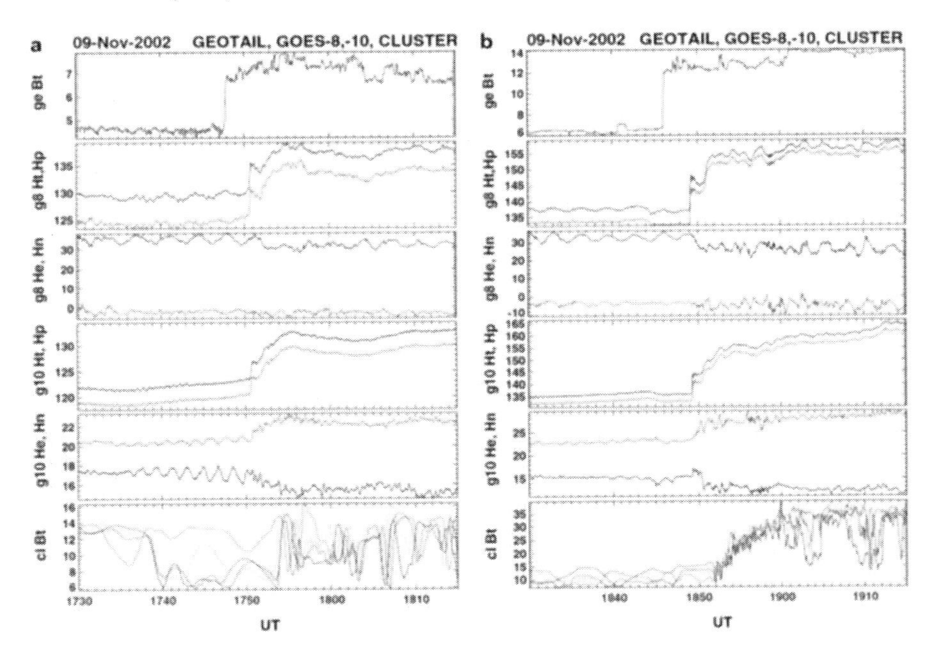

Figure 5.1.7. November 9th, 2002 events in detail. (a) Shock at 17:50 UT, (b) Shock at 18:50 UT. The top panel shows the magnitude of IMF as measured by Geotail. The next two panel pairs depict the GOES-8 and GOES-10 high-resolution magnetic field data in local S/C coordinates (Ht magnitude, Hp northward component (in red), He earthward component, Hn normal to both Hp and He, eastward (in red)). The bottom panel plots the magnitude of magnetic fields measured by the Cluster SC1 (black), SC2 (red), SC3 (green), and SC4 (blue). All panels in nT units. According to Andréeová and Přech (2007). Permission from COSPAR.

5.1.8. Effect of Interplanetary Shocks on the AL and Dst Indices: Influence on the Solar Wind-Magnetosphere-Ionosphere System

As noted Mays et al. (2007), interplanetary coronal mass ejections (ICMEs) are the interplanetary counterparts of coronal mass ejections (CMEs) at the Sun and are observed as enhanced magnetic structures in the solar wind lasting on the order of a day (Zurbuchen and Richardson, 2006). Magnetic clouds (MCs) are a subclass of ICMEs with above-average strength magnetic fields which rotate smoothly through a large angle in a low beta plasma. Earth-directed Halo ICMEs often trigger geomagnetic storms such as the storms of 3–6 October 2000 and 15–24 April 2002. Interplanetary shocks and their resulting geomagnetic activity are usually caused by Halo ICMEs and their associated dynamic interaction regions, also known as 'sheath regions' (Gosling et al., 1990). These sheath regions are accelerated due to the momentum exchange from the fast CME, and they have enhanced densities and temperatures, since they have interacted with the shock. Solar wind velocity and magnetic field strength variation across interplanetary shocks are correlated with the *Dst* index (Echer et al., 2004). Shock effects on the aurora as measured by the FAST and DMSP satellites have been studied by Zhou et al. (2003). It was found that there was a significant increase in electron precipitation the dawnside and duskside auroral oval zone after the shock/pressure pulse arrivals.

In order to understand the effect of interplanetary shock/sheath events on geomagnetic activity, Mays et al. (2007) used the WINDMI model which uses the solar wind driving dynamo voltage $U_{SW}(t)$ derived from Advanced Composition Explorer (ACE) satellite data (Stone et al., 1998) as input and outputs a predicted westward auroral electrojet index (*AL*) and disturbance storm time index (*Dst*). Mays et al. (2007) construct analytic solar wind plasma fields from ACE data for the 3–6 October 2000 event and derive an analytic input driving voltage. There are three basic phenomena that can lead to perturbations in the *AL* and *Dst*: 1) the CME, as defined by its composition or magnetic field configuration, 2) the sheath compressed solar wind, and 3) the shock itself. The role of the shock events are examined in Mays et al. (2007) by removing both the shock and sheath features individually from each analytic plasma field: solar wind density, velocity, and IMF magnitude, then examining the change in the WINDMI output of *AL* and *Dst*.

Mays et al. (2007) used the low dimensional WINDMI physics model of eight coupled ODE's which conserves energy and charge in the solar-wind driven magnetosphere-ionosphere system (Horton and Doxas, 1998). WINDMI outputs a predicted *AL* and *Dst* index and with the solar wind driving and ionospheric damping, even at constant solar wind dynamo voltage there is a rich spectrum of possible magnetosphere-ionosphere states. Measurements of solar wind proton density, solar wind velocity and the IMF in GSM coordinates for the two geomagnetic storm periods are available from the ACE satellite. Mays et al. (2007) used these quantities to derive the input dynamo driving voltage for the WINDMI model. The dynamo driving voltage $U_{SW}(t)$ was calculated from the analytic data using a formula given by Siscoe et al. (2002a) and Ober et al. (2003) for the coupling of the solar wind to the magnetopause using the solar wind dynamic pressure P_{SW} to determine the standoff distance. The formula for U_{SW} is given by

$$U_{SW}\,[\mathrm{kV}] = 30.0\,[\mathrm{kV}] + 57.6E_{SW}\,[\mathrm{mV/m}](P_{SW}\,[\mathrm{nPa}])^{-1/6}, \quad (5.1.1)$$

where

$$E_{SW} = V_{SW}\left(B_y^2 + B_z^2\right)^{1/2} \sin(\theta/2) \tag{5.1.2}$$

is the solar wind electric field with respect to the magnetosphere and the dynamic solar wind pressure

$$P_{SW} = n_{SW} m_p V_{SW}^2 . \tag{5.1.3}$$

Here m_p is the mass of proton and only the proton density contribution has been included in n_{SW}, even though the He can provide important contributions to the dynamic pressure of the plasma. The IMF clock angle θ is given by arctan (B_y/B_z) and u_{SW} is the solar wind flow velocity. The *AL* index is derived from measurements of the horizontal component of the Earth's magnetic field at stations located along the auroral oval in the Northern hemisphere (Rostoker, 1972). The minimum values are taken to be the strongest activity of the westward auroral electrojet which is given by the region 1 field aligned current in the model, that closes in the nightside magnetosphere through the nightside auroral ionosphere. The *Dst* index is obtained from the measurement of the Earth's magnetic field from observatories that are sufficiently distant from the auroral and equatorial electrojets (Sugiura, 1964). The *Dst* index is compared to the output from the WINDMI model through the ring current energy W_{rc} using the Dessler-Parker-Schopke relation (Dessler and Parker, 1959).

According to ACE data during this period was seen three fast forward shock events which signal the arrival at Earth of CMEs from solar eruptions on 15, 17, and 21 April. ACE IMF data and compositional signatures (elevated oxygen charge states O^{7+}/O^{6+} and unusually high Fe charge states) were used to identify the signatures of the ICME in the data. The first shock event (S1) was observed by ACE at 10:20 UT on 17 April moving at the calculated shock speed of 480 km/s and is associated with a halo CME with brightness asymmetry observed by SOHO/LASCO at 03:50 UT on 15 April moving at the plane-of-sky speed of 720 km/s away from the Sun (Manoharan et al., 2004). The CME driving the shock is observed by ACE as a MC beginning at the start of 18 April and continuing until approximately 19:00 UT. The shock and sheath features in the data are taken from 10:20 UT (S1) to 14:50 UT on 17 April. Seven sawtooth oscillations were observed on 18 April from about 02:00 UT to 21:00 UT whose signature can be seen in the *AL* shown in the bottom panels of Figure 1 as the shaded region. The *Dst*, also shown in this figure, reaches a −127 nT during this time. The second shock event (S2) was observed at 08:01 UT on 19 April with a speed of 650 km/s and is associated with a halo CME with outline asymmetry which left the Sun at 08:26 on 17 April moving with the plane-of-sky speed of 1240 km/s (Cane and Richardson, 2003). The shock on April 19 was followed by a more complicated solar wind disturbance observed by ACE from 15:00–20:00 UT 19 April and 10:00 UT 20 April to 12:00 UT 21 April likely resulting from a subsequent CME which dynamically interacts with the perturbation ahead. The interacting signatures looked qualitatively comparable to the well-documented case of October–November 2003 (Zurbuchen et al., 2004), but with clear signatures of solar wind between the two interacting CMEs. The shock/sheath features are taken from 08:01 UT (S2) to 13:00 UT on 19 April. This solar wind disturbance triggered a magnetic storm with *Dst* minima of −126 nT and −124 nT building up in the main phase and −148 nT and −149 nT at storm peak. The third shock event (S3) arrived during the recovery phase at 04:13 UT on 23 April with a speed of 680 km/s and is associated with an X-class flare and partial halo CME with outline asymmetry leaving the Sun at 01:27 UT on 21 April with the plane-of-sky speed of 2393 km/s. The magnetosphere was clipped by the shock/sheath region rather than the ICME, producing a weak magnetic stormwith minimum *Dst* of only −56 nT. Halo CMEs experience maximum projection effects in coronagraph images and therefore the plane-of-sky speeds should be taken as a lower limit of the actual speed.

Mays et al. (2007) compare WINDMI results from runs using both data and analytic input fields from which the shock/sheath feature has been removed from $B_\perp$. The analytic shock field $B_\perp$ without all three $\delta B_\perp$ shock/sheath features, and ACE data for the u_{sw} and n_{sw} parameters were used to derive the input solar wind dynamo voltage. WINDMI–*AL* and *Dst* results for this input are calculated. When the $\delta B_\perp$ shock/sheath feature is removed there is a significant decrease of 50% in the *AL* peaks −1600 nT (17 April 11:00 UT), −1824 nT and −1851 nT (19 April 16:48 UT and 20 April 04:51 UT), and −1297 nT (23 April 07:41 UT) associated with these shocks. Model results for the *Dst* show a –Dst decrease of 10–20% for roughly 12 hours after the first shock (17 April 11:20 UT to 18 April 07:00 UT) and a decrease of 20%–30% after the second shock (19 April 09:00 UT to 20 April 04:00 UT). As noted Mays et al. (2007), the jump $\delta B_\perp$ has the most significant impact on the *AL* and *Dst* compared to the other parameters. Removing the shock/sheath features from u_{SW} produces a slight decrease of 15%, 25% and 10% in the first, second, and third *AL* peaks respectively. There is only a slight increase of 10% and 5% of the first and third *AL* peaks when the shock/sheath features are removed from n_{SW}. The compressional jump δn_{SW} is only ~2 cm^{-3} for the second shock event. When the shock/sheath features are removed from all of the plasma fields ($B_\perp^{IMF}$, V_{SW}, n_{SW}) the –*AL* peaks decrease by a similar amount as when the $\delta B_\perp$ features are removed only. The jump $\delta B_\perp$ has the most impact on producing the three –*AL* peaks during this storm. The second shock/sheath combination on 19 April at 08:01 UT which produced *AL* peaks of −1824 nT and −1851 nT is the most effective of three shocks.

According to Mays et al. (2007), an unusual feature of the 3–6 October 2000 solar wind driver was the appearance of a fast forward shock advancing into a preceding magnetic cloud (Wang et al., 2003b). ACE data shows a magnetic cloud from 3 October at 10:18 UT through 5 October at 05:34 UT lasting about 42 hours. The signature of the magnetic cloud can be seen from the sinusoid-like waveforms of IMF B_y and B_z as the IMF clock angle changes linearly through an angle of 180° during this period. The fast forward shock occurs at 02:40 UT on 5 October with a calculated shock speed of 534 km/s and compression ratio of 2.3. There are jumps in the velocity from 364 km/s to 460 km/s, in the proton density from 7 cm^{-3} to 16 cm^{-3}, and in perpendicular magnetic field from 7 nT to 16 nT across the shock front. The *AL* data shows a first large spike with a peak of −1938 nT occurring at 06:51 UT on 5 October 2000. A second, larger spike of approximately –2790 nT in the *AL* index occurs at 1210 UT on 5 October 2000 initiated by a strong southward IMF excursion detected at ACE about an hour earlier. Periodic substorms occur in the interval of 06:00–12:00 UT 4 October and have been identified as sawtooth oscillations by Huang et al. (2003) and Reeves et al. (2003). The *Dst* minimum of −180 nT is reached on 5 October slightly after the strong southward IMF surge. Consistent with April 2002 analysis, when the shock/sheath feature is removed from $B_\perp$ the first *AL* peak of −1938 nT occurring at 07:20 UT 5 October 2000 decreased by ~50%. There is also a decrease of $-Dst$ by ~25% after the shock arrival time. The AL peak only decreases by 10% when the shock/sheath is removed from V_{SW} and the removal of the feature from n_{SW} produces an increase of 10% in the AL peak. Again, when the shock is dropped from all three plasma fields the result is similar to removing the $\delta B_{\perp}$ shock only. These results demonstrate that the first large AL peak was triggered by the shock/sheath front, and most strongly by the $\delta B_{\perp}$ jump.

Mays et al. (2007) summarize obtained results as following: **(i)** The question of how much interplanetary shock/sheath events contribute to the geoeffectiveness of solar wind drivers was examined based on a series of numerical experiments with WINDMI using observed solar wind drivers for the 15–24 April 2002 and 3–6 October 2000 events, each of which had interesting shock features. **(ii)** In these experiments, analytic fits to solar wind input parameters ($B_{\perp}$, v_{sw}, and n_{sw}) allowed shock/sheath features to be easily removed while leaving other features of the solar wind driver undisturbed. **(iii)** Percent changes in WINDMI-derived AL and Dst indices between runs with and without the observed shock/sheath feature were taken as a measure of its relative contribution to the geoeffectiveness. **(iv)** The interplanetary shock/sheath events during these storm periods are strongly related to geomagnetic activity predicted by the WINDMI model. **(v)** The $\delta B_{\perp}$ jumps at the shocks/sheath have a strong impact on the three AL peaks during the April 2002 storm. **(vi)** During the October 2000 storm the first large AL spike was triggered by the shock/sheath feature in $B_{\perp}$. **(vii)** The Siscoe et al. (2002a) solar wind dynamo voltage includes contributions from the number density, clock angle, and B_y, but which are not included in dynamo voltage more typically used. This is particularly important for the April 2002 shocks in which, for example, the second shock had a $\delta B_z < 1$ nT while $\delta B_y \sim 10$ nT therefore producing dynamo voltage $U_{SW} = 600$ kV while the rectified voltage is only 200 kV. **(viii)** The solar wind-magnetosphere coupling dynamics is most sensitive to variations in the solar wind velocity and IMF. This can be seen from the equation for the input Siscoe et al. (2002a) solar wind dynamo voltage where the input

$$U_{SW} \propto V_{SW}^{2/3} n_{SW}^{-1/6} B_{\perp}^{1/2}, \tag{5.1.4}$$

so it is expected that the removal of the shock compressional feature in the velocity and magnetic field parameters to decrease the driving voltage U_{SW}, and in the number density to increase U_{SW}. **(ix)** During these storms the magnetic field components have a 1.5–3 times increase across the shock front while the velocity does not increase by more than 1.5 times. **(x)** The jump in the number density can be as high as 4 times the upstream value, however, the $n_{SW}^{-1/6}$ dependence in the calculated U_{SW} hides this effect according to Eq. 5.1.4. **(xi)** Also shock features in the velocity and number density increase the solar wind dynamic pressure which causes the magnetopause to move close to the Earth and produces stronger coupling.

5.1.9. MHD Simulation for the Interaction of an Interplanetary Shock with the Earth's Magnetosphere

According to Samsonov et al. (2007), the global BATS-R-US MHD code is used to simulate the interaction of a moderately strong interplanetary shock with the Earth's magnetosphere. The model predicts the propagation of a transmitted fast shock through the magnetosheath and magnetosphere and the reflection of this shock from the inner numerical boundary. The reflected fast shock propagates sunward through the dayside magnetosphere and magnetosheath. The passage of the transmitted shock causes the bow shock and magnetopause to move inward, while the passage of the reflected fast shock causes these boundaries to move outward, consistent with previously reported in situ observations. Samsonov et al. (2007) noted that solar wind conditions play a dominant role in determining the level of geomagnetic activity. High-speed solar wind streams and coronal mass ejections launch interplanetary shocks which propagate antisunward through the solar wind with supersonic velocities. When the interplanetary shocks reach the magnetosphere, they initiate global magnetospheric disturbances. Geomagnetic sudden impulse or sudden commencement signatures are very clear global phenomena caused by interplanetary shocks (e.g., Nishida, M1978; Araki, 1994). Interplanetary shocks are usually fast forward MHD shocks. The interaction of an interplanetary shock with the magnetosphere follows several particular phases, including the interaction of the interplanetary shock with the bow shock, the interaction of the interplanetary shock with the magnetopause, the transmission of the interplanetary shock into the magnetosphere as a fast mode wave, modifications of the field-aligned and ionospheric current systems, and magnetic disturbances observed on the ground. As outlined Samsonov et al. (2007), the interaction of interplanetary shocks with the Earth's bow shock has long been a topic for theoretical, laboratory, and observational study (Ivanov, 1964; Dryer et al., 1967; Shen and Dryer, 1972; Dryer, 1973; Grib et al., 1979; Zhuang et al., 1981; Grib, 1982; Pushkar et al., 1991; Grib and Pushkar, 2006). More recently, one- and three-dimensional numerical MHD simulations for the interplanetary-bow shock interaction have been employed (Yan and Lee, 1996; Samsonov et al., 2006). The interaction generates a fast shock (FS) into the magnetosheath which propagates at a speed lower than that of the solar wind interplanetary shock (Samsonov et al., 2006), in agreement with predictions from the Rankine-Hugoniot conditions and observations (Koval et al., 2005, 2006a,b). The bow shock, a reverse fast shock, moves earthward with a velocity ~100 km/s. Three new discontinuities appear downstream from the bow shock in addition to the FS: a forward slow expansion wave (SEW), a contact discontinuity (CD), and a reverse slow shock (SS). Because the propagation velocities of these three discontinuities are very similar, they cannot be distinguished in the results of 3-D simulations. Instead, they take the form of a single discontinuity where the magnetic field strength and density increase, the temperature decreases, and the velocity remains unchanged (Samsonov et al., 2006).

Samsonov et al. (2007) outlined that Rankine-Hugoniot conditions can also be used to study the interaction of an interplanetary shock with a tangential discontinuity (TD) magnetopause. The interaction for typical shock conditions sets the magnetopause moving inward at speeds greater than 200 km/s and launches a transmitted fast shock into the magnetosphere that propagates with a velocity on the order of 1500 km/s. Grib (1972, 1973) predicted that the interaction would also launch an outward-propagating fast expansion wave (FEW) into the magnetosheath. Ridley et al. (2006) presented the results of a global simulation employing the BATS-R-US MHD code for the interaction of an extremely strong interplanetary shock with the magnetosphere. The transmitted interplanetary shock propagates through the magnetosheath and magnetosphere, then reflects from the inner numerical boundary and moves outward. Ridley et al. (2006) assumed that the reflection boundary corresponded to the plasmapause. The transmitted interplanetary shock propagated around the flanks of the magnetosphere toward the magnetotail. The dayside bow shock moved inward until it interacted with the reflected shock. Guo et al. (2005) used a PPMLR-MHD code to study the interaction of interplanetary shocks with the magnetosphere. They considered two cases: one when the shock normal lies along the Sun-Earth line, the other when the angle between the shock normal and the Sun-Earth line is 60°. Despite differing transitions, the system evolved to nearly the same final quasi-steady state configuration in both cases. Lee and Hudson (2001) used a 3-D dipole model for the magnetosphere to study a similar problem, namely the propagation of a sudden impulse associated with the interplanetary shock inside the magnetosphere. To simulate the response, they invoked an abrupt variation in the electric field at the outer boundary of their model, the magnetopause. Their results indicate that most of the impulse energy penetrates the plasmapause to excite strong low-frequency pulsations in the plasmasphere, but that a small portion of the initial impulse is reflected from the plasmapause and returns to the outer boundary. As noted Samsonov et al. (2007), despite these studies, many aspects of the interaction of solar wind shocks with the magnetosphere remain unclear. They used existing global MHD models to predict the motion of the bow shock (Šafránková et al., 2007) and corresponding signatures at geosynchronous orbit (Andréeová and Přech, 2007). However, there are some questions that contemporary global MHD models cannot address. Only a model employing a realistic plasmasphere can be used to estimate the energy fluxes transmitted into the plasmasphere and reflected back into the outer magnetosphere when an interplanetary shock strikes.

The work of Samsonov et al. (2007) employs the global BATS-R-US code to simulate the solar wind-magnetosphere interaction (Powell et al., 1999). The BATS-R-US code solves the MHD equations with a finite volume discretization in a 3-D block-adaptive Cartesian grid using conservative variables. As usual for global codes, the supersonic solar wind conditions are imposed on a plane perpendicular to the Sun-Earth line (X axis) upstream from the bow shock. There is an outflow boundary far downstream in the magnetotail. The inner numerical boundary is located about 3 r_E from the Earth. The boundary conditions allow no mass flux through this inner boundary and reflective boundary conditions are used for the mass density and the thermal pressure (Song et al., 1999). The magnetic field near the inner boundary is determined primarily by the imposed terrestrial magnetic field (Gombosi et al., 2003). As a basic approach, Samsonov et al. (2007) use a version of the BATS-RUS code that is coupled to the Rice Convection Model (RCM) (De Zeeuw et al., 2004). The latter model includes a representation of the inner magnetosphere and its connection to the ionosphere. Samsonov et al. (2007) compared results from the BATSR-US code with and without the RCM coupling, but found no significant differences. The resolution of the numerical grid has been enhanced in the dayside region near the Sun-Earth line, with the smallest computational cell $0.125\times0.125\times0.125\ R_E^3$. The region with the finest resolution is a square with dimensions of $15\times15\ R_E^2$ in the Y-Z plane situated between the terminator plane (X = 0) and the solar wind inflow boundary (X = 20 R_E). The grid spacing gradually increases with increasing |Y| and |Z|. Samsonov et al. (2007) summarize main obtained results as following: **(i)** There are used MHD simulations to study the interaction of interplanetary shocks with the Earth's magnetosphere. Like Samsonov et al. (2006), there are considered a moderately strong interplanetary shock with a density compression ratio ρ_2/ρ_1 = 2.84 propagating strictly antisunward in a solar wind flowing radially outward from the Sun. **(ii)** The initial interaction of the interplanetary shock with the Earth's bow shock sets the bow shock moving inward and launches a FS into the magnetosheath and a set of SEW CD-SS discontinuities. **(iii)** Because the velocity of the FS in the magnetosheath is less than that in the upstream solar wind, the originally planar shock front becomes distorted. The SEW-CD-SS discontinuities essentially move Earthward with the magnetosheath flow velocity. **(iv)** Once the transmitted FS interacts with the MP, a forward FS appears in the magnetosphere and a reflected FEW appears in the magnetosheath. In addition, the MP begins to move Earthward. **(v)** The inner numerical boundary of the global magnetospheric BATS-R-US code is a sphere with a radius of about 3 R_E. The FS reflects from this boundary, and a RFS propagates outward through the dayside magnetosphere and magnetosheath. The reflected shocks terminate the inward motion of the MP and BS. Because the global magnetospheric code does not incorporate a self-consistent plasmasphere, the results obtained from the simulation may not describe the real situation well. **(vi)** Therefore there are presented a supplementary 1-D MHD model for conditions along the Sun-Earth line to simulate the interaction of the FS with a more realistic plasmapause. It is find that most of the FS energy penetrates into the plasmasphere and that only a relatively small fraction (~30% of the incoming FS energy) is reflected. This is consistent with the FS reaching the ionosphere and producing the well-known sudden impulse in ground observations. **(vii)** However, in situ observations provide increasing evidence for a strong reflected fast mode wave (RFS). Following the arrival of interplanetary shocks, the bow shock moves inward and then outward, and there is a two step response in the geosynchronous magnetic field strength accompanied by a dawnward electric field pulse. The calculations and timing considerations suggest that the FS reflects from the dayside ionosphere and that the RFS propagates through the magnetosphere and magnetosheath to reach the bow shock, in qualitative agreement with predictions of the global MHD code. **(viii)** A joint analysis of high-resolution data at the subsolar bow shock and magnetopause, and in the subsolar magnetosphere (particularly flow velocities and electric fields) in conjunction with low-latitude ground observations of sudden impulse signatures would help to confirm or disprove described suggestions.

5.1.10. Case Study of Nightside Magnetospheric Magnetic Field Response to Interplanetary Shocks

As outlined Wang et al. (2010), observations show that the geosynchronous magnetic field in midnight sector sometimes decreases when an interplanetary fast forward shock (FFS) passes Earth, even though the magnetosphere is always compressed. Wang et al. (2010) perform case studies of the response observed by the GOES spacecraft at geosynchronous orbit near midnight to two interplanetary shocks passing Earth. One shock produces a decrease in B_Z (a negative response) and the other an increase in B_Z (a positive response). A global 3D MHD code is run to reproduce the responses at geosynchronous orbit, and to further provide information on the initiation and development of B_Z variations in the entire magnetosphere. The model reveals that when a FFS sweeps over the magnetosphere, there exist mainly two regions, a positive response region caused by the compressive effect of the shock and a negative response region which is probably associated with the temporary enhancement of earthward convection in the nightside magnetosphere. The spacecraft may observe an increase or decrease of the magnetic field depending on which region it is in. The numerical results reproduce the main characters of the geosynchronous magnetic field response to interplanetary shocks for these two typical cases.

5.1.11. Displacement of Large-Scale Open Solar Magnetic Fields from the Zone of Active Longitudes and the Heliospheric Storm of November 3–10, 2004: 1. The Field Dynamics and Solar Activity

The dynamics of the large-scale open field and solar activity at the second stage of the MHD process, including the origination and disappearance of the four-sector structure during the decline phase of cycle 23 (the stage when the blocking field is displaced from the main zone of active longitudes), has been considered by Ivanov (2010). Extremely fast changes in the scales of one of new sectors (from an extremely small sector ('singularity') to a usual sector that originated after the uniform expansion ('explosion') of singularity with a 'kick' into the zone of active longitudes, westward motion of the MHD disturbance front in the direction of solar rotation, and formation of an active quasi-rigidly corotating sector boundary responsible for the heliospheric storm of November 2004 have been detected in the field dynamics. It has been indicated that a very powerful group of sunspots AR 10656 (which disappeared after the explosion) with an area of up to 1540 ppmh (part per million hemisphere), a considerable deficit of the external energy release, and zero geoeffectiveness in spite of the closeness to the Earth helioprojection existed within singularity. It has been assumed that the energy escaped from this group with effort owing to the interaction between coronal ejections and narrow sector walls (singularity), and a considerable part of the energy was released in the outer layers of the convective zone, as a result of which singularity exploded and this explosion was accompanied by the above effects in the large-scale field and solar activity.

5.1.12. Geomagnetic Activity Triggered by Interplanetary Shocks

As noted Yue et al. (2010), interplanetary shocks can greatly disturb the Earth's magnetosphere, causing the global dynamic changes in the electromagnetic fields and the plasma. In order to investigate this, Yue et al. (2010) have systematically analyzed 106 interplanetary shock events based on OMNI data, GOES, and Los Alamos National Laboratory satellite observations during 1997–2007. It is revealed that the median value of IMF B_Z keeps negative/positive prior to shock arrival and becomes more negative/positive following the shock arrival. The statistical analysis shows that interplanetary shocks with southward IMF (46%) are likely to increase AE (AL, AU) and PC indices significantly. The amplitude of AE index increases from 200 to 600 nT, AU from 100 to 200 nT, AL from 50 to 400 nT, and PC from 1.5 to 3 approximately in 10 min, which could be a signature of geomagnetic activity/substorms onset (or substorm further intensification). Meanwhile, there is a strong injection of energetic electrons in the dawn region following the shock arrival and a strong depletion in the dusk region 30 min later, showing a clear dawn-dusk asymmetry. On the other hand, there is only the typical shock compression effect for interplanetary shocks with northward IMF (54%). The median value of AE index increased from 80 to 150 nT, AU from 50 to 90 nT, AL index decreased from −30 to −40 nT, and PC index increased from 0.6 to 1.2 in ~10 min following the shock arrival. Both individual cases and statistical studies indicate that the magnetosphere-ionosphere system must be preconditioned for a substorm-like geomagnetic activity to be triggered by an interplanetary shock with southward IMF impact, whereas interplanetary shock with northward IMF precondition shows only compression effect.

5.1.13. Inner Magnetosphere Plasma Characteristics in Response to Interplanetary Shock Impacts

In order to characterize plasma (0.03–45 keV) properties in response to interplanetary shock impact at the geosynchronous orbit, Yue et al. (2011) have examined 95 shock events from 1997 to 2004. These shock events have been categorized into two groups: shock fronts associated with southward IMF (47 cases) and with northward IMF (48 cases). Obtained results show that under southward IMF, the plasma becomes denser and hotter following the interplanetary shock arrival. The proton (0.1–45 keV) and electron (0.03–45 keV) number densities have peaks of 1.8 and of 2.5 cm^{-3} at the duskside, respectively (the typical tail plasma sheet density is about 0.7 cm^{-3}). After the interplanetary shock impact, the plasma (proton and electron) temperature anisotropy increases remarkably at the noon sector, decreases toward dawn and dusk, and minimizes at midnight, suggesting that both electromagnetic ion cyclotron and whistler waves can be stimulated mainly at the dayside magnetosphere. In addition, there are more oxygen ions injecting into the inner magnetosphere, and the density of ionospheric oxygen ions is comparable to proton density. However, for interplanetary shocks associated with northward IMF, the plasma density and temperature increases are insignificant, while slight enhancements of the plasma temperature anisotropy are distributed globally.

5.1.14. MHD Analysis of Propagation of an Interplanetary Shock across Magnetospheric Boundaries

As outlined Němeček et al. (2011), an important problem of the Space Weather Program is the interaction of interplanetary shocks with Earth's magnetosphere because their interaction often (but not always) leads to major geomagnetic storms. Since the huge interaction region can be covered by simultaneous spacecraft observations only sporadically, global MHD modeling can help in our understanding of the interaction process. Němeček et al. (2011) have developed a procedure that clearly distinguished magnetospheric boundaries in an output of the global MHD model and compare its results to spacecraft observations. Using one interplanetary shock observed by Geotail, it was compare its passage through the magnetosphere with predictions of two modifications of the global BATS-R-US MHD code. Němeček et al. (2011) demonstrate the complexity of the shock interaction with the bow shock, magnetopause, and oscillations of the whole system toward a new equilibrium state with a duration of 10-12 min. Moreover, based on the changes of the magnetopause and bow shock locations in the nightside region, Němeček et al. (2011) suggest that the information about the interplanetary shock hitting the subsolar magnetopause reaches the nightside magnetopause earlier than the interplanetary shock can arrive. This information is, probably, mediated by fast magnetosonic waves in the inner magnetosphere. Although it was found generally a good agreement, Němeček et al. (2011) discuss possible sources of deviations of modeled and observed locations of the boundaries.

5.1.15. Nightside Geosynchronous Magnetic Field Response to Interplanetary Shocks: Model Results

Inspired by the fact that spacecraft at geosynchronous orbit may observe an increase or decrease in the magnetic field in the midnight sector caused by interplanetary fast forward shocks (FFS), Sun et al. (2011) perform global MHD simulations of the nightside magnetospheric magnetic field response to interplanetary shocks. The model reveals that when a FFS sweeps over the magnetosphere, there exist mainly two regions: a positive response region caused by the compressive effect of the shock and a negative response region which is probably associated with the temporary enhancement of earthward convection in the nightside magnetosphere. Interplanetary shocks with larger upstream dynamic pressures have a higher probability of producing a decrease in B_Z that can be observed in the midnight sector at geosynchronous orbit, and other solar wind parameters such as the interplanetary magnetic field (IMF) B_Z and interplanetary shock speed do not seem to increase this probability. Nevertheless, the southward IMF B_Z leads to a stronger and larger negative response region, and a higher interplanetary shock speed results in stronger negative and positive response regions. Finally, a statistical survey of nightside geosynchronous B_Z response to interplanetary shocks between 1998 and 2005 is conducted to examine these model predictions.

5.1.16. Propagation of Inclined Interplanetary Shock through the Magnetosheath

As noted Samsonov (2011), normal of most interplanetary shocks are nearly aligned with the Sun-Earth line. But some shocks, especially those connected with corotating interaction regions, are sufficiently diverted from the typical orientation near 1 AU. It was obtain that shocks with normal lying in the XY plane and inclined at an angle about 40° or more from the Sun-Earth line can result in sudden impulse variations of different magnitudes in the dawn and dusk magnetosphere. Using the Rankine-Hugoniot equations, Samsonov (2011) calculate the downstream velocity in dependence on the interplanetary shock orientation. It was found for given upstream parameters that the downstream velocity $V_Y \cong 0.2V_X$, when $n_y \cong n_x$ and the upstream velocity is directed exactly along the Sun-Earth line. For more inclined shocks, the ratio V_Y/V_X may exceed 30 percent. Numerical three-dimensional (3-D) MHD simulations predict a set of MHD discontinuities propagating through the magnetosheath after interaction between an inclined shock and the bow shock. It is shown a clear difference between variations in the dusk magnetosheath downstream of the quasi-perpendicular bow shock (the region passed first by the inclined interplanetary shock) and in the dawn magnetosheath downstream of the quasi-parallel bow shock. In the dusk flank, the predicted variations are mainly similar to those obtained previously for a radially propagating shock at the Sun-Earth line. In the dawn flank, the forward fast shock with a small variation of the magnetic field magnitude is followed by another compound discontinuity bringing an increase of the density and magnetic field, but a decrease of the velocity and temperature. Samsonov (2011) suppose that this discontinuity consists of several basic MHD discontinuities moving with close velocities, therefore its composition cannot be determined exactly in 3-D simulations. Using an estimation of the Alfvén velocity in the magnetosphere, we find the transit time of the fast shock from the first impact at the bow shock to the ionosphere. This transit time is obtained to be 0.5-1 min longer for the inclined shock than for a radially propagating shock with a similar amplitude.

5.1.17. Proton Auroral Intensification Induced by Interplanetary Shock on 7 November 2004

Su et al. (2011) report a shock-induced auroral intensification event observed by the IMAGE spacecraft on 7 November 2004. The comparison of simultaneous auroral snapshots, obtained from FUV-SI12 and FUV-SI13 cameras onboard IMAGE spacecraft, indicates the dominance of proton precipitation (rather than electron precipitation) throughout the auroral oval region. The proton aurora in the postnoon sector showed the most significant intensification, with luminosity increasing by 5 times or more. Su et al. (2011) describe the main characteristics of interplanetary parameters observed by the ACE and Geotail satellites and plasma parameters within the mapped precipitation region detected by the Los Alamos National Laboratory 1990–1995 satellite. The generation mechanism of postnoon proton auroral intensification is further investigated on the basis of these observations. The estimated increase of loss cone size was not enough to produce the required proton auroral precipitation enhancement. The expected oxygen band electromagnetic ion cyclotron waves (no available observation), in the highly fluctuating density region during the shock period, might contribute to the enhanced precipitation of auroral protons. The finding of Su et al. (2011) is that the shock-driven buildup of 1–10 keV proton fluxes could account for the observed proton auroral intensification.

5.1.18. Different Bz Response Regions in the Nightside Magnetosphere after the Arrival of an Interplanetary Shock: Multipoint Observations Compared with MHD Simulations

Sun et al. (2012) present different magnetic field changes in the nightside magnetosphere in response to the interplanetary shock on 17 December 2007, using multiple spacecraft observations and global MHD simulations. The coexistence of two distinct B_Z response regions in the nightside magnetosphere in a single event has been observationally identified for the first time. From the inner magnetosphere to the tail, they are the positive response (B_Z increase) and the negative response (B_Z decrease). This scenario reasonably agrees with the

MHD model prediction. Moreover, the analysis of the response delay time shows that, for the three satellites which observed the negative responses of B_Z, the one closest to Earth was the last to respond. This phenomenon can also be understood based on the model prediction that the negative response region develops toward Earth after its formation. In addition, the temporarily enhanced earthward flows in the negative response region, which were suggested to be responsible for the formation of this region by previous model studies, were also supported by the observation. At last, a global view of the B_Z response processes in the nightside magnetosphere is presented based on MHD simulations.

5.1.19. Impact of the Rippling of a Perpendicular Shock Front on Ion Dynamics

According to Yang et al. (2012), both hybrid/full particle simulations and recent experimental results have clearly evidenced that the front of a supercritical quasi-perpendicular shock can be rippled. Recent two-dimensional simulations have focused on two different types of shock front rippling: (1) one characterized by a small spatial scale along the front is supported by lower hybrid wave activity, (2) the other characterized by a large spatial scale along the front is supported by the emission of large amplitude nonlinear whistler waves. These two rippled shock fronts are self-consistently observed when the static magnetic field is perpendicular to (so called "B_0-OUT" case) or within (so called "B_0-IN" case) the simulation plane, respectively. On the other hand, several studies have been made on the reflection and energization of incoming ions with a shock but most have been restricted to a one dimensional shock profile only (no rippling effects). Herein, two-dimensional test particle simulations based on strictly perpendicular shock profiles chosen at a fixed time in two-dimensional Particle-in-cell (PIC) simulations, are performed in order to investigate the impact of the shock front ripples on incident ion (H^+) dynamics. The acceleration mechanisms and energy spectra of the test-ions (described by shell distributions with different initial kinetic energy) interacting with a rippled shock front are analyzed in detail. Both "B_0-OUT" and "B_0-IN" cases are considered separately; in each case, y-averaged (front rippling excluded) and non-averaged (front rippling included) profiles will be analyzed. Present results show that: (1) the incident ions suffer both shock drift acceleration (SDA) and shock surfing acceleration (SSA) mechanisms. Moreover, a striking feature is that SSA ions not only are identified at the ramp but also within the foot which confirms previous 1-D simulation results; (2) the percentage of SSA ions increases with initial kinetic energy, a feature which persists well with a rippled shock front; (3) furthermore, the ripples increase the porosity of the shock front, and more directly transmitted (DT) ions are produced; these strongly affect the relative percentage of the different identified classes of ions (SSA, SDA and DT ions), their average kinetic energy and their relative contribution to the resulting downstream energy spectra; (4) one key impact of the ripples is a strong diffusion of ions (in particular through the frontiers of their injection angle domains and in phase space which are blurred out) which leads to a mixing of the different ion classes. This diffusion increases with the size of the spatial scale of the front ripples; (5) through this diffusion, an ion belonging to a given category (SSA, SDA, or DT) in y-averaged case changes class in non-averaged case without one-to-one correspondence.

5.1.20. Magnetospheric Responses to the Passage of the Interplanetary Shock on 24 November 2008

According to Kim et al. (2012), the passage of an interplanetary shock was detected by Wind, ACE, Geotail, and THEMIS-B in the solar wind on 24 November 2008. From the propagation time of the interplanetary shock at the spacecraft, it is expected that the interplanetary shock front is aligned with the Parker spiral and strikes the postnoon dayside magnetopause first. Using multipoint observations of the sudden commencement (SC) at THEMIS probes, GOES 11, and ETS in the dayside magnetosphere, Kim et al. (2012) confirmed that the magnetospheric response to the interplanetary shock starts earlier in the postnoon sector than in the prenoon sector. It's found that the estimated normal direction of the SC front is nearly aligned with the estimated interplanetary shock normal. It was also found that the SC front normal speed is much slower than the fast mode speed and is about 22–56% of the interplanetary shock speed traveling in the solar wind. Thus, Kim et al. (2012) suggest that the major field changes of the SC in the dayside magnetosphere are not due to the magnetic flux carried by hydromagnetic waves but to the increased solar wind dynamic pressure behind the shock front sweeping the magnetopause. The SC event appears as a step-like increase in the H component at the low-latitude Bohyun station and a negative-then-positive variation in the H component at the high-latitude Chokurdakh station in the morning sector. During the negative perturbation at CHD, the SuperDARN Hokkaido radar detected a downward motion in the ionosphere, implying westward electric field enhancement. Using the THEMIS electric field data, it is confirmed that the westward electric field corresponds to the inward plasma motion in the dayside magnetosphere due to the magnetospheric compression.

5.1.21. A Sunward Propagating Fast Wave in the Magnetosheath Observed after the Passage of an Interplanetary Shock

As noted Pallocchia (2013), the interaction of an interplanetary shock with the terrestrial magnetosphere causes some noticeable effects on the global scale structure of the magnetosphere itself. One of these effects is given by an earthward motion of the bow shock followed, some minutes later, by a sunward displacement. As demonstrated by past observational studies, in agreement with the theory of the shock-shock collision, the earthward motion of the bow shock is due directly to the interplanetary shock impact on it. Differently, the origin of the sunward motion of the bow shock is still not well understood. In this regard, on the basis of some issues of the present observational study, Pallocchia (2013) suggest a possible mechanism to account for the aforementioned outward displacement. The event, observed by the Cluster SC3 spacecraft on day 7 November 2004, is related to a magnetosheath perturbation triggered by an interplanetary shock impact. The main result of the data analysis is the first identification of a reverse (i.e., sunward directed) fast magnetosonic wave just after the interplanetary shock passage. The identification has been performed via a direct comparison of the observations with the MHD equations of nonlinear magnetosonic waves. The reverse fast wave consists of a smooth compression, approximately 2 min long, which is most of the magnetosheath plasma density increase produced by the entire perturbation. It was point out that signatures of this reverse wave are observable in two other events as well as in a three-dimensional MHD simulation reported in the literature on the topic. Moreover, it was provide a possible interpretation of this reverse wave in terms of a transient pressure build-up in the magnetosheath due to the post-shock reconfiguration of the plasma flow around the magnetopause obstacle.

5.1.22. Nonlinear Phenomena Related to the Solar Shock Motion in the Earth's Magnetosphere

The motion of the MHD nonlinear shock in the Earth's magnetosphere is considered by Grib (2013) in the scope of magnetic hydrodynamics. This wave comes from the solar wind and is refracted into the magnetosphere, generating a fast return rarefaction wave. It has been indicated that a wave refracted into the magnetosphere is a weak fast dissipative shock, propagating in magnetospheric plasma at a velocity higher than its propagation velocity in a solar wind stream. The wave motion near the Earth-Sun line with regard to the effect of the geomagnetic field transverse component is described. In this case, shock damping follows the generalized Crussard-Landau law and a wave retains its shock character up to the plasmapause, interacting with this region when an arbitrary MHD discontinuity is disintegrated. It is stated that an MHD shock loses its shock character when moving in a strongly inhomogeneous plasma within the plasmasphere and a weak shock reflected from the plasmapause can combine with a return secondary shock in the magnetosheath, promoting the experimentally observed backward motion of the bow shock front.

5.1.23. Sudden Impulse Observations in the Dayside Magnetosphere by THEMIS

Samsonov et al. (2014) present a study of the magnetospheric response to interplanetary shocks. It was analyzed eight events with simultaneous observations of sudden impulses in the dayside magnetosphere and interplanetary shocks in the solar wind. The spacecraft measurements in the equatorial plane, even those very close to the Earth, can be interpreted in terms of the vortices predicted by previous studies employing global MHD models. In fact, these vortices are velocity oscillations with the properties of Alfvén waves. The amplitude and frequency of the oscillations depend on radial distance from the Earth. The amplitude of the velocity perturbations decreases with increase of the density and magnetic field magnitude, but the velocity amplitude shows no dependence on magnitude of the solar wind dynamic pressure change attending the interplanetary shocks. The oscillations are observed both in the outer magnetosphere and the plasmasphere, but they become less sinusoidal near the plasmapause, i.e., in the region with a large-density gradient. The amplitude of the magnetic field enhancement in the sudden impulses also depends on the radial distance. The MHD simulations successfully predict the amplitudes of magnetic field increase and the first cycles of the velocity oscillations in these events.

5.1.24. The Chain Response of the Magnetospheric and Ground Magnetic Field to Interplanetary Shocks

As outlined Sun et al. (2015), in response to interplanetary shocks, magnetic field may decrease/increase (negative/positive response) in nightside magnetosphere, while at high latitudes on the ground it has two-phase bipolar variations: preliminary impulse and main impulse (MI). Using global MHD simulations, Sun et al. (2015) investigate the linkage between the MI phase variation on the ground and the magnetospheric negative response to an interplanetary shock. It is revealed that although the two phenomena occur at largely separated locations, they are physically related and form a response chain. The velocity disturbances near the flanks of the magnetopause cause the magnetic field to decrease, resulting in a dynamo which thus powers the transient field-aligned currents (FACs). These FACs further generate a pair of ionospheric current vortex, leading to MI variations on the ground. Therefore, Sun et al. (2015) report here the intrinsic physically related chain response of the magnetospheric and ground magnetic field to interplanetary shocks, and thus link the magnetospheric sudden impulse (SI) and ground SI together.

5.1.25. Impact Angle Control of Interplanetary Shock Geoeffectiveness: A Statistical Study

Oliveira and Raeder (2015) present a survey of interplanetary (IP) shocks using Wind and ACE satellite data from January 1995 to December 2013 to study how IP shock geoeffectiveness is controlled by IP shock impact angles. A shock list covering one and a half solar cycle is compiled. The yearly number of IP shocks is found to correlate well with the monthly sunspot number. Oliveira and Raeder (2015) use data from SuperMAG, a large chain with more than 300 geomagnetic stations, to study geoeffectiveness triggered by IP shocks. The SuperMAG SML index, an enhanced version of the familiar AL index, is used in our statistical analysis. The jumps of the SML index triggered by IP shock impacts on the Earth's magnetosphere are investigated in terms of IP shock orientation and speed. It’s find that, in general, strong (high speed) and almost frontal (small impact angle) shocks are more geoeffective than inclined shocks with low speed. The strongest correlation (correlation coefficient $R = 0.78$) occurs for fixed IP shock speed and for varied IP shock impact angle. Oliveira and Raeder (2015) attribute this result, predicted previously with simulations, to the fact that frontal shocks compress the magnetosphere symmetrically from all sides, which is a favorable condition for the release of magnetic energy stored in the magnetotail, which in turn can produce moderate to strong auroral substorms, which are then observed by ground-based magnetometers.

5.1.26. Rapid Enhancement of Low-Energy (< 100eV) Ion Flux in Response to Interplanetary Shocks Based on Two Van Allen Probes Case Studies: Implications for Source Regions and Heating Mechanisms

As outlined Yue et al. (2016), interactions between interplanetary shocks (IP) and the Geomagnetosphere manifest many important space physics phenomena including low-energy ion flux enhancements and particle acceleration. In order to investigate the mechanisms driving shock-induced enhancement of low-energy ion flux, Yue et al. (2016) have examined two IP shock events that occurred when the Van Allen Probes were located near the equator while ionospheric and ground observations were available around the spacecraft footprints. They have found that, associated with the shock arrival, electromagnetic fields intensified, and low-energy ion fluxes, including H^+, He^+, and O^+, were enhanced dramatically in both the parallel and perpendicular directions. During the 2 October 2013 shock event, both parallel and perpendicular flux enhancements lasted more than 20 min with larger fluxes observed in the perpendicular direction. In contrast, for the 15 March 2013 shock event, the low-energy perpendicular ion fluxes increased only in the first 5 min during an impulse of electric field, while the parallel flux enhancement lasted more than 30 min. In addition, ionospheric outflows were observed after shock arrivals. From a simple particle motion calculation, Yue et al. (2016) found that the rapid response of low-energy ions is due to drifts of plasmaspheric population by the enhanced electric field. However, the fast acceleration in the perpendicular direction cannot solely be explained by $\mathbf{E} \times \mathbf{B}$ drift, but betatron acceleration also plays a role. Adiabatic acceleration may also explain the fast response of the enhanced parallel ion fluxes, while ion outflows may contribute to the enhanced parallel fluxes that last longer than the perpendicular fluxes.

5.1.27. Statistical Study of Polar Negative Magnetic Bays Driven by Interplanetary Fast-Mode Shocks

As note Liou et al. (2017), the question of whether an interplanetary (IP) fast-mode shock can trigger the substorm expansion is still not resolved. Some scientists believe that the substorm expansion can be triggered by a sudden compression of the magnetosphere if a substorm growth phase is in progress (e.g., the magnetosphere precondition hypothesis). Such a hypothesis has not been rigorously tested. Liou et al. (2017) study 258 positive sudden impulse (SI^+) events associated with interplanetary fast-mode shocks observed by the ACE spacecraft from 1998 to 2009. The westward auroral electrojet SML index, derived from the SuperMAG network geomagnetic field data, is used to check for negative magnetic bays. It is found that the probability of having a magnetic bay onset that follows a shock impact peaks (~27%) sharply and immediately (within 3 min) after the shock-induced SI^+. This indicates that the majority of the shock events do not lead to magnetic bays. A superposed epoch analysis of the shock events associated with (Type-Y) and without (Type-N) negative bays indicates that the Type-Y events had stronger (~40%) solar wind driving than the Type-N ones. This result generally supports the magnetosphere precondition hypothesis. Liou et al. (2017) also found that (1) Type-Y events are associated with even stronger (~56%) solar wind driving downstream of the shock than Type-N events and (2) there is a good correlation between the SML and the polar cap magnetic PC indices both upstream and downstream of the shock. It is suggested by Liou et al. (2017) that sharp decreases in SML immediately after a shock impact are not associated with the substorm-driven DP-1 current system but with the convection-driven DP-2 current system.

5.2. Coupling of Coronal Mass Ejections and Magnetic Clouds with Geomagnetosphere

5.2.1. Stream-Interacting Magnetic Clouds Causing Very Intense Geomagnetic Storms

Dal Lago et al. (2002) calculate the magnetic field parameters for two stream-interacting magnetic clouds using Burlaga's (1988) model and compare with their observed magnetic field behavior. Presumably the model does not take into account the effects of interaction between the magnetic cloud and the high speed stream, so they could estimate quantitatively the increase in the peak Dst value due to this interaction. For this purpose, the observed Dst was used together with the Dst values calculated with the modified Burton formula, using as input the observed parameters and the parameters obtained by Burlaga's (1988) model. Obtained results show that the peak Dst value can be increased from (at least) 60 to 100 nT due to the interaction between the magnetic cloud and the high speed stream overtaking it. Main results are illustrated by Figures 5.2.1 – 5.2.5.

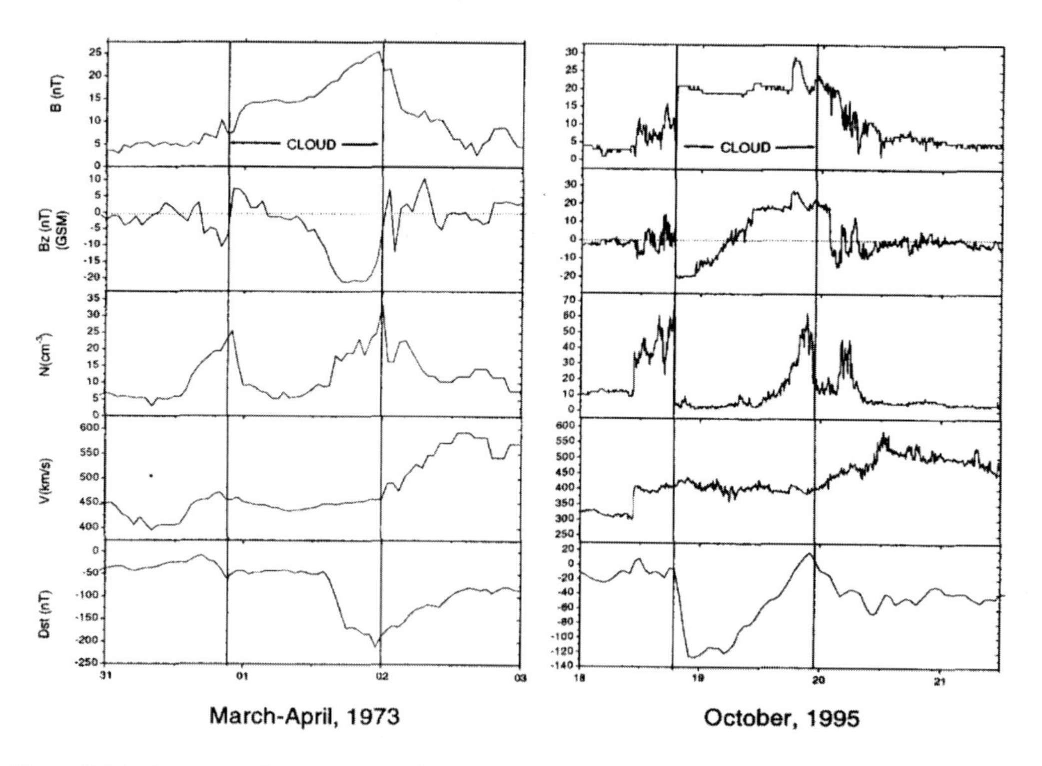

Figure 5.2.1. From top to bottom: magnetic field intensity, Bz component (in GSM coordinate system) of the IMF, density, velocity and the *Dst* index for the magnetic clouds observed in March-April 1973 and October 1995, respectively (adapted from Klein and Burlaga, 1982 and Lepping et al., 1997). According to Dal Lago et al. (2002). Permission from COSPAR.

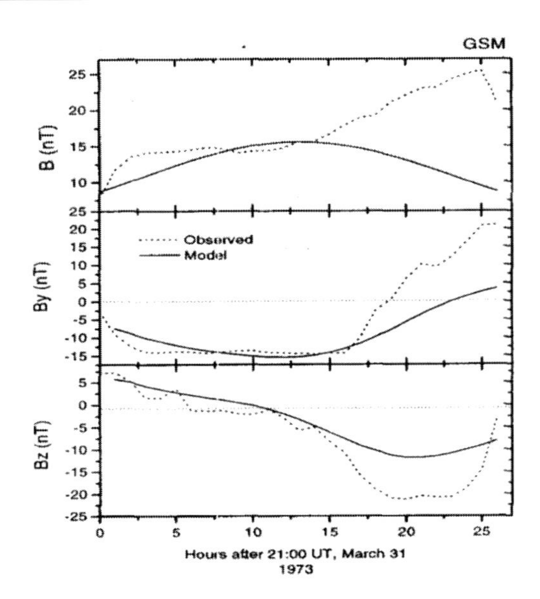

Figure 5.2.2. Comparison of the observed magnetic field of the cloud (dotted curve) with the magnetic field calculated by Burlaga's (1988) model (solid curve) for the magnetic cloud event of March-April 1973. A clear discrepancy occurs in the rear portion of the cloud due to the interaction with the high speed stream overtaking the cloud. According to Dal Lago et al. (2002). Permission from COSPAR.

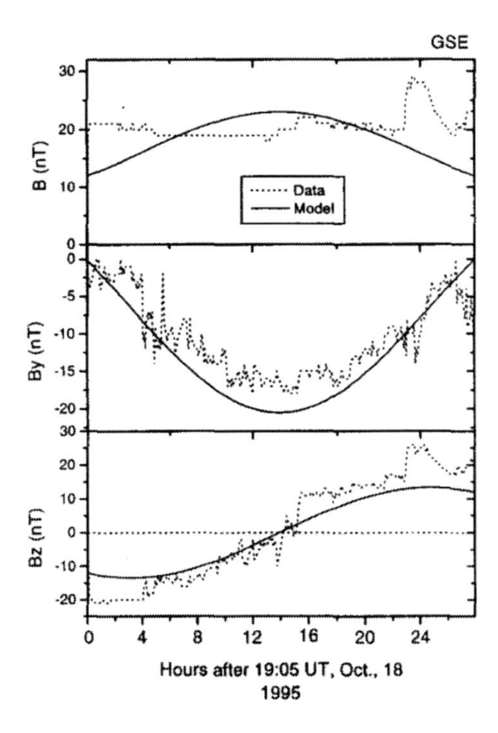

Figure 5.2.3. Same as in Figure 5.2.2, but for the magnetic cloud observed in October 1995. The discrepancies between the modeled and observed curves appear in the extremities due to interaction effects (the fitting is similar to that done by Lepping et al., 1997). According to Dal Lago et al. (2002). Permission from COSPAR.

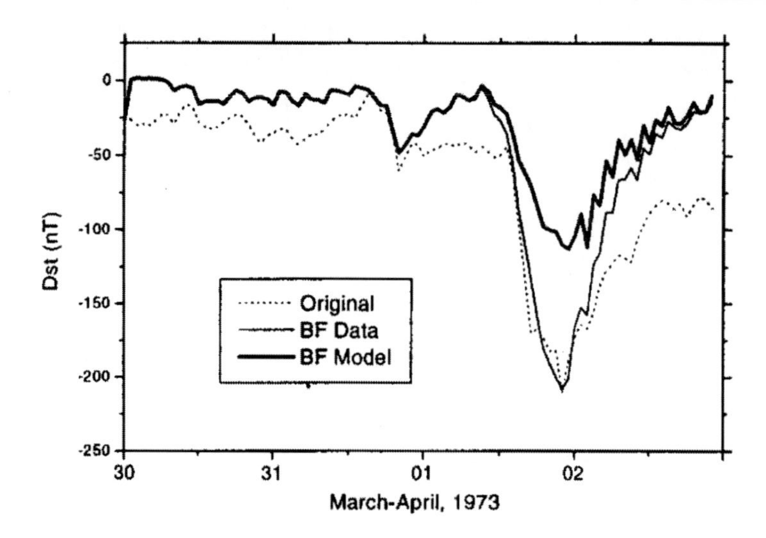

Figure 5.2.4. Original *Dst* index (dotted curve), *Dst* calculated with the modified Burton formula, using as input the observed parameters (solid thin curve), and *Dst* calculated with the modified Burton formula using as input the magnetic field parameters obtained by the Burlaga's (1988) model (solid thick curve) for the magnetic cloud observed in March-April 1973. The latter curve (solid thick) gives an estimation of what the geomagnetic effect would be without the interaction of the cloud with the high speed stream. According to Dal Lago et al. (2002). Permission from COSPAR.

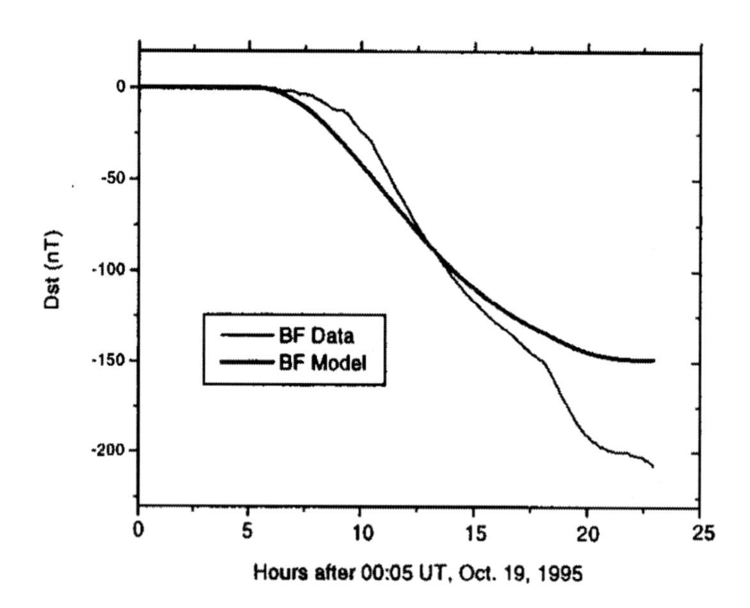

Figure 5.2.5. *Dst* calculated by the modified Burton formula using as input the observed parameters, using a reverse B_z component (thin curve) and the *Dst* calculated using as input the magnetic field parameters obtained by the model (also using a reverse B_z component) (thick curve). The difference between the two peaks is almost 60 nT. According to Dal Lago et al. (2002). Permission from COSPAR.

5.2.2. Modeling the Sun-to-Earth Propagation of a Very Fast CME

Manchester IV et al. (2006) present a three-dimensional (3-D) numerical ideal MHD model describing the time-dependent propagation of a CME from the solar corona to Earth in just 18 h. The simulations are performed using the BATS-R-US (Block Adaptive Tree Solarwind Roe Upwind Scheme) code. Manchester IV et al. (2006) begin by developing a global steady-state model of the corona that possesses high-latitude coronal holes and a helmet streamer structure with a current sheet at the equator. The Archimedean spiral topology of the IMF is reproduced along with fast and slow speed solar wind. Within this model system, Manchester IV et al. (2006) drive a CME to erupt by the introduction of a Gibson Low magnetic flux rope that is embedded in the helmet streamer in an initial state of force imbalance. The flux rope rapidly expands, driving a very fast CME with an initial speed of in excess of 4000 km/s and slowing to a speed of nearly 2000 km/s at Earth (Figure 5.2.6).

It was found: the developed model predicts a thin sheath around the flux rope, passing the Earth in only 2 h. Shocked solar wind temperatures at 1 AU are in excess of 10 million degrees. Physics based on adaptive mesh refinement (AMR) allows to capture the structure of the CME focused on a particular Sun-Earth line with high spatial resolution given to the bow shock ahead of the flux rope.

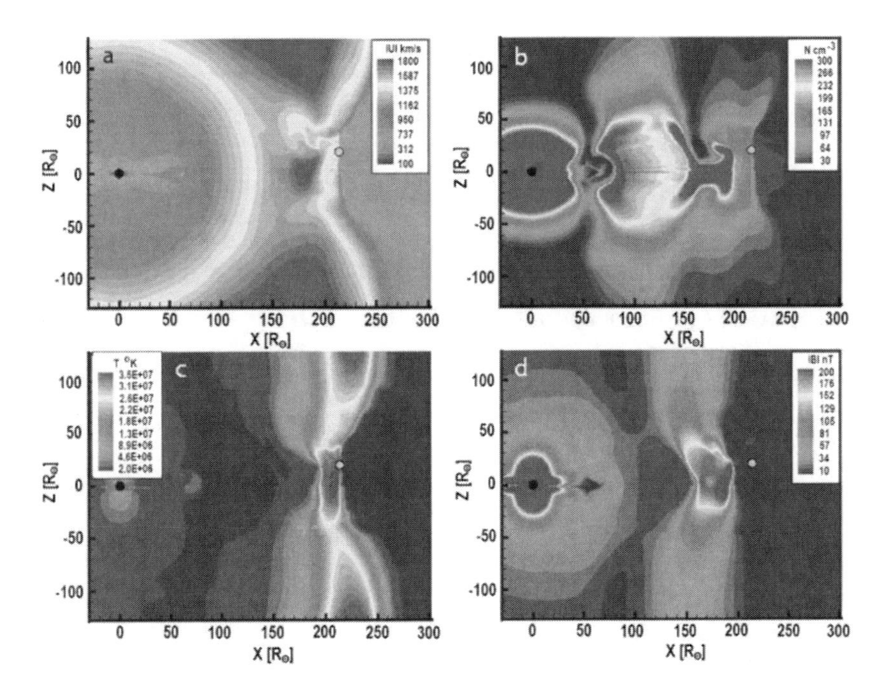

Figure 5.2.6. The properties of the CME and perturbed solar wind are shown in a series of images at t = 18 h when the shock front reaches the Earth located at X = 214 $R_\odot$, Z = 20 $R_\odot$. The color images on the meridional (X–Z) plane show the velocity magnitude, density and temperature in (a–c), respectively, and the magnetic field strength in (d). According to Manchester IV et al. (2006). Permission from COSPAR.

5.2.3. Altered Solar Wind-Magnetosphere Interaction at Low Mach Numbers: Coronal Mass Ejections

Lavraud and Borovsky (2008) illustrate some fundamental alterations of the solar wind-magnetosphere interaction that occur during low Mach number solar wind. It was first shown that low Mach number solar wind conditions are often characteristic of coronal mass ejections (CMEs) and magnetic clouds in particular. Lavraud and Borovsky (2008) then illustrate the pivotal role of the magnetosheath. This comes from the fact that low Mach number solar wind leads to the formation of a low thermal β magnetosheath downstream of the bow shock. This property influences magnetic forces and currents, in particular, and in turn alters magnetosheath-magnetosphere coupling. The implications of this unusual regime of interaction have generally been overlooked. Potentially affected phenomena include the following: (1) asymmetric magnetosheath flows (with substantial enhancements), (2) asymmetric magnetopause and magnetotail shapes, (3) changes in the development of the Kelvin-Helmholtz instability and giant spiral auroral features, (4) variations in the controlling factors of dayside magnetic reconnection, (5) cross polar cap potential saturation and Alfvén wings, and (6) global sawtooth oscillations. Lavraud and Borovsky (2008) was examine these phenomena, primarily by use of global MHD simulations, and discuss the mechanisms that rule such an altered interaction. Lavraud and Borovsky (2008) emphasize the fact that all these effects tend to occur simultaneously so as to render the solar wind-magnetosphere interaction drastically different from the more typical high Mach number case. In addition to the more extensively studied inner magnetosphere and magnetotail processes, these effects may have important implications during CME-driven storms at Earth, as well as at other astronomical bodies such as Mercury.

5.2.4. Effects on the Distant Geomagnetic Tail of a Fivefold Density Drop in the Inner Sheath Region of a Magnetic Cloud: A Joint Wind–ACE Study

As noted Farrugia et al. (2009), much attention has been directed at impulsive changes in solar wind dynamic pressure because they are known to cause important deformations of, and wave modes inside, the magnetosphere. Their effect is complicated because these pressure changes interact first with the bow shock where they can produce a number of other wave modes in the magnetosheath (Neubauer, 1975; Wu et al., 1993). These wave modes, which 'share' the oncoming density decrease with the initial discontinuity, travel ahead of it and thus they are the first triggers to which the magnetosphere reacts. Some of these effects at the dayside and the magnetosphere flanks have been studied (Kaufmann and Konradi, 1969; Sibeck, 1990; Fairfield et al., 2003). Farrugia et al. (2009) outlined that no study has tried to isolate effects as far downtail as ~ – 220 R_E. The aim of their paper is to discuss one such example. The interplanetary cause was a planar magnetic structure plastered against the front boundary of the magnetic cloud which reached Earth on November 20, 2003. Before Farrugia et al., (2008) studied this feature of the innermost sheath in the context of ICME sheaths consisting of accreted solar wind material, a theoretical standpoint advanced by Siscoe and Odstrcil (2008). Of direct relevance here is that at the forward edge of this filament there was a fivefold, abrupt drop of the dynamic pressure, one of the largest drops ever recorded in the solar wind. This, along with the unusually strong fields and lateral flows in the filament, make its interaction with the magnetosphere of particular interest, more so because spacecraft Wind was in the distant magnetosheath ~ – 220 R_E, and close enough to the geomagnetic tail (~ 40.5 R_E) to monitor any large deformation/motions. Indeed, during a 30-min time interval Wind entered the geomagnetic tail twice.

Using a serendipitous configuration of the ACE and Wind spacecraft, Farrugia et al. (2009) monitor the response of the distant geomagnetic tail ~ – 220 R_E) to an abrupt, approximately fivefold pressure drop (from ~19.0 to ~3.5 nPa) at the front boundary of a magnetic cloud on November 20, 2003. The interplanetary data are from ACE in orbit around the L1 point. The far-tail observations are from Wind, which was nominally in the magnetosheath, separated from the Sun–Earth line by ~ 40 R_E. The magnetic field in the innermost sheath region of the magnetic cloud had a large B_y (~ 30 nT) and substantial and variable flows lateral to the Sun–Earth line. There was also a significant northward field (~35 nT), unique in the vicinity of this magnetic cloud. These extreme values are reached in a filament forming the earliest relic of material accreted by the magnetic cloud en route to Earth. The effects resulting from these on the far geomagnetic tail are: (1) expansion, (2) tail twisting, and (3) tail tilting. These extreme conditions were in part responsible for a crossing by Wind of a neutral sheet which is tilted by ~85° to the ecliptic. Further, Wind made two successive excursions deep into the geomagnetic tail, in the first of which a tailward flow burst of ~1200 km/s was observed. The dayside part of the interaction of the sudden and large dynamic pressure drop with the bow shock is studied with a local 3D MHD simulation. Obtained results are illustrated by Figures 5.2.7 - 5.2.13.

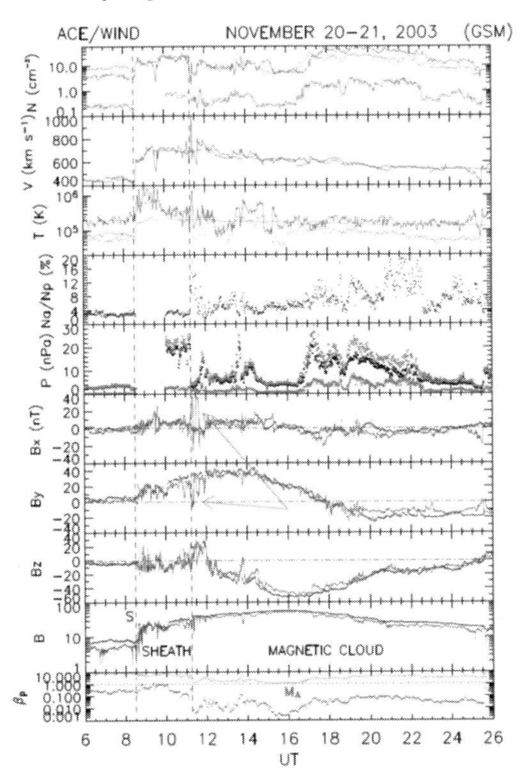

Figure 5.2.7. Overlaid ACE and Wind (blue trace) plasma and field data. The shocks seen by the two spacecraft have been aligned. Vertical guidelines indicate the times of passage at Wind of the shock ('S'), sheath and magnetic cloud. According to Farrugia et al. (2009). Permission from COSPAR.

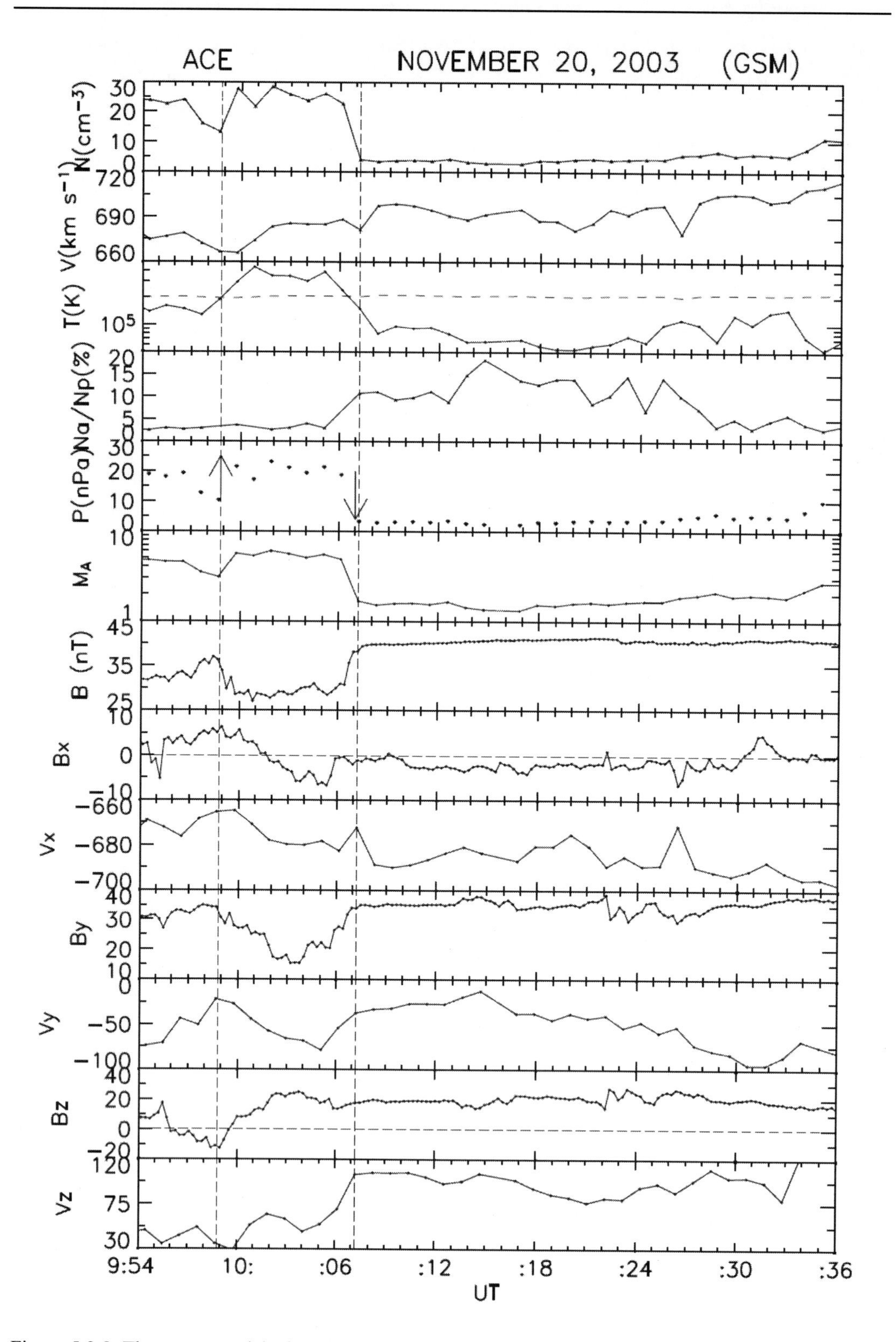

Figure 5.2.8. The structure picked out by ACE at the inner most region of the cloud sheath (bracketed by the vertical guidelines). According to Farrugia et al. (2009). Permission from COSPAR.

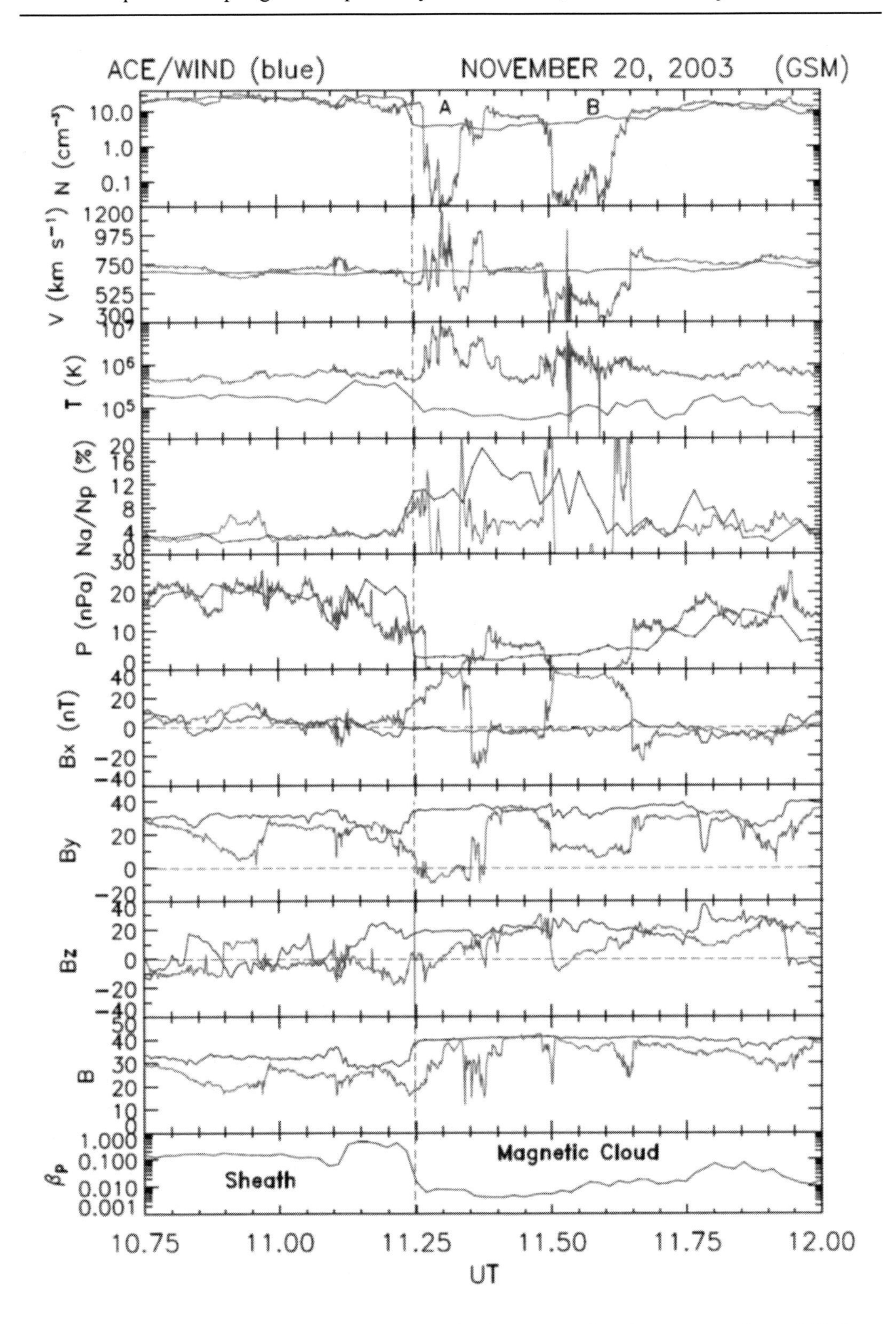

Figure 5.2.9. An overlay of Wind and ACE data at the time surrounding the two excursions of Wind into the tail lobes and plasma sheet. According to Farrugia et al. (2009). Permission from COSPAR.

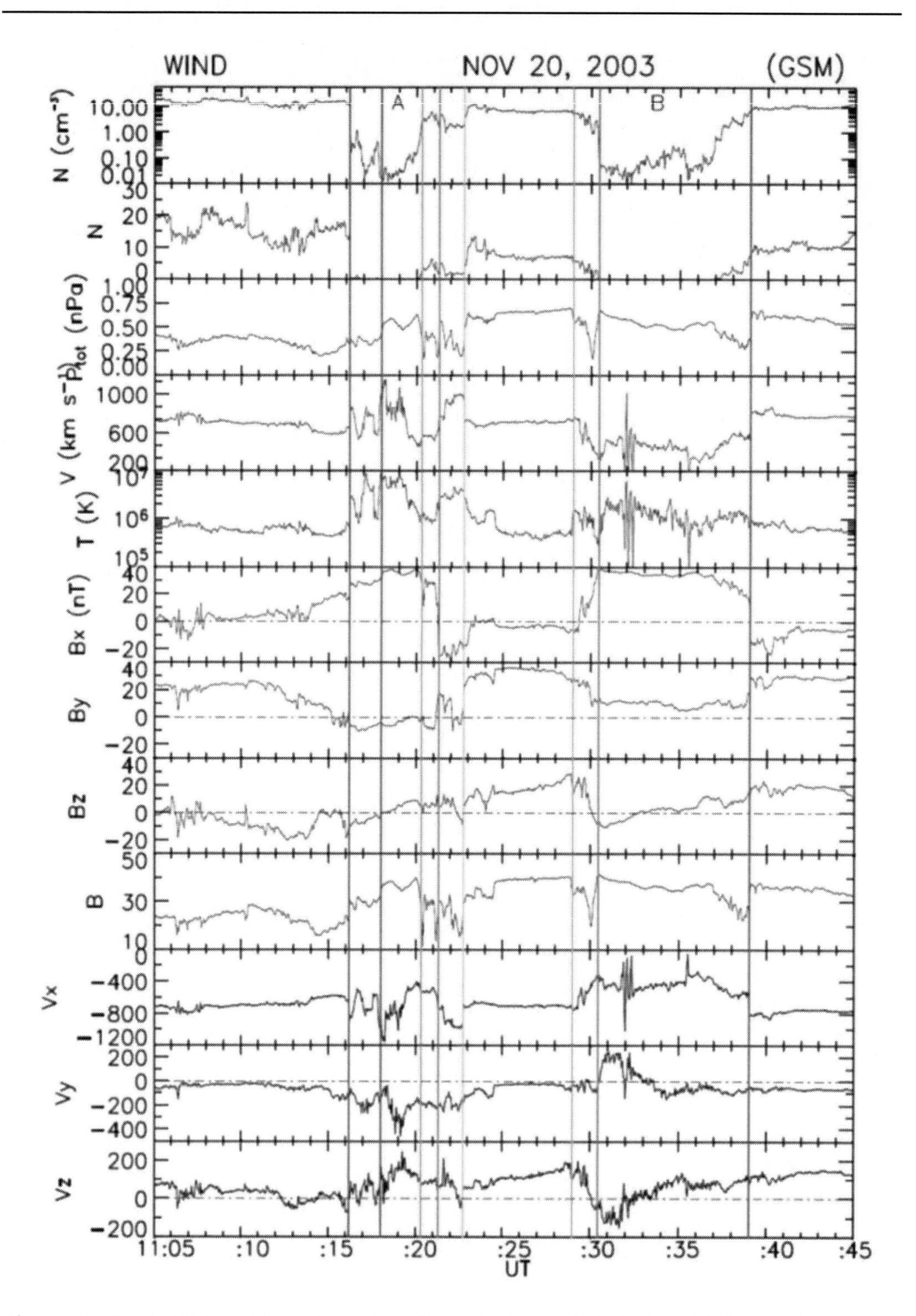

Figure 5.2.10. Wind field and flow observations from 11:05 to 11:45 UT. The guidelines mark the entry times of Wind into various regions of the tail. From top to bottom: plasma density (linear and log scale), total pressure, bulk speed and temperature, and field and flow components (GSM). According to Farrugia et al. (2009). Permission from COSPAR.

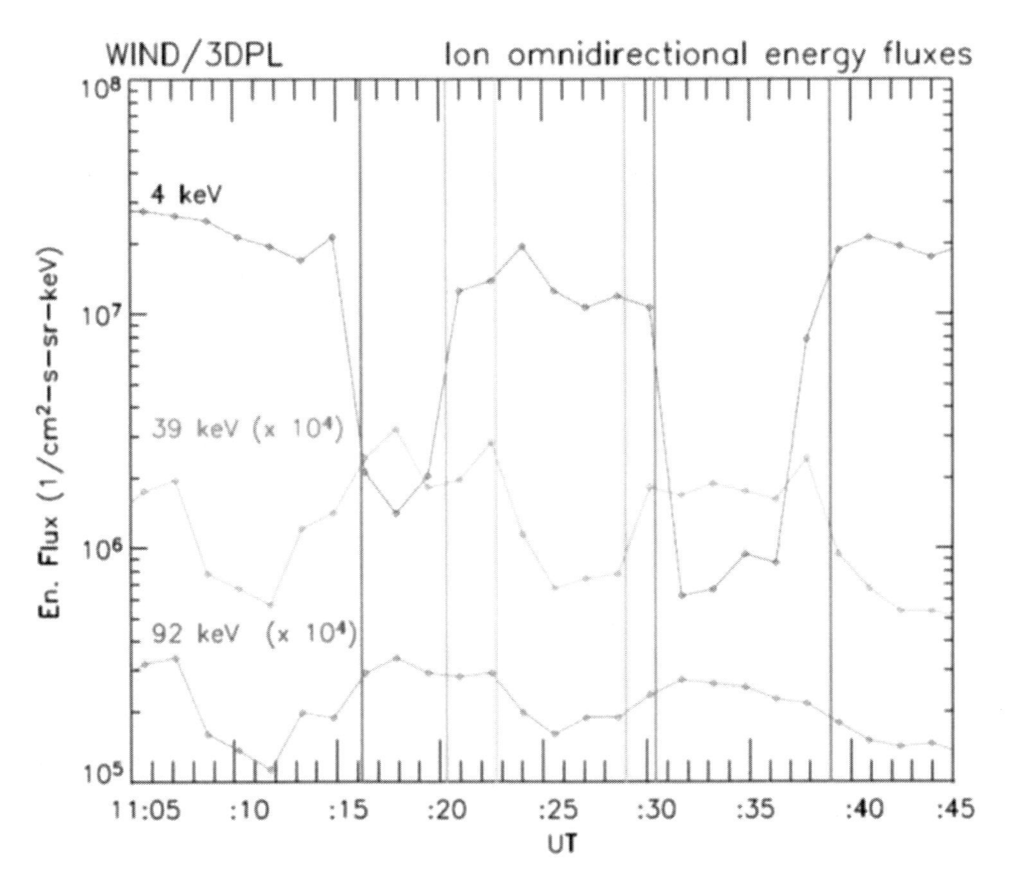

Figure 5.2.11. Omnidirectional ion energy fluxes for three energy ranges (4, 39 and 92 keV). Same color coding as in the previous figure. According to Farrugia et al. (2009). Permission from COSPAR.

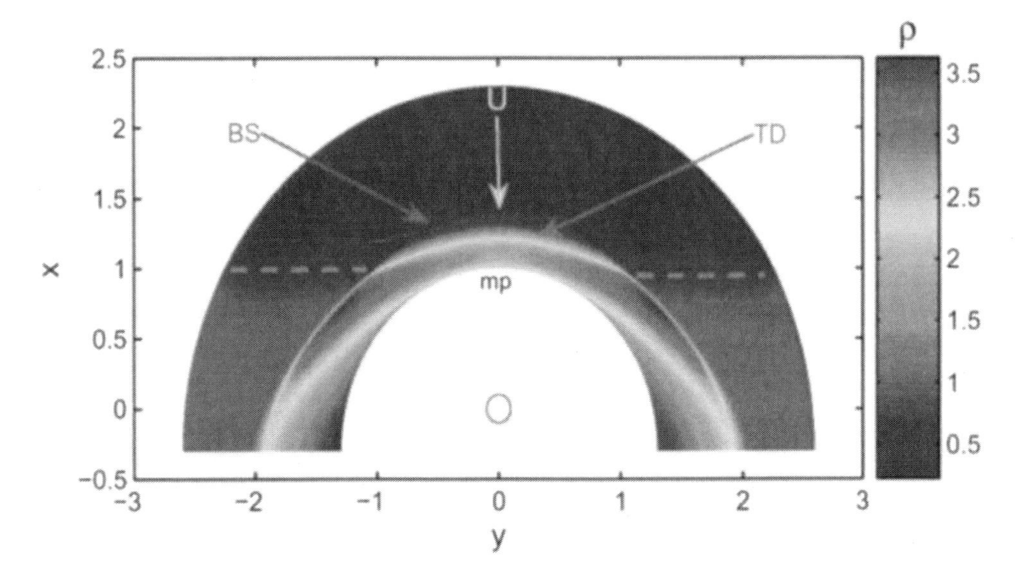

Figure 5.2.12. Normalized density distribution soon after the pressure front hits the bow shock. An (XY) cut is shown, where positive X is up and positive Y is to the right. According to Farrugia et al. (2009). Permission from COSPAR.

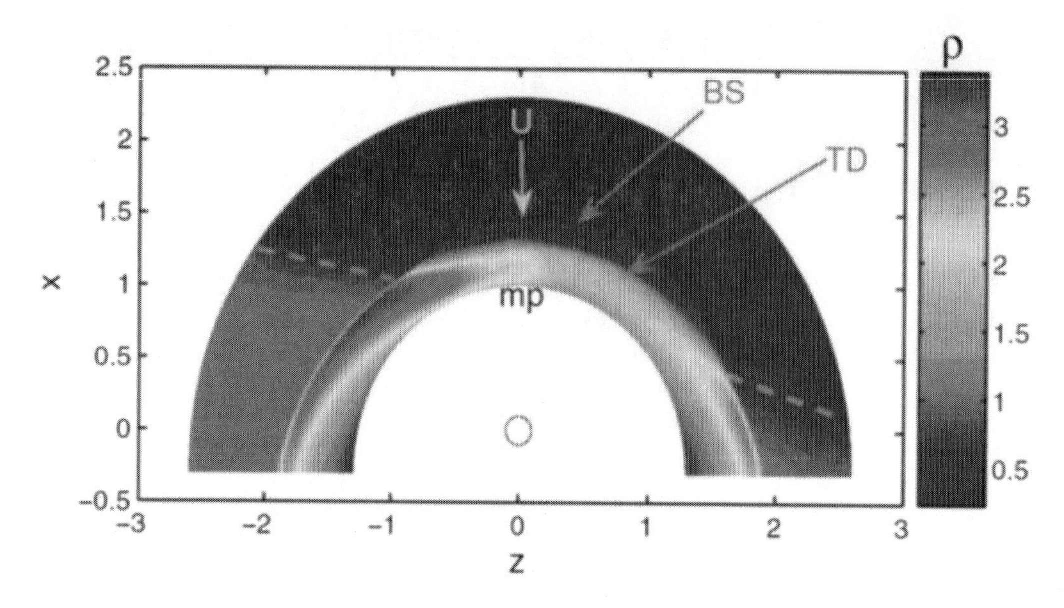

Figure 5.2.13. Same as Figure 5.2.12, but for the case when the normal to the planar structure has both X- and Z-components. Positive X is up and positive Z (north) is to the right. According to Farrugia et al. (2009). Permission from COSPAR.

5.2.5. Interplanetary Origins of November 2004 Superstorms

As outlined Echer et al. (2010), the Sun was very active in the declining phase of solar cycle 23. Large sunspot active regions gave origin to multiple flare and coronal mass ejection (CME) activity in the interval 2003–2005. On November 2004, the active region AR10696 was the origin of dozens of flares and many CMEs. Some events of this solar activity region resulted in two large geomagnetic storms, or superstorms (Dst $\leq$ 250 nT) on November 8, peak Dst = –373 nT, and on November 10, peak Dst = –289 nT. It is the purpose of Echer et al. (2010) article to identify the interplanetary origins of these two superstorms. The south ward-directed interplanetary magnetic fields (IMF B_S) that caused the two superstorms were related to a magnetic cloud (MC) field for the first superstorm, and a combination of sheath and MC fields for the second superstorm. However, this simple, classic picture is complicated by the presence of multiple shock sand waves. Six fast-forward shocks and, at least, two reverse waves were observed in the period of the two superstorms. A detailed analysis of these complex interplanetary features is performed in Echer et al. (2010).

5.2.6. Moderate Geomagnetic Storm (21–22 January 2005) Triggered by an Outstanding CME Viewed via Energetic Neutral Atoms

As outlined McKenna-Lawlor et al. (2010), an outstanding flare on 20 January 2005 was accompanied by a CME which arrived at the magnetopause at ~1712 UT, 21 January, and produced a strong compression-pressure pulse. Enhanced magnetospheric activity was stimulated. The associated development between <1800 UT and >02.19 UT on 21–22 January, 2005 of a ring current disturbance in energetic neutral atom (ENA) data, recorded aboard both the Double Star and the IMAGE (Imager for Magnetopause-to-Aurora Global Exploration) spacecraft, is described. A magnetic storm from ~1712 UT, 21 January, reached minimum Dst = ~−101 nT at ~0600 UT, 22 January, and its recovery phase endured until 27 January. ENA data indicate that the ring current experienced a deep injection of H^+ and O^+ ions at ~1830 UT when IMF B_Z was oriented southward. At this time, the ring current was strongly asymmetric, although later it became more symmetric. B_Z turned northward at 1946 UT. From ~0224 to ~0612 UT on 22 January, B_Z fluctuated such that it intermittently pointed southward (±10 nT). The moderate but extended response of the Geomagnetosphere to the strong pressure pulse is explained by a slow evolution in the orientation of B_Z under conditions of enhanced plasma sheet density. Modeling of dynamical parameters that represent various current systems that contributed to Dst revealed their individual characteristics. The changing geomagnetic field was also modeled. Comparisons with ENA data show that early asymmetric enhancements recorded in hydrogen and oxygen were accompanied by intensified external current systems that produced a magnetic field related compression of the magnetosphere. The gradual reduction in ring current asymmetry was complemented by the largely symmetrical configuration displayed by the corresponding, still intensified, modeled magnetic field.

5.2.7. Relationship between Dst(min) Magnitudes and Characteristics of ICMEs

As outlined Kane (2010), all interplanetary disturbances having shocks and directed towards the Earth are geoeffective, giving at least a storm sudden commencement (SSC) and giving Dst depressions [Dst(min)] in a wide range -10 to -500 nT, actual magnitudes roughly proportional to the magnitudes of negative B_Z(min) of interplanetary magnetic field. During 1965-1996, the ejecta and shock events not accompanied with magnetic clouds (MCs) had only ~17% intense storms [Dst(min) ≤ −200 nT] but ejecta and shocks accompanied with magnetic clouds (MCs) had ~40% intense storms. For solar cycle 23, ejecta and shocks without MCs had only ~3% intense storms; but ejecta and shocks accompanied by MCs had 11% intense storms. Thus, events accompanied with MCs gave larger percentage of intense storms. Events related to corotating interaction regions (CIRs) led to only weak and moderate storms [Dst(min): −20 nT to −140 nT]. For cycle 23, the plots of –Bz(min) vs –Dst(min) for ejecta and shocks without MCs showed a large scatter for the ranges B_Z(min) –3 to –20 nT vs Dst(min) −5 to −170 nT. Thus, analysis in this region would give confusing and uncertain results. Concentrating on intense events [Dst (min) ≤ –200 nT], the B_Z(min) vs Dst(min) plot showed a correlation of only +0.77±0.12, with considerable scatter in the Dst(min) range −200 to −300 nT [for the same value of B_Z(min), Dst(min) would have an uncertainty of about ± 50 nT]. The correlation did not improve when B_Z(min) was substituted by the product B_ZV, or by cumulative negative B_Z from the start to the peak of negative B_Z(min).

5.2.8. Statistical Properties and Geoefficiency of ICME and Their Sheaths during Intense Geomagnetic Storms

Guo et al. (2010) examine and compare the statistical properties of interplanetary coronal mass ejections (ICMEs) and their sheath regions in the near-Earth space, mainly focusing on the distributions of various physical parameters and their geoefficiency. The 53 events studied are a subset of events responsible for intense (Dst ≤ -100 nT) geomagnetic storms during the time period from 1996 to 2005. These events all fall into the single-type category in which each of the geomagnetic storms was caused by a well-isolated single ICME, free of the complexity of the interaction of multiple ICMEs. For both sheaths and ICMEs, we find that the distributions of the magnetic field strength, the solar-wind speed, the density, the proton temperature, the dynamic pressure, the plasma beta, and the Alfvén Mach number are approximately lognormal, while those of the B_Z component and the Y component of the electric field are approximately Gaussian. On the average, the magnetic field strengths, the B_Z components, the speeds, the densities, the proton temperatures, the dynamic pressures, the plasma betas, and the Mach numbers for the sheaths are 15, 80, 4, 60, 70, 62, 67, and 30% higher than the corresponding values for ICMEs, respectively, whereas the Y component of the electric field for the sheaths is almost 1 s of that for ICMEs. The two structures have almost equal energy transfer efficiency and comparable Newell functions (Newell et al., 2007), whereas they show statistically meaningful differences in the dayside reconnection rate, according to the Borovsky function (Borovsky, 2008).

5.2.9. Features of the Interaction of ICMEs/MCs with the Earth's Magnetosphere

As outlined Farrugia et al. (2013), the interaction of interplanetary coronal mass ejections (ICMEs) and magnetic clouds (MCs) with the Earth's magnetosphere exhibits various interesting features principally due to interplanetary parameters which change slowly and reach extreme values of long duration. These, in turn, allow to explore the geomagnetic response to continued and extreme driving of the magnetosphere. Farrugia et al. (2013) discuss elements of the following: **(i)** anomalous features of the flow in the terrestrial magnetosheath during ICME/MC passage and **(ii)** large geomagnetic disturbances when total or partial mergers of ICMEs/MCs pass Earth. In **(i)** it was emphasized two roles played by the upstream Alfvén Mach number in solar wind-magnetosphere interactions: (1) It gives rise to wide plasma depletion layers (by plasma depletion layer it is mean a magnetosheath region adjacent to the magnetopause where magnetic forces dominate over hydrodynamic forces), (2) It enhances the magnetosheath flow speed on draped magnetic field lines. In **(ii)** it was stress that the ICME mergers elicit geoeffects over and above those of the individual members. In addition, features of the non-linear behavior of the magnetosphere manifest themselves.

5.2.10. Steep Plasma Depletion in Dayside Polar Cap during a CME-Driven Magnetic Storm

According to Sakai et al. (2013), a series of steep plasma depletions was observed in the dayside polar cap during an interval of highly enhanced electron density on 14 October 2000 through EISCAT Svalbard Radar (ESR) field-aligned measurements and northward-directed low-elevation measurements. Each depletion started with a steep drop off to as low as $10^{11}m^{-3}$ from the enhanced level of $3\times10^{12}m^{-3}$ at F2 region altitudes, and it continued for 10–15 min before returning to the enhanced level. These depletions moved poleward at a speed consistent with the observed ion drift velocity. DMSP spacecraft observations over an extended period of time which includes the interval of these events indicate that a region of high ion densities extended into the polar cap from the equatorward side of the cusp, i.e., a tongue of ionization existed, and that the ion densities were very low on its prenoon side. Solar wind observations show that a sharp change from IMF $B_Y > 0$ to $B_Y < 0$ is associated with each appearance of the ESR electron density dropoff. From this unprecedented clear correlation Sakai et al. (2013) present a specific scenario: the series of plasma density depletions observed using the ESR is a result of the poleward drift of the undulating boundary of the tongue of ionization; this undulation is created in the cusp roughly 20 min before the ESR observation by the azimuthal intrusion, in response to the rapid prenoon shift of the footprint of the reconnection line, of the low-density plasmas originating in the morning sector.

5.2.11. An Extreme CME and Consequences for the Magnetosphere and Earth

According to Tsurutani and Lakhina (2014), a "perfect" interplanetary coronal mass ejection could create a magnetic storm with intensity up to the saturation limit (Dst ~ −2500 nT), a value greater than the Carrington storm. Many of the other space weather effects will not be limited by saturation effects, however. The interplanetary shock would arrive at Earth within ~12 h with a magnetosonic Mach number ~45. The shock impingement onto the magnetosphere will create a sudden impulse of ~234 nT, the magnetic pulse duration in the magnetosphere will be ~22 s with a dB/dt of ~30 nT/s, and the magnetospheric electric field associated with this dB/dt is about 1.9 V/m, creating a new relativistic electron radiation belt. The magnetopause location of 4 R_E from the Earth's surface will allow expose of orbiting satellites to extreme levels of flare and ICME shock-accelerated particle radiation. The results of these calculations are compared with current observational records. Comments are made concerning further data analysis and numerical modeling needed for the field of space weather.

5.2.12. Storm Time Evolution of ELF/VLF Waves Observed by DEMETER Satellite

Zhima et al. (2014), using the data of Sun-synchronous satellite (Detection of Electro-Magnetic Waves Transmitted from Earthquake Regions) DEMETER, investigate the storm time variations of ELF/VLF waves during the intense coronal mass ejections (CME)-driven storms from 2005 to 2009. The results show that there is a good correlation between the enhancement of ELF/VLF waves and the CME events. Immidately following the enhanced wave activity driven by CMEs at the initial phase, the wave intensity decreases temporarily at the beginning of storm main phase. The strongest waves predominantly occur from the late main phase to early recovery phase. The ELF waves below 3 kHz are significantly intensified during the whole storm time, while the high-frequency waves above 3 kHz seem strengthened predominantly during the late main and early recovery phase. The ELF waves below 3 kHz can exist in a wide L shell range, with the intensity peaking at L ~ 3 and 4. High-frequency waves at $f > 9$ kHz exist mostly outside the plasmapause. The stronger ELF/VLF waves on the dayside can last longer time than those on the nightside.

5.2.13. Solar Wind-Magnetosphere Coupling Efficiency during Ejecta and Sheath-Driven Geomagnetic Storms

Myllys et al. (2016) have investigated the effect of key solar wind driving parameters on solar wind-magnetosphere coupling efficiency during sheath and magnetic cloud-driven storms. The particular focus of the study was on the coupling efficiency dependence with Alfvén Mach number (M_A). The efficiency has been estimated using the dawn-dusk component of the interplanetary electric field (E_Y), Newell and Borovsky functions as a proxy for the energy inflow and the polar cap potential (PCN), and auroral electrojet (*AE*) and *SYM-H* indices as the measure of the energy output. Myllys et al. (2016) have also performed a time delay analysis between the input parameters and the geomagnetic indices. The optimal time lag and smoothing window length depend on the coupling function used and on the solar wind driver. For example, turbulent sheaths are more sensitive to the time shift and the averaging interval than smoother magnetic clouds. The results presented in this study show that the solar wind-magnetosphere coupling efficiency depends strongly on the definition used, and it increases with increasing M_A. Myllys et al. (2016) demonstrate that the PCN index distinctively shows both a Mach number dependent saturation and a Mach number independent saturation, pointing to the existence of at least two underlying physical mechanisms for the saturation of the index. By contrast, Myllys et al. (2016) show that the *AE* index saturates but that the saturation of this index is independent of the solar wind Mach number. Finally, it's find that the *SYM-H* index does not seem to saturate and that the absence of saturation is independent of the Mach number regime. Myllys et al. (2016) highlight the difference between the typical M_A conditions during sheath regions and magnetic clouds. The lowest M_A values are related to the magnetic clouds. As a consequence, sheaths typically have higher solar wind-magnetosphere coupling efficiencies than magnetic clouds.

5.3. Coupling of Corotating Interaction Regions with Geomagnetosphere

5.3.1. Empirical Modeling of a CIR-Driven Magnetic Storm

The empirical analysis of structure and evolution of the geomagnetic field and underlying electric currents during the 8–11 March 2008 magnetospheric storm, driven by a corotating interaction region (CIR), is presented by Sitnov et al. (2010). It is based on the high-resolution geomagnetic field model TS07D (http://geomag_field.jhuapl.edu/model/) and the low-altitude mapping of field-aligned currents obtained using Iridium satellites. Compared to storms, driven by coronal mass ejections, the equatorial currents are found to be overall more dawn-dusk symmetric. Only in the early main phase a moderate hook-like westward current flowing from the Region-2 inflow area on the dawn side to the dusk/afternoon magnetopause is detected, along with an eastward current near the pre-noon magnetopause. New tail-type currents are found to dominate the storm-time magnetosphere in early main and recovery phases at the moments of strong peaks of the solar wind dynamic pressure. Similar effects, found in CME-driven storms, coincide in time with the plasma sheet density bursts. A clear distinction between the periods dominated by the partial ring current and the tail-type currents seen in the model agrees with the Iridium data that indicate a significant reduction and contraction of the field-aligned current pattern in the latter case. Overall small magnitudes of the total field-aligned currents, with rather transient enhancements, appear to be the most distinctive feature of CIR-driven storms. Comparison between the model and observations made by five THEMIS probes shows good agreement on storm timescales. Moreover, every deviation arising during a substorm reveals characteristic signatures of the tail current sheet thinning and dipolarization.

5.3.2. Geomagnetic and Auroral Activity Driven by Corotating Interaction Regions during the Declining Phase of Solar Cycle 23

A superposed epoch analysis is performed by Luan et al. (2013) to investigate the relative impact of the solar wind/interplanetary magnetic field (IMF) on geomagnetic activity, auroral hemispheric power, and auroral morphology during corotating interaction regions (CIRs) events between 2002 and 2007, when auroral images from Thermosphere Ionosphere Mesosphere Energetics and Dynamics/Global Ultraviolet Imager were available. Four categories of CIRs have been compared. These were classified by the averaged IMF B_Z and the time of maximum solar wind dynamic pressure around the CIR stream interface or onset time. It is found that during CIR events: (1) The peaks of auroral power and Kp were largely associated with dominant southward B_Z, whereas auroral activity also became stronger with increases of solar wind speed, density, and dynamic pressure. (2) The percentage and absolute increases of auroral hemispheric power with solar wind speed were much greater under dominantly northward B_Z conditions than under dominantly southward B_Z conditions. (3) The enhancement of the auroral power and Kp with increasing solar wind speed followed the same pattern, for both dominantly southward and northward B_Z conditions, regardless of the behavior of solar wind density and dynamic pressure. These results suggest that, during CIR events, southward B_Z played the most critical role in determining geomagnetic and auroral activity, whereas solar wind speed was the next most important contributor. The solar wind dynamic pressure was the less important factor, as compared with B_Z and solar wind speed. Relatively strong auroral precipitation energy flux (> ~3 mW/m^2) occurred in a wider auroral oval region after the stream interface than before it for both dominantly northward and southward B_Z conditions. These conditions enhanced the auroral hemispheric power after the stream interface. Intense auroral precipitation (> ~4 mW/m^2) generally occurred widely at night under dominantly southward B_Z conditions, but the location of this precipitation in the auroral oval was different when it was associated with different solar wind density and speed conditions.

5.3.3. The Influence of Corotating Interaction Region (CIR) Driven Geomagnetic Storms on the Development of Equatorial Plasma Bubbles (EPBs) over Wide Range of Longitudes

As outlined Tulasi Ram et al. (2015), recurrent high speed solar wind streams from coronal holes on the Sun are more frequent and geoeffective during the declining phase of solar cycle which interact with the ambient solar wind leading the formation of corotating interaction regions (CIRs) in the interplanetary medium. These CIR-High Speed Stream (HSS) structures of enhanced density and magnetic fields, when they impinge up on the Earth's magnetosphere, can cause recurrent geomagnetic storms in the Geospace environment. Tulasi Ram et al. (2015) investigate the influence of two CIR-driven recurrent geomagnetic storms on the equatorial and low-latitude ionosphere in the context of the development of equatorial plasma bubbles over Indian and Asian longitudes.

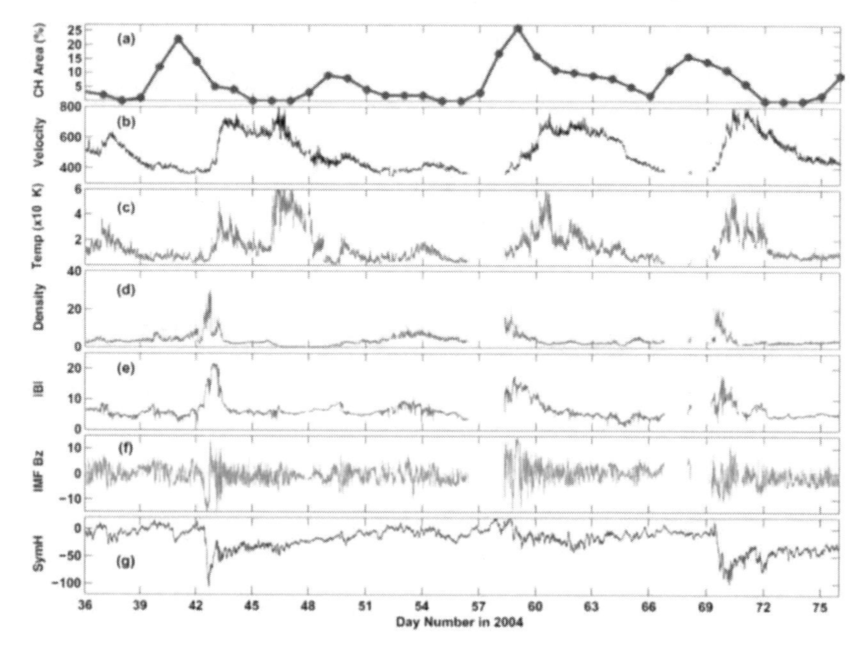

Figure 5.3.1. Daily variations of (a) fractional CH area (%), (b) solar wind proton velocity (V_P in km/s), (c) temperature (T_P in K), (d) density (N_P in particles/cm^3), (e) total magnetic field strength (|B| in nT), (f) z-component of the magnetic field (IMF B_z in nT) in GSM coordinates, and (g) low-latitude geomagnetic activity index SymH (in nT) during the day numbers from 36 to 76 in declining phase of the solar cycle year 2004. According to Tulasi Ram et al. (2015). Permission from COSPAR.

The results consistently indicate that prompt penetration of eastward electric fields into equatorial and low-latitudes under southward IMF B_Z can occur even during the CIR-driven storms. Further, the penetration of eastward electric fields augments the evening pre-reversal enhancement and triggers the development of EPBs over wide longitudinal sectors where the local post-sunset hours coincide with the main phase of the storm. Similar results that are consistently observed during both the CIR-driven geomagnetic storms are reported and discussed also by Tulasi Ram et al. (2015). Obtained results are illustrated by Figures 5.3.1 - 5.3.3.

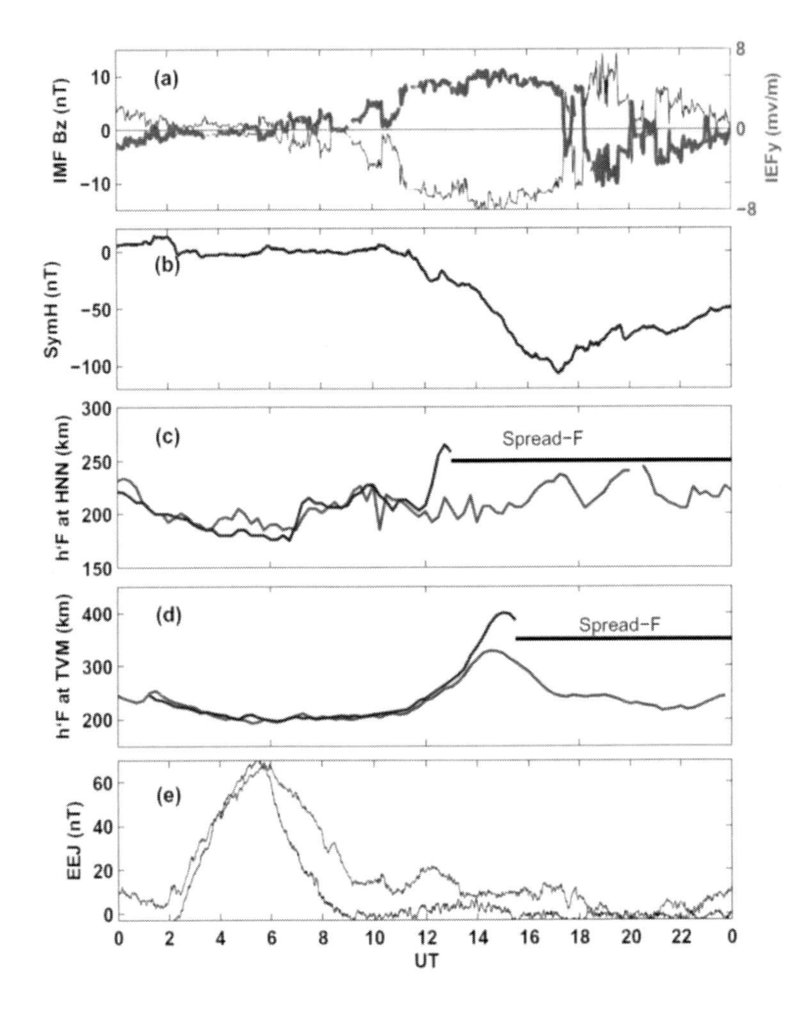

Figure 5.3.2. UT variations of (a) z-component of the interplanetary magnetic field, IMF B_z (blue curve) and dawn-dusk interplanetary electric field, IEFy (red curve), (b) SymH index, (c) virtual base height of the F-layer, h'F at Hainan Is, China, (d) h'F at Trivandrum, India and (e) equatorial electrojet (EEJ) strength derived from magnetometer observations at Tirunelveli and Alibagh during a CIR-driven geomagnetic storm on February 11, 2004. According to Tulasi Ram et al. (2015). Permission from COSPAR.

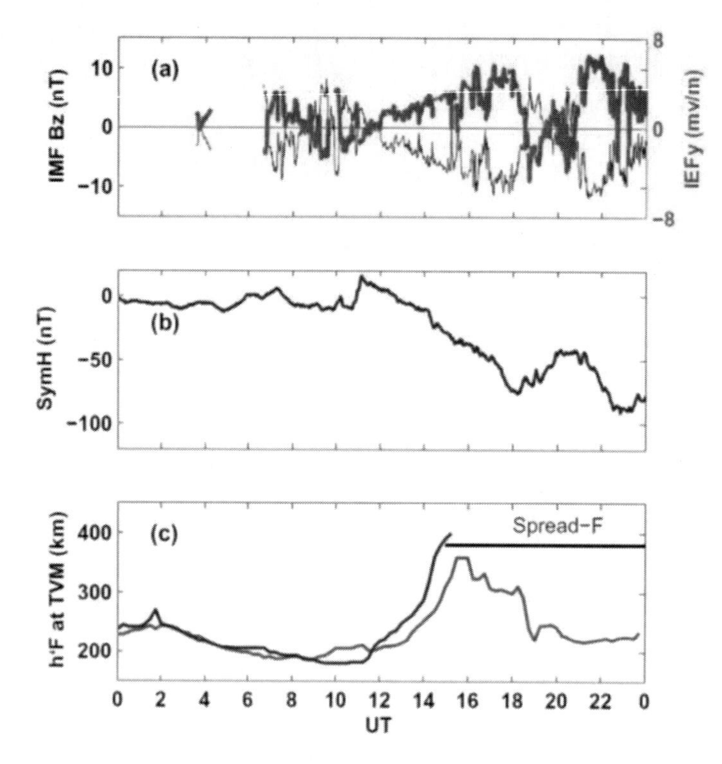

Figure 5.3.3. UT variations of (a) z-component of the interplanetary magnetic field, IMF B_z (blue curve) and dawn-dusk interplanetary electric field, IEFy (red curve), (b) SymH index (nT), (c) virtual base height of the F-layer, h'F at Trivandrum, India during a recurrent CIR-driven geomagnetic storm on March 09, 2004. According to Tulasi Ram et al. (2015). Permission from COSPAR.

5.4. Coupling of Solar Wind Streams with Geomagnetosphere

5.4.1. Variations in the Polar Precipitation Equatorward Boundary during the Interaction between the Earth's Magnetosphere and Solar Wind Streams from Isolated Solar Sources

The position of the auroral luminosity equatorward boundary during the interaction between the Earth's magnetosphere and isolated solar wind streams from different solar sources has been statistically studied by Zverev and Hviuzova (2008) based on the ground and satellite observations of auroras. These studies continue the series of the works performed in order to develop the technique for predicting auroras based on the characteristics of the interplanetary medium and auroral disturbances. The dependences of the minimal position of the auroral luminosity equatorward boundary (Φ') on the values of the azimuthal component of the interplanetary electric field (E_Y) and AL indices of magnetic activity, averaged over 6 and 24 h, are presented. The distribution limits for each type of isolated solar wind streams on the Φ'–E_Y and Φ'–AL planes have been determined.

5.4.2. The Properties of Two Solar Wind High Speed Streams and Related Geomagnetic Activity during the Declining Phase of Solar Cycle 23

Two high speed stream (HSS) solar wind intervals (days 283–294 and 314–318, 2003, hereafter called Events 1 and 2) during the declining phase of solar cycle 23 have been examined by Tsurutani et al. (2011a) in detail for their interplanetary characteristics and their resultant geomagnetic activity. Event 1 had an associated storm initial phase with peak Dst = +9 nT. This was caused by a high plasma density heliospheric plasma sheet (HPS) which impacted the magnetosphere. The southward component of IMF B_Z fluctuations in the corotating interaction regions (CIRs) of both Events 1 and 2 led to peak storm main phases of Dst = −85 and −62 nT, respectively. The extended storm "recovery" phases were associated with $\Delta\mathbf{B}/B_0$ ~1–2 Alfvénic fluctuations in the HSS proper. High-intensity, long-duration, continuous AE (HILDCAA) intervals were present, presumably due to the southward component of the Alfvén waves. The IMF B_X–V_X 4-h cross-correlation values were > 0.8 in Event 2, and lower, > 0.6, in Event 1. The difference in Alfvénicity between the two HSS events is not understood at present. The IMF B_Z 10-min to 3-h variances $\left(\sigma_z^2\right)$ and are highest during the CIRs. The normalized variances $\left(\sigma_z^2/B_0^2\right)$ during the HSS proper are approximately the same as those for the CIRs. For Event 1, the 1-h IMF $\left(\sigma_z^2\right)$ and $\left(\sigma_z^2/B_0^2\right)$ are 5.0 nT^2 and 1.1×10^{-1}, respectively. The IMF B_Z -AE cross-correlation (CC) coefficients during the storm main phase of Event 1 and for 24-h of the HSS of Event 2 give similar results. For the Event 1, a peak CC of − 0.4 occurred with a lag of 103 min, and for Event 2 a peak CC of −0.38 with a lag of 67 min was obtained. Both CC results were sharply peaked. The decay-portion of a HSS prior to Event 1 was characterized by low Np, low Bo and low Alfvén wave amplitudes. The 1-h IMF $\left(\sigma_z^2\right)$ and $\left(\sigma_z^2/B_0^2\right)$ were 0.84 nT^2 and 2.9×10^{-2}, respectively. This quiet interplanetary interval led to a quiet geomagnetic activity period (AE < 100 nT, Dst ~+5 nT). On the other hand, what is quite surprising is that this region was the most purely "Alfvénic" interval studied (CC of B_X–V_X = 0.95). The ε parameter was calculated using both GSE and GSM coordinates. It was found that ε is ~30% larger for GSM coordinates. Thus, the major cause of geomagnetic activity during HSSs is the large amplitude Alfvén waves but not coordinate transformations. Sector polarities (IMF B_Y values) may be a secondary factor. However, other models, like the tilted solar dipole, should be considered as well.

5.4.3. Response of the Polar Magnetic Field Intensity to the Exceptionally High Solar Wind Streams in 2003

As outlined Lukianova et al. (2012), the exceptionally high solar wind stream activity in 2003 caused a record intensity in the auroral electrojet currents, leading to a major reduction of the horizontal field at auroral latitudes and to a notable strengthening of the vertical geomagnetic field in the polar cap. This strengthening is clearly visible in the observatory annual values as a significant deflection in the corresponding secular variation. A similar but weaker deflection also occurs during the strongest high speed stream years of the earlier solar cycles, e.g., in 1983 and 1994. It is also found that, in addition to the disturbed times, the westward electrojet was often enhanced even during the most quiet times of the strongest high speed stream years. The quiet time level was more disturbed in 2003 than in other high speed stream years, when an exceptionally clear signal was seen in the polar cap Z intensity even in the annual mean curve in this year. Lukianova et al. (2012) exclude other current systems like the ring current or the DPY current as possible explanations.

5.4.4. Inner Magnetospheric Heavy Ion Composition during High-Speed Stream-Driven Storms

Ion composition data, taken by the Combined Release and Radiation Effects Satellite Magnetospheric Ion Composition Spectrometer instrument, are investigated by Forster et al. (2013) across eight high-speed solar wind-stream-driven storms (HSSs) during 1991. The HSSs are identified using solar wind data from OMNI alongside geomagnetic indices, and the behavior of ions in the energy range 31.2–426.0 keV is investigated. A case study of the single HSS event that occurred on 30 July 1991 is performed, and superposed epoch analyses of five events are conducted. The data show evidence of a local minimum (dropout) in the flux and partial number density of ionic species H^{+}, He^{+}, He^{++}, and O^{+} close to the onset of magnetospheric convection. The flux and number density rapidly fall and then recover over a period of hours. The initial rapid recovery in number density is observed to consist primarily of lower-energy ions. As the number density reaches its maximum, the ions show evidence of energization. Heavy ion-to-proton ratios are observed to decrease substantially during these HSS events.

5.4.5. The Response of the Inner Magnetosphere to the Trailing Edges of High-Speed Solar-Wind Streams

The effects of the leading edge stream interface of high-speed solar-wind streams (HSSs) upon the Earth's magnetosphere have been extensively documented by Denton and Borovsky (2017). The arrival of HSSs leads to significant changes in the plasmasphere, plasma sheet, ring current, and radiation belts, during the evolution from slow solar wind to persistent fast solar wind. Studies have also documented effects in the lower ionosphere and the neutral atmosphere. However, only cursory attention has been paid to the trailing-edge stream interface during the transition back from fast solar wind to slow solar wind. Here Denton and Borovsky (2017) report on the statistical changes that occur in the plasmasphere, plasma sheet, ring current, and electron radiation belt during the passage of the trailing-edge stream interface of HSSs, when the magnetosphere is in most respects in an extremely quiescent state. Counterintuitively, the peak flux of ~1 MeV electrons is observed to occur at this interface. In contrast, other regions of the magnetosphere demonstrate extremely quiet conditions. As with the leading-edge stream interface, the occurrence of the trailing-edge stream interface has a periodicity of 27 days, and hence, understanding the changes that occur in the magnetosphere during the passage of trailing edges of HSSs can lead to improved forecasting and predictability of the magnetosphere as a system.

5.4.6. Cosmic Radio Noise Absorption in the High-Latitude Ionosphere during Solar Wind High-Speed Streams

The effect of solar wind high-speed streams (HSSs) on energetic particle precipitation at auroral and subauroral latitudes (L = 3.8–5.7) is studied in Grandin et al. (2017) by using cosmic noise absorption (CNA) data measured by the Finnish riometer chain during 95 HSS events occurring between 2006 and 2008. The data are divided into "long" and "short" HSS events, depending on whether the maximum solar wind speed is reached within 24 h or later after the arrival of the corotating interaction region (CIR) at the bow shock. Grandin et al. (2017) find that CNA is more frequent during long events and extends to subauroral latitudes, unlike during short events. CNA is observed for at least 4 days after the CIR arrival during long events, and only 3 days during short events. In addition, CNA is divided into three categories, depending on whether it is associated with local substorm activity, with ultralow frequency (ULF) wave activity, or neither of these. Substorm-type CNA dominates in the 21–06 magnetic local time (MLT) sector, while ULF-type CNA dominates from morning to afternoon and follows the daily SYM-H index variations. This indicates the importance of having energetic (E >30 keV) electrons available for interacting with the very low frequency (VLF) chorus waves, which may be modulated by ULF waves. Finally, Grandin et al. (2017) find a correlation between ULF-type CNA on the dayside and substorm activity on the nightside with a 30 min delay, suggesting that substorm injections energize electrons, which then drift to the morning and dayside, where they precipitate into the ionosphere.

5.5. Coupling of Interplanetary Shocks and Coronal Mass Ejections with Geomagnetosphere

5.5.1. The Magnetospheric and Ionospheric Response to a Very Strong Interplanetary Shock and Coronal Mass Ejection

Ridley et al. (2006) present results from a coupled magnetospheric and ionospheric simulation of a very strong solar wind shock and coronal mass ejection (CME). The solar wind drivers that are used for this simulation were output from the Sun-to-Earth MHD simulation of the Carrington-like CME reported in Manchester et al. (2006). It was used the University of Michigan's BATS-R-US MHD code to model the global magnetosphere and coupled height integrated ionosphere. As the interplanetary shock swept over the magnetosphere, a wave is observed to propagate through the system. This is evident both in the magnetosphere and ionosphere. On the dayside, the magnetospheric bowshock is shown to bifurcate. The inner shock is pushed close to the inner boundary, where it "bounces" and propagates back outwards to meet the outer bowshock, which is propagating inwards. The inward and outward motion of the bowshocks can be observed propagating down the flanks of the magnetosphere. In the ionosphere, the wave is manifested as two pairs of field-aligned currents moving antisunward. The first pair is opposite of the normal region-1 current system, while the second pair is in the same sense as the normal region-1 system. The ionospheric potential shows a behavior consistent with the field-aligned current pattern, given the strong gradient in the conductance from the dayside to the nightside. As the magnetic cloud flows over the system, the entire magnetopause boundary is observed to move inside of geosynchronous orbit (6.6 R_E). At the time of the most extreme solar wind conditions, the magnetopause boundary encounters the inner edge of the magnetospheric simulation domain. During the magnetic cloud, the ionospheric cross-polar cap potential is shown to match by the Siscoe et al. (2002a,b,c) formulation relating the ionospheric potential to the solar wind and IMF conditions. It is shown that by using this formulation, the extremely large potentials observed in the MHD simulation are most likely saturated. Obtained results are illustrated by Figures 5.5.1 - 5.5.6.

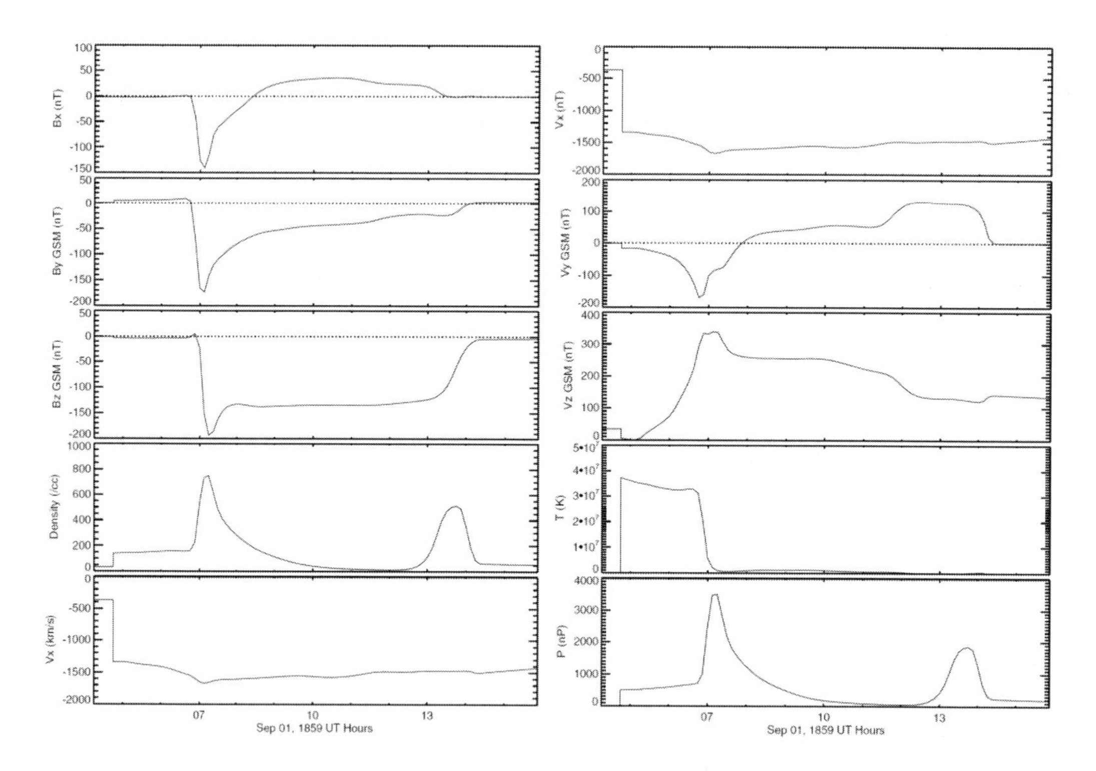

Figure 5.5.1. The IMF and solar wind conditions used to simulate the Carrington event. These are similar to the conditions reported by Manchester et al. (2006). The left figure contains, from top to bottom: B_x, B_y, B_z, the solar wind number density, and V_x. The right figure contains, from top to bottom: V_x, V_y, V_z, T, and the ram pressure of the solar wind. According to Ridley et al. (2006). Permission from COSPAR.

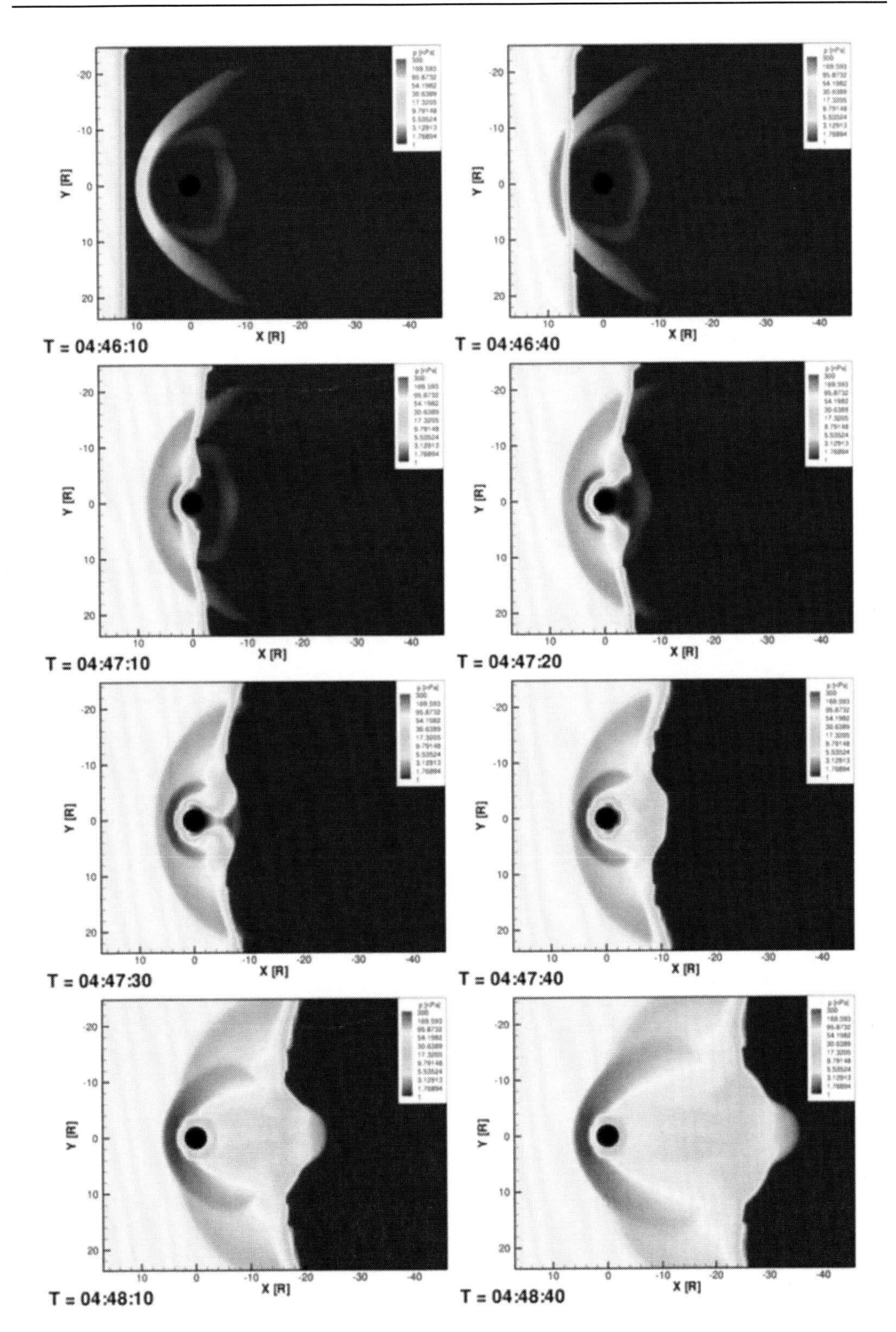

Figure 5.5.2. Plots of the pressure (in nPa) in the z = 0 plane at a number of different times during the shock encounter with the magnetosphere. The times are indicated on the lower left of each plot. According to Ridley et al. (2006). Permission from COSPAR.

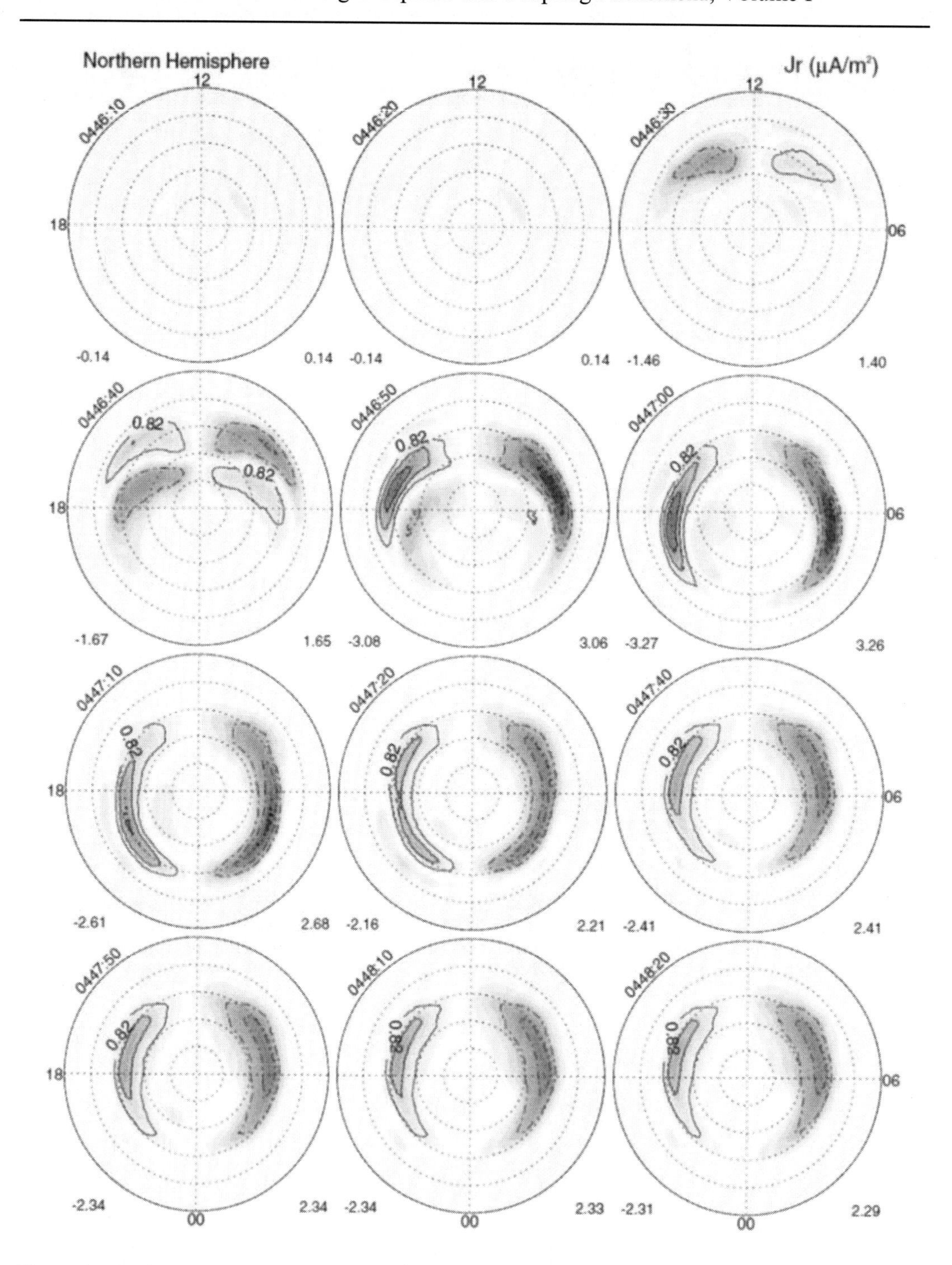

Figure 5.5.3. Plots of radial current (in $\mu A/m^2$) into the Northern ionosphere during the shock passage. The center of each plot is the magnetic pole, while the outer ring is 40° magnetic latitude. The right side of each plot is dawn while the top is noon (or towards the Sun). The maximum and minimum currents are indicated at the bottom of each plot, while the times are indicated in the upper left corner of each plot. According to Ridley et al. (2006). Permission from COSPAR.

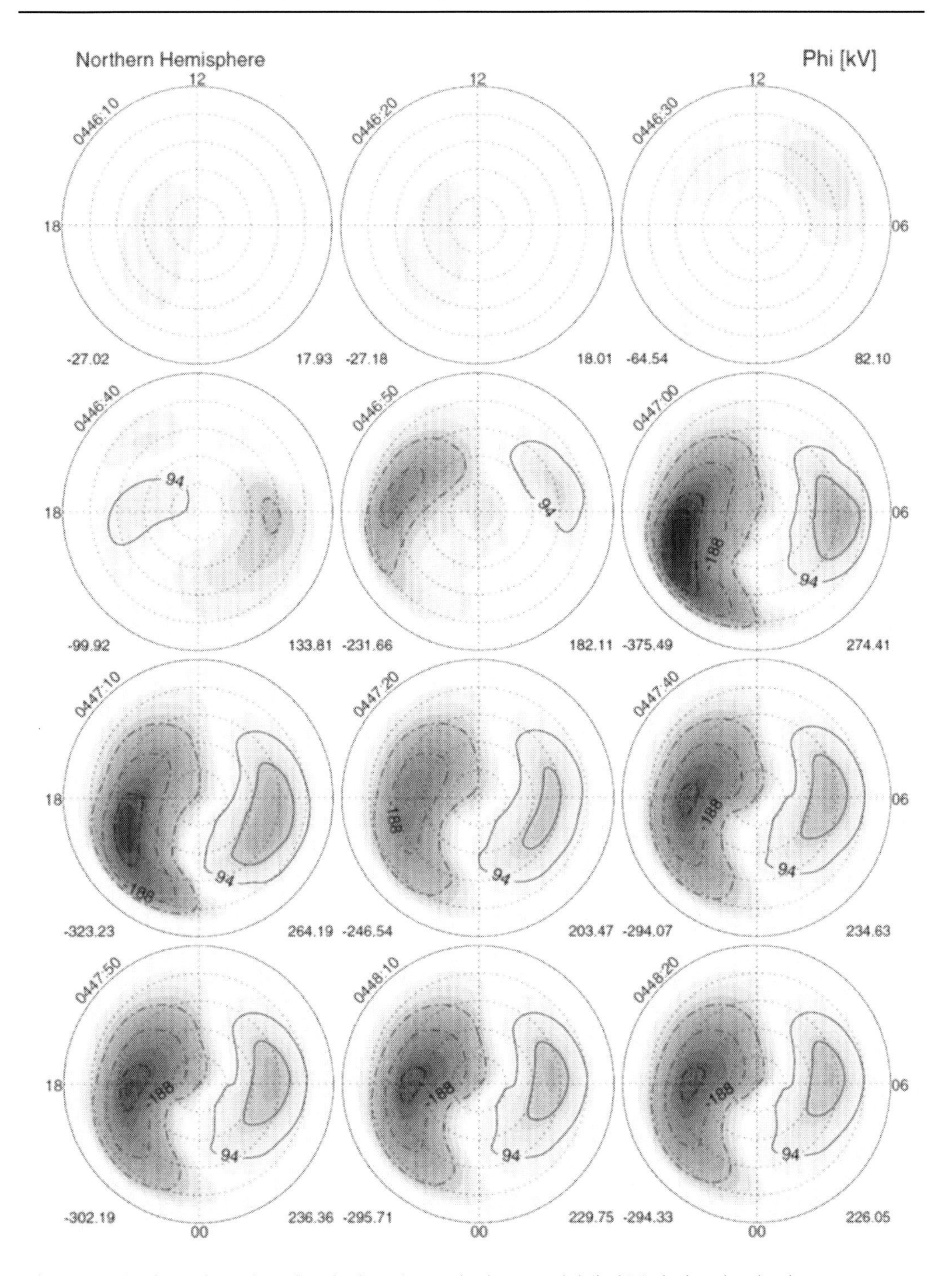

Figure 5.5.4. Plots of Northern hemisphere ionospheric potential (in kV) during the shock passage. These plots are in the same format as Figure 5.5.3. According to Ridley et al. (2006). Permission from COSPAR.

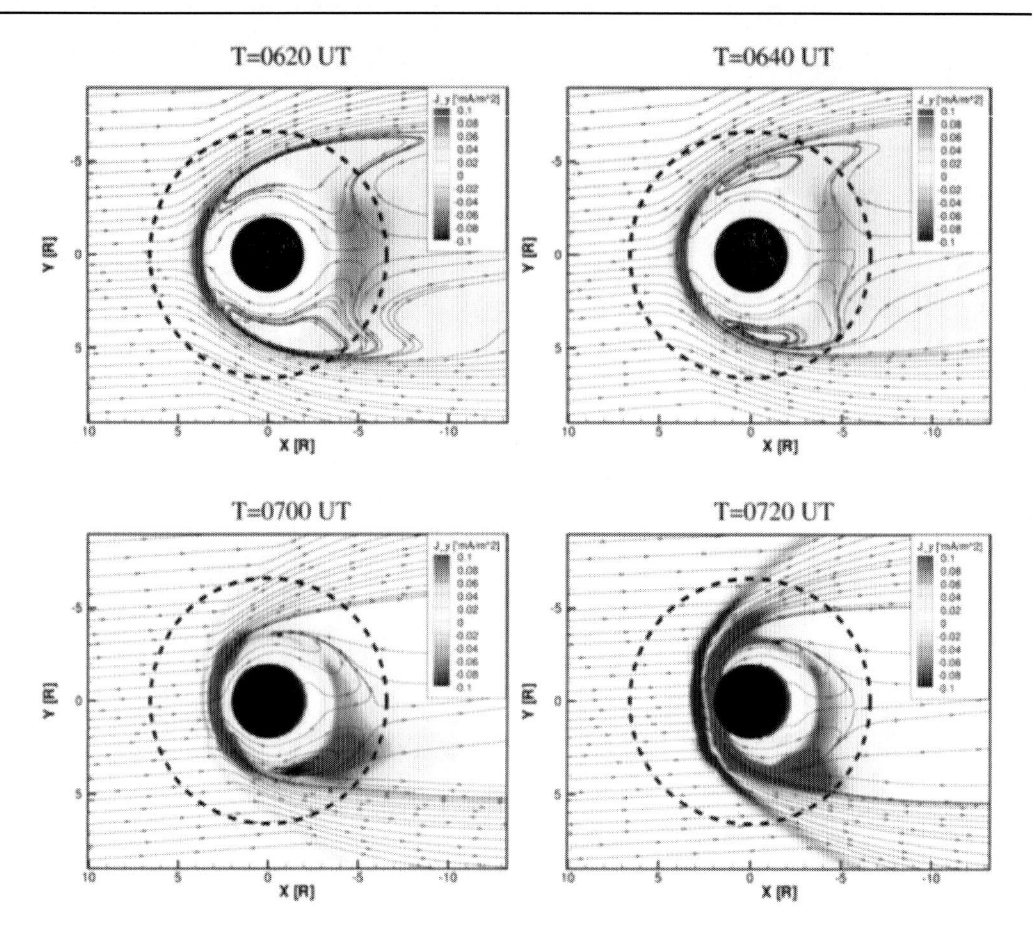

Figure 5.5.5. Plots of J_y (in μA/m^2) in the $z = 0$ plane at a number of different times during the CME encounter with the magnetosphere. Geosynchronous orbit (6.6 R_E) is shown as a dashed circle, while the inner boundary of the code (2 R_E) is shown as a filled black circle. According to Ridley et al. (2006). Permission from COSPAR.

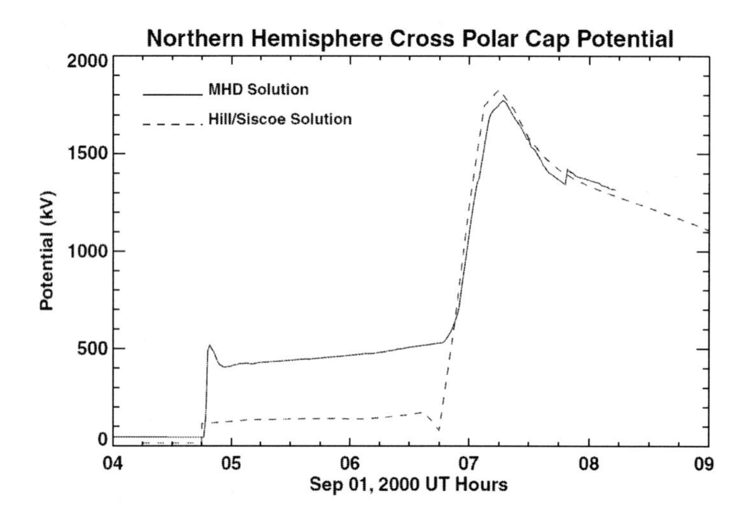

Figure 5.5.6. The ionospheric Northern hemisphere cross-polar cap potential as a function of time through the Carrington storm. According to Ridley et al. (2006). Permission from COSPAR.

5.5.2. Low-Latitude Geomagnetic Response to the Interplanetary Conditions during Very Intense Magnetic Storms

In Rawat et al. (2009) the variations in the horizontal and declination components of the geomagnetic field in response to the interplanetary shocks driven by fast halo coronal mass ejections, fast solar wind streams from the coronal hole regions and the dynamic pressure pulses associated with these events are studied. Close association between the field-aligned current density ($j_{||}$) and the fluctuations in the declination component (ΔD_{ABG}) at Alibag (geographic latitude 18.63°N, longitude 72.87°E) is found for intense storm conditions. Increase in the dawn-dusk interplanetary electric field (E_y) and ΔD_{ABG} are generally in phase.

However, when the magnetospheric electric field is directed from dusk to dawn direction, a prominent scatter occurs between the two. It is suggested that low-latitude ground magnetic data may serve as a proxy for the interplanetary conditions in the solar wind. Obtained results are illustrated by Figures 5.5.7 - 5.5.14.

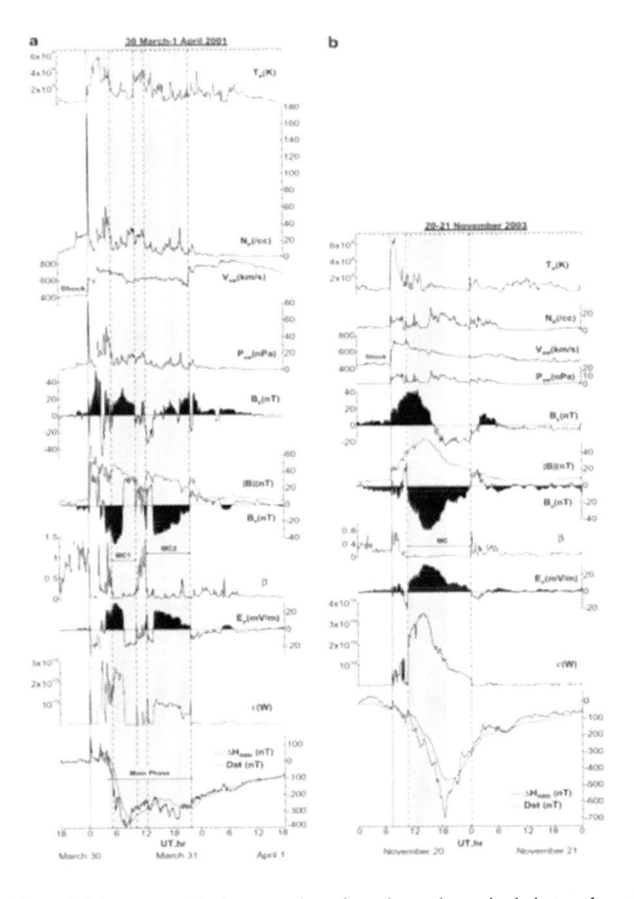

Figure 5.5.7. Panels (a) and (b), respectively, are showing the solar wind, interplanetary parameters, T_p, N_p, V_{sw}, P_{sw}, B_y, B_z, $|B|$, β, E_y, the rate of energy transferred into the magnetosphere (ε), the corresponding ground magnetic variation (ΔH_{ABG}) and the *Dst* for two intense storm events of 31 March 2001 and 20 November 2003. Alibag data have been shifted in accordance to the ACE data. Vertical shaded (light) strips depict the main phase period used and dark shaded portions in B_y, E_y indicate dusk-ward interval and in B_z show southward interval. According to Rawat et al. (2009). Permission from COSPAR.

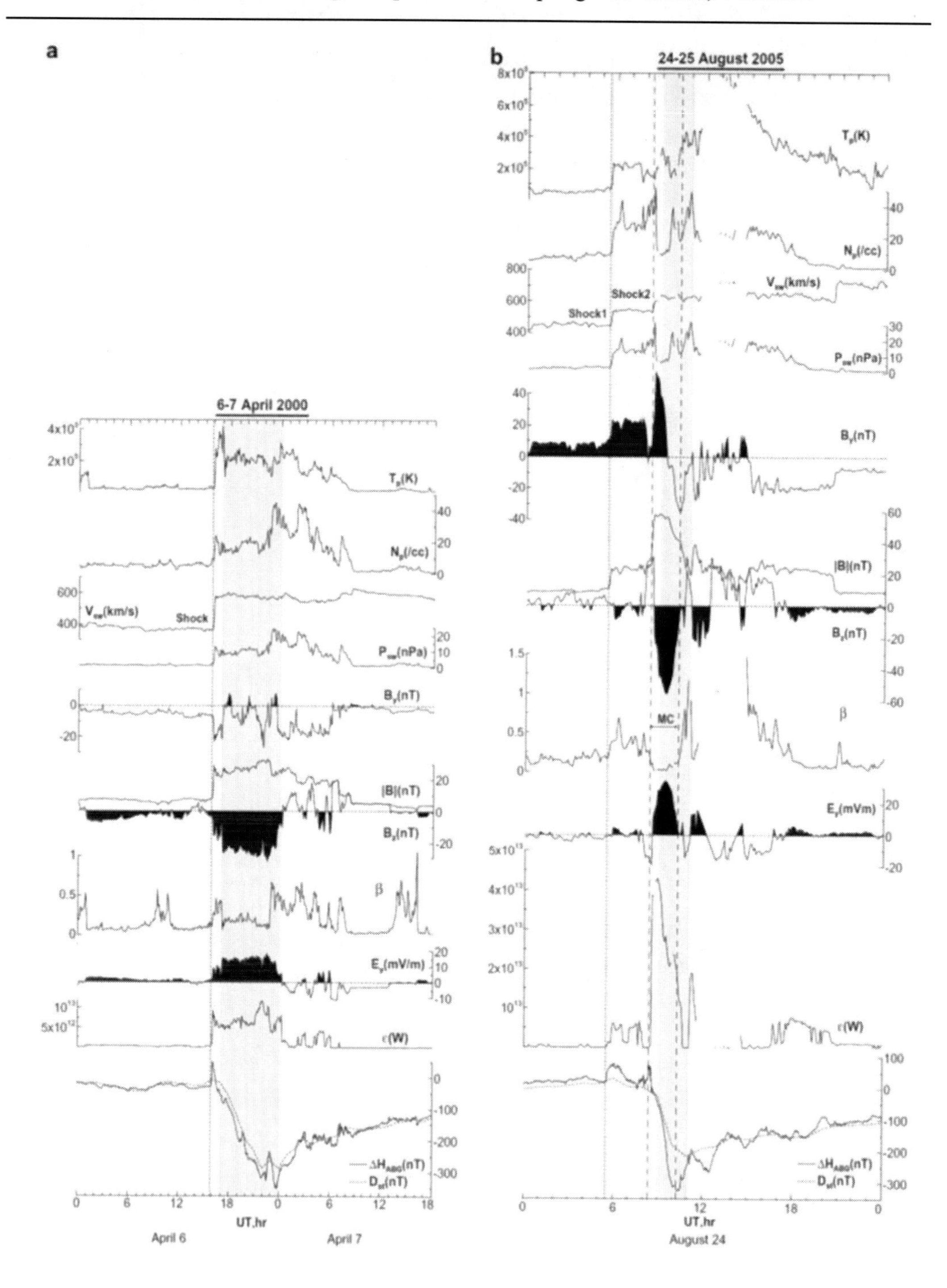

Figure 5.5.8. The same as in Figure 5.5.7, but for two major storm events of 6 April 2000 (Panel a) and 24 August 2005 (Panel b). According to Rawat et al. (2009). Permission from COSPAR.

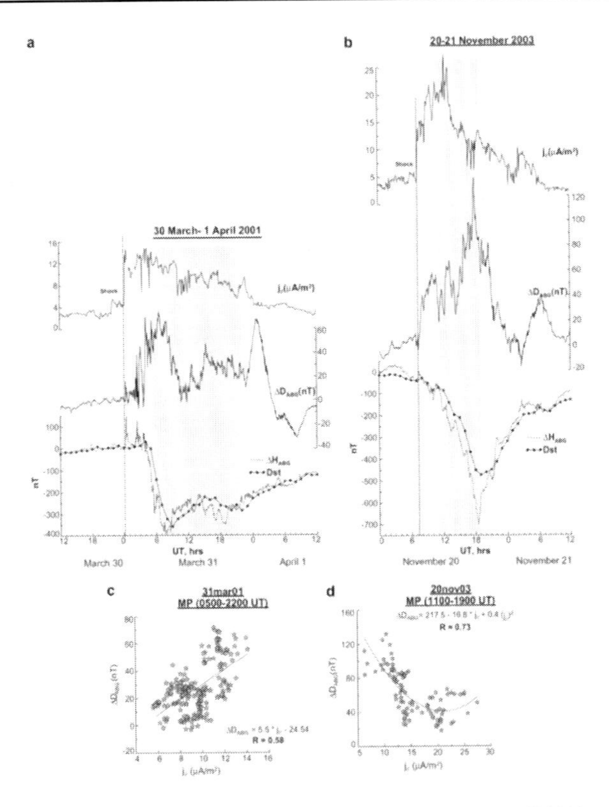

Figure 5.5.9. Variation of FAC ($j_{\|}$), ΔD_{ABG} and ΔH_{ABG} with time for (a) 30 March–1 April 2001 and (b) 20–21 November 2003. Shaded portion represents period of main phase used. Alibag data have been shifted in accordance to the ACE data (for $j_{\|}$). Panels (c) and (d), respectively, show scatter plots between $j_{\|}$ and ΔD_{ABG} for 30 March–1 April 2001 and 20–21 November 2003. According to Rawat et al. (2009). Permission from COSPAR.

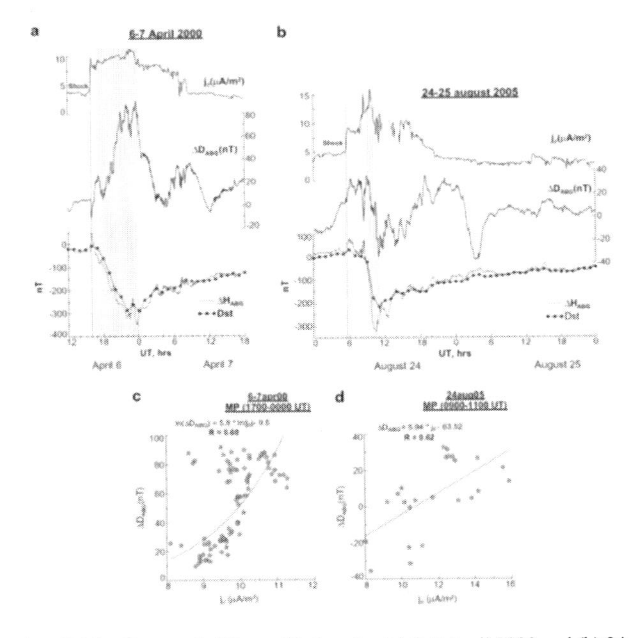

Figure 5.5.10. Variation of FAC ($j_{\|}$), ΔD_{ABG} and ΔH_{ABG} with time for (a) 6–7 April 2000 and (b) 24–25 August 2005. Shaded portion represents period of main phase used. Alibag data have been shifted in accordance to the ACE data (for $j_{\|}$). Panels (c) and (d), respectively, show scatter plots between $j_{\|}$ and ΔD_{ABG} for 6–7 April 2000 and 24–25 August 2005. According to Rawat et al. (2009). Permission from COSPAR.

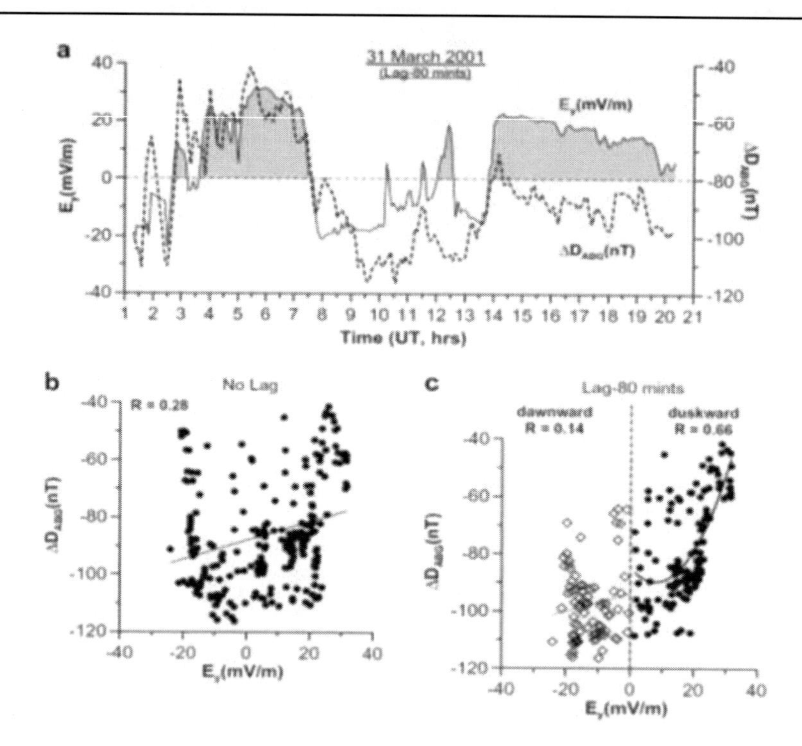

Figure 5.5.11. (a) Lagged variation of dawn-dusk electric field (E_y) and ΔD_{ABG} with time for 31 March 2001 during the main phase. Shaded portion mark duskward orientation of E_y. Alibag data (ΔD_{ABG}) has been shifted in accordance to the ACE data (for Ey). Panels (b) and (c) represent correlation between E_y and ΔD_{ABG}. Without lag (b) and with 80 min lag (c). Correlation more conspicuous in the dusk-ward E_y. According to Rawat et al. (2009). Permission from COSPAR.

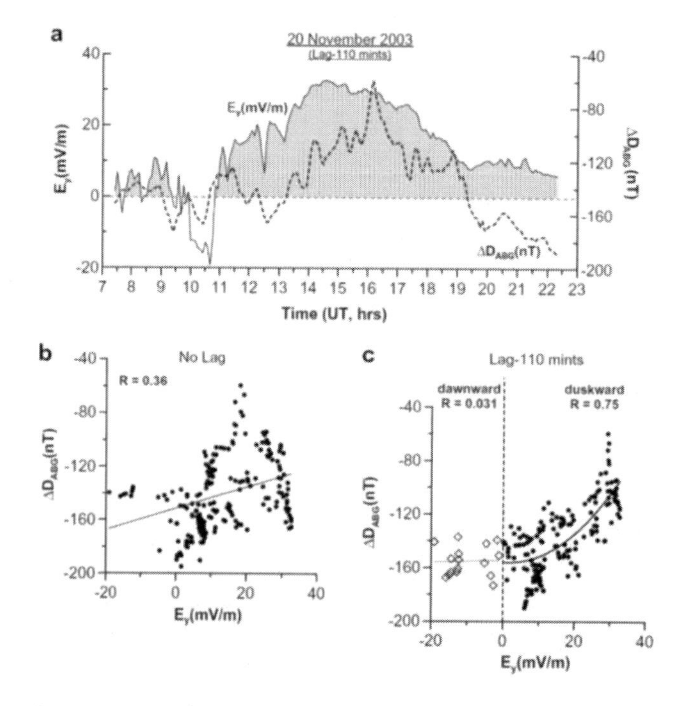

Figure 5.5.12. (a) Lagged variation of dawn-dusk electric field (E_y) and ΔD_{ABG} with time for 20 November 2003 during the main phase. Shaded portion mark dusk-ward orientation of E_y. Alibag data (ΔD_{ABG}) has been shifted in accordance to the ACE data (for E_y). Panels (b) and (c) represent correlation between E_y and ΔD_{ABG}. (b) Without lag (b) and with 110 min lag (c). Correlation more conspicuous in the dusk-ward E_y. According to Rawat et al. (2009). Permission from COSPAR.

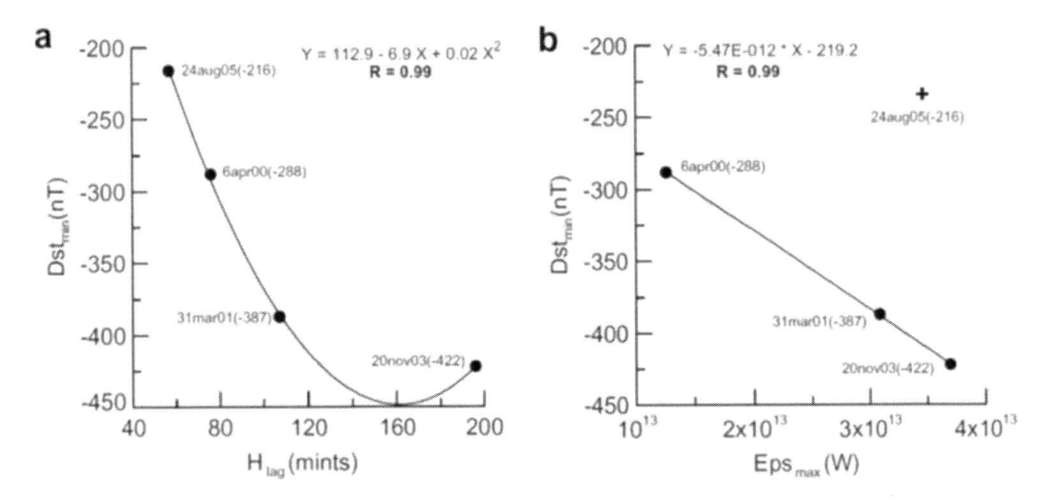

Figure 5.5.13. (a) Correspondence between H_{lag} (expressed as the time when ΔH_{ABG} attained its minimum value during the main phase with respect to the B_{zmin}) and storm intensity (Dst_{min}). (b) Epsilon maximum and Dst_{min}. According to Rawat et al. (2009). Permission from COSPAR.

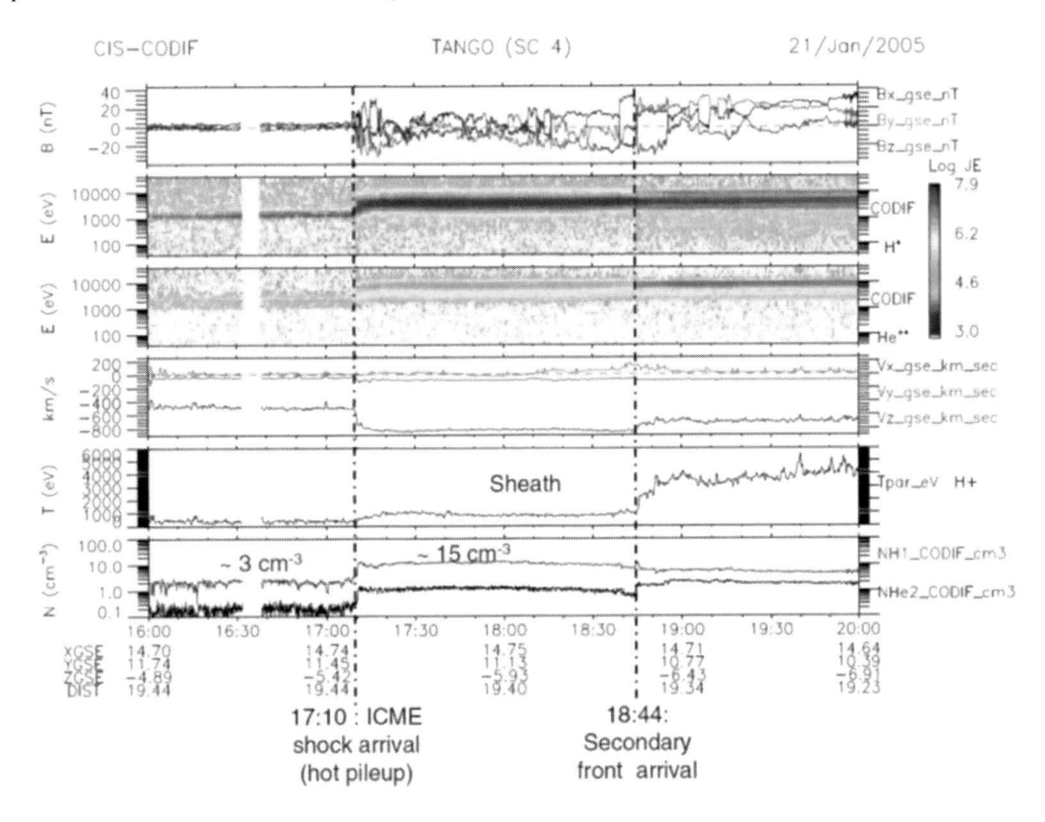

Figure 5.5.14. Cluster spacecraft SC 4 magnetic field and ion data for the January 21, 2005 event. From top to bottom: magnetic field, CIS-CODIF energy-time spectrograms separately for H^+ and He^{++} (data of energy and flux are in units: keV and $cm^{-2} s^{-1} sr^{-1} keV^{-1}$); H^+ ion bulk velocity; H^+ parallel temperature, H^+ and He^{++} densities, spacecraft coordinates (GSE system) and geocentric distance in R_E. According to Dandouras et al. (2009). Permission from COSPAR.

5.5.3. Magnetosphere Response to the 2005 and 2006 Extreme Solar Events as Observed by the Cluster and Double Star Spacecraft

As noted Dandouras et al. (2009), the four identical Cluster spacecraft, launched in 2000, orbit the Earth in a tetrahedral configuration and on a highly eccentric polar orbit (4–19.6 r_E). This allows the crossing of critical layers that develop as a result of the interaction between the solar wind and the Earth's magnetosphere. Since 2004 the Chinese Double Star TC-1 and TC-2 spacecrafts, whose payload comprise also backup models of instruments developed by European scientists for Cluster, provided two additional points of measurement, on a larger scale: the Cluster and Double Star orbits are such that the spacecrafts are almost in the same meridian, allowing conjugate studies. The Cluster and Double Star observations during the 2005 and 2006 extreme solar events are presented, showing uncommon plasma parameters values in the near-Earth solar wind and in the magnetosheath. Dandouras et al. (2009) came to the following conclusions: **(i)** Solar wind velocities up to ~900 km/s were measured during an ICME shock arrival, accompanied by a sudden increase in the density by a factor of ~5 (21 January 2005 event). **(ii)** During the secondary front of this ICME an enrichment in He^{++} was

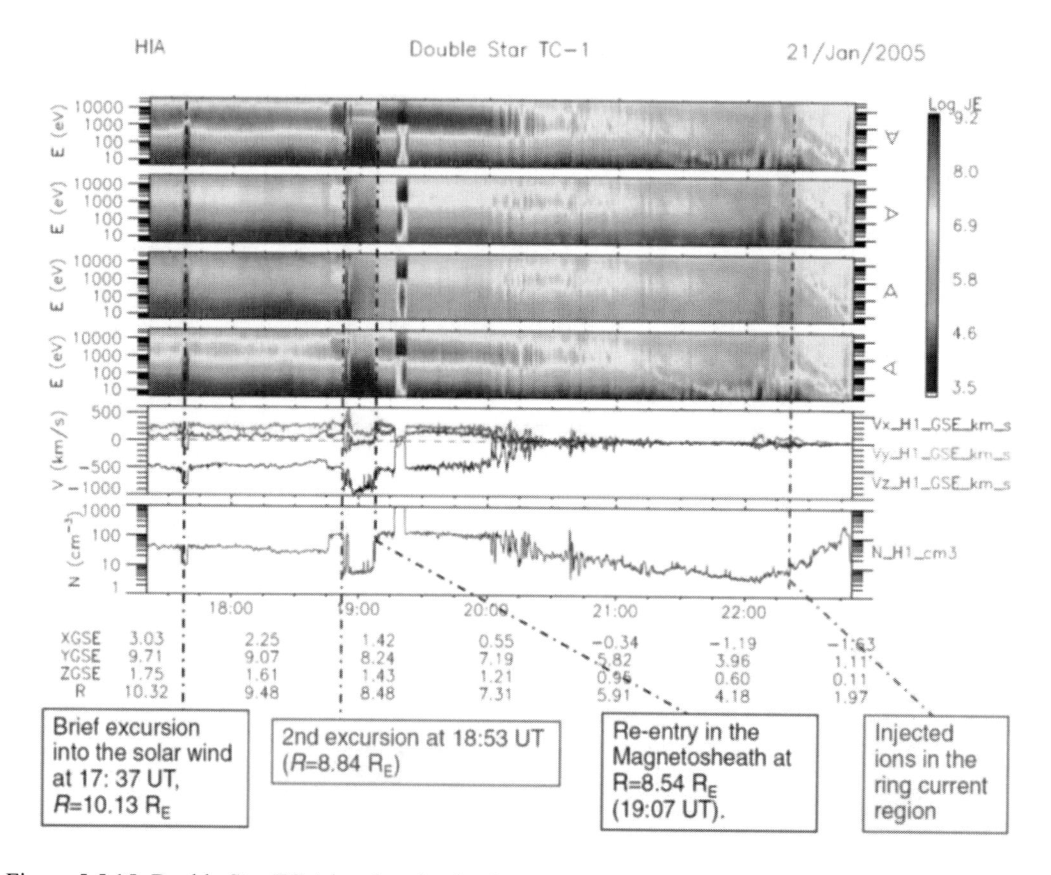

Figure 5.5.15. Double Star TC-1 ion data for the January 21, 2005 event. From top to bottom: HIA energy-time spectrograms for ions arriving in the 90° × 180° sector with a field-of-view pointing in the sun, dusk, tail, and dawn direction, respectively; data of energy and flux are in units: keV and $cm^{-2}\,s^{-1}\,sr^{-1}\,keV^{-1}$; ion bulk velocity; ion density, spacecraft coordinates (GSE system) and geocentric distance in R_E. According to Dandouras et al. (2009). Permission from COSPAR.

observed, probably indicating the arrival of the flare driver gas. **(iii)** The ICME resulted in a very strong magnetospheric compression. In the magnetosheath ion density values as high as 130 cm^{-3} were observed, and the plasma flow velocity values measured in this extreme magnetosheath regime reached 630 km/s, which is even higher than the typical solar wind velocity. **(iv)** Ring current development was monitored (14 December 2006 event). A 'nose-like' ion structure, previously formed in the ring current region and simultaneously detected by the Cluster and Double Star spacecraft in opposite MLT sectors, was 'washed out' by several successive injections of energetic particles. These injections resulted in a much harder ring current energy spectrum. **(v)** During this event the arrival of penetrating SEPs (Solar Energetic Particles) was recorded inside the pre-midnight magnetotail, about 5 hours after the impact of the CME ejecta on the Earth's magnetosphere. Obtained results are illustrated by Figures 5.5.15 – 5.5.16.

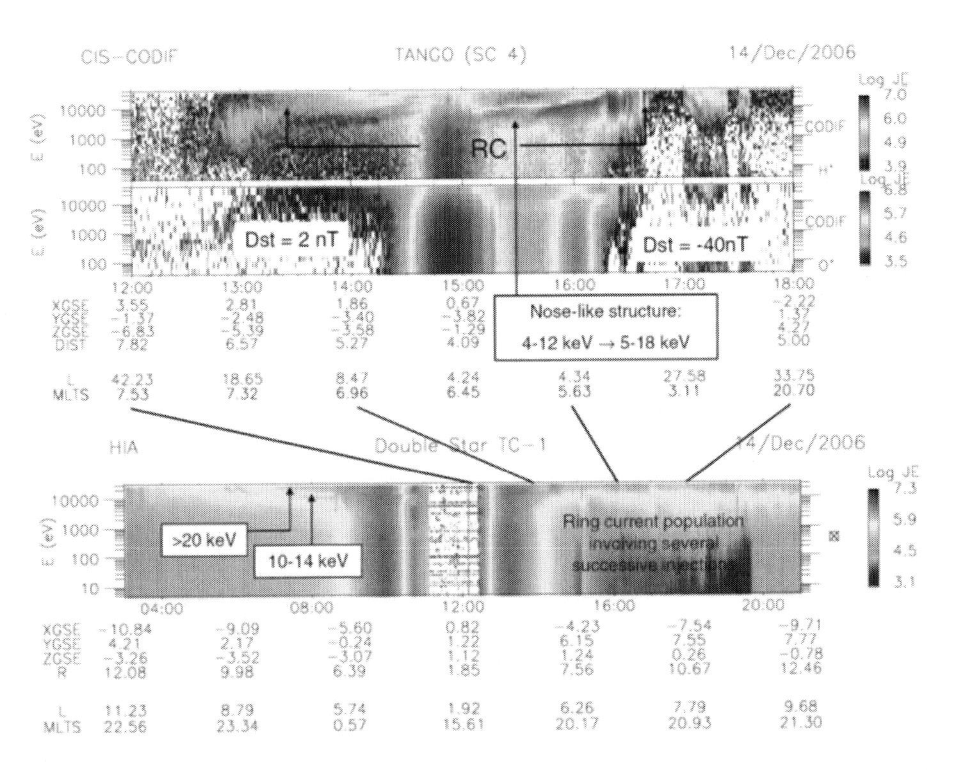

Figure 5.5.16. Combined Cluster SC 4 (CODIF H$^+$ and O$^+$ data) and Double Star TC-1 (HIA ion data) for the December 14, 2006 perigee pass. According to Dandouras et al. (2009). Permission from COSPAR.

5.5.4. Geomagnetic Storms Caused by Shocks and ICMEs

Ontiveros and Gonzalez-Esparza (2010) performed an event-by-event study of 47 geomagnetic storms (GSs) that occurred during the ascending phase of solar cycle 23. All the GSs are associated with the passage of a shock and an interplanetary coronal mass ejection (ICME). For each event, it was identified the section in the interplanetary medium causing the GS (the sheath behind the shock, the main body of the ICME or the combination of both). On average, the most intense GSs are caused by sheaths, followed by sheath-ICME combinations and by ICMEs. Ontiveros and Gonzalez-Esparza (2010) obtained the correlation coefficients between the intensity of each GS (minimum Dst) and several solar wind parameters. It was found that the well-known correlation between the GS intensity and the solar wind convected electric field, E_Y, stands for the GSs caused by ICMEs (CC = −0.88) and sheath-ICME combinations (CC = −0.95), but it is very low for the GSs caused by sheaths (CC = −0.44). In contrast, it was found a very good correlation between the GSs caused by sheaths and the total convected electric field (ΣE_Y) (CC = −0.89). On the other hand, it was estimated the total perpendicular pressure (Pt) for each interplanetary event associated with the GSs and identified the three different types of Pt profiles. The most intense GSs are related with interplanetary events with Pt = 1, but moderate and less intense storms are associated with the three Pt profiles. The correlations between the Dst and the solar wind parameters results that the CCs decrease significantly for interplanetary events having a Pt profile of 3.

5.6. Coupling of Coronal Mass Ejections and Corotating Interaction Regions with Geomagnetosphere

5.6.1. Motivation and Short History of the Problem

Motivated by recent observations and simulations of the formation of a cold and dense plasma sheet in the tail of the magnetosphere under northward IMF and of the direct influence of the plasma sheet density on the ring current strength, the paper of Lavraud et al. (2006b) aims at: **(1)** highlighting how the coupling of these effects may lead to a preconditioning of the magnetosphere under northward IMF, **(2)** performing first tests of the validity of this hypothesis. Lavraud et al. (2006b) outlined that the investigation of the geoeffectiveness of coronal mass ejections (CMEs) and corotating interaction regions (CIRs) has recently gained much interest in the context of space weather studies and the upcoming launch of missions aimed at studying such structures and their impact on the Earth (Zhang and Burlaga, 1988; Gosling et al., 1991; Gosling, 1993; Kamide et al., 1998; Gonzalez et al., 2002a,b; Cane and Richardson, 2003; Li and Luhmann, 2004; Wu and Lepping, 2002, 2005; Borovsky and Steinberg, 2006). It is known that a portion of CMEs have a well-defined embedded magnetic flux rope-like structure; these are called magnetic clouds. As noted Lavraud et al. (2006b), magnetic clouds often have a bipolar structure in their magnetic field. When the orientation of their main axis has a significant horizontal component, they are usually referred to as north-south or south-north polarity magnetic clouds. Magnetic clouds are often geoeffective structures as they usually possess a relatively strong and persistent southward magnetic field component either in the leading or trailing portion of the structure. However, the geoeffectiveness of CMEs can also be driven by the preceding sheath region of the CMEs owing to its compressed and fluctuating nature (Gosling et al., 1991; Huttunen et al., 2002). CIRs are similarly important geoeffective structures (Kamide et al., 1998; Gonzalez et al., 2002b; Borovsky and Steinberg, 2006; Denton et al., 2006) as the increased speed during such events is often accompanied by a southward component of the IMF, leading to a strong coupling through a large solar wind electric field imposed on the magnetosphere. The geoeffectiveness of solar wind structures (either CMEs or CIRs) is thus to first order attributable to the presence of a large solar wind electric field and a large dynamic pressure (e.g., Burton et al., 1975; Akasofu, 1981).

Lavraud et al. (2006b) noted that a commonly used index for assessment of geomagnetic activity is *Dst*, which is a measure of the strength of the ring current. *Dst* has been shown to correlate strongly with the magnitude of the solar wind electric field and dynamic pressure. These correlations allowed Burton et al. (1975) to construct a semi-empirical model of the *Dst* response based on inputs from the solar wind parameters. This predictive model was further modified and adapted successively by O'Brien and McPherron (2000) and Wang et al. (2003a). It is known that contributions to *Dst* come from other (e.g., tail) current systems as well (e.g., Hakkinen et al., 2003; Ganushkina et al., 2004). These currents may not be separated in Lavraud et al. (2006a). Their contribution, however, may be viewed as part of the global geo-effectiveness of the events under consideration in Lavraud et al. (2006a). As mentioned Lavraud et al. (2006b), from the study of the statistical significance of the difference between their predicted and the measured *Dst*, O'Brien and McPherron (2000) concluded that the errors in their model were caused by a few rather than many sources. An obvious source of errors is that of the measurements themselves (O'Brien and McPherron, 2000). In addition, because of the sole dependence of the model on the solar wind electric field and dynamic pressure, the effects of other operative physical processes ought to constitute additional sources of discrepancy. One such mechanism, investigated in Lavraud et al. (2006b), is a preconditioning of the magnetosphere under northward IMF; i.e., the possibility that the magnetosphere configures itself in a specific state as a function of the properties of the preceding solar wind conditions, which in turn may affect the strength of ensuing storms. Of particular importance in this context is the formation of a colder and denser plasma sheet during northward IMF (Terasawa et al., 1997; Fujimoto et al., 1998).

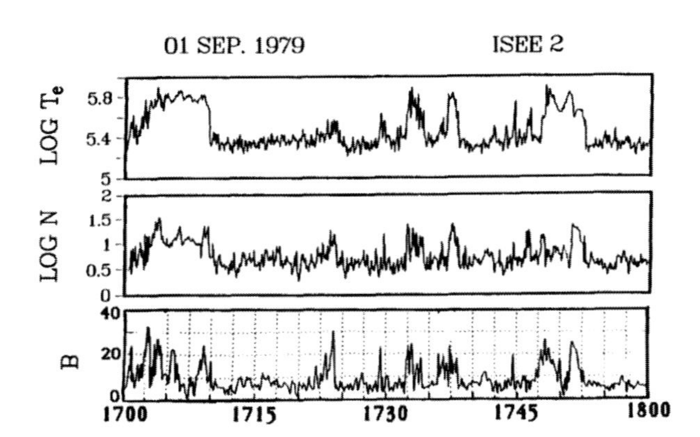

Figure 5.6.1. Plasma and field properties measured by ISEE-2 for a one-hour interval on 1 September 1979 containing a number of magnetic pulsation events. From top to bottom: Electron temperature and density (3-second measurements obtained every 3 seconds) and magnetic field strength (12-second averages plotted on a running 4-second center). According to Thomsen et al. (1998). Permission from COSPAR.

Both observations (Thomsen et al., 1998; see Figures 5.6.1- 5.6.5) and modeling efforts (Jordanova et al., 1998; Kozyra et al., 1998) have shown evidence for the role of the plasma sheet density in storm-time ring current strength, with larger plasma sheet density leading to larger negative *Dst* excursions. A lower plasma sheet temperature may add to the ring current strength as colder populations may convect deeper Earthward than their hotter counterpart (Ebihara and Ejiri, 2000). Lavraud et al. (2006b) argue that the formation of a cold, dense plasma sheet (CDPS) may lead to an enhanced *Dst* during the main phase of a storm as that denser plasma is being pushed inward during the strong convection associated with the storm main phase.

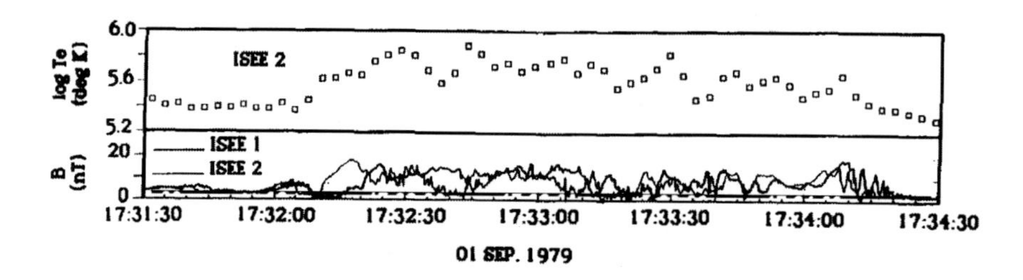

Figure 5.6.2. Temporal profile of high-resolution (16 vectors/sec) magnetic field strength observed by ISEE-1 (heavy trace) overlaid on that observed by ISEE-2 (light trace) for a pulsation event on 1 September 1979. The upper panel shows the 3-second measurements of the electron heating associated with the pulsation. The two-spacecraft timing shows a convective signature for the pulsation. According to Thomsen et al. (1998). Permission from COSPAR.

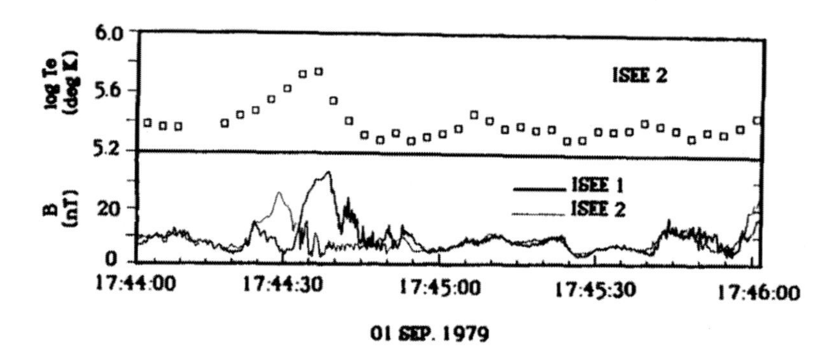

Figure 5.6.3. Same as Figure 5.6.2 but for a brief pulsation event at 1744:30 UT on 1 September 1979. Again the signature is convective, with ISEE-2 (light trace), which is the more sunward spacecraft, seeing both entrance and exit first. According to Thomsen et al. (1998). Permission from COSPAR.

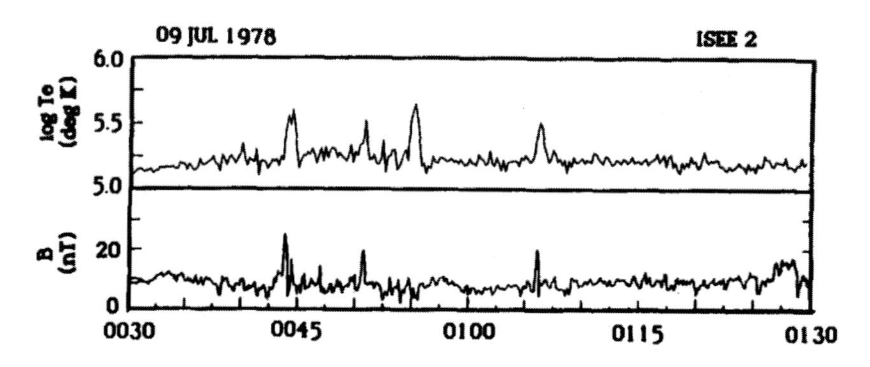

Figure 5.6.4. Electron temperature and magnetic field strength observed at ISEE-2 from 0030 to 0130 UT on 9 July 1978 showing several isolated magnetic pulsations. The nearest shock crossing occurred at ~ 0242 UT. According to Thomsen et al. (1998). Permission from COSPAR.

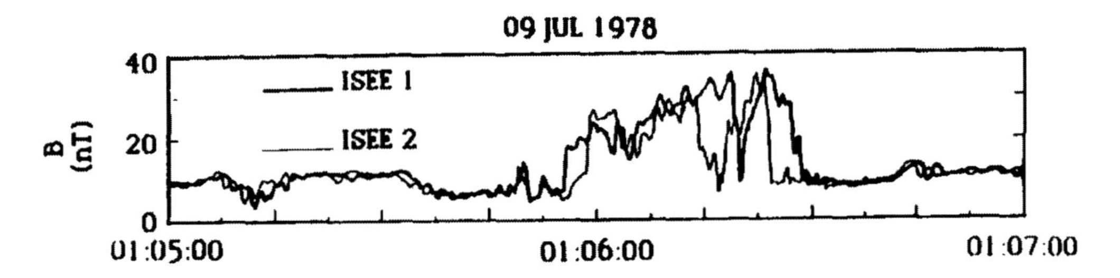

Figure 5.6.5. High time-resolution magnetic field strength profiles observed by ISEE-l (heavy trace) and ISEE-2 (light trace) for a magnetic pulsation at ~ 0106 UT on 9 July 1978. The profiles are clearly nested, with ISEE-1, the earthward spacecraft, entering the pulsation first and leaving it last. According to Thomsen et al. (1998). Permission from COSPAR.

As Lavraud et al. (2006b) outlined, a number of studies have investigated a possible dependence of CMEs geoeffectiveness on the polarity of magnetic clouds (Zhang and Burlaga, 1988; Fenrich and Luhmann, 1998; Li and Luhmann, 2004; Wu and Lepping, 2002, 2005). However, a relation between CME magnetic polarity and their absolute geoeffectiveness (in terms of *Dst* in particular) has not been fully pinned down. This is in part due to the lack of extended databases of magnetic clouds, i.e., only a subset of all CMEs. It is further complicated by the common presence of a sheath ahead of the CME itself, which often is the first geoeffective part of CMEs (Gosling et al., 1991). Periods of calms (defined as a *Kp* index lower than 1^+ for at least 6 hours) before CIR-driven storms have been noted and investigated recently by Borovsky and Steinberg (2006). They showed that such calm periods occur more frequently prior to CIR-driven storms than during more typical solar wind conditions. In their discussion, they suggested that such periods of calm may lead to a preconditioning of the magnetosphere by, in particular: **(1)** a possible mass loading after the buildup of a dense plasmasphere during quiet (low *Kp*) intervals, and **(2)** the formation of a cold and dense plasma sheet, in response to presumably northward IMF (as also suggested by Thomsen et al., 2003). As noted Lavraud et al. (2006b), there is ample evidence that both double high-latitude reconnection (Song and Russell, 1992; Øieroset et al., 2005; Li et al., 2005a; Lavraud et al., 2005a, 2006a) and the Kelvin- Helmholtz instability (Fujimoto et al., 1998; Hasegawa et al., 2004) may participate in the formation of the cold, dense plasma sheet (CDPS) under northward IMF. Lavraud et al. (2006b) do not focus on the relative importance of these processes, they only rely on the fact that the presence of the CDPS following extended periods of northward IMF is well established (e.g., Terasawa et al., 1997) and that this material has access to the inner magnetosphere under certain conditions (Thomsen et al., 2003; Lavraud et al., 2005b).

According to Lavraud et al. (2006b), any signature of the dependence of storm strength on the preceding solar wind conditions (IMF direction in particular) is necessarily embedded in the measured *Dst* trend. Primarily, *Dst* responds to coupling through the solar wind electric field and dynamic pressure. Therefore, to assess the role of secondary effects such as a preconditioning of the magnetosphere under northward IMF, one must either (1) compare the magnetospheric response of CMEs or CIRs which have similar solar wind characteristics apart from their magnetic structure (resulting in a limited number of suitable events, if any) or (2) statistically compare the actual magnetospheric response to that predicted by models which are developed to take into account the effects of the solar wind electric field and dynamic pressure. In the latter case, systematic differences observed between different sets of storms selected for specific preceding IMF conditions can be interpreted to be due to other coupling processes.

5.6.2. Used Data

Lavraud et al. (2006b) use solar wind data from the OMNI-2 data set (King and Papitashvili, 2005). This data set extends from 1963 with hourly averages of solar wind data from various solar wind monitors. Lag times from the observing spacecraft to Earth are already applied in the data set. The OMNI-2 data set has the same time resolution as the Dst index and therefore was used by O'Brien and McPherron (2000) for the construction of the Dst model described below. Data on Dst and Kp are also taken from the OMNI-2 data set. Lavraud et al. (2006b) also make use data of measurements from the Los Alamos National Laboratory MPA (Magnetospheric Plasma Analyzer) instruments onboard geosynchronous satellites. The MPA instruments are electrostatic analyzers. They measure the three-dimensional energy-per-charge distributions of both ions and electrons (see description of MPA in Bame et al., 1993; McComas et al., 1993). Data are taken from six different satellites in geosynchronous orbit covering the period between 1994 and 2003.

5.6.3. Used Dst Model

The *Dst* model used in Lavraud et al. (2006b) comes from Wang et al. (2003a). It is a modified version of the O'Brien and McPherron (2000) model that is designed to better represent the *Dst* dependence on the solar wind dynamic pressure. These models are based on that originally developed by Burton et al. (1975). The measured *Dst* index is known to have a substantial contribution coming from the magnetopause current system (i.e., the currents over the entire magnetopause induce deflections of the magnetic field at the surface of the Earth that are necessarily included as part of the *Dst* index). This contribution ought to be removed prior to analyzing *Dst* as an actual index representing the strength of the intra-magnetospheric currents, and, in particular, of the ring current itself. Contribution from the magnetopause currents mainly depends on the solar wind dynamic pressure P_{SW} and, in the model used here, is given by (e.g., O'Brien and McPherron, 2000):

$$Dst^* = Dst - 7.26\sqrt{P_{SW}} + 11, \tag{5.6.1}$$

where P_{SW} is the solar wind dynamic pressure, *Dst* is the measured index and *Dst** is that corrected for the contribution of magnetopause currents. In the models, *Dst** variations are assumed to be the result of an injection term and a decay term:

$$\frac{d}{dt} Dst^*(t) = Q(t) - \frac{Dst^*(t)}{\tau}, \tag{5.6.2}$$

where the first right hand term is injection, the second right hand term is decay, and τ is a decay timescale in hours. The statistical analysis of the response of *Dst** to solar wind parameters allowed Burton et al. (1975), and later O'Brien and McPherron (2000) and Wang et al. (2003a), to find the functional dependence of each term in the *Dst** variations as a function of the solar wind electric field (E_y or VB_Z parameter, in mV/m) and the solar wind dynamic pressure (P_{SW} in nPa). For the injection term Q (nT·h^{-1}), this parameterization is as follows:

$$Q = \begin{cases} 0, & \text{if } VB_z \le 0.49 \text{ mV/m}, \\ -4.4(VB_z - 0.49)(P_{SW}/P_0)^{\gamma}, & \text{if } VB_z > 0.49 \text{ mV/m}, \end{cases} \quad (5.6.3)$$

where the best fit parameters P_0 and γ were found to be 3.0 nPa and 0.2, respectively. The decay timescale τ (in hours) from Eq. 5.6.2 has the following parameterization:

$$\tau = \begin{cases} 8.70\exp(6.66/(6.04 + P_{SW})), & \text{for } B_z \ge 0, \\ 2.40\exp(9.74/(4.69 + VB_z)), & \text{for } B_z < 0, \end{cases} \quad (5.6.4)$$

The implementation of this *Dst** model thus requires that solar wind data be available. In the analysis of Lavraud et al. (2006b), from a given start time forward, the *Dst** index is calculated at time $t + \Delta t$ (Δt is 1 hour) by straight integration of Eq. 5.6.2 using the solar wind data at time t as input.

5.6.4. Events Selection

Lavraud et al. (2006b) study both CME- and CIR-driven storms. The list of CMEs used in this study comes from Cane and Richardson (2003). The list of CIRs comes from Borovsky and Steinberg (2006). As one aim of Lavraud et al. (2006b) study is to compare measured and modeled Dst indices, only CME- and CIR-driven storms having good solar wind data coverage were selected. In addition, Lavraud et al. (2006b) require that the storms be sufficiently, but not too strong, i.e., that during the 12 hours following storm onset, (1) the Kp index increases to a value of at least 4^+, (2) the difference between the storm onset Dst value and the minimum Dst value is at least 40 nT, (3) there is neither obvious strong activity before the storm onset nor a large multiple peak main phase, and (4) the minimum Dst is in the range (–30, –150) nT. This last criterion is required to hold for the entire storm duration and comes from the fact that the O'Brien and McPherron (2000) model has been derived from the analysis of storms having minimum Dst values larger than –150 nT, and it is thus valid only in that range. The storm onset times were determined by visual inspection of the data (from the Dst and VB_Z values in particular). This set of criteria led to a total of 60 CME-driven storms and 38 CIR-driven storms.

5.6.5. Illustration from Two Selected CME-Driven Storms

As the *Dst** model used in Lavraud et al. (2006b) study comes from a simple integration of Eq. 5.6.2 forward in time, the choice of the start time for integration may be critical. For consistency Lavraud et al. (2006b) begin the integration at a similar relative time for each storm. For every event in this study, the start time for the *Dst** model is taken to be 12 hours before the storm onset times. Then, Lavraud et al. (2006b) construct the model *Dst** traces for the next 24 hours, i.e., taking into account the 12 first hours of the storm main phase. They compare those to the measured *Dst** traces. Lavraud et al. (2006b) first illustrate the analysis and goal of the study by showing two sample CME-driven storm events. The first CME occurred on 22 January 2000 and the second on 25 June 1998. The two events were chosen for their clear difference in terms of the prevailing IMF direction prior to storm onset. Also, the other main parameters were quite similar during the first hours of the storm main phase (VB_Z and P_{SW} in particular). The 22 January 2000 event corresponds to a case with steady horizontal IMF for more than 6 hours preceding storm onset, with an IMF absolute clock angle of CA ~ 90°. By contrast, the 25 June 1998 event was preceded by a strong northward IMF (CA ~ 30°) for over 6 hours. For the 12 hours preceding the 22 January 2000 storm, the *y* component of the solar wind electric field E_Y was slightly positive, leading to a *Dst** value of the order –10 to –20 nT. The *Dst** value during the 12 hours preceding the 25 June 1998 was slightly higher (10–20 nT) owing to a negative E_Y (northward IMF). At storm onset, corresponding to a large increase in E_Y, both cases show a large decrease in *Dst**. In terms of the model *Dst**, that decrease lasts for as long as the injection term in Eq. 5.6.2 is large enough to overcome the decay term. As E_Y resumes lower values more quickly during the main phase for the 25 June 1998 than for the 22 January 2000 event, both the measured and modeled *Dst** values start to increase 5–6 hours after storm onset for that event. On 22 January 2000 the *Dst** values kept decreasing until the end of the 12 hour interval after storm onset owing to a more sustained positive E_Y.

5.6.6. Analysis of CME-Driven Storms

To search for a systematic discrepancy between the modeled and measured *Dst** as a function of the prevailing IMF direction during the period preceding storm onset, Lavraud et al. (2006b) conduct a superposed epoch analysis of the set of CME-driven storms. Figure 5.6.2 shows superposed epoch averages for the 24 hours surrounding storm onset for the 60 CME-driven storms. The 60 events were divided into two sets based on the preceding IMF conditions.

The observed difference is of the order of the average absolute deviation of the mean (i.e., the average absolute deviation from the mean divided by the square root of the number of events) shown as the statistical error bars on both the measured and modeled *Dst**, and is therefore not dramatic. However, it corresponds to 15% of the average *D*st* at this time; this trend will be confirmed by the separate study of CIR-driven storms (see below Section 5.6.7). Panel **a** shows that the dynamic pressure of the solar wind P_{dyn} is large for both sets of events. The discrepancy observed before storm onset between the modeled and measured *Dst** for both cases (Panel **d**) suggests that the pressure correction is not large enough. The discrepancy is slightly larger for the northward IMF cases (which have larger solar wind pressure associated with them). If an additional pressure correction were applied to the high-Pram main phase interval after onset, the corrected *Dst** would fall even further below the modeled value for the northward IMF case, therefore increasing the discrepancy for these conditions.

5.6.7. Analysis of CIR-Driven Storms

Lavraud et al. (2006b) present the superposed epoch results from the analysis of the 38 CIR-driven storms. The analysis performed and the event selection criteria are exactly the same as those used in the previous section for CME-driven storms. The total number of events corresponding to the horizontal/southward (northward) IMF case is 24 (14). It was observed that (1) the Kp values are similar for both sets of events (horizontal/southward and northward IMF cases), (2) the VB_Z parameter (or E_Y) and absolute IMF CA are more negative and lower, respectively, in the interval prior to storm onset for the events preceded by northward IMF, again as a result of the selection criteria, and (3) after storm onset, Ey is somewhat larger in the case of the events preceded by horizontal/southward IMF (black curves). In this case, this latter characteristic does not result in any clearly faster Dst* decrease in the first hours following storm onset as far as the measured Dst* is concerned (Panel d). The modeled Dst* depends directly on this parameter (VB_z or E_Y), however, and it therefore decreases slightly faster for the horizontal/southward IMF case. As discussed for CME-driven storms in Section 5.6.6, a possible misrepresentation of the magnetopause current system correction may be indicated as the slight difference between the modeled and observed Dst* in the immediate pre-storm hours.

In terms of the general trends, for CIR-driven storms the model Dst* again globally underestimates the decrease in measured Dst* when the IMF has been northward in the 6 hours preceding storm onset. Compared to the model predictions, storms with preceding northward IMF intervals are stronger than those without. The difference between the model and observations is even more significant than for the case of CME-driven storms when compared to the statistical error bars. The average underestimation is more than 10 nT for the northward IMF case a few hours after storm onset. It corresponds to 20% of the average Dst* at this time. This separate study of CIR-driven storms thus confirms the global underestimation of the Dst* magnitude by the model when the IMF has been northward for a substantial interval before storm onset.

Lavraud et al. (2006b) performed similar analyses using different criteria for the selection of northward IMF cases. The superposed epoch results for both CMEs and CIRs show a consistently larger Dst* magnitude than the model when one requires at least 3 and 4 hours of preceding northward IMF. Other analyses based on the IMF direction over the 12 hours preceding storm onset also confirmed this trend. This trend holds true when other sensible criteria are used for the definition of northward and horizontal/southward IMF cases. This underestimation of storm strength by the model may be attributed to additional physical processes not taken into account in the model formulation. Lavraud et al. (2006b) propose that such an additional coupling mechanism may be the occurrence of a preconditioning of the magnetosphere under northward IMF through the formation of a cold, dense plasma sheet in the mid-tail of the magnetosphere. To test this hypothesis, Lavraud et al. (2006b) study Los Alamos geosynchronous data for both sets of events. Geosynchronous spacecraft have previously been shown to constitute good monitors of plasma sheet access to the inner magnetosphere (Korth et al., 1999; Thomsen et al., 2003; Denton et al., 2005; Lavraud et al., 2005b).

5.6.8. Geosynchronous Data and Combined Results

Lavraud et al. (2006b) obtained superposed epoch results for the combined set of CME- and CIR-driven storms, zoomed on the early storm main phase. All available measurements during the time intervals of interest in the nightside region (18:00–06:00 LT) of geosynchronous orbit are used; the trends are similar when only considering measurements closer to midnight, but the statistics decrease. The global trends in the measured and modeled *Dst** for the combined set of events follow those from the separate sets discussed earlier. It was obtained that the model slightly underestimates storm strength when all events are taken into account and that in the cases preceded by southward/horizontal IMF the model better reproduces the measured *Dst**. By contrast, the model underestimates even more the strength of storms preceded by northward IMF. The underestimation attains 5–10 nT which is 10–20% of the *Dst** value at that time. For isolated cases with strong and sustained northward IMF preceding the storm, such as for the north-south polarity magnetic cloud of 25 June 1998, the underestimation may even be greater (e.g., 30%). Los Alamos data show a clear tendency for the plasma sheet accessing geosynchronous orbit in the midnight sector around storm onset to have a much larger density and a slightly lower temperature in the case of storms preceded by intervals of northward IMF than in the case of storms preceded by horizontal/southward IMF. This ensemble of results is compatible with the hypothesis that a preceding northward IMF interval leads to the formation of a cold, dense plasma sheet, which in turn is pushed inward by the increased convection at storm onset. This larger plasma sheet density seems to lead to a larger ring current than a more tenuous plasma sheet for similar solar wind driving conditions (within error uncertainties), as was suggested in the study by Thomsen et al. (1998) and several simulation studies (Jordanova et al., 1998; Kozyra et al., 1998). The colder nature of the plasma sheet also has a potential geoeffective role, as colder plasma can be convected further inward than hot plasma that curvature and gradient drifts more readily out of the ring current region (Ebihara and Ejiri, 2000).

5.6.9. Summary of Main Results of Lavraud et al. (2006b)

Lavraud et al. (2006b) summarized obtain main results as following: **(i)** It was studied the possible preconditioning effect of the IMF history on the geoeffectiveness of both CMEs (60 events) and CIRs (38 events). **(ii)** It was analyzed the magnetospheric response to those two types of solar wind structures (1) in terms of both measured and modeled *Dst** index and (2) in terms of the associated geosynchronous observations of the density and temperature of the plasma sheet accessing the inner magnetosphere at storm onset. **(iii)** It was first focused on the difference between measured and modeled (Wang et al., 2003a) *Dst** signatures during those events. **(iv)** It was shown that, for both cases separately as well as combined, the model *Dst** tends to underestimate the actual storm strength for events that are preceded by northward IMF intervals. Although the differences observed are of the order of the error bars, this trend was confirmed separately for the two sets of storms, i.e., CME- and CIR-driven storms. The average underestimation is of the order of 5–10 nT, which corresponds to 10–20% of the average *Dst** value for these events. **(v)** This result suggests that, for similar solar wind electric field and dynamic pressure profiles, CME or CIR structures that are preceded by strongly northward IMF are more geoeffective than those preceded by horizontal/southward

IMF. **(vi)** An example is the storm driven by the north-south polarity magnetic cloud on 25 June 1998, which shows an underestimation of ~30% in the magnitude of *Dst**. **(vii)** It was further analyzed the available Los Alamos geosynchronous data for the events under consideration and demonstrated that the plasma sheet density and temperature in the midnight sector around storm onset are significantly larger and slightly lower, respectively, for the events preceded by northward IMF intervals. **(viii)** This fact shows the presence of a colder and denser plasma sheet before storm onset in the mid-tail region and formed under the prevailing northward IMF conditions. **(ix)** This cold, dense plasma sheet apparently is conducive of an increased ring current owing to potentially both its larger density and colder nature. The formation of a cold, dense plasma sheet under northward IMF thus seems to precondition the magnetosphere by leading to the creation of a stronger ring current during the early main phase of the ensuing storm. **(x)** These results highlight the necessity for further studies of this, as well as other, potential preconditioning processes. Future models need to better represent the effects of such preconditioning mechanisms through the addition of appropriate formulations in the *Dst* models. **(xi)** The current number of CME and CIR events of substantial strength and with good solar wind data coverage (and steady preceding conditions) may not be extensive enough to perform such a task. However, the fact that north-south polarity magnetic clouds are expected to become prevalent during solar cycle 24 starting in 2007 (Mulligan et al., 1998; Li and Luhmann, 2004) may enable: (1) to further confirm, (2) to better quantify this mechanism, and (3) to potentially implement it in more elaborate schemes (e.g., Temerin and Li, 2002) of *Dst* prediction.

5.7. COUPLING DIFFERENT SOLAR WIND DISCONTINUITIES WITH GEOMAGNETOSPHERE

5.7.1. Geomagnetic Activity Associated with Magnetic Clouds, Magnetic Cloud-Like Structures and Interplanetary Shocks for the Period 1995–2003

Using nine years (1995–2003) of solar wind plasma and magnetic field data, solar sunspot number, and geomagnetic activity data, Wu and Lepping (2008) investigated the geomagnetic activity associated with magnetic clouds (MCs), magnetic cloud-like structures (MCLs), and interplanetary shock waves. Eighty-two MCs and one hundred and twenty-two MCLs were identified by using solar wind and magnetic field data from the WIND mission, and two hundred and sixty-one interplanetary shocks were identified over the period of 1995–2003 in the vicinity of Earth. It is found that MCs are typically more geoeffective than MCLs or interplanetary shocks. The occurrence frequency of MCs is not well correlated with sunspot number. By contrast, both occurrence frequency of MCLs and sudden storm commencements (SSCs) are well correlated with sunspot number. Obtained results are illustrated by Tables 5.7.1 – 5.7.2 and Figure 5.7.1.

Table 5.7.1. The magnitude of yearly averages of solar wind parameters for MCLs/MCs and Dst/SSCs. According to Wu and Lepping (2008). Permission from COSPAR

	Year									
	1995	1996	1997	1998	1999	2000	2001	2002	2003	Average
	MCLs									
No.	6	1	6	16	16	24	21	21	11	13.9
Dst_{min} (nT)	−49	−41	−59	−61	−31	−46	−36	−46	−40	−45
	MCs									
No.	8	4	17	11	4	14	10	10	4	9.5
Dst_{min} (nT)	−56	−27	−75	−75	−116	−134	−112	−92	−112	−91
	SSCs[a]									
No.	23	9	28	35	31	44	42	36	13	29
Dst_{min}[b] (nT)	−51	−22	−64	−65	−64	−91	−88	−60	−159	−74.6
	SSCs(no MCs/MCLs)[c]									
No.	17	8	15	25	25	31	26	24	11	20.2
Dst_{min}[d] (nT)	−46	−23	−51	−52	−51	−82	−87	−48	−168	−66

[a] Yearly occurrence rate of SSCs, [b] Dstmin caused by an SSC, [c] Yearly occurrence rate of SSCs without MCs/MCLs, [d] Dst_{min} without MCs/MCLs.

Table 5.7.2. Correlation coefficients between yearly occurrence rate of sunspot numbers (SN), MCs, MCLs, joint seta, SSCs and SSCs/without MCs/MCLs. According to Wu and Lepping (2008). Permission from COSPAR

	SN[b]	MCs	MCLs	Joint set[a]	SSCs	SSCs(no MC/MCL)[c]
SN[b]	–	0.12	0.97	0.81	0.77	0.82
MCs	–	–	0.28	0.67	0.61	0.41
MCLs	–	–	–	0.90	0.81	0.90
Joint Set[c]	–	–	–	–	0.95	0.89

[a] Joint set is MCs + MCLs, [b] Sunspot numbers, [c] SSCs without MCs or MCLs.

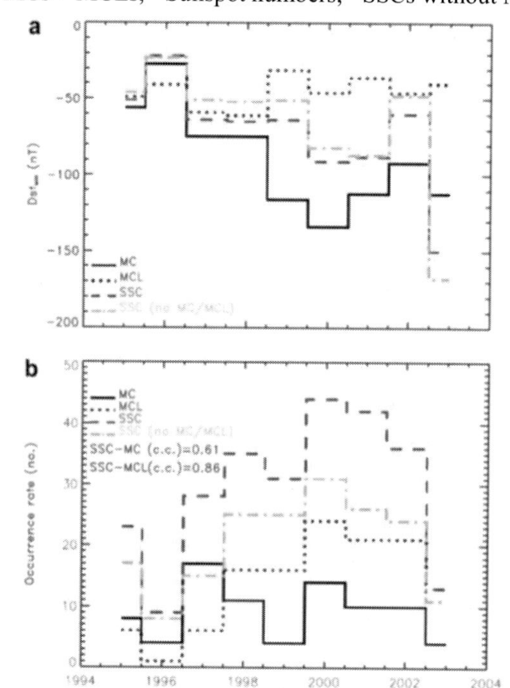

Figure 5.7.1. Yearly occurrence rate and Dstmin for MC, MCL, SSC, and SSC without MC/MCL. According to Wu and Lepping (2008). Permission from COSPAR.

5.7.2. Statistical Study of Interplanetary Condition Effect on Geomagnetic Storms

Based on the archive OMNI data for the period 1976-2000 an analysis has been made by Yermolaev et al. (2010) of 798 geomagnetic storms with Dst < –50 nT and their interplanetary sources-large-scale types of the solar wind: CIR (145 magnetic storms), Sheath (96), magnetic clouds MC (62), and Ejecta (161). The remaining 334 magnetic storms have no well-defined sources. For the analysis, Yermolaev et al. (2010) applied the double method of superposed epoch analysis in which the instants of the magnetic storm beginning and minimum of Dst index are taken as reference times. The well-known fact that, independent of the interplanetary source type, the magnetic storm begins in 1-2 h after a southward turn of the IMF ($B_Z < 0$) and both the end of the main phase of a storm and the beginning of its recovery phase are observed in 1-2 h after disappearance of the southward component of the IMF is confirmed. Also confirmed is the result obtained previously that the most efficient generation of magnetic storms is observed for Sheath before MC. On the average parameters B_Z and E_Y slightly vary between the beginning and end of the main phase of storms (minimum of Dst and Dst* indices), while Dst and Dst* indices decrease monotonically proportionally to integral of B_Z and E_Y over time. Such a behavior of the indices indicates that the used double method of superposed epoch analysis can be successfully applied in order to study dynamics of the parameters on the main phase of magnetic storms having different duration.

5.7.3. Review of Interplanetary Discontinuities and their Geomagnetic Effects

Interplanetary discontinuities and their geomagnetic effects are reviewed by Tsurutani et al. (2011b) for magnetospheric/space weather researchers. Discontinuities are particularly useful as diagnostics since they are clearly identifiable in interplanetary data and their geomagnetic effects are unambiguous most of the time. Directional discontinuities (DDs) are abrupt changes in the interplanetary magnetic field direction and plasma parameters. DDs may be rotational discontinuities (RDs), tangential discontinuities (TDs) contact discontinuities (CDs) or shocks (fast (FS), intermediate (IS) and slow (SS)). Shocks can propagate in the direction of the driver (forward shocks or FSs) or opposite to the driver (reverse shocks of RSs). Discontinuities interacting with other discontinuities may create new discontinuities. Fast forward shocks (FFSs) are shown to energize trapped particles by compressive effects, cause dayside aurora, lead to the creation of new radiation belts and to trigger nightside sector magnetospheric substorms. Fast reverse shocks (FRSs) or reverse waves (RWs) lead to magnetospheric expansions and the cessation of geomagnetic activity. TD-bow shock interactions create hot flow anomalies (HFAs) which then lead to outward expansions of the local magnetopause and dayside auroral enhancements. Some DD crossings may cause sudden southward IMF turnings. These cause magnetic reconnection and energy input into the magnetosphere–ionosphere–magnetotail system. Substorms sometimes occur thereafter. DDs that entail northward IMF turnings may lead to the triggering of substorms.

5.7.4. Geoefficiency of Solar Wind Discontinuities

Andreeova et al. (2011) have investigated the solar wind-magnetosphere coupling efficiency in response to solar wind dynamic pressure enhancements. It was investigate fast forward shocks, corotating interaction regions (CIRs), and pressure pulses separately, by conducting a statistical driver-response analysis. Andreeova et al. (2011) complement the observations by results from the GUMICS-4 global MHD simulation, which in all cases is in good agreement with the observational results. Analysis of shock propagation inside the magnetosphere shows that the events with higher speed are associated with higher ionospheric activity. The coupling efficiency depends on the disturbance characteristics: it is higher for small solar wind E_Y and smaller for high E_Y. The coupling efficiency is highest for small values of IMF B_Z and increases with increasing solar wind speed. These results confirm that pressure changes drive dynamic changes in the ionosphere independent of the changes associated with the driving reconnection electric field.

5.7.5. Interplanetary Origin of Intense, Superintense and Extreme Geomagnetic Storms

Gonzalez et al. (2011) present a review on the interplanetary causes of intense geomagnetic storms (Dst ≤ -100 nT), that occurred during solar cycle 23 (1997-2005). It was reported that the most common interplanetary structures leading to the development of intense storms were: magnetic clouds, sheath fields, sheath fields followed by a magnetic cloud and corotating interaction regions at the leading fronts of high speed streams. However, the relative importance of each of those driving structures has been shown to vary with the solar cycle phase. Superintense storms (Dst ≤ -250 nT) have been also studied in more detail for solar cycle 23, confirming initial studies done about their main interplanetary causes. The storms are associated with magnetic clouds and sheath fields following interplanetary shocks, although they frequently involve consecutive and complex ICME structures. Concerning extreme storms (Dst ≤ -400 nT), due to the poor statistics of their occurrence during the space era, only some indications about their main interplanetary causes are known. For the most extreme events, it was reviewed the Carrington event and also discuss the distribution of historical and space era extreme events in the context of the sunspot and Gleissberg solar activity cycles, highlighting a discussion about the eventual occurrence of more Carrington-type storms.

5.7.6. On Magnetospheric Response to Solar Wind Discontinuities

As outlined Zong and Hui (2011), the interaction of the Earth's magnetosphere with the solar wind is a fundamental problem in space plasma physics. Sudden changes in the Interplanetary IMF/solar wind conditions, e.g., coronal mass ejections (CMEs), solar energetic particles (SEPs), and interplanetary shocks, provide excellent opportunities to study the complex response of the Earth's magnetosphere and ionosphere to the solar wind.

The interaction of solar wind discontinuities with the magnetosphere includes several phases, including interaction with the bow shock, propagation in the magnetosheath, interaction with the magnetopause, transmission into the magnetosphere and the ionosphere. The studies of magnetospheric phenomena in conjunction with solar wind discontinuities are tractable from at least two obvious vantage points: First, unlike other magnetospheric phenomena, e.g., substorms, the specific energy source for the solar wind discontinuity related magnetospheric phenomena is clear. Second, the response of the magnetospheric system to solar wind discontinuities usually yield a significant and easily identified electromagnetic signal. Thus, there is no temporal ambiguity for the solar wind discontinuity related phenomena.

5.7.7. Dependence of Geomagnetic Activity during Magnetic Storms on the Solar Wind Parameters: Main Phase of Storm

Nikolaeva et al. (2012a) analyzed the development of the main phase of magnetic storms with Dst ≤ -50 nT, the interplanetary source of which consists of eight types of solar wind streams: magnetic clouds (MC, 17 storms); corotating interaction regions (CIR, 49 storms); Ejecta (50 storms); compressed region (Sheath) before Ejecta Sh_E (34 storms); the Sheath before a magnetic cloud Sh_{MC} (6 storms); all Sheath before all ICME, $Sh_E + Sh_{MC}$ (40 storms); all ICME, MC + Ejecta (67 storms); and an indeterminate type of stream IND (34 storms).Threshold value estimates of the integral electric field sumEy to attain an intensity level of moderate (Dst ≤ -50 nT) and strong (Dst ≤ -100 nT) magnetic storms indicate a tendency toward their dependence on the type of magnetic storm source. It can be assumed that, on average, Sheath before ICME have threshold values 1.5 times less that ICME themselves.

5.7.8. Dependence of Geomagnetic Activity during Magnetic Storms on the Solar Wind Parameters: Development of Storm

The paper of Nikolaeva et al. (2012b) analyzes the development of the main phase of 190 magnetic storms with Dst ≤ -50 nT depending on the type of source in the solar wind (magnetic clouds, MC; corotating interaction regions, CIR; Ejecta; Sheath before them, ShE; Sheath before MC, ShMC; all Sheath regions before ICME, $Sh_E + Sh_{MC}$; all ICME, MC + Ejecta; and an indeterminate type of solar wind stream, IND). It is shown that at the main phase of all types of magnetic storms, the Dst index decreases with increasing integral electric field ΣE_Y. The closeness of the relationship between these parameters (correlation coefficient CC) is higher for magnetic storms caused by Sheath before ICME than by MC and CIR. It is possible to assume that a high dynamic pressure intensifies the electric field effectiveness for four types of streams: Sheath (Sh_E, $Sh_E + Sh_{MC}$), CIR, and IND). Apparently, the Dst index does not depend on the IMF fluctuation level σB for almost all types of streams (differences within error limits) against the background of the Dst dependence on ΣE_Y of the main phase of magnetic storm.

5.7.9. Impact of Solar Wind Tangential Discontinuities on the Earth's Magnetosphere

The collision of a solar wind tangential discontinuity with the bow shock and magnetopause is considered by Grib (2012) in the scope of an MHD approximation. Using MHD methods of trial calculations and generalized shock polars, it has been indicated that a fast shock refracted into the magnetosheath originates when density increases across a tangential discontinuity and a fast rarefaction wave is generated when density decreases at this discontinuity. It has been indicated that a shock front shift under the action of collisions with a tangential discontinuity is experimentally observed and a fast bow shock can be transformed into a slow shock. Using a specific event as an example, it has been demonstrated that solar wind tangential discontinuity affects the geomagnetic field behavior.

5.7.10. Interplanetary Origins of Moderate (–100 nT < Dst ≤–50 nT) Geomagnetic Storms during Solar Cycle 23 (1996–2008)

The interplanetary causes of 213 moderate intensity (–100 nT < peak Dst ≤ –50 nT) geomagnetic storms that occurred in solar cycle 23 (1996-2008) are identified by Echer et al. (2013). Interplanetary drivers such as corotating interaction regions (CIRs), pure high speed streams (HSSs), interplanetary coronal mass ejections (ICMEs) of two types: those with magnetic clouds (MCs) and those without (non-magnetic cloud or ICME_nc), sheaths (compressed and/or draped sheath fields), as well as their combined occurrence, were identified as causes of the storms. The annual rate of occurrence of moderate storms had two peaks, one near solar maximum and the other in the descending phase, around 3 years later. The highest rate of moderate storm occurrence was found in the declining phase (25 storms/year). The lowest occurrence rate was 5.7 storms/year and occurred at solar minimum. All moderate intensity storms were associated with southward IMF, indicating that magnetic reconnection was the main mechanism for solar wind energy transfer to the magnetosphere. Most of these storms were associated with CIRs and pure HSSs (47.9%), followed by MCs and non-cloud ICMEs (20.6%), pure sheath fields (10.8%), and sheath and ICME combined occurrence (9.9%). In terms of solar cycle dependence, CIRs and HSSs are the dominant drivers in the declining phase and at solar minimum. CIRs and HSSs combined have about the same level of importance as ICMEs plus their sheaths in the rising and maximum solar cycle phases. Thus CIRs and HSSs are the main driver of moderate storms throughout a solar cycle, but with variable contributions from ICMEs, their shocks (sheaths), and combined occurrence within the solar cycle. This result is significantly different than that for intense (Dst ≤ –100 nT) and superintense (Dst ≤ – 250 nT) magnetic storms shown in previous studies. For superintense geomagnetic storms, 100% of the events were due to ICME events, while for intense storms, ICMEs, sheaths and their combination caused almost 80% of the storms. CIRs caused only 13% of the intense storms. The typical interplanetary electric field (E_Y) criteria for moderate magnetic storms were identified. It was found that ~80.1% of the storms follow the criteria of $E_Y \geq 2$ mV/m for intervals longer than 2 hours. It is concluded that southward directed IMF within CIRs/HSSs may be the main energy source for long-term averaged geomagnetic activity at Earth.

5.7.11. Very Intense Geomagnetic Storms and Their Relation to Interplanetary and Solar Active Phenomena

Szajko et al. (2013) revisit previous studies in which the characteristics of the solar and interplanetary sources of intense geomagnetic storms have been discussed. In this particular analysis, using the Dst time series, Szajko et al. (2013) consider the very intense geomagnetic storms that occurred during Solar Cycle 23 by setting a value of $Dst_{min} \leq -200$ nT as threshold. After carefully examining the set of available solar and in situ observations from instruments aboard the Solar and Heliospheric Observatory (SOHO) and the Advanced Composition Explorer (ACE), complemented with data from the ground, it was identified and characterized the solar and interplanetary sources of each storm. That is to say, Szajko et al. (2013) determine the time, angular width, plane-of-the-sky, lateral expansion, and radial velocities of the source coronal mass ejection (CME), the type and heliographic location of the CME solar source region (including the characteristics of the sunspot groups), and the time duration of the associated flare. After this, it was investigate the overall characteristics of the interplanetary main-phase storm driver, including the time arrival of the shock/disturbance at 1 AU, the type of associated interplanetary structure/ejecta, the origin of a prolonged and enhanced southward component (B_S of the interplanetary field, and other characteristics related to the energy injected into the magnetosphere during the storm (i.e., the solar wind maximum convected electric field, E_Y). The analyzed set consists of 20 events, some of these are complex and present two or more Dst minima that are, in general, due to consecutive solar events. The 20 storms are distributed along Solar Cycle 23 (which is a double-peak cycle) in such a way that 15% occurs during the rising phase of the cycle, 45% during both cycle maxima, and, surprisingly, 40% during the cycle descending phase. This latter set includes half of the superstorms and the only cycle extreme event. 85% of the storms are associated to full halo CMEs and 10% to partial halo events. One of the storms occurred at the time contact with SOHO was lost. The CME solar sources of all analyzed storms, but one, are active regions (ARs). The source of the remaining CME is a bipolar low-field region where a long and curved filament erupts. The ARs where the CMEs originate show, in general, high magnetic complexity; delta spots are present in 74% of the ARs, 10% are formed by several bipolar sunspot groups, and only 16% present a single bipolar sunspot group. All CMEs are associated to long duration events (LDEs), exceeding 3 h in all cases, with around 75% lasting more than 5 h. The associated flares are, in general, intense events, classified as M or X in soft X-rays; only 3 of them fall in the C class, with the one happening in the bipolar low field region hardly reaching the C level. It was calculate the lateral expansion velocity for most of the CMEs. The values found exceed in all cases but one the fast solar wind speed ($\approx$750 km/s). The average lateral expansion velocity is 2400 km/s. The spatial distribution of the solar CME sources on the solar disk shows an evident asymmetry; while there are no sources located more eastward than 12° in longitude, there are 7 events more westward than 12°. Nevertheless, the bulk of the solar sources are located near Sun center, i.e., at less than 20° in longitude or latitude. Considering the interplanetary structures responsible for a long and enhanced B_S, it is find that 35% correspond to magnetic clouds (MCs) or ICME fields, 30% to sheath fields, and 30% to combined sheath and MC or ICME fields. For only one storm the origin of B_S is related to the back compression of an ICME by a high speed stream coming from a coronal hole in the neighborhood of the corresponding CME source region. It was also found that for this particular set of storms the linear relation between E_Y and the storm intensity holds (with a correlation coefficient of 0.73). Obtained results are illustrated by Table 5.7.3 and Figures 5.7.2 – 5.7.4.

Table 5.7.3. Very intense geomagnetic storms during Solar Cycle 23. According to Szajko et al. (2013). Permission from COSPAR

#	Date and time Dst_{min}	Dst_{min} (nT)
1	04 May 1998-05:00 UT	−205
2	25 Sep 1998-09:00 UT	−207
3	22 Oct 1999-06:00 UT	−237
4	06 Apr 2000-23:00 UT	−287
5	16 Jul 2000-00:00 UT	−301
6	12 Aug 2000-09:00 UT	−235
7	17 Sep 2000-23:00 UT	−201
8	31 Mar 2001-08:00 UT	−387
9	31 Mar 2001-21:00 UT	−284
10	11 Apr 2001-23:00 UT	−271
11	06 Nov 2001-06:00 UT	−292
12	24 Nov 2001-16:00 UT	−221
13	30 Oct 2003-00:00 UT	−353
14	30 Oct 2003-22:00 UT	−383
15	20 Nov 2003-20:00 UT	−422
16	27 Jul 2004-13:00 UT	−197
17	08 Nov 2004-06:00 UT	−373
18	10 Nov 2004-10:00 UT	−289
19	15 May 2005-08:00 UT	−263
20	24 Aug 2005-11:00 UT	−216

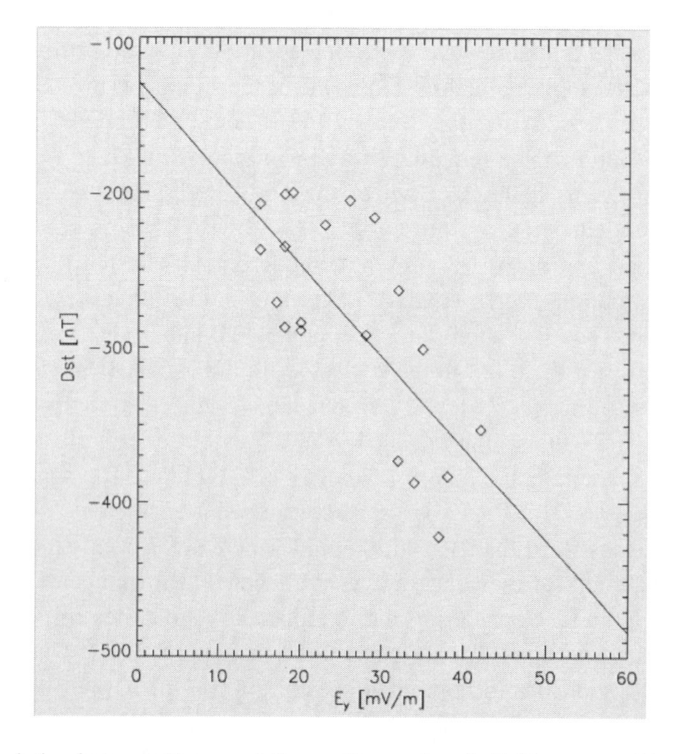

Figure 5.7.2. Correlation between Dst_{min} and the maximum electric field convected by the solar wind (E_y). According to Szajko et al. (2013). Permission from COSPAR.

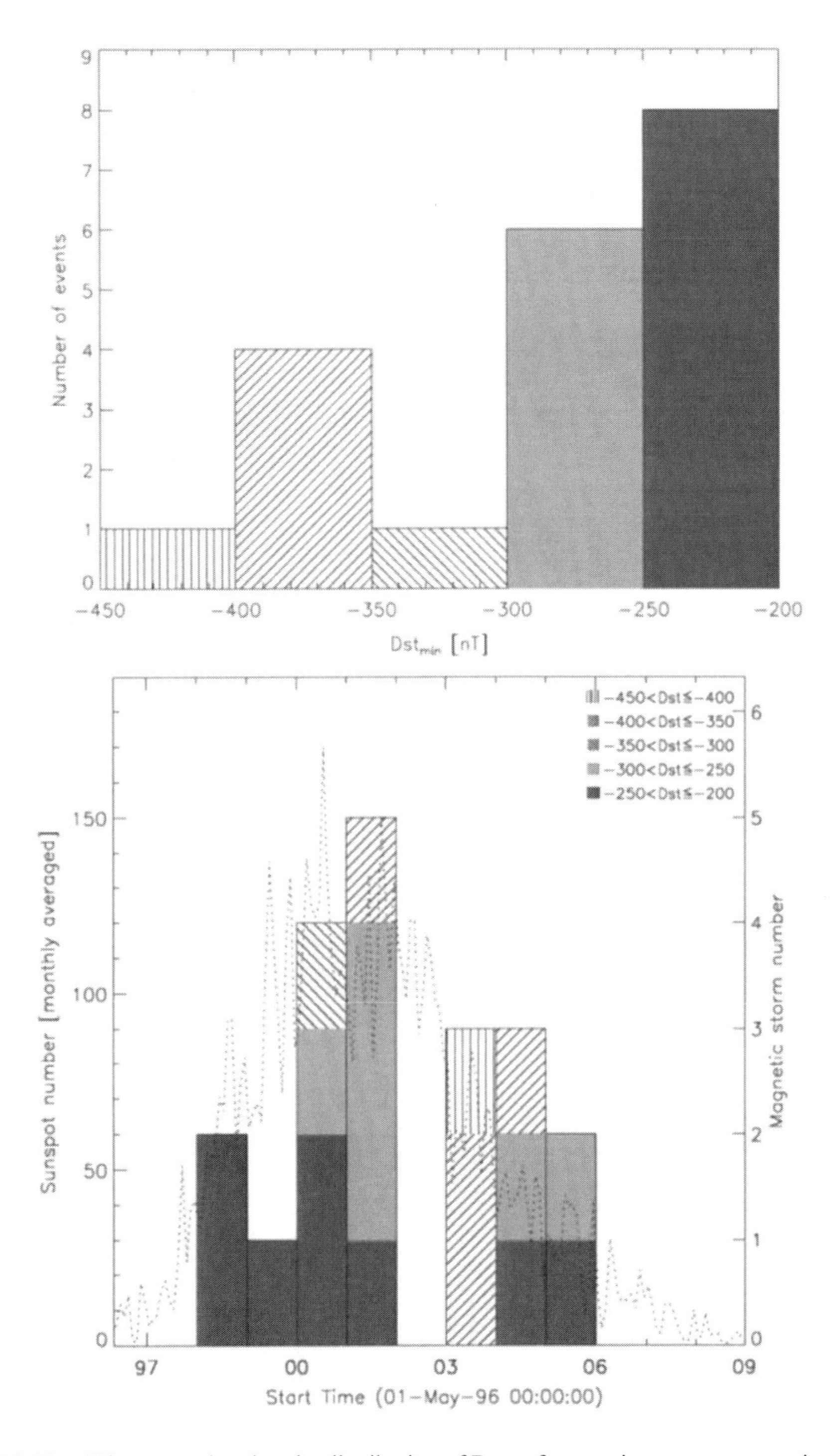

Figure 5.7.3. Top: Histogram showing the distribution of Dst_{min} for very intense geomagnetic storms (in 50 nT bin sizes). Bottom: Storm occurrence rate per year (vertical left-hand axis). The overlaid dotted line shows the monthly averaged sunspot numbers, as indicated in the right-hand vertical axis. In both figures, different shadings and hatchings correspond to Dst_{min} values separated in 50 nT ranges. The storm with $Dst_{min} = -197$ nT will be included, from now on, in the set that ranges between –250 and –200 nT. According to Szajko et al. (2013). Permission from COSPAR.

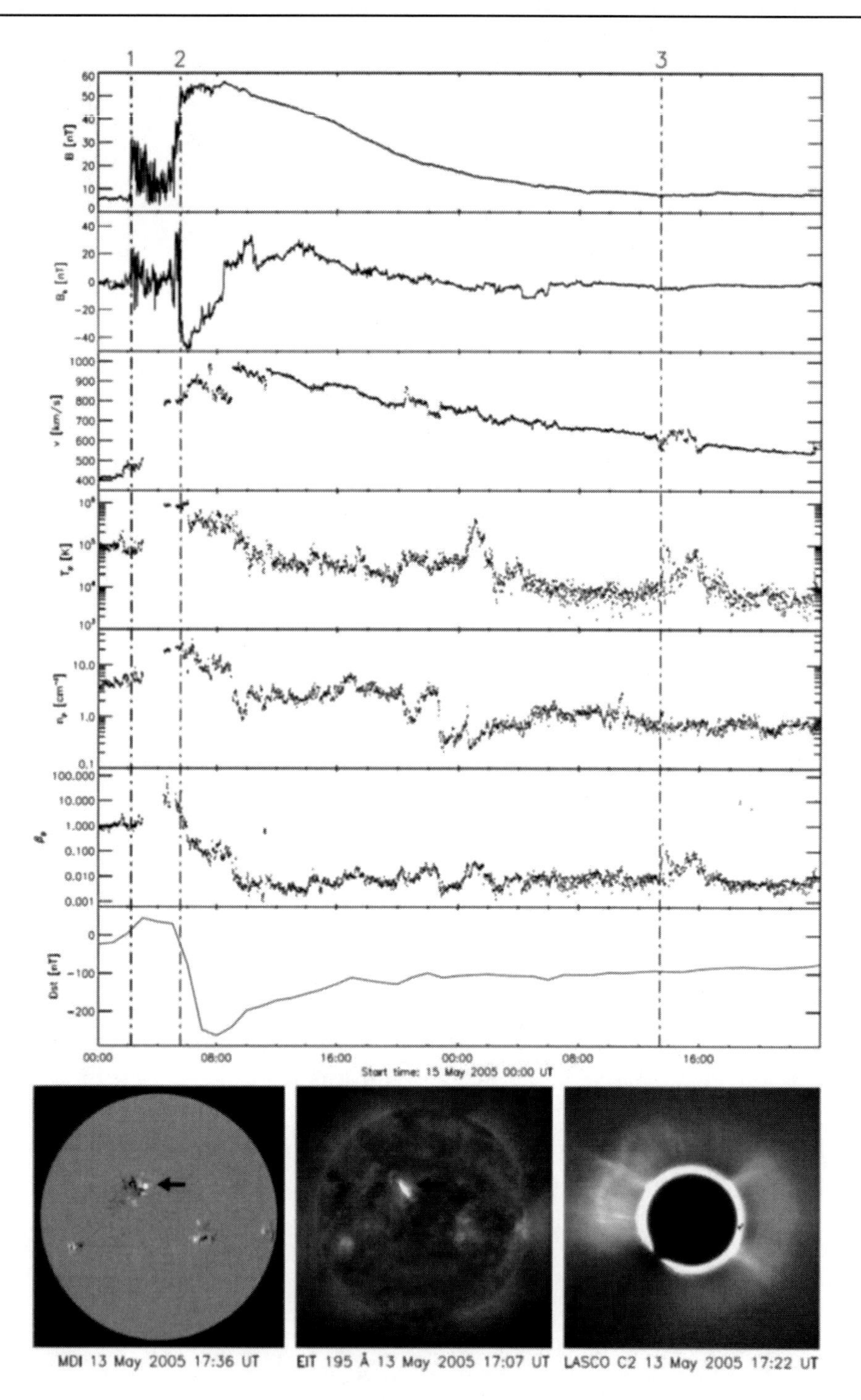

Figure 5.7.4. Interplanetary, geomagnetic, and solar data related to event number 19 in Table 5.7.3, $Dst_{min} = -263$ nT on 15 May 2005 at 08:00 UT. The figure shows, from top to bottom, the IP magnetic field intensity, the southward component of the IP field in GSM coordinates, the solar wind radial velocity, the proton temperature and density, the proton plasma β parameter, and the geomagnetic storm profile. The three images at the bottom illustrate the location of the source AR (pointed with an arrow) in an MDI magnetogram, the EUV flare in EIT, and the LASCO C2 field of view at 17:22 UT. According to Szajko et al. (2013). Permission from COSPAR.

5.7.12. Solar Transients Disturbing the Terrestrial Magnetic Environment at Higher Latitudes

Geomagnetic field variations during five major Solar Energetic Particle (SEP) events of solar cycle 23 have been investigated by Khan et al. (2014). The SEP events of 1 October 2001, 4 November 2001, 22 November 2001, 21 April 2002 and 14 May 2005 have been selected to study the geomagnetic field variations at two high-latitude stations, Thule (77.5° N, 69.2° W) and Resolute Bay (74.4° E, 94.5° W) of the northern polar cap. It was used the GOES proton flux in seven different energy channels (0.8–4 MeV, 4–9 MeV, 9–15 MeV, 15–40 MeV, 40–80 MeV, 80–165 MeV, 165–500 MeV). All the proton events were associated with geoeffective or Earth directed CMEs that caused intense geomagnetic storms in response to geospace. It was taken high-latitude indices, AE and PC, under consideration and found fairly good correlation of these with the ground magnetic field records during the five proton events. The departures of the H component during the events were calculated from the quietest day of the month for each event and have been represented as ΔH_{THL} and ΔH_{RES} for Thule and Resolute Bay, respectively. The correspondence of spectral index, inferred from event integrated spectra, with ground magnetic signatures ΔH_{THL} and ΔH_{RES} along with Dst and PC indices have been brought out. From the correlation analysis it was found a very strong correlation to exist between the geomagnetic field variation (ΔHs) and high-latitude indices AE and PC. To find the association of geomagnetic storm intensity with proton flux characteristics Khan et al. (2014) derived the correspondence between the spectral indices and geomagnetic field variations (ΔH_S) along with the Dst and AE index. It was found a strong correlation (CC = 0.88) to exist between the spectral indices and ΔH_S and also between spectral indices and AE and PC.

5.7.13. Statistical Analysis of the Geomagnetic Response to Different Solar Wind Drivers and the Dependence on Storm Intensity

As noted Katus et al. (2015), geomagnetic storms start with activity on the Sun that causes propagation of magnetized plasma structures in the solar wind. The type of solar activity is used to classify the plasma structures as being either interplanetary coronal mass ejection (ICME) or corotating interaction region (CIR) driven. The ICME-driven events are further classified as either magnetic cloud (MC) driven or sheath (SH) driven by the geoeffective structure responsible for the peak of the storm. The geoeffective solar wind flow then interacts with the magnetosphere producing a disturbance in near-Earth space. It is commonly believed that a SH-driven event behaves more like a CIR-driven event than a MC-driven event; however, in present analysis this is not the case. In this study, geomagnetic storms are investigated statistically with respect to the solar wind driver and the intensity of the events. it was used the Hot Electron and Ion Drift Integrator (HEIDI) model to simulate the inner magnetospheric hot ion population during all of the storms classified as intense ($Dst_{min} \leq -100$ nT) within solar cycle 23 (1996–2005). HEIDI is configured four different ways using either the Volland-Stern or self-consistent electric field and either event-based Los Alamos National Laboratory (LANL) magnetospheric plasma analyzer (MPA) data or a reanalyzed lower resolution version of the data that provides spatial resolution. Presenting the simulation results, geomagnetic data, and solar wind data along a normalized epoch timeline shows the average behavior throughout a typical storm of each classification. The error along the epoch timeline for each HEIDI configuration is used to rate the model's performance. It was also subgrouped the storms based on the magnitude of the minimum Dst. It was found that typically the LANL MPA data provide the best outer boundary condition. Additionally, the self-consistent electric field better reproduces SH- and MC-driven events throughout most of the storm timeline, but the Volland-Stern electric field better reproduces CIR-driven events. Contrary to what it was expect, examination of the HEIDI model results and solar wind data shows that SH-driven events behave more like MC-driven events than CIR-driven storms.

References

Akasofu S.-I., 1981. "Energy coupling between the solar wind and the magnetosphere", *Space Sci. Rev.*, **28**, No. 2, 121-190.

Andréeová K. and L. Přech, 2007. "Propagation of interplanetary shocks into the Earth's magnetosphere", *Adv. Space Res.*, **40**, No. 12, 1871-1880.

Andreeova K., T.I. Pulkkinen, M. Palmroth, and R. McPherron, 2011. "Geoefficiency of solar wind discontinuities", *J. Atmos. Sol. Terr. Phys.*, **73**, No. 1, 112-122.

Antonova E.E., 2002. "Magnetostatic Equilibrium and Turbulent Transport in Earth's Magnetosphere: A Review of Experimental Observation Data and Theoretical Approach", *Int. J. Geomagn. Aeron.*, **3**, No. 2, 117–130.

Araki T., 1994. "A physical model of geomagnetic sudden commencement", in *Solar Wind Sources of Magnetospheric Ultra-Low-Frequency Waves*, Geophys. Monogr. Ser., **81**, edited by M.J. Engebretson, K. Takahashi, and M. Scholer, AGU, Washington, D.C., 183-200.

Bale S.D., M.A. Balikhin, T.S. Horbury, V.V. Krasnoselskikh, H. Kucharek, E. Möbius, S.N. Walker, A. Balogh, D. Burgess, B. Lembège, E.A. Lucek, M. Scholer, S.J. Schwartz, and M.F. Thomsen, 2005. "Quasi-perpendicular Shock Structure and Processes", *Space Sci. Rev.*, **118**, No 1-4, 161-203.

Bame S.J., D.J. McComas, M.F. Thomsen, B.L. Barraclough, R.C. Elphic, J.P. Glore, J.T. Gosling, J.C. Chavez, E.P. Evans, and J.F. Wymer, 1993. "Magnetospheric plasma analyzer for spacecraft with restrained resources", *Rev. Sci. Instrum.*, **64**, No. 4, 1026-1033.

Blanc M. and Richmond A.D., 1980. "The Ionospheric disturbance dynamo". *J. Geophys. Res.*, **85**, No., 1669-.

Borovsky J. E., 2008. "The rudiments of a theory of solar wind/magnetosphere coupling derived from first principles", *J. Geophys. Res.*, **113**, A08228, doi:10.1029/2007 JA012646.

Borovsky J.E. and H.E. Funsten (2003), "MHD Turbulence in the Earth's Plasma Sheet: Dynamics, Dissipation and Driving," J. Geophys. Res. 107A, JA009601.

Borovsky J.E. and J.T. Steinberg, 2006. "The 'calm before the storm' in CIR/magnetosphere interactions: Occurrence statistics, solar wind statistics, and magnetospheric preconditioning", *J. Geophys. Res.*, **111**, No. A7, A07S10, doi:10.1029/2005JA011397, 1-29.

Bruinsma S.L. and Forbes J.M., 2007. Global observations of travelling atmospheric disturbances (TADs) in the thermosphere. Geophys. Res. Lett. 34, L14103.

Burgess D., 1989. "Cyclic behavior at quasi-parallel collisionless shocks", *Geophys. Res. Lett.*, **16**, 345–348.

Burgess D., E.A. Lucek, M. Scholer, S.D. Bale, M.A. Balikhin, A. Balogh, T.S. Horbury, V.V. Krasnoselskikh, H. Kucharek, B. Lembège, E. Möbius, S.J. Schwartz, M.F. Thomsen, and S.N. Walker, 2005. "Quasi-parallel Shock Structure and Processes", *Space Sci. Rev.*, **118**, No 1-4, 205-222.

Burke W.J., 2008. Stormtime energy budgets of the global thermosphere. In: Kintner, P.M., Coster, A.J., Fuller-Rowell, T., Mannucci, A.J., Mendillo, M., Heelis, R.A. (Eds.), Mid-Latitude Ionospheric Dynamics and Disturbances. Washington, DC, pp. 235–246 (American Geophysical Monograph 181).

Burke W.J., Huang C.Y., Marcos F.A., and Wise J.O., 2007. Interplanetary control of thermospheric densities during large magnetic storms. *J. Atmos. Sol.-Terr. Phys*. **69**, 279.

Burlaga L.F., 1988. "Magnetic Clouds and Force-Free with Constant Alpha", *J. Geophys. Res.*, **93**, 7217-7224.

Burton R.K., R.L. McPherron, and C.T. Russell, 1975."An empirical relationship between interplanetary conditions and *Dst*", *J. Geophys. Res.*, **80**, No. 31, 4204-4214.

Cane H.V. and I.G. Richardson, 2003. "Interplanetary coronal mass ejections in the near-Earth solar wind during 1996-2002", *J. Geophys. Res.*, **108**, No. A4, 1156, doi:10.1029/2002JA009817.

Carovillano R.L. and Siscoe G.L., 1973. Energy and momentum theorems in magnetospheric processes. Rev. Geophys. 11, 289.

Carrington R.C., 1860. "Description of a Singular Appearance Seen in the Sun on September 1, 1859," Mon. Not. R. Astron. Soc. 20, 13–15.

Dal Lago A., W.D. Gonzalez, A.L.C. de Gonzalez, and L.E.A. Vieira, 2002. "Stream-interacting magnetic clouds causing very intense geomagnetic storms", *Adv. Space Res.*, **30**, No. 10, 2225-2229.

Dandouras I.S., H. Rème, J. Cao, and P. Escoubet, 2009. "Magnetosphere response to the 2005 and 2006 extreme solar events as observed by the Cluster and Double Star spacecraft", *Adv. Space Res.*, **43**, No. 4, 618–623.

De Zeeuw D.L., S. Sazykin, R.A. Wolf, T.I. Gombosi, A.J. Ridley, and G. Toth, 2004. "Coupling of a global MHD code and an inner magnetospheric model: Initial results", *J. Geophys. Res.*, **109**, No. A12, A12219, doi:10.1029/2003JA010366, 1-14.

Denton M.H., and J.E. Borovsky, 2017. "The response of the inner magnetosphere to the trailing edges of high-speed solar-wind streams", *J. Geophys. Res. Space Physics*, **122**, 501–516, doi:10.1002/2016JA023592.

Denton M.H., M.F. Thomsen, H. Korth, S. Lynch, J.-C. Zhang, and M.W. Liemohn, 2005. "Bulk plasma properties at geosynchronous orbit", *J. Geophys. Res.*, **110**, No. A7, A07223, doi:10.1029/2004JA010861, 1-17.

Denton M.H., J.E. Borovsky, R.M. Skoug, M.F. Thomsen, B. Lavraud, M.G. Henderson, R.L. McPherron, J.C. Zhang, and M.W. Liemohn, 2006. "Geomagnetic storms driven by ICME- and CIR-dominated solar wind", *J. Geophys. Res.*, **111**, No. A7, A07S07, doi:10.1029/2005JA011436, 1-12.

Despirak I.V., A.A. Lubchich, H.K. Biernat, and A.G. Yahnin, 2008. "Poleward Expansion of the Westward Electrojet Depending on the Solar Wind and IMF Parameters," *Geomagn. Aeron.* **48** (3), 297–305 (In Russian: *Geomagn. i Aeron.* **48**, 282–293).

Dessler A.J. and E.N. Parker, 1959. "Hydromagnetic theory of geomagnetic storms", *J. Geophys. Res.*, **64**, No. 12, 2239-2252.

Dryer M., 1973. "Bow shock and its interaction with interplanetary shocks", *Radio Sci.*, **8**, No. 11, 893-901.

Dryer M., D.L. Merritt, and P.M. Aronson, 1967. "Interaction of a plasma cloud with the Earth's magnetosphere", *J. Geophys. Res.*, **72**, No. 11, 2955-2962.

Dubouloz N. and M. Scholer, 1995. "Two-dimensional simulations of magnetic pulsations upstream of the Earth's bow shock", *J. Geophys. Res.*, **100**, No. A6, 9461–9474.

Eastwood J., E.A. Lucek, C. Mazelle, K. Meziane, Y. Narita, J. Pickett, and R. Treumann, 2005. "The Foreshock". *Space Sci. Rev.*, **118**, No 1-4, 41-94.

Ebihara Y. and M. Ejiri, 2000. "Simulation study on fundamental properties of the storm-time ring current", *J. Geophys. Res.*, **105**, No. A7, 15843-15859.

Echer E., M.V. Alves, and W.D. Gonzalez, 2004. "Geoeffectiveness of interplanetary shocks during solar minimum (1995–1996) and solar maximum (2000)", *Sol. Phys.*, **221**, No. 2, 361-380.

Echer E., B.T. Tsurutani, F.L. Guarnieri, 2010. "Interplanetary origins of November 2004 superstorms", *J. Atmos. Sol. Terr. Phys.*, **72**, No. 4, 280–284.

Echer E., B.T. Tsurutani and W.D. Gonzalez, 2013. "Interplanetary origins of moderate (100 nT $<$ Dst$<$50 nT) geomagnetic storms during solar cycle 23 (1996-2008)", *J. Geophys. Res. Space Physics*, **118**, No. 1, doi: 10.1029/2012JA018086, 385-392.

Fairfield D.H., C.J Farrugia., T. Mukai, T. Nagai, and A. Fedorov, 2003. "Motion of the dusk flank boundary layer caused by solar wind pressure changes and the Kelvin–Helmholtz instability: 10–11 January 1997", *J. Geophys. Res.*, **108**, No. 12, 1460, doi:10.1029/2003JA010134.

Farrugia C.J., Erkaev N.V., Taubenschuss U., Shaidurov V.A., Smith C.W., and Biernat H.K., 2008. "A slow mode transition region adjoining the front boundary of a magnetic cloud as a relic of a convected solar wind feature: observations and MHD simulation". *J. Geophys. Res.* **113**, A00B01, doi:10.1029/2007JA012953.

Farrugia C.J., N.V. Erkaev, N.C. Maynard, I.G. Richardson, P.E. Sandholt, D. Langmayr, K.W. Ogilvie, A. Szabo, U. Taubenschuss, R.B. Torbert, and H.K. Biernat, 2009. "Effects on the distant geomagnetic tail of a fivefold density drop in the inner sheath region of a magnetic cloud: A joint Wind–ACE study", *Adv. Space Res.*, **44**, No. 11, 1288–1294.

Farrugia C.J., N.V. Erkaev, V.K. Jordanova, N. Lugaz, P.E. Sandholt, S. Mühlbachler, and R.B. Torbert, 2013. "Features of the interaction of interplanetary coronal mass ejections/magnetic clouds with the Earth's magnetosphere", *J. Atmos. Sol. Terr. Phys.*, **99**, 14-26.

Fenrich F.R. and J.G. Luhmann, 1998.. "Geomagnetic response to magnetic clouds of different polarity", *Geophys. Res. Lett.*, **25**, No. 15, 2999-3002.

Forster D.R., M.H. Denton, M. Grande, and C.H. Perry, 2013. "Inner magnetospheric heavy ion composition during high-speed stream-driven storms", *J. Geophys. Res. Space Physics*, **118**, No. 7, doi:10.1002/jgra.50292, 4066-4079.

Fujimoto M., T. Terasawa, T. Mukai, Y. Saito, T. Yamamoto, and S. Kokubun, 1998. "Plasma entry from the flanks of the near-Earth magnetotail: Geotail observations", *J. Geophys. Res.*, **103**, No. A3, 4391-4408.

Ganushkina N.Y., T.I. Pulkkinen, M.V. Kubyshkina, H.J. Singer, and C.T. Russell, 2004. "Long-term evolution of magnetospheric current systems during storm periods", *Ann. Geophys.*, **22**, No. 4, 1317-1334.

Giacalone J., S.J. Schwartz, and D. Burgess, 1993. "Observations of suprathermal ions in association with SLAMS", *Geophys. Res. Lett.*, **20**, No. 2, 149–152.

Giacalone J., S.J. Schwartz, and D. Burgess, 1994. "Artificial spacecraft in hybrid simulations of the quasi-parallel Earth's bow shock: Analysis of time series versus spatial profiles and a separation strategy for Cluster", *Ann. Geophys.*, **12**, No. 7, 591–601.

Golovchanskaya I.V., A.A. Ostapenko, and B.V. Kozelov (2006), "Relationship between the High-Latitude Electric and Magnetic Turbulence and the Birkeland Field-Aligned Currents," J. Geophys. Res., A12301.

Gombosi T.I., D.L. de Zeeuw, K.G. Powell, A.J. Ridley, I.V. Sokolov, Q.F. Stout and G. Tóth, 2003. "Adaptive Mesh Refinement for Global Magnetohydrodynamic Simulation", in *Space Plasma Simulation*, Lecture Notes in Physics, **615** (edited by J. Buchner, C. Dum, and M. Scholer), Springer-Verlag, Berlin/Heidelberg, 247-274.

Gonzalez W.D., B.T. Tsurutani, R.P. Lepping, and R. Schwenn, 2002a. "Interplanetary phenomena associated with very intense geomagnetic storms", *J. Atmos. Sol. Terr. Phys.*, **64**, No. 2, 173-181.

Gonzalez W.D., B.T. Tsurutani, and A.L. Clua de Gonzalez, 2002b. "Geomagnetic storms contrasted during solar maximum and near solar minimum", *Adv. Space Res.*, **30**, No. 10, 2301-2304.

Gonzalez W.D., E. Echer, B.T. Tsurutani, A.L. Clúa de Gonzalez, and A. Dal Lago, 2011. "Interplanetary Origin of Intense, Superintense and Extreme Geomagnetic Storms", *Space Sci. Rev.*, **158**, No. 1, 69-89.

Gosling J.T., 1993. "The solar flare myth", *J. Geophys. Res.*, **98**, No. A11, 18937-18949.

Gosling J.T., S.J. Bame, D.J. McComas, and J.L. Phillips, 1990. "Coronal mass ejections and large geomagnetic storms", *Geophys. Res. Lett.*, **17**, No. 7, 901-904.

Gosling J.T., D.J. McComas, J.L. Phillips, and S.J. Bame, 1991. "Geomagnetic activity associated with Earth passage of interplanetary shock disturbances and coronal mass ejections", *J. Geophys. Res.*, **96**, No. A5, 7831-7839.

Grandin M., A.T. Aikio, A. Kozlovsky, T. Ulich, and T. Raita, 2017. "Cosmic radio noise absorption in the high-latitude ionosphere during solar wind high-speed streams", *J. Geophys. Res. Space Physics*, **122**, 5203–5223, doi:10.1002/2017JA023923.

Greenstadt E.W., M.M. Hoppe, and C.T. Russell, 1982., "Large-amplitude magnetic variations in quasi-parallel shocks: correlation lengths measured by ISEE 1 and 2", *Geophys. Res. Lett.*, **9**, No. 7, 781–784.

Grib S.A., 1972. "Interaction of the solar-wind shock wave with the Earth's magnetosphere", *Dokl. Akad. Nauk BSSR.*, **16**, No. 6, 493-496. In Russian.

Grib S.A., 1973. "Some questions of the interaction of shock waves of the solar wind with the Earth's magnetosphere", *Geomagn. Aeron.*, **13**, No. 5, 669-673.

Grib S.A., 1982. "Interaction of non-perpendicular/parallel solar wind shock waves with the Earth's magnetosphere", *Space Sci. Rev.*, **32**, No.1-2, 43-48.

Grib S.A., 2012. "Impact of solar wind tangential discontinuities on the Earth's magnetosphere", *Geomagnetism and Aeronomy*, **52**, No. 8, 1113-1116.

Grib S.A., 2013. "Nonlinear phenomena related to the solar shock motion in the Earth's magnetosphere", *Geomagnetism and Aeronomy*, **53**, No. 4, 424-429. Russian Text:*Geomagnetizm i Aeronomiya*, **53**, No. 4, 451-456.

Grib S.A. and E.A. Pushkar, 2006. "Asymmetry of nonlinear interactions of solar MHD discontinuities with the bow shock", *Geomagn. Aeron.*, **46**, No. 4, 417-423.

Grib S.A., B.E. Briunelli, M. Dryer, and W.-W. Shen, 1979. "Interaction of interplanetary shock waves with the bow shock-magnetopause system", *J. Geophys. Res.*, **84**, No. A10, 5907-5922.

Guo X.C., Y.-Q. Hu, and C. Wang, 2005. "Earth's magnetosphere impinged by interplanetary shocks of different orientations", *Chin. Phys. Lett. (China)*, **22**, No. 12, doi:10.1088/0256-307X/22/12/067, 3221-3224.

Guo J., X. Feng, J. Zhang, P. Zuo, and C. Xiang, 2010. "Statistical properties and geoefficiency of interplanetary coronal mass ejections and their sheaths during intense geomagnetic storms", *J. Geophys. Res.*, **115**, No. A9, A09107, doi:10.1029/2009JA 015140, 1-8.

Hakkinen L.V.T., T.I. Pulkkinen, R.J. Pirjola, H. Nevanlinna, E.I. Tanskanen, and N.E. Turner, 2003. "Seasonal and diurnal variation of geomagnetic activity: Revised *Dst* versus external drivers", *J. Geophys. Res.*, **108**, No. A2, 1060, doi:10.1029/2002JA 009428, SMP2-1-13.

Hasegawa H., M. Fujimoto, T.-D. Phan, H. Rème, A. Balogh, M.W. Dunlop, C. Hashimoto, and R. Tandokoro, 2004. "Transport of solar wind into Earth's magnetosphere through rolled-up Kelvin-Helmholtz vortices, *Nature*, **430**, No. 7001, doi:10.1038/nature02799, 755-758.

Horton W. and I. Doxas, 1998. "A low-dimensional dynamical model for the solar wind driven geotail-ionosphere system", *J. Geophys. Res.*, **103**, No. A3, 4561-4572.

Huang C.Y. and Burke W.J., 2004. Transient sheets of field-aligned current observed by DMSP during the main phase of a magnetic storm. J. Geophys. Res. 109, A06303.

Huang C.-S., Foster J., Reeves G., Le G., Frey H., Pollock C., and Jahn J.-M., 2003. "Periodic magnetospheric substorms: Multiple space-based and ground-based instrumental observations", *J. Geophys. Res.*, **108**, No. A11, 1411, doi:10.1029/2003JA009992, SMP16-1-17.

Huttunen K.E.J., H.E.J. Koskinen, and R. Schwenn, 2002."Variability of magnetospheric storms driven by different solar wind perturbations", *J. Geophys. Res.*, **107**(A7), 1121, doi:10.1029/2001JA900171.

Ivanov K.G., 1964. "Interaction between advancing shock waves and strong discontinuities in space in the Earth's neighborhood", *Geomagn. Aeron.*, **4**, No. 4, 626-629.

Ivanov K.G., 2010. "Displacement of large-scale open solar magnetic fields from the zone of active longitudes and the heliospheric storm of November 3-10, 2004: 1. The field dynamics and solar activity", *Geomagnetism and Aeronomy*, **50**, No. 3, 285-297. Russian Text: *Geomagnetizm i Aeronomiya*, **50**, No. 3, 298-310.

Jordanova V.K., C.J. Farrugia, L. Janoo, J.M. Quinn, R.B. Torbert, K.W. Ogilvie, R.P. Lepping, J.T. Steinberg, D.J. McComas, and R.D. Belian, 1998. "October 1995 magnetic cloud and accompanying storm activity: Ring current evolution", *J. Geophys. Res.*, **103**, No. A1, 79-92.

Kamide Y., W. Baumjohann, I.A. Daglis, W.D. Gonzalez, M. Grande, J.A. Joselyn, R.L. McPherron, J.L. Phillips, E.G.D. Reeves, G. Rostoker, A.S. Sharma, H.J. Singer, B.T. Tsurutani, and V.M. Vasyliunas, 1998. "Current understanding of magnetic storms: storm-substorm relationships", *J. Geophys. Res.,* **103**, No. A8, 17705-17728.

Kane R.P., 2010. "Relationship between Dst(min) magnitudes and characteristics of ICMEs", *Indian J. Radio and Space Phys*., **39**, No. 4, 177-183.

Katus R.M., M.W. Liemohn, E.L. Ionides, R. Ilie, D. Welling, and L.K. Sarno-Smith, 2015. "Statistical analysis of the geomagnetic response to different solar wind drivers and the dependence on storm intensity", *J. Geophys. Res. Space Physics*, **120**, No. 1, doi:10.1002/2014JA020712, 310-327.

Kaufmann R.L. and Konradi A., 1969. "Explorer 12 magnetopause observations: large-scale nonuniform motions", *J. Geophys. Res*. **74**, No. 14, 3609–3627.

Khan P.A., S.C. Tripathi, O.A. Troshichev, M.A. Waheed, A.M. Aslam, and A.K. Gwal, 2014. "Solar transients disturbing the terrestrial magnetic environment at higher latitudes", *Astrophys. Space Sci*., **349**, No. 2, 647-656.

Kim K.-H., Lee D.-H., Shiokawa K., Lee E., Park J.-S., Kwon H.-J., Angelopoulos V., Park Y.-D., Hwang J., Nishitani N., Hori T., Koga K., Obara T., Yumoto K., and Baishev D.G., 2012. "Magnetospheric responses to the passage of the interplanetary shock on 24 November 2008", *J. Geophys. Res*., **117**, No. A10, A10209, doi:10.1029/2012JA017871, 1-11.

King J.H. and N.E. Papitashvili, 2005. "Solar wind spatial scales in and comparisons of hourly Wind and ACE plasma and magnetic field data", *J. Geophys. Res*., **110**, No. A2, A02209, doi:10.1029/2004JA010804, 1-8.

Kleimenova N.G. and O.V. Kozyreva (2005), "Spatial–Temporal Dynamics of Pi3 and Pc5 Geomagnetic Pulsations during the Extreme Magnetic Storms in October 2003," Geomagn. Aeron. 45 (1), 75–83 [Geomagn. Aeron. 45, 71–79 (2005)].

Klein L.W. and L.F. Burlaga, 1982. "Interplanetary Magnetic Clouds at 1 AU", *J. Geophys. Res*., **87**, 613-624.

Knipp D.J., Tobiska W.K., and Emery B.A., 2005. Direct and indirect thermospheric heating sources for solar cycles 21–23. Sol. Phys. 224, 495.

Kornilov I.A. (2009), "Localization of the Precipitating Electron Source in Active Longitudes during Breakup and Pulsating Auroras," Geomagn. Aeron. 49 (3).

Kornilova T.A., I.A. Kornilov, M.I. Pudovkin, and O.I. Kornilov (2003), "Two Types of Aurora Development during the Substorm Expansion Phase," Geomagn. Aeron. 43 (1), 40–49 [Geomagn. Aeron. 43, 36–45 (2003)].

Korth H., M.F. Thomsen, J.E. Borovsky, and D.J. McComas, 1999. "Plasma sheet access to geosynchronous orbit", *J. Geophys. Res*., **104**, No. A11, 25047-25062.

Koval A., J. Šafránková, Z. Němeček, L. Prech, A.A. Samsonov, and J.D. Richardson, 2005. "Deformation of interplanetary shock fronts in the magnetosheath", *Geophys. Res. Lett*., **32**, No. 15, L15101, doi:10.1029/2005GL023009, 1-4.

Koval A., J. Šafránková, Z. Němeček, and L. Přech, 2006a. "Propagation of interplanetary shocks through the solar wind and magnetosheath", *Adv. Space Res*., **38**, No. 3, 552-558.

Koval A., J. Šafránková, Z. Němeček, A.A. Samsonov, L. Přech, J.D. Richardson, and M. Hayosh, 2006b. "Interplanetary shock in the magnetosheath: Comparison of experimental data with MHD modeling", *Geophys. Res. Lett*., 33, L11102, doi:10.1029/2006 GL025707, 1-5.

Kozyra J.U., V.K. Jordanova, J.E. Borovsky, M.F. Thomsen, D.J. Knipp, D.S. Evans, D.J. McComas, and T.E. Cayton, 1998. "Effects of a high-density plasma sheet on ring current development during the November 2–6, 1993, magnetic storm", *J. Geophys. Res.*, **103**, No. A11, 26285-26306.

Kozyreva O.V., N.G. Kleimenova, T.A. Kornilova et al., 2006. "Unusual Spatial–Temporal Dynamics of Geomagnetic Disturbances during the Main Phase of the Extremely Strong Magnetic Storm of November 7–8, 2004", *Geomagn. Aeron.* **46** (5), 614–626. Russian text: *Geomagn. i Aeron.* **46**, 580–592.

Kupfer A.T., 1855. "Instructions Governing Magnetic and Meteorological Observations, Compiled by a Director of the Main Physical Observatory for Magnetic Observatories of the Mining Department, 2nd ed., Addition to the Code of Observations for 1852 (Korp. Gorn. Inzh., St. Petersburg, 1855) [in Russian].

Lavraud B., and J.E. Borovsky, 2008. "Altered solar wind-magnetosphere interaction at low Mach numbers: Coronal mass ejections", *J. Geophys. Res.*, **113**, No. A9, A00B08, doi:10.1029/2008JA013192, 1-25.

Lavraud B., M.F. Thomsen, M.G.G.T. Taylor, Y.L. Wang, T.D. Phan, S.J. Schwartz, R.C. Elphic, A. Fazakerley, H. Rème, and A. Balogh, 2005a "Characteristics of the magnetosheath electron boundary layer under northward interplanetary magnetic field: Implications for high-latitude reconnection", *J. Geophys. Res.*, **110**, No. A6, A06209, doi:10.1029/2004JA010808.

Lavraud B., M.H. Denton, M.F. Thomsen, J.E. Borovsky, and R.H.W. Friedel, 2005b. "Superposed epoch analysis of dense plasma access to geosynchronous orbit, *Ann. Geophys.*, **23**, No. 7, 2519-2529.

Lavraud B., M.F. Thomsen, B. Lefebvre, S.J. Schwartz, K. Seki, T.D. Phan, Y.L. Wang, A. Fazakerley, H. Rème, and A. Balogh, 2006a. "Evidence for newly closed magnetosheath field lines at the dayside magnetopause under northward IMF", *J. Geophys. Res.*, **111**, No. A5, A05211, doi:10.1029/2005JA011266, 1-10.

Lavraud B., M.F. Thomsen, J.E. Borovsky, M.H. Denton, and T.I. Pulkkinen, 2006b. "Magnetosphere preconditioning under northward IMF: Evidence from the study of coronal mass ejection and corotating interaction region geoeffectiveness", *J. Geophys. Res.*, **111**, A09208, doi:10.1029/2005JA011566, 1-10.

Lazutin L.L. and S.N. Kuznetsov, 2008. "Nature of Sudden Auroral Activations at the Beginning of Magnetic Storms", *Geomagn. Aeron.* **48** (2), 173–182. Russian text: *Geomagn. i Aeron.* **48**, 165–174.

Le G. and C.T. Russell, 1990. "A study of the coherence length of ULF waves in the Earth's foreshock", *J. Geophys. Res.*, **95**, No. A7, 10703–10706.

Le G. and C.T. Russell, 1992. "A study of ULF wave foreshock morphology ? II: Spatial variation of ULF waves", *Planet. Space Sci.*, **40**, No. 9, 1215–1225.

Lee D.-H. and M.K. Hudson, 2001. "Numerical studies on the propagation of sudden impulses in the dipole magnetosphere", *J. Geophys. Res.*, **106**, No. A5, 8435-8445.

Lepping R.P., L.F. Burlaga, A. Szabo, K.W. Ogilvie, W.H. Mish, D. Vassiliadis, A.J. Lazarus, J.T. Steinberg, C.J Farrugia, L. Janoo, and F. Mariani, 1997. "The Wind Magnetic Cloud and Events of October 18-20, 1995: Interplanetary Properties and as Triggers for Geomagnetic Activity", *J. Geophys. Res.*, **102**, 14049-14063.

Li Y. and J. Luhmann, 2004. "Solar cycle control of the magnetic cloud polarity and the geoeffectiveness", *J. Atmos. Sol. Terr. Phys.*, **66**, No. 3/4, 323-331.

Li X., D.N. Baker, S. Elkington, M. Temerin, G.D. Reeves, R.D. Belian, J.B. Blake, H.J. Singer, W. Peria, and G. Parks, 2003. "Energetic particle injections in the inner magnetosphere as a response to an interplanetary shock", *J. Atmos. Sol. Terr. Phys.*, **65**, No. 2, 233-244.

Li W.H., J. Raeder, J. Dorelli, M. Oieroset, and T.D. Phan, 2005a. "Plasma sheet formation during long period of northward IMF", *Geophys. Res. Lett.*, **32**, No. 12, L12S08, doi:10.1029/2004GL021524, 1-4.

Li X., D.N. Baker, M. Temerin, G.D. Reeves, R. Friedel, and C. Shen, 2005b. "Energetic electrons, 50 keV–6 MeV, at geosynchronous orbit: Their responses to solar wind variations", *Space Weather*, **3**, No. 4, S04001, doi:10.1029/2004SW000105.

Lin C.H., Richmond A.D., Liu J.Y., Yeh H.C., Paxton L.J., Tsai H.F., and Su S.–Y., 2005. Large-scale variations of the low latitude ionosphere during the October–November 2003 superstorm: observational results. J. Geophys. Res. 110, A09S28.

Liou K., T. Sotirelis, and J. Gjerloev, 2017. "Statistical study of polar negative magnetic bays driven by interplanetary fast-mode shocks", *J. Geophys. Res. Space Physics*, **122**, 7463–7472, doi:10.1002/2017JA024465.

Luan X., W. Wang, J. Lei, A. Burns, X. Dou, and J. Xu, 2013. "Geomagnetic and auroral activity driven by corotating interaction regions during the declining phase of Solar Cycle 23)", *J. Geophys. Res. Space Physics*, **118**, No. 3, doi:10.1002/jgra.50195, 1255-1269.

Lukianova R., K. Mursula, and A. Kozlovsky, 2012. "Response of the polar magnetic field intensity to the exceptionally high solar wind streams in 2003", *Geophys. Res. Lett.*, **39**, No. 4, L04101, doi:10.1029/2011GL050420, 1-5.

Manchester IV W.B., A.J. Ridley, T.I. Gombosi, and D.L. DeZeeuw, 2006. "Modeling the Sun-to-Earth propagation of a very fast CME", *Adv. Space Res.*, **38**, No. 2, 253-262.

Manoharan P.K., N. Gopalswamy, S. Yashiro, A. Lara, G. Michalek, and R.A. Howard, 2004. "Influence of coronal mass ejection interaction on propagation of interplanetary shocks", *J. Geophys. Res.*, **109**, No. A6, A06109, doi:10.1029/2003JA010300, 1-9.

Maruyama N., Richmond A.D., Fuller-Rowell T.J., Codrescu M.V., Sazykin S., Toffoletto F.R., Spiro R.W., and Millward G.H., 2005. Interaction between direct penetration and disturbance dynamo electric fields in the storm-time equatorial ionosphere. Geophys. Res. Lett. 32, L17105.

Mays M.L., W. Horton, J. Kozyra, T.H. Zurbuchen, C. Huang, and E. Spencer, 2007. "Effect of Interplanetary Shocks on the AL and Dst Indices", *Geophys. Res. Lett.*, **34**, L11104, doi:10.1029/2007GL029844, 1-5.

McComas D.J., S.J. Bame, B.L. Barraclough, J.R. Donart, R.C. Elphic, J.T. Gosling, M.B. Moldwin, K.R. Moore, and M.F. Thomsen, 1993. "Magnetospheric Plasma Analyzer: Initial three-spacecraft observations from geosynchronous orbit", *J. Geophys. Res.*, **98**, No. A8, 13453-13465.

McKenna-Lawlor S., L. Li, I. Dandouras, P.C. Brandt, Y. Zheng, S. Barabash, R. Bucik, K. Kudela, J. Balaz, and I. Strharsky, 2010. "Moderate geomagnetic storm (21-22 January 2005) triggered by an outstanding coronal mass ejection viewed via energetic neutral atoms", *J. Geophys. Res.*, **115**, No. A8, A08213, doi:10.1029/2009JA014663, 1-22.

Mulligan T., C.T. Russell, and J.G. Luhmann, 1998. "Solar cycle evolution of the structure of magnetic clouds in the inner heliosphere", *Geophys. Res. Lett.*, **25**, No. 15, 2959-2962.

Myllys M., E.K.J. Kilpua, B. Lavraud, and T.I. Pulkkinen, 2016. “Solar wind-magnetosphere coupling efficiency during ejecta and sheath-driven geomagnetic storms”, *J. Geophys. Res. Space Physics*, **121**, 4378–4396, doi:10.1002/2016JA022407.

Němeček Z., J. Safránková, A. Koval, J. Merka, and L. Prech, 2011. “MHD analysis of propagation of an interplanetary shock across magnetospheric boundaries”, *J. Atmos. Sol. Terr. Phys.*, **73**, No. 1, 20-29.

Neubauer F.M., 1975. “Nonlinear oblique interaction of interplanetary tangential discontinuities with magnetogasdynamic shocks”, *J. Geophys. Res.*, **80, No.** 10, 1213–1222.

Newell P.T., T. Sotirelis, K. Liou, C.-I. Meng, and F.J. Rich, 2007. “A nearly universal solar wind-magnetosphere coupling function inferred from 10 magnetospheric state variables”, *J. Geophys. Res.*, **112**, A01206, doi:10.1029/2006JA012015.

Nikolaeva N.S., Yu.I. Yermolaev, and I.G. Lodkina, 2012a. “Dependence of geomagnetic activity during magnetic storms on the solar wind parameters for different types of streams: 2. Main phase of storm”, *Geomagnetism and Aeronomy*, **52**, No. 1, 28-36. Russian Text: *Geomagnetizm i Aeronomiya*, **52**, No. 1, 31-40.

Nikolaeva N.S., Yu.I. Yermolaev, and I.G. Lodkina, 2012b. “Dependence of geomagnetic activity during magnetic storms on the solar wind parameters for different types of streams: 3. Development of storm”, *Geomagnetism and Aeronomy*, **52**, No. 1, 37-48. Russian Text: *Geomagnetizm i Aeronomiya*, **52**, No. 1, 41-52.

O’Brien T.P. and R.L. McPherron, 2000. “An empirical phase space analysis of ring current dynamics: Solar wind control of injection and decay”, *J. Geophys. Res.*, **105**, No. A4, 7707-7719.

Ober D.M., N.C. Maynard, and W.J. Burke, 2003. “Testing the Hill model of transpolar potential saturation”, *J. Geophys. Res.*, **108**, No. A12, 1467, doi:10.1029/2003JA010154, SMP27-1-7.

Øieroset M., J. Raeder, T.D. Phan, S. Wing, J.P. McFadden, W. Li, M. Fujimoto, H. Rème, and A. Balogh, 2005. "Global cooling and densification of the plasma sheet during an extended period of purely northward IMF on October 22–24, 2003", *Geophys. Res. Lett.*, **32**, No. 12, L12S07, doi:10.1029/2004GL021523, 1-4.

Oliveira D.M., and J. Raeder, 2015. “Impact angle control of interplanetary shock geoeffectiveness: A statistical study”, *J. Geophys. Res. Space Physics*, **120**, No. 6, doi:10.1002/2015JA021147, 4313–4323.

Ontiveros V., and J.A. Gonzalez-Esparza, 2010. “Geomagnetic storms caused by shocks and ICMEs”, *J. Geophys. Res.*, **115**, No. A10, A10244, doi:10.1029/2010JA015471, 1-9.

Pallocchia G., 2013. “A sunward propagating fast wave in the magnetosheath observed after the passage of an interplanetary shock”, *J. Geophys. Res. Space Physics*, **118**, No. 1, doi:10.1029/2012JA017851, 331-339.

Panasyuk M.I., S.N. Kuznetsov, L.L. Lazutin et al., 2004. “Magnetic Storms in October 2003”, *Kosm. Issled.* **42** (5), 509–554.

Powell K.G., P.L. Roe, T.J. Linde, T.I. Gombosi, and D.L. De Zeeuw, 1999. “A solution-adaptive upwind scheme for ideal magnetohydrodynamics”, *J. Comput. Phys. (USA)*, **154**, No. 2, 284-309.

Pushkar E.A., A.A. Barmin, and S.A. Grib, 1991. “Investigation in the MHD approximation of the incidence of the solar-wind shock wave on the near-Earth bow shock”, *Geomagnetism and Aeronomy*, **31**, No. 3, 410-412.

Rawat R., S. Alex, and G.S. Lakhina, 2009. "Low-latitude geomagnetic response to the interplanetary conditions during very intense magnetic storms", *Adv. Space Res.*, **43**, No. 10, 1575–1587.

Reeves G.D., K.L. McAdams, R.H.W. Friedel, and T.P. O'Brien, 2003. "Acceleration and loss of relativistic electrons during geomagnetic storms", *Geophys. Res. Lett.*, **30**, No. 10, 1529, doi:10.1029/2002GL016513, 6-1-4.

Richmond A.D. and Kamide Y., 1988. Mapping electrodynamic features of the high- latitude ionosphere from localized observations: technique. J. Geophys. Res. 93, 5741.

Richmond A.D. and Thayer J.P., 2000. Ionospheric Electrodynamics: A Tutorial, in Magnetospheric Current Systems. American Geophysical Union, Washington, D.C. (Geophysical Monograph 118) p. 131.

Ridley A.J., D.L. De Zeeuw, W.B. Manchester IV, and K.C. Hansen, 2006. "The magnetospheric and ionospheric response to a very strong interplanetary shock and coronal mass ejection", *Adv. Space Res.*, **38**, No. 2, 263-272.

Roldugin V.K., 1974. "Optical Effects in SC and SI at Loparskaya and Mirny Observatories", *Geomagn. Aeron.* **14** (1), 93–95.

Rostoker G., 1972. "Geomagnetic indices", *Rev. Geophys., Space Phys.,* **10**, No. 4, 935-950.

Rykachev M.A., 1899. "Historical Essay of the Main Physical Observatory for 50 Years of Its Activity", In *Imper. Akad. Nauk*, St. Petersburg, 1899, Part I [in Russian].

Ŝafránková J., Z. Němeček, L. Přech, A.A. Samsonov, A. Koval, and K. Andréeová. 2007. "Modification of interplanetary shocks near the bow shock and through the magnetosheath", *J. Geophys. Res.*, **112**, A08212, doi:10.1029/2007JA012503, 1-9.

Sakai J., S. Taguchi, K. Hosokawa, and Y. Ogawa, 2013. "Steep plasma depletion in dayside polar cap during a CME-driven magnetic storm", *J. Geophys. Res. Space Physics*, **118**, No. 1, doi:10.1029/2012JA018138, 462-471.

Samsonov A.A., 2011. "Propagation of inclined interplanetary shock through the magnetosheath", *J. Atmos. Sol. Terr. Phys.*, **73**, No. 1, 30-39.

Samsonov A.A., Z. Němeček, and J. Šafránková, 2006. "Numerical MHD modeling of propagation of interplanetary shock through the magnetosheath", *J. Geophys. Res.*, **111**, No. A8, A08210, doi:10.1029/2005JA011537, 1-9.

Samsonov A.A., D.G. Sibeck, and J. Imber, 2007. "MHD simulation for the interaction of an interplanetary shock with the Earth's magnetosphere", *J. Geophys. Res.*, **112**, A12220, doi:10.1029/2007JA012627, 1-9.

Samsonov A.A., D.G. Sibeck, B.M. Walsh, and N.V. Zolotova, 2014. "Sudden impulse observations in the dayside magnetosphere by THEMIS", *J. Geophys. Res. Space Physics*, **119**, No. 12, doi:10.1002/2014JA020012, 9476-9496.

Schwartz S.J. and D. Burgess, 1991. "Quasi-parallel shocks: A patchwork of three-dimensional structures". *Geophys. Res. Lett.*, **18**, 373–376.

Shen W.-W. and M. Dryer, 1972. "Magnetohydrodynamic theory for the interaction of an interplanetary double-shock ensemble with the Earth's bow shock", *J. Geophys. Res.*, **77**, No. 25, 4627-4644.

Sibeck D.G., 1990. "A model for the transient magnetospheric response to sudden solar wind dynamic pressure changes", *J. Geophys. Res.* **95**, No. A4, 3755–3771.

Siscoe G.L. and Odstrcil D., 2008. "Ways in which ICME sheaths differ from magnetosheaths", *J. Geophys. Res.*, **113**, No. A9, A00B07, doi:10.1029/2008JA013142.

Siscoe G.L., G.M. Erickson, B.U.Ö. Sonnerup, N.C. Maynard, J.A. Schoendorf, K.D. Siebert, D.R. Weimer, W.W. White, and G.R. Wilson, 2002a. “Hill model of transpolar potential saturation: Comparisons with MHD simulations”, *J. Geophys. Res.*, **107**, No. A6, 1075, doi:10.1029/2001JA000109, SMP8-1-8.

Siscoe G. L., N. U. Crooker, and K. D. Siebert, 2002b. “Transpolar potential saturation: Roles of region 1 current system and solar wind ram pressure”, *J. Geophys. Res.*, **107**, No. A10, 1321, doi:10.1029/2001JA009176, SMP 21-1-5.

Siscoe G.L., N.U. Crooker, G.M. Erickson, B.U.Ö. Sonnerup, N.C. Maynard, J.A. Schoendorf, K.D. Siebert, D.R. Weimer, W.W. White, and G.R. Wilson, 2002c. “MHD properties of magnetosheath flow”, *Planet. Space Sci.*, **50**, No. 5-6, 461-471.

Sitnov M.I., N.A. Tsyganenko, A.Y. Ukhorskiy, B.J. Anderson, H. Korth, A.T.Y. Lui, and P.C. Brandt, 2010. “Empirical modeling of a CIR–driven magnetic storm”, *J. Geophys. Res.*, **115**, No. A7, A07231, doi:10.1029/2009JA015169, 1-20.

Song P. and C.T. Russell, 1992. "Model of the formation of low-latitude boundary layer for strongly northward interplanetary magnetic field", *J. Geophys. Res.*, **97**, No. A2, 1411-1420.

Song P., D.L. DeZeeuw, T.I. Gombosi, C.P.T. Groth, and K.G. Powell, 1999. “A numerical study of solar wind-magnetosphere interaction for northward interplanetary magnetic field”, *J. Geophys. Res.*, **104**, No. A12, 28361-28378.

Stone E.C., A.M. Frandsen, R.A. Mewaldt, E.R. Christian, D. Margolies, J.F. Ormes, and F. Snow, 1998. “The Advanced Composition Explorer”, *Space Sci. Rev.*, **86**, No. 1-4, 1-22.

Su Z., Q.–G. Zong, C. Yue, Y. Wang, H. Zhang, and H. Zheng, 2011. “Proton auroral intensification induced by interplanetary shock on 7 November 2004”, *J. Geophys. Res.*, **116**, No. A8, A08223, doi:10.1029/2010JA016239, 1-8.

Sugiura M., 1964. “Hourly values of equatorial *Dst* for the IGY”, *Ann. Int. Geophys. Year*, **35**, No. 1, Pergamon, New York, 9-45.

Sun T.R., C. Wang, H. Li, and X.C. Guo, 2011. “Nightside geosynchronous magnetic field response to interplanetary shocks: Model results”, *J. Geophys. Res.*, **116**, No. A4, A04216, doi:10.1029/2010JA016074, 1-10.

Sun T.R., C. Wang, and Y. Wang, 2012. “Different B_Z response regions in the nightside magnetosphere after the arrival of an interplanetary shock: Multipoint observations compared with MHD simulations”, *J. Geophys. Res.*, **117**, No. A5, A05227, doi:10.1029/2011JA017303, 1-11.

Sun T.R., C. Wang, J.J. Zhang, V.A. Pilipenko, Y. Wang, and J.Y. Wang, 2015. “The chain response of themagnetospheric and ground magnetic field to interplanetary shocks”, *J. Geophys. Res. Space Physics*, **120**, No. 1, doi:10.1002/2014JA020754, 157-165.

Szajko N.S., G. Cristiani, C.H. Mandrini, and A. Dal Lago, 2013. “Very intense geomagnetic storms and their relation to interplanetary and solar active phenomena”, *Adv. Space Res.*, **51**, No. 10, 1842-1856.

Temerin M. and X. Li, 2002. "A new model for the prediction of Dst on the basis of the solar wind", J. Geophys. Res., 107, No. A12, 1472, doi:10.1029/2001JA007532, SMP31-1-8.

Terasawa T., M. Fujimoto, T. Mukai, I. Shinohara, Y. Saito, T. Yamamoto, S. Machida, S. Kokubun, A.J. Lazarus, J.T. Steinberg, and R.P. Lepping, 1997. "Solar wind control of density and temperature in the near-Earth plasma sheet: WIND/GEOTAIL collaboration", *Geophys. Res. Lett.*, **24**, No. 8, 935-938.

Thomsen M.F., J.T. Gosling, and C.T. Russell, 1988. "ISEE studies of the quasi-parallel bow shock", *Adv. Space. Res.*, **8**, No. 9-10, 175–178.

Thomsen M.F., J.T. Gosling, S.J. Bame, and T.G. Onsager, 1990. "Two-state heating at quasiparallel shocks", *J. Geophys. Res.*, **95**, No. A5, 6363–6374.

Thomsen M.F., J.E. Borovsky, D.J. McComas, and M.R. Collier, 1998. "Variability of the ring current source population", *Geophys. Res. Lett.*, **25**, No. 18, 3481-3484.

Thomsen M.F., J.E. Borovsky, R.M. Skoug, and C.W. Smith, 2003. "Delivery of cold, dense plasma sheet material into the near-Earth region", *J. Geophys. Res.*, **108**, No. A4, 1151, doi:10.1029/2002JA009544.

Tsubouchi K. and B. Lembège, 2004. "Full particle simulations of short large-amplitude magnetic structures (SLAMS) in quasi-parallel shocks", *J. Geophys. Res.*, **109**, A02114, doi:10.1029/2003JA010014.

Tsurutani B.T. and G.S. Lakhina, 2014. "An extreme coronal mass ejection and consequences for the magnetosphere and Earth", *Geophys. Res. Lett.*, **41**, No. 2, doi:10.1002/2013GL058825, 287-292.

Tsurutani B.T. and Zhou, 2003. "Interplanetary shock triggering of substorms: WIND and Polar", *Adv. Space Res.*, **31**, No. 4, 1063-1067.

Tsurutani B.T., E. Echer, F.L. Guarnieri, and W.D. Gonzalez, 2011a. "The properties of two solar wind high speed streams and related geomagnetic activity during the declining phase of solar cycle 23", *J. Atmos. Sol. Terr. Phys.*, **73**, No. 1, 164-177.

Tsurutani B.T., G.S. Lakhina, O.P. Verkhoglyadova, W.D. Gonzalez, E. Echer, and F.L. Guarnieri, 2011b. "A review of interplanetary discontinuities and their geomagnetic effects", *J. Atmos. Sol. Terr. Phys.*, **73**, No. 1, 5-19.

Tulasi Ram S., S. Kumar, S.-Y. Su, B. Veenadhari, and S. Ravindran, 2015. "The influence of Corotating Interaction Region (CIR) driven geomagnetic storms on the development of equatorial plasma bubbles (EPBs) over wide range of longitudes", *Adv. Space Res.*, **55**, No. 2, 535-544.

Tverskaya L.V., M.V. Tel'tsov, and S.I. Shkol'nikova (1989), "Dynamics of the Nightside Auroral Oval Related to Substorm Activity during Magnetic Storms," Geomagn. Aeron. 29 (2), 321–323.

Villante U, S Lepidi, P Francia, and T Bruno, 2004. "Some aspects of the interaction of interplanetary shocks with the Earth's magnetosphere: an estimate of the propagation time through the magnetosheath", *J. Atmos. Sol. Terr. Phys.*, **66**, No. 5, 337-341.

Vorobjev V.G., 1974. "Auroral Effects Related to SC", *Geomagn. Aeron.* **14** (1), 90–92.

Vorobjev V.G. and O.I. Yagodkina, 2007. "Auroral Precipitation Dynamics during Strong Magnetic Storms", *Geomagn. Aeron.* **47** (2), 198–205. Russian text: *Geomagn. i Aeron.* **47**, 185–192.

Wang C.B., J.K. Chao, and C.-H. Lin, 2003a. "Influence of the solar wind dynamic pressure on the decay and injection of the ring current", *J. Geophys. Res.*, **108**, No. A9, 1341, doi:10.1029/2003JA009851, SMP5-1-11.

Wang Y., P. Ye, S. Wang, and X. Xue, 2003b. "An interplanetary cause of large geomagnetic storms: Fast forward shock overtaking preceding magnetic cloud", *Geophys. Res. Lett.*, **30**, No. 13, 1700, doi:10.1029/2002GL016861, 33-1-4.

Wang C., T.R. Sun, X.C. Guo, and J.D. Richardson, 2010. "Case study of nightside magnetospheric magnetic field response to interplanetary shocks", *J. Geophys. Res.*, **115**, No. A10, A10247, doi:10.1029/2010JA015451, 1-8.

Weimer D.R., 2005. Improved ionospheric electrodynamic models and application to calculating Joule heating rates. J. Geophys. Res. 110, A05306.

Wilson G.R., Weimer D.R., Wise J.O., and Marcos F., 2006. Response of the thermosphere to Joule heating and particle precipitation. J. Geophys. Res. 111, A10314.

Wu C.-C. and R.P. Lepping, 2002. "Effects of magnetic clouds on the occurrence of geomagnetic storms: The first 4 years of Wind", *J. Geophys. Res.*, **107**, No. A10, 1314, doi:10.1029/2001JA000161, SMP19-1-8.

Wu C.-C. and R.P. Lepping, 2005. "Relationships for predicting magnetic cloud-related geomagnetic storm intensity", *J. Atmos. Sol. Terr. Phys.*, **67**, No. 3, 283-291.

Wu C.-C. and R.P. Lepping, 2008. "Geomagnetic activity associated with magnetic clouds, magnetic cloud-like structures and interplanetary shocks for the period 1995-2003", *Adv. Space Res.*, **41**, No. 2, 335-338.

Wu B.H., Mandt M.E., Lee L.C., and Cao J.K., 1993. "Magnetospheric response to solar wind dynamic pressure variations: interaction of interplanetary tangential discontinuities with the bow shock", *J. Geophys. Res.* **98**, No. A12, 21297–21311.

Wüest M.P., W. Lennartsson, and P.D.Craven, 2002. "Response of the dawn-side, high-altitude, high-latitude magnetosphere to the arrival of an interplanetary shockwave", *Adv. Space Res.*, **30**, No. 10, 2189-2194.

Yan M. and L.C. Lee, 1996. "Interaction of interplanetary shocks and rotational discontinuities with the Earth's bow shock", *J. Geophys. Res.*, **101**, No. A3, 4835-4848.

Yang Z.W., B. Lembège, and Q.M. Lu, 2012. "Impact of the rippling of a perpendicular shock front on ion dynamics", *J. Geophys. Res.*, **117**, No. A7, A07222, doi:10.1029/2011JA017211, 1-24.

Yermolaev Yu.I., I.G. Lodkina, N.S. Nikolaeva, and M.Yu. Yermolaev, 2010. "Statistical study of interplanetary condition effect on geomagnetic storms", *Cosmic Res.*, **48**, No. 6, 485-500. Russian text: *Kosmicheskie Issledovaniya*, **48**, No. 6, 499-515.

Yue C., Q.G. Zong, H. Zhang, Y.F. Wang, C.J. Yuan, Z.Y. Pu, S.Y. Fu, A.T.Y. Lui, B. Yang, and C.R. Wang, 2010. "Geomagnetic activity triggered by interplanetary shocks", *J. Geophys. Res.*, **115**, No. A5, A00I05, doi:10.1029/2010JA015356, 1-13.

Yue C., Q. Zong, Y. Wang, I.I. Vogiatzis, Z. Pu, S. Fu, and Q. Shi, 2011. "Inner magnetosphere plasma haracteristics in response to interplanetary shock impacts", *J. Geophys. Res.*, **116**, No. A11, A11206, doi:10.1029/2011JA016736, 1-12.

Yue C., W. Li, Y. Nishimura et al., 2016. "Rapid enhancement of low-energy (<100 eV) ion flux in response to interplanetary shocks based on two Van Allen Probes case studies: Implications for source regions and heating mechanisms", *J. Geophys. Res. Space Physics*, **121**, 6430–6443, doi:10.1002/2016JA022808.

Zhang G. and L.F. Burlaga, 1988. "Magnetic clouds, geomagnetic disturbances, and cosmic-ray decreases", *J. Geophys. Res.*, **93**, No. A4, 2511-2518.

Zhima Z., J.B. Cao, W.L. Liu, H.S. Fu, T.Y. Wang, X.M. Zhang, and X.H. Shen, 2014. "Storm time evolution of ELF/VLF waves observed by DEMETER satellite", *J. Geophys. Res. Space Physics*, **119**, No. 4, doi:10.1002/2013JA019237, 2612-2622.

Zhou X.-Y., R.J. Strangeway, P.C. Anderson, D.G. Sibeck, B.T. Tsurutani, G. Haerendel, H.U. Frey, and J.K. Arballo, 2003. "Shock aurora: FAST and DMSP observations", *J. Geophys. Res.*, **108**, No. A4, 8019, doi:10.1029/2002JA009701.

Zhuang H.C., C.T. Russell, E.J. Smith, and J.T. Gosling, 1981. "Three dimensional interaction of interplanetary shock waves with the bow shock and magnetopause: A comparison of theory with ISEE observations", *J. Geophys. Res.*, **86**, No. A7, 5590-5600.

Zong Q.-G. and Z. Hui, 2011. "On magnetospheric response to solar wind discontinuities", *J. Atmos. Sol. Terr. Phys.*, **73**, No. 1, 1-4.

Zurbuchen T.H. and I.G. Richardson, 2006. "In-situ solar wind and magnetic field signatures of interplanetary coronal mass ejections", *Space Sci. Rev.*, **123**, No. 1-3, 31-43.

Zurbuchen T.H., G. Gloeckler, F. Ipavich, J. Raines, C.W. Smith, and L.A. Fisk, 2004. "On the fast coronal mass ejections in October/November 2003: ACE-SWICS results", *Geophys. Res. Lett.*, **31**, No. 11, L11805, doi:10.1029/2004GL019461, 1-4.

Zverev V.L. and T.A. Hviuzova, 2008. "Variations in the polar precipitation equatorward boundary during the interaction between the Earth's magnetosphere and solar wind streams from isolated solar sources", *Geomagnetism and Aeronomy*, **48**, No. 1, 28-35. Russian Text: *Geomagnetizm i Aeronomiya*, **48**, No. 1, 32-39.

Chapter 6

COUPLING OF GEOMAGNETOSPHERE-IONOSPHERE SYSTEM WITH PROCESSES IN SPACE AND IN ATMOSPHERE

6.1. COUPLED MAGNETOSPHERE-IONOSPHERE SYSTEM: IONOSPHERIC CONDUCTANCE, CURRENTS, AND PLASMA CONVECTION IN THE INNER MAGNETOSPHERE

6.1.1. The Matter and Short History of the Problem

The paper of Hurtaud et al. (2007) focuses on a modeling of the seasonal and diurnal effects on the dynamics of the coupled magnetosphere-ionosphere system under different solar illumination conditions, to try to reproduce some of the observations concerning the region 2 (R2) field-aligned currents (FAC). This is performed by introducing in the Ionosphere Magnetosphere Model (IMM) the Earth's rotation axis tilt, the magnetic dipole axis tilt and an eccentric magnetic dipole. As noted Hurtaud et al. (2007), the Earth's magnetosphere is a complex system where numerous physical phenomena take place. It is strongly coupled to the solar wind and to the Earth's upper atmospheric layers and evolves depending on their relative state. In particular, seasonal effects have been observed. The geomagnetic activity tends to be higher around the equinoxes (Russell and McPherron, 1973; Mayaud, 1978, M1980; Crooker and Siscoe, 1986b; Crooker et al., 1992; Cliver et al., 2000, 2001) and the solar-produced conductivities are enhanced in the summer hemisphere. The conductivities depend not only on season, but also on the spatially varying geomagnetic field strength (e.g., Wagner et al., 1980; Gasda and Richmond, 1998). Seasonal variations also modify the distribution of the field-aligned currents (FAC). Lu et al. (1994, 1995) used the Assimilative Mapping of Ionospheric Electrodynamics (AMIE) procedure to derive conjugate patterns of the convection electric potential and of the region 1 (R1) and region 2 (R2) FAC densities to stress the inter-hemispheric differences. On January 28–29, 1992, under southward IMF condition, and for a total cross-polar-cap potential drop ranging from 85 to 94 kV, the total downward current (i.e., the radial component of the inward FAC integrated over the area poleward of ±50° magnetic latitude) was found greater in the summer hemisphere

(~6.3 MA) than in the winter one (~4.4 MA). Globally, the current densities were found higher in the summer hemisphere and the R2 FAC flowing in the winter hemisphere were roughly bounded between 15:00 MLT and 10:00 MLT (i.e., mainly on the night-side). For solstices, from statistical studies based on data from Ørsted and MAGSAT, Papitashvili et al. (2002) showed that the ratio of the total summer downward FAC over the total winter downward FAC depends on the orientation of the IMF and varies between 0.92 and 1.57. For the total upward currents, this ratio takes values between 1.31 and 1.62. On average the total downward or upward FAC are larger by a factor 1.35 in summer than in winter. These results are in agreement with Ridley et al. (2004), who showed in a MHD simulation study that solar-produced conductances promote the closure of FAC in sunlight. For equinox conditions, Papitashvili et al. (2002) found the total currents to be symmetrically distributed between the two hemispheres, and Ohtani et al. (2005a) showed the FAC to be more intense around the equinoxes than around the solstices (consistent with the idea that the geomagnetic activity tends to be higher around the equinoxes than around the solstices). Ridley et al. (2004) and Ridley (2007b) studied the seasonal asymmetry of the FAC with a combination of MHD and empirical models, and found that the modeled ratio of total R1 plus R2 FAC in summer to those in winter can be consistent with observations if the statistically observed seasonal variation of auroral particle fluxes (winter 20% larger than summer) is taken into account.

Hurtaud et al. (2007) outlined, that the seasonal effect on the FAC is in fact complex and depends on the location of FAC, with dayside FAC reacting differently from nightside FAC. Fujii et al. (1981) and Fujii and Iijima (1987) have shown that the dayside FAC increase by a factor of 2 and their latitude increases by 1° to 3° in the summer hemisphere with respect to the winter hemisphere. These results are corroborated by recent statistical studies based on data from Ørsted and MAGSAT (Christiansen et al., 2002; Papitashvili et al., 2002), DMSP (Ohtani et al., 2005a,b), Polar and IMAGE (Østgaard et al., 2005) and CHAMP (Wang et al., 2005) The seasonal variations of the FAC seem more complex on the nightside (Fujii et al., 1981; Fujii and Iijima, 1987). Indeed, whereas some of these observations reported no significant seasonal variation of the R1 and R2 current intensity on the nightside (Fujii et al., 1981; Fujii and Iijima, 1987; Christiansen et al., 2002; Wang et al., 2005), the others showed a preference for the large-scale field-aligned currents to flow in the winter hemisphere (Ohtani et al., 2005a,b; Østgaard et al., 2005). The latter results are supported by the observations of intense auroras occurring preferentially under dark conditions, where field-aligned electric fields are found (Newell et al., 1996, 2005). Ohtani et al. (2005a,b) showed an equatorward displacement of the nightside FAC in the summer hemisphere and a poleward displacement in the winter hemisphere, the reverse of the dayside FAC. They explained that this displacement is most likely due to the inter-hemispheric asymmetry of the magnetospheric configuration and, to a smaller extent, to the higher ionospheric conductivities on the dayside.

According to the opinion of Hurtaud et al. (2007), parts of these seasonal effects on the FAC can be due to the interaction between the magnetosphere and the ionosphere. In the inner magnetosphere, plasma motion results from the combined action of the convection electric field and plasma pressure gradients. The latter effect is responsible for a charge separation and the appearance of a partial ring current. The R2 FAC that originate there close in the ionosphere and, together with energetic precipitating particles, modify the electrical properties of the ionosphere. The modified ionospheric electric potential is then transmitted to the magnetosphere along the highly conductive magnetic lines of force, and in turn affects the plasma transport. The Ionosphere Magnetosphere Model (IMM) (Peymirat and Fontaine, 1994) describes the magnetospheric and ionospheric plasma dynamics, taking feedback into account. From an initial state, the IMM computes the temporal evolution of the ion and electron density, temperature and pressure, and also the electric potential in the magnetospheric equatorial plane. Furthermore, it calculates the particle precipitations along with the distributions of R2 field-aligned currents and ionospheric electric potential. Among other things, the magnetic field is assumed dipolar, and in previous studies the dipole axis was assumed to be aligned with the Earth's rotation axis and perpendicular to the Sun-Earth line, so that both hemispheres were equally illuminated and symmetric. As such, it could therefore not simulate seasonal effects on the dynamics of the coupled magnetosphere-ionosphere system. Another model of inner-magnetospheric plasma convection is the Rice Convection Model (RCM): see Wolf et al. (2006) and references therein. The RCM differs from the IMM in that it uses a kinetic rather than fluid formalism for the description of magnetospheric particle convection, and it allows for distortion of the geomagnetic field from a dipole in the magnetosphere. Similar to the IMM, the RCM has until now assumed that the Earth's internal magnetic field is a centered dipole oriented perpendicular to the Sun-Earth line, and that the ionospheric electric potential and field-aligned currents are symmetric about the dipole equator. Wolf et al. (2006) developed a formalism based on Euler potentials (Stern, 1970) to overcome these limitations. Hurtaud et al. (2007) consider cases where the dipole axis is either aligned with the Earth's rotation axis or tilted relative to it, and cases where the dipole is either at or displaced from the Earth's center (hereafter respectively referred to as the centered dipole and the eccentric dipole). In all cases the IMM coordinates are defined relative to the location and orientation of the dipole. As outlined Hurtaud et al. (2007), the R1 currents and cross-polar-cap potential depend in a complex manner on the ionospheric conductances and are coupled to the R2 currents (e.g., Fedder and Lyon, 1987; Christiansen et al., 2002; Ridley et al., 2004). The IMM can make predictions only about the region-2 currents and electric potential equatorward of its high-latitude boundary, for a given potential across this boundary. Thus the study of Hurtaud et al. (2007) is limited to examining seasonal effects only equatorward of its high-latitude boundary (or inside the outer boundary in the magnetospheric equatorial plane), for a given potential across this boundary. The value of this is to understand better how the effectiveness of shielding depends on season. A more realistic evaluation of this effect could be performed coupling the IMM with an MHD code as has been done with the RCM (De Zeeuw et al., 2004).

6.1.2. Extended Ionosphere-Magnetosphere Model (IMM)

In Hurtaud et al. (2007), the calculations are made on two independent grids, one ionospheric and the other magnetospheric, both in a solar magnetic (SM) frame (the Z axis is antiparallel to the magnetic dipole, the X-Z plane contains the direction of the Sun, and the Y axis, perpendicular to the Sun-Earth line, points toward dusk). This system rotates with both a yearly and a daily period with respect to the geocentric equatorial inertial (GEI) system (the Z axis is parallel to the Earth's rotation axis and the X axis is directed toward the Sun at March equinox). The magnetospheric grid lies in the magnetic equatorial plane (i.e., the X-Y plane).

It has 31×40 points (radius and longitude, respectively) distributed evenly in radius between 1.074 and 10.44 R_E, which correspond to 15.20° and 71.97° invariant latitude, and evenly in magnetic longitude between 00:00 and 24:00 magnetic local time (MLT). The ionospheric grid has 25×80 points (magnetic latitude and longitude, respectively) distributed unevenly in invariant latitude between 15.20° and 71.97° but evenly in magnetic longitude. The latitudinal grid spacing is smaller near the regions of interest like the auroral zone and the poleward boundary of the equatorial electrojet. The low-latitude boundary has been chosen so that in the case of the eccentric dipole all lines of force in the computational domain extend beyond the Earth's surface.

6.1.3. Computation of Height-Integrated Conductivities

According to Hurtaud et al. (2007), in the IMM, two sources of ionization and conductivity are considered: solar EUV radiation and precipitating electrons from the magnetosphere (Fontaine et al., 1985). It is assumed that the respective contributions to the Pedersen and Hall conductances (height-integrated conductivities) combine quadratically:

$$\Sigma_{P,H}^{dip} = \sqrt{\left(\Sigma_{P,H}^{S}(\chi)\right)^2 + \left(\Delta\Sigma_{P,H}(E,\Gamma)\right)^2}, \tag{6.1.1}$$

where $\Sigma_{P,H}^{dip}$ are the total conductances for a dipolar geomagnetic field, $\Sigma_{P,H}^{S}(\chi)$ are the solar components, χ is solar zenith angle, and $\Delta\Sigma_{P,H}(E,\Gamma)$ are the auroral particle components. $\Delta\Sigma_P$ and $\Delta\Sigma_H$ depend on the mean energy E and the energy flux Γ of precipitating electrons (Harel et al., 1981):

$$\Delta\Sigma_P = 5.2\Gamma^{1/2}, \quad \Delta\Sigma_H = 0.55\,\Delta\Sigma_P\,E^{3/5}, \tag{6.1.2}$$

where the conductances are expressed in Siemens (S), the electron mean energy in keV, and the electron energy flux in mW.m^{-2}.

6.1.4. Computation of the Ionospheric Electric Potential

Hurtaud et al. (2007) used the equation of the ionospheric electric potential solved in the IMM, which was introduced by Fontaine et al. (1985), who assumed a thin shell approximation. The use of a more realistic magnetic field like an eccentric dipole requires changing the equation for the ionospheric potential. For this aim Hurtaud et al. (2007) make use of the dipolar coordinates α, φ, β defined by (e.g., Richmond, 1973; Gagnepain et al., 1975):

$$\alpha = r/(\sin\theta)^2, \quad \varphi = \varphi, \quad \beta = \cos\theta/r^2, \tag{6.1.3}$$

where r, θ, φ are the usual spherical coordinates. Gagnepain et al. (1975) used dipolar coordinates to study the longitudinal gradients in the equatorial electrojet. Hurtaud et al. (2007) used a similar formalism except that their calculations extend from the polar cap to the equatorial electrojet, and they into account pressure driven currents from the magnetosphere.

6.1.5. Summary of Main Results of Hurtaud et al. (2007)

Hurtaud et al. (2007) summarize main obtained results as following:

(i) The tilt of the magnetic dipole toward or away from the Sun, due both to a seasonal tilt of the Earth's axis and to a tilt of the dipole with respect to the Earth's axis, as well as a shift of the dipole away from the center of the Earth, have been introduced in the IMM to investigate the response of the magnetosphere-ionosphere system to seasonal and diurnal variations of solar illumination.

(ii) When the dipole is aligned with the Earth's axis (case NT), the IMM predicts:

1) seasonal variations of the Pedersen conductances and FAC by 28.5–52.8%;
2) seasonal variations of the mid-latitude ionospheric electric field of about 19–29%;
3) seasonal variations of the ion and electron maximum pressure by 0.9–3.9% and of the ion and electron maximum energy flux by 2.9–3.6%;

(iii) This illustrates the fact that the seasons affect in an important way the distributions of the Pedersen conductances and the FAC but that such effects do not have much influence on the distribution of magnetospheric plasma.

(iv) The dipole tilt with respect to the rotation axis (case WT) induces diurnal variations that can be summarized as follows:

1) diurnal variations of the Pedersen conductances and FAC by about 5.5–29%;
2) diurnal variations of 1.5–4.2% for the ion maximum pressure and 0.4–2.1% for the electron maximum energy flux.

(v) These variations are about the same order as those induced by the seasons in case NT, illustrating the fact that the diurnal and seasonal effects have to be considered simultaneously. The diurnal variations modify the inter-hemispheric asymmetry of the FAC by 26–59% with respect to case NT.

(vi) The shift of the dipole with respect to the center of the Earth (case WS) induces some modifications compared to the WT case that can be summed up as follows:

1) an increase (decrease) of the daily variations of the conductances and the R2 FAC by factors between 1.2 and 2.3 in the Southern (Northern) Hemisphere, irrespective of the season, which contributes to increase the asymmetry between the Southern and Northern Hemispheres;
2) an increase of the daily variations of the ion maximum pressure and of the electron maximum energy flux at the December solstice and at the March and September equinoxes, but a decrease at the June solstice (this introduces an asymmetry between the two solstices which is not predicted when the shift is disregarded);
3) at December solstice, the increase of the pressure and the energy flux amounts to 6.7%.

(vii) These results underline the importance of considering the three different variations of the conductances induced by the Earth's rotation axis tilt, the dipole axis tilt and a more realistic magnetic field like the eccentric dipole.

(viii) The simulated patterns of the R2 field-aligned currents agree rather well with the observations, but a better agreement might be obtained by including additional physical effects in the IMM, like solar-activity dependence of the conductance model, a more realistic magnetic field model for the magnetosphere, and field-aligned potential drops.

6.1.6. A Fast, Parameterized Model of Upper Atmospheric Ionization Rates, Chemistry, and Conductivity

As outlined McGranaghan et al. (2015), rapid specification of ionization rates and ion densities in the upper atmosphere is essential when many evaluations of the atmospheric state must be performed, as in global studies or analyses of on-orbit satellite data. Though many models of the upper atmosphere perform the necessary specification, none provide the flexibility of computational efficiency, high accuracy, and complete specification. McGranaghan et al. (2015) introduce a parameterized, updated, and extended version of the GLobal AirglOW (GLOW) model, called GLOWfast, that significantly reduces computation time and provides comparable accuracy in upper atmospheric ionization, densities, and conductivity. McGranaghan et al. (2015) extend GLOW capabilities by (1) implementing the nitric oxide empirical model, (2) providing a new model component to calculate height-dependent conductivity profiles from first principles for the 80–200 km region, and (3) reducing computation time. The computational improvement is achieved by replacing the full, two-stream electron transport algorithm with two parameterizations: (1) photoionization and (2) electron impact ionization. It is found that GLOWfast accurately reproduces ionization rates, ion and electron densities, and Pedersen and Hall conductivities independent of the background atmospheric state and input solar and auroral activity. Obtained results suggest that GLOWfast may be even more appropriate for low characteristic energy auroral conditions. It is demonstrate in a suite of 3028 case studies that GLOWfast can be used to rapidly calculate the ionization of the upper atmosphere with few limitations on background and input conditions. McGranaghan et al. (2015) support these results through comparisons with electron density profiles from COSMIC.

6.1.7. Conductivities Consistent with Birkeland Currents in the AMPERE-Driven TIE-GCM

As outlined Marsal (2015), the AMPERE satellite mission has offered for the first time global snapshots of the geomagnetic field-aligned currents with unprecedented space and time resolution, thus providing an opportunity to feed an acknowledged first-principles model of the Earth's upper atmosphere such as the National Center for Atmospheric Research Thermosphere-Ionosphere-Electrodynamics General Circulation Model (NCAR TIE-GCM). In the first step, Marsal et al. (2012) used AMPERE data in the current continuity equation between the magnetosphere and the ionosphere to drive the TIE-GCM electrodynamics. In the work of Marsal (2015), ionospheric conductivities have been made consistent with enhanced upward field-aligned currents, which are assumed to correspond to electrons plunging as a result of downward acceleration by electric fields built up along the geomagnetic field lines. The resulting conductance distribution is reasonably commensurate with an independent model that has tried to quantify the ionizing effect of precipitating particles onto the auroral ionosphere. On the other hand, comparison of geomagnetic observatory data with the ground magnetic variations output by the model only shows a modest improvement with respect to Marsal et al. (2012) approach.

6.1.8. MICA Sounding Rocket Observations of Conductivity Gradient-Generated Auroral Ionospheric Responses: Small-Scale Structure with Large-Scale Drivers

A detailed, in situ study of field-aligned current (FAC) structure in a transient, substorm expansion phase auroral arc is conducted by Lynch et al. (2015) using electric field, magnetometer, and electron density measurements from the Magnetosphere-Ionosphere Coupling in the Alfvén Resonator (MICA) sounding rocket, launched from Poker Flat, AK. These data are supplemented with larger-scale, contextual measurements from a heterogeneous collection of ground-based instruments including the Poker Flat incoherent scatter radar and nearby scanning doppler imagers and filtered all-sky cameras. An electrostatic ionospheric modeling case study of this event is also constructed by using available data (neutral winds, electron precipitation, and electric fields) to constrain model initial and boundary conditions. MICA magnetometer data are converted into FAC measurements using a sheet current approximation and show an up-down current pair, with small-scale current density and Poynting flux structures in the downward current channel. Model results are able to roughly recreate only the large-scale features of the field-aligned currents, suggesting that observed small-scale structures may be due to ionospheric feedback processes not encapsulated by the electrostatic model. The model is also used to assess the contributions of various processes to total FAC and suggests that both conductance gradients and neutral dynamos may contribute significantly to FACs in a narrow region where the current transitions from upward to downward. Comparison of Poker Flat Incoherent Scatter Radar versus in situ electric field estimates illustrates the high sensitivity of FAC estimates to measurement resolution.

6.2. Magnetosphere-Ionosphere Coupling: Using Euler Potentials

6.2.1. The Matter and Short History of the Problem

Wolf et al. (2006) present a general formulation of the basic equations of large-scale magnetosphere-ionosphere coupling in terms of Euler potentials. They note that Vasyliunas (1970) was first who presented a computational scheme for self consistently calculating the electrodynamics and plasma processes coupling the inner magnetosphere and ionosphere. Expressed in general form, this scheme consists of three sets of relations:

1. The first relation is an expression relating plasma pressure gradients in Earth's magnetic field to field-aligned currents linking the magnetosphere and ionosphere. Neglecting inertial drift currents compared to gradient/curvature drift currents and assuming an isotropic plasma pitch angle distribution within a magnetic flux tube, this relation (termed the Vasyliunas equation) can be written as

$$\frac{J_{\|N}}{B_N} - \frac{J_{\|S}}{B_S} = \frac{\mathbf{b} \cdot \nabla V \times \nabla P}{B}, \quad (6.2.1)$$

where $J_{\|N}$ and $J_{\|S}$ are the density of Birkeland current along the magnetic field direction just above the Northern and Southern ionospheres, respectively, **b** is a unit vector along the magnetic field, $V = \int ds/B$ is the volume of a tube of unit magnetic flux, and P is magnetospheric particle pressure (e.g., Wolf, 1983; Heinemann and Pontius, 1990). The right side of Eq. 6.2.1 can be calculated anywhere on the field line.

2. The second set of relations consists of expressions for ionospheric current continuity. If one makes the conventional assumption that the induction electric field in the ionosphere is negligible, so that the electric field there is simply $\mathbf{E} = -\nabla\Phi$, then the standard expression for ionospheric conduction current is

$$\mathbf{J} = \sigma_P(-\nabla\Phi + \mathbf{v} \times \mathbf{B}) - \sigma_H(-\nabla\Phi + \mathbf{v} \times \mathbf{B}) \times \mathbf{b}, \quad (6.2.2)$$

where **b** is a unit vector in the direction of the magnetic field, and **v** is the neutral wind velocity. Setting the divergence in **J** equal to the computed field-aligned current yields an elliptic partial differential equation that can be solved for the electric potential distribution in the ionosphere, given the electrical conductance and neutral wind fields. Integrating Eq. 6.2.2 over the conducting region of each field line and requiring that the divergence of horizontal ionospheric conduction current be balanced by the field-aligned current from the magnetosphere leads to the relation

$$\nabla_h \cdot \left[\hat{\Sigma} \cdot (-\nabla\Phi)\right] + w = J_{\|} \sin I, \quad (6.2.3)$$

where $\hat{\Sigma}$ is the conductance tensor representing single-hemisphere field-line-integrated conductivities, w represents the field line integrals of products of conductivity and $\mathbf{v} \times \mathbf{B}$ (see, e.g., Forbes and Harel (1989) for explicit approximate forms) and I is the magnetic dip angle.

3. The third set of relations consists of expressions describing the bounce-averaged, energy-dependent adiabatic drift of magnetospheric plasma. In convection and ring current models, the bounce-averaged drift equation is usually written in a form like

$$\mathbf{v}_D = \frac{\mathbf{E}_{ind} \times \mathbf{B}}{B^2} + \frac{\mathbf{B} \times \nabla H}{qB^2}. \quad (6.2.4)$$

Here $\mathbf{E}_{ind}$ is the induction electric field, the total electric field is $\mathbf{E} = \mathbf{E}_{ind} - \nabla\Phi$, and q is the charge. Hamiltonian H is given by

$$H = q\Phi + W_K \tag{6.2.5}$$

where W_K is the kinetic energy of a particle written as a function of position, time, and the appropriate adiabatic invariants; for the case where the model particle distribution is organized in terms of the first two adiabatic invariants, as in most ring current-particle models, including specifically the Comprehensive Ring Current Model (CRCM) (e.g., Fok et al., 2001), W_K is taken to be a function of the first and second adiabatic invariants μ and J, as well as position and time. The plasma distribution is advanced in time using

$$\left(\frac{\partial}{\partial t} + v_D \cdot \nabla\right) f = S - L\,, \tag{6.2.6}$$

where S and L represent source and loss terms for the magnetospheric flux tube (e.g., up-flowing particles, precipitation, and charge exchange), and f is a function of appropriate adiabatic invariants, position, and time. Wolf et al. (2006) note that the derivation of Eq. 6.2.1 requires neglect of inertial currents compared to gradient/curvature drift currents, an assumption that is valid only for timescales that are long compared to Alfvén wave travel times along field lines (typically a few minutes). The time dependence introduced in Eq. 6.1.9 occurs on a drift timescale, which is assumed to be much longer.

6.2.2. Using of the Euler potentials

Wolf et al. (2006) outlined that early implementations of the Vasyliunas (1970) magnetosphere-ionosphere coupling scheme made many simplifying assumptions. Over the years many of the assumptions and limitations originally built into the Rice Convection Model (Harel et al., 1981) and other early convection models (e.g., Senior and Blanc, 1984) have been eliminated (e.g., Toffoletto et al., 2003). Wolf et al. (2006) formulate the magnetosphere-ionosphere coupling relations described by Eqs. 6.2.1, 6.2.3, 6.2.4, and 6.2.6 in terms of Euler potentials in order to remove two important restrictions remaining from the original convection model implementations, namely, (1) the assumption that Earth's intrinsic magnetic field is a dipole with axis along the rotation axis, perpendicular to the solar wind flow, and (2) the assumption of simple conjugacy between northern and southern hemispheres: a field line that crosses the northern ionosphere reference altitude at magnetic latitude Λ and longitude ϕ crosses the southern ionosphere reference altitude at magnetic latitude $-\Lambda$ and longitude ϕ. Elimination of restrictions 1 and 2 requires a basic change in computational coordinates, which, in the past, has always been a fixed, spherical-polar grid on an ionospheric reference sphere that does not rotate with the Earth. Wolf's et al. (2006) new non-orthogonal mesh is fixed in the northern ionosphere, rotates with the Earth, and is based on Euler potentials, which are constant along field lines. The southern ionosphere grid is the mapping along magnetic field lines of the northern one, and, thus, necessarily varies in time as the planet rotates through non-symmetric and possibly time-varying magnetic field configurations.

According to Wolf et al. (2006), substituting the definition of Euler potentials in terms of the magnetic field (Stern, 1970),

$$\mathbf{B} = \nabla\alpha \times \nabla\beta, \tag{6.2.7}$$

into Eq. 6.2.1, and using $\nabla V = (\partial V/\partial\alpha)\nabla\alpha + (\partial V/\partial\beta)\nabla\beta$ and a similar expression for ∇P leads to the particularly elegant form

$$\frac{J_{\|N}}{B_N} - \frac{J_{\|S}}{B_S} = \frac{\partial V}{\partial\alpha}\frac{\partial P}{\partial\beta} - \frac{\partial P}{\partial\alpha}\frac{\partial V}{\partial\beta}, \tag{6.2.8}$$

where the subscript S refers to values evaluated just above the Southern ionosphere, and N refers to the same height above the northern ionosphere. The right side of Eq. 6.2.8 can be evaluated anywhere along the field line.

6.2.3. Summary of Main Obtained Results in Wolf et al. (2006)

Wolf et al. (2006) came to following conclusions:

(i) The new formulation of the basic equations of magnetosphere-ionosphere coupling allows inner-magnetosphere convection models to be generalized beyond the earlier simplifying assumption of an aligned dipole, zero-tilt representation of Earth's internal magnetic field.

(ii) This formulation facilitates proper consideration of seasonal and IMF-B_y effects in the magnetosphere as well as seasonal and longitude effects in the ionosphere.

(iii) The new formulation utilizes a computational mesh based on Euler potentials, allowing the adiabatic drift and Birkeland current equations to be written in an elegant form.

(iv) The expressions for conservation of ionospheric currents also are most naturally derived and expressed in terms of Euler potentials.

(v) The developed procedure has the disadvantage that it does not produce a grid in the polar cap region of the southern ionosphere, but that does not limit its usefulness for models of inner-magnetospheric convection and the ring current.

6.3. Global and Dynamical Magnetosphere-Ionosphere Coupled System

6.3.1. Alfvénic-Coupling Algorithm for Global and Dynamical Magnetosphere-Ionosphere Coupled System

Yoshikawa et al. (2010) proposes a new formula that describes a dynamical magnetosphere-ionosphere coupling system through the field-aligned current (FAC) closure and electrostatic potential connectivity. In the past, magnetosphere-ionosphere coupling processes were described as either ‘inductive’ or ‘static.’ The inductive coupling scheme is based on the reflection of MHD waves at the ionosphere, whereas in the static coupling scheme the electrostatic potential is determined through the FAC closure. In contrast, in the new formulation these two schemes are combined by the ‘Alfvénic-coupling’ algorithm. The concept of the Alfvénic coupling is as follows. When a distribution of ionospheric current is changed from the background condition through the mapping of magnetospheric disturbances and/or by the change of ionospheric conductivity, an ionospheric reflection electric filed is instantaneously generated to satisfy the current continuity condition. The electrostatic potential of this reflection field also feeds back to the magnetosphere and excites shear Alfvénic disturbances in the magnetosphere. Therefore, the current continuity condition has to include the FAC of Alfvénic disturbances. In this sense, the reflection potential is not only a modification of the ionospheric potential but also a source of inductive Alfvénic disturbances. Quantitative estimation of feedback components to the magnetosphere reveals that the new formulation is more suitable to global magnetosphere-ionosphere simulations than the static magnetosphere-ionosphere coupling, which has been the only scheme used in the past.

6.3.2. Exploring the Influence of Ionospheric O+ Outflow on Magnetospheric Dynamics: The Effect of Outflow Intensity

The ionospheric O^+ outflow varies dramatically during geomagnetic activities, but the influence of its initial characteristics on the magnetospheric dynamics has not been well established. To expand a previous study on the impact of ionospheric heavy ions outflow originating from different source regions on the magnetotail dynamics and dayside reconnection rate, the study of Yu and Ridley (2013) conducts two idealized numerical experiments with different O^+ outflow densities to examine the consequent change in the magnetosphere system, especially on the solar wind-magnetosphere coupling efficiency. Results indicate that a larger O^+ outflow is capable of triggering the Kelvin-Helmholtz instability (KHI) on the magnetopause flanks. The subsequent surface waves enhance the solar wind-magnetosphere coupling efficiency by transmitting more solar wind energy into the magnetosphere-ionosphere system, increasing the cross polar cap potential index. This index is initially reduced after the ionospheric mass loading owing to the direct depression in the dayside reconnection rate as commonly reported from earlier literature. The above KHI is generated under steady state solar wind conditions, suggesting that besides the commonly recognized cause, the elevated solar wind speed, ionospheric heavy ions outflow is another potential factor in disturbing the boundary by enhancing the mass density near the magnetopause and thus lowering the threshold for generating KHI. During storms, the increased ionospheric mass source causes an increased probability of KHI, which allows more solar wind plasma into the magnetosphere. This implies there is a possibility of even further nonlinear coupling between the magnetosphere and solar wind.

6.3.3. Global Empirical Model of TEC Response to Geomagnetic Activity

A global total electron content (TEC) model response to geomagnetic activity described by the Kp index is built by Mukhtarov et al. (2013) using the Center for Orbit Determination of Europe (CODE) TEC data for a full 13 years, January 1999 to December 2011. The model describes the most probable spatial distribution and temporal variability of the geomagnetically forced TEC anomalies assuming that these anomalies at a given modified dip latitude depend mainly on the Kp index, local time (LT), and longitude. The geomagnetic anomalies are expressed by the relative deviation of TEC from its 15 day median and are denoted as rTEC. The rTEC response to the geomagnetic activity is presented by a sum of two responses with different time delay constants and different signs of the cross-correlation function. It has been found that the mean dependence of rTEC on Kp index can be expressed by a cubic function. The LT dependence of rTEC is described by Fourier time series which includes the contribution of four diurnal components with periods 24, 12, 8, and 6 h. The rTEC dependence on longitude is presented by Fourier series which includes the contribution of zonal waves with zonal wave numbers up to 6. In order to demonstrate how the model is able to reproduce the rTEC response to geomagnetic activity, three geomagnetic storms at different seasons and solar activity conditions are presented. The model residuals clearly reveal two types of the model deviation from the data: some underestimation of the largest TEC response to the geomagnetic activity and randomly distributed errors which are the data noise or anomalies generated by other sources. The presented TEC model fits to the CODE TEC input data with small negative bias of −0.204, root mean squares error RMSE = 4.592, and standard deviation error STDE = 4.588. The model offers TEC maps which depend on geographic coordinates (5° × 5° in latitude and longitude) and universal time (UT) at given geomagnetic activity and day of the year. It could be used for both science and possible service (nowcasting and short-term prediction); for the latter, a detailed validation of the model at different geophysical conditions has to be performed in order to clarify the model predicting quality.

6.3.4. Simulation of O+ Upflows Created by Electron Precipitation and Alfvén Waves in the Ionosphere

A two-dimensional model of magnetosphere-ionosphere coupling is presented by Sydorenko and Rankin (2013). It includes Alfvén wave dynamics, ion motion along the geomagnetic field, chemical reactions between ions and neutrals, collisions between different species, and a parametric model of electron precipitation. Representative simulations are presented, along with a discussion of the physical mechanisms that are important in forming oxygen ion field-aligned plasma flows. In particular, it is demonstrated that ion upwelling is strongly affected by the ponderomotive force of standing Alfvén waves in the ionospheric Alfvén resonator, and by enhanced electric fields that are produced when electrons are heated by soft electron precipitation. It is verified that the simulations are in qualitative agreement with available theoretical predictions. In the resonator, in addition to the ponderomotive force, a contribution to the upflow comes from centrifugal acceleration. Heating by the current of standing waves increases parallel electric fields and ion pressure gradients only at low altitudes where they are easily balanced by friction with neutrals. This prevents development of fast field-aligned ion flows in the E-layer and lowers F-layer.

6.3.5. Alfvén Wave Boundary Condition for Responsive Magnetosphere-Ionosphere Coupling

As note Wright and Russell (2014), the solution of electric fields and currents in a height-resolved ionosphere is traditionally solved as an elliptic equation with Dirichlet or Neumann boundary condition in which the magnetosphere is represented as an unresponsive (prescribed) voltage generator or current source. Wright and Russell (2014) derive an alternative boundary condition based upon Alfvén waves in which only the Alfvén wave from the magnetosphere that is incident upon the ionosphere is prescribed. For a uniform magnetosphere the new boundary condition is evaluated at the magnetosphere-ionosphere interface. The resulting solution is interpreted as a responsive magnetosphere and establishes a key stage in the full self-consistent and nonlinear coupling of the magnetosphere and ionosphere.

6.3.6. Some Aspects of Magnetosphere-Ionosphere Relations

The paper of Petrukovich et al. (2015) reviews the characteristics of plasma-wave perturbations produced by wave-particle interactions in the magnetosphere-ionosphere system. These perturbations may, in particular, be due to lightning discharges and to the radiation of high-power, low frequency transmitters. These can form waveguide channels, i.e., density inhomogeneities, which originate in the ionosphere above the radiation source and extend along geomagnetic field lines in the magnetosphere. Although different in nature, the natural and manmade radiations may have similar effects on the processes in the near-earth plasma, causing the excitation of a variety of emissions in it, and stimulating the precipitation of charged particles from the magnetosphere into the ionosphere.

6.3.7. Topside Ionospheric Vary-Chap Scale Height Retrieved from the COSMIC/FORMOSAT-3 Data at Midlatitudes

As note Wang et al. (2015), an α-Chapman function with a continuously varying scale height can be used to describe the topside F_2 vertical electron density profile that seamlessly connects the ionosphere with the plasmasphere. The local time, seasonal, latitudinal and longitudinal variations of the Vary-Chap scale heights (VCSHs) in 2007 at midlatitudes are investigated using the electron density profiles retrieved from the COSMIC/FORMOSAT-3 ionospheric radio occultation observations. The results show that the VCSHs at 400–550 km have prominent altitudinal dependence and diurnal, seasonal variations, and are not tightly correlated with the neutral or plasma-scale heights. The values of the VCSHs in the daytime are larger than those in the nighttime in spring, autumn and winter, comparable and even larger nighttime ones can be detected in summer. The VCSHs change slowly from the midnight to predawn and change abruptly around the sunrise, and the daytime maximum are in winter and the minimum are in summer. In the midlatitudes, the daytime VCSHs decrease with latitude, while at nighttime, the latitudinal changes of VCSHs are not so distinct. The daytime VCSHs in northern hemisphere are larger than those in southern hemisphere. The VCSHs have complicated longitudinal variations and no obvious wavelike structures. What is more, the IRI-2012 model is capable of reproducing the trend of the diurnal, seasonal and latitudinal VCSH variations. But the values have large differences with those retrieved from the COSMIC/FORMOSAT-3 data. Obtained results are illustrated by Figs. 6.3.1 – 6.3.4.

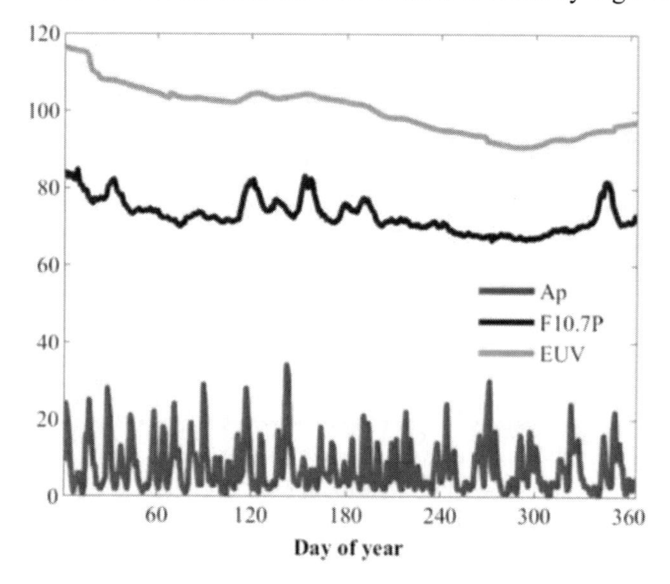

Figure 6.3.1. The solar flux index F10.7P (F10.7P = (F10.7 + F10.7A)/2, in units of 10^{-22} W/m^2/Hz, blue line), 81-day average SOHO/SEM 0.1–50 nm EUV (in units of 2×10^{12} photons/m^2/s for better comparison, green line) and daily geomagnetic Ap index (red line) during 2007. According to Wang et al. (2015). Permission from COSPAR.

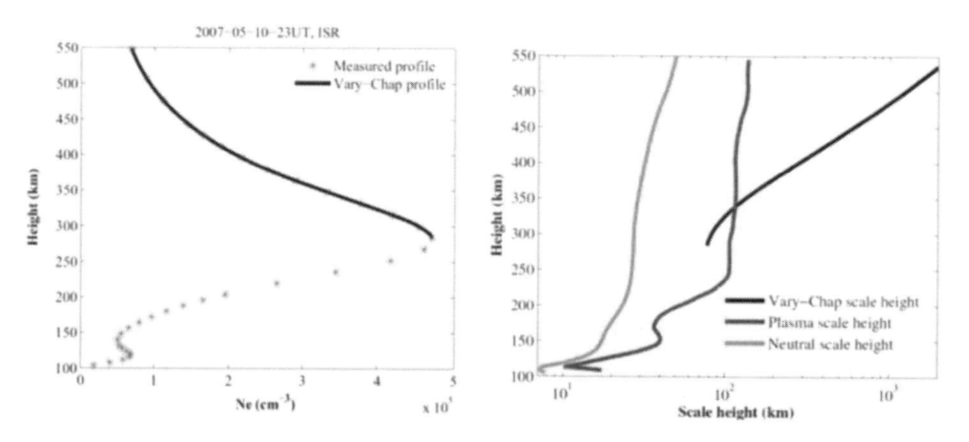

Figure 6.3.2. Left panel: Millstone Hill ISR observed electron density profile (red asterisk). Right panel: calculated Vary-Chap scale height (VCSH) profile from the electron density profile shown on the left. The Vary-Chap profile (blue line) calculated with VCSH shown on the right is superimposed on the measured profile on the left. The neutral scale height (green line) and plasma scale height (red line) are also showed in the right panel. According to Wang et al. (2015). Permission from COSPAR.

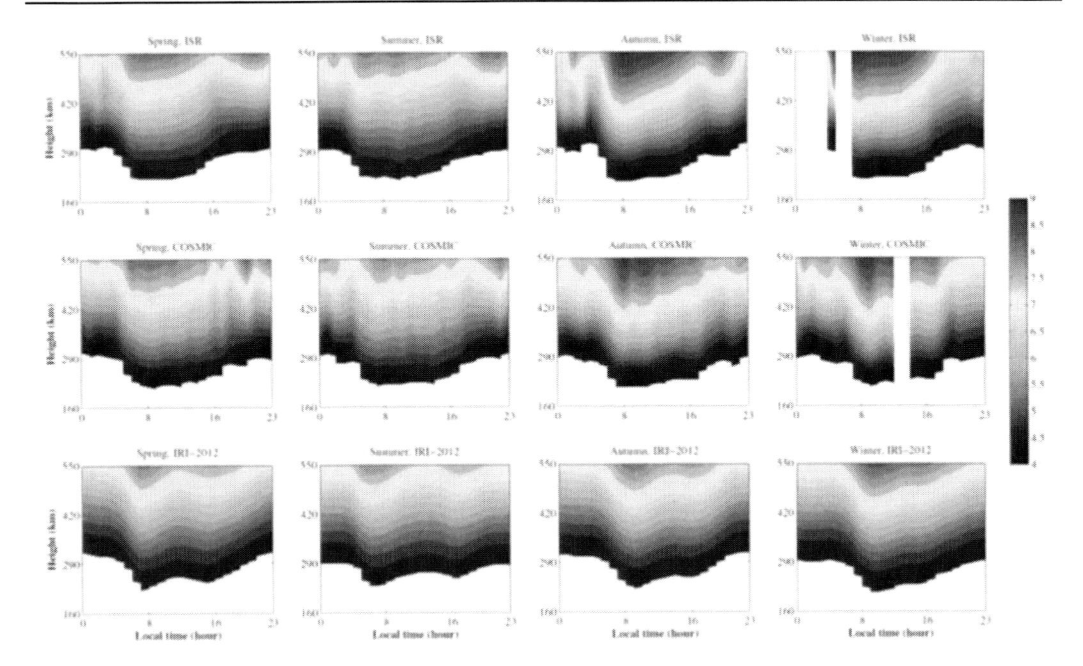

Figure 6.3.3. Local time variation of VCSH in four seasons retrieved from the Millstone Hill ISR data (upper panel), COSMIC data (middle panel) and IRI-2012 model (lower panel) at geographical coordinate (42.8°N, 71.5°W). Note that the VCSH value is logarithmic. According to Wang et al. (2015). Permission from COSPAR.

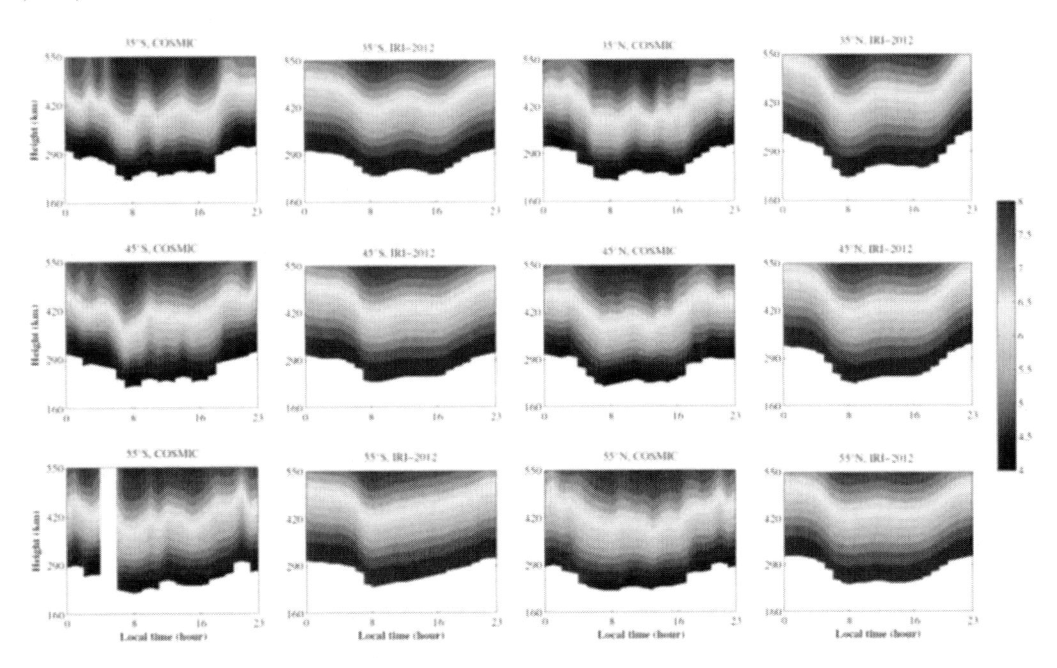

Figure 6.3.4. Local time variation of VCSH in northern spring at magnetic latitude 35°, 45° and 55° of both hemispheres from the COSMIC data and IRI-2012 model, the longitude is 120°E. Note that the VCSH value is logarithmic. According to Wang et al. (2015). Permission from COSPAR.

6.3.8. Dominant Modes of Variability in Large-Scale Birkeland Currents

Properties of variability in large-scale Birkeland currents are investigated by Cousins et al. (2015) through empirical orthogonal function (EOF) analysis of 1 week of data from AMPERE. Mean distributions and dominant modes of variability are identified for both the Northern and Southern Hemispheres. Differences in the results from the two hemispheres are observed, which are attributed to seasonal differences in conductivity (the study period occurred near solstice). A universal mean and set of dominant modes of variability are obtained through combining the hemispheric results, and it is found that the mean and first three modes of variability (EOFs) account for 38% of the total observed squared magnetic perturbations (δB2) from both hemispheres. The mean distribution represents a standard Region 1/Region 2 (R1/R2) morphology of currents and EOF 1 captures the strengthening/weakening of the average distribution and is well correlated with the north-south component of the IMF. EOF 2 captures a mixture of effects including the expansion/contraction and rotation of the (R1/R2) currents; this mode correlates only weakly with possible external driving parameters. EOF 3 captures changes in the morphology of the currents in the dayside cusp region and is well correlated with the dawn-dusk component of the IMF. The higher-order EOFs capture more complex, smaller-scale variations in the Birkeland currents and appear generally uncorrelated with external driving parameters. The results of the EOF analysis described here are used for describing error covariance in a data assimilation procedure utilizing AMPERE data.

6.3.9. The Anomaly of Plasmapause and Ionospheric through Positions from DEMETER data

The paper of Ruzhin et al. (2014) addresses the study of the specific pattern of the subauroral ionosphere marked with the anomalous positions of the plasmapause, the equatorial boundary of the mid-latitude (main) ionospheric trough, and the light-ion trough under quiet solar and geophysical conditions near the magnetospheric shell with the McIlwain parameter $L = 3$. The anomaly was identified on the base of data of active experiments with the SURA heating facility on October 2, 2007, which was conducted as part of the SURA-International Space Station (SURA-ISS) program in the framework of the DEMETER satellite mission. Joint analysis of the orbital data from DEMETER and ISS, together with the results of the complex ground-based measurements, shows that the revealed effect, which is characteristic of the premidnight sector north of the Moscow-SURA satellite path, is not local. It is observed in a vast territory, extending from the west to the east along the same L-shell, from at least Sweden to Kamchatka. The conclusions suggested by the DEMETER data are supported by analysis of the meridional distributions of the $F2$-peak plasma frequencies provided by GPS radio probing of the ionosphere. Comparison of these results with the model latitudinal-longitudinal and meridional distributions of the $F2$-peak plasma density provided by the IRI 2007 and SMI (Russian standard model of the ionosphere) models shows that the model predictions are at odds with the empirical data

6.3.10. Magnetosphere-Ionosphere Coupling Scales and Energy Dumping

Haerendel (2014) reviews three magnetosphere-ionosphere coupling scales characterizing (1) inverted-V auroral arcs, (2) so-called Alfvénic arcs, and (3) dense plasma clouds artificially injected into the magnetosphere. The three scale-breaking processes are different but follow all from the principle of perfect matching of the wave impedance of the energy and momentum carrier with the effective resistance of the energy dump. As a consequence, wave reflections from the ionosphere are absent or present only in a short initial phase and lead to quasi-stationary wave fields. In inverted-V auroral arcs the energy conversion occurs in the low magnetosphere by post acceleration of the current-carrying hot electrons. For the Alfvénic arc the energy is transferred first to dispersive Alfvén waves which are then damped in the topside ionosphere by accelerating electrons along B and ions transversely to B. Barium plasma clouds break into narrow striations mapping to ionospheric scales that experience a strongly reduced Pedersen conductivity and thus achieve the perfect matching.

6.3.11. RCM-E and AMIE Studies of the Harang Reversal Formation during a Steady Magnetospheric Convection Event

Yang et al. (2014) presente the results of a modeling study on the formation of the Harang reversal (HR) during a steady magnetospheric convection event. The Harang reversal is identified as the boundary of the northward and southward electric field in the nightside auroral zone using the Assimilative Mapping of Ionospheric Electrodynamics (AMIE) procedure. Yang et al. (2014) simulate the event with the Rice Convection Model-Equilibrium (RCM-E) by adjusting its boundary conditions to approximately match THEMIS and GOES observations in the nightside magnetosphere. Obtained results show that the HR is collocated with an upward region 1 field-aligned current, where converging ionospheric currents cause a southward/northward electric field on the poleward/equatorward side of the HR. These results also indicate that the electric field reversal is slightly poleward of the ionospheric east–west current reversal and is to the northeast of the ground magnetic reversal, which is consistent with previous observations. We also test the sensitivity of the HR formation to a variety of parameters in the RCM-E simulations. Yang et al. (2014) find that (1) the reduction of the flux tube entropy parameter $PV^{5/3}$ near the midnight sector plays a major role in the formation of the HR; (2) a run carried out assuming uniform conductance produced the same major features as the run with more realistic precipitation-enhanced conductance; and (3) the detailed pattern of the polar cap potential distribution plays a minor role, but its dawn-dusk asymmetry significantly controls the location of the HR with respect to midnight. The RCM-E simulations also predict $PV^{5/3}$ and flow distributions associated with the magnetospheric source of the HR in the plasma sheet, which can be further tested against observations.

6.3.12. Response of the Equatorial Ionosphere to the Geomagnetic DP 2 Current System

The response of equatorial ionosphere to the magnetospheric origin DP 2 current system fluctuations is examined by Yizengaw et al. (2016) using ground-based multi-instrument observations. The interaction between the solar wind and magnetosphere generates a convection electric field that can penetrate to the ionosphere and cause the DP 2 current system. The quasiperiodic DP 2 current system, which fluctuates coherently with fluctuations of the IMF B_z, penetrates nearly instantaneously to the dayside equatorial region at all longitudes and modulates the electrodynamics that governs the equatorial density distributions. Using magnetometers at high and equatorial latitudes, Yizengaw et al. (2016) demonstrate that the quasiperiodic DP 2 current system penetrates to the equator and causes the dayside equatorial electrojet (EEJ) and the independently measured ionospheric drift velocity to fluctuate coherently with the high-latitude DP 2 current as well as with the IMF B_z component. At the same time, radar observations show that the ionospheric density layers move up and down, causing the density to fluctuate up and down coherently with the EEJ and IMF B_z. Main conclusions are following: (i) The solar wind-magnetosphere interaction generates DP 2 current fluctuation. (ii) The DP 2 current fluctuations penetrate to the equator and cause the equatorial electrodynamics to fluctuate, () It also causes the equatorial density to fluctuate which might affect the communication and navigation systems.

6.3.13. An Empirical Model of Ionospheric Total Electron Content (TEC) Near the Crest of the Equatorial Ionization Anomaly (EIA)

Hajra et al. (2016) present a geomagnetic quiet time (Dst > –50 nT) empirical model of ionospheric total electron content (TEC) for the northern equatorial ionization anomaly (EIA) crest over Calcutta, India. The model is based on the 1980–1990 TEC measurements from the geostationary Engineering Test Satellite-2 (ETS-2) at the Haringhata (University of Calcutta, India: 22.58° N, 88.38° E geographic; 12.09° N, 160.46° E geomagnetic) ionospheric field station using the technique of Faraday rotation of plane polarized VHF (136.11 MHz) signals. The ground station is situated virtually underneath the northern EIA crest. The monthly mean TEC increases linearly with F10.7 solar ionizing flux, with a significantly high correlation coefficient (r = 0.89–0.99) between the two. For the same solar flux level, the TEC values are found to be significantly different between the descending and ascending phases of the solar cycle. This ionospheric hysteresis effect depends on the local time as well as on the solar flux level. On an annual scale, TEC exhibits semiannual variations with maximum TEC values occurring during the two equinoxes and minimum at summer solstice. The semiannual variation is strongest during local noon with a summer-to-equinox variability of ~50–100 TEC units. The diurnal pattern of TEC is characterized by a pre-sunrise (0400–0500 LT) minimum and near-noon (1300–1400 LT) maximum. Equatorial electrodynamics is dominated by the equatorial electrojet which in turn controls the daytime TEC variation and its maximum. Hajra et al. (2016) combine these long-term analyses to develop an empirical model of monthly mean TEC. The model is validated using both ETS-2 measurements and recent GNSS measurements. It is found that the developed model efficiently estimates the TEC values within a 1-r range from the observed mean values.

6.3.14. Comparative Quality Analysis of Models of Total Electron Content in the Ionosphere

Ivanov et al. (2016) present a brief description and comparative analysis of the Klobuchar, GEMTEC, and NTCM-GL models of total electron content in the ionosphere. The quality of model performance against experimental data on the total electron content is compared. Statistical estimates for the residual positioning error are obtained for each of these models on the basis of the international Global Navigation Satellite Systems (GNSS) Service data. The GEMTEC and NTCM-GL models are shown to have a higher positioning accuracy than the Klobuchar model. The best results of the ionospheric error correction are provided by the GEMTEC model.

6.4. Plasma Bubbles, Fluctuations, Pulsations, Inertial Alfvén and Ion Acoustic Waves

6.4.1. Continuous Generation and Two-Dimensional Structure of Equatorial Plasma Bubbles Observed by High-Density GPS Receivers in Southeast Asia

High-density GPS receivers located in Southeast Asia (SEA) were utilized by Buhari et al. (2014) to study the two-dimensional structure of ionospheric plasma irregularities in the equatorial region. The longitudinal and latitudinal variations of tens of kilometer-scale irregularities associated with equatorial plasma bubbles (EPBs) were investigated using two-dimensional maps of the rate of total electron content change index (ROTI) from 127 GPS receivers with an average spacing of about 50–100 km. The longitudinal variations of the two-dimensional maps of GPS ROTI measurement on 5 April 2011 revealed that 16 striations of EPBs were generated continuously around the passage of the solar terminator. The separation distance between the subsequent onset locations varied from 100 to 550 km with 10 min intervals. The lifetimes of the EPBs observed by GPS ROTI measurement were between 50 min and over 7 h. The EPBs propagated 440–3000 km toward the east with velocities of 83–162 m/s. The longitudinal variations of EPBs by GPS ROTI keogram coincided with the depletions of 630 nm emission observed using the airglow imager. Six EPBs were observed by GPS ROTI along the meridian of Equatorial Atmosphere Radar (EAR), while only three EPBs were detected by the EAR. The high-density GPS receivers in SEA have an advantage of providing time continuous descriptions of latitudinal/longitudinal variations of EPBs with both high spatial resolution and broad geographical coverage. The spatial periodicity of the EPBs could be associated with a wavelength of the quasiperiodic structures on the bottomside of the F region which initiate the Rayleigh-Taylor instability.

6.4.2. Density Cavity Formation through Nonlinear Interaction of 3-D Inertial Alfvén Wave and Ion Acoustic Wave

Density cavities having transverse-scale size of the order of electron inertial length have been observed in auroral ionosphere by Freja spacecraft. In Sharma et al. (2014), density cavity formation in auroral regions has been investigated using numerical simulation techniques. Nonlinear interaction of inertial Alfvén waves and ion acoustic waves has been suggested as a possible mechanism to apprehend density cavity formation in auroral regions. In the proposed mechanism, strong modification in the density profile takes place due to the ponderomotive force of inertial Alfvén waves and accounts for these density depleted regions. In literature also, it has been demonstrated that these cavities are associated with the ponderomotive forces of inertial Alfvén wave. The presented model also attempts to understand the salient features of auroral density cavities which are reported in literature from the analysis of data available from FAST spacecraft. Relevance of obtained results with observations in auroral ionosphere has also been discussed.

6.4.3. Geomagnetic Control of Equatorial Plasma Bubble Activity Modeled by the TIEGCM with Kp

Describing the day-to-day variability of Equatorial Plasma Bubble (EPB) occurrence remains a significant challenge. Carter et al. (2014) use the Thermosphere-Ionosphere Electrodynamics General Circulation Model (TIEGCM), driven by solar (F10.7) and geomagnetic (Kp) activity indices, to study daily variations of the linear Rayleigh-Taylor (R-T) instability growth rate in relation to the measured scintillation strength at five longitudinally distributed stations. For locations characterized by generally favorable conditions for EPB growth (i.e., within the scintillation season for that location), we find that the TIEGCM is capable of identifying days when EPB development, determined from the calculated R-T growth rate, is suppressed as a result of geomagnetic activity. Both observed and modeled upward plasma drifts indicate that the prereversal enhancement scales linearly with Kp from several hours prior, from which it is concluded that even small Kp changes cause significant variations in daily EPB growth.

6.4.4. Nonlinear Growth, Bifurcation, and Pinching of Equatorial Plasma Bubble Simulated by Three-Dimensional High-Resolution Bubble Model

A new three-dimensional high-resolution numerical model to study equatorial plasma bubble (EPB) has been developed by Yokoyama et al. (2014). The High-Resolution Bubble (HIRB) model is developed in a magnetic dipole coordinate system for the equatorial and low-latitude ionosphere with a spatial resolution of as fine as 1 km. Adopting a higher-order numerical scheme than those used in the existing models, the HIRB model is capable of reproducing the bifurcation, pinching, and turbulent structures of EPB. From a seeding perturbation resembling large-scale wave structure (LSWS), EPB grows nonlinearly from the crest of LSWS upwelling, bifurcates at the top of EPB, then becomes turbulent at the topside of the F region. One of the bifurcated EPB is pinched off from the primary EPB and stops growing after pinching. The narrow channel of EPB tends to have a wiggle due to the secondary instability along the wall of EPB. Because of the fringe field effect above and below the EPB, upward drifting low-density plasma converges toward the F peak altitude, forming a narrow-depleted channel, and diverges above the peak, forming a flattened top of the EPB. The flattened top which has a steep upward density gradient is so unstable that bifurcation can easily occur even from a very small thermal perturbation. A higher density region between the bifurcated EPB moves downward due to westward polarization electric field. The EPB is pinched off when it reaches the wall of the primary EPB. It is concluded that turbulent plume-like irregularities can be spontaneously generated only from large-scale perturbation at the bottomside F region.

6.4.5. Magnetosphere-Ionosphere Coupling of Global Pi2 Pulsations

As note Keiling et al. (2014), global Pi2 pulsations have mainly been associated with either low/middle latitudes or middle/high latitudes and, as a result, have been treated as two different types of Pi2 pulsations, either the plasmaspheric cavity resonance or the transient response of the substorm current wedge, respectively. However, in some reports, global Pi2 pulsations have a single period spanning low/middle/high latitudes. This "super" global type has not yet been satisfactorily explained. In particular, it has been a major challenge to identify the coupling between the source region and the ground. Keiling et al. (2014) report two consecutive super global Pi2 events which were observed over a wide latitudinal and longitudinal range. Using four spacecraft that were azimuthally spread out in the nightside and one spacecraft in the tail lobe, it was possible to follow the Pi2 signal along various paths with time delays from the magnetotail to the ground. Furthermore, it was found that the global pulsations were a combination of various modes including the transient Alfvén and fast modes, field line resonance, and possibly a forced cavity-type resonance. As for the source of the Pi2 periodicity, oscillatory plasma flow inside the plasma sheet during flow braking (e.g., interchange oscillations) is a likely candidate. Such flow modulations, resembling the ground Pi2 pulsations, were recorded for both events.

6.4.6. Nighttime Magnetic Field Fluctuations in the Topside Ionosphere at Midlatitudes and Their Relation to Medium-Scale Traveling Ionospheric Disturbances: The Spatial Structure and Scale Sizes

As note Park et al. (2015b), previous studies suggested that electric and/or magnetic field fluctuations observed in the nighttime topside ionosphere at midlatitudes generally originate from quiet time nocturnal medium-scale traveling ionospheric disturbances (MSTIDs). However, decisive evidences for the connection between the two have been missing. Park et al. (2015b) make use of the multispacecraft observations of midlatitude magnetic fluctuations (MMFs) in the nighttime topside ionosphere by the Swarm constellation. The analysis results show that the area hosting MMFs is elongated in the NW-SE (NE-SW) direction in the Northern (Southern) Hemisphere. The elongation direction and the magnetic field polarization support that the area hosting MMFs is nearly field aligned. All these properties of MMFs suggest that they have close relationship with MSTIDs. Expectation values of root-mean-square field-aligned currents associated with MMFs are up to about 4 nA/m^2. MMF coherency significantly drops for longitudinal distances of $\geq 1°$.

6.4.7. Dual Radar Investigation of E Region Plasma Waves in the Southern Polar Cap

Origins and characteristics of small-scale plasma irregularities in the polar ionosphere are investigated by Forsythe and Makarevich (2015) using a dual radar setup in which the E region is probed from opposite directions by two Super Dual Auroral Radar Network facilities at the McMurdo and Dome Concordia Antarctic stations. In certain time intervals, velocity agreement is observed when velocities are compared at the same physical location in the horizontal plane. Such an agreement is widely expected if velocity at a given location is largely controlled by the convection electric field. In other cases, however, velocity agreement is unexpectedly observed when measurements are considered at the same slant range (distance along the radar beam) for both radars. This implies that it is not the electric field at a given location that is a controlling factor. Raytracing results show that the same range agreement may be explained for certain E region density conditions when echo altitude increases with radar range. Backscatter observations under generally unfavorable conditions for irregularity generation and the critical role of propagation conditions in the polar cap are discussed. The observed E region velocity in the polar cap is demonstrated to depend indirectly on the plasma density distribution, which is important for establishing the fundamental dependence on the convection electric field.

6.4.8. Enhanced N2 and O2 Densities Inferred from EISCAT Observations of Pc5 Waves and Associated Electron Precipitation

An advanced two-dimensional numerical model of the coupled ionosphere and magnetosphere is used by Sydorenko et al. (2016) to analyze EISCAT observations of ULF waves that are accompanied by electron precipitation with a wide energy spectrum. The observations show columns of significantly enhanced electron density produced by pulsating precipitation at altitudes between 150 km and 300 km. After each precipitation pulse, the plasma density returns to its initial value within 2 min. Simulations reveal that such a high-density decay rate cannot be reproduced with the composition of neutrals corresponding to a quiet time provided by the Mass Spectrometer Incoherent Scatter model. To explain the rapid density decay rate using the model of the coupled ionosphere and magnetosphere, the density of nitrogen and oxygen molecules was increased, while the density of oxygen atoms was decreased. The modified neutral densities improved not only the decay rate but also the altitude profile of plasma density which had no F_2 layer maximum before the wave and the pulsating precipitation started.

6.4.9. The Distribution of Spectral Index of Magnetic Field and Ion Velocity In Pi2 Frequency Band in BBFs: THEMIS Statistics

A statistical study of the THEMIS FGM and ESA data is performed by Wu et al. (2016) on turbulence of magnetic field and velocity for 218 selected 12 min intervals in bursty bulk flows (BBFs). The spectral index α in the frequency range of 0.005–0.06 Hz are Gaussian distributions. The peaks indexes of total ion velocity V_i and parallel velocity $V_{\|}$ are 1.95 and 2.07 nearly the spectral index of intermittent low frequency turbulence with large amplitude. However, most probable α of perpendicular velocity $V_{\perp}$ is about 1.75. It is a little bigger than 5/3 of Kolmogorov (1941). The peak indexes of total magnetic field B_T is 1.70 similar to $V_{\perp}$. Compression magnetic field $B_{\|}$ are 1.85 which is smaller than 2 and bigger than 5/3 of Kolmogorov (1941). The most probable spectral index of shear $B_{\perp}$ is about 1.44 which is close to 3/2 of Kraichnan (1965). Max $V_{\perp}$ have little effect on the power magnitude of V_T and $V_{\|}$ but is positively correlated to spectral index of $V_{\perp}$. The spectral power of B_T, $B_{\|}$ and $B_{\perp}$ increase with max perpendicular velocity but spectral indexes of them are negatively correlated to $V_{\perp}$. As outlined Wu et al. (2016), the spectral index and the spectral power of magnetic field over the frequency interval 0.005–0.06 Hz is very different from that over 0.08–1 Hz. Some obtained results are illustrated by Fig. 6.4.1 – 6.4.3.

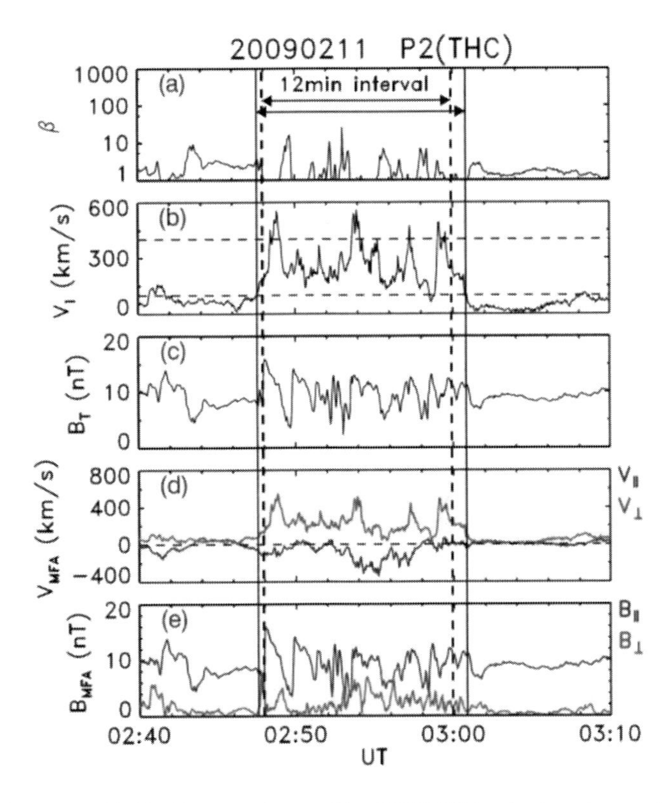

Figure 6.4.1. 12 min interval in BBF event observed by THEMIS P3 in the plasma sheet on 11 February 2009. This figure shows the (a) β, (b) total ion velocity V_i, (c) total magnetic field B_T, (d and e) MFA parallel and perpendicular components of the velocity and magnetic field. The two vertical solid lines identify the starting and the ending times of BBF. The two vertical dashed lines starting and ending times of 12 min interval. According to Wu et al. (2016). Permission from COSPAR.

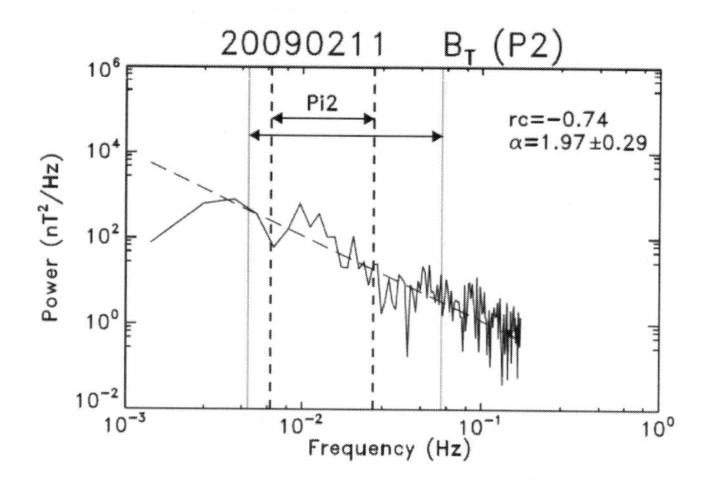

Figure 6.4.2. The power spectral of total magnetic field B_T observed by P3 on 11 February 2009. The linear fit is made over the frequency range 0.005– 0.06 Hz (long arrow). The traditional Pi2 range is contained in this range (short arrow). α given in the figure is the spectral index for the fitted region. rc indicates the cross correlation coefficient of the linear fit. According to Wu et al. (2016). Permission from COSPAR.

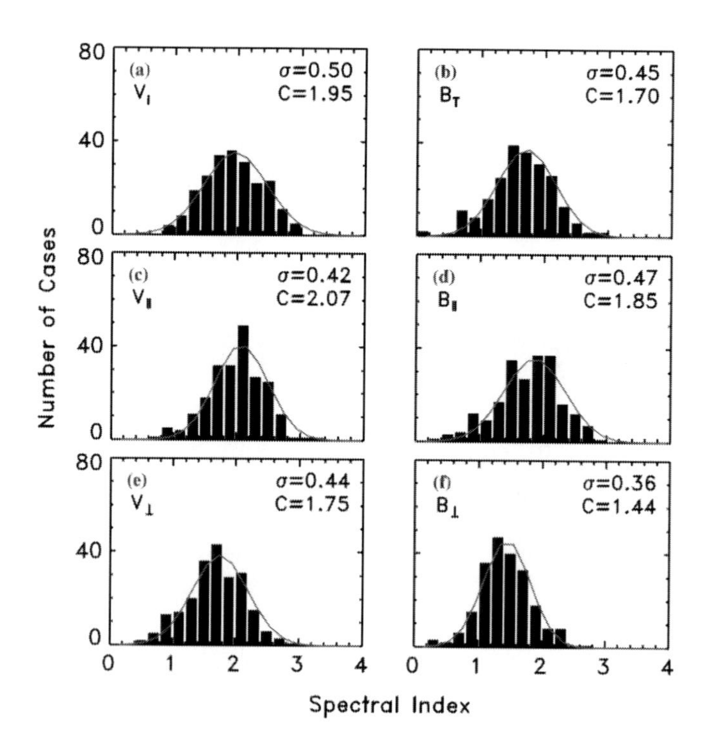

Figure 6.4.3. Histograms of spectral index of all 218 cases. The red lines are the Gaussian curve fitting. C is the most probable spectral index, σ denotes the standard deviation. This figure shows the spectral index distribution of (a–f) the total ion velocity V_i, the total magnetic field B_T, the parallel component of velocity $V_{\|}$, the parallel component magnetic field $B_{\|}$, the perpendicular component velocity $V_{\perp}$ and the perpendicular component magnetic field $B_{\perp}$, separately. According to Wu et al. (2016). Permission from COSPAR.

6.4.10. The Origin of Infrasonic Ionosphere Oscillations over Tropospheric Thunderstorms

As note Shao and Lay (2016), thunderstorms have been observed to introduce infrasonic oscillations in the ionosphere, but it is not clear what processes or which parts of the thunderstorm generate the oscillations. Shao and Lay (2016) present a new technique that uses an array of ground-based GPS total electron content (TEC) measurements to locate the source of the infrasonic oscillations and compare the source locations with thunderstorm features to understand the possible source mechanisms. The location technique utilizes instantaneous phase differences between pairs of GPS-TEC measurements and an algorithm to best fit the measured and the expected phase differences for assumed source positions and other related parameters. In this preliminary study, the infrasound waves are assumed to propagate along simple geometric ray paths from the source to the measurement locations to avoid extensive computations. The located sources are compared in time and space with thunderstorm development and lightning activity. Sources are often found near the main storm cells, but they are more likely related to the downdraft process than to the updraft process. The sources are also commonly found in the convectively quiet stratiform regions behind active cells and are in good coincidence with extensive lightning discharges and inferred high-altitude sprites discharges.

6.4.11. Dust Acoustic Dromions in a Magnetized Dusty Plasma with Superthermal Electrons and Ions

An investigation of dust acoustic (DA) dromions in a magnetized dusty plasma consisting of inertial dust fluid, kappa-type distributed electrons, and ions is presented by Saini et al. (2016). Using reductive perturbation technique, Saini et al. (2016) have derived coupled nonlinear evolution equations of (2 + 1) dimensions (called Davey-Stewartson (DS-I) equations). Hirota bilinear method has been employed to derive the analytical solution of DS-I equations. The solutions of such equations are exponentially localized and are called dromions. The combined effects of various physical parameters such as superthermality of charged particles, strength of magnetic field, and dust concentration have been studied on the existence regions and propagation properties of DA dromions in context with observations of POLAR satellite in the presence of superthermal particles in polar cap boundary layer region of Earth's atmosphere.

6.4.12. First Three Dimensional Wave Characteristics in the Daytime Upper Atmosphere Derived from Ground-Based Multiwavelength Oxygen Dayglow Emission Measurements

First results on the three-dimensional wave characteristics in the daytime upper atmosphere have been derived by Pallamraju et al. (2016) using measurements of oxygen dayglow emissions at 557.7, 630.0, and 777.4 nm that originate at around 130, 230, and 300 km (peak of the F region). The horizontal scale sizes of gravity waves (GWs), their time periods, phase propagation angle (counterclockwise from east), and phase speeds are found to vary in the range of 27–227 km, 32–70 min, 207°–253°, and 6–76 ms^{-1}, respectively. Two-dimensional measurements on the horizontal scale sizes in the daytime have not been reported before. Further, using Hines' (1960) dispersion relation for GWs, vertical scale sizes and phase angles have also been derived. This technique opens up new possibilities in the investigations of daytime wave dynamics in three dimensions in the upper atmosphere.

6.4.13. Predawn Plasma Bubble Cluster Observed in Southeast Asia

According to Watthanasangmechai et al. (2016), predawn plasma bubble was detected as deep plasma depletion by GNU Radio Beacon Receiver (GRBR) network and in situ measurement onboard Defense Meteorological Satellite Program F15 (DMSPF15) satellite and was confirmed by sparse GPS network in Southeast Asia. In addition to the deep depletion, the GPS network revealed the coexisting submesoscale irregularities. A deep depletion is regarded as a primary bubble. Submesoscale irregularities are regarded as secondary bubbles. Primary bubble and secondary bubbles appeared together as a cluster with zonal wavelength of 50 km. An altitude of secondary bubbles happened to be lower than that of the primary bubble in the same cluster. The observed pattern of plasma bubble cluster is consistent with the simulation result of the recent high-resolution bubble (HIRB) model. This event is only a single event out of 76 satellite passes at nighttime during 3–25 March 2012 that significantly shows plasma depletion at plasma bubble wall. The inside structure of the primary bubble was clearly revealed from the in situ density data of DMSPF15 satellite and the ground-based GRBR total electron content.

6.4.14. Zakharov Simulations of Beam-Induced Turbulence in the Auroral Ionosphere

As note Akbari et al. (2016), recent detections of strong incoherent scatter radar echoes from the auroral F region, which have been explained as the signature of naturally produced Langmuir turbulence, have motivated to revisit the topic of beam-generated Langmuir turbulence via simulation. Results from one-dimensional Zakharov simulations are used by Akbari et al. (2016) to study the interaction of ionospheric electron beams with the background plasma at the F region peak. A broad range of beam parameters extending by more than 2 orders of magnitude in average energy and electron number density is considered. A range of wave interaction processes, from a single parametric decay, to a cascade of parametric decays, to formation of stationary density cavities in the condensate region, and to direct collapse at the initial stages of turbulence, is observed as Akbari et al. (2016) increase the input energy to the system. The effect of suprathermal electrons, produced by collisional interactions of auroral electrons with the neutral atmosphere, on the dynamics of Langmuir turbulence is also investigated. It is seen that the enhanced Landau damping introduced by the suprathermal electrons significantly weakens the turbulence and truncates the cascade of parametric decays.

6.5. GEOMAGNETIC ACTIVITY AND VARIATIONS OF IONOSPHERIC PEAK HEIGHT

6.5.1. Empirical Orthogonal Function Analysis and Modeling of the Ionospheric Peak Height during the Years 2002–2011

As outlined Lin et al. (2014), the ionospheric peak electron density height (h_m) is one of the most important ionospheric parameters characterizing high-frequency radio wave propagation conditions. In Lin et al. (2014), a global h_m model based on the empirical orthogonal function (EOF) analysis method is constructed by using Global Navigation Satellite Systems ionospheric radio occultation measurements from COSMIC and CHAMP as well as global ionosonde data during the years 2002–2011. The variability of h_m can be well represented by the several EOF base functions Ek and the corresponding coefficients Pk. The rapid convergence of EOF decomposition makes it possible to use only the first four EOFs components, which express 99.133% of total variance in this study, to construct the empirical model. The variations of hm with respect to the magnetic latitude, local time, season, and solar cycle have been studied, and the EOF-based hm model has been validated through comparisons with the International Reference Ionosphere (IRI) model and other observation. The evaluations indicate that the EOF and IRI model give better h_mF_2 at middle and high latitudes than those at low latitudes. Since the limited data were used in the EOF model during high solar activity years, its accuracy degrades to some extent. During nighttime of spring, summer, and winter in the auroral zone, the hm derived from the EOF model may range from 90 to 150 km because of the reduction of hm, which is due to particle precipitation, whereas the IRI model does not include this reduction. During the periods of low solar activity, the F_2 peak heights (h_mF_2) from the EOF model are in good agreement with the observed data, while the IRI model tends to overestimate the h_mF_2 in the middle and high latitudes.

6.5.2. Geomagnetic Activity Effect on the Global Ionosphere during the 2007–2009 Deep Solar Minimum

In Chen et al. (2014) the significant effect of weaker geomagnetic activity during the 2007–2009 deep solar minimum on ionospheric variability on the shorter-term time scales of several days was highlighted via investigating the response of daily mean global electron content (GEC, the global area integral of total electron content derived from ground-based GPS measurements) to geomagnetic activity index Ap. Based on a case during the deep solar minimum, the effect of the recurrent weaker geomagnetic disturbances on the ionosphere was evident. Statistical analyses indicate that the effect of weaker geomagnetic activity on GEC variations on shorter-term time scales was significant during 2007–2009 even under relatively quiet geomagnetic activity condition; daily mean GEC was positively correlated with geomagnetic activity. However, GEC variations on shorter-term time scales were poorly correlated with geomagnetic activity during the solar cycle descending phase of 2003–2005 except under strong geomagnetic disturbance condition. Statistically, the effects of solar EUV irradiance, geomagnetic activity, and other factors (e.g., meteorological sources) on GEC variations on shorter-term time scales were basically equivalent during solar minimum.

6.6. Influence of Annual, Solar Cycle, and Secular Variations on Magnetosphere-Ionosphere Coupling

6.6.1. Electron–Ion–Neutral Temperatures and Their Ratio Comparisons over Low Latitude Ionosphere

Annual variations in temperatures of plasma (electron– T_e, ion–T_i) using SROSS-C2 satellite and neutral (T_n) using NRLMSIS-00 (neutral atmospheric model) have been investigated and their ratios have been compared over half solar cycle (years 1995–2000) by Bardhan et al. (2015). The region under consideration spans over 5–35°N geog. latitude and 65–95°E geog. longitude in the Indian sector, at an average altitude of 500 km. T_e and T_i exhibit similar, while T_n show completely different diurnal features. During nighttime T_e, T_i and T_n attain equilibrium with each other, but plasma cools faster than neutrals. Magnitude of T_e and T_i reduces, while that of T_n increases with increasing solar activity. Ratio comparisons (T_e/T_n, T_i/T_n and T_e/T_i) show higher sensitivity of electrons compared to ions and neutrals. T_e/T_n, T_i/T_n exhibits linear/direct relationship with solar flux, while T_e/T_i doesn't. The deviation of T_e, T_i and T_n from equilibrium temperature decreases with increasing solar activity. Obtained results are illustrated by Figures 6.6.1 – 6.6.4.

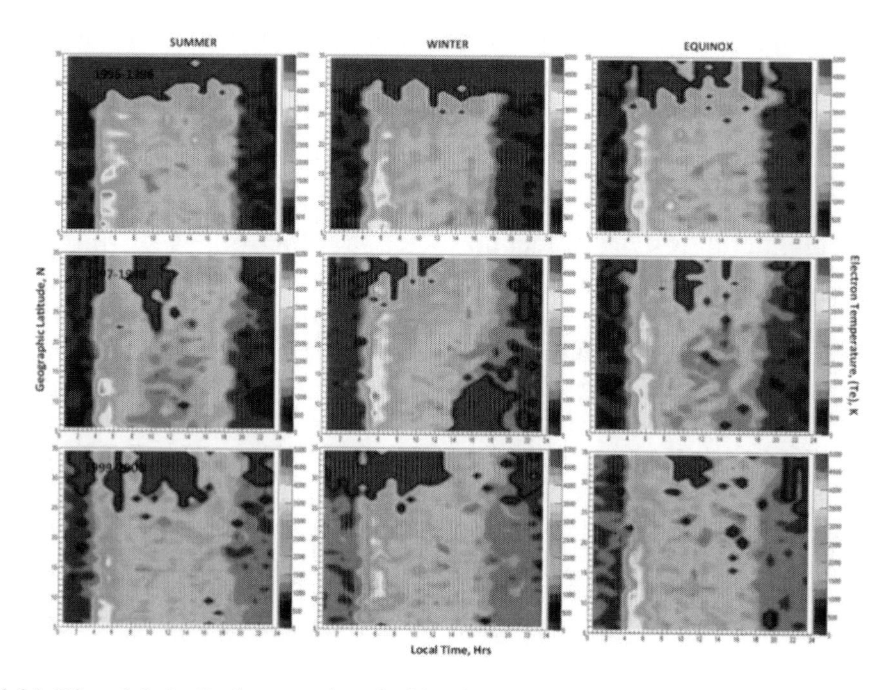

Figure 6.6.1. Diurnal, latitudinal, seasonal, and with solar activity variability of electron temperature (Te) during low (1995–1996; F10.7 ~ 74.5, upper panel), moderate (1997–1998; F10.7 ~99.5, middle panel) and high (1999–2000; F10.7 ~ 166, lower panel) solar activity years. According to Bardhan et al. (2015). Permission from COSPAR.

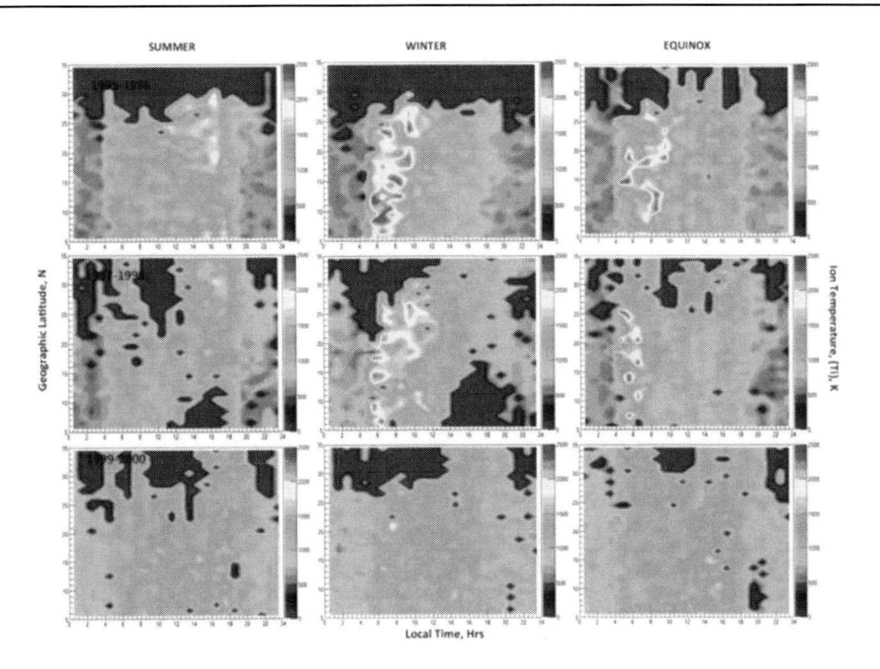

Figure 6.6.2. The same as Figure 6.6.1, but for ion temperature (Ti). According to Bardhan et al. (2015). Permission from COSPAR.

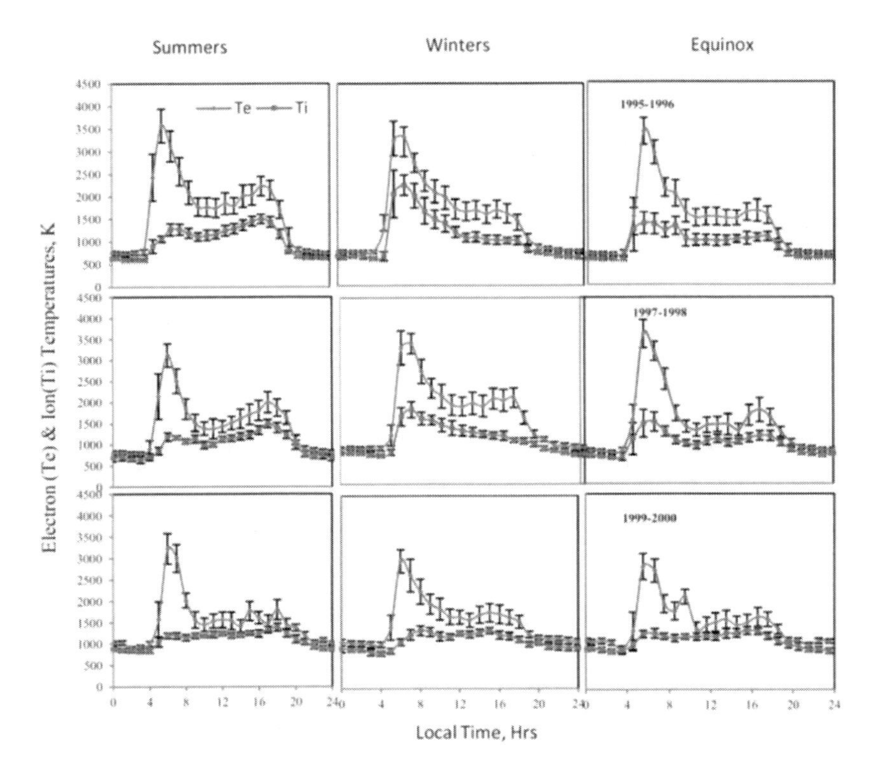

Figure 6.6.3. Diurnal and seasonal (summer-left column; winter – middle column and equinox – right column) variations of electron (Te) and ion (Ti) temperatures during low (1995–1996; F10.7 ~ 74.5, upper panel), moderate (1997–1998; F10.7 ~99.5, middle panel) and high (1999–2000; F10.7 ~ 166, lower panel) solar activity years. According to Bardhan et al. (2015). Permission from COSPAR.

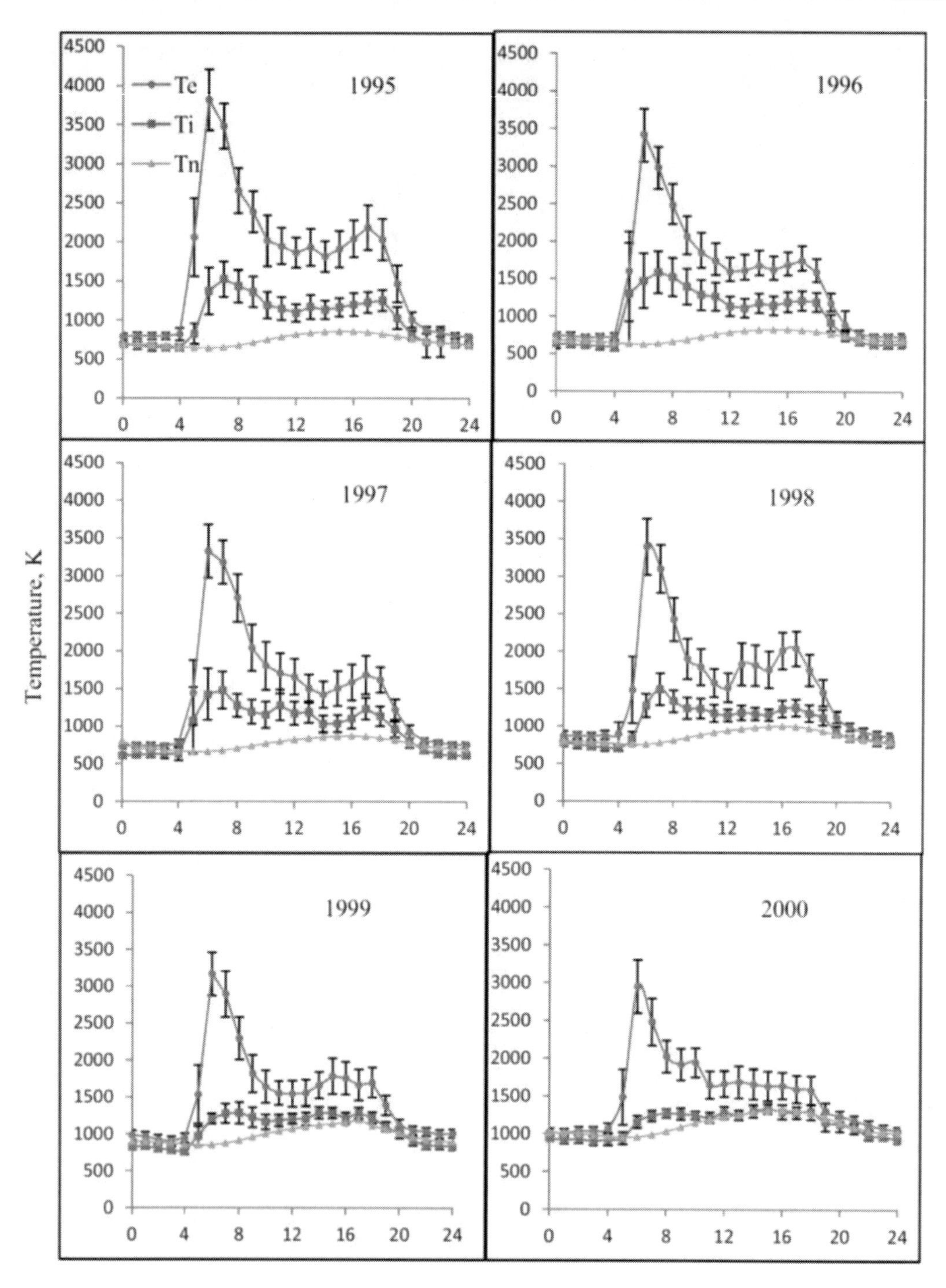

Figure 6.6.4. Diurnal variation of electron (Te-red), ion (Ti-blue) and neutral (Tn-green) temperatures over half solar cycle (1995–2000). According to Bardhan et al. (2015). Permission from COSPAR.

6.6.2. Solar Control of F Region Radar Backscatter: Further Insights from Observations in the Southern Polar Cap

The role of solar wind and illumination in production of small-scale F region plasma irregularities is investigated by Lamarche and Makarevich (2015) using a 4 year data set collected by SuperDARN facility at the McMurdo station, Antarctica (MCM). Statistical analysis of ionospheric echoes detected by MCM shows that radar backscatter from the polar F region occurs in wide and persistent bands in range that exhibit systematic changes with local time, season, and solar cycle. It is demonstrated that all variations considered together form a distinct pattern. A comparison with the F region model densities and ray tracing simulations shows that this pattern is largely controlled by the F region solar-produced ionization during the day. During the night, however, MCM observations reveal a significant additional source of plasma density in the polar cap as compared with the model. An example of conjugate radar observations is presented that supports the idea of polar patches being an additional source of ionization on the nightside. Echo occurrence exhibits a clear peak near the solar terminator, which suggests that small-scale irregularities form in turbulent cascade from large scales. Further, echo occurrence is enhanced for particular IMF orientations during the night. Observations indicate that solar illumination control of irregularity production is strong and not restricted to the nightside. Indirect solar wind control is also exerted by the IMF-dependent convection pattern, since the gradient-drift instability favors certain orientations between the plasma density gradients and convection velocity.

6.6.3. Diurnal Variation of Winter F Region Ionosphere for Solar Minimum at Both Zhongshan Station, Antarctica, and Svalbard Station, Arctic

Diurnal variation features of wintertime F_2 peak electron density (N_mF_2) representative for solar minimum at both Zhongshan station, Antarctica, and Svalbard station are compared and analyzed by Zhang et al. (2015b). Both stations are located around cusp latitude and are almost on the same geomagnetic meridian plane in both hemispheres. For quiet time period, typical N_mF_2 diurnal variation features at Svalbard station show double peaks with a decrease of N_mF_2 around magnetic local noon (~UT + 3 h); N_mF_2 diurnal variation at Zhongshan station shows one major peak around magnetic local noon (~UT + 1.75 h), followed by a sharp decrease of N_mF_2, and a subpeak around 1500 UT. Simulation results of the high-latitude ionospheres in both hemispheres agree well with observations at both stations. It is found that the major difference of N_mF_2 variation between both stations can be explained by the unique location of each station relative to the sunlit demarcation line during the day. For quiet time period, photoionization from lower latitude contributes to the major peak of N_mF_2 in the diurnal variation at Zhongshan station, while the interaction between horizontal convection and auroral precipitation is the main cause for N_mF_2 variation at Svalbard station. For active time period, both stations show the increase of N_mF_2 due to transportation of higher plasma density from lower latitudes on the dayside with the expansion of the polar cap and the additional ionization from soft-precipitating electrons.

6.6.4. f_0F_2 Long-Term Trend Linked to Earth's Magnetic Field Secular Variation

Long-term trend of the critical frequency of the F_2 ionospheric region, f_oF_2, at Phu Thuy station (21.03°N, 105.96°E), Vietnam, located under the northern crest of the equatorial ionization anomaly (EIA) is studied by Pham Thi Thu et al. (2016). Annual mean data are analyzed at 04 LT and 12 LT for the period 1962–2002 using monthly median values and monthly mean values during magnetically quiet days ($am < 20$). In both cases it's obtain similar trends at 4 LT and 12 LT, which Pham Thi Thu et al. (2016) interpret as an absence of geomagnetic activity effect over trends. The positive trends obtained are not consistent with the negative values expected from greenhouse gases effect at this layer of the upper atmosphere. The increasing trend observed at 12 LT is qualitatively in agreement with the expected effect of the secular displacement of the dip equator over the EIA latitudinal profile. At 04 LT, when the EIA is absent, the positive trend is in qualitative agreement with the secular variation of the Earth's magnetic field inclination, I, and the consequent increase of the $\sin(I)\cos(I)$ factor at the corresponding location.

6.6.5. The Persistence of the NWA Effect during the Low Solar Activity Period 2007–2009

As note Jakowski et al. (2015), the ionospheric Nighttime Winter Anomaly (NWA) was first reported more than three decades ago based on total electron content (TEC) and vertical sounding data. The aim of Jakowski et al. (2015) paper is to provide further evidence that the NWA effect is a persistent feature in the Northern Hemisphere at the American and in the Southern Hemisphere at the Asian longitude sector under low solar activity conditions. The analysis of ground-based GPS derived TEC and peak electron density data from radio occultation measurements on Formosat-3/COSMIC satellites confirms and further supports the findings published in earlier NWA papers. So it has been confirmed and further specified that the NWA appears at longitude sectors where the displacement between the geomagnetic and the geographic equator maximizes. Here NWA peaks at around 40°–50° geomagnetic midlatitude supporting the idea that wind-induced plasma uplifting in the conjugated summer hemisphere is the main driving force for the accumulation of ionospheric plasma in the topside ionosphere and plasmasphere. In parallel, the midsummer nighttime anomaly (MSNA) is caused at the local ionosphere. Simultaneously, interhemispheric coupling causes severe downward plasma fluxes in the conjugated winter hemisphere during night causing the NWA at low solar activity. With increasing solar activity, the downward plasma fluxes lose their impact due to the much stronger increasing background ionization that masks the NWA. It is assumed that MSNA and related special anomalies such as the Weddell Sea Anomaly and the Okhotsk Sea Anomaly are closely related to the NWA via enhanced wind-induced uplifting of the ionosphere.

6.6.6. Spherical Cap Harmonic Analysis of the Arctic Ionospheric TEC for One Solar Cycle

As outlined Liu et al. (2014a), precise knowledge of the Arctic ionosphere total electron content (TEC) and its variations has scientific relevance due to the unique characteristics of the polar ionosphere. Understanding the Arctic TEC is also important for precise positioning and navigation in the Arctic. The study of Liu et al. (2014a) utilized the spherical cap harmonic analysis (SCHA) method to map the Arctic TEC for the most recent solar cycle from 2000 to 2013 and analyzed the distributions and variations of the Arctic TEC at different temporal and spatial scales. Even with different ionosphere conditions during the solar cycle, the results showed that the existing International Global Navigation Satellite Systems Service stations are sufficient for mapping the Arctic TEC. The SCHA method provides adequate accuracy and resolution to analyze the spatiotemporal distributions and variations of the Arctic TEC under different ionosphere conditions and to track ionization patches in this polar region (e.g., the ionization event of 26 September 2011). The results derived from the SCHA model were compared to direct observations using the Super Dual Auroral Radar Network radar. The SCHA method is able to predict the TEC in the long and short terms. This paper presented a long-term prediction with a relative uncertainty of 75% for a latency of one solar cycle and a short-term prediction with errors of ±2.2 total electron content units (TECUs, 1 TECU = 10^{16} el. m^{-2}), ±3.8 TECU, and ±4.8 TECU for a latency of 1, 2, and 3 days, respectively. The SCHA is an effective method for mapping, predicting, and analyzing the Arctic TEC.

6.7. Modeling of Magnetosphere-Ionosphere Coupling

6.7.1. The Comprehensive Inner Magnetosphere-Ionosphere (CIMI) Model

As outlined Fok et al. (2014), simulation studies of the Earth's radiation belts and ring current are very useful in understanding the acceleration, transport, and loss of energetic particles. Recently, the Comprehensive Ring Current Model (CRCM) and the Radiation Belt Environment (RBE) model were merged to form a Comprehensive Inner Magnetosphere-Ionosphere (CIMI) model. CIMI solves for many essential quantities in the inner magnetosphere, including ion and electron distributions in the ring current and radiation belts, plasmaspheric density, Region 2 currents, convection potential, and precipitation in the ionosphere. It incorporates whistler mode chorus and hiss wave diffusion of energetic electrons in energy, pitch angle, and cross terms. CIMI thus represents a comprehensive model that considers the effects of the ring current and plasmasphere on the radiation belts. Fok et al. (2014) have performed a CIMI simulation for the storm on 5–9 April 2010 and then compared obtained results with data from the Two Wide-angle Imaging Neutral-atom Spectrometers and Akebono satellites. Fok et al. (2014) identify the dominant energization and loss processes for the ring current and radiation belts. Fok et al. (2014) find that the interactions with the whistler mode chorus waves are the main cause of the flux increase of MeV electrons during the recovery phase of this particular storm. When a self-consistent electric field from the CRCM is used, the enhancement of MeV electrons is higher than when an empirical convection model is applied. Fok et al. (2014) also demonstrate how CIMI can be a powerful tool for analyzing and interpreting data from Van Allen Probes mission.

6.7.2. A New Interhemispheric 16-Moment Model of the Plasmasphere-Ionosphere System: IPIM

Marchaudon and Blelly (2015) present a new interhemispheric numerical model: the IRAP plasmasphere-ionosphere model (IPIM). This model describes the transport of the multispecies ionospheric plasma from one hemisphere to the other along convecting and corotating magnetic field lines, taking into account source processes at low altitude such as photoproduction, chemistry, and energization through the coupling with a kinetic code solving the transport of suprathermal electron along the field line. Among the new developments, a 16-moment-based approach is used for the transport equations in order to allow development of strong temperature anisotropy at high altitude and it's consider important but often neglected effects, such as inertial acceleration (centrifugal and Coriolis). In Marchaudon and Blelly (2015), after presenting in detail the principle of the model, it's focus on preliminary results showing the original contribution of this new model. For these first runs, Marchaudon and Blelly (2015) simulate the convection and corotation transport of closed flux tubes in the plasmasphere for tilted/eccentric dipolar magnetic field configuration in solstice and equinox conditions. Marchaudon and Blelly (2015) follow different flux tubes between 1.2 and 6 R_E and demonstrate the capability of the model to describe a wide range of density (above 15 orders of magnitude). The relevance of the mathematical approach used is highlighted, as anisotropies can develop above 3000 km in the plasmasphere as a result of the mirroring effect related to the anisotropic pressure tensor. Moreover, Marchaudon and Blelly (2015) show that the addition of inertial acceleration may become critical to describe plasma interhemispheric transport above $4R_E$. The ability of the model to describe the external plasmasphere is demonstrated, and innovative studies are foreseen, regarding the dynamics of the plasma along the magnetic field lines (in particular interhemispheric exchanges and "opening"/"closure" of a flux tube).

6.7.3. A Regional Ionospheric TEC Mapping Technique over China and Adjacent Areas on the Basis of Data Assimilation

A regional total electron content (TEC) mapping technique over China and adjacent areas (70°E–140°E and 15°N–55°N) is developed by Aa et al. (2015) on the basis of a Kalman filter data assimilation scheme driven by Global Navigation Satellite Systems (GNSS) data from the Crustal Movement Observation Network of China and International GNSS Service. The regional TEC maps can be generated accordingly with the spatial and temporal resolution being 1°×1° and 5 min, respectively. The accuracy and quality of the TEC mapping technique have been validated through the comparison with GNSS observations, the International Reference Ionosphere model values, the global ionosphere maps from Center for Orbit Determination of Europe, and the Massachusetts Institute of Technology Automated Processing of GPS TEC data from Madrigal database. The verification results indicate that great systematic improvements can be obtained when data are assimilated into the background model, which demonstrates the effectiveness of this technique in providing accurate regional specification of the ionospheric TEC over China and adjacent areas.

6.7.4. Assessment of IRI and IRI-Plas Models over the African Equatorial and Low-Latitude Region

As outlined Adebiyi et al. (2016), a reliable ionospheric specification by empirical models is important to mitigate the effects of the ionosphere on the operations of satellite-based positioning and navigation systems. The study of Adebiyi et al. (2016) evaluates the capability of the International Reference Ionosphere (IRI) and IRI extended to the plasmasphere (IRI-Plas) models in predicting the total electron content (TEC) over stations located in the southern hemisphere of the African equatorial and low-latitude region. TEC derived from Global Positioning System (GPS) measurements were compared with TEC predicted by both the IRI-Plas 2015 model and the three topside options of the IRI 2012 model. Generally, the diurnal and the seasonal structures of modeled TEC follow quite well with the observed TEC in all the stations, although with some upward and downward offsets observed during the daytime and nighttime. The prediction errors of both models exhibit latitudinal variation and these showed seasonal trends. The values generally decrease with increase in latitude. The TEC data-model divergence of both models is most significant at stations in the equatorial region during the daytime and nighttime. Conversely, both models demonstrate most pronounced convergence during the nighttime at stations outside the equatorial region. The IRI-Plas model, in general, performed better in months and seasons when the three options of the IRI model underestimate TEC. Factors such as the height limitation of the IRI model, the inaccurate predictions of the bottom-side and topside electron density profiles were used to explain the data-model discrepancies.

6.7.5. Can an Electron Gun Solve the Outstanding Problem of Magnetosphere-Ionosphere Connectivity?

As outlined Delzanno et al. (2016), determining the magnetic connectivity between magnetospheric phenomena and ionospheric phenomena is an outstanding problem of magnetospheric and ionospheric physics. Accurately establishing this connectivity could answer a variety of long-standing questions. The most viable option to solve this is by means of a high-power electron beam fired from a magnetospheric spacecraft and spotted at its magnetic footpoint in the ionosphere. This has technical difficulties. Progress has been made on mitigating the major issue of spacecraft charging. The remaining physics issues are identified, together with the need for a synergistic effort in modeling, laboratory experiments, and, ultimately, testing in space. The goal of this commentary is to stimulate awareness and interest on the magnetosphere-ionosphere connectivity problem and possibly accelerate progress toward its solution.

6.7.6. Can Atomic Oxygen Production Explain the Ionospheric Annual Asymmetry?

As outlined Lei et al. (2016), the causes of the ionospheric annual asymmetry, which refers to a larger averaged electron density at geomagnetic conjugate latitudes in December than in June, remain an unresolved problem that still generates considerable interest. According to Lei et al. (2016), the ionospheric annual asymmetry in the peak electron density of the F_2 layer (N_mF_2) is typically 20–40%, which cannot be explained by the 7% annual asymmetry in photoionization caused by the shorter Sun-Earth distance in December. Mikhailov and Perrone (2011, 2015) suggested that the annual asymmetry in atomic oxygen production due to O2 dissociation is sufficient to explain the ionospheric annual asymmetry at middle latitudes. In the study of Lei et al. (2016), a series of the Global Mean Model (GMM) simulations have been conducted to test this hypothesis. Although O_2 dissociation and eddy diffusion processes are included in the GMM, the simulated annual asymmetry of N_mF_2 is only 13%. Furthermore, the annual asymmetry increase in neutral composition in the GMM simulations can only explain about one fifth of the ionospheric annual asymmetry. Therefore, the atomic oxygen production mechanism is unlikely to be a major contributor to the ionospheric annual asymmetry.

6.7.7. Coherent Seasonal, Annual, and Quasi-Biennial Variations in Ionospheric Tidal/SPW Amplitudes

Chang et al. (2016) examine the coherent spatial and temporal modes dominating the variation of selected ionospheric tidal and stationary planetary wave (SPW) signatures from 2007 to 2013 FORMOSAT-3/COSMIC (Constellation Observing System for Meteorology, Ionosphere, and Climate) total electron content observations using multidimensional ensemble empirical mode decomposition (MEEMD) from the Hilbert-Huang Transform. Chang et al. (2016) examine the DW1, SW2, DE3, and SPW4 components, which are driven by a variety of in situ and vertical coupling sources. The intrinsic mode functions (IMFs) resolved by MEEMD analysis allows for the isolation of the dominant modes of variability for prominent ionospheric tidal/SPW signatures in a manner not previously used, allowing the effects of specific drivers to be examined individually. The time scales of the individual IMFs isolated for all tidal/SPW signatures correspond to a semiannual variation at equatorial ionization anomaly (EIA) latitudes maximizing at the equinoxes, as well as annual oscillations at the EIA crests and troughs. All tidal/SPW signatures show one IMF isolating an ionospheric quasi-biennial oscillation (QBO) in the equatorial latitudes maximizing around January of odd-numbered years. This total electron content QBO variation is in phase with a similar QBO variation isolated in both the Global Ultraviolet Imager (GUVI) zonal mean column O/N_2 density ratio ($\Sigma O/N_2$) and the F10.7 solar radio flux index around solar maximum, while showing temporal variation more similar to that of GUVI $\Sigma O/N_2$ during the time around the 2008/2009 extended solar minimum. These results point to both quasi-biennial variations in solar irradiance and thermosphere/ionosphere composition as a generation mechanism for the ionospheric QBO.

6.7.8. Day-to-Day Variability of Midlatitude Ionospheric Currents Due to Magnetospheric and Lower Atmospheric Forcing

As note Yamazaki et al. (2016), from previous studies on the solar quiet (Sq) variation of the geomagnetic field, is known that the strength and pattern of ionospheric dynamo currents change significantly from day to day. The study of Yamazaki et al. (2016) investigates the relative importance of two sources that contribute to the day-to-day variability of the ionospheric currents at middle and low latitudes. One is high-latitude electric fields that are caused by magnetospheric convection, and the other is atmospheric waves from the lower atmosphere. Global ionospheric current systems, commonly known as Sq current systems, are simulated using the National Center for Atmospheric Research thermosphere-ionosphere-mesosphere-electrodynamics general circulation model. Simulations are run for 1–30 April 2010 with a constant solar energy input but with various combinations of high-latitude forcing and lower atmospheric forcing. The model well reproduces geomagnetic perturbations on the ground, when both forcings are taken into account. The contribution of high-latitude forcing to the total Sq current intensity (J_{total}) is generally smaller than the contribution of wave forcing from below 30 km, except during active periods (Kp $\geq$ 4), when J_{total} is enhanced due to the leakage of high-latitude electric fields to lower latitudes. It is found that the penetration electric field drives ionospheric currents at middle and low latitudes not only on the dayside but also on the nightside, which has an appreciable effect on the Dst index. It is also found that quiet time day-to-day variability in J_{total} is dominated by symmetric-mode migrating diurnal and semidiurnal tidal winds at 45–60° latitude at ~110 km.

6.7.9. Effects of the Interplanetary Magnetic Field on the Location of the Open-Closed Field Line Boundary

Using global magnetohydrodynamic (MHD) simulation, Wang et al. (2016) investigate the effect of the interplanetary magnetic field (IMF) on the location of the open-closed field line boundary (OCB), in particular the duskside and dawnside OCB and their asymmetry. They first model the typical OCB-crossing events on 22 October 2001 and 24 October 2002 observed by DMSP. The MHD model presents a good estimate of OCB location under quasi-steady magnetospheric conditions. Wang et al. (2016) then systemically study the location of the OCB under different IMF conditions. The model results show that the dawnside and duskside OCBs respond differently to IMF conditions when B_Y is present. An empirical expression describing the relationship between the OCB latitudes and IMF conditions has been obtained. It is found that the IMF conditions play an important role in determining the dawn-dusk OCB asymmetry, which is due to the magnetic reconnection at the dayside magnetopause. The differences between the dawn and dusk OCB latitudes from MHD predictions are in good agreement with the observations.

6.7.10. Electrostatic Analyzer Measurements of Ionospheric Thermal Ion Populations

Fernandes and Lynch (2016) define the observational parameter regime necessary for observing low-altitude ionospheric origins of high-latitude ion upflow/outflow. They present measurement challenges and identify a new analysis technique which mitigates these impediments. To probe the initiation of auroral ion upflow, it is necessary to examine the thermal ion population at 200–350 km, where typical thermal energies are tenths of eV. Interpretation of the thermal ion distribution function measurement requires removal of payload sheath and ram effects. Fernandes and Lynch (2016) use a 3-D Maxwellian model to quantify how observed ionospheric parameters such as density, temperature, and flows affect in situ measurements of the thermal ion distribution function. It's define the viable acceptance window of a typical top hat electrostatic analyzer in this regime and show that the instrument's energy resolution prohibits it from directly observing the shape of the particle spectra. To extract detailed information about measured particle population, Fernandes and Lynch (2016) define two intermediate parameters from the measured distribution function, then use a Maxwellian model to replicate possible measured parameters for comparison to the data. Liouville's theorem and the thin-sheath approximation allow to couple the measured and modeled intermediate parameters such that measurements inside the sheath provide information about plasma outside the sheath. Fernandes and Lynch (2016) apply this technique to sounding rocket data to show that careful windowing of the data and Maxwellian models allows for extraction of the best choice of geophysical parameters. More widespread use of this analysis technique will help the science community expand its observational database of the seed regions of ionospheric outflows.

6.7.11. Equatorial Ionization Anomaly in the Low-Latitude Topside Ionosphere: Local Time Evolution and Longitudinal Difference

As note Chen et al. (2016), although the equatorial ionization anomaly (EIA) has been widely studied, it was seldom investigated from observations for the topside ionosphere. In paper of Chen et al. (2016) the climatology characteristics of the latitudinal structure of topside ion density (N_i) were investigated in detail using ROCSAT-1 and DMSP N_i observations. EIA double-peak structure may exist at 600 km, depending on longitude, local time, season, and solar activity, while it is not a prevalent characteristic at 840 km even at solar maximum sunset. Local time evolution of the EIA at 600 km was presented. The double-peak structure begins to appear at noontime, being later than the appearance of the EIA in F_2 peak region. The pronounced EIA induced by the strong prereversal enhancement at solar maximum begins to appear at 19:00 LT and can last to premidnight; EIA crest-to-trough ratio is largest (lowest) at March equinox (June solstice) and reaches a maximum at 20:00 LT in all seasons. EIA structure shows evident longitudinal difference. Pronounced EIA exists around about 100°E at 13:00 LT at the two equinoxes and June solstice, while it exists at more extensive longitudes (about 90°E to 240°E) at December solstice. The trans-equator plasma transport induced by neutral winds can weaken the double-peak structure in the topside ionosphere. The longitudinal difference in the EIA structure at 600 km is related to the longitudinal variations of equatorial upward plasma drift and geomagnetic declination.

6.7.12. Geomagnetic Control of the Midlatitude f_oF_1 and f_oF_2 Long-Term Variations: Recent Observations in Europe

A new very simple method, allowing an easy control, has been applied by Perrone and Mikhailov (2016) to extract long-term (11 year) $(\delta f_oF_2)_{11y}$ and $(\delta f_oF_1)_{11y}$ variations from June f_oF_2 and f_oF_1 monthly median observations at European Slough/Chilton and Juliusruh stations, including recent data until 2015. The aim of the analysis was to check the validity of the geomagnetic control of f_oF_2 and f_oF_1 long-term variations in the 21st century with the main accent on the period including the last deep solar minimum in 2008–2009. The geomagnetic control was shown to be valid. Moreover, the dependence on geomagnetic activity has become more pronounced and explicit after 1990. A simultaneous analysis of f_oF_2 and f_oF_1 long-term variations improves the reliability of the obtained conclusions and helps understand the physical mechanism of these variations. Due to common neutral composition and the similarity of photochemical processes noontime f_oF_2 and f_oF_1 demonstrate similar long-term variations: the correlation coefficient between $(\delta f_oF_2)_{11y}$ and $(\delta f_oF_1)_{11y}$ is 0.834 at Slough/Chilton and 0.884 at Juliusruh with the 99% confidence level according to Fisher's F criterion. Midnight long-term $(\delta f_oF_2)_{11y}$ variations also manifest a pronounced dependence on Ap_{11y} which may be interpreted in the framework of the geomagnetic control concept.

6.7.13. Geomagnetic Control of the Midlatitude Daytime f_oF_1 and f_oF_2 Long-Term Variations: Physical Interpretation Using European Observations

As note Mikhailov and Perrone (2016), morphological analysis of Slough/Chilton and Juliusruh f_oF_2 and f_oF_1 long-term variations for the period including recent observations made in the previous paper of authors (Perrone and Mikhailov, 2016) has shown that the geomagnetic control is valid in the 21st century, moreover, the dependence on geomagnetic activity has become more pronounced and explicit after 1990. A new method to retrieve thermospheric neutral composition (O, O_2, and N_2), exospheric temperature T_{ex}, and the total solar EUV flux with $\lambda < 1050$ Å from routine f_oF_1 ionosonde observations has been developed to understand the mechanism of this geomagnetic control. The method was tested using CHAMP/STAR neutral gas density measurements. The retrieved for the first time thermospheric parameters at Slough/Chilton and Juliusruh over the period of ~ 5 solar cycles were used to analyze the mechanism of f_oF_1 and f_oF_2 long-term variations in the light of the geomagnetic control concept. It was shown that the control was provided via two channels: [O] and [O]/[N_2] variations. Geomagnetic activity presented by 11 year running mean weighted index Ap_{11y} controls the $(O/N_2)_{11y}$ ratio variations, while solar activity presented by $(F10.7)_{11y}$ controls atomic oxygen $[O]_{11y}$ variations. Atomic oxygen, the main aeronomic parameter controlling daytime f_oF_1 and f_oF_2 variations, manifests solar cycle and long-term (for some solar cycles) variations with the rising phase in 1965–1985 and the falling phase in 1985–2008. These long-term [O] variations are reflected in f_oF_2 and f_oF_1 long-term variations.

The origin of these long-term variations is in the Sun. The empirical thermospheric model Mass Spectrometer Incoherent Scatter-86 driven by Ap and F10.7 indices manifests $[O]_{11y}$ and $(O/N_2)_{11y}$ variations similar to the retrieved ones including the period of deep solar minimum with a very low atomic oxygen concentration in 2008. As underlined Mikhailov and Perrone (2016), this confirms the basic idea of the geomagnetic control concept that ionospheric long-term variations have a natural (not anthropogenic) origin related to long-term variations in solar and geomagnetic activity.

6.7.14. Signal Propagation Time from the Magnetotail to the Ionosphere: OpenGGCM Simulation

As outlined Ferdousi and Raeder (2016), distinguishing the processes that occur during the first 2 min of a substorm depends critically on the correct timing of different signals between the plasma sheet and the ionosphere. To investigate signal propagation paths and signal travel times, Ferdousi and Raeder (2016) use a MHD global simulation model of the Earth magnetosphere and ionosphere, OpenGGCM-CTIM model. By creating single impulse or sinusoidal pulsations in various locations in the magnetotail, the waves are launched, and Ferdousi and Raeder (2016) investigate the paths taken by the waves and the time that different waves take to reach the ionosphere. They find that it takes approximately about 27, 36, 45, 60, and 72 s for waves to travel from the tail plasma sheet at $x = -10, -15, -20, -25$, and -30 R_E, respectively, to the ionosphere, contrary to previous reports. Ferdousi and Raeder (2016) also find that waves originating in the plasma sheet generally travel faster through the lobes than through the plasma sheet. It's found also that MHD waves travel faster through the lobes, and they reach higher latitudes earlier than lower latitudes.

6.7.15. The Importance of Neutral Hydrogen for the Maintenance of the Midlatitude Winter Nighttime Ionosphere: Evidence from IS Observations at Kharkiv, Ukraine, and Field Line Interhemispheric Plasma Model Simulations

Kotov et al. (2016) investigate the causes of nighttime enhancements in ionospheric density that are observed in winter by the incoherent scatter (IS) radar at Kharkiv, Ukraine. Calculations with a comprehensive physical model reveal that large downward ion fluxes from the plasmasphere are the main cause of the enhancements. These large fluxes are enabled by large upward H^+ fluxes into the plasmasphere from the conjugate summer hemisphere during the daytime. The nighttime downward H^+ flux at Kharkiv is sensitive to the thermosphere model H density, which had to be increased by factors of 2 to 3 to obtain model-data agreement for the topside H^+ density. Other studies support the need for increasing the thermosphere model H density for all seasons at solar minimum. It was found that neutral winds are less effective than plasmaspheric fluxes for maintaining the nighttime ionosphere. This is partly because increased equatorward winds simultaneously oppose the downward H^+ flux. The model calculations also reveal the need for a modest additional heat flow from the plasmasphere in the afternoon. This source could be the quiet time ring current.

Kotov et al. (2016) outlined that: (i) Hydrogen density is important in the development of nighttime N_mF_2 enhancements at midlatitudes, (ii) NRLMSISE-00 model H density should be increased by factors of 2 to 3 at solar minimum, () The increase of h_mF_2 has a small effect on the nighttime N_mF_2.

6.7.16. Toward An Integrated View of Ionospheric Plasma Instabilities: 2. Three Inertial Modes of a Cubic Dispersion Relation

The cubic dispersion relation describing inertial modes of fundamental ionospheric instabilities at arbitrary altitude is demonstrated to yield three distinct solutions, and their stability is analyzed by Makarevich (2016) using a combination of numerical and analytic techniques. A robust numerical method is developed for obtaining all three solutions for arbitrary altitude, and analytic expressions are developed for two solutions. In the E region, one unstable and two stable modes are found for strong electric fields, with the unstable mode being the Farley-Buneman instability. In the F region, zero, one, or two unstable modes are found depending on the plasma density gradient. The first unstable mode represents a finite-temperature generalization of the inertial mode of the gradient drift instability (GDI) in the F region that has been previously considered for cold-plasma case. The instability cone width and the gradient strength cutoff values are analyzed analytically and inertial effects are shown to drastically alter their behavior for decameter-scale waves. In particular, progressively stronger gradients are required to excite the instability with an increasing electric field. Another strongly unstable mode is found at high altitudes and for sufficiently sharp gradients, although the applicability of this solution is limited due to its high-frequency nature. The results strengthen the case for analyzing different ionospheric instabilities within the same formalism and provide an additional framework for interpreting the experimentally observed irregularity formation times that are inconsistent with those predicted by the standard GDI theory.

6.7.17. Tucumán Ionospheric Model (TIM): Initial Results for STEC Predictions

As outlined Scidá et al. (2016), most ionospheric models can calculate vertical total electron content (VTEC) predictions, but only a few are suitable for calculating slant total electron content (STEC). This ionospheric magnitude is generally measured for electron content determinations, with VTEC particularly corresponding to an elevation of 90°. This is generally obtained by applying a mapping function to STEC measurements, which leads to important calculation errors. Moreover, the equatorial region has unique characteristics, such as the fountain effect and the equatorial electrojet, which lead to significant errors in the model's calculations. In Scidá et al. (2016), the Tucumán ionospheric model (TIM) is presented as a novel alternative for calculating the STEC in low-latitude regions (–24° to 24° dip latitude). The model is based on spatial geometry where the considered trajectory is segmented, and the corresponding electron density calculations for the resulting segment end points are determined using the semi-empirical low-latitude ionospheric model (SLIM) with reference to their corresponding magnetic coordinates and height. Finally, the electron density values are integrated along the path to obtain the STEC. The work of Scidá et al. (2016) describes the TIM and tests their STEC predictions for five ray paths around the world (totaling 16 cases under study), which are compared with experimental data from satellites and with those calculated by the NeQuick model. Moreover, the TIM performance for VTEC predictions is also checked and compared with VTEC data obtained from Global Positioning System (GPS) signals, IRI model, and NeQuick model predictions, for six GPS receiver stations during the equinox and solstice (totaling 12 cases studied). Comparisons of the TIM predictions with experimental data show that 53% of the calculation has, in general, deviations <30%. For the considered cases, TIM reproduces the experimental data better than the other models.

6.8. Energetics of Magnetosphere-Ionosphere Coupling

6.8.1. Magnetosphere-Ionosphere Energy Interchange in the Electron Diffuse Aurora

As outlined Khazanov et al. (2014), the diffuse aurora has recently been shown to be a major contributor of energy flux into the Earth's ionosphere. Therefore, a comprehensive theoretical analysis is required to understand its role in energy redistribution in the coupled ionosphere-magnetosphere system. In previous theoretical descriptions of precipitated magnetospheric electrons ($E\sim 1$ keV), the major focus has been the ionization and excitation rates of the neutral atmosphere and the energy deposition rate to thermal ionospheric electrons. However, these precipitating electrons will also produce secondary electrons via impact ionization of the neutral atmosphere. The paper of Khazanov et al. (2014) presents the solution of the Boltzman-Landau kinetic equation that uniformly describes the entire electron distribution function in the diffuse aurora, including the affiliated production of secondary electrons ($E< 600$ eV) and their ionosphere-magnetosphere coupling processes. In this article, Khazanov et al. (2014) discuss for the first time how diffuse electron precipitation into the atmosphere and the associated secondary electron production participate in ionosphere-magnetosphere energy redistribution.

6.8.2. ULF Wave Electromagnetic Energy Flux into the Ionosphere: Joule Heating Implications

As outlined Hartinger et al. (2015), ultralow-frequency (ULF) waves—in particular, Alfvén waves–transfer energy into the Earth's ionosphere via Joule heating, but it is unclear how much they contribute to global and local heating rates relative to other energy sources. Hartinger et al. (2015) use THEMIS satellite data to investigate the spatial, frequency, and geomagnetic activity dependence of the ULF wave Poynting vector (electromagnetic energy flux) mapped to the ionosphere. Its use these measurements to estimate Joule heating rates, covering latitudes at or below the nominal auroral oval and below the open/closed field line boundary. Hartinger et al. (2015) find ULF wave Joule heating rates (integrated over 3–30 mHz frequency band) typically range from 0.001 to 1 mW/m^2. Its compare these rates to empirical models of Joule heating associated with large-scale, static (on ULF wave timescales) current systems, finding that ULF waves nominally contribute little to the global, integrated Joule heating rate. However, there are extreme cases with ULF wave Joule heating rates of $\geq$ 10 mW/m^2 - in these cases, which are more likely to occur when Kp $\geq$ 3, ULF waves make significant contributions to the global Joule heating rate. Hartinger et al. (2015) also find ULF waves routinely make significant contributions to local Joule heating rates near the noon and midnight local time sectors, where static current systems nominally contribute less to Joule heating; the most important contributions come from lower frequency (<7 mHz) waves.

6.8.3. Ionosphere-Magnetosphere Energy Interplay in the Regions of Diffuse Aurora

As outlined Khazanov et al. (2016), both electron cyclotron harmonic (ECH) waves and whistler mode chorus waves resonate with electrons of the Earth's plasma sheet in the energy range from tens of eV to several keV and produce the electron diffuse aurora at ionospheric altitudes. Interaction of these superthermal electrons with the neutral atmosphere leads to the production of secondary electrons (E<500–600 eV) and, as a result, leads to the activation of lower energy superthermal electron spectra that can escape back to the magnetosphere and contribute to the thermal electron energy deposition processes in the magnetospheric plasma. The ECH and whistler mode chorus waves, however, can also interact with the secondary electrons that are coming from both of the magnetically conjugated ionospheres after they have been produced by initially precipitated high-energy electrons that came from the plasma sheet. After their degradation and subsequent reflection in magnetically conjugate atmospheric regions, both the secondary electrons and the precipitating electrons with high (E>600 eV) initial energies will travel back through the loss cone, become trapped in the magnetosphere, and redistribute the energy content of the magnetosphere-ionosphere system. Thus, scattering of the secondary electrons by ECH and whistler mode chorus waves leads to an increase of the fraction of superthermal electron energy deposited into the core magnetospheric plasma. Khazanov et al. (2016) outlined also that: (i) Magnetosphere-ionosphere coupling is a key in forming the diffuse auroral electron distribution function (ii) It is present the solution of the Boltzmann-Landau kinetic equation () It is important to consider the coupled ionosphere-magnetosphere system.

6.9. Energetic Electrons in the Geomagnetosphere – Ionosphere Coupled System

6.9.1. Superthermal Electron Magnetosphere-Ionosphere Coupling in the Diffuse Aurora in the Presence of ECH Waves

As note Khazanov et al. (2015a), there are two main theories for the origin of the diffuse auroral electron precipitation: first, pitch angle scattering by electrostatic electron cyclotron harmonic (ECH) waves, and second, by whistler mode waves. Precipitating electrons initially injected from the plasma sheet to the loss cone via wave-particle interaction processes degrade in the atmosphere toward lower energies and produce secondary electrons via impact ionization of the neutral atmosphere. These secondary electrons can escape back to the magnetosphere, become trapped on closed magnetic field lines, and deposit their energy back to the inner magnetosphere. ECH and whistler mode waves can also move electrons in the opposite direction, from the loss cone into the trap zone, if the source of such electrons exists in conjugate ionospheres located at the same field lines as the trapped magnetospheric electron population. Such a situation exists in the simulation scenario of superthermal electron energy interplay in the region of diffuse aurora presented and discussed by Khazanov et al. (2014) and quantified in Khazanov et al. (2015a) by taking into account the interaction of secondary electrons with ECH waves.

6.9.2. Electron Distribution Function Formation in Regions of Diffuse Aurora

As note Khazanov et al. (2015b), the precipitation of high-energy magnetospheric electrons (E ~ 0.6 – 10 KeV) in the diffuse aurora contributes significant energy flux into the Earth's ionosphere. To fully understand the formation of this flux at the upper ionospheric boundary, ~700–800 km, it is important to consider the coupled ionosphere-magnetosphere system. In the diffuse aurora, precipitating electrons initially injected from the plasma sheet via wave-particle interaction processes degrade in the atmosphere toward lower energies and produce secondary electrons via impact ionization of the neutral atmosphere. These precipitating electrons can be additionally reflected upward from the two conjugate ionospheres, leading to a series of multiple reflections through the magnetosphere. These reflections greatly influence the initially precipitating flux at the upper ionospheric boundary (700–800 km) and the resultant population of secondary electrons and electrons cascading toward lower energies. Khazanov et al. (2015) present the solution of the Boltzman-Landau kinetic equation that uniformly describes the entire electron distribution function in the diffuse aurora, including the affiliated production of secondary electrons (E < 600 eV) and its energy interplay in the magnetosphere and two conjugated ionospheres. This solution takes into account, for the first time, the formation of the electron distribution function in the diffuse auroral region, beginning with the primary injection of plasma sheet electrons via both electrostatic electron cyclotron harmonic waves and whistler mode chorus waves to the loss cone, and including their subsequent multiple atmospheric reflections in the two magnetically conjugated ionospheres. It is demonstrated that magnetosphere-ionosphere coupling is key in forming the electron distribution function in the diffuse auroral region.

6.9.3. A Unified Model of Auroral Arc Growth and Electron Acceleration in the Magnetosphere-Ionosphere Coupling

A unified physical model of auroral arc growth and electron acceleration is constructed by Watanabe (2014) from the gyrokinetic and two-fluid equations for the magnetosphere-ionosphere (M-I) coupling system. The present theory describes destabilization of kinetic Alfvén waves (KAWs) in the M-I coupling, where development of upward and downward field-aligned currents carried by the KAWs leads to ionospheric density enhancement and depletion, respectively. The feedback M-I coupling through KAWs elucidates growth of auroral arcs and field-aligned acceleration of electrons self-consistently and provides a possible explanation to the Alfvénic auroras observed by the FAST spacecraft.

6.9.4. Energetic Electron Precipitation into the Middle Atmosphere – Constructing the Loss Cone Fluxes from MEPED POES

As outlined Nesse et al. (2016), the impact of energetic electron precipitation (EEP) on the chemistry of the middle atmosphere (50–90 km) is still an outstanding question as accurate quantification of EEP is lacking due to instrumental challenges and insufficient pitch angle coverage of current particle detectors. The Medium Energy Proton and Electron Detectors (MEPED) instrument on board the NOAA/Polar Orbiting Environmental Satellites (POES) and MetOp spacecraft has two sets of electron and proton telescopes pointing close to zenith (0°) and in the horizontal plane (90°). Using measurements from either the 0° or 90° telescope will underestimate or overestimate the bounce loss cone flux, respectively, as the energetic electron fluxes are often strongly anisotropic with decreasing fluxes toward the center of the loss cone. By combining the measurements from both telescopes with electron pitch angle distributions from theory of wave-particle interactions in the magnetosphere, a complete bounce loss cone flux is constructed for each of the electron energy channels >50 keV, >100 keV, and >300 keV. Nesse et al. (2016) apply a correction method to remove proton contamination in the electron counts. It is also account for the relativistic (>1000 keV) electrons contaminating the proton detector at subauroral latitudes. This gives the full range coverage of electron energies that will be deposited in the middle atmosphere. Finally, Nesse et al. (2016) demonstrate the method's applicability on strongly anisotropic pitch angle distributions during a weak geomagnetic storm in February 2008. Nesse et al. (2016) compare the electron fluxes and subsequent energy deposition estimates to OH observations from the Microwave Limb Sounder on the Aura satellite substantiating that the estimated fluxes are representative for the true precipitating fluxes impacting the atmosphere.

6.10. Ionospheric Outflow in the Geomagnetosphere – Ionosphere Coupled System

6.10.1. Asymmetric Ionospheric Outflow Observed at the Dayside Magnetopause

As outlined Lee et al. (2015), an important source of the terrestrial magnetospheric plasma is the Earth's ionospheric outflows from the high-latitude regions of both hemispheres. The ionospheric ion outflows have rarely been observed at the dayside magnetopause. Lee et al. (2015) report Cluster observations of the ionospheric ion outflows observed at the dayside magnetopause. The low-energy (up to 1.5 keV) electrons are detected with bidirectional pitch angle distributions indicating that the magnetic field lines are closed. The unidirectional cold ions (< 200 eV) are observed in the magnetosphere by both C1 and C3 satellites. The pitch angle distributions (0°–75°) of the cold ions (< 1 keV) at the dayside magnetopause indicate that these cold ions are the ionospheric outflows coming only from the Southern Hemisphere. The cold ions (< 200 eV) fluxes are modulated by the ULF wave electric field. Two different species (possibly H^+ and He^+) are observed in the magnetosphere. Obtained results suggest that the ionospheric outflows can directly reach the dayside magnetopause region and may participate in the reconnection process.

6.10.2. The Role of the Ionosphere in Providing Plasma to the Terrestrial Magnetosphere

As outlined Chappell (2015), through the more than half century of space exploration, the perception and recognition of the fundamental role of the ionospheric plasma in populating the Earth's magnetosphere has evolved dramatically. A brief history of this evolution in thinking is presented. Both theory and measurements have unveiled a surprising new understanding of this important ionosphere-magnetosphere mass coupling process. The highlights of the mystery surrounding the difficulty in measuring this largely invisible low energy plasma are also discussed. This mystery has been solved through the development of instrumentation capable of measuring these low energy positively-charged outflowing ions in the presence of positive spacecraft potentials. This has led to a significant new understanding of the ionospheric plasma as a significant driver of magnetospheric plasma content and dynamics.

6.10.3. Modeling the Effects of Ionospheric Oxygen Outflow on Bursty Magnetotail Flows

Using a global multifluid MHD model, Garcia-Sage et al. (2015) demonstrate the effects of magnetospheric O^+ on bursty magnetotail flows. Garcia-Sage et al. (2015) carry out two simulations without ionospheric outflow to use as baseline, one driven by real solar wind data and one driven by idealized solar wind. Solar wind data from 1 October 2001 are used as a storm time solar wind driver. During this event, the plasma sheet was observed to be rich in O^+, making the event of interest for a model analysis of the effects of ionospheric origin O^+ on magnetospheric dynamics. Garcia-Sage et al. (2015) carry out outflow comparison simulations for both the realistic and idealized solar wind drivers using a simple empirical model that places auroral outflow in regions where downward propagating Poynting flux and electron precipitation are present, combined with a low-flux thermal energy O^+ outflow over the entire polar region. Garcia-Sage et al. (2015) demonstrate the effects of O^+ on magnetotail structure and the occurrence rate and strength of bursty, fast earthward flows. The addition of

O^+ to the magnetotail stretches the tail and increases the velocity of bursty earthward flows. This increase is shown to be produced by reconnection events in an extended current sheet created by tail stretching.

6.10.4. Driving Ionospheric Outflows and Magnetospheric O+ Energy Density with Alfvén Waves

Chaston et al. (2016) show how dispersive Alfvén waves observed in the inner magnetosphere during geomagnetic storms can extract O^+ ions from the topside ionosphere and accelerate these ions to energies exceeding 50 keV in the equatorial plane. This occurs through wave trapping, a variant of "shock" surfing, and stochastic ion acceleration. These processes in combination with the mirror force drive field-aligned beams of outflowing ionospheric ions into the equatorial plane that evolve to provide energetic O^+ distributions trapped near the equator. These waves also accelerate preexisting/injected ion populations on the same field lines. Chaston et al. (2016) show that the action of dispersive Alfvén waves over several minutes may drive order of magnitude increases in O^+ ion pressure to make substantial contributions to magnetospheric ion energy density. These wave accelerated ions will enhance the ring current and play a role in the storm time evolution of the magnetosphere.

6.10.5. Evidence and Effects of the Sunrise Enhancement of the Equatorial Vertical Plasma Drift in the F Region Ionosphere

As note Zhang et al. (2016), recent studies based on the satellite observations demonstrated that the equatorial vertical plasma drift can also enhance near sunrise in a way similar to the prereversal enhancement. However, it is not clear whether the signature of this sunrise enhancement appears in observations with other sounding techniques. Zhang et al. (2016) explore the Jicamarca (12°S, 283.2°E) incoherent scatter radar measurements to present the evidence of sunrise enhancement in vertical plasma drift on 12 May and 10 June 2004, which are under magnetically quiet and solar minimum conditions. The effects of the sunrise enhancement on the ionosphere are, for the first time, investigated by analyzing the ionograms recorded by the Digisonde Portable Sounder at Jicamarca and conducting the Theoretical Ionospheric Model of the Earth in Institute of Geology and Geophysics, Chinese Academy of Sciences. The observations showed that, during the sunrise enhancement, the F_2 layer peak height is lifted remarkably, and the F_2 layer peak density and bottom-side electron density tend to decrease compared to the days without sunrise enhancements. The simulations indicated that the sunrise enhancement drift can lift the equatorial ionosphere to higher heights and distort the equatorial electron density profiles. What is more, the simulations display an F_3 layer in the equatorial F region during the sunrise enhancement, and a new F_2 layer develops at lower altitudes under the jointed control of the usual photochemical and dynamical processes.

6.11. Subauroral Ion Drifts in the Geomagnetosphere – Ionosphere Coupled System

6.11.1. Hemispheric Asymmetry of Subauroral Ion Drifts: Statistical Results

A large database of more than 18,000 subauroral ion drift (SAID) events from DMSP observations from 1987 to 2012 is used by Zhang et al. (2015a) to systematically investigate the features of SAID. SAID occurs mostly at ~62°/−60° magnetic latitude (MLAT) and ~22:15/22:45 magnetic local time (MLT) for geomagnetically quiet conditions and at ~58°/−56° MLAT and ~22:15/22:45 MLT for geomagnetically disturbed conditions in the North Hemisphere (NH)/South Hemisphere (SH), respectively. Significant north-south asymmetries in SAID occurrence, shape, and geomagnetic activity variations are found in this statistical study. The latitudinal width of a SAID is larger in the NH than in the SH. An interesting finding of this work is that the SAID occurrence probability peaks have a ~180° difference in longitude between the two hemispheres in the geographic coordinates for both geomagnetically quiet and disturbed conditions. The SAID width peaks in almost the same geomagnetic meridian zone with a geomagnetic longitude of ~80°–120° in both hemispheres. Significant hemispheric asymmetries and spike signatures with sharp dips are found in all the latitudinal profiles of the horizontal velocities of SAIDs. The SAID is highly correlated to geomagnetic activity, indicating that the location and evolution of the SAID might be influenced by global geomagnetic activity, auroral dynamics, and the dynamics of ring currents.

6.11.2. Daytime Plasma Drifts in the Equatorial Lower Ionosphere

Hui and Fejer (2015) have used extensive radar measurements from the Jicamarca Observatory during low solar flux periods to study the quiet time variability and altitudinal dependence of equatorial daytime vertical and zonal plasma drifts. The daytime vertical drifts are upward and have largest values during September–October. The day-to-day variability of these drifts does not change with height between 150 and 600 km, but the bimonthly variability is much larger in the F region than below about 200 km. These drifts vary linearly with height generally increasing in the morning and decreasing in the afternoon. The zonal drifts are westward during the day and have largest values during July–October. The 150 km region zonal drifts have much larger day-to-day, but much smaller bimonthly variability than the F region drifts. The daytime zonal drifts strongly increase with height up to about 300 km from March through October, and more weakly at higher altitudes. The December solstice zonal drifts have generally weaker altitudinal dependence, except perhaps below 200 km. Current theoretical and general circulation models do not reproduce the observed altitudinal variation of the daytime equatorial zonal drifts.

6.12. Geomagnetosphere – Ionosphere Coupled System and Ionospheric Density Cavities and Ducts

6.12.1. Magnetospheric Signatures of Ionospheric Density Cavities Observed by Cluster

Russell et al. (2015) present Cluster measurements of large amplitude electric fields correlated with intense downward field-aligned currents, observed during a nightside crossing of the auroral zone. The data are reproduced by a simple model of magnetosphere-ionosphere coupling which, under different conditions, can also produce a divergent electric field signature in the downward current region, or correlation between the electric and perturbed magnetic fields. It does conclude that strong electric field associated with intense downward field-aligned current, such as this observation, is a signature of ionospheric plasma depletion caused by the downward current. It is also shown that the electric field in the downward current region correlates with downward current density if a background field is present, e.g., due to magnetospheric convection.

6.12.2. Real-Time Imaging of Density Ducts between the Plasmasphere and Ionosphere

As note Loi et al. (2015), ionization of the Earth's atmosphere by sunlight forms a complex, multilayered plasma environment within the Earth's magnetosphere, the innermost layers being the ionosphere and plasmasphere. The plasmasphere is believed to be embedded with cylindrical density structures (ducts) aligned along the Earth's magnetic field, but direct evidence for these remains scarce. Loi et al. (2015) report the first direct wide-angle observation of an extensive array of field-aligned ducts bridging the upper ionosphere and inner plasmasphere, using a novel ground-based imaging technique. Loi et al. (2015) establish their heights and motions by feature tracking and parallax analysis. The structures are strikingly organized, appearing as regularly spaced, alternating tubes of over-densities and under-densities strongly aligned with the Earth's magnetic field. These findings represent the first direct visual evidence for the existence of such structures.

6.12.3. Density Duct Formation in the Wake of a Travelling Ionospheric Disturbance: Murchison Widefield Array Observations

As outlined Loi et al. (2016), geomagnetically aligned density structures with a range of sizes exist in the near-Earth plasma environment, including 10–100 km wide VLF/HF wave-ducting structures. Their small diameters and modest density enhancements make them difficult to observe, and there is limited evidence for any of the several formation mechanisms proposed to date. Loi et al. (2016) present a case study of an event on 26 August 2014 where a travelling ionospheric disturbance (TID) shortly precedes the formation of a complex collection of field-aligned ducts, using data obtained by the Murchison Widefield

Array (MWA) radio telescope. Their spatiotemporal proximity leads to suggest a causal interpretation. Geomagnetic conditions were quiet at the time, and no obvious triggers were noted. Growth of the structures proceeds rapidly, within 0.5 h of the passage of the TID, attaining their peak prominence 1–2 h later and persisting for several more hours until observations ended at local dawn. Analyses of the next 2 days show field-aligned structures to be preferentially detectable under quiet rather than active geomagnetic conditions. Loi et al. (2016) used a raster scanning strategy facilitated by the speed of electronic beam-forming to expand the quasi-instantaneous field of view of the MWA by a factor of 3. These observations represent the broadest angular coverage of the ionosphere by a radio telescope to date.

6.12.4. Vertical Structure of Medium-Scale Traveling Ionospheric Disturbances

Ssessanga et al. (2015) develop an algorithm of computerized ionospheric tomography (CIT) to infer information on the vertical and horizontal structuring of electron density during nighttime medium-scale traveling ionospheric disturbances (MSTIDs). To facilitate digital CIT Ssessanga et al. (2015) have adopted total electron contents (TEC) from a dense Global Positioning System (GPS) receiver network, GEONET, which contains more than 1000 receivers. A multiplicative algebraic reconstruction technique was utilized with a calibrated IRI-2012 model as an initial solution. The reconstructed F2 peak layer varied in altitude with average peak-to-peak amplitude of ~52 km. In addition, the F2 peak layer anticorrelated with TEC variations. This feature supports a theory in which nighttime MSTID is composed of oscillating electric fields due to conductivity variations. Moreover, reconstructed TEC variations over two stations were reasonably close to variations directly derived from the measured TEC data set. This tomographic analysis may thus help understand three-dimensional structure of MSTIDs in a quantitative way.

6.13. Dipolarization Fronts and Equatorial Electrojets in Geomagnetosphere – Ionosphere System

6.13.1. Three-Dimensional Current Systems and Ionospheric Effects Associated with Small Dipolarization Fronts

Palin et al. (2015) present a case study of eight successive plasma sheet (PS) activations (usually referred to as bursty bulk flows or dipolarization fronts), associated with small individual inline image increases on 31 March 2009 (0200–0900 UT), observed by the THEMIS mission. This series of events happens during very quiet solar wind conditions, over a period of 7 h preceding a substorm onset at 1230 UT. The amplitude of the dipolarizations increases with time. The low-amplitude dipolarization fronts are associated with few (1 or 2) rapid flux transport events (RFT, Eh>2 mV/m), whereas the large-amplitude ones encompass many more RFT events. All PS activations are associated with small and localized substorm current wedge (SCW)-like current system signatures, which seems to be the consequence of RFT arrival in the near tail. The associated ground magnetic perturbations affect a larger part of the contracted auroral oval when, in the magnetotail, more RFT are embedded in PS activations (>5). Dipolarization fronts with very low amplitude, a type usually not included in statistical studies, are of particular interest because we found even those to be associated with clear small SCW-like current system and particle injections at geosynchronous orbit. This exceptional data set highlights the role of flow bursts in the magnetotail and leads to the conclusion that it may be observing the smallest form of a substorm or rather its smallest element. This study also highlights the gradual evolution of the ionospheric current disturbance as the plasma sheet is observed to heat up.

6.13.2. Day-to-Day Variability of Equatorial Electrojet and Its Role on the Day-to-Day Characteristics of the Equatorial Ionization Anomaly over the Indian and Brazilian Sectors

As note Venkatesh et al. (2015), the equatorial electrojet (EEJ) is a narrow band of current flowing eastward at the ionospheric *E* region altitudes along the dayside dip equator; mutually perpendicular electric and magnetic fields over the equator results in the formation of equatorial ionization anomaly (EIA), which in turn generates large electron density variabilities. Simultaneous study on the characteristics of EEJ and EIA is necessary to understand the role of EEJ on the EIA variabilities. This is helpful for the improved estimation of total electron content (TEC) and range delays required for satellite-based communication and navigation applications. Venkatesh et al. (2015) reports simultaneous variations of EEJ and GPS-TEC over Indian and Brazilian sectors to understand the role of EEJ on the day-to-day characteristics of the EIA. Magnetometer measurements during the low solar activity year 2004 are used to derive the EEJ values over the two different sectors. The characteristics of EIA are studied using two different chains of GPS receivers along the common meridian of 77°E (India) and 45°W (Brazil). The diurnal, seasonal, and day-to-day variations of EEJ and TEC are described simultaneously. Variations of EIA during different seasons are presented along with the variations of the EEJ in the two hemispheres. The role of EEJ variations on the characteristic features of the EIA such as the strength and temporal extent of the EIA crest has also been reported. Further, the time delay between the occurrences of the day maximum EEJ and the well-developed EIA is studied and corresponding results are presented in the paper of Venkatesh et al. (2015).

6.13.3. Characteristics of Equatorial Electrojet Derived from Swarm Satellites

The vector magnetic field measurements from three satellite constellation, Swarm mission (Alpha ‘Swarm-A’, Bravo ‘Swarm-B’, and Charlie ‘Swarm-C’) during the quiet days (daily $\Sigma Kp \leq 10$) of the years 2014–2015 are used by Thomas et al. (2017) to study the characteristic features of equatorial electrojet (EEJ). A program is developed to identify the EEJ signature in the X (northward) component of the magnetic field recorded by the satellite.

An empirical model is fitted into the observed EEJ signatures separately for both the hemispheres, to obtain the parameters of electrojet current such as peak current density, total eastward current, the width of EEJ, position of the electrojet axis, etc. The magnetic field signatures of EEJ at different altitudes are then estimated. Swarm B and C are orbiting at different heights (separation ~50 km) and during the month of April 2014, both the satellites were moving almost simultaneously over nearby longitudes. The magnetic field estimates at the location of Swarm-C obtained using the observations of Swarm B are compared with the actual observations of Swarm-C. A good correlation between the actual and the computed values (correlation coefficient = 0.98) authenticates the method of analysis. The altitudinal variation of the amplitude and the width of the EEJ signatures are also depicted. The ratio of the total eastward flowing forward to westward return currents is found to vary between 0.1 and 1.0. The forward and return current values in the northern hemisphere are found to be ~0.5 to 2 times of those in the southern hemisphere, thereby indicating the hemispheric asymmetry. The latitudinal extents of the forward and return currents are found to have longitudinal dependence similar to that of the amplitude and the width of EEJ showing four peak structures. Local time dependence of EEJ parameters has also been investigated. In general, the results are found to be consistent with previous studies. In order to examine the existence of the EEJ associated meridional currents, Thomas et al. (2017) have estimated the vertical current density using combination of two satellites separated in longitude.

6.13.4. Energy Conversion at Dipolarization Fronts

Khotyaintsev et al. (2017) use multispacecraft observations by Cluster in the Earth's magnetotail and 3-D particle-in-cell simulations to investigate conversion of electromagnetic energy at the front of a fast plasma jet. They find that the major energy conversion is happening in the Earth (laboratory) frame, where the electromagnetic energy is being transferred from the electromagnetic field to particles. This process operates in a region with size of the order several ion inertial lengths across the jet front, and the primary contribution to $E \times j$ is coming from the motional electric field and the ion current. In the frame of the front Khotyaintsev et al. (2017) find fluctuating energy conversion with localized loads and generators at sub-ion scales which are primarily related to the lower hybrid drift instability excited at the front; however, these provide relatively small net energy conversion.

6.13.5. Enhancement of Oxygen in the Magnetic Island Associated with Dipolarization Fronts

According to Wang et al. (2017), a significant enhancement of O^+ is observed by Cluster inside an earthward propagating magnetic island behind a dipolarization front (DF). Such enhancement, from 0.005 to 0.03 cm^{-3}, makes the O^+ flux inside the magnetic island ~20 times larger than that outside the magnetic island. In the meantime, the H^+ density is nearly a constant, 0.1 cm^{-3}, during the magnetic-island encounter. This results in a dramatic increase of the density ratio, n_{O+}/n_{H+}, from 0.05 to 0.3 (about 10 times as large as the average value in the plasma sheet) and a dramatic decrease of the local Alfvén speed from $V_A \approx 770$ km/s to

$V_A \approx 430$ km/s inside the magnetic island. The decrease of Alfvén speed indicates an asymmetric reconnection and a slow magnetic reconnection rate near the secondary X line. Since the reconnection rates at the primary X line and secondary X line are imbalanced, the DFs and magnetic islands are pushed to propagate earthward by the outflow of the primary reconnection, as demonstrated in recent simulations.

6.13.6. Particle Energization by a Substorm Dipolarization

As outlined Kabin et al. (2017), magnetotail dipolarizations, often associated with substorms, produce significant energetic particle enhancements in the nighttime magnetosphere. Kabin et al. (2017) apply recently developed by them magnetotail dipolarization model (Kabin et al., 2011) to the problem of energizing electrons and ions. This model is two-dimensional in the meridional plane and is characterized by the ability to precisely control the location of the transition from the dipole-like to tail-like magnetic fields. Both magnetic and electric fields are calculated, self-consistently, as the transition zone retreats farther into the tail and the area around the Earth occupied by dipole-like lines increases in size. These fields are used to calculate the motion of electrons and ions and changes in their energies. Kabin et al. (2017) consider the energizing effects of the fields restricted to ±15° and ±30° sectors around the midnight meridian, as well the axisymmetric case. Energies of some electrons increase by a factor of 25, which is more than enough to produce observable ionospheric signatures. Electrons are treated using the Guiding Center approximation, while protons and heavier particles generally require description based on the Lorentz equations.

6.14. Checking and Using International Reference Ionosphere (IRI) Model

6.14.1. Diurnal variations of the ionospheric electron density height profiles over Irkutsk: Comparison of the incoherent scatter radar measurements, GSM TIP simulations and IRI predictions

According to Zherebtsov et al. (2017), the long-duration continuous Irkutsk incoherent scatter radar (ISR) measurements allowed to obtain the monthly averaged height-diurnal variations of the electron density in the 180–600 km altitudinal range for four seasons (winter, spring, summer, autumn) and for two solar activity levels (low and moderate). Considering these electron density variations as "quiet ionosphere patterns" Zherebtsov et al. (2017) compared them with the Global Self-consistent Model of the Thermosphere, Ionosphere and Protonosphere (GSM TIP) simulations and the International Reference Ionosphere (IRI) predictions. It was found that some observational features revealed from the ISR measurements are reproduced nicely by both the theoretical and empirical models, and some features agree better with the GSM TIP than with IRI. None of the models is able to reproduce a detailed multi-peak behavior of the electron density observed by ISR at ~300 km and above for the spring and autumn under low solar activity, while for the spring the GSM TIP tends to reproduce the morning and daytime peaks at the same local times as they are seen from the ISR observations.

6.14.2. Seasonal and solar cycle effects on TEC at 95°E in the ascending half (2009–2014) of the subdued solar cycle 24: Consistent underestimation by IRI 2012.

TEC measured at Dibrugarh (27.5°N, 94.9°E, 17.5°N Geomag.) from 2009 to 2014 is used by Kakoti et al. (2017) to study its temporal characteristics during the ascending half of solar cycle 24. The measurements provide an opportunity to assess the diurnal, seasonal and longterm predictability of the IRI 2012 (with IRI Nequick, IRI01-corr, IRI 2001 topside options) during this solar cycle which is distinctively low in magnitude compared to the previous cycles. The low latitude station Dibrugarh is normally located at the poleward edge of the northern EIA. A semi-annual variation in GPS TEC is observed with the peaks occurring in the equinoxes. The peak in spring (March, April) is higher than that in autumn (September, October) irrespective of the year of observation. The spring autumn asymmetry is also observed in IRI TEC. In contrast, the winter (November, December, January, February) anomaly is evident only in high activity years. TEC bears a distinct nonlinear relationship with 10.7 cm solar flux (F10.7). TEC increases linearly with F10.7 up to about 125 sfu beyond which it tends to saturate. The correlation between TEC and solar flux is found to be a function of local time and peaks at 10:00 LT. TEC varies nonlinearly with solar EUV flux similar to its variation with F10.7. The nonlinearity is well captured by the IRI. The saturation of TEC at high solar activity is attributed to the inability of the ionosphere to accommodate more ionization after it reaches the level of saturation ion pressure. Annual mean TEC increased from the minimum in 2009 almost linearly till 2012, remains at the same level in 2013 and then increased again in 2014. IRI TEC shows a linear increase from 2009 to 2014. IRI01-corr and IRI-NeQuick TEC are nearly equal at all local times, season and year of observation while IRI-2001 simulated TEC are always higher than that simulated by the other two versions. As outlined Kakoti et al. (2017), the IRI 2012 underestimates the TEC at about all local times except for a few hours in the midday in all season or year of observation. The discrepancy between model and measured TEC is high in spring and in the evening hours. The consistent underestimation of the TEC at this longitude by the IRI may be attributed to the inadequate ingestion of F region data from this longitude sector into the model and exclusion of the plasmaspheric content.

6.14.3. Seasonal characteristics of COSMIC measurements over Indian sub-continent during different phases of solar activity

The seasonal characteristics of F2 region is investigated by Aggarwal and Sharma (2017) using peak electron density (NmF2) and corresponding altitude (hmF2) measurements obtained by COSMIC observations over Indian sub-continent (Geog. 5–40°N and 60–100°E) for the complete mission during the solar cycle 24 (Apr, 2006–Dec, 2013). A stronger EIA is observed in summer and winter during high solar activity whereas occurs in equinoxes during all levels of solar activity. The noontime winter anomaly in NmF2 is absent/weaker during low (2006–2009)/high (2010–2013) solar activity over equator whereas is stronger in low and moderate solar activity (2006–2007 and 2010–2011) but absent in minimum and maximum (2008–2009 and 2012–2013) solar activity over low-latitude respectively. The higher hmF2 is observed over the equator throughout the day (equator: 230–415 km and low-latitude: 220–340 km). The observed parameters, NmF2 and hmF2 are compared with the IRI-modeled values using CCIR and URSI options. As outlined Aggarwal and Sharma (2017), the comparison exhibits a higher discrepancy in NmF2 in nighttime and morning for equatorial and lower discrepancy for the whole day over the low-latitude whereas smaller (<20%) discrepancy exists in the hmF2 throughout the day using both IRI options. In conclusion, Aggarwal and Sharma (2017) found CCIR option in more agreement to the observed values for both equatorial and low-latitude regions respectively.

6.14.4. Solar cycle variation of ionospheric parameters over the low latitude station Hainan during 2002–2012 and its comparison with IRI-2012 model

The low latitude ionospheric data observed by digisonde in Chaina at Hainan station (19.5°N, 109.1°E) in a whole solar activity cycle period from 2002 to 2012 within Ap < 20 have been analyzed by Wang et al. (2017) to explore the diurnal, seasonal, annual variations and solar activity dependences of the ionospheric peak parameters (foF2, hmF2, and Chapman scale height Hm), as well as some quantitative comparison with IRI- 2012 modeling predictions. The results show that the winter anomaly in the daytime foF2 appears at different levels of solar activity. The semiannual anomaly in the daytime and nighttime foF2 with two maxima in equinox seasons is present. The foF2 have a close correlation with a solar activity factor F10.7P = (F10.7 + F10.7A)/2 and the correlation coefficients CC in their diurnal variation are around 0.7. The slope of foF2 varying with F10.7P in daytime is usually smaller than in nighttime. The afternoon and evening hmF2 show good correlation with F10.7P (their CC values exceed 0.6), but hmF2 at other time are low or poor related to F10.7P. The prominent character of hmF2 in equinox and summer seasons is its strong increase at sunset in high solar activity period, which may be due to pre-reversal enhancement (PRE) of local electric field. Wang et al. (2017) also note that hmF2 values around midnight slightly decrease with increasing F10.7P index in equinox seasons. The diurnal variation of Hm usually has two peaks around noontime and pre-sunrise. The daytime Hm has an annual variation with maximum in summer and minimum in winter. Moreover, the dependence of the daytime Hm on solar activity is not strong due to meridional wind and other factors. As outlined Wang et al. (2017), the above results over Hainan are considerably different from those reported over Millstone Hill, which is attributed to their different geomagnetic locations. According to Wang et al. (2017), the quantitative results compared between IRI-2012 model predictions and observations show that the predicted foF2 values are basically underestimated and the magnitude of their deviations obviously increases with increasing solar activity. The predicted hmF2 obtained with measured M(3000)F2 inputs in low and moderate solar activity agree well with the observed ones. However, their deviations in high solar activity are significantly magnified.

6.14.5. Variability of Ionospheric Parameters during Solar Minimum and Maximum Activity and Assessment of IRI Model

The ionospheric parameters (electron and ion plasma temperatures Te and Ti and total ion density Ni) as obtained by the Indian SROSS-C2 satellite (altitude ~500 km) have been investigated by Sharma et al. (2017) during low (year 1995, F10.7 ~ 77 sfu) and high (year 2000, F10.7 ~ 177 sfu) solar activity periods. The region under study spans over 5°S-30°N geomag. latitude and 60–100°E geog. longitude over the Indian sector. The observations are compared with the modelled values using IRI-2007 and IRI-2012 versions to assess model predictability. Sharma et al. (2017) found that minimum plasma temperatures (Te and Ti) in nighttime gets twice hotter whereas maximum temperatures in early morning gets reduced by half (cooler) when the solar flux gets doubled indicating a direct relation of Te and Ti with solar flux, F10.7 in nighttime but inverse in the morning hours. The ion density (Ni) exhibits solar activity dependence throughout the day and increases by one order when solar activity gets doubled. As outlined Sharma et al. (2017), the modelled Te and Ti are found in agreement to the observed values for high solar activity over both the regions. Whereas the discrepancy exists during low solar activity period over both the regions with over-/under-estimated values in nighttime/morning and noontime respectively. The latest IRI-2012 model improves the nighttime Te and Ti whereas the modelled Ni is found in complete agreement to the observations.

6.15. Variable Pixel Size Ionospheric Tomography

A novel ionospheric tomography technique based on variable pixel size was developed by Dunyong et al. (2017) for the tomographic reconstruction of the ionospheric electron density (IED) distribution. In variable pixel size computerized ionospheric tomography (VPSCIT) model, the IED distribution is parameterized by a decomposition of the lower and upper ionosphere with different pixel sizes. Thus, the lower and upper IED distribution may be very differently determined by the available data. The variable pixel size ionospheric tomography and constant pixel size tomography are similar in most other aspects. There are some differences between two kinds of models with constant and variable pixel size respectively, one is that the segments of GPS signal pay should be assigned to the different kinds of pixel in inversion; the other is smoothness constraint factor need to make the appropriate modified where the pixel change in size. For a real dataset, the variable pixel size method distinguishes different electron density distribution zones better than the constant pixel size method. Furthermore, it can be non-chided that when the effort is spent to identify the regions in a model with best data coverage. As outlined Dunyong et al. (2017), the variable pixel size method can not only greatly improve the efficiency of inversion, but also produce IED images with high fidelity which are the same as a used uniform pixel size method. In addition, variable pixel size tomography can reduce the underdetermined problem in an ill-posed inverse problem when the data coverage is irregular or less by adjusting quantitative proportion of pixels with different sizes. In comparison with constant pixel size tomography models, the variable pixel size ionospheric tomography technique achieved relatively good results in a numerical simulation. A careful validation of the reliability and superiority of variable pixel size ionospheric tomography was performed. Finally, according to the results of the statistical analysis and quantitative comparison, the proposed method offers an improvement of 8% compared with conventional constant pixel size tomography models in the forward modeling.

REFERENCES

Aa E., W. Huang, S. Yu, S. Liu, L. Shi, J. Gong, Y. Chen, and H. Shen, 2015. "A regional ionospheric TECmapping technique over China and adjacent areas on the basis of data assimilation", *J. Geophys. Res. Space Physics*, **120**, No. 6, doi:10.1002/2015JA021140, 5049–5061.

Adebiyi S.J., I.A. Adimula, O.A. Oladipo, and B.W. Joshua, 2016. "Assessment of IRI and IRI-Plas models over the African equatorial and low-latitude region", *J. Geophys. Res. Space Physics*, **121**, doi:10.1002/2016JA022697.

Aggarwal M. and D.K. Sharma, 2017. "Seasonal characteristics of COSMIC measurements over Indian sub-continent during different phases of solar activity", *Advances in Space Research*, **59**, 2279–2294.

Akbari H., P. Guio, M.A. Hirsch, and J.L. Semeter, 2016. "Zakharov simulations of beam-induced turbulence in the auroral ionosphere", *J. Geophys. Res. Space Physics*, **121**, 4811–4825, doi:10.1002/2016JA022605.

Bardhan A., D.K. Sharma, M.S. Khurana, M. Aggarwal, and S. Kumar, 2015. "Electron–ion–neutral temperatures and their ratio comparisons over low latitude ionosphere", *Adv. Space Res.*, **56**, No. 10, 2117–2129.

Buhari S.M., M. Abdullah, A.M. Hasbi, Y. Otsuka, T. Yokoyama, M. Nishioka, and T. Tsugawa, 2014. "Continuous generation and two-dimensional structure of equatorial plasma bubbles observed by high-density GPS receivers in Southeast Asia", *J. Geophys. Res. Space Physics*, **119**, No. 12, doi:10.1002/2014JA020433, 10569-10580.

Carter B.A., J.M. Retterer, E. Yizengaw, K. Groves, R. Caton, L. McNamara, C. Bridgwood, M. Francis,M. Terkildsen, R. Norman, and K. Zhang, 2014. "Geomagnetic control of equatorial plasma bubble activity modeled by the TIEGCM with Kp", *Geophys. Res. Lett.*, **41**, No. 15, doi:10.1002/2014GL060953, 5331-5339.

Chang L.C., Y.-Y. Sun, J. Yue, J.C. Wang, and S.-H. Chien, 2016. "Coherent seasonal, annual, and quasi-biennial variations in ionospheric tidal/SPW amplitudes", *J. Geophys. Res. Space Physics*, **121**, 6970–6985, doi:10.1002/2015JA022249.

Chappell C.R., 2015. "The Role of the Ionosphere in Providing Plasma to the Terrestrial Magnetosphere-An Historical Overview", *Space Sci. Rev.*, **192**, No. 1, doi: 10.1007/s11214-015-0168-5, 5-25.

Chaston C.C., J.W. Bonnell, G.D. Reeves, and R.M. Skoug, 2016. "Driving ionospheric outflows and magnetospheric O^+ energy density with Alfvén waves", *Geophys. Res. Lett.*, **43**, 4825–4833, doi:10.1002/2016GL069008.

Chen Y., L. Liu, H. Le, and W. Wan, 2014. "Geomagnetic activity effect on the global ionosphere during the 2007–2009 deep solar minimum", *J. Geophys. Res. Space Physics*, **119**, No. 5, doi:10.1002/2013JA019692, 3747-3754.

Chen Y., L. Liu, H. Le, W. Wan, and H. Zhang, 2016. "Equatorial ionization anomaly in the low-latitude topside ionosphere: Local time evolution and longitudinal difference", *J. Geophys. Res. Space Physics*, **121**, doi:10.1002/2016JA022394.

Christiansen F., V.O. Papitashvili, and T. Neubert, 2002. "Seasonal variations of high-latitude field-aligned currents inferred from Ørsted and Magsat observations", *J. Geophys. Res.*, **107**, No. A2, 1029, doi:10.1029/2001JA900104, SMP5-1-5-13.

Cliver E.W., Y. Kamide, and A.G. Ling, 2000. "Mountains versus valleys: Semiannual variation of geomagnetic activity", *J. Geophys. Res.*, **105**, No. A2, 2413-2424.

Cliver E.W., Y. Kamide, and A.G. Ling, and N. Yokoyama, 2001 "Semiannual variation of the geomagnetic *Dst* index: Evidence for a dominant nonstorm component", *J. Geophys. Res.*, **106**, No. A10, 21297-21304.

Cousins E.D.P., T. Matsuo, A.D. Richmond, and B.J. Anderson, 2015. "Dominant modes of variability in large-scale Birkeland currents", *J. Geophys. Res. Space Physics*, **120**, No. 8, doi:10.1002/2014JA020462, 6722–6735.

Crooker N.U. and G.L. Siscoe, 1986b. "The effect of the solar wind on the terrestrial environment", in *Physics of the Sun*, edited by P.A. Sturrock, D. Reidel, Hingham, Mass., 193-249.

Crooker N.U., E.W. Cliver, and B.T. Tsurutani, 1992. "The semiannual variation of great geomagnetic storms and the post-shock Russel-McPherron effect preceding coronal mass ejecta", *Geophys. Res. Lett.*, **19**, No. 5, 429-432.

Delzanno G.L., J.E. Borovsky, M.F. Thomsen, B.E. Gilchrist, and E. Sanchez, 2016. "Can an electron gun solve the outstanding problem of magnetosphere-ionosphere connectivity?",
J. Geophys. Res. Space Physics, **121**, doi:10.1002/2016JA022728, 6769–6773.

De Zeeuw D.L., S. Sazykin, R.A. Wolf, T.I. Gombosi, A.J. Ridley, and G. Toth, 2004. "Coupling of a global MHD code and an inner magnetospheric model: Initial results", *J. Geophys. Res.*, **109**, No. A12, A12219, doi:10.1029/2003JA010366, 1-14.

Dunyong Z., Z. Hongwei, W. Yanjun, N. Wenfeng, L. Chaokui, A. Minsi, H. Wusheng, and Z. Wei, 2017. "Variable pixel size ionospheric tomography", *Advances in Space Research*, **59**, 2969–2986.

Fedder J.A. and J.G. Lyon, 1987. "The solar wind-magnetosphere-ionosphere current-voltage relationship", *Geophys. Res. Lett.*, **14**, No. 8, 880-883.

Ferdousi B. and J. Raeder, 2016. "Signal propagation time from the magnetotail to the ionosphere: OpenGGCM simulation", *J. Geophys. Res. Space Physics*, **121**, 6549–6561, doi:10.1002/2016JA022445.

Fernandes P.A. and K.A. Lynch, 2016. "Electrostatic analyzer measurements of ionospheric thermal ion populations", *J. Geophys. Res. Space Physics*, **121**, doi:10.1002/2016JA022582, 1-10.

Fok M.-C., R.A. Wolf, R.W. Spiro, and T.E. Moore, 2001. "Comprehensive computational model of Earth's ring current", *J. Geophys. Res.*, **106**, No. A5, 8417-8424.

Fok M.-C., N. Y. Buzulukova, S.-H. Chen, A. Glocer, T. Nagai, P. Valek, and J. D. Perez , 2014. "The Comprehensive Inner Magnetosphere-Ionosphere Model", *J. Geophys. Res. Space Physics*, **119**, No. 9, doi:10.1002/2014JA020239, 7522-7540.

Fontaine D., M. Blanc, L. Reinhart, and R. Glowinski, 1985. "Numerical simulations of the magnetospheric convection including the effects of electron precipitation", *J. Geophys. Res.*, **90**, No. A9, 8343-8360.

Forbes J.M. and M. Harel, 1989. "Magnetosphere-thermosphere coupling: An experiment in interactive modeling", *J. Geophys. Res.*, **94**, No. A3, 2631-2644.

Forsythe V.V. and R.A. Makarevich, 2015. "Dual radar investigation of E region plasma waves in the southern polar cap", *J. Geophys. Res. Space Physics*, **120**, No. 10, doi:10.1002/2015JA021664, 9132–9147.

Fujii R. and T. Iijima, 1987. "Control of ionospheric conductivities on large-scale Birkeland current intensities under geomagnetic quiet conditions", *J. Geophys. Res.*, **92**, No. A5, 4505-4513.

Fujii R., T. Iijima, T.A. Potemra, and M. Sugiura, 1981. "Seasonal dependence of large-scale Birkeland currents", *Geophys. Res. Lett.*, **8**, No. 10, 1103-1106 (1981).

Gagnepain J., M. Crochet, and A.D. Richmond, 1975. "Theory of longitudinal gradients in the equatorial electrojet", *J. Atmos. Terr. Phys.*, **38**, No. 3, 279-286.

Garcia-Sage K., T.E. Moore, A. Pembroke, V.G. Merkin, and W.J. Hughes, 2015. "Modeling the effects of ionospheric oxygen outflow on bursty magnetotail flows", *J. Geophys. Res. Space Physics*, **120**, No. 10, doi:10.1002/2015JA021228, 8723–8737.

Gasda S. and A.D. Richmond, 1998. "Longitudinal and interhemispheric variations of auroral ionospheric electrodynamics in a realistic geomagnetic field", *J. Geophys. Res.*, **103**, No. A3, 4011-4021.

Haerendel G., 2014. "M-I coupling scales and energy dumping", *Geophys. Res. Lett.*, **41**, No. 6, doi:10.1002/2014GL059582, 1846-1853.

Hajra R., S.K. Chakraborty, B.T. Tsurutani et al., 2016. "An empirical model of ionospheric total electron content (TEC) near the crest of the equatorial ionization anomaly (EIA)", *J. Space Weather Space Clim.*, **6**, A29, 1-9. DOI: 10.1051/swsc/2016023.

Harel M., R.A. Wolf, P.H. Reiff, R.W. Spiro, W.J. Burke, F.J. Rich, and M. Smiddy, 1981. "Quantitative simulation of a magnetospheric substorm: 1. Model logic and overview", *J. Geophys. Res.*, **86**, No. A4, 2217-2241.

Hartinger M.D., M.B. Moldwin, S. Zou, J.W. Bonnell, and V. Angelopoulos, 2015. "ULF wave electromagnetic energy flux into the ionosphere: Joule heating implications", *J. Geophys. Res. Space Physics*, **120**, No. 1, doi:10.1002/2014JA020129, 494-510.

Heinemann M. and D.H. Pontius Jr., 1990. "Representations of currents and magnetic fields in isotropic magnetohydrostatic plasma", *J. Geophys. Res.*, **95**, No. A1, 251-257.

Hines C.O., 1960. "Internal atmospheric gravity waves at ionospheric heights", *Can. J. Phys.*, **38**, 1441–1481.

Hui D. and B.G. Fejer, 2015. "Daytime plasma drifts in the equatorial lower ionosphere", *J. Geophys. Res. Space Physics*, **120**, No. 11, doi:10.1002/2015JA021838, 9738–9747.

Hurtaud Y., C. Peymirat, and A.D. Richmond, 2007. "Modeling seasonal and diurnal effects on ionospheric conductances, region-2 currents, and plasma convection in the inner magnetosphere", *J. Geophys. Res.*, **112**, A09217, doi:10.1029/2007JA012257, 1-17 (2007).

Ivanov V.B., O.A. Gorbachev, and A.A. Kholmogorov, 2016. "Comparative Quality Analysis of Models of Total Electron Content in the Ionosphere", *Geomagnetism and Aeronomy*, **56**, No. 3, 318–322. Original Russian Text published in *Geomagnetizm i Aeronomiya*, **56**, No. 3, 340–344.

Jakowski N., M.M. Hoque, M. Kriegel, and V. Patidar, 2015. "The persistence of the NWA effect during the low solar activity period 2007–2009", *J. Geophys. Res. Space Physics*, **120**, No. 10, doi:10.1002/2015JA021600, 9148–9160.

Kabin K., E. Spanswick, R. Rankin, E. Donovan, and J. C. Samson, 2011. "Modeling the relationship between substorm dipolarization and dispersionless injection", *J. Geophys. Res.*, *116*, A04201, doi:10.1029/2010JA015736.

Kabin K., G. Kalugin, E. Donovan, and E. Spanswick, 2017. "Particle energization by a substorm dipolarization", *J. Geophys. Res. Space Physics*, **122**, 349–367, doi:10.1002/2016JA023459.

Kakoti G., P.K. Bhuyan, and R. Hazarika, 2017. "Seasonal and solar cycle effects on TEC at 95°E in the ascending half (2009–2014) of the subdued solar cycle 24: Consistent underestimation by IRI 2012", *Advances in Space Research*, **60**, 257–275.

Keiling A., O. Marghitu, J. Vogt, O. Amm, C. Bunescu, V. Constantinescu, H. Frey, M. Hamrin, T. Karlsson, R. Nakamura, H. Nilsson, J. Semeter, and E. Sorbalo, 2014. "Magnetosphere-ionosphere coupling of global Pi2 pulsations", *J. Geophys. Res. Space Physics*, **119**, No. 4, doi:10.1002/2013JA019085, 2717-2739.

Khazanov G.V., A. Glocer, and E.W. Himwich, 2014. "Magnetosphere-ionosphere energy interchange in the electron diffuse aurora", *J. Geophys. Res. Space Physics*, **119**, No.1, doi:10.1002/2013JA019325, 171-184.

Khazanov G.V., A.K. Tripathi, R.P. Singhal, E.W. Himwich, A. Glocer, and D.G. Sibeck, 2015a. "Superthermal electron magnetosphere-ionosphere coupling in the diffuse aurora in the presence of ECH waves", *J. Geophys. Res. Space Physics*, **120**, No. 1, doi:10.1002/2014JA020641, 445-459

Khazanov G.V., A.K. Tripathi, D. Sibeck, E. Himwich, A. Glocer, and R.P. Singhal, 2015b. "Electron distribution function formation in regions of diffuse aurora", *J. Geophys. Res. Space Physics*, **120**, No. 11, doi:10.1002/2015JA021728, 9891–9915.

Khazanov G.V., A. Glocer, D.G. Sibeck, A.K. Tripathi, L.G. Detweiler, L.A. Avanov, and R.P. Singhal, 2016. "Ionosphere-magnetosphere energy interplay in the regions of diffuse aurora", *J. Geophys. Res. Space Physics*, **121**, 6661–6673, doi:10.1002/2016JA022403.

Khotyaintsev Y.V., A. Divin, A. Vaivads, M. André, and S. Markidis, 2017. "Energy conversion at dipolarization fronts", *Geophys. Res. Lett.*, **44**, 1234–1242, doi:10.1002/2016GL071909.

Kolmogorov A.N., 1941. "The local structure of turbulence in incompressible viscous fluid for very large Reynolds numbers". *Dokl. Akad. Nauk SSSR*, 299–303.

Kotov D.V., P.G. Richards, O.V. Bogomaz, L.F. Chernogor, V. Truhlik, L.Y. Emelyanov, Y.M. Chepurnyy, and I.F. Domnin, 2016. "The importance of neutral hydrogen for the maintenance of the midlatitude winter nighttime ionosphere: Evidence from IS observations at Kharkiv, Ukraine, and field line interhemispheric plasma model simulations", *J. Geophys. Res. Space Physics*, **121**, doi:10.1002/2016JA022442.

Kraichnan R.H., 1965. "Inertial-range spectrum of hydromagnetic turbulence". *Phys. Fluids*, **8**, 1385–1387.

Lamarche L.J., and R.A. Makarevich, 2015. "Solar control of F region radar backscatter: Further insights from observations in the southern polar cap", *J. Geophys. Res. Space Physics*, **120**, No. 11, doi:10.1002/2015JA021663, 9875–9890.

Lee S.H., H. Zhang, Q.-G. Zong, Y. Wang, A. Otto, H. Rиme, and K.-H. Glassmeier, 2015. "Asymmetric ionospheric outflow observed at the dayside magnetopause", *J. Geophys. Res. Space Physics,* **120**, No. 5, doi:10.1002/2014JA020943, 3564-3573.

Lei J., W. Wang, A.G. Burns, X. Luan, and X. Dou, 2016. "Can atomic oxygen production explain the ionospheric annual asymmetry?", *J. Geophys. Res. Space Physics*, **121**, doi:10.1002/2016JA022648.

Lin J., X. Yue, Z. Zeng, Y. Lou, X. Shen, Y. Wu, W.S. Schreiner, and Y.-H. Kuo, 2014. "Empirical orthogonal function analysis and modeling of the ionospheric peak height during the years 2002–2011", *J. Geophys. Res. Space Physics*, **119**, No. 5, doi:10.1002/2013JA019626, 3915-3929.

Liu J., R. Chen, J. An, Z. Wang, and J. Hyyppa, 2014a. "Spherical cap harmonic analysis of the Arctic ionospheric TEC for one solar cycle", *J. Geophys. Res. Space Physics*, **119**, No.1, doi:10.1002/2013JA019501, 601-619.

Loi S.T., T. Murphy, I.H. Cairns et al., 2015. "Real-time imaging of density ducts between the plasmasphere and ionosphere", *Geophys. Res. Lett.*, **42**, No. 10, doi:10.1002/2015GL063699, 3707-3714.

Loi S.T., I.H. Cairns, T. Murphy, P.J. Erickson, M.E. Bell, A. Rowlinson, B.S. Arora, J. Morgan, R.D. Ekers, N. Hurley-Walker, and D.L. Kaplan, 2016. "Density duct formation in the wake of a travelling ionospheric disturbance: Murchison Widefield Array observations", *J. Geophys. Res. Space Physics*, **121**, No. 2, doi:10.1002/2015JA022052, 1569–1586.

Lu G., A.D. Richmond, B.A. Emery, P.H. Reiff, O. de la Beaujardière, F.J. Rich, W.F. Denig, H.W. Kroehl, L.R. Lyons, J.M. Ruohoniemi, E. Friis-Christensen, H. Opgenoorth, M.A.L. Persson, R.P. Lepping, A.S. Rodger, T. Hughes, A. McEwin, S. Dennis, R. Morris, G. Burns, and L. Tomlinson, 1994. "Interhemispheric Asymmetry of the High-Latitude Ionospheric Convection Pattern", *J. Geophys. Res.*, **99**, No. A4, 6491-6510.

Lu G., L.R. Lyons, P.H. Reiff, W.F. Denig, O. de la Beaujardière, H.W. Kroehl, P.T. Newell, F.J. Rich, H. Opgenoorth, M.A.L. Persson, J.M. Ruohoniemi, E. Friis-Christensen, L. Tomlinson, R. Morris, G. Burns, and A. McEwin, 1995. "Characteristics of ionospheric convection and field-aligned currents in the dayside cusp region", *J. Geophys. Res.*, **100**, No. A7, 11845-11861.

Lynch K.A., D.L. Hampton, M. Zettergren, T.A. Bekkeng, M. Conde, P.A. Fernandes, P. Horak, M. Lessard, R. Miceli, R. Michell, J. Moen, M. Nicolls, S.P. Powell, and M. Samara, 2015. "MICA sounding rocket observations of conductivity gradient-generated auroral ionospheric responses: Small-scale structure with large-scale drivers", *J. Geophys. Res. Space Physics*, **120**, No. 11, doi:10.1002/2014JA020860, 9661–9682.

Makarevich R.A., 2016. "Toward an integrated view of ionospheric plasma instabilities: 2. Three inertial modes of a cubic dispersion relation", *J. Geophys. Res. Space Physics*, **121**, doi:10.1002/2016JA022864.

Marchaudon A. and P.-L. Blelly, 2015. "A new interhemispheric 16-moment model of the plasmasphere-ionosphere system: IPIM", *J. Geophys. Res. Space Physics*, **120**, No. 7, doi:10.1002/2015JA021193, 5728–5745.

Marsal S., 2015. "Conductivities consistent with Birkeland currents in the AMPERE-driven TIE-GCM", *J. Geophys. Res. Space Physics*, **120**, No. 9, doi:10.1002/2015JA021385, 8045–8065.

Marsal S., A.D. Richmond, A. Maute, and B.J. Anderson, 2012. "Forcing the TIEGCM model with Birkeland currents from the active magnetosphere and planetary electrodynamics response experiment", *J. Geophys. Res.*, **117**, A06308, doi:10.1029/2011JA017416.

Mayaud P.N., 1978. "The annual and daily variations of the *Dst* index", *Geophys. J. R. Astron. Soc. (UK)*, **55**, No. 1, 193-201.

McGranaghan R., D.J. Knipp, S.C. Solomon, and X. Fang, 2015. "A fast, parameterized model of upper atmospheric ionization rates, chemistry, and conductivity", *J. Geophys. Res. Space Physics*, **120**, No. 6, doi:10.1002/2015JA021146, 4936–4949.

Mikhailov A.V. and L. Perrone, 2011. "On the mechanism of seasonal and solar cycle N_mF_2 variations: A quantitative estimate of the main parameters contribution using incoherent scatter radar observations", *J. Geophys. Res.*, **116**, A03319, doi:10.1029/2010JA016122.

Mikhailov A.V. and L. Perrone, 2015. "The annual asymmetry in the F_2 layer during deep solar minimum (2008–2009): December anomaly", *J. Geophys. Res. Space Physics*, **120**, 1341–1354, doi:10.1002/2014JA020929.

Mikhailov A.V. and L. Perrone, 2016. "Geomagnetic control of the midlatitude daytime f_oF_1 and f_oF_2 long-term variations: Physical interpretation using European observations", *J. Geophys. Res. Space Physics*, **121**, doi:10.1002/2016JA022716.

Mukhtarov P., B. Andonov, and D. Pancheva, (2013). "Global empirical model of TEC response to geomagnetic activity", *J. Geophys. Res. Space Physics*, **118**, No. 10, doi:10.1002/jgra.50576, 6666-6685.

Nesse T.H., M.I. Sandanger, L.-K.G. Ødegaard, J. Stadsnes, A. Aasnes, and A.E. Zawedde, 2016. "Energetic electron precipitation into the middle atmosphere – Constructing the loss cone fluxes from MEPED POES", *J. Geophys. Res. Space Physics*, **121**, 5693–5707, doi:10.1002/2016JA022752.

Newell P.T., C.-I. Meng, and K.M. Lyons, 1996. "Suppression of discrete aurorae by sunlight", *Nature*, **381**, No. 6585, 766-767.

Newell P.T., S. Wing, T. Sotirelis, and C.-I. Meng, 2005. "Ion aurora and its seasonal variations", *J. Geophys. Res.*, **110**, No. A1, A01215, doi:10.1029/2004JA010743, 1-10.

Ohtani S., G. Ueno, T. Higuchi, and H. Kawano, 2005a. "Annual and semiannual variations of the location and intensity of large-scale field-aligned currents", *J. Geophys. Res.*, **110**, No. A1, A01216, doi:10.1029/2004JA010634, 1-15.

Ohtani S., G. Ueno, and T. Higuchi, 2005b. "Comparison of large-scale field-aligned currents under sunlit and dark ionospheric conditions", *J. Geophys. Res.*, **110**, No. A9, A09230, doi:10.1029/2005JA011057, 1-14.

Østgaard N., N.A. Tsyganenko, S.B. Mende, H.U. Frey, T.J. Immel, M. Fillingim, L.A. Frank, and J.B. Sigwarth, 2005. "Observations and model predictions of substorm auroral asymmetries in the conjugate hemispheres", *Geophys. Res. Lett.*, **32**, No. 5, L05111, doi:10.1029/2004GL022166, 1-4.

Palin L., C. Jacquey, H. Opgenoorth, M. Connors, V. Sergeev, J.-A. Sauvaud, R. Nakamura, G.D. Reeves, H.J. Singer, V. Angelopoulos, and L. Turc, 2015. "Three-dimensional current systems and ionospheric effects associated with small dipolarization fronts", *J. Geophys. Res. Space Physics,* **120**, No. 5, doi:10.1002/2015JA021040, 3739-3757.

Pallamraju D., D.K. Karan, and K.A. Phadke, 2016. "First three dimensional wave characteristics in the daytime upper atmosphere derived from ground-based multiwavelength oxygen dayglow emission measurements", *Geophys. Res. Lett.*, **43**, 5545–5553, doi:10.1002/2016GL069074.

Papitashvili V.O., F. Christiansen, and T. Neubert, 2002. "A new model of field-aligned currents derived from high-precision satellite magnetic field data", *Geophys. Res. Lett.*, **29**, No.14, 1683, doi:10.1029/2001GL014207, 28-1-4.

Park J., H. Lühr, G. Kervalishvili, J. Rauberg, I. Michaelis, C. Stolle, and Y.-S. Kwak, 2015b. "Nighttime magnetic field fluctuations in the topside ionosphere at midlatitudes and their relation to medium-scale traveling ionospheric disturbances: The spatial structure and scale sizes", *J. Geophys. Res. Space Physics*, **120**, No. 8, doi:10.1002/2015JA021315, 6818–6830.

Perrone L. and A.V. Mikhailov, 2016. "Geomagnetic control of the midlatitude f_oF_1 and f_oF_2 long-term variations: Recent observations in Europe", *J. Geophys. Res. Space Physics*, **121**, 7183– 7192, doi:10.1002/2016JA022715.

Petrukovich A.A., M.M. Mogilevsky, A.A. Chernyshov, and D.R. Shklyar, 2015. "Some aspects of magnetosphere-ionosphere relations", *Physics-Uspekhi*, **58**, No. 6, 606–611. Russian Text: *Uspekhi Fizicheskikh Nauk*, **185**, No. 6, 649–654.

Peymirat C. and D. Fontaine, 1994. "Numerical simulation of magnetospheric convection including the effect of field-aligned currents and electron precipitation", *J. Geophys. Res.*, **99**, No. A6, 11155-11176.

Pham Thi Thu H., C. Amory-Mazaudier, M. Le Huy, and A.G. Elias, 2016. "f_oF_2 long-term trend linked to Earth's magnetic field secular variation at a station under the northern crest of the equatorial ionization anomaly", *J. Geophys. Res. Space Physics*, **121**, No. 1, doi:10.1002/2015JA021890, 719–726.

Richmond A.D., 1973. "Equatorial electrojet – I. Development of a model including winds and instabilities", *J. Atmos. Terr. Phys.*, **35**, No. 6, 1083-1103.

Ridley A.J., 2007b. "Effect of seasonal changes in the ionospheric conductances on magnetospheric field-aligned currents", *Geophys. Res. Lett.*, **34**, No. 5, L05101, doi:10.1029/2006GL028444, 1-5.

Russell C.T. and R.L. McPherron, 1973. "Semiannual variations of geomagnetic activity", *J. Geophys. Res.*, **78**, No. 1, 92-108.

Russell A.J.B., T. Karlsson, and A.N. Wright, 2015. "Magnetospheric signatures of ionospheric density cavities observed by Cluster", *J. Geophys. Res. Space Physics*, **120**, No. 3, doi:10.1002/2014JA020937, 1876-1887.

Ruzhin Yu.Ya., M. Parrot, V.M. Smirnov, and V.Kh. Depuev, 2014. "The anomaly of plasmapause and ionospheric trough positions from DEMETER data", *Geomagnetism and Aeronomy*, **54**, No. 6, 763-772. Russian Text: *Geomagnetizm i Aeronomiya*, **54**, No. 6, 780-788.

Saini N.S., Y. Ghai, and R. Kohli, 2016. "Dust acoustic dromions in a magnetized dusty plasma with superthermal electrons and ions", *J. Geophys. Res. Space Phys.*, **121**, 5944–5958, doi:10.1002/2015JA022138.

Senior C. and M. Blanc, 1984. "On the control of magnetospheric convection by the spatial distribution of ionospheric conductivities", *J. Geophys. Res.*, **89**, No. A1, 261-284.

Shao X.-M. and Lay E.H., 2016. "The origin of infrasonic ionosphere oscillations over tropospheric thunderstorms", *J. Geophys. Res. Space Physics*, **121**, 6783–6798, doi:10.1002/ 2015JA022118.

Sharma R.P., N. Yadav, and N. Pathak, 2014. "Density cavity formation through nonlinear interaction of 3-D inertial Alfvén wave and ion acoustic wave", *J. Geophys. Res. Space Physics*, **119**, No. 12, doi:10.1002/2014JA020249, 10561-10568.

Sharma D.K., M. Aggarwal, and A. Bardhan, 2017. "Variability of ionospheric parameters during solar minimum and maximum activity and assessment of IRI model", *Advances in Space Research*, **60**, 435–443.

Scidá L.A., R.G. Ezquer, M.A. Cabrera, C. Jadur, and A.M. Sfer, 2016. "Tucumán ionospheric model (TIM): Initial results for STEC predictions", *Advances in Space Research*, **58,** 821–834.

Ssessanga N., Y.H. Kim, and E. Kim, 2015. "Vertical structure of medium-scale traveling ionospheric disturbances", *Geophys. Res. Lett.*, **42**, No. 21, doi:10.1002/2015GL066093, 9156–9165.

Stern D.P., 1970. "Euler Potentials", *Am. J. Phys. (USA)*, **38**, No. 4, 494-501.

Sydorenko D. and R. Rankin, (2013). "Simulation of O^+ upflows created by electron precipitation and Alfvén waves in the ionosphere", *J. Geophys. Res. Space Physics*, **118**, No. 9, doi:10.1002/jgra.50531, 5562-5578.

Sydorenko D., R. Rankin, and A.W. Yau, 2016. "Enhanced N_2 and O_2 densities inferred fromEISCAT observations of Pc5 waves and associated electron precipitation", *J. Geophys. Res. Space Physics*, **121**, No. 1, doi:10.1002/2015JA021508, 549–566.

Thomas N., G. Vichare, and A.K. Sinha, 2017. "Characteristics of equatorial electrojet derived from Swarm satellites", *Advances in Space Research*, **59**, 1526–1538.

Toffoletto F., S. Sazykin, R. Spiro, and R. Wolf, 2003. "Inner magnetospheric modeling with the Rice Convection Model", *Space Sci. Rev.*, **107**, No. 1-2, 175-196.

Vasyliūnas V.M., 1970. "Mathematical models of magnetospheric convection and its coupling to the ionosphere", in *Particles and Fields in the Magnetosphere*, edited by B.M. McCormac, Springer, New York, 60-71.

Venkatesh K., P.R. Fagundes, D.S.V.V.D. Prasad, C.M. Denardini, A.J. de Abreu, R. de Jesus, and M. Gende, 2015. "Day-to-day variability of equatorial electrojet and its role on the day-to-day characteristics of the equatorial ionization anomaly over the Indian and Brazilian sectors", *J. Geophys. Res. Space Physics*, **120**, No. 10, doi:10.1002/2015JA021307, 9117–9131.

Wagner C.-U., D. Möhlmann, K. Schäfer, V.M. Mishin, and M.I. Matveev, 1980. "Large-scale electric fields and currents and related geomagnetic variations in the quiet plasmasphere", *Space Sci. Rev.*, **26**, No. 4, 391-446.

Wang H., H. Lühr, and S.Y. Ma, 2005. "Solar zenith angle and merging electric field control of field-aligned currents: A statistical study of the Southern Hemisphere", *J. Geophys. Res.*, **110**, No. A3, A03306, doi:10.1029/2004JA010530, 1-15.

Wang S., S. Huang, and H. Fang, 2015. "Topside ionospheric Vary-Chap scale height retrieved from the COSMIC/FORMOSAT-3 data at midlatitudes", *Adv. Space Res.*, **56**, No. 5, 893–899.

Wang C., J.Y. Wang, R.E. Lopez, L.Q. Zhang, B.B. Tang, T.R. Sun, and H. Li, 2016. "Effects of the interplanetary magnetic field on the location of the open-closed field line boundary", *J. Geophys. Res. Space Physics*, **121**, 6341–6352, doi:10.1002/2016JA022784.

Wang J., J.B. Cao, H.S. Fu, W.L. Liu, and S. Lu, 2017a. "Enhancement of oxygen in the magnetic island associated with dipolarization fronts", *J. Geophys. Res. Space Physics*, **122**, 185–193, doi:10.1002/2016JA023019.

Wang G.J., J.K. Shi, Z. Wang, X. Wang, E. Romanova, K. Ratovsky, and N.M. Polekh, 2017b. "Solar cycle variation of ionospheric parameters over the low latitude station Hainan, China, during 2002–2012 and its comparison with IRI-2012 model", *Advances in Space Research*, **60**, 381–395.

Zherebtsov G.A., K.G. Ratovsky, M.V. Klimenko, V.V. Klimenko, A.V. Medvedev, S.S. Alsatkin, A.V. Oinats, and R.Yu. Lukianova, 2017. "Diurnal variations of the ionospheric electron density height profiles over Irkutsk: Comparison of the incoherent scatter radar measurements, GSM TIP simulations and IRI predictions", *Advances in Space Research*, **60**, 444–451.

Watanabe T.-H., 2014. "A unified model of auroral arc growth and electron acceleration in the magnetosphere-ionosphere coupling", *Geophys. Res. Lett.*, **41**, No. 17, doi:10.1002/2014GL061166, 6071-6077.

Watthanasangmechai K., M. Yamamoto, A. Saito, R. Tsunoda, T. Yokoyama, P. Supnithi, M. Ishii, and C. Yatini, 2016. "Predawn plasma bubble cluster observed in Southeast Asia", *J. Geophys. Res. Space Physics*, **121**, 5868–5879, doi:10.1002/2015JA022069.

Wolf R.A., 1983. "The quasi-static (slow-flow) region of the magnetosphere", in *Solar Terrestrial Physics*, edited by R.L. Carovillano and J.M. Forbes, Springer, New York, 303-368.

Wolf R.A., R.W. Spiro, S. Sazykin, F.R. Toffoletto, P. Le Sager, and Huang T.S., 2006. "Use of Euler potentials for describing magnetosphere-ionosphere coupling", *J. Geophys. Res.*, **111**, No. A7, A07315, doi:10.1029/2005JA011558, 1-7.

Wright A.N. and A.J.B. Russell, 2014. "Alfvén wave bounary condition for responsive magnetosphere-ionosphere coupling", *J. Geophys. Res. Space Physics*, **119**, No. 5, doi:10.1002/2014JA019763, 3996-4009.

Wu Q., A.M. Du, M. Volwerk, and G.Q. Wang, 2016. "The distribution of spectral index of magnetic field and ion velocity in Pi2 frequency band in BBFs: THEMIS statistics", *Advances in Space Research*, **58**, 847–855.

Yamazaki Y., K. Häusler, and J.A. Wild, 2016. "Day-to-day variability of midlatitude ionospheric currents due to magnetospheric and lower atmospheric forcing", *J. Geophys. Res. Space Physics*, **121**, 7067–7086, doi:10.1002/2016JA022817.

Yang J., F. Toffoletto, G. Lu, and M. Wiltberger, 2014. "RCM-E and AMIE studies of the Harang reversal formation during a steady magnetospheric convection event", *J. Geophys. Res. Space Physics*, **119**, No. 9, doi:10.1002/2014JA020207, 7228-7242.

Yizengaw E., M.B. Moldwin, E. Zesta, M. Magoun, R. Pradipta, C.M. Biouele, A.B. Rabiu, O.K. Obrou, Z. Bamba, and E.R. de Paula, 2016. "Response of the equatorial ionosphere to the geomagnetic DP 2 current system", *Geophys. Res. Lett.*, **43**, 7364–7372, doi:10.1002/2016GL070090.

Yokoyama T., H. Shinagawa, and H. Jin, 2014. "Nonlinear growth, bifurcation, and pinching of equatorial plasma bubble simulated by three-dimensional high-resolution bubble model", *J. Geophys. Res. Space Physics*, **119**, No. 12, doi:10.1002/2014JA020708, 10474-10482.

Yoshikawa A., H. Nakata, A. Nakamizo, T. Uozumi, M. Itonaga, S. Fujita, K. Yumoto, and T. Tanaka, 2010. "Alfvénic coupling algorithm for global and dynamical magnetosphere-ionosphere coupled system", *J. Geophys. Res.*, **115**, A04211, doi:10.1029/2009JA014924.

Yu Y. and A.J. Ridley, (2013). "Exploring the influence of ionospheric O+ outflow on magnetospheric dynamics: The effect of outflow intensity", *J. Geophys. Res. Space Physics*, **118**, No. 9, doi:10.1002/jgra.50528, 5522-5531.

Zhang X.-X., F. He, W. Wang, and B. Chen, 2015a. "Hemispheric asymmetry of subauroral ion drifts: Statistical results", *J. Geophys. Res. Space Physics*, **120**, No. 6, doi:10.1002/2015JA021016, 4544–4554.

Zhang B.-C., S.-G. Yang, S. Xu, R.-Y. Liu, I. Häggström, Q.-H. Zhang, Z.-J. Hu, D.-H. Huang, and H.-Q. Hu, 2015b. "Diurnal variation of winter F region ionosphere for solar minimum at both Zhongshan Station, Antarctica, and Svalbard Station, Arctic", *J. Geophys. Res. Space Physics*, **120**, No. 11, doi:10.1002/2015JA021465, 9929–9942.

Zhang R., L. Liu, H. Le, and Y. Chen, 2016. "Evidence and effects of the sunrise enhancement of the equatorial vertical plasma drift in the *F* region ionosphere", *J. Geophys. Res. Space Physics*, **121**, 4826–4834, doi:10.1002/2016JA022491.

REFERENCES FOR MONOGRAPHS AND BOOKS

Afanasieva V.I. (Ed.), M1954. *Manual of Variable Magnetic Field of USSR*, Gidrometeoizdat, Leningrad (in Russian).

Akasofu S.I., M1968. *Polar and Magnetospheric Substroms*, D. Reidel Publ. Co., Dordrecht, Holland.

Akasofu S.I., M1977. *Physics of Magnetospheric Substorms*, D. Reidel Publ. Co., Dordrecht-Boston.

Akasofu S.I. and S. Chapman, M1972. *Solar-Terrestrial Physics*, Clarendon Press, Oxford. In Russian: M1974, Mir, Moscow.

Akasofu S.I. and J.R. Kan (Eds.), M1981. *Physics of Auroral Arc Formation*, Geophys. Monogr. Ser., **25**, AGU, Washington, D.C.

Akhiezer I.A., R.V. Polovin, A.G. Sitenko, and K.N. Stepanov, M1974. *Plasma Electrodynamics*, Nauka, Moscow (in Russian).

Alania M.V. and L.I. Dorman, M1981. *Cosmic Ray Distribution in the Interplanetary Space,* METSNIEREBA, Tbilisi (in Russian).

Alania M.V., L.I. Dorman, R.G. Aslamazashvili, R.T. Gushchina, and T.V. Dzhapiashvili, M1987. *Galactic Cosmic Ray Modulation by Solar Wind*, Metsniereba, Tbilisi (in Russian).

Alfvén H., M1950. *Cosmical Electrodynamics*, Oxford.

Alfvén H., M1981. *Cosmic Plasma*, Springer, New York.

Alfvén H. and C.-G. Fälthammar, M1963. *Cosmical Electrodynamics*, Oxford.

Armstrong T.P. and R. Sahi, M1996. *HISCALE Data Analysis Handbook*, Dep. of Phys. and Astron., Univ. of Kansas, Lawrence.

Artsymovich L.A and R.Z. Sagdeev, M1979. *Plasma Physics for Physicists*, Atomizdat, Moscow (in Russian).

Aschwanden M., M2011. *Self-organized Criticality in Astrophysics,* Springer, Berlin.

Bagenal F., T. Dowling, and W. Mckinnon (Eds.), M2004. *Jupiter: The Planet, Satellites and Magnetosphere*, Cambridge Univ. Press, New York, M2004.

Bak P., M1996. *How Nature Works,* Copernicus, New York.

Baranov V.B. and K.V. Krasnobaev, M1977. *Hydromagnetic Theory of Cosmic Plasma*, Nauka, Moscow (in Russian).

Bateman H. and A. Erdelyi, M1953. *Higher Transcendental Functions*, McGraw-Hill Book, Co. Inc. N.Y.-Toronto-London (In Russian: Inostrannaja Literatura, Moscow, Vol. 1 – M1965, Vol. 2 – M1966).

Bateman H. and A. Erdelyi, M1954. *Tables of Integral Transforms*, McGraw-Hill, Maidenhead, U.K..

Battrick B. (Ed.), M1979. *Magnetospheric Boundary Layers*, Eur. Space Agency, Paris, ESA SP-148.

Battrick et al. (Eds.), M2001. *Solar Encounter: Proceedings of the First Solar Orbiter Workshop*, Eur. Space Agency, ESA-493.

Berezhko E.G., V.K. Elshin, G.F. Krymsky, and G.F. Petukhov, M1988. *Cosmic Ray Generation by Shock Waves*, Nauka, Novosibirsk (in Russian).

Berezin I.S. and I.P. Zhidkov, M1959. *Methods of Computations*, Physmatgiz, Moscow (in Russian).

Berezinsky V.S., S.V. Bulanov, V.L. Ginzburg, V.A. Dogiel, and V.S. Ptuskin (ed. V.L. Ginzburg), M1990. *Astrophysics of Cosmic Rays*, Physmatgiz, Moscow (in Russian). English translation: North-Holland, Amsterdam, M1990.

Bleeker J.A.M., J. Geiss, and M.C.E. Huber (Eds.), M2002. *The Century of Space Science*, Kluwer Academic Publishers, Dordrecht.

Boguslavsky S.A., M1929. *Paths of Electrons in the Electromagnetic Fields*, ONTI, Moscow-Leningrad (in Russian).

Bohr H., M1950. *Passage of Atomic Particles Through Matter*. Inostr. Literatura, Moscow (in Russian).

Bonch-Bruevich V.L. and S.V. Tyablikov, M1961. *Method of Green's Functions in Statistical Mechanics*, Fizmatgiz, Moscow (in Russian).

Buchner J., C.T. Dum, and M. Scholer (Eds.), M2003. *Space Plasma Simulations*, Lecture Notes in Physics, **615,** Springer-Verlag, Berlin/Heidelberg, M2003.

Burch J.L. (Ed.), M2000. *The IMAGE Mission*, Kluwer Acad., Norwell.

Burch J.L. and J.H. Waite, Jr. (Eds.), M1994. *Solar System Plasmas in Space and Time*, Geophys. Monogr. Ser., **84**, AGU, Washington D.C.

Burch J.L., R.L. Carovillano, and S.K. Antiochos (Eds.), M1998. *Sun-Earth Plasma Connections*, Geophys. Monogr. Ser., **109**, AGU, Washington, D.C.

Burch J., M. Schulz, and H. Spence (Eds.), M2005. *Inner Magnetosphere Interactions: New Perspectives From Imaging*, Geophys. Monogr. Ser., **159**, AGU, Washington, D.C.

Burke W. and T.-D. Guyenne (Eds.), M1996. *Environmental Modelling for Space-based applications*, ESA Symposium Proceedings, ESA SP-392, ESA Publ. Div., Noordwijk.

Buti B. (Ed.), M1985. *Advances in Space Plasma Physics*, World Sci., Hackensack, N.J.

Butler R.F., M1992. *Paleomagnetism: Magnetic Domains to Geologic Terraines*, Blackwell Scientific Publications, Oxford.

Case K. and P. Zweifel, M1967. *Linear Theory of Transport*, Addison-Wesley Publ. Co., Massachusetts.

Chamberlain J.W., M1961. *Physics of the Aurora and Airflow*, Academic Press, New York and London.

Chandrasekhar S., M1960. *Plasma Physics*, Chicago Univ. Press, Chicago.

Chang T. and J.R. Jasperse (Eds.), M1996. *Physics of Space Plasmas*, **14**, MIT Centre for Theor. Geo/Cosmo Plasma Phys., Cambridge, MA.

Chang T., M.K. Hudson, J.R. Jasperse, R.G. Johnson, P.M. Kintner and M. Schulz (Eds.), M1986. *Ion Acceleration in the Magnetosphere and Ionosphere*, Geophys. Monogr. Ser., **38**, AGU, Washington, D.C.

Chang T., G.B. Crew, and J.R. Jasperse (Eds.), M1988. *Physics of Space Plasmas*, SPI Conf. Proc. and Reprint Ser., No. 8, Scientific Publishers, Inc., Cambridge, MA.

Channell J.E.T., D.V. Kent, W. Lowrie, and J.G. Meert (Eds.), M2004. *Timescales of the Paleomagnetic Field*, Geophysical Monograph Series, Volume 145, pp 328.

Chapman S. and J. Bartels, M1962. *Geomagnetism*, Clarendon Press, Oxford.

Corovillano R.L., Z.F. McClay, and H.F. Padoski (Eds.), M1968. *Physics of the Magnetosphere*, D. Reidel Publ. Co., Dordrecht.

Daglis I.A. (ed.), M2001. *Space Storms and Space Weather Hazards*, Proc. NATO Advanced Study Institute, 19-29 June 2000, Hersonissos, Crete, NATO Science Series, Vol. 38, Kluwer Academic Publishers, Dordrecht.

Demidovich B.P., I.A. Maron, and É.Z. Shuvalova, M1962: *Numerical Methods of Analysis*, Physmatgiz, Moscow (in Russian).

Dessler A.J. (Ed.), M1983. *Physics of the Jovian Magnetosphere*, Cambridge Univ. Press, New York.

DeWitt C., J. Hiebolt, and A. Lebeau (Eds.), M1963. *Geophysics: The Earth's Environment*, Proceedings of the 1962 Les Houches Summer School, Gordon and Breach, New York.

Dorman L.I., M1957. *Cosmic Ray Variations*, Gostekhteorizdat, Moscow (in Russian). English translation: US Department of Defense, Ohio Air-Force Base, M1958.

Dorman Lev I., M1963a. *Geophysical and Astrophysical Aspects of Cosmic Rays,* North-Holland, Amsterdam, in series 'Progress in Physics of Cosmic Ray and Elementary Particles', Eds. J.G. Wilson and S.A. Wouthuysen, Vol. 7, pp. 1-324.

Dorman Lev I., M1963b. *Cosmic Ray Variations and Space Research,* NAUKa, Moscow (in Russian).

Dorman Lev I., M1972a. *Acceleration Processes in Space*, VINITI, Moscow (in Russian).

Dorman Lev I., M1972b. *Meteorological Effects of Cosmic Rays*, NAUKA, Moscow (in Russian); In English: NASA, Washington DC, M1973.

Dorman Lev I., M1974. *Cosmic Rays: Variations and Space Exploration,* North-Holland, Amsterdam.

Dorman Lev I., M1975a. *Experimental and Theoretical Principles of Cosmic Ray Astrophysics,* PHYSMATGIZ, Moscow (in Russian).

Dorman Lev I., M1975b. *Variations of Galactic Cosmic Rays,* Moscow State University Press, Moscow (in Russian).

Dorman Lev I., M1978. *Cosmic Rays of Solar Origin*, VINITI, Moscow (in series "Summary of Science", Space Investigations, Vol.12), in Russian.

Dorman Lev I., M2004. *Cosmic Rays in the Earth's Atmosphere and Underground,* Kluwer Academic Publishers, Dordrecht/Boston/London.

Dorman Lev I., M2006. *Cosmic Ray Interactions, Propagation, and Acceleration in Space Plasmas*, Springer, Dordrecht.

Dorman Lev I., M2009. *Cosmic Rays in Magnetospheres of the Earth and other Planets,* Springer, Heidelberg.

Dorman Lev I., M2010. *Solar Neutrons and Related Phenomena,* Springer, Heidelberg.

Dorman Lev I., M2017a. *Plasmas and Energetic Processes in the Geomagnetosphere. Volume I. Internal and Space Sources, Structure, and Main Properties of Geomagnetosphere*, Nova Science Publishers, New York.

Dorman Lev I., M2017b. *Plasmas and Energetic Processes in the Geomagnetosphere. Volume II. Magnetospheric Sheets, Reconnections, Particle Acceleration, and Substorms,* Nova Science Publishers, New York.

Dorman Lev I. and Irina V. Dorman, M2014. *Cosmic Ray History*, Nova Science Publishers, New York.

Dorman L.I. and E.V. Kolomeets, M1968. *Solar Diurnal and Semidiurnal Cosmic Ray Variations from Maxima to Minima of Solar Activity.* Nauka, Moscow (in Russian); In English: NASA, Washington, M1972.

Dorman L.I. and I.D. Kozin, M1983. *Cosmic Radiation in the Upper Atmosphere.* PHYSMATGIZ, Moscow (in Russian).

Dorman L.I. and L.I. Miroshnichenko, M1968. *Solar Cosmic Rays*, PHYSMATGIZ, Moscow (in Russian). In English: NASA, Washington DC, M1976.

Dorman L.I., V.S. Smirnov and M.I. Tyasto, M1971. *Cosmic Rays in the Earth's Magnetic Field,* PHYSMATGIZ, Moscow (in Russian); Translation in English: NASA, Washington DC, M1973.

Dorman L.I., R.T. Gushchina, D.F. Smart, and M.A. Shea, M1972. *Effective Cut-Off Rigidities of Cosmic Rays,* NAUKA, Moscow (in Russian and in English).

Dorman L.I., I.A. Pimenov and V.S. Satsuk, M1978. *Mathematical Service of Geophysical Investigations on the Example of Cosmic Ray Variations,* Nauka, Moscow (in Russian).

Dorman L.I., I.Ya. Libin, and Ya.L. Blokh, M1979. *Scintillation Method of Cosmic Ray Investigations,* Nauka, Moscow (in Russian).

Dungey J.W., M1954. *Electrodynamics of the Outer Atmospheres*, vol. 69, Ionos. Sci. Rep., Cambridge, Pa.

Dyer E.R. (Ed.), M1972. *Critical Problems of Magnetospheric Physics*, Proceedings of the Symposium, 11-13 May, 1972 in Madrid, Spain, Washington, D.C., National Acad. of Sci.

Egeland A., O. Holter, and A. Omholt (Eds.), M1973. *Cosmical Geophysics*, Universitetsforlaget, Oslo, Norway.

Engebretson M.J., K. Takahashi and M. Scholer (Eds.), M1994. *Solar Wind Sources of Magnetospheric Ultra- Low-Frequency Waves*, Geophys. Monogr. Ser., **81**, AGU, Washington D.C.

Foing B. and B. Battrick (Eds.), M2002. *Earth-Like Planets and Moons*, Proceedings of the 36th ESLAB Symposium, ESA SP-514.

Formisano V. (Ed.), M1975. *Magnetospheres Earth and Jupiter*, Proc. Neil Brice Memor. Symp., Frascati, Italy, 1974, Dordrecht-Boston, D. Reidel Publishing Co..

Friedberg J.P., M1987. *Ideal Magnetohydrodynamics*, Springer, New York.

Fritz T.A. and S.F. Fung (Eds.), M2006. *The Magnetospheric Cusps: Structure and Dynamics,* Netherlands, Springer.

Gehrels T. (Ed.), M1976. *Jupiter: Studies of the Interior, Atmosphere, Magnetosphere, and Satellites*, Proceedings of the Colloquium, Tucson, Ariz., 1975, University of Arizona Press, Tucson, Arizona.

Gehrels T. and M.S. Matthews (Eds.), M1984. *Saturn*, Univ. of Ariz. Press, Tucson.

Ginzburg V.L. and A.A. Rukhadze, M1970. *Waves in Magneto-Active Plasma*, NAUKA, Moscow (in Russian).

Gombosi T.I., M1998. *Physics of Space Environment*, Cambridge University Press, Cambridge.

Hasegawa A., M1975. *Plasma Instabilities and Non-Linear Effects*, Springer, New York.

Hatton C.J. and D.A. Carswell, M1963. *Asymptotic Directions of Approach of Vertically Incident Cosmic Rays for 85 Neutron Monitor Stations*, Report No. CRGR-1165, Atomic Energy of Canada Ltd.

Hayakawa M. and Y. Fujinawa (Eds.), M1994. *Electromagnetic Phenomena Related to Earth-quake Prediction*. TERRAPUB, Tokyo.

Hayakawa M. and O.A. Molchanov (Eds.), M2002. *Seismo Electromagnetics Lithosphere-Atmosphere-Ionosphere Coupling*. TERRAPUB, Tokyo.

Helliwell R.A., M1965. *Whistlers and Related Ionospheric Phenomena*, Stanford Univ. Press, Stanford, Calif.

Hess V.N., Ml972. *Radiation Belts and the Magnetosphere*, Gosatomizdat, Moscow (in Russian).

Hoshino M., Y. Omura and L.J. Lanzerotti, (Eds.), M2005. *Frontiers in Magnetospheric Plasma Physics*, COSPAR Colloq. Ser., **16**, Elsevier Ltd., Oxford, UK, M2005.

Hultqvist B., M. Øieroset, G. Paschmann, and R. Treumann (Eds.), M1999. *Magnetospheric Plasma Sources and Losses*, Springer, New York.

Hultqvist B. and L. Stenflo (Eds.), M1975. *Physics of the hot plasma in the magnetosphere*; Proceedings of the Thirtieth Nobel Symposium, Kiruna, Sweden, April 2-4, 1975, Plenum Press, New York.

Hundhausen R.I., M1972. *Coronal Expansion and Solar Wind*, Springer-Verlag, Berlin-Heidelberg-New York.

Isaev S.I. and M.I. Pudovkin, M1972. *Auroras and Magnetospheric Processes*, NAUKA, Leningrad (in Russian).

Jackson J.D., M1975. *Classical Electrodynamics*, John Wiley, Hoboken, N.J.

Jeffrey A. and T. Taniuti, M1964. *Non-linear Wave Propagation*, Elsevier, New York.

Jones A.V., M1974. *Aurora*, D. Reidel Publ. Co., Dordrecht.

Johnson R.E., M1990. *Energetic Charged Particle Interactions with Atmospheres and Surfaces*, Springer-Verlag, New York.

Kadomtsev B.B., M1988. *Collective Phenomena in Plasma*, NAUKA, Moscow (in Russian).

Kamide Y. and W. Baumjohann, M1993. *Magnetosphere–Ionosphere Coupling*, Springer-Verlag.

Kamide Y. and J.A. Slavin (Eds.), M1986. *Solar Wind–Magnetosphere Coupling*, Terra Sci., Tokyo.

Kantrowitz A. and H. Petschek, M1966. *Plasma Physics in Theory and Application*, McGraw-Hill, New York.

Kelley M.C. and Heelis R.A., M1989. *The Earth's Ionosphere: Plasma Physics and Electrodynamics*, Academic Press, San Diego.

Kennel C.F., M1995. *Convection and Substorms* (*Paradigms of Magnetospheric Phenomenology*), Oxford University Press, New York.

Khorosheva O.V., M1967. *Spatial and Temporal Distribution of Auroras*, Moscow, Nauka. In Russian.

Kikuchi H. (Ed.), M1994. *Dusty and Dirty Plasmas, Noise, and Chaos in Space and in the Laboratory*, Plenum Press, New York.

Kivelson M.G. and C.T. Russell (Eds.), M1995. *Introduction to Space Physics*, Cambridge Univ. Press, New York.

Klimontovich Yu.L., M1990. *Turbulent Motion and the Structure of Chaos. The New Approach to the Statistical Theory of Open Systems*, Kluwer Academic Publishers, Dordrecht.

Klimontovich Yu.L., M1999. *Statistical Theory of Open Systems*, vol. 1, 2, Yanus-K, Moscow (in Russian).

Klyatskin V.I., M1975. *Statistical Description of Dynamical Systems with Fluctuating Parameters*, NAUKA, Moscow (in Russian).

Knott K. and B. Battrick (Eds.), M1976. *Scientific Satellite Program during the International Magnetospheric Study*, D. Reidel Publ. Co., Dordrecht.

Kokubun S. and Y. Kamide (Eds.), M1998. *Proceedings of the International Conference on Substorms-4*, Astrophysics and Space Science Library, **238**, Terra Sci. Pub. Co., Kluwer Academic Publishers.

Komarov I.V., L.I. Ponomarev, and S.Yu. Slavyanov, M1976. *Spherical and Colon Spherical Functions*, NAUKA, Moscow (in Russian).

Krall N.A. and A.W. Trivelpiece, M1973. *Principles of Plasma Physics*, McGraw-Hill, New York.

LaBelle J.W. and R.A. Treumann (Eds.), M2006. *Geospace Electromagnetic Waves and Radiation*, Lect. Not. in Phys., **687**, Springer, Berlin, New York,

Landau L.D. and E.M. Lifshitz, M1957. *Electrodynamics of Continuous Matter*, GOSTEKHIZDAT, Moscow (in Russian).

Lans D.H., M1962. *Numerical Methods for Computers*, Foreign Literature Press (IL), Moscow (in Russian).

Lemaire J., M1985. *Frontiers of the Plasmasphere (Theoretical Aspects)*, Université Catholique de Louvain, Faculté des Sciences, Louvain-la-Neuve, ISBN 2-87077-310-2

Lemaire J.F. and K.I. Gringauz, M1998. *The Earth's Plasmasphere*, Cambridge Univ. Press, New York.

Lemaire J.F., D. Heynderickx, and D.N. Baker (Eds.), M1997. *Radiation Belts: Models and Standards*, AGU, Geophys. Monogr, **97**.

Leroy C. and P.G. Rancoita, M2004. *Principles of Radiation Interaction in Matter and Detection*, World Sci., Hackensack, N.J.

Longmair K., M1966. *Plasma Physics*, Atomizdat, Moscow (in Russian).

Lui A. (Ed.), M1987. *Magnetotail Physics*, Johns Hopkins Univ. Press, Baltimore, Md.

Lyatsky V.B., M1978. *Current Systems of Magnetospheric-Ionospheric Disturbances*, Nauka, Moscow (in Russian).

Lyons L.R. and D.J. Williams, M1984. *Quantitative Aspects of Magnetospheric Physics*, Springer, New York.

Lyons L. and D. Willer, M1987. *Magnetospheric Physics*. Mir, Moscow (in Russian).

Maltsev Yu.P. (Ed.), M1993. *Magnetospheric–Ionospheric Physics*: A Handbook, Nauka, St. Petersburg (in Russian).

Mandelbrot B.B., M2002, *Gaussian Self-Affinity and Fractals: Globality, the Earth, 1/f Noise and R/S*, Springer-Verlag, NY.

Mathews J. and R.L. Walker, M1971. *Mathematical Methods of Physics*, Addison Wesley, Boston, Mass.

Matsumo H. and Y. Omura (Eds.), M1993. *Computer Space Plasma Physics: Simulation Techniques and Software*, Terra Sci., Tokyo.

Mayaud P.N. (Ed.), M1980. *Derivation, Meaning and Use of Geomagnetic Indices*, Geophys. Monogr. Ser., vol. 22, AGU, Washington, D.C.

McCormac B.M. (Ed.), M1966. *Radiation Trapped in the Earth's Magnetic Field*, Astrophysics and Space Science Library, **5**, Springer, New York.

McCormac B.M. (Ed.), M1968. *Earth's Particles and Fields*, Springer, New York.

McCormac B.M. (Ed.), M1970. *Particles and Fields in the Magnetosphere*, D. Reidel Publ. Co., Dordrecht.

McCormac B.M., (Ed.), M1972. *Earth's Magnetospheric Processes*, Proc. Symp. Cortina (1971), D. Reidel Publ. Co., Dordrecht.

McCormac B.M. (Ed.), M1974. *Magnetospheric Physics*, D. Reidel Publ. Co., Hongham, Mass.

McCormac B.M. (Ed.), M1976. *Magnetospheric particles and fields*, Astrophysics and Space Science Library, **58**, Dordrecht, D. Reidel Publishing Co.

McCormack B.M. and A. Renzini (Eds.), M1970. *Particles and Fields in the Magnetosphere,* Proc. Symposium, Santa Barbara, Calif., 1969, Astrophysics and Space Science Library, **17**, Springer-Verlag New York Inc.

McElhinny M.W. and P.L. McFadden, M2000. *Paleomagnetism*, Academic Press, New York.

McFadden L.-A., P. Weissman, and T. Johnson (Eds.), M2007: *Encyclopedia of the Solar System* (Second Edition), Elsevier.

Melrose D.B., M1980. *Plasma Astrophysics*, Gordon and Breach, Newark, N.J.

Melrose D.B., M1986. *Instabilities in Space and Laboratory Plasmas*, Cambridge Univ. Press, Cambridge.

Meng C.-I., M.J. Rycroft, and L.A. Frank (Eds.), M1991. *Auroral Physics*, Cambridge Univ. Press, New York.

Merrill R.T. and M.W. McElhinny, M1983. *The Earth's Magnetic Field*, Academic Press, London.

Merrill R.T., M.W. McElhinny, and P.L. McFadden, M1997. *The Magnetic Field of the Earth: Paleomagnetism, the Core and the Deep Mantle*, Academic Press, San Diego, CA.

Miroshnichenko L.I., M2001. *Solar Cosmic Rays*, Kluwer Acad. Publishers, Dordrecht.

Miroshnichenko L.I. and V.M. Petrov, M1985. *Dynamics of Radiation Conditions in Space*, Energoatomizdat, Moscow (in Russian).

Mitra S.K., M1955. *The Upper Atmosphere*, Foreighn Literature, Moscow (in Russian).

Moffatt H.K., M1978. *Magnetic Field Generation in Electrically Conducting Fluids*, Cambridge Univ. Press, New York.

Monin A.S. and A.M. Yaglom, M1965. *Statistical Hydromechanics*, Vol. 1, Nauka, Moscow (in Russian).

Monin A.S. and A.M. Yaglom, M1967. *Statistical Hydromechanics*, Vol. 2, Nauka, Moscow (in Russian).

Moon P. and D.E. Spencer, M1971. *Field Theory Handbook*, 2nd ed., Springer, New York.

Morse P.M. and H. Feshbach, M1953. *Methods of Theoretical Physics*, Vol. 1, 2, McGraw-Hill Book Co Inc., N.Y.-Toronto-London. In Russian: Inostrannaja Literatura, Moscow, M1958.

Mysovskikh I.P., M1962. *Lectures on Computation Methods*, Moscow, FIZMATGIZ (in Russian).

Nishida A., M1978. *Geomagnetic Diagnosis of the Magnetosphere*, Series Physics and Chemistry in Space, Springer, New York.

Nishida A., D.N. Baker, and S.W.H. Cowley (Eds.), M1998. *New Perspectives on the Earth's Magnetotail*, Geophys. Monogr. Ser., **105**, AGU, Washington, D.C.

Ohtani S.-I., R. Fujii, M. Hesse and R.L. Lysak (Eds.), M2000. *Magnetospheric Current Systems*, Geophys. Monogr. Ser., **118**, AGU, Washington, D.C.

Øksendal B., M1998. *Stochastic Differential Equations*, 5th edition, Springer, Berlin, Germany.

Panasyuk M.I. (Ed.), M2007. *Model of Cosmos,* Vol. 1, 8th edition, Moscow (in Russian).

Pandalai (Ed.), M2000. *Recent Research Developments in Plasmas*, Transworld Res. Network, Trivandrum, India, **1**.

Parker E.N., M1963. *Interplanetary Dynamical Processes*, John Wiley and Sons, New York-London. In Russian (transl. by L.I. Miroshnichenko, Ed. L.I. Dorman): Inostrannaja Literatura, Moscow, M1965.

Parker E.N., M2007. *Conversations on Electric and Magnetic Fields in the Cosmos*, Princeton Univ. Press, Princeton, NJ.

Parks G., M1991. *Physics of Space Plasmas*, Addison-Wesley-Longman, Reading, Mass.

Paschmann G. and P.W. Daly (Eds.), M1998. *Analysis Methods for Multi-Spacecraft Data*, ISSI Sci. Rep. SR-001, Eur. Space Agency, Paris.

Paschmann G. and P.W. Daly (Eds.), M2000. *Analysis Methods for Multi-Spacecraft Data*, ISSI Sci. Rep. SR-001, Springer, New York.

Petviashvili V.I. and O.A. Pokhotelov, M1992. *Solitary Waves in Plasmas and in the Atmosphere*, Gordon and Breach, New York.

Piddington J.H., M1969. *Cosmic Electrodynamics*, Wiley Interscience, New York.

Pikelner S.B., M1961. *Principles of Cosmic Electrodynamics*, PHYSMATGIZ, Moscow (in Russian).

Pikelner S.B., M1966. *Principles of Cosmic Electrodynamics*, PHYSMATGIZ, Moscow (in Russian).

Ponomarev E.A., M1985. *Magnetospheric Substorm Mechanism*, NAUKA, Moscow. In Russian.

Potemra T.A. (Ed.), M1983. *Magnetospheric Currents*, Geophys. Monogr. Ser., **28**, AGU, Washington, D.C.

Press W.H., B.P. Flannery, S.A. Teulolsky, and W.T. Wetterling, M1999. *Numerical Recipes in C*, Cambridge Univ. Press, New York.

Pudovkin M.I. and V.S. Semenov, M1985. *The Reconnection Theory and Interaction of the Solar Wind with the Earth's Magnetosphere*, in series "Results of Researches on the International Geophysical Projects", NAUKA, Moscow (in Russian).

Pudovkin M.I. and A.D. Shevnin (Eds.), M1978. *Geomagnetic Activity and its Prediction*, Nauka, Moscow.

Pudovkin M.I., Koselov V.P., Lazutin L.L., Troshichev O.A., and Chertkov A.D., M1977, *The physical base for the prediction of the magnetospheric disturbances*. Nauka, Leningrad (in Russian).

Pulinets S.A. and Boyarchuk K.A., M2004. *Ionospheric Precursors of Earthquakes*. Springer, Berlin, Germany.

Pulkkinen T.I., N.A. Tsyganenko, and R.H.W. Friedel, (Eds.), M2005. *The Inner Magnetosphere: Physics and Modeling*, Geophys. Monogr. Ser., **155**, AGU, Washington, D.C.

Reeves G.D. (Ed.), M1996. *Workshop on the Earth's Trapped Particle Environment*, AIP, Woodbury, New York.

Roederer J.G., M1970. *Dynamics of Geomagnetically Trapped Radiation*, Springer-Verlag, New York.

Rolfe E.J. and B. Kaldeich (Eds.), M1996. *International Conference on Substorms*, Proc. of the 3rd Intern. Conf., Versailles, ICS-3, ESA SP-389, Paris, European Space Agency.

Rucker H.O., S.J. Bauer, and B.M. Pedersen (Eds.), M1988. *Planetary Radio Emissions II*, Austrian Acad. of Sci. Press, Vienna.

Rucker H.O., S.J. Bauer, and B.M. Pedersen (Eds.), M1992. *Planetary Radio Emissions III*, Austrian Acad. of Sci. Press, Vienna.

Rucker H.O., S.J. Bauer, and A. Lecacheux (Eds.), M1997. *Planetary Radio Emissions IV*, Aust. Acad. of Sci. Press, Vienna.

Rucker H.O., M.L. Kaiser, and Y. Leblanc (Eds.), M2001. *Planetary Radio Emissions V*, Aust. Acad. of Sci. Press, Vienna.

Russell C.T., E.R. Priest, and L.C. Lee (Eds.), M1990. *Physics of Magnetic Flux Ropes*, Geophys. Monogr. Ser., **58**, AGU, Washington, DC.

Rytov S.M., Yu.A. Kravtsov, and V.I. Tatarsky, M1977. *Introduction to Statistical Radiophysics*, Nauka, Moscow (in Russian).

Ryzhik I.M. and S.M. Gradstein, M1971. *Tables of Integrals, Sums, Series and Products*, Nauka, Moscow (in Russian).

Sagdeev R.Z. and M.N. Rozenblud (Eds.), M1984. *Base of Plasma Physics, Vol. 2*, Energoatomizdat, Moscow (in Russian).

Sagdeev R.Z., D.A. Usikov, and G.M. Zaslavsky, M1998. *Nonlinear Physics*, Harwood Academic Publishers, London.

Sandholt P.E., Carlson H.E., and Egeland A., M2002. *Dayside and Polar Cap Aurora*. Kluwer Academic Publishers, Dordrecht.

Sanz-Serna J.M. and M.P. Calvo, M1994. *Numerical Hamiltonian Problems*, Chapman & Hall, London.

Sauvaud J.-A. and Z. Němeček (Eds.), M2005. *Multiscale Processes in the Earth's Magnetosphere: From Interball to Cluster*, NATO Sci. Ser., **178**, Springer, New York.

Schindler K., M2007. *Physics of Space Plasma Activity*, Cambridge Univ. Press, New York.

Schulz M. and L. Lanzerotti, M1974. *Particle Diffusion in the Radiation Belts*, Springer, New York.

Schunk R.W. (Ed.), M1996. *Handbook of Ionospheric Models*, Scientific Committee on Solar-Terrestrial Physics.

Schunk R.W. and A.F. Nagy, M2000. *Ionospheres*, Cambridge Univ. Press, New York.

Sellers W.D., M1965. *Physical Climatology*, University of Chicago Press, Chicago,IL, USA, ISBN-10: 0226746992/ISBN-13: 978-0226746999.

Sergeev V.A. and N.A. Tsyganenko, M1980. *The Earth's Magnetosphere*, Nauka, Moscow (in Russian).

Shabansky V.P., M1972. *Electrodynamical Phenomena in the Earth's Environment*, Nauka, Moscow (in Russian).

Sharma A.S., Y. Kamide, and G.S. Lakhina (Eds.), M2004. *Disturbances in Geospace: The Storm-Substorm Relationship*, Geophys. Monogr. Ser., **142**, AGU, Washington, DC.

Shkarofsky I., T. Johnston, and M. Bachynsky, M1966. *The Particle Kinetics of Plasmas*, Addison-Wesley Publ. Co., Massachusetts.

Song P., B.U.O. Sonnerup, and M.F. Thomsen (Eds.), M1995. *Physics of the Magnetopause*, Geophys. Monogr. Ser., **90**, AGU, Washington, DC.

Spitzer L., M1956. *Physics of Fully Ionized Gasses*, New York.

Stix T.H., M1962. *The Theory of Plasma Waves*, McGraw-Hill, New York.

Stix T.H., M1992. *Waves in Plasmas*, Am. Inst. of Phys., College Park, Md.

Stix M., M2002. *The Sun: an Introduction* (second edition), Springer-Verlag, New York.

Stone R.G. and B.T. Tsurutani (Eds.) M1985. *Collisionless Shocks in the Heliosophere: A Tutorial Review*, Geophys. Monogr. Ser., **34**, AGU, Washington, D.C.

Störmer C., M1955. *The Polar Aurora*, Oxford University Press, London and New York.

Tatarsky V.I., M1959. *Theory of Fluctuation Events during Wave Propagation in Turbulent Atmosphere*, Acad. of Sci. USSR, Moscow (in Russian).

Tatarsky V.I., M1967. *Wave Propagation in Turbulent Atmosphere*, Nauka, Moscow (in Russian).

Titchmarsh E.C., M1958. *Eigenfunction Expansion of Differential Equations*, Vol. 1, 2, Clarendon Press, Oxford.

Tsurutani B.T.,.W. D. Gonzalez, Y. Kamide, and J.K. Arballo (Eds.), M1997. *Magnetospheric Storms*, Geophys. Monogr. Ser., vol. 98, AGU, Washington, D.C.

Tsurutani B.T., R.L. McPherron, W.D. Gonzalez, G. Lu, J.H.A. Sobral and N. Gopalswarmy (Eds.), M2006. *Recurrent Magnetic Storms: Corotating Solar Wind Streams*, Geophys Monogr. Ser., **167**, AGU, Washington, D.C.

Tverskoy B.A., M1968. *Radiation Belt Dynamics*, Nauka, Moscow (in Russian).

Tsitovich V.N., M1971. *Theory of Turbulent Plasma*, Atomizdat, Moscow (in Russian).

Van Allen J.A., M2004. *Origins of Magnetospheric Physics*. University of Iowa Press, Iowa City, USA.

Van Dyke M., M1964. *Perturbation methods in fluid mechanics*, Academic Press, New York.

Velinov P., G. Nestorov, and L. Dorman, M1974. *Cosmic Ray Influence on the Ionosphere and Radio Wave Propagation*, Bulgarian Ac. Sci. Press, Sofia (in Russian).

Vernov S.N. and L.I. Dorman (Eds.), M1965. *Variations of Cosmic Rays and Solar Corpuscular Streams: Collection of Articles,* National Aeronautics and Space Administration, Washington, D.C.

Volland H., M1984. *Atmospheric Electrodynamics*, Springer-Verlag, New York.

Walker A.D.M., M2005. *Magnetohydrodynamic Waves in Geospace: The Theory of ULF Waves and Their Interaction with Energetic Particles in the Solar-Terrestrial Environment*, Inst. of Phys. Publ., Bristol, U.K.

Walt M., M1994. *Introduction to Geomagnetically Trapped Radiation*, Cambridge Univ. Press.

Watson G.N., M1949. *A Treatise on the Theory of Bessel Functions*, Inostr. Literatura, Moscow (in Russian).

Weissman P., L. McFadden, and T. Johnson (Eds.), M1999. *Encyclopedia of the Solar System*, Academic Press, London.

Williams D.J. (Ed.), M1976. *Physics of Solar Planetary Environments*, AGU, Washington, D.C.

Williams J. and G.D. Mead (Eds.), M1969. *Magnetospheric Physics*, D. Reidel Publ. Co., Dordrecht.

Winglee R.M. (Ed.), M2002. *Proceedings of the Sixth International Conference on Substorms*, Univ. of Wash. Press, Seattle.

Zeldovich Ya.B. and Yu.P. Raizer, M1966. *The Physics of Shock Waves and High Temperature Hydro-Dynamical Phenomena,* Nauka, Moscow (in Russian).

Zeleny L.M. and I.S. Veselovsky (Eds.), M2008. *Plasmas Heliogeophysics*, Vol. 1 and Vol. 2. Physmatlit, Moscow (in Russian).

Zombeck M., M1982. *Handbook of Space Astronomy and Astrophysics,* Cambridge University Press.

AUTHOR'S CONTACT INFORMATION

Professor, Dr. of Phys.-Math. Science Lev I. Dorman

Founder and Head of Israel Cosmic Ray & Space Weather Centre
and Israeli-Italian Emilio Ségre Observatory
(Mt. Hermon), affiliated to Tel Aviv University, Shamir Research Institute,
and Israel Space Agency, ISRAEL;

Founder and Chief Scientist of Cosmic Ray Department of N.V. Pushkov Institute of
Terrestrial Magnetizm, Ionosphere, and Radio-Wave Propagation (IZMIRAN),
Russian Academy of Science, Moscow, RUSSIA

Email: lid010529@gmail.com, lid@physics.technion.ac.il

Subject Index

D

E

F

G

H

I

J

K

L

M

N

O

P

Q

R

S

T

U

V

Z

AUTHOR INDEX

A

B

C

D

E

F

G

H

I

J

K

L

M

N

O

P

R

S

T

U

V

W

Y

Z